AF510738

Recent Titles in This Series

48 **Walter Gautschi, editor,** Mathematics of Computation 1943–1993: A half century of computational mathematics (Vancouver, British Columbia, August 1993)

47 **Ingrid Daubechies, editor,** Different perspectives on wavelets (San Antonio, Texas, January 1993)

46 **Stefan A. Burr, editor,** The unreasonable effectiveness of number theory (Orono, Maine, August 1991)

45 **De Witt L. Sumners, editor,** New scientific applications of geometry and topology (Baltimore, Maryland, January 1992)

44 **Béla Bollobás, editor,** Probabilistic combinatorics and its applications (San Francisco, California, January 1991)

43 **Richard K. Guy, editor,** Combinatorial games (Columbus, Ohio, August 1990)

42 **C. Pomerance, editor,** Cryptology and computational number theory (Boulder, Colorado, August 1989)

41 **R. W. Brockett, editor,** Robotics (Louisville, Kentucky, January 1990)

40 **Charles R. Johnson, editor,** Matrix theory and applications (Phoenix, Arizona, January 1989)

39 **Robert L. Devaney and Linda Keen, editors,** Chaos and fractals: The mathematics behind the computer graphics (Providence, Rhode Island, August 1988)

38 **Juris Hartmanis, editor,** Computational complexity theory (Atlanta, Georgia, January 1988)

37 **Henry J. Landau, editor,** Moments in mathematics (San Antonio, Texas, January 1987)

36 **Carl de Boor, editor,** Approximation theory (New Orleans, Louisiana, January 1986)

35 **Harry H. Panjer, editor,** Actuarial mathematics (Laramie, Wyoming, August 1985)

34 **Michael Anshel and William Gewirtz, editors,** Mathematics of information processing (Louisville, Kentucky, January 1984)

33 **H. Peyton Young, editor,** Fair allocation (Anaheim, California, January 1985)

32 **R. W. McKelvey, editor,** Environmental and natural resource mathematics (Eugene, Oregon, August 1984)

31 **B. Gopinath, editor,** Computer communications (Denver, Colorado, January 1983)

30 **Simon A. Levin, editor,** Population biology (Albany, New York, August 1983)

29 **R. A. DeMillo, G. I. Davida, D. P. Dobkin, M. A. Harrison, and R. J. Lipton,** Applied cryptology, cryptographic protocols, and computer security models (San Francisco, California, January 1981)

28 **R. Gnanadesikan, editor,** Statistical data analysis (Toronto, Ontario, August 1982)

27 **L. A. Shepp, editor,** Computed tomography (Cincinnati, Ohio, January 1982)

26 **S. A. Burr, editor,** The mathematics of networks (Pittsburgh, Pennsylvania, August 1981)

25 **S. I. Gass, editor,** Operations research: mathematics and models (Duluth, Minnesota, August 1979)

24 **W. F. Lucas, editor,** Game theory and its applications (Biloxi, Mississippi, January 1979)

23 **R. V. Hogg, editor,** Modern statistics: Methods and applications (San Antonio, Texas, January 1980)

22 **G. H. Golub and J. Oliger, editors,** Numerical analysis (Atlanta, Georgia, January 1978)

21 **P. D. Lax, editor,** Mathematical aspects of production and distribution of energy (San Antonio, Texas, January 1976)

20 **J. P. LaSalle, editor,** The influence of computing on mathematical research and education (University of Montana, August 1973)

19 **J. T. Schwartz, editor,** Mathematical aspects of computer science (New York City, April 1966)

(*Continued in the back of this publication*)

Proceedings of Symposia in
APPLIED MATHEMATICS

Volume 48

Mathematics of Computation 1943–1993: A Half-Century of Computational Mathematics

Mathematics of Computation
50th Anniversary Symposium
August 9–13, 1993
Vancouver, British Columbia

Walter Gautschi
Editor

American Mathematical Society
Providence, Rhode Island

Proceedings of the Mathematics of Computation
50th Anniversary Symposium
held in Vancouver, British Columbia
August 9–13, 1993

with the support of the National Science Foundation
Grant No. DMS-9219447, the U.S. Department of Energy
Grant No. DE-FG02-93ER25152, the National Security Agency
Grant No. MDA904-93-3008, and NEC Research Institute.

1991 *Mathematics Subject Classification.* Primary 65-06, 11-06, 11Yxx.

Library of Congress Cataloging-in-Publication Data

Mathematics of Computation 50th Anniversary Symposium (1993: Vancouver, B.C.)
Mathematics of computation, 1943–1993: a half-century of computational mathematics: Mathematics of Computation 50th Anniversary Symposium, August 9–13, 1993, Vancouver, British Columbia / Walter Gautschi, editor.
p. cm. – (Proceedings of symposia in applied mathematics, ISSN 0160-7634; v. 48)
Includes bibliographical references and indexes.
ISBN 0-8218-0291-7 (acid-free).
1. Numerical analysis—Congresses. 2. Number theory—Data processing—Congresses. I. Gautschi, Walter. II. Mathematics of computation. III. Title. IV. Series.
QA297.M385 1993
519.5–dc20 94-31835
CIP

Copying and reprinting. Individual readers of this publication, and nonprofit libraries acting for them, are permitted to make fair use of the material, such as to copy an article for use in teaching or research. Permission is granted to quote brief passages from this publication in reviews, provided the customary acknowledgment of the source is given.

Republication, systematic copying, or multiple reproduction of any material in this publication (including abstracts) is permitted only under license from the American Mathematical Society. Requests for such permission should be addressed to the Manager of Editorial Services, American Mathematical Society, P.O. Box 6248, Providence, Rhode Island 02940-6248. Requests can also be made by e-mail to reprint-permission@math.ams.org.

The appearance of the code on the first page of an article in this book indicates the copyright owner's consent for copying beyond that permitted by Sections 107 or 108 of the U.S. Copyright Law, provided that the fee of $1.00 plus $.25 per page for each copy be paid directly to the Copyright Clearance Center, Inc., 222 Rosewood Drive, Danvers, Massachusetts 01923. This consent does not extend to other kinds of copying, such as copying for general distribution, for advertising or promotional purposes, for creating new collective works, or for resale.

© Copyright 1994 by the American Mathematical Society. All rights reserved.
The American Mathematical Society retains all rights
except those granted to the United States Government.
Printed in the United States of America.
∞ The paper used in this book is acid-free and falls within the guidelines
established to ensure permanence and durability.
♲ Printed on recycled paper.

Portions of this volume were typeset by the authors using AMS-TEX and AMS-LATEX,
the American Mathematical Society's TEX macro systems.
Papers in this volume are final versions, which will not be published elsewhere,
unless otherwise indicated by a footnote.

10 9 8 7 6 5 4 3 2 1 99 98 97 96 95 94

Contents

Contributed Papers

Part II. Minisymposium on Computational Number Theory Dedicated to the memory of Derrick Henry Lehmer

Invited Papers

Contributed Papers

Preface

The year 1993 marked an important milestone in the history of the journal *Mathematics of Computation*: the 50th anniversary of its founding. It began publication in 1943 as an archival journal under the name *Mathematical Tables and Other Aids to Computation*, but gradually changed character and became an influential — and probably the oldest — research journal in the area of computational mathematics. (For a brief history of the journal, see the article by E. Isaacson following this Preface.) To commemorate the event, an International Conference was held in Vancouver from August 9 to 13, 1993. It consisted of a Symposium on Numerical Analysis and a Minisymposium on Computational Number Theory. The latter was dedicated to the memory of Derrick Henry Lehmer, one of the founders of the journal, who passed away in 1991. The present volume is the proceedings of the conference, containing in Part I the papers of the Symposium, and in Part II those of the Minisymposium.

The Organizing Committee consisted of James H. Bramble, Walter Gautschi (Chairman), Eugene Isaacson, Vidar Thomée, and Hugh C. Williams. The first four members of the committee organized the Symposium; Professor Williams organized the Minisymposium.

This volume contains fourteen invited papers of which all but two were presented at the conference. Those not presented are the historical essay of H.C. Williams and J.O. Shallit on integer factorization, and the paper on componentwise perturbation bounds in linear algebra by N.J. Higham, which was solicited for these proceedings only after the conclusion of the conference. Three invited papers were presented but are not included here. They are: G.W. Stewart, "On the early history of the singular value decomposition", which had already been committed for *SIAM Review* and has since appeared there in vol. 35 (1993), pp. 551–566; B. Engquist, "Numerical approximations of hyperbolic differential equations", and L.R. Petzold, "A brief history of numerical methods for differential-algebraic equations", for which manuscripts have not been received.

It was envisaged that the invited papers present surveys — with ample bibliographic documentation — on the various subdisciplines covered by *Mathematics of Computation*, in a historical perspective and in a language accessible to as wide an audience as possible. Contemporary specialized work, on the other hand, was to be addressed in the (refereed and space-limited) contributed papers, of which forty-six have been accepted and appear in these proceedings.

We are pleased that all invited speakers who submitted manuscripts have admirably followed our intent, and we are able, therefore, to present in these proceedings informative surveys on such topics as the numerical solution of ordinary and partial differential equations, numerical linear algebra, approximation and optimization, numerical integration, special functions, and computational number theory. A brief outline of the topics covered follows.

Linear algebraic systems arising in either finite difference or finite element approximations to elliptic boundary value problems are typically very large, sparse, and ill conditioned. Direct methods of solution are out of the question, and even the standard iterative methods either converge too slowly or are applicable only to problems with special geometries. A major advance occurred in the early 1960s, when Fedorenko systematically pursued the idea of multilevel methods. This led to the development of what are now called *multigrid methods.* Bramble in his paper traces the history of these methods, as applied to second-order elliptic boundary value problems, and highlights the advances made in the last ten years in understanding their convergence properties. A related idea — *multiresolution* — decomposes unknowns into a hierarchy of subsets which are given individual treatments. The paper by Temam studies this idea in the context of nonlinear evolution equations and nonlinear dynamics, with a view toward the approximation of attractors. Here, Fourier-type expansions suggest differentiated treatments of high-frequency and low-frequency components, whereas finite difference approximations indicate the introduction of "incremental unknowns" associated with different grid levels. The former idea is illustrated by Temam for abstract evolution equations in Hilbert space, the latter for a reaction-diffusion equation on the square.

A different approach to solving partial differential equations is via *integral equations.* This is the topic of Nédélec's paper, in which this method is illustrated on Helmholtz and Laplace equations, the Maxwell equation for electric and magnetic fields, and problems in elasticity. The integral operators involved are of the single-layer potential and double-layer potential type, and the paper surveys the modern theory of such operators, including singular and hypersingular ones. Many real-world applications, for example in geophysics and medical imaging, require the solution of *inverse problems*, where typically information in inaccessible regions of space is to be reconstructed from measurements made in accessible regions. Problems of this type, especially nonlinear multidimensional ones involving partial differential equations, are discussed in the paper by M. Cheney.

The numerical solution of (initial value problems for) *ordinary differential equations* has had a long history, but has matured into a firm branch of mathematics only in the early 1960s through the work of Dahlquist and Henrici. The paper by Stetter shows how the field has developed from this point on. It deals not only with the basic principles underlying discrete-variable methods and analytical techniques for stability analysis, but also illuminates the difficulties encountered, and special techniques called for, in a variety of specialized problems such as stiff problems, highly oscillatory problems, Hamiltonian problems, and differential equations with delays. Questions of computer implementation and testing are also discussed, and there is an interesting account of error estimation, including the problem of validation, i.e., the construction of a set-valued trajectory guaranteed to contain the exact solution trajectory. Codes for validated solution of ordinary differential equations indeed have become important tools for providing computational proof of mathematical properties of differential systems, such as the existence of periodic solutions or of chaotic motion in dynamical systems.

Error and perturbation analyses in *linear algebra* problems are traditionally phrased in terms of matrix norms. There is, however, a loss of information inherent in the use of norms, as they ignore the relative magnitudes of matrix elements and sparsity patterns. A componentwise analysis, on the other hand, allows one to take such information into account and accordingly leads to sharper and more informative estimates. Both forward and backward componentwise perturbation bounds are proved and discussed in the paper by N. J. Higham in connection with such problems as linear algebraic systems, matrix inversion and factorization, the least squares problem, and eigenvalue and singular value problems. Most of the results on this last topic indeed are quite recent, dating from the last five years.

Extremely useful tools in approximation are *spline functions*, i.e., piecewise (relative to a given partition of the domain) polynomial functions enjoying global smoothness of a certain degree. While the theory of univariate splines is highly developed and well understood by now, the analogous theory for multivariate splines, in spite of great strides made since the early 1970s, has still many unanswered questions and is currently the subject of intense research. The paper by Schumaker gives an informative survey of bivariate spline spaces and goes on to review applications of bivariate splines to problems of surface interpolation and scattered data fitting, computer-aided geometric design, and image compression.

Recent advances in problems of *unconstrained optimization*, specifically the problem of finding a local minimum of a smooth nonlinear function, are reviewed in the paper by Goldfarb. When first and second derivatives are available analytically, Newton's method, based on second-order Taylor series approximation, still plays a central role but requires modifications (a number of which are described) to make it well defined, globally convergent, efficient and practical. In the absence of analytic second derivatives, one can resort to quasi-Newton methods in which the Hessian matrix is updated in various ways from one iteration step to the next. There is an interesting discussion on how these updated methods perform in practice and how they can be adapted to large sparse problems. More recent extensions of Newton's method, based on fourth-order tensor approximations of Taylor's series, are also discussed as well as variants of the conjugate gradient method.

The age-old subject of *numerical integration* is represented by a paper by Lyness and Cools, which treats a rather specialized problem —

numerical cubature over triangles — but one that is potentially useful in finite element applications and in adaptive cubature using successive triangulations. Among the techniques discussed are transformations to a square domain, polynomial moment fitting (taking advantage of the symmetries of the triangle), and extrapolation techniques suggested by appropriate Euler-Maclaurin expansions. Integrands that are regular as well as those exhibiting corner or line singularities are analyzed.

Tables of *special functions* and their documentation have been the prime motivation for starting our journal. In spite of the rapid advances in computer technology during the intervening fifty years, the need for special functions still persists, but has shifted from tables to numerical approximations and software. It is therefore particularly fitting to include in these proceedings a detailed documentation of mathematical techniques for generating approximations to functions, and of software packages, libraries, and interactive systems dealing with the whole arsenal of special functions. This is provided in the paper by Lozier and Olver and in its extensive bibliography of nearly 500 items, covering functions both in the real and complex domain, and involving one or several variables.

The field of *computational number theory* — another subject area that has been an integral part of our journal since its inception — is being addressed in four invited papers. The one by Williams and Shallit is entirely historical and deals with factoring and primality testing before the age of electronic computers, covering the two hundred years from 1750 to about 1950. The paper pays respect to the tireless efforts of the early pioneers of computational number theory, who, in the absence of powerful computing machinery, had to invent ingenious mechanical and other devices to help them along. Yet many of their ideas carried in themselves the germs of today's modern techniques of factorization. An account of some of these modern techniques, culminating in the number field sieve of Pollard, is presented in the paper by Pomerance. It examines the various mathematical ideas that, in combination, led to the impressive recent successes of factoring general 120-digit numbers and special numbers of even larger size.

Other aspects of computational number theory are dealt with in the two remaining papers. The one by Odlyzko is devoted to analytic functions important in number theory and to their zeros, notably the Riemann zeta function and Dirichlet series, and their relevance to prime number distribution. In particular, a brief history is given on the progress made in trying to establish the Riemann Hypothesis and on the computational techniques employed. The paper by Biehl and Buchmann, finally, discusses deterministic algorithms and their complexity in algebraic number theory. The problems discussed include the computation of fundamental units and regulators, principal ideal testing, solving prime norm equations, computing the structure of a class group, computing the order of an ideal class, and determining discrete logarithms in the class group.

It is a pleasure to acknowledge the support and encouragement given by Jim Maxwell to organize this conference and his efforts to secure funding for it, and the logistics and technical support provided by Wayne Drady, Donna Harmon, and the staff at AMS. I am particularly thankful to my colleagues on the Organizing Committee for their help in putting this conference together, and above all to the speakers and authors for their great effort in producing a collection of outstanding papers. I also wish to acknowledge the assistance of the referees — too many to name individually — for their review of contributed papers and their helpful comments on invited papers as well. Finally, I gratefully acknowledge financial support from the National Science Foundation, the National Security Agency, the Department of Energy, and the NEC Research Institute. Last, but not least, my thanks go to the members of the Proceedings of Symposia in Applied Mathematics Committee for allowing these proceedings to appear in their series.

Walter Gautschi
Department of Computer Sciences
Purdue University
West Lafayette, IN 47907-1398
June 1994

Proceedings of Symposia in Applied Mathematics
Volume **48**, 1994

Mathematics of Computation: A Brief History

EUGENE ISAACSON

1. Editors and Policies

Mathematics of Computation (MoC) is published quarterly by the American Mathematical Society (AMS). The journal first appeared in January 1943 as *Mathematical Tables and Aids to Computation* (MTAC) under the auspices of the Committee for MTAC (CMTAC) of the National Research Council (NRC) of the National Academy of Sciences (NAS) of the United States. During 1943, the sole EDITOR of MTAC was Raymond Clare Archibald (R.C.A.). Archibald was also the chairman of the CMTAC, which had 24 members and constituted the editorial board of the journal. The purpose of MTAC and the committee is explained in the following excerpt from the "Introductory" in volume I (1943):

> This Quarterly Journal, a new publication of the National Research Council, is to serve as a clearing-house for information concerning mathematical tables and other aids to computation. ...
>
> ... Up to the present the chief duty of the Committee has been to prepare a series of comprehensive Reports on mathematical tables, valuable for various types of research in different fields. D. H. Lehmer's *Guide to Tables in the Theory of Numbers* was published in February, 1941. It is expected that a more elaborate Report, in another field, may be ready for publication during 1943. Since most members of the Committee are at the present time deeply involved in national service, the completion of still further Reports in the near future is likely to be very difficult to achieve.
>
> Meanwhile, it is hoped this new periodical may render notable current service, and that in years to come it, and the Reports, may be regarded as the standard sources to which one may naturally turn for guidance in connection with all mathematical tables of importance in contemporary research.
>
> On behalf of the Committee,
>
> R. C. A.

1991 *Mathematics Subject Classification*. Primary 65-03, 01-08; Secondary 65A05.
This paper is in final form and no version of it will be submitted elsewhere.

© 1994 American Mathematical Society
0160-7634/94 $1.00 + $.25 per page

Derrick Henry Lehmer was a member of the CMTAC. Beginning in 1944, his appointment as EDITOR, along with Archibald, assured that a wide field of mathematical interests would be covered in MTAC-MoC and that these would be unique among numerical analysis journals in that they would become "the journal of choice for almost all authors of papers on computational number theory" (see [6]). In a letter [2] of March 24, 1989, Lehmer responds to a request from Walter Gautschi for information on the early history of MTAC with:

"... History is better served by the back numbers of the quarterly. ... "

Throughout the early years, Emma Lehmer, Derrick's wife, helped to edit the journal, but had no editorial title. She is a mathematician and has published in MTAC-MoC.

From an examination of the cover pages, the name *Mathematical Tables and Other Aids to Computation* seems to have been chosen without editorial comment, beginning with the October 1943 issue. Archibald and Lehmer continued as EDITORS of MTAC with the *cooperation* of L. J. Comrie and S. A. Joffe from 1944 until the end of 1949, when Archibald retired. For the one issue of January 1946, Harry Bateman was listed along with Archibald and Lehmer as EDITOR. Unfortunately, Bateman died in January, 1946 (see reference in [1]). The EDITOR(S) managed the journal.

Beginning with the January 1950 issue, the journal was managed by the CHAIRMAN of the editorial committee. Lehmer continued as the first Chairman. The following are the names of the chairmen and the years that they served:

D. H. Lehmer	1950-1954
C. B. Tompkins	1955-1958
Harry Polachek	1959-1965
Eugene Isaacson	1966-1974
J. H. Bramble	1975-1983
Walter Gautschi	1984-present

Starting with the January 1979 issue, the Chairman of the editorial committee was given the title MANAGING EDITOR.

2. Change of Name

In the mid 1940s, electro-mechanical computers produced a flurry of many volumes of tables of special functions. These machines were soon replaced by electronic computers, and the subsequent nature of articles submitted to MTAC changed rapidly. Table-making was replaced by algorithm-making with its attendant analysis of all aspects of computation (see [3]). The editorial committee of MTAC changed the name of the journal to MoC, beginning with the January 1960 issue. Harry Polachek explained the rationale in an introduction to that number [4].

3. Change of Publisher

In the late 1950s, the NRC-NAS was interested in finding a competent organization to assume responsibility for publishing MTAC. Polachek and his editorial committee contacted the American Mathematical Society (AMS), the Association for Computing Machinery, and the Society for Industrial and Applied Mathematics. All three organizations offered to take control of the journal. The editorial board of MoC recommended that the AMS be given the opportunity. The AMS proposed that the journal would be managed as it had been in the past — that is, by the editorial committee of MoC (which until then had been chosen by the NRC-NAS).

Thus, in January 1962 MoC was published by the AMS under the editorial control of the NRC-NAS (which selected the editorial committee). This began a four-year trial period for changing publishers. At the start, the NRC-NAS gave the AMS the balance of funds that were left from the original founding contribution for MTAC received from the Rockefeller Foundation (see [5]). By the July 1965 issue, the AMS assumed full control of both the publication and the selection of the editorial committee for MoC. The management of the journal is still in the hands of its managing editor and editorial committee. The AMS appoints the MANAGING EDITOR and three other members that together form the EDITORIAL COMMITTEE; the managing editor then appoints the rest of the editorial board (with approval of the AMS). In July 1993, the editorial board of MoC had 27 members.

Acknowledgment

In addition to the cited references, the author had access to letters in the MoC files from past and present members of MTAC-MoC editorial boards (i.e., F. J. Murray, Daniel Shanks, and John Todd). Financial support was received from a grant made to the AMS by the National Science Foundation.

References

1. Isaacson, Eugene, *The Origin of Mathematics of Computation and Some Personal Recollections*, A History of Scientific Computing, ed. Stephen G. Nash, ACM Press, Addison-Wesley Publishing Co., New York, 1990, 211-216.

2. Lehmer, D. H., Unpublished letter to Walter Gautschi, in MoC files, March 24,1989, 1 p.

3. Nash, S. G., ed., *A History of Scientific Computing,* ACM Press, Addison-Wesley Publishing Co., New York, 1990. (ACM Conference on the History of Scientific and Numeric Computation, 1987, Princeton, NJ), 359 p.

4. Polachek, Harry, *Mathematics of Computation,* Math Comp. **14** (1960), 1-2.

5. ______________, *History of Mathematics of Computation, 1959-1965,* Unpublished manuscript in MoC files, received Jan. 3, 1988, 11 p.

6. Williams, H. C., *Dedication*, Math Comp. **61** (1993), v-viii.

Courant Institute of Mathematical Sciences, New York University, New York, NY 10012; :: *E-mail address* :: isaacson@acfcluster.nyu.edu ::

Part I

Symposium on Numerical Analysis

Invited Papers

Proceedings of Symposia in Applied Mathematics
Volume 48, 1994

ON THE DEVELOPMENT OF MULTIGRID METHODS AND THEIR ANALYSIS

JAMES H. BRAMBLE

ABSTRACT. In this paper we discuss the development of multigrid methods. In particular we concentrate on multigrid as applied to second-order elliptic boundary value problems. The aim is to trace the development of techniques for proving convergence estimates. This is done by highlighting and describing some of the important papers in the literature.

1. INTRODUCTION

The purpose of this paper is to describe the development of the multigrid method as it is applied mainly to the solution of finite difference and finite element approximations to second-order elliptic boundary value problems. There has been an enormous amount of work on the solution of the typical linear systems which result from such approximations, and no attempt is made here to cover this fully. Only certain papers concerning the development of the multigrid approach will be discussed. Many interesting papers which are not directly referred to in this note may be found in the bibliography. For other works on the subject the reader is referred to the bibliographies in the book of Hackbusch [49] and the survey article of Bank, Mandel and McCormick [8].

We shall start by giving a little motivation for the multigrid approach. For this purpose we note that the systems resulting from the abovementioned approximations are very large and very ill-conditioned. For example, the difference equation resulting from the usual 5-point approximation to the Poisson equation in the plane with mesh size h has a condition number proportional to h^{-2}, with the dimension of solution space also proportional to h^{-2}. For many large-scale scientific applications the resulting system is too large to be dealt with by direct methods. Consequently, the usual approach to obtaining the solution of such a system is via some systematic iterative method. Among the earliest such methods are the Jacobi and Gauss-Seidel methods (cf. [87]). The convergence rates for these methods are very poor because of the ill-conditioning. An important advance in developing methods which have better convergence properties was the successive overrelaxation method

1991 *Mathematics Subject Classification.* Primary 65N30; Secondary 65F10.

This manuscript has been authored under contract number DE-AC02-76CH00016 with the U.S. Department of Energy. Accordingly, the U.S. Government retains a non-exclusive, royalty-free license to publish or reproduce the published form of this contribution, or allow others to do so, for U.S. Government purposes. This work was also supported in part under the National Science Foundation Grant No. DMS-9007185 and by the U.S. Army Research Office through the Mathematical Sciences Institute, Cornell University, Contract No. DAAL03-91-C-0027.

This paper is in final form, and no version of it will be submitted for publication elsewhere.

© 1994 American Mathematical Society
0160-7634/94 $1.00 + $.25 per page

(SOR) (cf. Young [95] or Varga [87]). This method still suffers from a substantial deterioration in its rate of convergence for small mesh sizes h. Also a parameter has to be chosen in order that the theoretical improvement over the Gauss-Seidel method be achieved.

Other developments in the progress to obtain efficient methods for such problems are the alternating direction implicit methods (ADI) (cf. the papers by Peaceman and Rachford [78] and Douglas [38]). Generally, these methods apply only to separable equations on special domains, e.g. the Poisson equation on a rectangular domain. With a proper choice of a set of iteration parameters, the increase in rate of convergence over the abovementioned methods is dramatic. The Jacobi and Gauss-Seidel methods give rise to iteration matrices with spectral radii proportional to $1-ch^2$, and that of the SOR method behaves like $1-ch$, where c is some constant independent of h. The contraction number corresponding to the ADI method (for this example) is bounded below one, uniformly in h, provided that the number of properly chosen iteration parameters is proportional to $|\log h|$. This means that the total cost of solving the relevant system by means of the ADI method is much less than the cost by means of the Jacobi or Gauss-Seidel methods.

The conjugate gradient method (CG) (cf. Hestenes and Stiefel [53] or Lanczos [56]) was developed in the 1950s, but its importance as an iterative method for elliptic difference equations was not recognized for some time. Wachspress [89] considered using the ADI method for a model problem as a preconditioner for a more general system in a CG iteration. The problem of constructing good preconditioners for elliptic problems is very important but will not be specifically discussed here. It should be mentioned, however, that the multigrid technique provides an excellent means of constructing effective preconditioners.

The first papers on the multigrid method were those of Fedorenko [39], where he considered a two-level method, and [40], where he demonstrated that a multilevel method could be constructed with particularly good theoretical convergence properties. He showed that under certain conditions for a model problem, an iterative procedure could be developed for which the spectral radius of the error operator was bounded away from one independently of the dimension or mesh size. This was a major breakthrough in the development of iteration methods for such problems.

The main idea is to notice that simple methods like the Jacobi or Gauss-Seidel methods have the very important property that the high-frequency components of the error are rapidly reduced. Hence, with one or more iterations of such a simple method, a new problem can be formulated whose solution is smoother and hence can be well approximated on a mesh of, say, twice the size. This reduces the dimension of the problem which needs to be solved for a good correction, i.e., reducing also the low-frequency components in the error. Fedorenko noticed that the new problem was of the same form as the original but relative to a coarser grid, and hence had the idea of proceeding in some systematic recursive manner. This is the basic multigrid idea.

In the remainder of this paper I will touch on some of the highlights in the development of the multigrid method and the evolution of techniques for its analysis and consequent understanding of how this powerful tool should be applied. The book by Hackbusch [49] gives a quite thorough treatment of the state of development up until 1985, and, therefore, only certain papers up until that time will be discussed here. For specific details of some of the later work we refer to the book by the author [12].

In the second section we describe a model approximate problem and somewhat general multigrid iteration procedures which apply to this situation. I have chosen to use a finite element formulation since, in this framework it is easy to describe the essential points of the various results. I will refer to some results, which are specifically about difference equations, in the finite element context since, at least in particular cases, these can be put into this setting.

The later sections will be devoted to summarizing some results from the literature. After 1970 quite a few papers on the subject appeared in the journal *Mathematics of Computation*, and, since this survey material is being presented in celebration of its fiftieth anniversary, I will specifically note those papers with the symbol MOC.

2. The model problems and multigrid methods

We shall consider as our example the Dirichlet problem for domains $\Omega \subset R^2$ which may have curved boundaries. For simplicity we let $\partial\Omega$ be piecewise smooth and in fact, in many of the works discussed here, it will be polygonal.

Let

$$A(v,w) = \sum_{i,j=1}^{2} \int_{\Omega} a_{ij} \frac{\partial v}{\partial x_i} \frac{\partial w}{\partial x_j}\, dx + \int_{\Omega} avw\, dx$$

on $H_0^1(\Omega) \times H_0^1(\Omega)$, where the 2×2 symmetric matrix with entries a_{ij} is uniformly positive definite and the function a is nonnegative.

The weak formulation of the Dirichlet problem is the following: Given f in an appropriate Sobolev space (e.g. $L_2(\Omega)$), find $u \in H_0^1(\Omega)$ such that

$$A(u,w) = (f,w) \text{ for all } w \in H_0^1(\Omega). \tag{2.1}$$

Here $(\cdot,\cdot)$ is the usual L_2-inner product. Once we define a subspace $M \subset H_0^1(\Omega)$, the Galerkin approximation in M is: Find $U_h \in M$ such that

$$A(U_h,\chi) = (f,\chi) \text{ for all } \chi \in M. \tag{2.2}$$

We assume that a_{ij} and a are smooth enough so that the following a priori estimate is valid:

$$\|u\|_{1+\alpha} \leq C\|f\|_{-1+\alpha} \tag{2.3}$$

for some $\alpha \in (0,1]$. The norm $\|\cdot\|_{1+\alpha}$ is the norm on the space defined by interpolation between the spaces $H_0^1(\Omega)$ and $H^2(\Omega)$. The space with negative index is defined by duality. That is,

$$\|f\|_{-1+\alpha} = \sup_{\theta \in H_0^1(\Omega)} \frac{(f,\theta)}{\|\theta\|_{1-\alpha}},$$

where the norm $\|\cdot\|_{1-\alpha}$ is the norm on the space defined by interpolation between the spaces $L^2(\Omega)$ and $H_0^1(\Omega)$.

Now the simplest finite element method is obtained by making a special choice of M in (2.2). For the purpose of describing one of the most common setups for the multigrid method, we let Ω be a polygon and let τ_1 be a (coarse) triangulation of Ω consisting of, say, just a few triangles. We define M_1 to be the space of continuous piecewise linear functions on τ_1 which vanish on $\partial\Omega$. Let h_1 be the length of the side of maximum length in τ_1.

If the triangulation τ_k and the space M_k have been defined, define τ_{k+1} by subdividing each triangle of τ_k into four triangles by joining the midpoints of the sides. Define M_{k+1} analogously relative to τ_{k+1}. Thus $M_1 \subset M_2 \subset \cdots \subset M_J = M \subset H_0^1(\Omega)$. We define $h_{k+1} = \frac{1}{2}h_k$ and $h = h_J$. These are the multilevel spaces, which may be thought of as the prototype for multilevel procedures.

The finite element approximation is (2.2) with $M = M_J$, and it is in terms of equations of this type that we want to describe the work on multigrid methods. Though the above example is good to keep in mind, we will allow considerably more generality. For instance, nonconforming and mixed finite element methods lead to more general multigrid formulations.

We shall present multigrid algorithms in which we include the cases of nonnested spaces (those defined in the above example are nested), forms and inner products which may vary from level to level, nonsymmetric linear smoothers allowing the number of smoothing steps to vary from level to level, as well as the so-called V- and W-cycles.

The general setup consists of several ingredients. We list them as follows.

1. Let $M_1, \ldots, M_J$ be J finite-dimensional spaces with $M_J = M$.
2. Assume that we have $J-1$ linear operators $I_k : M_{k-1} \to M_k$, $k = 2, \ldots, J$ which connect the spaces.
3. Assume that we have symmetric positive definite forms $A_k(\cdot,\cdot)$ on $M_k \times M_k$, with $A_J(\cdot,\cdot) = A(\cdot,\cdot)$.
4. Assume that we have for each k an inner product $(\cdot,\cdot)_k$ on $M_k \times M_k$, with induced norm $\|\cdot\|_k$.
5. Define $A_k : M_k \to M_k$ by

$$(A_k w, \varphi)_k = A_k(w, \varphi) \text{ for all } \varphi \in M_k.$$

6. Define $P_{k-1} : M_k \to M_{k-1}$ and $Q_{k-1} : M_k \to M_{k-1}$ by

$$A_{k-1}(P_{k-1}w, \varphi) = A_k(w, I_k\varphi)$$

and

$$(Q_{k-1}w, \varphi)_{k-1} = (w, I_k\varphi)_k,$$

for all $\varphi \in M_{k-1}$.

7. Let $R_k : M_k \to M_k$ be a linear (smoothing) operator, and in addition define

$$R_k^{(\ell)} = \begin{cases} R_k & \text{if } \ell \text{ is odd,} \\ R_k^t & \text{if } \ell \text{ is even,} \end{cases}$$

where R_k^t is the adjoint with respect to the inner product $(\cdot,\cdot)_k$.

We are now ready to state the most general multigrid algorithm which we will consider here.

As described in [12], we will define a linear operator $B = B_J$ which completely characterizes the multigrid method. We shall do this by means of an induction process. Once B has been defined, the linear multigrid iteration is

$$x^{n+1} = x^n - B(Ax^n - f) \quad \text{for} \quad n \geq 0, \tag{2.4}$$

with x^0 arbitrary.

Algorithm MG: Let p be a positive integer, and let $m(k)$ be a positive integer depending on k.

0) $B_1 = A_1^{-1}$.

For $g \in M_k$ we define $B_k g$ in terms of B_{k-1}, $2 \leq k \leq J$.

1) $x^0 = 0$, $q^0 = 0$.

2) Define x^ℓ, $\ell = 1, \dots, m(k)$, by

$$x^\ell = x^{\ell-1} + R_k^{(\ell+m(k))}(g - A_k x^{\ell-1}).$$

3) $y^{m(k)} = x^{m(k)} + I_k q^p$, where q^i for $i = 1, \dots, p$ is defined by

$$q^i = q^{i-1} + B_{k-1}[Q_{k-1}(g - A_k x^{m(k)}) - A_{k-1} q^{i-1}].$$

4) Define y^ℓ for $\ell = m(k)+1, \dots, 2m(k)$ by

$$y^\ell = y^{\ell-1} + R_k^{(\ell+m(k))}(g - A_k y^{\ell-1}).$$

5) $B_k g = y^{2m(k)}$.

If we set $e^n = x^n - x$, where $x = A^{-1}f$, then (2.4) may be written

$$e^{n+1} = (I - BA)e^n. \tag{2.5}$$

Thus, the linear error operator corresponding to (2.4) is $I - BA$. In the paper by Maitre and Musy [59], for example, the recurrence relation defining $I - BA$ is given and is equivalent to the above definition.

We could have allowed for even more generality, but many of the main results in the literature can be adequately described by Algorithm MG. For the symmetric model problem, this algorithm always leads to a symmetric operator B.

As is customary in the literature, we will give names to certain special cases.

1. The case $p = 1$, with $m(k) = m$, is called the V-cycle.
2. The case $p = 2$, with $m(k) = m$, is called the W-cycle.
3. The case $p = 1$, with nonconstant $m(k)$, is called the variable V-cycle.
4. The case M_k contained in M_{k+1} is called nested.
5. The case $A_{k-1}(v, w) = A_k(I_k v, I_k w)$ for all $v, w \in M_{k-1}$ is called inherited.
6. The nested-inherited case, with I_k the natural imbedding of M_{k-1} in M_k, is called variational.

In describing results we will often refer to the various cases by name.

3. Early work

As pointed out in §1, Fedorenko [39] was the first to describe a systematic multigrid method. He formulated a two-level method for solving the difference equations corresponding to the Poisson equation on a rectangular domain in the plane. He then proposed in [40] the first recursive multigrid algorithm based on his two-level method. Although he did not provide a proof of the excellent convergence properties, he had the right idea.

Bakhvalov [3] generalized the method of Fedorenko to the case of difference approximations on the square for problems with variable coefficients which were not necessarily selfadjoint. He proved convergence of the W-cycle provided sufficiently many smoothing steps were performed. He also proposed what is called the "full multigrid" algorithm and showed that an approximation to the solution of the difference equations could be computed with optimal complexity. This means that the number of arithmetic operations required for a certain accuracy is proportional to the number of unknowns involved. This was a very interesting result, especially from a theoretical point of view.

A few years later, Astrakhantsev [1] carried out the program of Bakhvalov [3] for the third boundary value problem in an arbitrary smooth plane domain. In the third boundary value problem a linear combination of the solution and its conormal derivative are prescribed on the boundary. At this point, the Russians seem to have dropped work on the multigrid idea. I was told by a Russian colleague that the method was considered to be too complicated to be of general practical value.

4. Work in the 1970s

The work of Brandt was the first developmental work in the West on multilevel techniques. In the paper [28] in 1973 he described a so-called full multigrid algorithm. Although this work and his later paper [29] (MOC) did not contain complete proofs of convergence results, his insights and computational experience had a strong effect on subsequent interest in multigrid as a potentially powerful tool for solving discrete elliptic systems.

In a 1975 paper [70], Nicolaides considered multilevel methods. Then in 1977 [71] (MOC) he proved that the contraction number for the W-cycle with sufficiently many smoothing steps was uniformly bounded away from one. This study was with respect to the L_2-norm and made assumptions which were quite stringent. For example, he assumed that the problem (2.1) satisfied (2.3) with $\alpha = 1$. Hence, if the domain is polygonal, it must be convex. He also proposed the full multigrid algorithm and proved its optimal complexity. Studies requiring the W-cycle with sufficiently many smoothing steps were common to many subsequent works. This work was also in the category of variational, in particular only the case of nested subspaces was treated. He continued his studies in [72] (MOC), [73] (MOC) and [74] (MOC), where he considered practical and theoretical aspects of the multigrid approach to symmetric and nonsymmetric problems.

Hackbusch, in several papers [44], [45], [46] (MOC), [47] and [48] presented convergence results, primarily on the W-cycle with sufficiently many smoothing steps. He formulated abstract conditions, which he called the smoothing condition and the approximation property. In verifying the approximation property, at least in the finite element case, it seems to the author that it was also required that problem (2.1) satisfy (2.3) with $\alpha = 1$. Conditions of the type expressed by Hackbusch, as he pointed out, were at least implicit in all of the previous works on the subject.

So far as I can tell, all results to this time treated only the W-cycle with sufficiently many smoothing steps. Although computational experience had shown by then that the V-cycle with one smoothing step was successful, no proof of this was available. The variational case was still the one most commonly studied.

5. Work in the early 1980s

The paper of Bank and Dupont [6] (MOC), which appeared in 1981, was very important to the development of multigrid convergence theory. Their work was in the framework of finite element approximations. Although they specifically considered the case of the Neumann problem, their results cover our example (2.1) and (2.2) virtually without change. They studied the W-cycle in the nested-inherited case with piecewise linear continuous finite elements. Their analysis covered the very important, quite general case in which (2.3) is required only to be satisfied for some positive $\alpha \leq 1$. They proved that the W-cycle, with m sufficiently large, provided a uniform contraction number with respect to the norm defined by $A(\cdot, \cdot)$. They used specifically the smoother associated with the damped Jacobi iteration. This result shed new light on the meaning of the role that regularity plays with respect to multigrid convergence properties. Though a lot was done which extends their results, there are some situations for which their treatment has not been improved upon (cf. the paper by Brenner [33] (MOC)).

Another paper which has had an important effect in the development of the theory is that of Braess and Hackbusch [11]. This treated our model problem with conforming finite elements in a convex polygonal domain. The multigrid algorithm studied was the nested-inherited case with a damped Jacobi iteration as the smoother. The condition (2.3) was assumed with $\alpha = 1$. They showed that the contraction number δ satisfies

$$\delta \leq \frac{C}{C+m}, \tag{5.1}$$

where C is a constant independent of the number of levels, and m is the number of smoothing steps. Thus, for example, no more than one smoothing step was needed to guarantee a uniform contraction. This was the first time that a proof was given for the convergence of a multigrid algorithm which put no restrictions on the number of smoothing steps required. This paper was a major step in showing that, at least for regular symmetric problems, the multigrid algorithm is robust.

There were several other papers written during this period whose themes were close to those of these two works. In this connection we note here the papers of C. Douglas [37], Maitre and Musy [57] and [58], McCormick [64] and [65], McCormick and Ruge [66] and [67] (MOC), and the survey article by Bank, Mandel and McCormick [8]. Several of these works studied other smoothers, such as the smoother coming from the Gauss-Seidel iteration. In particular, the paper by McCormick [65] generalized the Braess-Hackbusch result to include smoothers which, in a sense, were comparable to one coming from a damped Jacobi iteration. He also allowed the use of nonlinear smoothers, such as the steepest descent or the conjugate gradient method. The estimate (5.1), for $m \geq 1$, was only obtained in the case of linear smoothers.

We note also the paper by Mandel [61] in which he extends the work of Braess and Hackbusch to the nonsymmetric or indefinite case. Another extension was given by Yserentant [96] (MOC), where he considers problems on nonconvex polygonal domains so that (2.3) is not satisfied with $\alpha = 1$. By introducing a graded mesh and a weighted norm, he shows that conditions sufficient for the Braess-Hackbusch result are satisfied, so that an estimate of the type (5.1) is valid.

6. Further results based on the two-level approach

All of the work discussed so far has been based on an analysis of a two-level scheme. That is, estimates are obtained relative to two successive spaces (levels). Assuming that such estimates hold uniformly on all such pairs, a convergence rate is then proved for the entire multilevel process. For a discussion of convergence proofs up until 1987, we refer the reader to the interesting paper of Parter [77].

The results of Braess and Hackbusch [11] required that $\alpha = 1$ in (2.3) and allowed any $m \geq 1$ for the V- and W-cycles. The W-cycle was treated in Bank and Dupont [6] (MOC) for any positive $\alpha \leq 1$ but required that m be sufficiently large. This kind of condition is quite nebulous and undesirable. The papers Mandel [62] and Bramble and Pasciak [16] (MOC) extended the abovementioned results. They provided uniform estimates for the contraction numbers for the W-cycle with any positive number of smoothing steps and any positive $\alpha \leq 1$. In both of these papers the estimates showed improvement when the number of smoothing steps is increased.

The papers by Bramble and Pasciak [16] (MOC) and Decker, Mandel and Parter [36] generalize [11] in another way. They both prove that, for the V-cycle (so far, always in the nested-inherited case) with any positive $\alpha \leq 1$, the contraction number δ_J is bounded by

$$\delta_J \leq 1 - cJ^{(\alpha-1)/\alpha}$$

for some constant c which is independent of the number of levels J. The bound in [16] (MOC) is

$$\delta_J \leq \frac{MJ^{\frac{1-\alpha}{\alpha}}}{MJ^{\frac{1-\alpha}{\alpha}} + m^\alpha}$$

for some constant M which is independent of J. This estimate explicitly displays an improvement when the number of smoothing steps is increased.

The paper by Bramble and Pasciak [16] (MOC) contains, in addition to the abovementioned results, the first systematic formulation of a V-cycle algorithm in which the number of smoothing steps is increased as we proceed to lower levels. For example, it is shown that if $m(J)$ is chosen to be m, and if $m(k-1) = 2m(k)$, then the contraction number δ for Algorithm MG, in the variational case with any positive $\alpha \leq 1$, satisfies

$$\delta \leq \frac{C}{C + m^\alpha} \tag{6.1}$$

for some constant C which is independent of the number of levels J.

In our example problem, consisting of nested spaces of piecewise linear finite elements, all of the previously mentioned multigrid methods can be shown to have the property that the cost of one multigrid cycle (the computation of B) is proportional to the number of unknowns, i.e., the dimension of M. These results all represent further evidence of the robustness of the multigrid approach.

7. Nonnested or noninherited multigrid

So far, our discussion has concerned entirely the case of nested spaces and inherited forms. It is of interest to lift such restrictions so as to cover some quite natural applications. In the case of three-dimensional problems with piecewise polynomials on simplices, nested spaces may be difficult to construct. Also in finite element methods with nonconforming elements, such as the Morley method for the first biharmonic boundary value problem, there is no known nested multigrid algorithm.

The paper by Bramble, Pasciak and Xu [25] (MOC) considers the more general framework. Nonnested spaces and noninherited forms are allowed. Referring to §2, we note that in the variational case the condition $A_{k-1}(v,w) = A_k(I_k v, I_k w)$ for all $v, w \in M_{k-1}$ is satisfied. It is shown that if this condition is replaced by the weaker condition

$$A_k(I_k v, I_k v) \leq A_{k-1}(v,v) \tag{7.1}$$

for all $v \in M_{k-1}$, then the results of Bramble and Pasciak [16] (MOC) for the V-cycle in the variational case, which were described in §6, carry over to the more general case. That is, the contraction number δ_J satisfies

$$\delta_J \leq \frac{MJ^{\frac{1-\alpha}{\alpha}}}{MJ^{\frac{1-\alpha}{\alpha}} + m^\alpha}$$

for some constant M which is independent of J.

Another result in [25] (MOC) concerns the W-cycle. If the above condition (7.1) is replaced by the even weaker condition

$$A_k(I_k v, I_k v) \leq 2A_{k-1}(v,v) \tag{7.2}$$

for all $v \in M_{k-1}$, then the results of [16] (MOC) for the W-cycle in the variational case, which were described in §6, carry over to the more general case. That is, (6.1) holds as well in the present case.

Perhaps the most novel of the results of Bramble, Pasciak and Xu [25] (MOC) concerns the variable V-cycle. For this algorithm there is no condition of the type of (7.1) or (7.2). It is shown that, under the usual kind of approximation assumption, although the operator B, generated with such a strategy, may not be a contraction, it is always a good preconditioner for A. That is to say, B and A^{-1} are spectrally equivalent, with constants of equivalence independent of the number of levels J. This means that B could naturally be used as a preconditioner in the CG algorithm (cf. [12]).

There are a few other papers dealing with specific problems for which nonnested multigrid methods are convenient. The objective of these papers is the construction of the operators I_k so that the approximation properties needed to apply the original techniques of Bank and Dupont [6] (MOC) are satisfied. We refer specifically to the papers of Brenner [31] (MOC), [32] and [33] (MOC). The paper [31] (MOC) considers nonconforming piecewise linear elements for second-order elliptic equations, while [32] treats the Morley nonconforming method for the first biharmonic boundary value problem. There is no known set of subspaces M_k corresponding to the Morley nonconforming method. The above papers were put into a unified framework in [34]. We remark that the papers [31] (MOC) and [32] prove that the conditions required for the variable V-cycle to produce a uniform preconditioner are satisfied, although this is not explicitly stated. For some other results concerning the multigrid approach to the biharmonic problem we refer the reader to the papers of Peisker [79] and Peisker and Braess [80].

In two papers by S. Zhang [100] (MOC) and [101] (MOC) a similar analysis is carried out. In the paper by Bramble, Pasciak and Xu [25] (MOC) an example was given in the case in which the spaces were defined on grids which were "loosely coupled". It was shown that I_k could be chosen to be the operator which interpolates functions from M_{k-1} into M_k in the two-dimensional case. This result was extended to the three-dimensional case by Scott and Zhang in [85] (MOC). The analysis required an interesting construction which was not trivial.

The thesis of Hanisch [50] contains a formulation and analysis of multigrid methods for the mixed finite element method of Ciarlet and Raviart for the biharmonic equation [35] and for the method of Raviart and Thomas [83] as well as others. In [50] it is shown that the conditions of [25] (MOC) concerning the variable V-cycle are satisfied and in some examples (7.2) is not. His computational examples show that the W-cycle fails unless enough smoothing steps are performed. One case required seven smoothing iterations per cycle, while the variable V-cycle with one smoothing on the finest level always produced a uniform preconditioner.

For some other works on nonnested or noninherited multigrid methods we refer to the papers Goldstein [41] (MOC), Pitkäranta and Saarinen [82] (MOC) and Verfürth [88].

8. V-cycle estimates with weak or no regularity assumptions

We shall return to the variational case since, in this situation, some very strong results have been recently obtained for the V-cycle with one or more smoothing steps. In describing some of these results, I will try to show how they have evolved.

The first result which did not require regularity of the underlying problem appears to be that of Mandel [60]. He proved that the two-level scheme had an associated contraction number less than one which was independent of the dimension of M. His result did not generalize to the multilevel case.

A step toward the multilevel result was taken in Bramble, Pasciak and Xu [24] (MOC), where an additive multilevel preconditioner was introduced and a bound on the condition number was given. Though this bound was not best possible, this work represented a substantial step forward in the development of multigrid theory.

The paper by Bramble, Pasciak, Wang and Xu [20] (MOC) was the first to provide a technique for analyzing so-called multiplicative-type algorithms such as the Schwarz alternating method with more than two subspaces. It was recognized then that the standard multigrid V- and W-cycle algorithms, in the variational case, had this character. That is, the associated error operator $I - BA$ in (2.5) took the form of a product of operators, each of which was defined on all of M. The second paper by Bramble, Pasciak, Wang and Xu [21] (MOC) was the first to make use of this important fact in the analysis of multigrid algorithms. These papers were a departure from the usual two-level analyses heretofore used in all previous multigrid convergence proofs. Already in [20] (MOC), the estimate

$$\delta_J \leq 1 - \frac{1}{CJ^2} \tag{8.1}$$

was given for the V-cycle with no requirement of regularity. By looking specifically at the multigrid algorithm, it was shown in [21] (MOC) that

$$\delta_J \leq 1 - \frac{1}{CJ}.$$

This was a substantial improvement over (8.1).

In 1991, Oswald [76] proved uniform bounds for the upper and lower eigenvalues in the additive preconditioner of Bramble, Pasciak and Xu [24] (MOC). At about the same time, X. Zhang [102] proved that the largest eigenvalue was uniformly bounded. These results were just what was needed to prove a uniform estimate for the V-cycle algorithm. This was done by Bramble and Pasciak in [19] (MOC). The previously mentioned upper bounds of Oswald and Zhang were based on a certain so-called strengthened Cauchy-Schwarz inequality.

Another uniform result for multigrid is given in Bramble and Pasciak [18]. Specifically, it is shown there that if the conditions of the Braess-Hackbusch result (5.1) of [11] are met, but with the relaxed condition that (2.3) need only hold for some positive $\alpha \leq 1$, then (5.1) is satisfied with $m = 1$. No use is made of the strengthened Cauchy-Schwarz inequality. None of the abovementioned new results, with very weak hypotheses, proves that the contraction numbers improve in the presence of an increased number of smoothing steps.

It should be mentioned that all of the works discussed in this section are strong enough that the case of approximations allowing very general local refinements are also included. Except for the very special situation in Yserentant [96] (MOC), and the hierarchical basis method of Yserentant [97] and Bank, Dupont and Yserentant [7], no results covering refinements had been proved previously. The first refinement situation studied, which was closely related to the multigrid approach, was done in the additive multilevel paper of Bramble, Pasciak and Xu [24] (MOC). The abovementioned subsequent papers [21] (MOC), [19] (MOC) and [18] all prove estimates which are valid with or without local refinements with smoothings required only on the refinement region.

These new results all show the robustness of the multigrid approach. In fact, results from the papers by Bramble and Pasciak [19] (MOC) and Bramble, Goldstein and Pasciak [13] can be combined to see that the multigrid V-cycle, with one smoothing step, provides a uniform contraction for irregular problems on piecewise smooth domains, approximated using quadratic forms coming from numerical integration. Local refinements may also be included.

In a recent survey article by Xu [93], he synthesizes some of the abovementioned results and points out connections between multigrid and domain decomposition methods. See also the survey article by Yserentant [98]. The book by the author [12] contains in detail most of the results discussed in this section.

REFERENCES

1. G.P. Astrakhantsev, *An iterative method for solving elliptic net problems*, U.S.S.R. Comput. Math. and Math. Phys. **11** (1971), 171–182.
2. O. Axelsson and I. Gutafsson, *Preconditoning and two-level multigrid methods of arbitrary degree of approximation*, Math. Comp. **40** (1983), 219–242.
3. N.S. Bakhvalov, *On the convergence of a relaxation method with natural constraints on the elliptic operator*, U.S.S.R. Comput. Math. and Math. Phys. **6** (1966), 101–135.
4. R. Bank, *A comparison of two multilevel iterative methods for nonsymmetric and indefinite elliptic finite element equations*, SIAM J. Numer. Anal. **18** (1981), 724–743.
5. R.E. Bank and C.C. Douglas, *Sharp estimates for multigrid rates of convergence with general smoothing and acceleration*, SIAM J. Numer. Anal. **22** (1985), 617–633.
6. R.E. Bank and T. Dupont, *An optimal order process for solving finite element equations*, Math. Comp. **36** (1981), 35–51.
7. R.E. Bank, T.F. Dupont, and H. Yserentant, *The hierarchical basis multigrid method*, Numer. Math. **52** (1988), 427–458.
8. R. Bank, J. Mandel and S. McCormick, *Variational multigrid theory*, Multigrid Methods, Ed. S. McCormick, SIAM, Philadelphia, 1987, pp. 131–178.
9. R.E. Bank and D.J. Rose, *Analysis of a multilevel iterative method for nonlinear finite element equations*, Math. Comp. **39** (1982), 453–465.
10. D. Braess, *The convergence rate of a multigrid method with Gauss-Seidel relaxation for the Poisson equation*, Math. Comp. **42** (1984), 505–519.
11. D. Braess and W. Hackbusch, *A new convergence proof for the multigrid method including the V-cycle*, SIAM J. Numer. Anal. **20** (1983), 967–975.
12. J.H. Bramble, *Multigrid Methods*, Longman Research Notes in Mathematics, London, 1993.
13. J.H. Bramble, C.I. Goldstein and J.E. Pasciak, *Analysis of V-cycle multigrid algorithms for forms defined by numerical quadrature*, Multigrid Methods, Proceedings of the Third Copper Mountain Conference (S. McCormick, ed.), Marcel Dekker, New York, 1992.
14. J.H. Bramble, Z. Leyk, and J.E. Pasciak, *Iterative schemes for nonsymmetric and indefinite elliptic boundary value problems*, Math. Comp. **60** (1993), 1–22.
15. J.H. Bramble, Z. Leyk, and J.E. Pasciak, *The analysis of multigrid algorithms for pseudo-differential operators of order minus one* (1992), Mathematical Sciences Institute Tech. Report 92-28, Cornell University.
16. J.H. Bramble and J.E. Pasciak, *New convergence estimates for multigrid algorithms*, Math. Comp. **49** (1987), 311–329.
17. J.H. Bramble and J.E. Pasciak, *The analysis of smoothers for multigrid algorithms*, Math. Comp. **58** (1992), 467–488.
18. J.H. Bramble and J.E. Pasciak, *Uniform convergence estimates for multigrid V-cycle algorithms with less than full elliptic regularity*, Brookhaven National Laboratory, Tech. Report BNL-47892 (1992).
19. J.H. Bramble and J.E. Pasciak, *New estimates for multigrid algorithms including the V-cycle*, Math. Comp. **60** (1993), 447–471.
20. J.H. Bramble, J.E. Pasciak, J. Wang, and J. Xu, *Convergence estimates for product iterative methods with applications to domain decomposition*, Math. Comp. **57** (1991), 1–21.
21. J.H. Bramble, J.E. Pasciak, J. Wang, and J. Xu, *Convergence estimates for multigrid algorithms without regularity assumptions*, Math. Comp. **57** (1991), 23–45.
22. J.H. Bramble and R. Scott, *Simultaneous approximation in scales of Banach spaces*, Math. Comp. **32** (1978), 947–954.
23. J.H. Bramble, J.E. Pasciak and J. Xu, *The analysis of multigrid algorithms for nonsymmetric and indefinite elliptic problems*, Math. Comp. **51** (1988), 389–414.

24. J.H. Bramble, J.E. Pasciak and J. Xu, *Parallel multilevel preconditioners*, Math. Comp. **55** (1990), 1–22.
25. J.H. Bramble, J.E. Pasciak and J. Xu, *The analysis of multigrid algorithms with nonnested spaces or noninherited quadratic forms*, Math. Comp. **56** (1991), 1–34.
26. J.H. Bramble and J. Xu, *Some estimates for weighted L^2 projections*, Math. Comp. **56** (1991), 463–476.
27. J.H. Bramble and M. Zlámal, *Triangular elements in the finite element method*, Math. Comp. **24** (1970), 809–820.
28. A. Brandt, *Multi-level adaptive technique (MLAT) for fast numerical solution to boundary value problems*, Proceedings of the 3rd International Conference on Numerical Methods in Fluid Mechanics, Springer-Verlag, Berlin and New York, 1973, pp. 82–89.
29. A. Brandt, *Multi-level adaptive solutions to boundary-value problems*, Math. Comp. **31** (1977), 333–390.
30. A. Brandt, *Algebraic multigrid theory: the symmetric case*, Appl. Math. Comput. **19** (1986), 23–56.
31. S.C. Brenner, *An optimal-order multigrid method for P1 nonconforming finite elements*, Math. Comp. **52** (1989), 1–16.
32. S.C. Brenner, *An optimal-order nonconforming multigrid method for the biharmonic equation*, SIAM J. Numer. Anal. **26** (1989), 1124–1138.
33. S.C. Brenner, *A nonconforming multigrid method for the stationary Stokes equations*, Math. Comp. **55** (1990), 411–437.
34. S.C. Brenner, *Multigrid methods for nonconforming finite elements*, Multigrid Methods, Proceedings of the Fourth Copper Mountain Conference (J. Mandel, S.F. McCormick, J.E. Dendy, Jr., C. Farhat, G. Londsdale, S.V. Parter, J.W. Ruge and K. Stüben, eds.), SIAM, Philadelphia, 1989, pp. 54–65.
35. P.G. Ciarlet and P.-A. Raviart, *A mixed finite element method for the biharmonic equation*, Mathematical Aspects of Finite Elements in Partial Differential Equations (C. de Boor, ed.), Academic Press, New York, 1974, pp. 125–145.
36. N. Decker, J. Mandel and S. Parter, *On the role of regularity in multigrid methods*, Multigrid Methods, Proceedings of the Third Copper Mountain Conference (S. McCormick, ed.), Marcel Dekker, New York, 1988.
37. C.C. Douglas, *Multi-grid algorithms with applications to elliptic boundary-value problems*, SIAM J. Numer. Anal. **21** (1984), 236–254.
38. J. Douglas, Jr., *On the numerical solution of heat conduction problems in two or three space dimensions*, Trans. Amer. Math. Soc. **82** (1956), 421–439.
39. R.P. Fedorenko, *A relaxation method for solving elliptic difference equations*, U.S.S.R. Comput. Math. and Math. Phys. **1, no. 4** (1962), 1092–1096.
40. R.P. Fedorenko, *The speed of convergence of one iterative process*, U.S.S.R. Comput. Math. and Math. Phys. **4, no. 3** (1964), 227–235.
41. C.I. Goldstein, *Multigrid analysis of finite element methods with numerical integration*, Math. Comp. **56** (1991), 409–436.
42. G.H. Golub and D.P. O'Leary, *Some history of the conjugate gradient and Lanczos algorithms: 1948-1976*, SIAM Rev. **31** (1989), 50–102.
43. M. Griebel and P. Oswald, *Remarks on the abstract theory of additive and multiplicative Schwarz algorithms*, Preprint.
44. W. Hackbusch, *On the convergence of a multigrid iteration applied to finite element equations*, Report 77-8, Universität zu Köln, 1977.
45. W. Hackbusch, *On the multi-grid method applied to difference equations*, Computing **20** (1978), 291–306.
46. W. Hackbusch, *Convergence of multi-grid iterations applied to difference equations*, Math. Comp. **34** (1980), 425–440.
47. W. Hackbusch, *Survey of convergence proofs for multi-grid iterations*, Special Topics of Applied Mathematics (J. Frehse, D. Pallaschke, U. Trottenberg, eds.), North-Holland, Amsterdam, 1980, pp. 152–164.
48. W. Hackbusch, *On the convergence of multi-grid iterations*, Beiträge Numer. Math. **9** (1980), 213–239.

49. W. Hackbusch, *Multi-Grid Methods and Applications*, Springer-Verlag, New York, 1985.
50. M.R. Hanisch, *Multigrid preconditioning for mixed finite element methods*, Cornell University (Thesis) (1991).
51. M.R. Hanisch, *Multigrid preconditioning for the biharmonic Dirichlet problem*, SIAM J. Numer. Anal. **30** (1993), 184–214.
52. P.W. Hemker and H. Schippers, *Multiple grid methods for the solution of Fredholm integral equations of the second kind*, Math. Comp. **36** (1981), 215–232.
53. M.R. Hestenes and E.S. Stiefel, *Methods of conjugate gradients for solving linear systems*, J. Res. Nat. Bur. Standards **49** (1952), 409–436.
54. L. Kronsö, *A note on the "nested iterations" method*, BIT **15** (1975), 107–110.
55. M. Kočvara and J. Mandel, *A multigrid method for three-dimensional elasticity and algebraic convergence estimates*, Appl. Math. Comput. **23** (1987), 121–135.
56. C. Lanczos, *Solution of systems of linear equations by minimized iterations*, J. Res. Nat. Bur. Standards **49** (1952), 33–53.
57. J.-F. Maitre and F. Musy, *Multigrid methods: convergence theory in a variational framework*, SIAM J. Numer. Anal. **21** (1984), 657–671.
58. J.-F. Maitre and F. Musy, *Algebraic formalisation of the multigrid method in the symmetric and positive definite case — a convergence estimation for the V-cycle*, Multigrid Methods for Integral and Differential Equations (D.J. Paddon and H. Holstien, eds.), Clarendon Press, Oxford, 1985.
59. J.-F. Maitre and F. Musy, *Multigrid methods for symmetric variational problems: a general theory and convergence estimates for usual smoothers*, Appl. Math. Comput. **21** (1987), 21–43.
60. J. Mandel, *Étude algébrique d'une méthode multigrille pour quelques problèmes de frontière libre*, C.R. Acad. Sci. Paris Sér. I. Math. **298** (1984), 469–472.
61. J. Mandel, *Multigrid convergence for nonsymmetric, indefinite variational problems and one smoothing step*, Appl. Math. Comput. **19** (1986), 201–216.
62. J. Mandel, *Algebraic study of multigrid methods for symmetric, definite problems*, Appl. Math. Comput. **25** (1988), 39–56.
63. J. Mandel, S.F. McCormick and J. Ruge, *An algebraic theory for multigrid methods for variational problems*, SIAM J. Numer. Anal. **25** (1988), 91–110.
64. S.F. McCormick, *Multigrid methods for variational problems: further results*, SIAM J. Numer. Anal. **21** (1984), 255–263.
65. S.F. McCormick, *Multigrid methods for variational problems: general theory for the V-cycle*, SIAM J. Numer. Anal. **22** (1985), 634–643.
66. S.F. McCormick and J. Ruge, *Multigrid methods for variational problems*, SIAM J. Numer. Anal. **19** (1982), 924–929.
67. S.F. McCormick and J. Ruge, *Unigrid for multigrid simulation*, Math. Comp. **41** (1983), 43–62.
68. S.F. McCormick and J. Thomas, *The fast adaptive composite grid (FAC) method for elliptic equations*, Math. Comp. **46** (1986), 439–456.
69. L.S.D. Morley, *The triangular equilibrium element in the solution of plate bending problems*, Aeronaut. Quart. **19** (1968), 149–169.
70. R.A. Nicolaides, *On multiple grid and related techniques for solving discrete elliptic systems*, J. Computational Phys. **19** (1975), 418–431.
71. R.A. Nicolaides, *On the l^2 convergence of an algorithm for solving finite element equations*, Math. Comp. **31** (1977), 892–906.
72. R.A. Nicolaides, *On the observed rate of convergence of an iterative method applied to a model elliptic difference equation*, Math. Comp. **32** (1978), 127–133.
73. R.A. Nicolaides, *On multigrid convergence in the indefinite case*, Math. Comp. **32** (1978), 1082–1086.
74. R.A. Nicolaides, *On some theoretical and practical aspects of multigrid methods*, Math. Comp. **33** (1979), 933–952.
75. J. Nitsche, *Ein Kriterium für die Quasi-Optimalität des Ritzschen Verfahrens*, Numer. Math. **11** (1968), 346–348.

76. P. Oswald, *On discrete norm estimates related to multilevel preconditioners in the finite element method*, Preprint.
77. S.V. Parter, *Remarks on multigrid convergence theorems*, Appl. Math. Comput. **23** (1987), 103–120.
78. D.W. Peaceman and H.H. Rachford, Jr., *The numerical solution of parabolic and elliptic differential equations*, SIAM J. Appl. Math. **3** (1955), 28–41.
79. P. Peisker, *A multilevel algorithm for the biharmonic problem*, Numer. Math. **46** (1985), 623–634.
80. P. Peisker and D. Braess, *A conjugate gradient method and a multigrid algorithm for Morley's finite element approximation of the biharmonic equation*, Numer. Math. **50** (1987), 567–586.
81. T. von Petersdorff and E.P. Stephan, *On the convergence of the multigrid method for a hypersingular integral equation of the first kind*, Numer. Math. **57** (1990), 379–391.
82. J. Pitkäranta and T. Saarinen, *A multigrid version of a simple finite element method for the Stokes problem*, Math. Comp. **45** (1985), 1–14.
83. P.-A. Raviart and J.M. Thomas, *A mixed finite element method for 2-nd order elliptic problems*, Mathematical Aspects of Finite Element Methods, Lecture Notes in Mathematics, vol. 606 (I. Galligani and E. Magenes, eds.), Springer-Verlag, New York, 1977, pp. 292–315.
84. A.H. Schatz, *An observation concerning Ritz-Galerkin methods with indefinite bilinear forms*, Math. Comp. **28** (1974), 959–962.
85. R. Scott and S. Zhang, *Higher-dimensional nonnested multigrid methods*, Math. Comp. **58** (1992), 457–466.
86. S. Spekreijse, *Multigrid solution of monotone second-order discretizations of hyperbolic conservation laws*, Math. Comp. **49** (1987), 135–155.
87. R.S. Varga, *Matrix Iterative Analysis*, Prentice-Hall, Englewood Cliffs, NJ, 1962.
88. R. Verfürth, *A multilevel algorithm for mixed problems*, SIAM J. Numer. Anal. **21** (1984), 264–284.
89. E.L. Wachspress, *Extended application of alternating direction implicit iteration model problem theory*, SIAM J. Appl. Math. **11** (1963), 994–1016.
90. J. Wang, *Convergence analysis without regularity assumptions for multigrid algorithms based on SOR smoothing*, Preprint.
91. J. Xu, *Convergence estimates for some multigrid algorithms*, Proc. 1989 Houston Domain Decomp. Methods Conf. (oral presentation).
92. J. Xu, *Theory of Multilevel Methods*, Penn. State Univ., Dept. Math. Rep AM-48, 1989.
93. J. Xu, *Iterative methods by space decomposition and subspace correction*, SIAM Rev. **34** (1992), 581–613.
94. J. Xu and X. Cai, *A preconditioned GMRES method for nonsymmetric and indefinite problems*, Math. Comp. **59** (1992), 311–319.
95. D.M. Young, *Iterative methods for solving partial difference equations of elliptic type*, Trans. Amer. Math. Soc. **76** (1954), 92–111.
96. H. Yserentant, *The convergence of multi-level methods for solving finite-element equations in the presence of singularities*, Math. Comp. **47** (1986), 399–409.
97. H. Yserentant, *On the multi-level splitting of finite element spaces*, Numer. Math. **49** (1986), 379–412.
98. H. Yserentant, *Old and new convergence proofs for multigrid methods*, Acta Numerica 1993, (A. Iserles, ed.), Cambridge University Press, 1993, pp. 285–326.
99. S. Zhang, *Multi-level Iterative Techniques*, Penn. State Univ. (Thesis), Math. Res. Rep. 88020, 1988.
100. S. Zhang, *Optimal-order nonnested multigrid methods for solving finite element equations I: On quasi-uniform meshes*, Math. Comp. **55** (1990), 23–36.
101. S. Zhang, *Optimal-order nonnested multigrid methods for solving finite element equations II: On non-quasi-uniform meshes*, Math. Comp. **55** (1990), 439–450.
102. X. Zhang, *Multilevel Schwarz methods*, Numer. Math. **63** (1992), 521–539.

DEPARTMENT OF MATHEMATICS, CORNELL UNIVERSITY, ITHACA, NY 14853
E-mail address: bramble@math.cornell.edu

Proceedings of Symposia in Applied Mathematics
Volume **48**, 1994

AN INTRODUCTION TO INVERSE PROBLEMS

MARGARET CHENEY

ABSTRACT. This paper gives an introduction to the field of inverse problems. It gives examples of some of the different types of inverse problems, with emphasis on those that involve partial differential equations. For certain of the nonlinear, multidimensional ones, a summary of some of the known reconstruction methods is given.

1. INTRODUCTION

The goal of this paper is to give an introduction to the field of inverse problems. This is a field with applications in many areas, such as geophysics, medical imaging, nondestructive testing, and remote sensing. To some extent the researchers in each area have developed their own nomenclature and their own techniques, many of which overlap. Because the field is so vast, it would be impossible to give a proper survey in a short paper. Instead, this paper will discuss some ways to classify inverse problems according to their mathematical structure. The eventual focus will be on the nonlinear, multidimensional inverse problems that arise in partial differential equations. For this type, the paper includes a summary of some of the known reconstruction methods. This paper represents a personal view in the sense that the problems and methods discussed are those that I happen to know and like. I apologize to those who find that I have omitted their favorite works.

2. EXAMPLES OF LINEAR INVERSE PROBLEMS

Inverse problems can be classified as nonlinear and linear. The simplest example of a linear problem that could be called an inverse problem is solving a system of linear equations. If we consider the system $Ax = b$, then the forward or direct problem is to use A and x to find b. An inverse problem is to use A and b to find x. This simple case already exhibits some of the difficulties that appear in solving inverse problems: first, if A is singular, either there may be many solutions, solutions may exist only for certain b, or there may be no solution at all. The problem can be either underdetermined or overdetermined. Second, if A is ill-conditioned, then the inverse problem is ill-posed: small errors in b can give rise to large errors in the solution x [EG, G].

1991 *Mathematics Subject Classification.* Primary 35R30.

This work was supported in part by the Office of Naval Research grant N-00014-93-1-0048 and by the National Science Foundation grant GER-9023630.

This paper is in final form and no version of it will be submitted for publication elsewhere.

© 1994 American Mathematical Society
0160-7634/94 $1.00 + $.25 per page

Solving a system of linear equations is not usually thought of as an inverse problem, even though the numerical treatment of linear inverse problems can be reduced to this case. Real-world inverse problems generally arise in trying to obtain information about an inaccessible region of space from measurements made in the accessible region [Sb].

An example of such a problem is the problem of X-ray Computed Tomography, in which one would like to find the internal density of a body from measurements of X-rays that are passed through it. From the measurements, one obtains the integral of the density along each line. Lines in two dimensions can be parametrized by two variables, and similarly, the density of a two-dimensional body also depends on two variables. Thus the inverse problem is to reconstruct a function of two variables from a function of two variables. It is a linear problem. One method for solving it, called the filtered backprojection method, depends on inversion formulas for the Radon transform [Dn] and has a nice geometrical interpretation [Dv, Nt]. Another method, called the Algebraic Reconstruction Technique or ART, is to expand everything in some basis and reduce the problem to that of solving a system of linear equations [Dn, Nt].

More often, problems of obtaining information about inaccessible regions involve partial differential equations. An example of a linear inverse problem arises in electrocardiology. By pasting electrodes to the surface of the skin, physicians can measure the electric potential on the surface of the body. From this, they would like to know what currents are flowing inside the heart. However, the electric potential on the surface of the body is a single function of two variables, whereas the current inside the heart is a vector function (i.e., three functions) of three variables. This is apparently a hopeless problem. To obtain a tractable problem, it is necessary to reduce the number of unknowns in some way. One way to do this is to give up the hope of finding the currents inside the heart, and instead look for the electric potential on the surface of the heart. A simple model is the following. Here u is the electric potential, σ is the electric conductivity, B is the body, ν is the outward unit normal vector, and H is the heart; we assume that the conductivity is known:

$$\text{(conservation of current)} \quad \nabla \cdot \sigma \nabla u = 0 \quad \text{in } B \setminus H,$$

$$\text{(no applied current)} \quad \sigma \frac{\partial u}{\partial \nu}\Big|_{\partial B} = 0,$$

$$\text{(measured)} \quad u|_{\partial B} = \text{given}.$$

We would like to find u on ∂H. Here the partial differential equation is elliptic, but the information we are given is Cauchy data. This is a classic example of an ill-posed problem. Methods for solving this problem have been discussed in [CF]. Because it is linear, the corresponding numerical problem can be reduced to the problem of solving a linear system.

However, many of the important inverse problems are nonlinear. Let's look at some examples before discussing reconstruction algorithms.

3. Examples of Nonlinear Inverse Problems

A nonlinear inverse problem arises in the problem of finding the electrical conductivity in the above example. This is called Electrical Impedance Imaging, Electrical

Impedance Tomography or Electric Current Computed Tomography. Here we apply currents to the surface of a body, measure the voltages that result, and try to reconstruct the electrical conductivity in the interior. A simple mathematical model is the following. As before, we use u for the electrical potential, σ for the electric conductivity, and B for the body. We denote the applied current density by j and the measured voltage by v. Then we have

$$\nabla \cdot \sigma \nabla u = 0 \quad \text{in the body } B, \tag{3.1}$$

$$\sigma \frac{\partial u}{\partial \nu}|_{\partial B} = j, \tag{3.2}$$

$$u|_{\partial B} = v. \tag{3.3}$$

The forward problem is to find v, given σ and j. This is a linear problem. The inverse problem is to find σ in the interior, given all pairs (j, v). This problem, which we refer to as an elliptic inverse problem, is nonlinear. We note that for a three-dimensional problem, σ depends on three variables, whereas the map from functions on the surface to functions on the surface can be thought of as being determined by an integral kernel, which is a function of four variables. However, not every function of four variables can arise as the kernel of a boundary map, so we now also have the problem of characterizing the functions of four variables that are admissible as data. The inverse problem of reconstructing σ from the boundary map arises in the inverse problem of electrocardiology, and also has potential applications to medical diagnosis, nondestructive testing, and process control.

Another example of a nonlinear inverse problem is the seismic prospecting or the ultrasound inverse problem. Here we want information about the speed of acoustic wave propagation $1/n(x, y, z)$. The excess pressure U satisfies

$$(\nabla^2 - n^2(x, y, z)\partial_t^2)U = 0. \tag{3.4}$$

We apply a known impulse

$$\partial_z U|_{z=0} = F(x, y, t) \tag{3.5}$$

and measure

$$U|_{z=0} = G(x, y, t). \tag{3.6}$$

In order to have a well-posed forward problem, we also need the conditions

$$U \quad \text{is initially quiescent}$$

and

$$U \text{ is outgoing.}$$

The forward problem is to find G, given knowledge of $F(x, y, t)$ and $n^2(x, y, z)$. The inverse problem is to find $n^2(x, y, z)$, given $F(x, y, t)$ and $G(x, y, t)$ on the surface. We note that the measured data depend on three variables (x, y, and t) and the unknown sound speed is also a function of three variables. We refer to this type of inverse problem, in which we have a nonspatial parameter t, as a hyperbolic inverse problem. This nonlinear inverse problem has applications to nondestructive testing, seismic prospecting, and to medical imaging.

For geophysical and nondestructive testing applications, a more accurate mathematical model would use the full equations of linear elasticity. There are similar nonlinear inverse problems involving electromagnetic waves, in which one uses Maxwell's equations; such problems arise in the use of radar for remote sensing, in the "eddy current" method of detecting flaws in metals, and in medical imaging using microwaves. Other, related inverse problems arise in the case when the medium is constant except for a perturbation that is far away from the measurement location. Then one measures the far-field asymptotics of the scattered wave, and the inverse problem is called an inverse scattering problem. These problems arise, for example, in the radar identification of aircraft and in the everyday problem of vision, in which light scattered into our eyes is used by our brains to give us a map of the three-dimensional world we live in [Sa].

The acoustic inverse problem can also be studied in the frequency domain rather than the time domain. Here one assumes the applied field is time-harmonic:

$$U(x,y,z,t) = u(x,y,z,k)e^{ikt},$$

$$F(x,y,t) = f(x,y,k)e^{ikt},$$

$$G(x,y,t) = g(x,y,k)e^{ikt}.$$

The hyperbolic wave equation then becomes an elliptic equation:

$$[\nabla^2 + n^2k^2]u = 0, \tag{3.7}$$

$$\partial_z u|_{z=0} = f, \tag{3.8}$$

$$u|_{z=0} = g. \tag{3.9}$$

One inverse problem would be to find $n^2(x,y,z)$ from knowledge of $f(x,y,k)$ and $g(x,y,k)$ for all k. Here we are still trying to determine a function of three variables from a function of three variables. Because the data depend on the "extra" parameter k, we still call this a hyperbolic inverse problem. We could, however, attempt to determine n^2 from data at only a single frequency. This problem, which we call an elliptic inverse problem, is underdetermined: we have only a function of two variables, from which we want to determine a function of three variables. To make up for the lack of temporal information, we need more spatial information: we instead assume that we know all pairs (f, g), or equivalently, we assume we know the map $R : f \mapsto g$, which is a function of four variables.

These nonlinear inverse problems are all multidimensional problems, and they are all ill-posed. They look very similar either to boundary value problems for hyperbolic equations or to Cauchy problems for elliptic equations.

4. Special Features of One-dimensional Problems

For the hyperbolic inverse problems, the initial work was done for the special case in which the medium parameters depend only on one dimension. This case is much less ill-posed than the general multidimensional case, because a simple interchange of the space and time variables converts a boundary value problem to a Cauchy problem. Moreover, a one-dimensional hyperbolic problem with a variable propagation speed can be converted to one with constant speed, by the following transformation to travel-time coordinates.

For the one-dimensional acoustic scattering problem,

$$(c^2(z)\partial_z^2 - \partial_t^2)w(t,z) = 0, \tag{4.1}$$

the change of variables is

$$x = \int_0^z \frac{dz'}{c(z')},$$

which transforms (4.1) to

$$(\partial_x^2 - \partial_t^2 - V(x))u(t,x) = 0, \tag{4.2}$$

where V is related to c by

$$V(x) = \partial_x \log c(z(x)).$$

Inverse problems for (4.2) are easier to study because the characteristics of (4.2) are straight lines, whereas the characteristics of (4.1) depend on the unknown $c^2(z)$.

The one-dimensional problems have been thoroughly studied (see, for example, [CS, SPSH, CKNS] and references therein). However, because most of the real-world problems are multidimensional, this survey will concentrate on methods that apply to the multidimensional problems.

5. Methods for Multidimensional Nonlinear Inverse Problems

The methods that apply to the above multidimensional nonlinear inverse problems fall into several categories: 1) optimization; 2) linearization or linearization plus iteration; 3) integral equation methods; and 4) layer-stripping methods.

These methods will be illustrated with two examples, the hyperbolic frequency-domain acoustic problem (3.7)–(3.9) and the elliptic impedance imaging problem (3.1)–(3.3).

5.1. Optimization. Optimization methods generally try to minimize some functional of the form

$$\text{Error}[(\text{predicted}) — (\text{measured})] + \alpha(\text{regularization term}).$$

The methods differ chiefly in what error function is used, what sort of regularization is used, and the method used for chosing the regularization parameter α. Optimization methods have the advantage that they are robust and usually work with real data. For many inverse problems, they are the only methods that have actually yielded numerical results. However, they often have the disadvantage that they can converge to answers that are not solutions; this happens when the functional being minimized has multiple local minima. They are also generally expensive because most (but not all [CK]) require solving the forward problem many times. Such a large amount of research has been done on optimization methods that the topic requires its own survey paper; I won't attempt such a survey here.

5.2. Linearization. The second way to try to solve a nonlinear inverse problem is to linearize it. Essentially, one assumes a known solution, and expands everything in sight about this solution. Linearizing the inverse problem usually amounts to assuming that the interior field arises from a known medium whereas the measurements are due to perturbations in the medium. In the context of quantum-mechanical scattering, replacing the true field by the incident field is often called the Born approximation [N]; replacing the true field by another known field is called the distorted wave Born approximation. Sometimes when this type of approximation is made, the resulting linear inverse problem can be solved easily, even by an explicit formula.

For the impedance imaging problem, linearization works like this. We assume a known conductivity σ_0 and a corresponding electric potential u_0:

$$\nabla \cdot \sigma_0 \nabla u_0 = 0, \tag{5.1}$$

$$\sigma_0 \partial_\nu u_0 |_{\partial B} = j. \tag{5.2}$$

We expand u and σ about u_0 and σ_0:

$$u = u_0 + \delta u,$$

$$\sigma = \sigma_0 + \delta\sigma.$$

We substitute these into (3.1) and (3.2) to obtain

$$\nabla \cdot \sigma_0 \nabla \delta u = -\nabla \cdot \delta\sigma \nabla u_0 + O(\delta^2), \tag{5.3}$$

$$\sigma_0 \partial_\nu \delta u|_{\partial B} = -\delta\sigma \partial_\nu u_0 |_{\partial B} + O(\delta^2). \tag{5.4}$$

The linearizing approximation is to drop the $O(\delta^2)$ terms. The unknowns $\delta\sigma$ and δu are then no longer multiplied together. To see how we can solve this linearized problem, we multiply (5.3) by $\tilde{u}_0$, which satisfies (5.1) and (5.2) but with j replaced by a current density $\tilde{j}$. We then integrate over B and use Green's theorem to obtain

$$\int_B \delta\sigma \nabla \tilde{u}_0 \cdot \nabla u_0 \approx \int_{\partial B} \delta u \tilde{j}.$$

Here the right side can be computed from the data, and everything on the left side is known except $\delta\sigma$, so we expect to be able to solve this equation. This is the method published in 1980 by Calderón [Ca], who also showed how to choose u_0 and $\tilde{u}_0$ so that the left side is a Fourier transform of $\delta\sigma$.

Calderón's scheme has not been used for computation. In 1984, Barber and Brown [BB] implemented a backprojection method that gives an approximate solution of the linearized inverse problem: they assumed that the interior electric potential corresponds to a constant conductivity distribution (the linearizing assumption) and backprojected the data along the equipotential lines. This procedure is very sensitive to perturbations in the measurement positions, which means that it is only successful in making images of the changes in the interior conductivity distribution. Our group at Rensselaer is currently using a method based on optimization and linearization to make images [Si, CINSG]. In the 1989 *Mathematics of Computation* issue dedicated to their father, David and Eli Isaacson [II] gave some explicit examples of exact solutions to the linearized inverse problem. This work showed that solutions to the linearized problem have more discontinuities than the true solution. Numerical analysis of the linearized problem has been carried out by a number of researchers [AS, IC, DS, SV].

The technique of linearization can also be used to study nonelliptic inverse problems [BlC, SS1, SS2, Sy2]. For example, a *Mathematics of Computation* paper that uses linearization is the 1988 work of Liu and Chen [LC] that proves convergence of a "linearization plus iteration" algorithm for an inverse problem for a one-dimensional diffusion equation.

The "linearization plus iteration" method, say, for the acoustic inverse problem (3.7), goes like this: 1) guess n^2; 2) solve the forward problem to obtain u; 3) find n^2 from knowledge of u and the data, generally by solving the linearized problem; 4) repeat, starting with step 2. In many cases step 3) can be done by means of an explicit formula [DT, St]; in the multidimensional case, for example, we have the formula [RC]

$$n^2(p) = \int_{\infty}^{\infty} \int_{S^2} (\text{scattering data})\ u(k, \hat{e}, p) d\hat{e} dk,$$

where $p = (x, y, z)$ and where the integration is over all incident angles $\hat{e}$ and frequencies k. In the 1989 issue of *Mathematics of Computation* dedicated to Eugene Isaacson, Bayliss, Li, and Morawetz [BLM] use a related method for reconstructing a medium from backscattering data. Although the methods of [RC] and [BLM] apply in principle to the multidimensional case, they have been tested only in the one-dimensional case. One reason for this is that these methods are expensive, because the forward problem must be solved many times. Another disadvantage is that in general the conditions for convergence of this type of iteration are unknown.

5.3. Integral equation methods. A third class of methods is those based on integral equations. These were first developed for the one-dimensional Schrödinger equation

$$(\partial_x^2 + k^2 - V(x))\psi(k, x) = 0,$$

which governs quantum-mechanical scattering; ψ is the probability density, $V(x)$ characterizes the interaction between particles, and k^2 is the energy of the incident particle. The relevant inverse problem is to find $V(x)$ from scattering data. Physicists have studied the inverse problem in order to obtain information about the interaction between elementary particles. In 1951 Gel'fand and Levitan [GL] and in 1955 Marchenko [M] developed methods for solving the inverse problem based on integral equations. These methods were turned into numerical algorithms by a number of workers in the late 1970s and 1980s; some review papers are [BBu] and [BK]. These methods essentially deal with the equation

$$[\partial_x^2 - \partial_t^2 - V(x)]u(t, x) = 0,$$

which is related to the Schrödinger equation by a Fourier transform. The hyperbolicity of this equation is used in some way by almost all the numerical methods; they are closely related to the layer-stripping schemes that will be discussed later. Reconstruction methods have also been developed for various other one-dimensional equations that arise in the study of solitons.

Since the work of Marchenko and Gel'fand and Levitan, researchers have been trying to extend these methods to higher dimensions. This has proved difficult, but progress was made by Kay and Moses [KM], Faddeev [F1, F2], Newton [N1, N2, N3], Melin [Me]. These methods met with varying degrees of theoretical success, but from a numerical point of view they are all similar in that they require solving a three-dimensional integral equation with the point p in space as a parameter; one must solve enough of these equations so that the results can be differentiated with respect to p. For a three-dimensional problem, a naive algorithm would require $O(N^{12})$ operations [YL].

There are a few other integral equation approaches to inverse scattering problems. One suggested by Prosser [P] involves solving convolution equations by iteration. Another is the $\bar{\partial}$-method, which uses analyticity properties and the Cauchy integral formula in several complex variables [BC, HN, NA]. As far as I know, neither method has been implemented numerically or analyzed for stability.

Integral equation methods have also been developed for fixed-frequency problems such as the impedance imaging problem. A breakthrough was the 1986 work of Sylvester and Uhlmann [SU], who introduced a new complex parameter ζ into the problem, defined certain special solutions $u(\zeta, p)$, and worked out their large-ζ behavior. This amounts to a sort of complex geometrical optics. Sylvester and Uhlmann used this technique to prove uniqueness of the inverse boundary value problem; in 1988 Nachman [Na1] combined the Sylvester-Uhlmann work with the d-bar ideas to develop a constructive method for solving the inverse problem. A two-dimensional version of this method [Na2] is as follows. First, starting with the boundary data, one solves a boundary integral equation for each ζ to obtain the boundary values of the special Sylvester-Uhlmann solutions $u(\zeta, p)$. These boundary values can then be used in an integral formula to compute an intermediate quantity for every ζ; this result is then used in another integral formula to reconstruct the conductivity for every p. This method has not yet been tried numerically, nor has anyone analyzed its stability.

5.4. Layer-stripping. The final class of methods is layer-stripping methods. The idea behind these methods is first to find the medium parameters on the top surface, then mathematically strip away the top layer, and repeat. These methods have the advantage that they seem computationally tractable. However, they tend to be unstable, which is to be expected in light of the fact that the multidimensional problems are ill-posed.

Layer-stripping methods in the one-dimensional case are closely related to fast methods of solving the Gel'fand-Levitan and Marchenko equations [BBu, BK]. One-dimensional layer-stripping methods based on invariant imbedding were developed by the group at Ames Labs in Iowa [CDK] and extended by them to one-dimensional dissipative media [KK] and viscoelastic media [CKa]. These methods, which depend on splitting the waves into upgoing and downgoing components, have been extended by Weston [W] to the three-dimensional case. Another approach was taken by Yagle and Levy [YL], who considered the hyperbolic equation

$$(\nabla^2 - \partial_t^2 - V)u = 0.$$

They write this equation as a system in which $\partial_y^2 - \partial_t^2$ has been split into an upgoing and downgoing part:

$$(\partial_y + \partial_t)w = (V - \partial_x^2)u, \tag{5.5}$$

$$(\partial_y - \partial_t)u = w. \tag{5.6}$$

They combine this with a geometrical-optics expansion that takes advantage of constant propagation speed:

$$u(x, y, t) = \delta(t - y) + p(x, y, t)H(t - y), \tag{5.7}$$

$$w(x, y, t) = q(x, y, t)H(t - y). \tag{5.8}$$

Here δ denotes the Dirac delta function and H the Heaviside function that is one when its argument is positive and zero otherwise. Substituting (5.7) and (5.8) into (5.5) and (5.6) gives rise to the system

$$(\partial_y + \partial_t)p = q,$$
$$(\partial_y - \partial_t)q = (V - \partial_x^2)p,$$
$$V(x, y) = -2q(x, y, t = y),$$

which can be solved by a layer-by-layer marching scheme. It is unstable because of the differentiation with respect to x, but can be made stable by filtering out the high frequencies in the x-direction [C]. This algorithm has been tested by Yagle and Raadhakrishnan [YR].

For the impedance imaging problem, my colleagues and I [SCII] have developed a somewhat different layer-stripping algorithm, which I will illustrate in the case when the body is the lower half-plane. We begin with the boundary map

$$R(y) : \sigma \partial_y u|_y \mapsto u|_y,$$

which is known at the surface $y = 0$. We first use this information to find the medium parameters on the surface by applying $R(0)$ to highly oscillatory current patterns. In particular, we use the formula

$$\frac{1}{\sigma(x, y)} = \lim_{n \to \infty} \frac{1}{n} e^{-inx} R(y) e^{inx}.$$

The idea here is that very little energy from a highly oscillatory pattern penetrates far into the body, so the corresponding measured voltage pattern gives information about the medium parameters near the boundary. The second part of the layer-stripping process is to synthesize the data we would have measured if the top layer were absent. In this case, we find a differential equation that tells us how the boundary map R depends on the depth y:

$$\partial_y R = \frac{1}{\sigma} + R \partial_x \sigma \partial_x R.$$

We use a simple Euler scheme to step into the medium. To stabilize the algorithm, we smooth in the x-direction as we proceed inward.

These ideas may work for many other problems as well.

6. Testing of Reconstruction Algorithms

Care is needed when testing reconstruction algorithms with synthetic data. In particular, the algorithm that generates the data should have no connection to the reconstruction algorithm [CK]. To expand upon this principle, Rainer Kress has defined the following "inverse crimes":

(1) using the same algorithm to generate data as is used to solve the inverse problem. An example is using a linearized algorithm to generate data and then using a linearized reconstruction algorithm;
(2) using the same mesh to generate the data as is used in the reconstruction algorithm;
(3) placing objects or discontinuities so that they fit precisely on the mesh used in the reconstruction algorithm;
(4) using inhomogeneities that are linear combinations of the basis functions used in the reconstruction algorithm.

The ultimate test of a reconstruction algorithm is whether it works on real data.

References

AS. A. Allers and F. Santosa, *Stability and resolution analysis of a linearized problem in electrical impedance tomography*, Inverse Problems **7** (1991), 515–533.

BB. D.C. Barber and B.H. Brown, *Applied potential tomography*, J. Phys. E. Sci. Instrum. **17** (1984), 723–733.

BLM. A. Bayliss, Y. Li, and C.S. Morawetz, *Scattering by a potential using hyperbolic methods*, Math. Comp. **52** (1989), 321–338.

BBu. K.P. Bube and R. Burridge, *The one-dimensional inverse problem of reflection seismology*, SIAM Rev. **25** (1983), 497–559.

BC. R. Beals and R.R. Coifman, *Multidimensional inverse scattering and nonlinear PDE*, Proc. Symp. Pure Math. **43** (1985), 45–70; *Linear spectral problems, nonlinear equations and the $\bar{\partial}$-method*, Inverse Problems **5** (1988), 87–130.

BlC. N. Bleistein and J.K. Cohen, *Velocity inversion procedure for acoustic waves*, Geophysics **44** (1979), 1077–1087.

BK. A.M. Bruckstein and T. Kailath, *Inverse scattering for discrete transmission-line models*, SIAM Rev. **29** (1987), 359–389.

Ca. A.P. Calderón, *On an inverse boundary value problem*, Seminar on Numerical Analysis and its Applications, ed. W.H. Meyer and M.A. Raupp, Brazilian Mathematics Society, Rio de Janeiro, 1980, pp. 65–73.

C. M. Cheney, *Stability analysis of the Yagle-Levy multidimensional inverse scattering algorithm*, J. Math. Phys. **31** (1990), 2141–2144.

CDK. J.P. Corones, M.E. Davison, and R.J. Krueger, *Direct and inverse scattering in the time domain via invariant imbedding equations*, J. Acoust. Soc. Amer. **75** (1983), 1535–1541.

CER. D. Colton, R. Ewing, and W. Rundell, eds., *Inverse Problems in Partial Differential Equations*, SIAM, Philadelphia, 1990.

CF. P. Colli Franzone, *Il problema inverso dell'elettrocardiografia*, Boll. Un. Mat. Ital. B **15** (1978), 30–51; and P. Colli Franzone, L. Guerri, B. Taccardi, and C. Viganotti, The direct and inverse potential problems in electrocardiology. Numerical aspects of some regularization methods and application to data collected in isolated dog heart experiments, Pub. 222, Laboratorio di Analisi Numerica del Consiglio Nazionale delle Ricerche, Pavia, 1979.

CINSG. M. Cheney, D. Isaacson, J.C. Newell, S. Simske, and J.C. Goble, *NOSER: An algorithm for solving the inverse conductivity problem*, Internat. J. Imaging Systems and Technology **2** (1991), 66–75.

CK. D. Colton and R. Kress, *Inverse Acoustic and Electromagnetic Scattering Theory*, Springer, New York, 1992.

CKa. J.P. Corones and A. Karlsson, *Transient direct and inverse scattering for inhomogeneous viscoelastic media: obliquely incident SH mode*, Inverse Problems **4** (1988), 643–660.

CKNS. J.P. Corones, G. Kristensson, P. Nelson, and D.L. Seth, eds., *Invariant Imbedding and Inverse Problems*, SIAM, Philadelphia, 1990.

CS. K. Chadan and P.C. Sabatier, *Inverse Problems in Quantum Scattering Theory*, Springer, New York, 1977.

Dn. S.R. Deans, *The Radon Transform and Some of Its Applications*, Wiley, New York, 1983.

DS. D.C. Dobson and F. Santosa, *Resolution and stability analysis of an inverse problem in electrical impedance tomography—dependence on the input current patterns*, Institute for Mathematics and its Applications preprint No. 1038 (1992) (to appear in: SIAM J. Appl. Math.).

DT. P. Deift and E. Trubowitz, *Inverse scattering on the line*, Comm. Pure Appl. Math. **32** (1979), 121–151.

Dv. A. J. Devaney, *A computer simulation study of diffraction tomography*, IEEE Trans. Biomed. Eng. **BME-30** (1983), 377–386.

EG. H.W. Engl and C.W. Groetsch, eds., *Inverse and Ill-Posed Problems*, Academic Press, New York, 1987.

F1. L.D. Faddeev, *Increasing solutions of the Schroedinger equation*, Dokl. Akad. Nauk SSSR **165** (1965), 514–517; translated in Soviet Phys. Dokl. **10** (1966), 1033–1035.

F2. ______, *Inverse problem of quantum scattering theory. II*, J. Soviet Math. **5** (1976), 334–396.

G. C.W. Groetsch, *Inverse Problems in the Mathematical Sciences*, Vieweg, Braunschweig, 1993.

GL. I.M. Gel'fand and B.M. Levitan, *On the determination of a differential equation by its spectral function*, Izv. Akad. Nauk SSSR, Ser. Mat. **15** (1951), 309–360; translated in Amer. Math. Soc. Transl. Ser. 2, **1** (1955), 253–304.

HN. G.M. Henkin and R.G. Novikov, *$\bar{\partial}$-equation in the multidimemsional inverse scattering problem*, Uspekhi Math. Nauk **42** (1987), 93–152; translated in Russian Math. Surveys **42** (1987), 109–180.

IC. D. Isaacson and M. Cheney, *Effects of measurement precision and finite numbers of electrodes on linear impedance imaging algorithms*, SIAM J. Appl. Math. **51** (1991), 1705–1731.

II. D. Isaacson and E.L. Isaacson, *Comment on Calderón's paper: 'On an inverse boundary value problem'*, Math. Comp. **52** (1989), 553–559.

KK. G. Kristensson and R.J. Krueger, *Direct and inverse scattering in the time domain for a dissipative wave equation, parts 1-4*, J. Math. Phys. **27** (1986), 1667–1682; ibid. **27** (1986), 1683–1693; ibid. **28** (1987), 360–370; Inverse Problems **5** (1989), 375–388.

KM. I. Kay and H.E. Moses, *The determination of the scattering potential from the spectral measure function V. The Gel'fand-Levitan equation for the three-dimensional scattering problem*, Nuovo Cimento **22** (1961), 689–705.

LC. X.Y. Liu and Y.M. Chen, *Convergence of a generalized pulse-spectrum technique (GPST) for inverse problems of 1-D diffusion equations in space-time domain*, Math. Comp. **51** (1988), 477–489.

M. V.A. Marchenko, *The construction of the potential energy from the phases of the scattered waves*, Dokl. Akad. Nauk SSSR **104** (1955), 695–698.

Me. A. Melin, *Intertwining methods in the theory of inverse scattering*, Internat. J. Quant. Chem. **31** (1987), 739–746.

MK. C. Morawetz and G. Kriegsmann, *The calculations of an inverse potential problem*, SIAM J. Appl. Math. **43** (1983), 844–854.

NA. A.I. Nachman and M. Ablowitz, *A multidimensional inverse scattering method*, Stud. Appl. Math. **71** (1984), 243–250.

Na1. A.I. Nachman, *Reconstruction from boundary measurements*, Ann. Math. **128** (1988), 531–576.

Na2. ———, *Global uniqueness for a two-dimensional inverse boundary value problem*, University of Rochester Department of Mathematics preprint no. 19 (1993).

N. R.G. Newton, *Scattering of Waves and Particles*, 2nd edition, Springer-Verlag, New York, 1982.

N1. ———, *The Gel'fand-Levitan method in the inverse scattering problem*, Scattering Theory in Mathematical Physics, ed. J.A. Levita and J.P. Marchand, Reidel, Dortrecht, 1974, pp. 193–235.

N2. ———, *Inverse scattering II. Three dimensions*, J. Math. Phys. **21** (1980), 1698–1715; ibid. **22** (1981), 631; ibid. **23** (1982), 693.

N3. ———, *Inverse Schrödinger Scattering in Three Dimensions*, Springer, New York, 1989.

Nt. F. Natterer, *Mathematics of Computed Tomography*, Wiley, New York, 1986.

P. R.T. Prosser, *Formal solutions of inverse scattering problems, parts 3 and 4*, J. Math. Phys. **21** (1980), 2648–2653; ibid. **23** (1982), 2127–2130.

RC. J.H. Rose and M. Cheney, *Self-consistent equations for variable-velocity three-dimensional inverse scattering*, Phys. Rev. Lett. **59** (1987), 954–957.

Sa. O. Sacks, *To see and not see*, May 10, 1993, New Yorker.

Sb. P.C. Sabatier, ed., *Applied Inverse Problems*, Springer, New York, 1978.

Si. S. Simske, Master's thesis, Rensselaer Polytechnic Institute, 1986.

SCII. E. Somersalo, M. Cheney, D. Isaacson, and E. Isaacson, *Layer stripping: a direct numerical method for impedance imaging*, Inverse Problems **7** (1991), 899–926.

SPSH. F. Santosa, Y.-H. Pao, W.W. Symes, and C. Holland, eds., *Inverse Problems of Acoustic and Elastic Waves*, SIAM, Philadelphia, 1984.

SS1. P. Sacks and W. W. Symes, *Recovery of the elastic parameters of a layered half-space*, Geophys. J. Royal Astronom. Soc. **88** (1987), 593–620.

SS2. ———, *Uniqueness and continuous dependence for a multidimensional hyperbolic inverse problem*, Comm. Partial Differential Equations **10** (1985), 635–676.

St. D.C. Stickler, *Inverse scattering in a stratified medium*, J. Acoust. Soc. Amer. **74** (1983), 994–1005.

SU. J. Sylvester and G. Uhlmann, *A uniquenes theorem for an inverse boundary value problem in electrical prospection*, Comm. Pure Appl. Math. **39** (1986), 91–112; *A global uniqueness theorem for an inverse boundary value problem in electrical prospection*, Ann. Math. **125** (1987), 153–169; *Inverse boundary value problems at the boundary—continuous dependence*, Comm. Pure Appl. Math. **41** (1988), 197–221.

SV. F. Santosa and M. Vogelius, *A backprojection algorithm for electrical impedance imaging*, SIAM J. Appl. Math. **50** (1990), 216–243.

Sy. W.W. Symes, *Numerical stability in an inverse scattering problem*, SIAM J. Numer. Anal. **17** (1980), 707–732.

Sy2. ———, *Linearization stability for an inverse problem in several-dimensional wave propagation*, SIAM J. Math. Anal. **17** (1986), 132–151.

W. V. Weston, *Wave splitting and the reflection operator for the wave equation*, Inverse Methods in Action, ed. P.C. Sabatier, Springer, New York, 1990, pp. 227–238.

YL. A.E. Yagle and B. Levy, *Layer-stripping solutions of multidimensional inverse scattering problems*, J. Math. Phys. **27** (1986), 1701–1710.

YR. A.E. Yagle and P. Raadhakrishnan, *Numerical performance of layer stripping algorithms for two-dimensional inverse scattering problems*, preprint 1992.

Mathematical Sciences Department, Rensselaer Polytechnic Institute, Troy, NY
E-mail address: chenem@rpi.edu

Proceedings of Symposia in Applied Mathematics
Volume 48, 1994

Algorithms for Unconstrained Optimization: A Review of Recent Developments

Donald Goldfarb*

Abstract

During the last decade there have been many important advances in methods for solving unconstrained optimization problems. We briefly review these developments in this paper, restricting our attention to local minimization of smooth nonlinear functions of a finite number of variables.

1. Introduction

In this paper we review advances that have taken place in the last decade in methods for solving unconstrained optimization problems; i.e., problems of the form

$$\text{given } f : \Re^n \to \Re,$$
$$\text{find } x^* \text{ such that } f(x^*) \leq f(x) \text{ for all } x \in D,$$

where D is an open neighborhood containing x^* and the nonlinear function $f(x)$ is assumed to be at least twice continuously differentiable. Thus we shall only cover algorithms for finding a *local minimum* of smooth functions f. Good surveys of *global* and *nondifferentiable* (*nonsmooth*) optimization can be found in Lemaréchal (1989) and Rinnooy Kan and Timmer (1989), respectively. More complete information on unconstrained optimization algorithms and more extensive lists of references can be found in the books by Dennis and Schnabel (1983), Fletcher (1987) and Gill, Murray and Wright (1981).

Our emphasis in this review paper will be on practical optimization algorithms — i.e., those that are effective and efficient in practice — and we will discuss what is currently known, both theoretically and empirically, about them. We will also discuss several algorithms that are not as efficient as their competitors but whose analysis or theoretical properties are important for understanding, developing, and improving practical algorithms.

1991 *Mathematics Subject Classification.* Primary 90C30; Secondary 65K05, 65K10, 49M07, 49M10, 45M15, 45M37.

*Research supported by National Science Foundation Grant DMS91-0619 and Department of Energy Grant DE-FG02-92ER25126.

This paper is in final form and no version of it will be submitted for publication elsewhere.

© 1994 American Mathematical Society
0160-7634/94 $1.00 + $.25 per page

2. Newton's Method

The most important method — from a fundamental point of view — for finding a local minimum of an unconstrained nonlinear function $f(x)$ is Newton's Method. Newton's method approximates $f(x)$ at the current iterate (estimate of the solution) x_k by the quadratic model provided by the second-order Taylor series approximation to $f(x)$ at x_k,

$$m(x) \equiv f(x_k) + g(x_k)^T(x - x_k) + \frac{1}{2}(x - x_k)^T G(x_k)(x - x_k), \tag{1}$$

where $g(x) \in \Re^n$ and $G(x) \in \Re^{n \times n}$ are, respectively, the gradient and Hessian of $f(x)$. It then computes a new estimate of x_{k+1} as that point $x \in \Re^n$ at which $m(x)$ is stationary. That is,

$$x_{k+1} = x_k + s_k^N,$$

where the Newton step s_k^N is obtained by solving

$$G(x_k)s_k^N = -g(x_k). \tag{2}$$

Although Newton's method is quadratically convergent in the neighborhood of a strict local minimum x^* of $f(x)$ (where $G(x^*)$ is positive definite), without modification it is not a practical method, since it is not globally convergent and not even defined if $G(x_k)$ is singular. Moreover, each iteration is expensive, requiring the computation of the full $n \times n$ matrix of second derivatives of $f(x)$ at each iteration, and $O(n^3)$ work to solve (2) for the Newton step.

To obtain a well-defined and globally convergent method, Gill and Murray (1974) proposed a modified Cholesky factorization algorithm which, given $G(x_k)$, produces its Cholesky factorization LL^T (or LDL^T) if $G(x_k)$ is positive definite and not too badly conditioned or, when this is not the case, produces the Cholesky factorization of $\hat{G}_k \equiv G(x_k) + E$, where E is a nonnegative diagonal matrix computed during the factorization process so that $\hat{G}_k$ is positive definite and reasonably well conditioned. The modified Hessian matrix $\hat{G}_k$ is then used in place of $G(x_k)$ to compute a *modified* Newton step, which is then combined with a standard line search (i.e., one which satisfies the so-called Wolfe (1969, 1971) conditions) to give a globally convergent method. Recently a new and improved modified Cholesky factorization algorithm which computes the elements of E using Gershgorin bounds has been developed by Schnabel and Eskow (1991).

Another way to ensure a descent step and global convergence of Newton's method is to combine it with a trust region approach. This approach was first suggested by Levenberg (1944) and Marquardt (1963) for solving least squares problems. Rather than modifying the Newton direction to make it a descent direction and then choosing an appropriate step length, a trust region algorithm based upon Newton's method first chooses an approximate step length Δ_k and then computes the next iterate x_{k+1} to be that value of x that approximately minimizes the quadratic $m(x)$ in (1) subject to the step size bound

$$\| x - x_k \| \leq \Delta_k. \tag{3}$$

The Newton step s_k^N in (2) solves this trust region problem if $G(x_k)$ is positive definite and the Newton step satisfies (3). Otherwise, if $\| \cdot \|$ is the l_2 norm, $x - x_k = d$ is the solution, where d satisfies

$$(G(x_k) + \mu I)d = -g(x_k)$$

for some $\mu \geq 0$ for which $G(x_k) + \mu I$ is at least positive semidefinite. The step size bound Δ_k is then adjusted (raised, lowered, or left unchanged) depending upon the ratio of the actual reduction of $f(x)$ to that predicted by the quadratic model $m(x)$. Gay (1981), Sorensen (1982b) and Moré and Sorensen (1983) have extended earlier methods of Hebden (1973) and Moré (1978) for efficient and approximate solution to the trust region problem, and very strong global convergence results for these methods have been given by Powell (1984b), Shultz, Schnabel and Byrd (1985) and Byrd, Schnabel and Shultz (1988), generalizing earlier work of Powell (1970, 1975) and Thomas (1975). These results show that approximately defined trust region methods generate iterates which always have an accumulation point at which the gradient vanishes and the Hessian is positive semidefinite, while for line search methods $\liminf_{k \to \infty} \| g(x_k) \| = 0$ is the best that has been proved. Excellent expositions on trust region methods can be found in Dennis and Schnabel (1983, 1989), Moré (1983) and Moré and Sorensen (1984).

In addition, advances in automatic differentiation and the development of software for carrying it out (e.g., see Bishof and Griewank (1992), Griewank (1989), Griewank and Corliss (1991) and Griewank, Juedes, Srinivasan and Tyner (1994)) and the introduction by Griewank and Toint (1982a) of *partial separability* (see §3) and its extension *group partial separability* by Conn, Gould, and Toint (1990, 1992) has made the computation of the Hessian matrix much less burdensome, so that modified Newton's methods are practical for problems with a small to moderate number of variables.

3. Quasi-Newton Methods

Quasi-Newton methods (also known as variable metric and secant methods) remain the methods of choice for solving unconstrained optimization problems when function evaluations are expensive and analytic derivatives are not available. The most commonly used quasi-Newton method is the BFGS method, named after Broyden (1970), Fletcher (1970), Goldfarb (1970), and Shanno (1970), all of whom proposed using the update formula

$$B_{k+1} = B_k - \frac{B_k s_k s_k^T B_k}{s_k^T B_k s_k} + \frac{y_k y_k^T}{y_k^T s_k}$$

to compute a new approximation B_{k+1} to the Hessian $G(x_{k+1})$ from the current approximation B_k, where $s_k = x_{k+1} - x_k$ and $y_k = g(x_{k+1}) - g(x_k)$. This update satisfies the secant equation

$$B_{k+1}s_k = y_k \tag{4}$$

and maintains positive definiteness of the Hessian approximations so long as $s_k^T y_k > 0$, which is easy to guarantee by a line search along the search direction $d_k = -B_k^{-1}g(x_k)$ used in the method.

These properties are shared by all members of the *restricted* Broyden class of methods — i.e., those which use the update

$$B_{k+1} = B_k - \frac{B_k s_k s_k^T B_k}{s_k^T B_k s_k} + \frac{y_k y_k^T}{y_k^T s_k} + \phi(s_k^T B_k s_k) v_k v_k^T, \tag{5}$$

where

$$v_k = \frac{y_k}{y_k^T s_k} - \frac{B_k s_k}{s_k^T B_k s_k}$$

and $\phi \in [0,1]$. Moreover, all of these methods are invariant under linear transformations of the variables and converge q-superlinearly to a strict local minimizer x^* of $f(x)$ if $\| x_1 - x_* \|$ and $\| B_1 - G(x^*) \|$ are sufficiently small. These results were proved by Ritter (1979, 1980), Stachurski (1981) and Griewank and Toint (1982b), extending the proofs of Broyden, Dennis and Moré (1973) for the BFGS ($\phi = 0$) and the DFP ($\phi = 1$) (Davidon (1959), Fletcher and Powell (1963)) — the first quasi-Newton method that was proposed for unconstrained optimization.

Although all of the restricted Broyden class of methods generate the same sequence of iterates if exact line searches are carried out (Dixon (1972)), their performance in practice with inexact line searches varies markedly, with that of the BFGS method far superior to that of the DFP method. An explanation for this fact was provided by Powell (1986), who studied the global behavior of the BFGS and DFP methods with unit step lengths applied to the two-dimensional quadratic function $f(x) = \frac{1}{2}x^T x, x \in \Re^2$, starting with an initial Hessian approximation

$$B_1 = \begin{bmatrix} 1 & 0 \\ 0 & \lambda_1 \end{bmatrix}. \tag{6}$$

Powell showed that

$$B_k = \begin{bmatrix} 1 & 0 \\ 0 & \lambda_k \end{bmatrix},$$

and that, if λ_1 is large and if a bad starting point $x_1 = (\cos\theta, \sin\theta)$ where $\theta = \arctan \lambda_1^{\frac{1}{2}}$ is chosen, then the DFP method reduces λ_k by about one at every iteration and requires approximately λ_1 iterations to converge, while the BFGS method requires roughly $2.4 \log_{10} \lambda_1$ iterations. When $\lambda \leq 1$, both methods are efficient for all starting points. The enormous difference in the performance of these two methods that is possible was further demonstrated by Byrd, Nocedal and Yuan (1987), who tested members of the restricted Broyden class of methods with a line search satisfying the Wolfe conditions on the quartic function $f(x) = \frac{1}{2}(x_1^2 + x_2^2) + 0.05(5x_1^2 + 2x_1x_2 + 3x_2^2)^2$ starting from $x_1 = (\cos 70°, \sin 70°)$ and $B_1 = \begin{pmatrix} 1 & 0 \\ 0 & 10^4 \end{pmatrix}$. The table below shows the number of iterations required to obtain $\| x_k \| \leq 10^{-4} \| x_1 \|$ for various values of $\phi \in [0, 1]$.

ϕ	0(BFGS)	0.2	0.4	0.6	0.8	0.9	0.99	0.999	1(DFP)
iterations	15	21	26	32	66	115	630	2233	4041

The DFP method took 3000 iterations to reduce the trace of B_k from 10^4 to 1100 while the BFGS method took only 10 iterations to reduce it from 10^4 to 3.

Byrd, Nocedal and Yuan (1987) also extended Powell's (1976) proof that the BFGS method with a line search satisfying the Wolfe conditions is globally convergent when applied to a conxex function $f(x)$, to the entire restricted Broyden class except the DFP method. The global convergence of the DFP method remains an open question. Byrd, Nocedal and Yuan's analysis shows that for $\phi < 1$ the members of the Broyden class have a self-correcting property (for large eigenvalues) that is strongest in the BFGS method and not present in the DFP method ($\phi = 1$).

Specifically, Byrd, Nocedal and Yuan first show that for any $\phi \in [0, 1]$

$$\det(B_{k+1}) \geq \det(B_k) \frac{y_k^T s_k}{s_k^T B_k s_k},$$

which, since $y_k^T s_k = s_k^T [\int_0^1 G(x_k + \tau s_k) d\tau] s_k$, ensures that small eigenvalues in B_k are efficiently corrected by all of the methods, and hence that the step lengths α_k taken do not tend to zero. Since a descent method that uses a line search satisfying the Wolfe conditions will be globally convergent if $\cos\theta_k \geq \delta > 0$ for all k, where θ_k is the angle between the step direction and the negative gradient, and since for all methods in the Broyden class it can be shown that

$$\cos\theta_k \geq \frac{\alpha_k}{\gamma \cdot \text{trace}(B_k)}$$

for some positive constant γ, one need only show that trace(B_k) is bounded above. Byrd, Nocedal and Yuan show that

$$\text{trace}(B_{k+1}) \leq \text{trace}(B_k) - \frac{(1-\phi)\alpha_k}{\gamma \cdot \cos^2\theta_k} + t(\phi, \alpha_k, \theta_k, B_k),$$

where $t(\phi, \alpha_k, \theta_k, B_k)$ represents terms that are less crucial to their analysis. This inequality shows that if, because of some excessively large eigenvalues in

B_k, $\cos\theta_k$ is very small, then as long as $\phi < 1$, trace(B_{k+1}) will be significantly less than trace(B_k). Unfortunately, this powerful self-correcting mechanism is not present in the DFP method. Another recent result is provided by Byrd, Tapia and Zhang (1990), who show that if the BFGS method converges, then it converges superlinearly. Moreover, this result does not require $f(x)$ to be convex.

Powell (1986) showed that if λ_1 is small in (6), then both the BFGS and DFP updates are very effective. In fact, his analysis suggests that the BFGS update is better at correcting underestimates rather than overestimates of the eigenvalues of $\nabla^2 f(x)$. This has motivated several researchers to investigate using members of the Broyden family of updates (5) corresponding to negative values of ϕ. Zhang and Tewarson (1988) and Lukšan (1993) report a consistent, though modest, improvement for this approach over the BFGS method. Zhang and Tewarson also prove global and linear convergence for values of ϕ_k that satisfy $(1-\nu)\phi_k^c \leq \phi_k \leq 0$, for all iterations k, where ν is a fixed constant in $(0,1)$ and ϕ_k^c is the value of ϕ_k for which the update yields a singular matrix. Byrd, Liu and Nocedal (1992) analyze the case $\phi_k < 0$ in some detail and show that Zhang and Tewarson's method is not superlinearly convergent.

The most successful attempts to adapt quasi-Newton methods for solving large sparse problems have been the development of *partitioned quasi-Newton* methods by Griewank and Toint (1982a, 1982b, 1982c and 1984) and *limited memory* BFGS methods by Nocedal (1980) and Buckley and LeNir (1983). Griewank and Toint base their method on the fact that every function $f(x)$ with a sparse Hessian is *partially separable*, i.e.,

$$f(x) = \sum_{i=1}^{k} f_i(x), \tag{7}$$

where each *element* function $f_i(x)$ depends on only a small subset of the variables (components of x). In Griewank and Toint's partitioned quasi-Newton method, an approximation B_i to the Hessian of each element function $f_i(x)$ is updated and $B = \sum_{i=1}^{k} B_i$ is used as an approximation to the Hessian of $f(x)$. To avoid factorizing B, Griewank and Toint use a truncated conjugate gradient method to determine a search direction.

In Nocedal's limited memory BFGS method, the approach taken is to store only a limited number, say at most m, of the most recent pairs $\{s_j, y_j\}$ that would have been used in the BFGS method to update B_k (or $H_k = B_k^{-1}$) rather than these matrices, and to compute $H_k \nabla f(x_k)$ by a recursive formula. The method of Buckley and LeNir, on the other hand, combines cycles of BFGS and conjugate gradient iterations.

The performance of the partitioned quasi-Newton method in practice is excellent if the number of variables that each element function depends on is fairly small. The performance of Nocedal's limited memory BFGS method is also quite good, and the method is especially useful for solving large unstructured problems. (For example, see Liu and Nocedal (1989) and Gilbert and Lemaréchal (1989).) Recently, Byrd, Nocedal and Schnabel (1994) derived compact representations of the BFGS and symmetric rank-one (see (9) below) matrices for use in limited memory methods.

A very promising quasi-Newton updating method was recently proposed by Fletcher (1992) for solving large sparse problems. Specifically, Fletcher determines a sparse Hessian approximation B_{k+1} from the current approximation B_k by solving the following problem:

$$\begin{array}{ll} \min\limits_{B>0} & \Psi(H_k B) \equiv \operatorname{trace}(H_k B) - \ln\det(H_k B) \\ & B = B^T \\ \text{s.t.} & B s_k = y_k \\ & B_{ij} = 0 \quad \text{for all } i, j \ni G_{ij}(x) = 0, \end{array} \tag{8}$$

where $H_k = B_k^{-1}$ and the notation $B > 0$ means that B is required to be positive definite. Since it can be shown (Sorenson (1982a)) that the requirements of positive definiteness and sparsity in a quasi-Newton update can be incompatible, Fletcher provides an ad hoc remedy when problem (8) has no solution. Notice that the function $\Psi(A)$, which was introduced by Byrd and Nocedal (1989) to analyze the convergence properties of quasi-Newton methods, has the form of a logarithmic barrier function of the eigenvalues of A. Fletcher's method is a sparse version of the BFGS method since it can be shown that the solution to the above problem with the last (i.e., sparsity) condition eliminated is the standard BFGS update (Fletcher (1991)). Fletcher has shown that his method requires low storage and computational effort and that it performs well in preliminary computational tests.

The simplest and one of the first quasi-Newton methods to have been proposed (Davidon (1959)) is the symmetric rank-one (SR1) update

$$B_{k+1} = B_k + \frac{(y_k - B_k s_k)(y_k - B_k s_k)^T}{s_k^T (y_k - B_k s_k)}. \tag{9}$$

If $s_k^T(y_k - B_k s_k) \neq 0$, this formula gives the unique solution to the problem of determining a symmetric matrix B_{k+1} that satisfies the secant equation (4) and is such that $B_{k+1} - B_k$ is a rank-one matrix, given the symmetric matrix B_k, y_k, and s_k. If $y_k = B_k s_k$, then $B_{k+1} = B_k$ solves this problem; however, if $y_k \neq B_k s_k$ and $s_k^T(y_k - B_k s_k) = 0$, then there is no solution.

When applied to a convex quadratic function $f(x)$, the SR1 update has long been known to have the important property that if the search directions are linearly independent and the denominator $s_k^T(y_k - B_k s_k)$ is always nonzero, then after n steps B_{n+1} equals the exact Hessian of $f(x)$. On the other hand, the denominator $s_k^T(y_k - B_k s_k)$ in the update formula can be zero or nearly zero, leading to either a breakdown in the update or numerical instabilities. Because of these difficulties and because the SR1 update does not generate positive definite matrices, which is important in line search algorithms, the update was rarely used in practice, until about ten years ago.

The recent interest in the SR1 update began with Griewank and Toint's (1982c) use of it to approximate the Hessians of each of the element functions $f_i(x)$ in (7), when the curvature condition $s_k^T y_k > 0$ does not hold and, hence, the BFGS update cannot be applied. More recently, Conn, Gould and Toint (1988a, 1988b, 1991) reported that the SR1 update outperformed the BFGS update when used in a trust region method for bound constrained problems. Further support for the SR1 update is provided by Khalfan, Byrd and Schnabel (1993) who found the SR1 to be somewhat better than the BFGS update in both line search and trust region implementations. Moreover, they proved $(n+1)$-step q-superlinear and $2n$-step q-quadratic convergence for a line search SR1 method, assuming that for all k

$$| s_k^T (y_k - B_k s_k) | \geq r \parallel s_k \parallel \parallel y_k - B_k s_k \parallel \tag{10}$$

for some $r \in (0,1)$, and $\parallel B_k \parallel \leq M$ for some $M > 0$, and that for all $k \geq k_0$, B_k is positive definite.

In their computational study, Khalfan, Byrd and Schnabel found that condition (10) was rarely violated (at most once in any of the problems that they solved) and that B_k was positive definite on most iterations. Although, as strong global convergence result as those that have been proved for BFGS update have not been proved for the SR1 update, the results described above indicate that, properly implemented, the SR1 update is a viable competitor to the widely used BFGS update.

Finally, we note that a modified version of the SR1 update has been proposed by Osborne and Sun (1988) in which B_k is scaled by a scale factor $\gamma > 0$, so that when the update formula is applied to γB_k it gives a positive definite matrix B_{k+1}, assuming that B_k is positive definite and $s_k^T y_k > 0$. Limited computational tests (Osborne and Sun (1988) and Schnabel (1989)) show that this method is promising.

The idea of scaling the matrix B_k in a variable metric updating formula was originally proposed by Oren and Luenberger (1974) as a way to ameliorate the difference between B_k and $\nabla^2 f(x_k)$ somewhat before applying the update. Unfortunately, this destroys the property

$$B_{k+1} s_j = y_j, \qquad j = 1, \ldots, k-1, \tag{11}$$

that the Broyden family (5) of updates has when $f(x)$ is quadratic and exact line searches are performed. Moreover, numerical tests of the approach have been disappointing (e.g., see Shanno and Phua (1980) for test results and Nocedal and Yuan (1991) for a possible explanation of them).

A recent method of Powell (1987), which allows for a partial scaling of B_k while maintaining (11), is based upon working with

$$B_k^{-1} = Z_k Z_k^T,$$

and updating Z_k to Z_{k+1}, so that B_{k+1} is the matrix corresponding to the BFGS update of B_k. This factorization is particularly useful in algorithms for linearly constrained problems (e.g., see Goldfarb and Idnani (1983)). Powell's updating method scales only the last $n-k+1$ columns of Z_k on iteration k before updating Z_k, and it can be shown that the conditions (11) continue to hold. Siegel (1991) and Lalee and Nocedal (1991) have proposed variations of this method which have performed well in preliminary tests.

4. Other Methods

In this section we discuss some recent results on tensor methods and the conjugate gradient method. Tensor methods, which were first developed by Schnabel and Frank (1984) for solving systems of nonlinear equations, have recently been proposed by Schnabel and Chow (1991) for solving constrained optimization problems. These methods base each iteration on the fourth-order tensor model

$$m(x_k + d) = f(x_k) + g(x_k)^T d + \frac{1}{2} d^T G(x_k) d + \frac{1}{6} T_k ddd + \frac{1}{24} V_k dddd, \qquad (12)$$

where $T_k \in \Re^{n\times n\times n}$ and $V_k \in \Re^{n\times n\times n\times n}$ are symmetric low-rank tensor approximations to $\nabla^3 f(x_k)$ and $\nabla^4 f(x_k)$, respectively. Schnabel and Chow show how to form and solve the tensor model (12) so that it interpolates $f(x)$ and $\nabla f(x)$ at the previous p most recent iterates. In practice the tensor method with $p = 1$ works well, outperforming a standard method based on a quadratic model, and requires only an additional $O(n)$ storage locations and $O(n^2)$ arithmetic operations per iteration over that needed by the standard method.

Although the conjugate gradient method for optimization has been known for nearly thirty years (Fletcher and Reeves (1964)), the first practical global convergence proof for a member of this class of methods was only fairly recently provided by Al-Baali (1985). Al-Baali proved that the Fletcher-Reeves method with a line search that satisfies the strong Wolfe conditions with appropriately chosen constants converges in the sense that $\liminf_{k\to\infty} g(x_k) = 0$ under the assumption that the level set $L \equiv \{x \mid f(x) \leq f(x_1)\}$ is bounded and that $f(x)$ is continuously differentiable and $\nabla f(x)$ is Lipschitz continuous in some neighborhood of L. In practice, the Fletcher-Reeves method is often much less efficient than the Polak-Ribière (1969) variant of the conjugate gradient method. But no practical global convergence results have been proved for the latter method and Powell (1984a) has shown that the Polak-Ribière method with exact line searches can cycle infinitely away from a local minimum of a twice continuously differentiable function. For an excellent review of what is currently known about conjugate gradient methods in particular, and unconstrained optimization algorithms in general, see Nocedal (1992).

References

[1] M. Al-Baali (1985), "Descent property and global convergence of the Fletcher-Reeves method with inexact line search", *IMA J. Numer. Anal.* **5**, 121–124.

[2] C. Bishof and A.O. Griewank (1992), "ADIFOR: A FORTRAN system for portable automatic differentiation", Technical Report MCS-P317-0792, Argonne National Laboratory (Argonne, IL).

[3] G.G. Broyden (1970), "The convergence of a class of double rank minimization algorithms", parts I and II, *J. Inst. Math. Appl.* **6**, 76–90, 222–236.

[4] G.G. Broyden, J.E. Dennis and J.J. Moré (1973), "On the local and superlinear convergence of quasi-Newton methods", *J. Inst. Math. Appl.* **12**, 223–246.

[5] A. Buckley and A. LeNir (1983), "QN-like variable storage conjugate gradients", *Math. Programming* **27**, 155–175.

[6] R.H. Byrd, D.C. Liu and J. Nocedal (1992), "On the behavior of Broyden's class of quasi-Newton methods", *SIAM J. Optim.* **2**, 533–557.

[7] R.H. Byrd and J. Nocedal (1989), "A tool for the analysis of quasi-Newton methods with application to unconstrained minimization", *SIAM J. Numer. Anal.* **26**, 727–739.

[8] R.H. Byrd, J. Nocedal and R.B. Schnabel (1994), "Representations of quasi-Newton matrices and their use in limited memory methods", *Math. Programming* **63**, 129–156.

[9] R.H. Byrd, J. Nocedal and Y.-X. Yuan (1987), "Global convergence of a class of quasi-Newton methods on convex problems", *SIAM J. Numer. Anal.* **24**, 1171–1190.

[10] R.H. Byrd, R.B. Schnabel and G.A. Shultz (1988), "Approximate solution of the trust region problem by minimization over two-dimentional subspaces", *Math. Programming* **40**, 247–263.

[11] R.H. Byrd, R.A. Tapia and Y. Zhang (1990), "An SQP augmented Lagrangian BFGS algorithm for constrained optimization", Tech. Report, University of Colorado (Boulder).

[12] A.R. Conn, N.I.M. Gould and Ph.L. Toint (1988a), "Global convergence of a class of trust region algorithms for optimization with simple bounds", *SIAM J. Numer. Anal.* **25**, 433–460.

[13] A.R. Conn, N.I.M. Gould and Ph.L. Toint (1988b), "Testing a class of methods for solving minimization problems with simple bounds on the variables", *Math. Comp.* **50**, 399–430.

[14] A.R. Conn, N.I.M. Gould and Ph.L. Toint (1990), "An introduction to the structure of large-scale nonlinear optimization problems and the LANCELOT project", in *Computing methods in applied science and engineering* (R. Glowinski and A. Lichnewski, eds.), SIAM (Philadelphia), pp. 42–54.

[15] A.R. Conn, N.I.M. Gould and Ph.L. Toint (1991), "Convergence of quasi-Newton matrices generated by the symmetric rank one update", *Math. Programming* **50**, 177–195.

[16] A.R. Conn, N.I.M. Gould and Ph.L. Toint (1992), "LANCELOT: a Fortran package for large-scale nonlinear optimization (release A)", *Springer Ser. Comput. Math.* **17**, Springer-Verlag (Berlin).

[17] W.C. Davidon (1959), "Variable metric method for minimization", Argonne National Lab Report (Argonne, IL). (Reprinted in *SIAM J. Optim.* **1** (1991), 1–17.)

[18] J.E. Dennis, Jr. and R.B. Schnabel (1983), *Numerical methods for unconstrained optimization and nonlinear equations*, Prentice-Hall, Inc. (Englewood Cliffs, NJ).

[19] J.E. Dennis, Jr. and R.B. Schnabel (1989), "A view of unconstrained optimization", in *Handbooks Oper. Res. Management Sci.* **1**, *Optimization* (G.L. Nemhauser, A.H.G. Rinnooy Kan and M.J. Todd, eds.), North-Holland (Amsterdam), pp. 1–72.

[20] L.C.W. Dixon (1972), "Quasi-Newton algorithms generate identical points", in *Math. Programming* **2**, 383–387.

[21] R. Fletcher (1970), "A new approach to variable metric methods", *Comput. J.* **13**, 317–322.

[22] R. Fletcher (1987), *Practical methods of optimization*, 2nd edition, John Wiley & Sons (New York).

[23] R. Fletcher (1991), "A new result for quasi-Newton formulae", *SIAM J. Optim.* **1**, 18–21.

[24] R. Fletcher (1992), "An optimal positive definite update for sparse Hessian matrices", Report NA/145 Univ. of Dundee, to appear in *SIAM J. Optim.*

[25] R. Fletcher and M.J.D. Powell (1963), "A rapidly convergent descent method for minimization", *Comput. J.* **6**, 163–168.

[26] R. Fletcher and C.M. Reeves (1964), "Function minimization by conjugate gradients", *Comput. J.* **7**, 149–154.

[27] D. Gay (1981), "Computing optimal locally constrained steps", *SIAM J. Sci. Statist. Comput.* **2**, 186–197.

[28] J.C. Gilbert and C. Lemaréchal (1989), "Some numerical experiments with variable storage quasi-Newton algorithms", *Math. Programming* **45**, 407–436.

[29] P.E. Gill and W. Murray (1974), "Newton type methods for unconstrained and linearly constrained optimization", *Math. Programming* **28**, 311–350.

[30] P.E. Gill, W. Murray and M.H. Wright (1981), *Practical optimization*, Academic Press (London).

[31] D. Goldfarb (1970), "A family of variable metric methods derived by variational means", *Math. Comp.* **24**, 23–26.

[32] D. Goldfarb and A. Idnani (1983), "A numerically stable dual method for solving strictly convex quadratic programs", *Math. Programming* **27**, 1–33.

[33] A. Griewank (1989), "On automatic differentiation", in *Mathematical programming: recent developments and applications* (M. Iri and K. Tanabe, eds.), Kluwer Academic Publishers (Dordrecht), pp. 83–108.

[34] A. Griewank and G.F. Corliss (1991), *Automatic differentiation of algorithms: theory, implementation, and application*, SIAM (Philadelphia).

[35] A. Griewank, D. Juedes, J. Srinivasan and C. Tyner (1994), "ADOL-C, a package for automatic differentiation of algorithms written in C/C++", *ACM Trans. Math. Software*, to appear.

[36] A. Griewank and Ph.L. Toint (1982a), "On the unconstrained optimization of partially separable objective functions", in *Nonlinear optimization 1981* (M.J.D. Powell, ed.), Academic Press (London), pp. 301–312.

[37] A. Griewank and Ph.L. Toint (1982b), "Local convergence analysis of partitioned quasi-Newton updates", *Numer. Math.* **39**, 429–448.

[38] A. Griewank and Ph.L. Toint (1982c), "Partitioned variable metric updates for large structured optimization problems", *Numer. Math.* **39**, 119–137.

[39] A. Griewank and Ph.L. Toint (1984), "Numerical experiments with partially separable optimization problems", in *Numerical analysis: proceedings Dundee 1983*, Lecture Notes in Math. 1066 (D.F. Griffiths, ed.), Springer-Verlag (Berlin), pp. 203–220.

[40] M.D. Hebden (1973), "An algorithm for minimization using exact second derivatives", Technical Report TP515, Atomic Energy Research Estab. (Harwell, England).

[41] H. Khalfan, R.H. Byrd and R.B. Schnabel (1993), "A theoretical and experimental study of the symmetric rank one update", *SIAM J. Optim.* **3**, 1–24.

[42] M. Lalee and J. Nocedal (1991), "Automatic column scaling strategies for quasi-Newton methods", Report No. NAM 04, EECS Department, Northwestern University (Evanston, IL).

[43] C. Lemaréchal (1989), "Nondifferentiable optimization", in *Handbooks Oper. Res. Management Sci.* **1**, *Optimization* (G.L. Nemhauser, A.H.G. Rinnooy Kan and M.J. Todd, eds.), North-Holland (Amsterdam), pp. 529–572.

[44] K. Levenberg (1944), "A method for the solution of certain problems in least squares", *Quart. Appl. Math.* **2**, 164–168.

[45] D.C. Liu and J. Nocedal (1989), "On the limited memory BFGS method for large scale optimization", *Math. Programming* **45**, 503–528.

[46] L. Lukšan (1993), "Computational experience with known variable metric updates", Institute of Computer & Information Sciences, Czechoslovak Academy of Sciences (Prague), working paper.

[47] D. Marquardt (1963), "An algorithm for least-squares estimation of nonlinear parameters", *SIAM J. Appl. Math.* **11**, 431–441.

[48] J.J. Moré (1978), "The Levenberg-Marquardt algorithm: implementation and theory", in *Numerical analysis: proceedings Dundee 1977* (G.A. Watson, ed.), Lecture Notes in Math. 630, Springer-Verlag (Berlin), pp. 105–116.

[49] J.J. Moré (1983), "Recent developments in algorithms and software for trust region methods", in *Mathematical programming: the state of the art* (A. Bachem, M. Grötschel and G. Korte, eds.), Springer-Verlag (Berlin), pp. 258–287.

[50] J.J. Moré and D.C. Sorensen (1983), "Computing a trust region step", *SIAM J. Sci. Statist. Comput.* **4**, 553–572.

[51] J.J. Moré and D.C. Sorensen (1984), "Newton's method", in *Studies in numerical analysis* (G.H. Golub, ed.), The Mathematical Association of America, pp. 29–82.

[52] J. Nocedal (1980), "Updating quasi-Newton matrices with limited storage", *Math. Comp.* **35**, 773–782.

[53] J. Nocedal (1992), "Theory of algorithms for unconstrained optimization", *Acta Numerica* **1**, 199–242.

[54] J. Nocedal and Y.-X. Yuan (1991), "Analysis of a self-scaling quasi-Newton method", Technical Report NAM-02, Department of Electrical Engineering and Computer Science, Northwestern University (Evanston, IL).

[55] S.S. Oren and D.G. Luenberger (1974), "Self-scaling variable metric (SSVM) algorithms I: criteria and sufficient conditions for scaling a class of algorithms", *Management Sci.* **20**, 845–862.

[56] M.R. Osborne and L.P. Sun (1988), "A new approach to the symmetric rank-one updating algorithm", Rept. NMO/01, Department of Statistics, IAS, Australian National University.

[57] E. Polak and G. Ribière (1969), "Note sur la convergence de méthodes de directions conjuguées", *Rev. Française Informat. Recherche Opérationnelle* **3**, no. 16, 35–43.

[58] M.J.D. Powell (1970), "A new algorithm for unconstrained optimization", in *Nonlinear programming* (J.B. Rosen, O.L. Mangasarian and K. Ritter, eds.), Academic Press (New York), pp. 31–65.

[59] M.J.D. Powell (1975), "Convergence properties of a class of minimization algorithms", in *Nonlinear programming 2* (O.L. Mangasarian, R.R. Meyer and S.M. Robinson, eds.), Academic Press (New York), pp. 1–27.

[60] M.J.D. Powell (1976), "Some global convergence properties of a variable metric algorithm for minimization without exact line searches", in *Nonlinear programming, SIAM-AMS Proc., Vol. IX* (R.W. Cottle and C.E. Lemke, eds.), Amer. Math. Soc. (Providence, RI).

[61] M.J.D. Powell (1984a), "Nonconvex minimization calculations and the conjugate gradient method", in *Numerical analysis: proceedings Dundee 1983* (D.F. Griffiths, ed.), Lecture Notes in Math. 1066, Springer-Verlag (Berlin), pp. 122–141.

[62] M.J.D. Powell (1984b), " On the global convergence of trust region algorithms for unconstrained minimization", *Math. Programming* **29**, 297–303.

[63] M.J.D. Powell (1986), "How bad are the BFGS and DFP methods when the objective function is quadratic?", *Math. Programming* **34**, 34–47.

[64] M.J.D. Powell (1987), "Updating conjugate directions by the BFGS formula", *Math. Programming* **38**, 29–46.

[65] A.H.G. Rinnooy Kan and G.T. Timmer (1989), " Global Optimization", in *Handbooks Oper. Res. Management Sci.* **1**, *Optimization* (G.L. Nemhauser, A.H.G. Rinnooy Kan and M.J. Todd, eds.), North-Holland (Amsterdam), pp. 631–662.

[66] K. Ritter (1979), "Local and superlinear convergence of a class of variable metric methods", *Computing* **23**, 287–297.

[67] K. Ritter (1980), "On the rate of superlinear convergence of a class of variable metric methods", *Numer. Math.* **35**, 293–313.

[68] R.B. Schnabel (1989), "Sequential and parallel methods for unconstrained optimization", in *Mathematical programming, recent developments and applications* (M. Iri and K. Tanabe, eds.), Kluwer Academic Publishers (Dordrecht), pp. 227–261.

[69] R.B. Schnabel and T.-T. Chow (1991), "Tensor methods for unconstrained optimization using second derivatives", *SIAM J. Optim.* **1**, 293–315.

[70] R.B. Schnabel and E. Eskow (1991), "A new modified Cholesky factorization", *SIAM J. Sci. Statist. Comput.* **11**, 1136–1158.

[71] R.B. Schnabel and P.D. Frank (1984), "Tensor methods for nonlinear equations ", *SIAM J. Numer. Anal.* **21**, 815–843.

[72] G.A. Shultz, R.B. Schnabel and R.H. Byrd (1985), "A family of trust-region-based algorithms for unconstrained minimization with strong global convergence properties", *SIAM J. Numer. Anal.* **22**, 47–67.

[73] D.F. Shanno (1970), "Conditioning of quasi-Newton methods for function minimization", *Math. Comp.* **24**, 647–657.

[74] D.F. Shanno and K.H. Phua (1980), "Remark on Algorithm 500: minimization of unconstrained multivariate functions", *ACM Trans. Math. Software* **6**, 618–622.

[75] D. Siegel (1991), "Modifying the BFGS update by a new column scaling technique", Technical Report DAMTP 1991/NA5, Department of Applied Mathematics and Theoretical Physics, University of Cambridge.

[76] D.C. Sorensen (1982a), "Collinear scaling and sequential estimation in sparse optimization algorithms", *Math. Programming Stud.* **18**, 135–159.

[77] D.C. Sorensen (1982b), "Newton's method with a model trust region modification", *SIAM J. Numer. Anal.* **19**, 409–426.

[78] A. Stachurski (1981), " Superlinear convergence of Broyden's bounded Θ-class of methods", *Math. Programming* **20**, 196–212.

[79] S.W. Thomas (1975), "Sequential estimation techniques for quasi-Newton algorithms", Technical Report TR75-227, Department of Computer Science, Cornell University.

[80] P. Wolfe (1969), "Convergence conditions for ascent methods", *SIAM Rev.* **11**, 226–235.

[81] P. Wolfe (1971), "Convergence conditions for ascent methods. II: Some corrections", *SIAM Rev.* **13**, 185–188.

[82] Y. Zhang and R.P. Tewarson (1988), "Quasi-Newton algorithms with updates from the pre-convex part of Broyden's family", *IMA J. Numer. Anal.* **8**, 487–509.

INDUSTRIAL ENGINEERING AND OPERATIONS RESEARCH DEPARTMENT, COLUMBIA UNIVERSITY, NEW YORK, NY 10027

E-mail address: goldfarb@cunixb.cc.columbia.edu

Proceedings of Symposia in Applied Mathematics
Volume **48**, 1994

A Survey of Componentwise Perturbation Theory in Numerical Linear Algebra

NICHOLAS J. HIGHAM

ABSTRACT. Perturbation bounds in numerical linear algebra are traditionally derived and expressed using norms. Norm bounds cannot reflect the scaling or sparsity of a problem and its perturbation, and so can be unduly weak. If the problem data and its perturbation are measured componentwise, much smaller and more revealing bounds can be obtained. A survey is given of componentwise perturbation theory in numerical linear algebra, covering linear systems, the matrix inverse, matrix factorizations, the least squares problem, and the eigenvalue and singular value problems. Most of the results described have been published in the last five years.

Our hero is the intrepid, yet sensitive matrix A.
Our villain is E, who keeps perturbing A.
When A is perturbed he puts on a crumpled hat: $\widetilde{A} = A + E$.
G. W. Stewart and J.-G. Sun, *Matrix Perturbation Theory* (1990)

1. Introduction

Matrix analysis would not have developed into the vast subject it is today without the concept of representing a matrix by a single symbol. Similarly, perturbation theory would not be such a rich and useful area if it were not for norms. A norm compresses the mn numbers in an $m \times n$ matrix into a single scalar measure of size, enabling a perturbation result to be presented in a form that is easy to interpret and gives insight. In the recent book *Matrix Perturbation Theory* [**71**] by Stewart and Sun, norms are exploited throughout.

1991 *Mathematics Subject Classification.* Primary 65F35, 65G05.
Key words and phrases. Componentwise perturbation bounds, componentwise backward error, linear systems, matrix inverse, matrix factorizations, least squares problem, underdetermined system, eigenvalue problem, singular value problem.
This work was supported by Science and Engineering Research Council grant GR/H52139.
This paper is in final form and no version of it will be submitted for publication elsewhere.

© 1994 American Mathematical Society
0160-7634/94 $1.00 + $.25 per page

Although a great deal can be achieved with norms, they have two main failings. First, norms ignore structure in the form of both scaling and sparsity. Norms can tell us the overall size of a perturbation but not how that size is distributed among the elements it perturbs, and this information can be important when the data is badly scaled or contains many zeros. The second failing is that norm bounds can lack sharpness (particularly in a rounding error analysis), because they are often the result of several applications of the triangle inequality and the submultiplicative inequality, in not all of which there can be equality simultaneously.

It is for these reasons that an alternative style of analysis has become popular, in which the use of norms is kept to a minimum (but not abandoned altogether); it is known as *componentwise analysis*. The idea is not new. Its history can be very roughly outlined as follows (more details are given in the following sections).

Wilkinson frequently obtained componentwise backward error results as intermediate steps in an error analysis, but usually stated the final result in normwise form. Some of the earliest examples are contained in his 1960 paper [**79**] and 1963 book [**80**]. In 1964, Oettli and Prager [**54**] obtained a formula for the componentwise backward error of an approximate solution to a linear system. Bauer [**8**] applied componentwise perturbation analysis to linear systems and to matrix inversion in 1966. Stoer and Bulirsch gave a componentwise error analysis of Gaussian elimination in the original 1972 German edition of *Introduction to Numerical Analysis* [**72**]. In a series of papers Skeel explored componentwise perturbation analysis for linear systems and componentwise error analysis for Gaussian elimination, obtaining new results about the stability of Gaussian elimination and the behaviour of iterative refinement [**63**], [**64**], [**65**]. Skeel's papers did not attract immediate attention in the numerical linear algebra community, but gradually came to be appreciated during the 1980s. Since the mid 1980s componentwise analysis has been widely used for linear systems and linear system solvers, and componentwise perturbation theory and error analysis has been developed for eigenproblems and eigensolvers. Perhaps the best indication of the utility of componentwise analysis is that it has been fully exploited in the development of LAPACK, the state-of-the-art package of Fortran programs for solving linear equation and eigenvalue problems [**1**].

Although the componentwise approach is now well known, at least among numerical analysts, it is hardly represented in current textbooks. For example, Golub and Van Loan [**34**] and Stewart and Sun [**71**] give componentwise error analysis for linear equations but not for eigenvalue problems, while Horn and Johnson [**47**], [**48**] give no componentwise perturbation results at all.

In this paper we survey componentwise perturbation theory in numerical linear algebra. We do not consider rounding error analysis, except briefly when it helps to illuminate the perturbation theory. No survey of componentwise rounding error analysis is currently available, but some relevant references for standard problems, which themselves include further references, are as follows:

- [**41**] for substitution for triangular systems,
- [**43**] for Gaussian elimination,
- [**20**] for Cholesky factorization,
- [**25**] for matrix inversion,
- [**46**] for stationary iterative methods,
- [**44**] for QR factorization and its application to least squares problems,
- [**22**] for underdetermined system solvers,
- [**23**] for the QR algorithm for the bidiagonal SVD problem,
- [**24**] for Jacobi's method for the symmetric eigenproblem and SVD.

In §2 we consider backward and forward perturbation bounds for square linear systems. The matrix inverse is considered in §3. Section 4 is concerned with the least squares problem, for which both normwise and componentwise results have recently been obtained; underdetermined systems are also discussed in this section. Matrix factorizations are discussed in §5. The active area of componentwise perturbation theory for eigenvalue and singular value problems is described in §6. Finally, some concluding remarks are given in §7.

It is possible to develop probabilistic perturbation theory for all the problems considered here, by making assumptions about the statistical distribution of the perturbations. We do not consider this approach, but refer the interested reader to the papers by Fletcher [**29**] and Stewart [**69**].

2. Linear Systems

Throughout this section we are concerned with a linear system $Ax = b$, where $A \in \mathbb{R}^{n\times n}$. In the context of uncertain data or inexact arithmetic there are two important questions:

(1) How much do we have to perturb the data A and b for an approximate solution y to be the exact solution of the perturbed system—in other words, what is the backward error of y?

(2) How much does x change if we perturb A and b, that is, how sensitive is the solution to perturbations in the data?

When we use norms to measure size, the answers are given by the following classical results. We denote by $\|\cdot\|$ any vector norm and the corresponding subordinate matrix norm, and $\kappa(A) = \|A\|\|A^{-1}\|$ is the matrix condition number. The matrix E and the vector f have nonnegative entries and represent tolerances against which the perturbations are measured (their role becomes clear when we consider componentwise results).

The first theorem is due to Rigal and Gaches [**58**], and was also given by Kovarik [**52**].

THEOREM 2.1 (RIGAL AND GACHES). *The* normwise backward error

$$\eta_{E,f}(y) := \min\{\epsilon : (A+\Delta A)y = b+\Delta b,\ \|\Delta A\| \le \epsilon\|E\|,\ \|\Delta b\| \le \epsilon\|f\|\} \tag{2.1}$$

is given by

$$\eta_{E,f}(y) = \frac{\|r\|}{\|E\|\|y\| + \|f\|}, \tag{2.2}$$

where $r = b - Ay$.

PROOF. It is straightforward to show that the right-hand side of (2.2) is a lower bound for $\eta(y)$. This lower bound is attained for the perturbations

$$\Delta A_{\min} = \frac{\|E\|\|y\|}{\|E\|\,\|y\| + \|f\|} rz^T, \qquad \Delta b_{\min} = -\frac{\|f\|}{\|E\|\,\|y\| + \|f\|} r, \tag{2.3}$$

where z is a vector dual to y, that is,

$$z^T y = \|z\|_D \|y\| = 1, \qquad \text{where} \qquad \|z\|_D = \max_{v \neq 0} \frac{|z^T v|}{\|v\|}. \quad \square$$

For the particular choice $E = |A|$ and $f = |b|$, the quantity $\eta_{E,f}(y)$ is called the *normwise relative backward error.*

The next result measures the sensitivity of the system.

THEOREM 2.2. *Let* $Ax = b$ *and* $(A + \Delta A)y = b + \Delta b$, *where* $\|\Delta A\| \leq \epsilon \|E\|$ *and* $\|\Delta b\| \leq \epsilon \|f\|$, *and assume that* $\epsilon \|A^{-1}\|\|E\| < 1$. *Then*

$$\frac{\|x - y\|}{\|x\|} \leq \frac{\epsilon}{1 - \epsilon \|A^{-1}\|\|E\|} \left(\frac{\|A^{-1}\|\|f\|}{\|x\|} + \|A^{-1}\|\|E\| \right), \tag{2.4}$$

and this bound is attainable to first order in ϵ.

PROOF. The bound (2.4) is straightforward to derive. It is attained to first order in ϵ for $\Delta A = \epsilon \|E\|\|x\| w v^T$ and $\Delta b = -\epsilon \|f\| w$, where $\|w\| = 1$, $\|A^{-1} w\| = \|A^{-1}\|$, and v is a vector dual to x. $\square$

Associated with the way of measuring perturbations used in these two theorems is the normwise condition number

$$\begin{aligned} \kappa_{E,f}(A, x) \quad := \quad \lim_{\epsilon \to 0} \sup \{ \; \frac{\|\Delta x\|}{\epsilon \|x\|} &: (A + \Delta A)(x + \Delta x) = b + \Delta b, \\ &\|\Delta A\| \leq \epsilon \|E\|, \quad \|\Delta b\| \leq \epsilon \|f\| \; \}. \end{aligned}$$

Because the bound of Theorem 2.2 is sharp, it follows that

$$\kappa_{E,f}(A, x) = \frac{\|A^{-1}\|\|f\|}{\|x\|} + \|A^{-1}\|\|E\|.$$

For the choice $E = |A|$ and $f = |b|$ (where $|A|$ denotes the matrix $(|a_{ij}|)$), we have $\kappa(A) \leq \kappa_{E,f}(A, x) \leq 2\kappa(A)$, and the bound (2.4) can be weakened slightly to yield the familiar form

$$\frac{\|x - y\|}{\|x\|} \leq \frac{2\epsilon \kappa(A)}{1 - \epsilon \kappa(A)}.$$

A numerical example illustrates the above results. Let A be the 8×8 Vandermonde matrix with (i,j) element $j^{2(i-1)}$, and let $b = e_1$, the first unit vector, so that x is the first column of A^{-1}. We take y to be the approximate solution to $Ax = b$ computed by Gaussian elimination with partial pivoting. All our experiments are performed in Matlab, for which the unit roundoff $u \approx 1.1 \times 10^{-16}$. We find that $\eta_\infty(y) = 3.05 \times 10^{-21}$ for $E = |A|$ and $f = |b|$, and $\kappa_\infty(A) = 1.68 \times 10^{13}$. This is an admirably small backward error, but it may be uninformative for two reasons. First, the elements of A vary over 12 orders of magnitude, so while our backward error perturbations are small compared with the large elements of A, we may be making large perturbations in the small elements (indeed we *are* in this particular example). Second, we are perturbing the zero elements of b (as can be seen from (2.3) together with the fact that for this example the residual r has no zero entries); this is unsatisfactory if we wish to regard y as the first column of the inverse of a perturbed matrix.

Next, let $b = Ae$, where $e = [1, 1, \ldots, 1]^T$, and let z be the solution to the perturbed system $(A + \Delta A)z = b + \Delta b$, where $\Delta A = \text{tol}|A|$ and $\Delta b = \text{tol}|b|$, with $\text{tol} = 8u$. We find that

$$\frac{\|x - z\|_\infty}{\|x\|_\infty} = 2.40 \times 10^{-12}, \tag{2.5}$$

while the corresponding bound from (2.4) with $\epsilon = \text{tol}$, $E = |A|$ and $f = |b|$ is 3.03×10^{-2}. Thus the normwise forward error bound is extremely pessimistic for this special choice of perturbation.

To obtain a more satisfactory backward error measure and a sharper perturbation bound, we need componentwise analysis.

The *componentwise backward error* is defined as

$$\omega_{E,f}(y) = \min\{\epsilon : (A + \Delta A)y = b + \Delta b,\ \ |\Delta A| \le \epsilon E,\ \ |\Delta b| \le \epsilon f\}, \tag{2.6}$$

where inequalities between matrices hold componentwise. In this definition each element of a perturbation is measured relative to its individual tolerance, so, unlike in the normwise definition, we are making full use of the $n^2 + n$ parameters in E and f.

How should E and f be chosen? The most common choice of tolerances is $E = |A|$ and $f = |b|$, which yields the *componentwise relative backward error*. For this choice

$$a_{ij} = 0 \Rightarrow \Delta a_{ij} = 0 \quad \text{and} \quad b_i = 0 \Rightarrow \Delta b_i = 0$$

in (2.6), and so if $\omega_{E,f}(y)$ is small then y solves a problem that is close to the original one in the sense of componentwise relative perturbations and has the same sparsity pattern. Another attractive property of the componentwise relative backward error is that it is insensitive to the scaling of the system: if $Ax = b$ is scaled to $(S_1 A S_2)(S_2^{-1}x) = S_1 b$, where S_1 and S_2 are diagonal, and y is scaled to $S_2^{-1}y$, then ω remains unchanged.

The choice $E = |A|ee^T$, $f = |b|$ gives a row-wise backward error. The constraint $|\Delta A| \le \epsilon E$ is now $|\Delta a_{ij}| \le \epsilon \alpha_i$, where α_i is the 1-norm of the ith row of A, so perturbations to the ith row of A are being measured relative to the norm of that row. A columnwise backward error can be formulated in a similar way.

The third natural choice of tolerances is $E = \|A\|ee^T$ and $f = \|b\|e$, for which $\omega_{E,f}(y)$ is the same as the normwise backward error $\eta_{E,f}(y)$, up to a constant.

As for the normwise backward error in Theorem 2.1, there is a simple formula for $\omega_{E,f}(y)$.

THEOREM 2.3 (OETTLI AND PRAGER [**54**]). *The componentwise backward error is given by*

$$\omega_{E,f}(y) = \max_i \frac{|r_i|}{(E|y| + f)_i}, \tag{2.7}$$

where $r = b - Ay$, and $\xi/0$ is interpreted as zero if $\xi = 0$ and infinity otherwise.

PROOF. It is easy to show that the right-hand side of (2.7) is a lower bound for $\omega(y)$, and that this bound is attained for the perturbations

$$\Delta A = D_1 E D_2, \qquad \Delta b = -D_1 f, \tag{2.8}$$

where $D_1 = \operatorname{diag}\big(r_i/(E|y| + f)_i\big)$ and $D_2 = \operatorname{diag}\big(\operatorname{sign}(y_i)\big)$. □

The next result gives a forward error bound corresponding to the componentwise backward error. It is a straightforward generalization of a result of Skeel [**63**, Theorems 2.1 and 2.2]. A monotonic norm is one for which $|x| \le |y|$ implies $\|x\| \le \|y\|$, which can be shown to be equivalent to $\| \, |x| \, \| = \|x\|$ [**71**, p. 52].

THEOREM 2.4. *Let $Ax = b$ and $(A + \Delta A)y = b + \Delta b$, where $|\Delta A| \le \epsilon E$ and $|\Delta b| \le \epsilon f$, and assume that $\epsilon \| \, |A^{-1}|E \, \| < 1$, where $\| \cdot \|$ is a monotonic norm. Then*

$$\frac{\|x - y\|}{\|x\|} \le \frac{\epsilon}{1 - \epsilon \| \, |A^{-1}|E \, \|} \frac{\| \, |A^{-1}|f + |A^{-1}|E|x| \, \|}{\|x\|}, \tag{2.9}$$

and for the ∞-norm this bound is attainable to first order in ϵ.

PROOF. The bound (2.9) is straightforward to derive. For the ∞-norm it is attained, to first order in ϵ, for $\Delta A = \epsilon D_1 E D_2$ and $\Delta b = -\epsilon D_1 f$, where $D_2 = \operatorname{diag}(\operatorname{sign}(x_i))$ and $D_1 = \operatorname{diag}(\xi_j)$, where $\xi_j = \operatorname{sign}(A^{-1})_{kj}$ and $\| \, |A^{-1}|f + |A^{-1}|E|x| \, \|_\infty = \big(|A^{-1}|f + |A^{-1}|E|x|\big)_k$. □

When the components of x vary widely in magnitude, the bound (2.9) can be weak for the smaller components. It is possible to obtain perturbation bounds for individual solution components by refraining from taking norms in the analysis. Chandrasekaran and Ipsen present this type of analysis [**14**]; see also the condition number (2.14) below. Componentwise solution bounds for Markov chains are obtained by Ipsen and Meyer [**49**], who measure the perturbation matrix normwise, and by O'Cinneide [**53**], who measures the perturbation matrix componentwise.

Theorem 2.4 implies that the condition number

$$\begin{aligned}\mathrm{cond}_{E,f}(A,x) \quad := \quad \lim_{\epsilon\to 0}\sup\{ \frac{\|\Delta x\|_\infty}{\epsilon\|x\|_\infty} &: (A+\Delta A)(x+\Delta x) = b+\Delta b, \\ &|\Delta A| \le \epsilon E, \quad |\Delta b| \le \epsilon f\,\}\end{aligned}$$

is given by

$$\mathrm{cond}_{E,f}(A,x) = \frac{\|\,|A^{-1}|f + |A^{-1}|E|x|\,\|_\infty}{\|x\|_\infty}. \tag{2.10}$$

For the special case $E = |A|$ and $f = |b|$ we have the condition numbers introduced by Skeel [**63**]:

$$\mathrm{cond}(A,x) := \frac{\|\,|A^{-1}||A||x|\,\|_\infty}{\|x\|_\infty},$$

which differs from $\mathrm{cond}_{|A|,|b|}(A,x)$ by at most a factor 2, and

$$\mathrm{cond}(A) := \mathrm{cond}(A,e) = \|\,|A^{-1}||A|\,\|_\infty \le \kappa_\infty(A). \tag{2.11}$$

Note that $\mathrm{cond}(A)$ is invariant under row scaling $Ax = b \to (DA)x = Db$, where D is diagonal, and for this reason it can be arbitrarily smaller than $\kappa_\infty(A)$. In fact, it is straightforward to show that

$$\min\{\,\kappa_\infty(DA) : D \text{ diagonal}\,\} = \mathrm{cond}(A), \tag{2.12}$$

where the optimal scaling D equilibrates the rows of A, that is, DA has rows of unit 1-norm.

How does cond compare with κ? Chandrasekaran and Ipsen [**14**] note the following inequalities. First, if D_R equilibrates the rows of A ($D_R|A|e = e$) then

$$\frac{\kappa_\infty(A)}{\kappa_\infty(D_R)} \le \mathrm{cond}(A) \le \kappa_\infty(A)$$

(these inequalities imply (2.12)). Thus $\mathrm{cond}(A)$ can be much smaller than $\kappa_\infty(A)$ only when the rows of A are badly scaled. Second, if D_C equilibrates the columns of A ($e^T|A|D_C = e^T$) then

$$\frac{\kappa_1(A)}{n\kappa_\infty(D_C)}\min_j \frac{\|A^{-1}e_j\|_\infty}{\|A^{-1}\|_1} \le \mathrm{cond}(A,x) \le \kappa_\infty(A).$$

These inequalities show that $\mathrm{cond}(A,x)$ can be much smaller than $\kappa_\infty(A)$ only when the columns of A are badly scaled or the columns of A^{-1} are badly scaled.

Returning to our numerical example, we find that $\omega_{E,f}(y) = 1.10 \times 10^{-12}$ for $E = |A|$ and $f = |b|$ or $f = 0$. This tells us that for y to be the first column of the inverse of a perturbed matrix we must make relative changes to A four orders of magnitude larger than the unit roundoff. For the perturbed system, Theorem 2.4 with $\epsilon = \text{tol}$, $E = |A|$ and $f = |b|$ gives the bound

$$\frac{\|x - z\|_\infty}{\|x\|_\infty} \le 4.08 \times 10^{-10},$$

which is eight orders of magnitude smaller than the normwise bound from Theorem 2.2, and only a factor 170 larger than the actual forward error (2.5).

A numerical example of Kahan [**51**] is also instructive. Let

$$A = \begin{bmatrix} 2 & -1 & 1 \\ -1 & \epsilon & \epsilon \\ 1 & \epsilon & \epsilon \end{bmatrix}, \quad b = \begin{bmatrix} 2(1+\epsilon) \\ -\epsilon \\ \epsilon \end{bmatrix},$$

where $0 < \epsilon \ll 1$, so that $x = [\epsilon, -1, 1]^T$. The normwise condition number $\kappa_\infty(A) = 2(1 + \epsilon^{-1})$, so the system is very sensitive to arbitrary perturbations in A and b. Moreover,

$$|A^{-1}||A| = \begin{bmatrix} 1 & \epsilon & \epsilon \\ \dfrac{2\epsilon + 1}{2\epsilon} & 1 & 1 \\ \dfrac{2\epsilon + 1}{2\epsilon} & 1 & 1 \end{bmatrix},$$

so $\text{cond}(A) = 3 + (2\epsilon)^{-1}$, which implies that the system is also very sensitive to componentwise perturbations for some right-hand sides. However, $\text{cond}(A, x) = 5/2 + \epsilon$, so *for this particular* b the system is very well-conditioned under componentwise perturbations.

We mention in passing another interesting role of the matrix $|A^{-1}||A|$. Recall from (2.12) that $\min_D \kappa_\infty(DA) = \| \, |A^{-1}||A| \, \|_\infty$. Bauer [**7**] shows that for two-sided scalings

$$\min\{\, \kappa_\infty(D_1 A D_2) : D_1,\, D_2 \text{ diagonal} \,\} \ge \rho(|A^{-1}||A|), \tag{2.13}$$

where ρ is the spectral radius, and he characterizes the minimizing D_1 and D_2 as $\text{diag}(x_1)^{-1}$ and $\text{diag}(x_2)^{-1}$, where x_1 and x_2 are Perron vectors of $|A||A^{-1}|$ and $|A^{-1}||A|$, respectively (thus the minimum is achieved whenever these Perron vectors have positive entries, which is guaranteed if A is irreducible—see Businger [**13**]). For the Kahan example,

$$\rho(|A^{-1}||A|) \approx 2.62 + 1.79\epsilon \ll 3 + (2\epsilon)^{-1} = \| \, |A^{-1}||A| \, \|_\infty,$$

and, in fact, $\kappa_\infty(DAD) = 3$ for $D = \text{diag}(\epsilon^{1/2}, \epsilon^{-1/2}, \epsilon^{-1/2})$, so a symmetric two-sided scaling is nearly optimal in this case.

2.1. Further notes and references. It is clear from Bauer's comments in [**8**] that the bound (2.9), with $E = |A|$ and $f = |b|$, was known to him, though he does not state the bound. This is the earliest reference we know in which componentwise analysis is used to derive forward perturbation bounds (for more details of Bauer's paper see §3).

Theorems 2.1 and 2.3 both remain valid when A is rectangular. Componentwise backward error for rectangular A was considered by Oettli, Prager and Wilkinson [**55**], but their results are subsumed by those of Oettli and Prager [**54**] and Rigal and Gaches [**58**].

The componentwise analyses can be generalized in three main ways:

(1) We can use more general measures of size for the data and the solution. Higham and Higham [**39**] measure ΔA, Δb and Δx by $\nu_p([(\Delta a_{ij}/e_{ij})\ (\Delta b_i/f_i)])$ and $\nu_p((\Delta x_i/g_i))$, where $\nu_p(A) = \left(\sum_{i,j} |a_{ij}|^p\right)^{1/p}$, $1 \le p \le \infty$, and the e_{ij}, f_i and g_i are tolerances. In [**39**] the corresponding backward error is shown to be given by the explicit formula

$$\left\| \left(\frac{r_j}{\| D_j \left[\begin{smallmatrix} y \\ -1 \end{smallmatrix} \right] \|_q} \right) \right\|_p,$$

where $r = b - Ay$, $D_j = \operatorname{diag}(e_{j1}, \ldots, e_{jn}, f_j)$ and $p^{-1} + q^{-1} = 1$; bounds for the corresponding condition number are also obtained. Theorem 2.3, and Theorem 2.4 with the ∞-norm, correspond to $p = \infty$ and $g_i \equiv \|x\|_\infty$. If we take $p = \infty$ and $g = |x|$, we are measuring the change in the solution in a componentwise relative sense, and the condition number is [**39**]

$$\| \operatorname{diag}(|x_i|)^{-1} |A^{-1}| (E|x| + f) \|_\infty. \tag{2.14}$$

This latter case has also been considered by Rohn [**59**] and Gohberg and Koltracht [**32**].

(2) The analysis can be extended to systems with multiple right-hand sides [**39**]. For the general ν_p measure described in (1), the backward error can be computed by finding the minimum p-norm solutions to n underdetermined linear systems.

(3) Structure in A and b can be preserved in the analysis. For example, if A is symmetric or Toeplitz then its perturbation can be forced to be symmetric or Toeplitz too, while still using componentwise measures. References include Higham and Higham [**38**] and Gohberg and Koltracht [**32**] for linear structure, and Bartels and D. J. Higham [**6**] for Vandermonde structure. A symmetry-preserving normwise backward error is explored by Bunch, Demmel and Van Loan [**12**], and symmetry-preserving normwise condition numbers are considered by D. J. Higham [**37**].

Practical issues associated with the use of componentwise backward error and perturbation bounds in the context of sparse matrices are considered by Arioli, Demmel and Duff [**2**]. In particular, they show how to estimate the condition number (2.10) cheaply, given a factorization of A. Their approach avoids explicit computation of A^{-1} by manipulating the condition number into a form that can be estimated using a matrix norm estimator of Hager [**36**] and Higham [**40**]. This approach can be adapted to estimate virtually any form of componentwise condition number for a linear system.

Finally, we mention how the analysis of this section is reflected in LAPACK. The LAPACK expert driver routine `xGESVX`[1] solves a linear system $Ax = b$ by Gaussian elimination with partial pivoting and then applies iterative refinement in fixed precision. The main termination criterion for the refinement is $\omega_{|A|,|b|}(\widehat{x}) \le u$, where $\widehat{x}$ is the current iterate and u is the unit roundoff. This routine returns as a forward error estimate the upper bound in

$$\frac{\|x-\widehat{x}\|_\infty}{\|x\|_\infty} \le \frac{\|\,|A^{-1}|(|\widehat{r}|+\xi)\,\|_\infty}{\|x\|_\infty},$$

where $\widehat{r} = fl(b - A\widehat{x})$ and $\xi = (n+1)u(|A||\widehat{x}| + |b|)$ bounds the rounding errors in forming $\widehat{r}$. This a posteriori bound is used in preference to (2.9) because it is expected to be smaller in general, since less inequalities are used in its derivation.

3. The Matrix Inverse

Much of the analysis in §2 can be adapted to the matrix inverse. If $A \in \mathbb{R}^{n\times n}$ and $A+\Delta A$ are nonsingular then, writing $X \equiv A^{-1}$ and $X+\Delta X \equiv (A+\Delta A)^{-1}$, we have $\Delta X = -A^{-1}\Delta A A^{-1} - A^{-1}\Delta A\Delta X$. Therefore derivation of perturbation bounds and condition numbers involves consideration of the first-order term $A^{-1}\Delta A A^{-1}$. For normwise measures it is well known that the condition number for inversion is the usual matrix condition number $\kappa(A)$. This result can be stated in a precise and general way as follows [**37**]: for the mixed subordinate norm $\|A\|_{\alpha,\beta} = \max \|Ax\|_\beta/\|x\|_\alpha$,

$$\mu_{\alpha,\beta}(A) := \lim_{\epsilon\to 0} \sup\{ \frac{\|\Delta X\|_{\beta,\alpha}}{\epsilon\|X\|_{\beta,\alpha}} : \|\Delta A\|_{\alpha,\beta} \le \epsilon\|A\|_{\alpha,\beta} \} = \|A\|_{\alpha,\beta}\|A^{-1}\|_{\beta,\alpha}.$$

Note that this definition uses the $\|\cdot\|_{\alpha,\beta}$ norm on the data space and the $\|\cdot\|_{\beta,\alpha}$ norm on the solution space, as is natural. A componentwise condition number is

$$\mu_E(A) := \lim_{\epsilon\to 0} \sup\{ \frac{\|\Delta X\|_\infty}{\epsilon\|X\|_\infty} : |\Delta A| \le \epsilon E \} \le \frac{\|\,|A^{-1}|E|A^{-1}|\,\|_\infty}{\|A^{-1}\|_\infty}.$$

In general, the inequality is strict, but there is equality when $|A^{-1}| = D_1A^{-1}D_2$ for D_i of the form $\mathrm{diag}(\pm 1)$, [**30**, Theorem 1.10], [**31**]. For perturbations satisfying $|\Delta A| \le \epsilon|A|$, Bauer [**8**, p. 413] obtains the bound

$$|\Delta X| \le (I - \epsilon|A^{-1}||A|)^{-1}|A^{-1}||A||A^{-1}|\epsilon,$$

and makes various deductions from it.

---

[1]The leading 'x' in routine names stands for the Fortran data type: single precision (`S`), double precision (`D`), complex (`C`) or double precision complex (`Z`).

Measuring ΔX componentwise relative to X, and taking $E = |A|$, gives the condition number

$$\mu_2(A) := \lim_{\epsilon\to 0} \sup\{ \max_{i,j} \frac{|\Delta x_{ij}|}{|x_{ij}|} : |\Delta A| \le \epsilon |A| \} = \max_{i,j} \frac{\big(|A^{-1}||A||A^{-1}|\big)_{ij}}{\big(|A^{-1}|\big)_{ij}}.$$

This condition number appears in a perturbation bound of Bauer [**8**, p. 413] and was derived by Rohn [**59**].

The natural definition of backward error of an approximate inverse Y of A is a smallest ΔA such that $(A + \Delta A)Y = I$. If Y is nonsingular there is a unique ΔA satisfying this equation and the problem of computing the backward error is trivial.

It is well known that the normwise matrix condition number is the reciprocal of the relative distance to singularity. Specifically, Kahan [**51**, pp. 775–76] shows that

$$\min\{ \frac{\|\Delta A\|_{\alpha,\beta}}{\|A\|_{\alpha,\beta}} : A + \Delta A \text{ singular} \} = \mu_{\alpha,\beta}^{-1}.$$

The componentwise distance to singularity,

$$d_E(A) = \min\{ \epsilon : A + \Delta A \text{ singular}, \ |\Delta A| \le \epsilon E \},$$

has been characterized by Rohn [**60**], [**61**] as

$$d_E(A) = \frac{1}{\max_{S_1,S_2} \rho_0(S_1 A^{-1} S_2 E)},$$

where the maximum is taken over all signature matrices $S_i = \operatorname{diag}(\pm 1)$ and where

$$\rho_0(X) = \max\{ |\lambda| : \lambda \text{ is a real eigenvalue of } A \}.$$

This formula involves 4^n eigenproblems, so is computationally intractable (in fact it is NP-hard [**57**]).

Demmel [**19**] shows by complexity arguments that there can be no simple relationship between $d_E(A)$ and the quantity $\| \, |A^{-1}|E \, \|_\infty$, which is an upper bound for $\mu_E(A)$. He also presents evidence for the conjecture that

$$\frac{1}{\rho(|A^{-1}||A|)} \le d_{|A|}(A) \le \frac{\gamma_n}{\rho(|A^{-1}||A|)}$$

for a constant γ_n. The lower bound always holds and Demmel identifies several classes of matrices for which the upper bound holds. This conjecture is both plausible and aesthetically pleasing because $d_{|A|}(A)$ is invariant under two-sided diagonal scalings of A and $\rho(|A^{-1}||A|)$ is the minimum condition number achievable by such scalings, as shown by (2.13).

4. The Least Squares Problem

In this section we consider the least squares (LS) problem $\min_x \|Ax - b\|_2$, where $A \in \mathbb{R}^{m\times n}$ $(m \geq n)$ has full rank. Normwise perturbation theory for the LS problem has been developed by various authors. The earliest bounds were obtained by Golub and Wilkinson [**33**]. The following theorem of Wedin is taken from [**78**, Theorem 5.1]. Here, $\kappa_2(A) = \|A\|_2\|A^+\|_2$, where A^+ is the pseudo-inverse.

THEOREM 4.1 (WEDIN). *Let $A \in \mathbb{R}^{m\times n}$ $(m \geq n)$ and $A + \Delta A$ both be of full rank, and let*

$$\|Ax - b\|_2 = \min, \quad r = b - Ax,$$
$$\|(A + \Delta A)y - (b + \Delta b)\|_2 = \min,$$
$$\|\Delta A\|_2 \leq \epsilon\|A\|_2, \quad \|\Delta b\|_2 \leq \epsilon\|b\|_2.$$

Provided that $\kappa_2(A)\epsilon < 1$,

$$\frac{\|x - y\|_2}{\|x\|_2} \leq \frac{\kappa_2(A)\epsilon}{1 - \kappa_2(A)\epsilon}\left(1 + \frac{\|b\|_2}{\|A\|_2\|x\|_2} + \kappa_2(A)\frac{\|r\|_2}{\|A\|_2\|x\|_2}\right). \quad \square \tag{4.1}$$

Bounds can also be given for the change in the residual, but we do not consider them here.

The bound (4.1) is usually interpreted as saying that the sensitivity of the LS problem is measured by $\kappa_2(A)$ when the residual is small or zero and by $\kappa_2(A)^2$ otherwise. For further discussion see standard textbooks (for example, [**34**, pp. 230–231]).

Surprisingly, it is easier to derive componentwise perturbation bounds than normwise ones for the LS problem. The key idea is to express the LS solution and its residual as the solution of the augmented system

$$\begin{bmatrix} I & A \\ A^T & 0 \end{bmatrix}\begin{bmatrix} r \\ x \end{bmatrix} = \begin{bmatrix} b \\ 0 \end{bmatrix}, \tag{4.2}$$

which is simply another way of writing the normal equations, $A^TAx = A^Tb$. This is a square nonsingular system, so standard techniques can be applied. The perturbed system of interest is

$$\begin{bmatrix} I & A + \Delta A \\ (A + \Delta A)^T & 0 \end{bmatrix}\begin{bmatrix} s \\ y \end{bmatrix} = \begin{bmatrix} b + \Delta b \\ 0 \end{bmatrix}, \tag{4.3}$$

where we assume that

$$|\Delta A| \leq \epsilon E, \quad |\Delta b| \leq \epsilon f. \tag{4.4}$$

From (4.2) and (4.3) we obtain

$$\begin{bmatrix} I & A \\ A^T & 0 \end{bmatrix}\begin{bmatrix} s - r \\ y - x \end{bmatrix} = \begin{bmatrix} \Delta b - \Delta A y \\ -\Delta A^T s \end{bmatrix}.$$

Premultiplying by the inverse of the matrix on the left gives

$$\begin{bmatrix} s-r \\ y-x \end{bmatrix} = \begin{bmatrix} I - AA^+ & (A^+)^T \\ A^+ & -(A^TA)^{-1} \end{bmatrix} \begin{bmatrix} \Delta b - \Delta A y \\ -\Delta A^T s \end{bmatrix}.$$

From the second block component we obtain

$$|y-x| \le \epsilon(|A^+|(f+E|y|) + |(A^TA)^{-1}|E^T|s|). \tag{4.5}$$

This yields the following theorem.

THEOREM 4.2. *Let $A \in \mathbb{R}^{m\times n}$ $(m \ge n)$ and $A + \Delta A$ be of full rank. For the perturbed LS problem described by* (4.3) *and* (4.4) *we have*

$$\frac{\|x-y\|}{\|x\|} \le \epsilon \frac{\|\,|A^{-1}|(f+E|y|)\,\| + \|\,|(A^TA)^{-1}|E^T|s|\,\|}{\|x\|}, \tag{4.6}$$

for any monotonic norm. □

For a square system, $s = 0$, and we essentially recover Theorem 2.4. Note, however, that the right-hand side of (4.6) contains the perturbed vectors y and s. For theoretical analysis it may be preferable to use an alternative bound in which x and r replace y and s and there is an extra factor

$$\left(1 - \epsilon \left\| \begin{bmatrix} |I - AA^+| & |(A^+)^T| \\ |A^+| & |(A^TA)^{-1}| \end{bmatrix} \begin{bmatrix} 0 & E \\ E^T & 0 \end{bmatrix} \right\| \right)^{-1},$$

where the term in parentheses is assumed to be positive. For practical computation (4.6) is unsatisfactory because we do not know $s = b + \Delta b - (A + \Delta A)y$. However, as Stewart and Sun observe [**71**, p. 159], $\widehat{r} = b - Ay$ is computable and

$$|s| \le |\widehat{r}| + \epsilon(f + E|y|),$$

and using this bound in (4.6) makes only a second-order change.

A componentwise bound of the form (4.6) was first derived by Björck in 1988 and variations of it have been given by Arioli, Duff and de Rijk [**3**], Björck [**11**] and Higham [**42**].

Apart from its increased sharpness over (4.1), the bound (4.6) has better scaling properties. It is not invariant under row or column scalings, but it is less sensitive to these scalings than (4.1). In [**42**] we examined the famous Longley test problem—a regression problem which has a notoriously ill-conditioned 16×7 coefficient matrix with $\kappa_2(A) \approx 5 \times 10^9$. We found that (4.5) gives tight bounds for the effect of random componentwise relative perturbations of the problem generated in experiments of Beaton, Rubin and Barone [**10**]. Thus componentwise perturbation bounds are potentially useful in regression analysis, as an alternative to the existing statistically-based techniques.

Although Wilkinson proved in the 1960s that a specific method for solving the LS problem, namely the QR factorization method, yields a small normwise backward error [**81**], it has long been an open problem to obtain a formula for the backward error of an arbitrary approximate solution. Stewart [**68**] discusses the problem and offers some backward perturbations that are candidates for being of minimal norm. Little progress had been made towards solving this problem until very recently when Waldén, Karlson and Sun [**77**] found an extremely elegant solution. We will denote by $\lambda_{\min}$ and $\sigma_{\min}$ the smallest eigenvalue of a symmetric matrix and the smallest singular value of a general matrix, respectively.

THEOREM 4.3 (WALDÉN, KARLSON AND SUN). *Let $A \in \mathbb{R}^{m\times n}$ $(m \geq n)$, $b \in \mathbb{R}^m$ and $r = b - Ay$. The normwise backward error*

$$\eta_F(y) = \min\{\, \|[\Delta A,\ \theta\Delta b]\|_F : \|(A+\Delta A)y - (b+\Delta b)\|_2 = \min\,\} \tag{4.7}$$

is given by

$$\eta_F(y) = \begin{cases} \dfrac{\|r\|_2}{\|y\|_2}\sqrt{\gamma}, & \lambda_* \geq 0, \\ \left(\dfrac{\|r\|_2^2}{\|y\|_2^2}\gamma + \lambda_*\right)^{1/2}, & \lambda_* < 0, \end{cases}$$

where

$$\lambda_* = \lambda_{\min}\Bigl(AA^T - \gamma\frac{rr^T}{\|y\|_2^2}\Bigr), \quad \gamma = \frac{\theta^2\|y\|_2^2}{1+\theta^2\|y\|_2^2}. \quad \Box$$

The backward error (4.7) is not a direct generalization of the one in (2.1) because it minimizes $\|[\Delta A,\ \theta\Delta b]\|_F$ instead of $\max\{\|\Delta A\|_2/\|E\|_2, \|\Delta b\|_2/\|f\|_2\}$. However, the parameter θ allows us some flexibility: taking the limit $\theta \to \infty$ gives the case where only A is perturbed.

Theorem 4.3 can be interpreted as saying that if $\lambda_* \geq 0$ then the backward error is essentially that given by Theorem 2.1 for a consistent system. If $\lambda_* < 0$, however, the nearest perturbed system of which y is the LS solution is inconsistent. A sufficient condition for $\lambda_* < 0$ is $r \notin \text{range}(A)$ (assuming $\gamma \neq 0$).

The formulae given in Theorem 4.3 are unsuitable for computation because they can suffer from catastrophic cancellation when $\lambda_* < 0$. Instead, the following alternative formula derived in [**77**] should be used:

$$\eta_F(y) = \min\left\{\frac{\|r\|_2}{\|y\|_2}\sqrt{\gamma},\ \sigma_{\min}\left(\left[A,\quad \sqrt{\gamma}\frac{\|r\|_2}{\|y\|_2}\left(I - \frac{r_y r_y{}^T}{\|r\|_2^2}\right)\right]\right)\right\}.$$

To illustrate Theorem 4.3, we consider an LS problem with a 25×15 Vandermonde matrix $A = (p_i^{j-1})$, where the p_i are equally spaced on $[0,1]$, and a right-hand side b with elements equally spaced on $[0,1]$. The condition number $\kappa_2(A) = 1.47 \times 10^9$. We solved the LS problem in Matlab in two different ways: by using the normal equations with Cholesky factorization, to give $\widehat{x}_{NE}$, and via Householder QR factorization, obtaining $\widehat{x}_{QR}$. We found that

$$\frac{\eta_F(\widehat{x}_{NE})|_{\theta=1}}{\|[A,b]\|_F} = 2.37 \times 10^{-10}, \qquad \frac{\eta_F(\widehat{x}_{NE})|_{\theta=\infty}}{\|A\|_F} = 2.56 \times 10^{-10},$$

$$\frac{\eta_F(\widehat{x}_{QR})|_{\theta=1}}{\|[A,b]\|_F} = 2.97 \times 10^{-17}, \qquad \frac{\eta_F(\widehat{x}_{QR})|_{\theta=\infty}}{\|A\|_F} = 3.22 \times 10^{-17}.$$

Thus, as would be expected from known backward error analysis, $\widehat{x}_{QR}$ is a backward stable solution but $\widehat{x}_{NE}$ is not. In this example it makes little difference whether or not we perturb b.

Componentwise backward error for the LS problem has been investigated by Arioli, Duff and de Rijk [**3**], Björck [**11**] and Higham [**42**]. The simplest approach is to apply the componentwise backward error $\omega_{E,f}(y)$ of (2.6) to the augmented system (4.2), setting

$$E = \begin{bmatrix} 0 & E_A \\ E_A^T & 0 \end{bmatrix}$$

so as not to perturb the diagonal blocks I and 0 of the augmented system coefficient matrix. However, this approach allows A and A^T to undergo different perturbations ΔA_1 and ΔA_2 with $\Delta A_1 \neq \Delta A_2^T$, so does not give a true backward error. This problem can be overcome by using the structured componentwise backward error of [**38**] to force symmetry of the perturbations; see [**38**] for details. One problem remains: as far as the backward error of y is concerned, the vector r in the augmented system is a vector of free parameters, so to obtain the true componentwise backward error we have to minimize the structure-preserving componentwise backward error over all r. This is a nonlinear optimization problem to which no closed-form solution is known. Experiments in [**38**] and [**44**] show that when y is a computed LS solution, $r = b - Ay$ is often a good approximation to the minimizing r.

4.1. Underdetermined systems. Perturbation theory is also available for the minimum 2-norm solution to an underdetermined system. Wedin's result, Theorem 4.1, is still valid when $m < n$ and x and y are interpreted as minimum 2-norm solutions of $Ax = b$ and $(A + \Delta A)y = b + \Delta b$. Of course, $r = 0$, so there is no $\kappa_2(A)^2$ term in the bound. A componentwise bound is given by Demmel and Higham [**22**].

THEOREM 4.4 (DEMMEL AND HIGHAM). *Let $A \in \mathbb{R}^{m\times n}$ $(m \leq n)$ be of full rank and $0 \neq b \in \mathbb{R}^m$. Suppose*

$$|\Delta A| \leq \epsilon E, \quad |\Delta b| \leq \epsilon f,$$

where $\epsilon\|E\|_2 < \sigma_{\min}(A)$. If x and y are the minimum norm solutions to $Ax = b$ and $(A + \Delta A)y = b + \Delta b$, respectively, then

$$\frac{\|x - y\|_2}{\|x\|_2} \leq \Big(\|\, |I - A^+A| \cdot E^T \cdot |A^{+^T} x| \,\|_2 + \|\, |A^+| \cdot (f + E|x|) \,\|_2 \Big) \frac{\epsilon}{\|x\|_2} + O(\epsilon^2).$$

The bound is attainable to within a constant factor depending on n. $\square$

Note that $\|I - A^+A\|_2 = \min(1, n - m)$, so again there is no $\kappa_2(A)^2$ effect in the bound. This theorem is used in [**22**], with different choices of E and f, to investigate the stability of standard methods for computing the minimal 2-norm solution to an underdetermined system.

5. Matrix Factorizations

Perturbation theory for matrix factorizations has been an active area of research in recent years. The aim of such analysis is to determine bounds for the changes in the factors of a matrix when the matrix is perturbed. Most of the existing results are norm-based. One of the first results was that of Stewart [**67**]. He showed that if $A \in \mathbb{R}^{m \times n}$ has rank n and

$$A = QR \quad \text{and} \quad A + \Delta A = (Q + \Delta Q)(R + \Delta R)$$

are QR factorizations, then, for sufficiently small ΔA,

$$\begin{aligned} \frac{\|\Delta R\|_F}{\|R\|_F} &\le c_n \kappa_F(A) \frac{\|\Delta A\|_F}{\|A\|_F}, \\ \|\Delta Q\|_F &\le c_n \kappa_F(A) \frac{\|\Delta A\|_F}{\|A\|_F}, \end{aligned}$$

where c_n is a modest constant.

Similar analyses for Cholesky, $\mathrm{LDL^T}$, LU and QR factorizations are given by Barrlund [**5**], Stewart [**70**], and Sun [**73**].

Componentwise analyses have been given by Zha [**82**] for the QR factorization and Sun [**74**] for the Cholesky, $\mathrm{LDL^T}$, LU and QR factorizations. Sun's results are rather complicated to state and interpret. Zha's bounds can be summarized as follows, with the same assumptions and notation as for Stewart's result above. Let $|\Delta A| \le \epsilon G|A|$, where G is nonnegative with $|g_{ij}| \le 1$. Then, for sufficiently small ϵ,

$$\begin{aligned} \frac{\|\Delta R\|_\infty}{\|R\|_\infty} &\le c_{m,n}\epsilon\big(\mathrm{cond}(R^{-1}) + \mathrm{cond}(R^T)\big) + O(\epsilon^2), \\ \|\Delta Q\|_\infty &\le c_{m,n}\epsilon\big(\mathrm{cond}(R^{-1}) + \mathrm{cond}(R^T)\big) + O(\epsilon^2), \end{aligned}$$

where cond is defined in (2.11). The quantity $\phi(A) = \mathrm{cond}(R^{-1}) + \mathrm{cond}(R^T)$ can therefore be thought of as a condition number for the QR factorization under the columnwise class of perturbations considered. Note that ϕ is independent of the column scaling of A. Zha [**82**] shows how this perturbation result can be used together with a componentwise rounding error analysis to bound the difference $Q - \widehat{Q}$ between the exact QR factor Q and the computed one from Householder QR factorization.

6. Eigenvalue and Singular Value Problems

Perturbation theory for eigenvalue problems is inherently more difficult than for linear equations because the mapping from data to solution is no longer a rational map. Normwise perturbation theory for eigenproblems is well developed (see, for example, Stewart and Sun's comprehensive presentation [**71**]) but componentwise results have been obtained only in the last five years or so.

Existing results are not easy to present in an organized fashion because they cover different problems (standard or generalized, symmetric positive definite or indefinite), make different assumptions, and measure perturbations in different ways. We give a selected summary, concentrating on the standard eigenvalue and singular value problems and noting which results have counterparts for the generalized problem. We do not necessarily state results in their most general form, or use their authors' original notation. In particular, the roles of A and H are reversed between our presentation and those of most of the references, because it seems natural to denote the original matrix, whose eigenvalues or singular values we seek, by A.

Most of the eigenvalue results are for symmetric matrices, but we begin with the nonsymmetric case. It is a standard result that if $A \in \mathbb{R}^{n\times n}$ has a simple eigenvalue λ with corresponding right and left eigenvectors x and y, then for sufficiently small $\|\Delta A\|$ there is an eigenvalue μ of $A+\Delta A$ with

$$\mu = \lambda + \frac{y^*\Delta A x}{y^* x} + O(\|\Delta A\|^2). \tag{6.1}$$

(See, for example, [**71**, Theorem 4.2.3] or [**81**, Chapter 2]). If we assume that $\|\Delta A\| \le \epsilon\|E\|$, then we have the perturbation bound

$$|\mu - \lambda| \le \epsilon\|E\|\frac{\|y\|_D\|x\|}{|y^* x|} + O(\epsilon^2), \tag{6.2}$$

which is sharp to first order (equality is attained for ΔA an appropriate multiple of uv^T, where u is dual to y and v is dual to x). On the other hand, if we assume that $|\Delta A| \le \epsilon E$, we obtain the componentwise perturbation bound

$$|\mu - \lambda| \le \epsilon\frac{|y^*|E|x|}{|y^* x|} + O(\epsilon^2) \tag{6.3}$$

of Geurts [**31**], for which equality to first order is attained for

$$\Delta A = \epsilon \operatorname{diag}(\operatorname{sign}(y)) E \operatorname{diag}(\operatorname{sign}(x)),$$

where, for possibly complex α, $\operatorname{sign}(\alpha) = \overline{\alpha}/|\alpha|$ (or 1 if $\alpha = 0$). The first-order terms in the bounds (6.2) and (6.3) yield condition numbers under the respective classes of perturbations. One interesting implication of (6.3) is that the Perron root of an irreducible nonnegative matrix is perfectly conditioned under componentwise relative perturbations ($|\Delta A| \le \epsilon E = \epsilon A$), for the eigenvectors corresponding to the Perron root have positive elements and so the bound is $|\mu - \lambda| \le \epsilon\lambda + O(\epsilon^2)$. In fact, this bound holds for any $\epsilon \le 1$ without the second-order term [**27**, Theorem 1].

Corresponding eigenvector perturbation results can also be obtained. We state condition numbers in the form given by Chatelin [**15**] and Frayssé [**30**], whose analysis extends earlier analysis by Wilkinson [**81**], Stewart [**66**] and Geurts [**31**]; see also the book by Saad [**62**, pp. 95–97]. With the same notation as above (with

λ simple), a condition number of x under perturbations measured normwise by $\|\Delta A\| \le \epsilon \|E\|$ is

$$\|E\| \|\mathcal{P}\|,$$

where $\mathcal{P} = X(B - \lambda I)^{-1} Y^*$ is a partial inverse for Y in the space range$(y)^\perp$, and where

$$Z^{-1}AZ = \begin{bmatrix} \lambda & c^T \\ 0 & B \end{bmatrix}, \quad Z = [\, x \quad X \,], \quad Z^{-1} = \begin{bmatrix} y^* \\ Y^* \end{bmatrix}.$$

Note that $\|\mathcal{P}\|$ depends on the particular choice of X and Y. This freedom stems from the nonuniqueness of an eigenvector up to scalar multiplication, which means that there is not a unique way to measure the perturbation in an eigenvector. What measure is appropriate depends on the application (see [**71**, p. 241] for a discussion and example). The minimum condition number in the 2-norm is obtained when X and Y are orthonormal and is

$$\|\mathcal{P}\|_2 = \|(B - \lambda I)^{-1}\|_2 = \mathrm{sep}(B, \lambda)^{-1}.$$

A condition number for perturbations measured componentwise by $|\Delta A| \le \epsilon E$, with the ∞-norm on the output space, is

$$\frac{\| \, |\mathcal{P}| E |x| \, \|_\infty}{\|x\|_\infty}.$$

In the special case where A is diagonalizable we can take B to be diagonal and X and Y to comprise right and left eigenvectors. Then it is clear that the size of $\mathcal{P}$ depends on both the separation of λ from the other eigenvalues and on the individual eigenvalue sensitivities.

Another classical eigenvalue perturbation result is a theorem of Bauer and Fike [**9**, Theorem 3a], which says that if μ is an eigenvalue of $A + \Delta A \in \mathbb{R}^{n \times n}$ and $X^{-1}AX = \mathrm{diag}(\lambda_1, \ldots, \lambda_n)$ then

$$\min_{\lambda \in \lambda(A)} |\lambda - \mu| \le \kappa_p(X) \|\Delta A\|_p,$$

for any Hölder p-norm. As Deif [**17**] shows, it is trivial to strengthen this result to

$$\min_{\lambda \in \lambda(A)} |\lambda - \mu| \le \| \, |X^{-1}| |\Delta A| |X| \, \|,$$

for any monotonic norm.

In the rest of this section we describe the new-style componentwise perturbation results, developed initially by Demmel and his co-workers. For the eigenvalue problem, most of these results require that A be symmetric positive definite. We therefore let $A \in \mathbb{R}^{n \times n}$ be symmetric positive definite and consider a symmetric perturbation $A + \Delta A$. Let the eigenvalues of A and $A + \Delta A$ be ordered $\lambda_n \le \cdots \le \lambda_1$ and $\lambda_n' \le \cdots \le \lambda_1'$, respectively. From classical perturbation results for the symmetric eigenproblem we know that

$$|\lambda_i - \lambda_i'| \le \|\Delta A\|_2$$

(specifically, this follows from Weyl's inequality [**71**, p. 203], which itself is a consequence of the Courant–Fischer minimax theorem [**71**, p. 201]). For $\lambda_1 = \|A\|_2$ this bound says that $|\lambda_1 - \lambda_1'|/\lambda_1 \le \|\Delta A\|_2/\|A\|_2$, so this largest eigenvalue undergoes a relative change no larger than that in A. But eigenvalues with $\lambda_i \ll \lambda_1$ can undergo a large relative change, and the best relative perturbation bound that can be deduced for all i is

$$\frac{|\lambda_i - \lambda_i'|}{\lambda_i} \le \frac{\|\Delta A\|_2}{\lambda_i} \le \frac{\|\Delta A\|_2}{\lambda_n} = \kappa_2(A)\frac{\|\Delta A\|_2}{\|A\|_2}. \tag{6.4}$$

This bound is sharp, in that for any A there is a perturbation ΔA for which equality is attained with $i = n$. However, by restricting the class of perturbations, tighter bounds can be obtained. Demmel and Veselić [**24**, Theorems 2, 3] derive a potentially much smaller bound for the case of componentwise relative perturbations.

THEOREM 6.1 (DEMMEL AND VESELIĆ). *Let $A \in \mathbb{R}^{n\times n}$ be symmetric positive definite and write $A = DHD$, where $D = \operatorname{diag}(A)^{1/2}$. Let the symmetric perturbation $\Delta A = D\Delta H D$ satisfy $\|\Delta H\|_2 \equiv \epsilon < \lambda_n(H)$. Then*

$$\frac{|\lambda_i - \lambda_i'|}{\lambda_i} \le \kappa_2(H)\epsilon. \tag{6.5}$$

In particular, if $|\Delta A| \le (\epsilon/n)|A|$, so that $|\Delta h_{ij}| \le \epsilon/n$, then $\|\Delta H\|_2 \le \epsilon$ and the bound (6.5) *holds if $\epsilon < \lambda_n(H)$.* □

The key idea in this theorem is to factor out the diagonal of A to leave a symmetric positive definite matrix H with unit diagonal. A famous result of van der Sluis [**75**, Theorem 4.1] states that

$$\kappa_2(H) \le n \min_{F \text{ diagonal}} \kappa_2(FAF),$$

so D is nearly a condition-minimizing diagonal scaling. It follows that $\kappa_2(H) \le n\kappa_2(A)$ and that $\kappa_2(H) \ll \kappa_2(A)$ is possible if A is badly scaled.

The proof of Theorem 6.1 in [**24**] is quite short and is based on the following lemma.

LEMMA 6.2. *Let $A \in \mathbb{R}^{n\times n}$ be symmetric positive definite and write $A = YHY$, where Y is any nonsingular matrix. Let the symmetric perturbation $\Delta A = Y\Delta HY$ satisfy $|z^T\Delta Hz| \le \eta\, z^THz$ for all z. Then*

$$1 - \eta \le \frac{\lambda_i'}{\lambda_i} \le 1 + \eta.$$

PROOF. Note first that

$$\frac{|x^T\Delta Ax|}{x^TAx} = \frac{|x^TY\Delta HYx|}{x^TYHYx} \equiv \frac{|z^T\Delta Hz|}{z^THz} \le \eta.$$

From the Courant–Fischer minimax theorem we have

$$\begin{aligned}\lambda_i' &= \max_{\dim(S)=i} \min_{x\in S} \frac{x^T(A+\Delta A)x}{x^Tx}\\ &= \max_{\dim(S)=i} \min_{x\in S} \frac{x^T(A+\Delta A)x}{x^TAx}\cdot\frac{x^TAx}{x^Tx}\\ &\le (1+\eta)\max_{\dim(S)=i} \min_{x\in S} \frac{x^TAx}{x^Tx} = (1+\eta)\lambda_i.\end{aligned}$$

The lower bound is proved similarly, starting with the max–min expression for λ_i. □

PROOF OF THEOREM 6.1. We have

$$\frac{|z^T\Delta Hz|}{z^THz} = \frac{|z^T\Delta Hz|}{z^Tz}\frac{z^Tz}{z^THz} \le \frac{\epsilon}{\lambda_n(H)}.$$

We obtain (6.5) on applying Lemma 6.2 with $Y = D$ and $\eta = \epsilon/\lambda_n(H) = \epsilon\|H^{-1}\|_2$, and using the inequality $\|H\|_2 \ge \max_{i,j}|h_{ij}| = 1$. □

A weakness of Theorem 6.1 is that it gives the same relative perturbation bound for each eigenvalue, yet some eigenvalues may be much less sensitive than others. Furthermore, λ_1 is insensitive to perturbations, yet the bound (6.5) can be large even for $i = 1$. Demmel and Veselić note that, under the same conditions as in Theorem 6.1, one can show, using (6.1), that

$$\frac{|\lambda_i - \lambda_i'|}{\lambda_i} \le \epsilon\frac{\|Dv_i\|_2^2}{\lambda_i} + O(\epsilon^2),$$

where $Av_i = \lambda_i v_i$ with $\|v_i\|_2 = 1$, and that this bound is attained, to first order, for $\Delta A = \epsilon D^2$. This bound depends on i, and can be much smaller than (6.5), but it is a first-order bound only.

Demmel and Veselić show, via rounding error analysis, that Jacobi's method (with a suitable stopping criterion), the bisection method, and inverse iteration (the latter two methods both applied to the original matrix) all compute the eigenvalues of a symmetric positive definite matrix to within the accuracy specified by the bound of Theorem 6.1 (with ϵ the machine precision). By contrast, any method that begins by tridiagonalizing the matrix (for example, the QR algorithm) cannot achieve this level of accuracy.

We turn now to eigenvectors. Standard perturbation theory (for which a classic reference is [**16**]) is expressed in terms of the absolute gap for eigenvalues, defined by

$$\text{absgap}_i = \min_{j\ne i}\frac{|\lambda_i - \lambda_j|}{\|A\|_2}.$$

Let y be a unit eigenvector of $A+\Delta A$ with Rayleigh quotient $\alpha = y^TAy$. Let λ_i be the eigenvalue of A closest to α and x_i the corresponding unit eigenvector.

Denoting by $\theta(x_i, y)$ the acute angle between x_i and y, we have (by rewriting the bound of [**56**, Theorem 11-7-1])

$$(6.6) \qquad |\sin\theta(x_i, y)| \le \frac{4}{\mathrm{absgap}_i} \frac{\|\Delta A\|_2}{\|A\|_2}.$$

Demmel and Veselić [**24**, Theorem 2.5] derive a potentially much smaller bound.

THEOREM 6.3 (DEMMEL AND VESELIĆ). *Let $A = DHD$ be as in Theorem 6.1, and define $A(\epsilon) = D(H + \epsilon E)D$, where E with $\|E\|_2 = 1$ is an arbitrary symmetric matrix. Let $\lambda_i(\epsilon)$ be the ith eigenvalue of $A(\epsilon)$ and assume that $\lambda_i(0)$ is simple, so that the corresponding unit eigenvector $v_i(\epsilon)$ is well defined for sufficiently small ϵ. Then*

$$(6.7) \qquad \|v_i(\epsilon) - v_i(0)\|_2 \le \frac{(n-1)^{1/2}\kappa_2(H)\epsilon}{\mathrm{relgap}_i} + O(\epsilon^2),$$

where the relative gap is defined by

$$\mathrm{relgap}_i = \min_{j\ne i} \frac{|\lambda_i - \lambda_j|}{|\lambda_i\lambda_j|^{1/2}}. \quad \square$$

The new bound (6.7) contains an extra factor $\kappa_2(H)$ compared with (6.6), but the important difference is that the relative gap in the denominator of (6.7) is much larger than the absolute gap in (6.6) for a small eigenvalue that is in a cluster of eigenvalues. For example, if $\lambda_1 = 1$, $\lambda_2 = 2\times 10^{-10}$ and $\lambda_3 = 10^{-10}$, then

$$\begin{aligned} \mathrm{absgap}(1\colon 3) &= 1 - 2\times 10^{-10},\ 10^{-10},\ 10^{-10}, \\ \mathrm{relgap}(1\colon 3) &= 2^{-1/2}(10^5 - 2\times 10^{-5}),\ 2^{-1/2},\ 2^{-1/2}. \end{aligned}$$

How sharp are these bounds? For any A, approximate equality is attained for some i in the bound (6.5) for a perturbation ΔA satisfying the conditions of Theorem 6.1 [**24**, Prop. 2.10]. This implies that the only symmetric positive definite matrices whose eigenvalues are determined to high relative accuracy by the matrix elements are those $A = DHD$ where H is well conditioned. A weaker sharpness result holds for (6.7) [**24**, Prop. 2.11].

We mention another interesting result that can be proved by Rayleigh quotient manipulations similar to those in the proof of Lemma 6.2. This result adds further information to the straightforward inequalities $\lambda_n \le a_{ii} \le \lambda_1$, by showing that the eigenvalues of A differ from the diagonal elements by at most a factor $\kappa_2(H)$.

THEOREM 6.4 (DEMMEL AND VESELIĆ [**24**, Prop. 2.7]). *Let $A \in \mathbb{R}^{n\times n}$ be symmetric positive definite and write $A = DHD$, where $D = \mathrm{diag}(A)^{1/2}$. Let $\alpha_{nn} \le \cdots \le \alpha_{11}$ be the diagonal entries of A arranged in decreasing order. Then*

$$\lambda_n(H) \le \frac{\lambda_i}{\alpha_{ii}} \le \lambda_1(H) \le n. \quad \square$$

Analogous results to Theorem 6.1 and 6.3 hold for singular values and singular vectors. Let $A \in \mathbb{R}^{m\times n}$, where there is no restriction on m and n. We denote the singular values of A and $A+\Delta A$ by $\sigma_{\min} \le \cdots \le \sigma_1$ and $\sigma'_{\min} \le \cdots \le \sigma'_1$, respectively.

THEOREM 6.5 (DEMMEL AND VESELIĆ [**24**, Theorem 2.14]). *Let $A \in \mathbb{R}^{m\times n}$ have full rank and write $A = BD$, where D is diagonal and the columns of B have unit 2-norm. Let $\Delta A = \Delta B D$ satisfy $\|\Delta B\|_2 \equiv \epsilon < \sigma_{\min}(B)$. Then*

$$\frac{|\sigma_i - \sigma'_i|}{\sigma_i} \le \kappa_2(B)\epsilon. \tag{6.8}$$

In particular, if $|\Delta A| \le (\epsilon/\sqrt{n})|A|$ then $\|\Delta B\|_2 \le \epsilon$ and the bound (6.8) *holds if $\epsilon < \sigma_{\min}(B)$.* □

We will not state the corresponding theorem giving a perturbation bound for the singular vectors; it involves a relative gap defined differently than for the eigenvalue case as $\mathrm{relgap}_i = \min_{j\ne i} |\sigma_i - \sigma_j|/(\sigma_i + \sigma_j)$ (see [**18**], [**24**]). As for Theorem 6.1, the bound in Theorem 6.5 is attainable, though not necessarily for a componentwise perturbation of B (that is, approximate equality is attained for a perturbation that satisfies $\|\Delta B\|_2 = \epsilon < \sigma_{\min}(B)$ but does not necessarily satisfy $|\Delta A| \le (\epsilon/\sqrt{n})|A|$) [**24**, Prop. 2.21].

The earliest componentwise eigenvalue perturbation results were obtained by Kahan in 1966 [**50**]. His results were slightly strengthened in [**4**] and [**23**].

THEOREM 6.6 (KAHAN). *Let $A \in \mathbb{R}^{n\times n}$ be a symmetric tridiagonal matrix with zero diagonal elements, and let the symmetric perturbation ΔA satisfy $|\Delta A| \le \epsilon|A|$. Then*

$$\frac{\lambda_i}{\tau^{2n-1}} \le \lambda'_i \le \tau^{2n-1}\lambda_i,$$

where $\tau = \max(1+\epsilon, 1/(1-\epsilon))$. □

Thus relative perturbations of size ϵ in the elements of A change the eigenvalues by relative amounts at most $(2n-1)\epsilon + O(\epsilon^2)$.

COROLLARY 6.7. *Let $B \in \mathbb{R}^{n\times n}$ be bidiagonal and let $|\Delta B| \le \epsilon|B|$. Then*

$$\frac{\sigma_i}{\tau^{2n-1}} \le \sigma'_i \le \tau^{2n-1}\sigma_i,$$

where $\tau = \max(1+\epsilon, 1/(1-\epsilon))$. □

PROOF. Use the facts that the eigenvalues of $C = \left[\begin{smallmatrix} 0 & B \\ B^T & 0 \end{smallmatrix}\right]$ are plus and minus the singular values of B, and that C can be permuted to a symmetric tridiagonal matrix with zero diagonal. □

Corollary 6.7 states that the singular values of a bidiagonal matrix are determined to approximately the same relative accuracy as the elements of the matrix. Demmel and Kahan [**23**] use this result, together with rounding error analysis, to show that a zero-shift version of the QR algorithm computes the singular values of a bidiagonal matrix to high relative accuracy. Fernando and Parlett [**28**] have developed a shifted quotient-difference algorithm that satisfies even smaller error bounds than the algorithm of Demmel and Kahan, and is at least as fast.

Barlow and Demmel [**4**] were interested in finding classes of matrices other than those in Theorem 6.6 or Corollary 6.7 that determine their eigenvalues or singular values to high relative accuracy. Their main result shows that such a class is the class of "scaled diagonally dominant" symmetric positive definite matrices A: those for which $\|H - I\|_2 = \gamma < 1$, where $A = DHD$ with $D = \mathrm{diag}(A)^{1/2}$. For this class the relative eigenvalue perturbations are bounded by a multiple of $(1-\gamma)^{-1}$. Barlow and Demmel also obtained an analogous result for the generalized symmetric positive definite eigenproblem. As Demmel and Veselić later showed, the assumption on diagonal dominance is unnecessary: Theorem 6.1 shows that it is only necessary that $\kappa_2(H)$ is not large for the eigenvalues to be accurately determined. Indeed, for a scaled diagonally dominant matrix, $\kappa_2(H) < (1+\gamma)/(1-\gamma)$. Barlow and Demmel also gave perturbation results for the eigenvectors of scaled diagonally dominant matrices; these revealed, for the first time, the role of the eigenvalue relative gap in measuring eigenvector sensitivity.

Demmel and Gragg [**21**] consider the question "for which sparsity patterns do small componentwise relative perturbations to the matrix cause small relative perturbations to the singular values?" From Corollary 6.7 we know that one allowable sparsity pattern is that of a bidiagonal matrix. Demmel and Gragg give a very elegant answer to the question: a sparsity pattern has this property if and only if its associated bipartite graph is acyclic. Examples of such acyclic sparsity patterns are, in addition to the bidiagonal pattern,

$$\begin{bmatrix} \times & & & & & \times \\ & \times & & & & \times \\ & & \times & & & \times \\ & & & \times & & \times \\ & & & & \times & \times \\ & & & & & \times \end{bmatrix}, \quad \begin{bmatrix} \times & & & \times & & \\ & \times & & & \times & \\ & & \times & & & \times \\ & & & \times & & \times \\ & & & & \times & \times \\ & & & & & \times \end{bmatrix}.$$

Eisenstat and Ipsen [**26**] answer an open question raised in [**21**] by obtaining, for the singular vectors of a matrix with an acyclic sparsity pattern, a perturbation bound that involves a relative gap for the singular values. They also use a theorem of Ostrowski to give new, elegant proofs of Theorem 6.6 and of Demmel and Gragg's componentwise perturbation bound for the singular values of a matrix with an acyclic sparsity pattern.

The results of Demmel and Veselić have been generalized to indefinite symmetric matrices by two pairs of authors. Veselić and Slapničar make use of the polar decomposition $A = US \in \mathbb{R}^{n\times n}$, where U is orthogonal and $S = (A^TA)^{1/2}$ is symmetric positive semidefinite.

THEOREM 6.8 (VESELIĆ AND SLAPNIČAR [**76**, Theorem 2.13]). *Let $A \in \mathbb{R}^{n\times n}$ be symmetric, with the polar decomposition $A = US$, and write $A = DHD$, where $D = \mathrm{diag}(S)^{1/2}$. Let the symmetric perturbation ΔA satisfy $|\Delta A| \le \epsilon|A|$ and assume that $\nu(A)\epsilon < 1$, where $\nu(A) = \|\,|H|\,\|_2\|DS^{-1}D\|_2$. Then*

$$|\lambda_i - \lambda_i'| \le \nu(A)\epsilon|\lambda_i|. \quad \square$$

If A is positive definite then $S = A$, so $\kappa_2(H) \le \nu(A) \le \sqrt{n}\kappa_2(H)$, and Theorem 6.8 essentially reduces to Theorem 6.1. Veselić and Slapničar also prove an eigenvector perturbation result and a generalization of Theorem 6.8 for the symmetric generalized eigenproblem.

Gu and Eisenstat [**35**] give results for indefinite matrices that have a form very different from those of Veselić and Slapničar. To state a sample result we need the notation that $\mu_0(A) = |\lambda|$, where λ is the eigenvalue of smallest absolute value among all eigenvalues of principle submatrices of A.

THEOREM 6.9 (GU AND EISENSTAT [**35**, Corollary 5]). *Let $A \in \mathbb{R}^{n\times n}$ be symmetric with $\mu_0(H) > 0$, where $A = DHD$ with D a positive definite diagonal matrix. Let the symmetric perturbation $\Delta A = D\Delta HD$ satisfy $\|\Delta H\|_2 \equiv \epsilon < \mu_0(H)$. Then*

$$|\lambda_i - \lambda_i'| \le |\lambda_i|\epsilon\frac{(2\mu_0(H) - \epsilon)}{(\mu_0(H) - \epsilon)^2}. \quad \square$$

If A is positive definite then $\mu_0(H) = \lambda_n(H)$, and if we take $D = \mathrm{diag}(A)^{1/2}$ the bound is

$$|\lambda_i - \lambda_i'| \le 2\epsilon\lambda_i/\lambda_n(H) + O(\epsilon^2) \le 2\epsilon\lambda_i\kappa_2(H) + O(\epsilon^2),$$

which is asymptotically the same as the bound (6.5) in Theorem 6.1, to within a factor 2. Gu and Eisenstat also give an eigenvector perturbation bound for indefinite A involving the relative gap, and a singular value perturbation result that allows two-sided diagonal scalings.

The componentwise bounds described in this section are reflected in several LAPACK routines. For example, the LAPACK routines `xSTEBZ` and `xPTEQR` for finding the eigenvalues of a symmetric (positive definite) tridiagonal matrix achieve the accuracy shown possible by Theorem 6.1. And the singular values of a bidiagonal matrix are computed to high relative accuracy by `xBDSQR` (cf. Corollary 6.7).

We conclude with a numerical example involving the "ipjfact" matrix from the Test Matrix Toolbox [**45**], for which $a_{ij} = (i+j)!$. This matrix is symmetric

α_{ii}	λ_i	α_{ii}/λ_i
2.00e+00	8.12e-04	2.46e+03
2.40e+01	3.32e-02	7.22e+02
7.20e+02	5.72e-01	1.26e+03
4.03e+04	1.28e+01	3.15e+03
3.63e+06	1.01e+03	3.59e+03
4.79e+08	2.21e+05	2.17e+03
8.72e+10	1.12e+08	7.79e+02
2.09e+13	1.25e+11	1.67e+02
6.40e+15	3.23e+14	1.98e+01
2.43e+18	2.44e+18	9.97e−01

TABLE 1. Sorted diagonal elements versus eigenvalues.

positive definite but badly scaled and ill conditioned. For $n = 10$, $\kappa_2(A) = 3.00 \times 10^{21}$, while $\kappa_2(H) = 6.04 \times 10^8$ and

$$\mathrm{absgap}_n = 1.33 \times 10^{-20}, \quad \mathrm{relgap}_n = 6.24.$$

Therefore, for perturbations satisfying $|\Delta A| \leq \epsilon|A|$, the new bound (6.5) is smaller than the traditional bound (6.4) by a factor of order 10^{12}, and is able to guarantee some unchanged digits in all the eigenvalues if $\epsilon < 10^{-10}$. Similarly, the new eigenvector perturbation bound (6.7) is of the order 10^{12} times smaller than traditional bound (6.6). Finally, the sorted diagonal elements α_{ii} of A and the eigenvalues λ_i of A are shown in Table 1. The maximum value of α_{ii}/λ_i is 3.59×10^3 (occurring for $i = 5$), which is less than $1.30 \times 10^8 = \lambda_n(H)^{-1}$, illustrating Theorem 6.4.

7. Concluding Remarks

Componentwise perturbation results are now available for most of the standard problems in numerical linear algebra. The forward perturbation results vary in their form and underlying assumptions, but most have the following property: when the problem data is badly scaled and the perturbations are measured in a componentwise relative fashion, the bounds can be much smaller than the traditional normwise bounds.

There are several reasons why componentwise bounds do not feature strongly in the literature of the 1960s and 1970s. In the first place, many (dense) problems that arise in practice are well scaled, and for these, normwise bounds are adequate. Second, badly scaled problems can be equilibrated prior to being solved, and various equilibration algorithms have been developed over the years. However, how best to scale a matrix is a tricky question for which there is no simple answer (see **[68]** for a discussion). A third reason is that normwise bounds tend to be "simpler" than componentwise ones, with the dependence on the matrix A often appearing only through $\kappa(A)$. A fourth reason is that the most impressive applications of componentwise analysis are in proving results that were not contemplated until relatively recently. Examples of such results are that to improve the stability of a solution to a linear system by iterative refinement it suffices to compute the residual in the working precision [**64**], and that the singular values of a bidiagonal matrix can be computed to the precision to which they are determined by the matrix entries [**23**].

Acknowledgements

I thank Des Higham for suggesting perturbations to the manuscript.

References

1. E. Anderson, Z. Bai, C. H. Bischof, J. W. Demmel, J. J. Dongarra, J. J. Du Croz, A. Greenbaum, S. J. Hammarling, A. McKenney, S. Ostrouchov, and D. C. Sorensen, *LAPACK users' guide*, Society for Industrial and Applied Mathematics, Philadelphia, 1992.
2. M. Arioli, J. W. Demmel, and I. S. Duff, *Solving sparse linear systems with sparse backward error*, SIAM J. Matrix Anal. Appl. **10** (1989), no. 2, 165–190.
3. M. Arioli, I. S. Duff, and P. P. M. de Rijk, *On the augmented system approach to sparse least-squares problems*, Numer. Math. **55** (1989), 667–684.
4. Jesse L. Barlow and James W. Demmel, *Computing accurate eigensytems of scaled diagonally dominant matrices*, SIAM J. Numer. Anal. **27** (1990), no. 3, 762–791.
5. Anders Barrlund, *Perturbation bounds for the LDL^H and LU decompositions*, BIT **31** (1991), 358–363.
6. Sven G. Bartels and Desmond J. Higham, *The structured sensitivity of Vandermonde-like systems*, Numer. Math. **62** (1992), 17–33.
7. F. L. Bauer, *Optimally scaled matrices*, Numer. Math. **5** (1963), 73–87.
8. ———, *Genauigkeitsfragen bei der Lösung linearer Gleichungssysteme*, Z. Angew. Math. Mech. **46** (1966), no. 7, 409–421.
9. F. L. Bauer and C. T. Fike, *Norms and exclusion theorems*, Numer. Math. **2** (1960), 137–141.
10. Albert E. Beaton, Donald B. Rubin, and John L. Barone, *The acceptability of regression solutions: Another look at computational accuracy*, J. Amer. Statist. Assoc. **71** (1976), no. 353, 158–168.
11. Å. Björck, *Component-wise perturbation analysis and error bounds for linear least squares solutions*, BIT **31** (1991), 238–244.
12. James R. Bunch, James W. Demmel, and Charles F. Van Loan, *The strong stability of algorithms for solving symmetric linear systems*, SIAM J. Matrix Anal. Appl. **10** (1989), no. 4, 494–499.
13. P. A. Businger, *Matrices which can be optimally scaled*, Numer. Math. **12** (1968), 346–348.
14. Shivkumar Chandrasekaran and Ilse C. F. Ipsen, *On the sensitivity of solution components in linear systems of equations*, SIAM J. Matrix Anal. Appl. (1994), To appear.
15. Françoise Chatelin, *Eigenvalues of matrices*, John Wiley, Chichester, 1993.
16. Chandler Davis and W. M. Kahan, *The rotation of eigenvectors by a perturbation. III*, SIAM J. Numer. Anal. **7** (1970), no. 1, 1–46.
17. A. S. Deif, *Realistic a priori and a posteriori error bounds for computed eigenvalues*, IMA J. Numer. Anal. **9** (1990), 323–329.
18. Percy Deift, James W. Demmel, Luen-Chau Li, and Carlos Tomei, *The bidiagonal singular value decomposition and Hamiltonian mechanics*, SIAM J. Numer. Anal. **28** (1991), 1463–1516.
19. J. W. Demmel, *The componentwise distance to the nearest singular matrix*, SIAM J. Matrix Anal. Appl. **13** (1992), no. 1, 10–19.
20. James W. Demmel, *On floating point errors in Cholesky*, LAPACK Working Note #14, Department of Computer Science, University of Tennessee, Knoxville, 1989.

21. James W. Demmel and William Gragg, *On computing accurate singular values and eigenvalues of matrices with acyclic graphs*, Linear Algebra Appl. **185** (1993), 203–217.
22. James W. Demmel and Nicholas J. Higham, *Improved error bounds for underdetermined system solvers*, SIAM J. Matrix Anal. Appl. **14** (1993), no. 1, 1–14.
23. James W. Demmel and W. Kahan, *Accurate singular values of bidiagonal matrices*, SIAM J. Sci. Statist. Comput. **11** (1990), no. 5, 873–912.
24. James W. Demmel and Krešimar Veselić, *Jacobi's method is more accurate than QR*, SIAM J. Matrix Anal. Appl. **13** (1992), no. 4, 1204–1245.
25. Jeremy J. Du Croz and Nicholas J. Higham, *Stability of methods for matrix inversion*, IMA J. Numer. Anal. **12** (1992), 1–19.
26. Stanley C. Eisenstat and Ilse C. F. Ipsen, *Relative perturbation techniques for singular value problems*, Research Report YALEU/DCS/RR-942, Department of Computer Science, Yale University, July 1993.
27. L. Elsner, I. Koltracht, M. Neumann, and D. Xiao, *On accurate computations of the Perron root*, SIAM J. Matrix Anal. Appl. **14** (1993), no. 2, 456–467.
28. K. Vince Fernando and Beresford N. Parlett, *Accurate singular values and differential qd algorithms*, Technical Report TR2/92, NAG Ltd., Oxford, July 1992, To appear in Numer. Math.
29. R. Fletcher, *Expected conditioning*, IMA J. Numer. Anal. **5** (1985), 247–273.
30. V. Frayssé, *Reliability of computer solutions*, Ph.D. thesis, L'Institut National Polytechnique de Toulouse, Toulouse, France, July 1992, CERFACS Report TH/PA/92/11.
31. A. J. Geurts, *A contribution to the theory of condition*, Numer. Math. **39** (1982), 85–96.
32. I. Gohberg and I. Koltracht, *Mixed, componentwise, and structured condition numbers*, SIAM J. Matrix Anal. Appl. **14** (1993), no. 3, 688–704.
33. G. H. Golub and J. H. Wilkinson, *Note on the iterative refinement of least squares solution*, Numer. Math. **9** (1966), 139–148.
34. Gene H. Golub and Charles F. Van Loan, *Matrix computations*, second ed., Johns Hopkins University Press, Baltimore, Maryland, 1989.
35. Ming Gu and Stanley C. Eisenstat, *Relative perturbation theory for eigenproblems*, Research Report YALEU/DCS/RR-934, Department of Computer Science, Yale University, February 1993.
36. William W. Hager, *Condition estimates*, SIAM J. Sci. Statist. Comput. **5** (1984), no. 2, 311–316.
37. Desmond J. Higham, *Condition numbers and their condition numbers*, Numerical Analysis Report NA/143, Department of Mathematics and Computer Science, University of Dundee, Scotland, May 1992, To appear in Linear Algebra Appl.
38. Desmond J. Higham and Nicholas J. Higham, *Backward error and condition of structured linear systems*, SIAM J. Matrix Anal. Appl. **13** (1992), no. 1, 162–175.
39. ———, *Componentwise perturbation theory for linear systems with multiple right-hand sides*, Linear Algebra Appl. **174** (1992), 111–129.
40. Nicholas J. Higham, *FORTRAN codes for estimating the one-norm of a real or complex matrix, with applications to condition estimation (Algorithm 674)*, ACM Trans. Math. Software **14** (1988), no. 4, 381–396.
41. ———, *The accuracy of solutions to triangular systems*, SIAM J. Numer. Anal. **26** (1989), no. 5, 1252–1265.
42. ———, *Computing error bounds for regression problems*, Statistical Analysis of Measurement Error Models and Applications, Contemporary Mathematics 112 (Philip J. Brown and Wayne A. Fuller, eds.), American Mathematical Society, Providence, Rhode Island, 1990, pp. 195–208.
43. ———, *How accurate is Gaussian elimination?*, Numerical Analysis 1989, Proceedings of the 13th Dundee Conference (D. F. Griffiths and G. A. Watson, eds.), Pitman Research Notes in Mathematics, vol. 228, Longman Scientific and Technical, Essex, UK, 1990, pp. 137–154.
44. ———, *Iterative refinement enhances the stability of QR factorization methods for solving linear equations*, BIT **31** (1991), 447–468.

45. ———, *The Test Matrix Toolbox for Matlab*, Numerical Analysis Report No. 237, University of Manchester, England, December 1993.
46. Nicholas J. Higham and Philip A. Knight, *Componentwise error analysis for stationary iterative methods*, Linear Algebra, Markov Chains, and Queueing Models (Carl D. Meyer and Robert J. Plemmons, eds.), IMA Volumes in Mathematics and its Applications, vol. 48, Springer-Verlag, New York, 1993, pp. 29–46.
47. Roger A. Horn and Charles R. Johnson, *Matrix analysis*, Cambridge University Press, 1985.
48. ———, *Topics in matrix analysis*, Cambridge University Press, 1991.
49. Ilse C. F. Ipsen and Carl D. Meyer, *Uniform stability of Markov chains*, SIAM J. Matrix Anal. Appl. (1994), To appear.
50. W. Kahan, *Accurate eigenvalues of a symmetric tri-diagonal matrix*, Technical Report No. CS41, Computer Science Department, Stanford University, 1966, Revised June 1968.
51. ———, *Numerical linear algebra*, Canad. Math. Bull. **9** (1966), 757–801.
52. Z. V. Kovarik, *Compatibility of approximate solutions of inaccurate linear equations*, Linear Algebra Appl. **15** (1976), 217–225.
53. Colm Art O'Cinneide, *Entrywise perturbation theory and error analysis for Markov chains*, Numer. Math. **65** (1993), 109–120.
54. W. Oettli and W. Prager, *Compatibility of approximate solution of linear equations with given error bounds for coefficients and right-hand sides*, Numer. Math. **6** (1964), 405–409.
55. W. Oettli, W. Prager, and J. H. Wilkinson, *Admissible solutions of linear systems with not sharply defined coefficients*, SIAM J. Numer. Anal. **2** (1965), no. 2, 291–299.
56. Beresford N. Parlett, *The symmetric eigenvalue problem*, Prentice-Hall, Englewood Cliffs, New Jersey, 1980.
57. Svatopluk Poljak and Jiří Rohn, *Checking robust nonsingularity is NP-hard*, Math. Control Signals Systems **6** (1993), 1–9.
58. J. L. Rigal and J. Gaches, *On the compatibility of a given solution with the data of a linear system*, J. Assoc. Comput. Mach. **14** (1967), no. 3, 543–548.
59. J. Rohn, *New condition numbers for matrices and linear systems*, Computing **41** (1989), 167–169.
60. Jiri Rohn, *Systems of linear interval equations*, Linear Algebra Appl. **126** (1989), 39–78.
61. ———, *Nonsingularity under data rounding*, Linear Algebra Appl. **139** (1990), 171–174.
62. Youcef Saad, *Numerical methods for large eigenvalue problems*, Manchester University Press, Manchester, and Halsted Press, New York, 1992.
63. Robert D. Skeel, *Scaling for numerical stability in Gaussian elimination*, J. Assoc. Comput. Mach. **26** (1979), no. 3, 494–526.
64. ———, *Iterative refinement implies numerical stability for Gaussian elimination*, Math. Comp. **35** (1980), no. 151, 817–832.
65. ———, *Effect of equilibration on residual size for partial pivoting*, SIAM J. Numer. Anal. **18** (1981), no. 3, 449–454.
66. G. W. Stewart, *Error and perturbation bounds for subspaces associated with certain eigenvalue problems*, SIAM Rev. **15** (1973), no. 4, 727–764.
67. ———, *Perturbation bounds for the QR factorization of a matrix*, SIAM J. Numer. Anal. **14** (1977), no. 3, 509–518.
68. ———, *Research, development, and LINPACK*, Mathematical Software III (John R. Rice, ed.), Academic Press, New York, 1977, pp. 1–14.
69. ———, *Stochastic perturbation theory*, SIAM Rev. **32** (1990), no. 4, 579–610.
70. ———, *On the perturbation of LU, Cholesky, and QR factorizations*, SIAM J. Matrix Anal. Appl. **4** (1993), no. 4, 1141–1145.
71. G. W. Stewart and Ji-guang Sun, *Matrix perturbation theory*, Academic Press, London, 1990.
72. J. Stoer and R. Bulirsch, *Introduction to numerical analysis*, Springer-Verlag, New York, 1980.
73. Ji-guang Sun, *Perturbation bounds for the Cholesky and QR factorizations*, BIT **31** (1991), 341–352.

74. ———, *Rounding-error and perturbation bounds for the Cholesky and LDL^T factorizations*, Linear Algebra Appl. **173** (1992), 77–97.
75. A. van der Sluis, *Condition numbers and equilibration of matrices*, Numer. Math. **14** (1969), 14–23.
76. Krešimar Veselić and Ivan Slapničar, *Floating-point perturbations of Hermitian matrices*, Linear Algebra Appl. **195** (1993), 81–116.
77. Bertil Waldén, Rune Karlson, and Ji-guang Sun, *Optimal backward perturbation bounds for the linear least squares problem*, Report LiTH-MAT-R-1992-06, Department of Mathematics, Linköping University, Sweden, February 1992, To appear in J Numer. Linear Algebra Appl.
78. Per-Åke Wedin, *Perturbation theory for pseudo-inverses*, BIT **13** (1973), 217–232.
79. J. H. Wilkinson, *Error analysis of floating-point computation*, Numer. Math. **2** (1960), 319–340.
80. ———, *Rounding errors in algebraic processes*, Notes on Applied Science No. 32, Her Majesty's Stationery Office, London, 1963.
81. ———, *The algebraic eigenvalue problem*, Oxford University Press, 1965.
82. Hongyuan Zha, *A componentwise perturbation analysis of the QR decomposition*, SIAM J. Matrix Anal. Appl. **4** (1993), no. 4, 1124–1131.

DEPARTMENT OF MATHEMATICS, UNIVERSITY OF MANCHESTER, MANCHESTER, M13 9PL, ENGLAND

E-mail address: na.nhigham@na-net.ornl.gov

Proceedings of Symposia in Applied Mathematics
Volume **48**, 1994

Numerical Evaluation of Special Functions

D. W. LOZIER AND F. W. J. OLVER

ABSTRACT. Higher transcendental functions continue to play varied and important roles in investigations by engineers, mathematicians, scientists and statisticians. The purpose of this paper is to assist in locating useful approximations and software for the numerical generation of these functions, and to offer some suggestions for future developments in this field.

CONTENTS

1991 *Mathematics Subject Classification.* Primary 65D20; Secondary 33-00.
The research of the second author has been supported by NSF Grant CCR 89-14933.
This paper is in final form and no version of it will be submitted for publication elsewhere.

© 1994 American Mathematical Society
0160-7634/94 $1.00 + $.25 per page

1. Introduction

When automatic computers began to appear in the 1950s various confident, and often incorrect, predictions were made concerning the impact of these devices on applied mathematics, science and engineering. One of these predictions was that the need for special functions, or higher transcendental functions (as they are also known), would disappear entirely. This was based on the observation that the main use of these functions in those days was to approximate the solutions of classical partial differential equations: with automatic computers it would become possible to solve these equations by direct numerical methods. This observation is indeed correct; nevertheless, a perusal of current computational journals in the sciences reveals a persistent need for numerical algorithms to generate Airy functions, Bessel functions, Coulomb wave functions, error functions and exponential integrals—to name but a few of the classical special functions. Equally significantly, the National Bureau of Standards' Handbook of Mathematical Functions [**AS64**][1] continues to be one of the best-selling mathematical books of all time[2].

The purpose of the present paper is to provide some assistance to those mathematicians, engineers, scientists, and statisticians who discover that they need to generate numerical values of the special functions in the course of solving their problems. "Generate" is the operative word here: we are thinking primarily of either software or numerical approximations that can be programmed fairly easily. Numerical tables are not covered in this survey. Furthermore, for the most part we shall concentrate on the functions themselves; only in certain cases do we include, for example, zeros, inverse functions or indefinite integrals. Elementary functions, also, are excluded[3]. Lastly, we believe that the majority of readers would prefer us to emphasize the more useful algorithms rather than make an attempt to be encyclopedic: algorithms or approximations that have clearly been superseded are omitted.

We identify three stages in the development of computational procedures for the special functions:

(i) Derivation of relevant mathematical properties.
(ii) Development of numerical approximations and algorithms.
(iii) Construction and testing of robust software.

Included in Stage (i) are asymptotic expansions, continued fractions, difference and differential equations, functional identities, integral representations, and Taylor-series expansions. Included in Stage (ii) are expansions in series of Chebyshev polynomials ("Chebyshev series"), minimax polynomial and ratio-

[1] An explanation of the scheme used for acronyms of references is given on p. 25.

[2] In 1988 the National Bureau of Standards became the National Institute of Standards and Technology.

[3] Methods for constructing and testing algorithms for generating elementary functions are surveyed in [**CW80**]. See also [**Bai93, Bre76, Bre78a, Bre78b, Cod93a, LCY65, MY91, Smi89, Smi91, Ziv91**].

nal approximations, Padé approximations, numerical quadrature, and numerical solution of difference and differential equations. **In this paper the emphasis will be on Stages (ii) and (iii), but in §2 we supply some general references for Stage (i).**

In §3 we make a general survey of software libraries and packages[4] that include collections of special functions.

In §§4 and 5 the functions are treated individually. We list software that is already available and readily programmable numerical approximations. In §6 we also include references to articles (or books) that may be useful for the testing or comparison of existing software, or in the construction of new libraries. We do not attempt what would be a herculean task of testing and comparing everything that is available. Our philosophy in this survey has to be that of *caveat emptor*: no algorithm, approximation or package that we mention should be relied upon to produce accurate output in the absence of evidence of independent and systematic checks.

As in the progress of other branches of numerical analysis, procedures used to evaluate special functions are influenced heavily by the computing equipment available at the time. In the era of desk-calculating machines the medium was a printed table of numerical values of the wanted function, or functions, generally for equispaced values of the arguments. Nontabular values were calculated by means of Lagrange's interpolation formula or central-difference interpolation formulas [**Fox56**]. In other words, local polynomial approximations were employed. These interpolation procedures were reasonably successful for functions of a single variable, but two-dimensional interpolation on desk-calculating machines was often a laborious computation that was prone to error. The daunting task that faced a (human) computer is summed up in the following quotation from the Introduction to Karl Pearson's tables of the incomplete gamma function [**Pea22**]:

"As a matter of fact, supposing the use of a machine, which every modern computer has at his command, no interpolation suggested ought to take more than an hour's work and many much less. If the user of these tables groans under that hour, let him compute *de novo* a function value, say $I(6.86877, 47.1813)$ —including of course $\Gamma(48.1813)$—to seven-figure accuracy, and when he has completed the task, we believe his feelings towards those who have provided him with these tables will be very sensibly modified."

With electronic computers, the number of arithmetic operations that could be contemplated for the generation of a single function value increased considerably. In consequence, high-degree global approximations appeared for functions of a single variable in the form of minimax polynomial or rational approximations,

[4]Certain company products are identified in the text. In no case does such identification imply recommendation or endorsement by the National Institute of Standards and Technology, nor does it imply that the products are necessarily the best available for the purpose.

or truncated Chebyshev-series expansions; see, for example, [**Cle62, HCL+68**] and [**Luk77b**].

Chebyshev series in two dimensions also became feasible [**CP66, Luk71a, Luk71b, Luk72a**]. However, because of their more complicated asymptotic behavior, special functions of two variables cannot be covered comprehensively simply by use of polynomial or rational approximations or Chebyshev series. The situation is cloudier still when the variables or parameters are complex, or of course when they are more than two in number. For this reason, to achieve maximum speed, a comprehensive software package for generating a function of two (or more) variables typically employs several different algorithms in addition to, or quite commonly in place of, polynomial or rational approximations or Chebyshev series. The construction and testing of such a package invariably entails prodigious effort.

Computers continue to increase in sophistication and power; in consequence we should expect further changes in the algorithms used to generate the special functions. So far, the potential offered by the introduction of vector and parallel computing machines has not been exploited to any great extent. It might well lead to simplifications in the algorithms needed for many functions, as well as to an increase in execution speeds. We refer again to this possibility in the concluding section (§7).

In assembling the bibliography of this paper we have been assisted by the references collected and classified by the late Dr. L. W. Fullerton in his 1980 Bell Laboratories report [**Ful80**], by access to Dr. N. M. Temme's private collection of references, and by GAMS, the Guide to Available Mathematical Software prepared by the National Institute of Science and Technology [**BHKS90**] and also available online[5].

We have searched through issues of the following journals of the past twenty-five years for relevant references:

Applied Statistics (Appl. Statist.), *Association for Computing Machinery Transactions on Mathematical Software* (ACM Trans. Math. Software), *BIT*, *Collected Algorithms from the Association for Computing Machinery* (CALGO), *Communications of the Association for Computing Machinery* (Comm. ACM), *Computer Physics Communications* (Comput. Phys. Comm.), *Computing*, *Journal of Computational and Applied Mathematics* (J. Comput. Appl. Math.), *Journal of Computational Physics* (J. Comput. Phys.), *Mathematical Reviews* (Math. Rev.), *Mathematics of Computation* (Math. Comp.), *Numerische Mathematik* (Numer. Math.), *U.S.S.R. Computational Mathematics and Mathematical Physics* (U.S.S.R. Comput. Math. and Math. Phys.)[6], *Zeitschrift für*

[5]To get started with the online GAMS system, `telnet gams.nist.gov` and login as `gams` or `xgams`. No password is required. Information on commercial and noncommercial software is provided, and electronic retrieval of noncommercial software is possible.

[6]In 1991 this journal was retitled *Computational Mathematics and Mathematical Physics.*

Angewandte Mathematik und Physik (Z. Angew. Math. Phys.), *Zentralblatt für Mathematik und ihre Grenzgebiete* (Zbl.).

However, because of the sheer volume and diversity of publications on special functions it is almost inevitable that we have overlooked some useful algorithms and important articles. In this event we tender, in advance, our apologies to the authors.

2. Mathematical Developments

Comprehensive compendia of mathematical properties of the special functions are provided by the National Bureau of Standards' Handbook of Mathematical Functions [**AS64**], published originally in 1964, and the 3-volume set that resulted from the Bateman Manuscript Project [**EMOT53a, EMOT53b, EMOT55**], published in 1953 and 1955. These references employ the same notation for the special functions, and we shall follow them. The NBS Handbook has been reprinted many times by the U. S. Government Printing Office and has also been issued in whole, or in part, by other publishers including Dover Publications, Moscow Nauka, Verlag Harri Deutsch and Wiley-Interscience.

The forerunner of [**AS64**] is the book of Jahnke and Emde [**JE45**], published originally in 1909, and still in print[7]. It continues to be useful, especially for its collection of graphs. Other useful compendia include those of Magnus, Oberhettinger and Soni [**MOS66**], and (from the standpoint of hypergeometric functions) Luke [**Luk69a**]. For an introductory compendium, see the recent "atlas" of Spanier and Oldham [**SO87**].

Books and articles that include descriptions or surveys of general methods for computing special functions include [**Bre78b, DKK81, Gau75, HCL$^+$68, Luk69b, Luk77b, PT84, PTVF92, Tem78, vdLT84**].

Other books and articles that provide indepth coverage of pertinent topics include:

[**Ask89**, survey of compendia], [**BG81a, BG81b**, Padé approximations], [**BH75**, asymptotic approximations], [**Bre91**, continued fractions, Padé approximations], [**BvI93**, Padé approximations], [**Cod70**, polynomial and rational approximations], [**Fik68**, polynomial and rational approximations], [**FP68**, Chebyshev polynomials], [**JT80**, continued fractions], [**Kar91**, power series], [**KG80**, statistical computations], [**Luk75**, supplement to **AS64**—especially for functions of hypergeometric type], [**Mor80**, power series], [**Olv74**, asymptotic approximations], [**Riv90**, Chebyshev polynomials], [**Tem77**, integral representations], [**Tem85**, asymptotic approximations], [**Wim84**, recurrence relations], [**Won89**, asymptotic approximations].

[7]A more recent edition, with F. Lösch added as author [**JEL60**], is no longer in print.

3. Packages, Libraries and Systems

This section reviews a selection of mathematical software with respect to its support for the numerical evaluation of special functions[8]. In some cases only a descriptive overview is given while in others cross references by individual function are given in the subsequent sections §4 and §5. The cross-referenced packages, libraries and systems are marked with a ♣. We used the following criteria in assigning the ♣ marks:

First, a marked item must be readily accessible. Often this means it is commercial software that is purchased or leased for a fee but we also include software that is distributed, sometimes over computer networks, by journals and research institutions.

Second, a marked item must have a significant following in North America. (In most cases the unmarked software is used widely elsewhere.)

Third, a marked item must be reasonably comprehensive in its coverage of special functions.

3.1. Software Packages. In this paper *software package* will mean a set of subroutines, or just a single subroutine, that addresses a subfield of numerical mathematics. There are three important series of software packages that include special functions. These packages are research contributions written in a variety of programming languages.

3.1.1. ♣ *ACM Algorithms.* These were published in the *Communications of the ACM*, Volumes 3–18 (1960–75) and since then in the *ACM Transactions on Mathematical Software* (*TOMS*). The transition between the two journals took place with Algorithm 493 in Volume 1, Number 2 of *TOMS*. Algol was required originally but Fortran and other languages were allowed after it became clear that this condition was too restrictive. A recent statement of the ACM Algorithms Policy appears in [**Kro91**]. The policy requires ACM Algorithms to be self-contained, adequately documented through comments in the code, and reasonably portable from one machine to another. A test program with sample output is also required. The policy provides for addenda to previously published algorithms. All ACM Algorithms that appear in *TOMS* are refereed. The ACM Algorithms Policy has been in effect, with little change, since 1975.

A compendium of all ACM Algorithms is maintained in the 4-volume publication [**ACM**], a useful feature of which is that all addenda to an algorithm are printed together with the algorithm. This makes it unnecessary to gather the information from different journal issues. A quarterly looseleaf service keeps [**ACM**] up-to-date. The software itself is available on tape or diskette from the ACM Algorithms Distribution Service. Order forms are printed in each issue of

[8]General reviews of mathematical software appear regularly in the Computers and Mathematics column of the *Notices of the American Mathematical Society*. A useful guide to these reviews is given in [**Dev93**].

TOMS. Alternatively, unofficial and possibly temporary electronic dissemination is being made available by Netlib [**DG87**][9].

For indexing purposes, each ACM Algorithm is assigned a symbol from a modification of the SHARE classification system; cf. [**ACM**] or [**ACM64**]. Cumulative indexes by SHARE classification for 1960–1980, 1981–1986 and 1987–1988 appear in [**ACM**]. Algorithm 620 [**Ham85, HM90b, RH84**] provides a data base and Fortran program for preparing a cumulative index by SHARE classification. This data base is updated periodically by the ACM Algorithms Distribution Service and Netlib.

3.1.2. ♣ *AS Algorithms.* A section for statistical subroutines in *Applied Statistics* was established in 1968 to "encourage the development of a published literature on statistical computing" as the specialized needs of statistical computing were "only partly met" by the algorithm sections of other journals [**AS68**]. The current version of detailed instructions and other information for authors of AS Algorithms, first published in 1968, can be found in [**RWGH87**]. All submissions adhere to a standard format and are refereed. A test program is required for the referee's use. Addenda to previously published algorithms are accepted and are subjected to the same refereeing process as original algorithms. An index appears at the end of every volume. A cumulative index of the first 251 AS Algorithms (1968–1989, with addenda) appears in [**HM90a**], organized according to the GAMS scheme [**BHK91**]. Corrected and improved versions of selected AS Algorithms appear in [**GH85**]. Instructions for obtaining AS Algorithms on computer diskette can be found in issues of *Applied Statistics* starting in 1993.

3.1.3. ♣ *CPC Programs.* The journal *Computer Physics Communications* was begun in 1969 to "facilitate the exchange of physics programs and of relevant information about the use of computers in the physics community". It publishes descriptions of CPC Programs and, in addition, general papers on the computational aspects of physics and physical chemistry. Programs and their descriptions are refereed.

Program descriptions consist of a Program Summary (a concise description in a standard format with keywords) and a Long Write-Up (a detailed description of the underlying physics and algorithms). A test program is required for each CPC Program, and each CPC Program is required to be well documented and as portable and self-contained as possible. An index of CPC Programs is printed at the end of every volume. Two cumulative indexes without Program Summaries [**CPC87, CPC90**] and one with Program Summaries [**CPC84**] exist. A more attractive alternative is the up-to-date cumulative index, with Program Summaries, that is accessible by electronic mail[10]. All these indexes are organized

[9]To get started with Netlib, send the message "help" to `netlib@ornl.gov`.

[10]To get started with the cumulative index, send the message "get cpc intro cpcindex" to `listral@earn-relay.ac.uk`, or see the instructions printed in every issue of *Computer Physics Communications*.

according to a physics-oriented classification scheme.

CPC Programs can be ordered individually or by subscription service. Ordering instructions and an order form are printed in the back of every issue of *Computer Physics Communications.*

3.2. Intermediate Libraries. Under this heading we place software that is in some sense intermediate between software packages, which embody original research contributions, and comprehensive libraries (§3.3 below). The libraries we consider here provide support only for mathematical functions. Furthermore, they are largely restricted to codification of existing algorithms with all their advantages—and limitations.

3.2.1. ♣ *C Mathematical Function Handbook.* This volume with diskette [**Bak92**] is keyed to the *NBS Handbook of Mathematical Functions* [**AS64**]. Most chapters of the *NBS Handbook* have a counterpart here in which brief introductory material is followed by C code listings. A complex arithmetic package is included since C supports only real and integer arithmetic. [**Bak92**] advocates the use of C because it "is rapidly becoming the *lingua franca* of the computer world" and "algorithms written in C should be very portable". He has written two other books on C programming for technical applications.

3.2.2. ♣ *Mathematical Function Library for Microsoft Fortran or C.* These volumes [**ULI90**] exist to "provide users with a comprehensive set of mathematical function routines to assist them in solving their mathematical problems on IBM PC/XT/AT or compatibles". Each consists of a looseleaf manual with diskettes. The documentation for each function gives usage instructions, input range, accuracy, definition of the function, algorithm, sample program and sample results. The functions are evaluated only for real arguments. The diskettes contain a compiled library in microprocessor assembly code for use with Microsoft compilers and the Fortran or C source code for use with other compilers.

3.2.3. ♣ *Methods and Programs for Mathematical Functions.* This volume with separate diskette of C programs [**Mos89**] presents a selection of special functions with real arguments and integer or real parameters. The programs are designed for double precision, and tables of test results included for every function typically show absolute or relative errors (whichever is appropriate) of the order of 10^{-16}. Where polynomial or rational approximations are used, the expansion coefficients were generated in multiple precision using C programs that are given in the book.

3.3. Comprehensive Libraries. When new algorithms are developed they tend to appear first as subroutines in software packages (§3.1 above). Later they may be assimilated into more complete software products such as intermediate libraries (§3.2 above). Even more useful are *comprehensive libraries* that integrate evaluation of special functions with other essential elements of numerical computing and offer additional advantages such as uniform documentation, style of usage, and handling of error conditions. Corrections and improvements, particularly in orienting the programming toward particular computer architectures, are often made.

3.3.1. *CERN Library.* The European Laboratory for Particle Physics maintains a comprehensive software library [**CER93**], mostly in Fortran but with a few routines in assembly language, in support of high-energy physics research. The coverage of special functions includes the error function of real and complex argument; the Dawson and Fresnel integrals; exponential, sine, cosine and arctangent integrals; the gamma and digamma functions of real and complex argument; incomplete gamma function; real dilogarithm and complex generalized polylogarithms; Bessel functions of real argument and orders 0, $\pm\frac{1}{4}$, $\pm\frac{1}{3}$, $\pm\frac{1}{2}$, $\pm\frac{2}{3}$, $\pm\frac{3}{4}$ and 1; Bessel functions of real order and real or complex argument; Bessel functions of complex order and argument; zeros of the Bessel functions J and Y and of their derivatives; Coulomb wave functions of complex order, argument and parameter; complete and incomplete elliptic integrals; Jacobi's elliptic functions (real and complex); Jacobi's theta functions (real); Bose-Einstein and Fermi-Dirac integrals ; Legendre and associated Legendre functions; conical functions of the first kind; Struve functions. The library is distributed, with some restrictions, to organizations outside CERN.

3.3.2. ♣ *IMSL Library.* International Mathematical and Statistical Libraries was incorporated in 1970 "with the intent of providing high-quality, supported Fortran subroutine libraries in mathematics and statistics" [**Air84**]. In its first ten years it produced libraries tailored to twelve different computer lines, providing an alternative to manufacturer-supplied libraries. Currently the company offers a wide range of products for large-scale scientific computing. At the center of its product line is the IMSL numerical subroutine library for mathematics and statistics, which includes an extensive coverage of real and complex special functions [**IMS91**]; this reference includes a GAMS index [**BHK91**] and a KWIC (keyword in context) index. The library is optimized, vectorized and parallelized where appropriate, depending on the target computer architecture, but it contains no vector or parallel support for special functions.

A subset of the IMSL library is offered also as a C library. This is an essential component of a powerful interactive system (§3.4 below) which has the capability of providing graphical and numerical computing with very large data sets. Optionally, Maple (§3.4.3 below) can be incorporated to provide for symbolic computing.

3.3.3. ♣ *NAG Library.* An overview of the development, structure and contents of the NAG Fortran library [**NAG91**] is given in [**FP84**]. After originating as a cooperative project among several British computing centers in 1970 with the purpose of providing "a balanced, general-purpose numerical algorithms library," the Numerical Algorithms Group formed a not-for-profit company in 1976 to provide for the wider distribution of the library. The library is organized around the ACM modification of the SHARE classification system (see §3.1.1 above) and is available for a wide cross-section of computing systems. A KWIC (keyword in context) index and an index in the GAMS classification scheme [**BHK91**] are provided in the library documentation.

Subsets of the full library are available in Ada, Algol 68, C, Fortran 90 and Pascal. NAG is actively developing and marketing an interactive system (§3.4 below) that integrates most of the numerical power of the full NAG library with online symbolic and graphical capabilities.

3.3.4. *NSWC Library.* In 1976 the Naval Surface Warfare Center, Dahlgren, Virginia, began development of "a [Fortran] library of general purpose subroutines that would provide a basic computational ability in a variety of mathematical activities" [**Mor93**]. The design goals stressed reliability, transportability, efficiency, ease of use, and generality. In 1993 the library contained 576 user-level subroutines, including ones in real arithmetic for the error function and its inverse; Dawson's integral and Fresnel integrals; exponential, sine and cosine integrals; gamma, psi and polygamma functions; dilogarithm; incomplete gamma function and its inverse; incomplete beta function; complete and incomplete elliptic integrals; Jacobi's and Weierstrass' elliptic functions; Bessel functions of real argument and order. It also contained Airy functions and complete elliptic integrals of complex argument, and Bessel functions of complex argument and integer or complex order.

3.3.5. *NUMAL Library.* In 1973 the Mathematisch Centrum, Amsterdam, introduced this library of numerical procedures in Algol-60 with " the aim ... to provide Algol-60 programmers with a high-level numerical library which contains a set of validated numerical procedures together with supporting documentation" [**Hem81**]. In 1979 it contained approximately 430 subroutines, including ones for the exponential, sine and cosine integrals; gamma function; incomplete gamma and beta functions; error and inverse error functions; Fresnel integrals; modified and unmodified Bessel functions of integer, half-integer or real order; Airy functions. All subroutines are for real variables.

3.3.6. ♣ *Numerical Recipes.* This partly pedagogical series of books offers "for each topic considered, a certain amount of general discussion, a certain amount of analytical mathematics, a certain amount of discussion of algorithmics, and (most important) actual implementations of these ideas in the form of working computer routines" [**PTVF92**]. Besides being listed fully in the text, the computer routines are available for purchase on diskette or under a variety of licensing arrangements, one of which is tailored to the needs of classroom instructors. Example books with test programs and diskettes are available also. Standard fields of numerical computation are covered, with approximation of functions and evaluation of special functions included. The book is published in four versions with the software coded in Basic, C, Fortran or Pascal; another volume for Basic is [**Spr91**].

3.3.7. *NUMPAC Library.* The Nagoya University Mathematical Package is used widely in Japan. It is a comprehensive Fortran library oriented toward Japanese computers, including vector supercomputers. Coverage includes Airy functions; error and inverse error functions; Dawson and Fresnel integrals; exponential, sine and cosine integrals; complex gamma function; digamma function; dilogarithm; Riemann's zeta function; Bessel functions of integer or real order and real or complex argument; zeros and integrals of Bessel functions; complete and incomplete elliptic integrals; Jacobi's elliptic functions; incomplete beta and gamma functions; Legendre polynomials and associated Legendre functions; classical orthogonal polynomials; Struve functions; Abramowitz, Debye and elliptic theta functions; solutions of the Blasius and Thomas-Fermi equations. Information can be obtained from Ichizo Ninomiya, Chubu University, Kasugai, Aichi, 487 Japan or Yasuyo Hatano, Chukyo University, Yagoto, Nagoya, 466 Japan.

3.3.8. *PORT Library.* This library [**FHS78b, Fox84**] is mentioned here because it provides a framework [**FHS78a**] for constructing portable Fortran libraries that has proven its utility. The framework supplies computer arithmetic parameters via Fortran function calls. Algorithms are coded so as to be valid for a range of values of the arithmetic parameters; actual values are substituted at run time. The PORT framework is used, for example, in the SLATEC library (§3.3.10 below). It is particularly valuable in the special function routines because of their sensitivity to precision, underflow and overflow. The PORT framework is available as ACM Algorithm 528 (§3.1.1 above).

3.3.9. ♣ *Scientific Desk Library.* C. Abaci offers the following products: (i) the Scientific Desk Library, a Fortran-based collection of numerical software; (ii) the Scientific Desk Analysis System, an interactive system (§3.4 below); (iii) software produced by others, including the ACM algorithms (§3.1.1 above). The library is available in object code for personal computers under a variety of Fortran compilers and in Fortran source code for other computers. The Analysis System, which is strongly oriented toward statistics, simplifies the programming burden and provides for simple graphical output. C. Abaci distributes the SPECFUN collection [**Cod93b**] of Fortran programs for special functions and the ELEFUNT, INTFUNT and CELEFUNT tests [**Cod93a, CW80**] for elementary functions. Inquire at *C. Abaci, Inc., P. O. Box 2626, Raleigh, NC 27602.*

3.3.10. ♣ *SLATEC Library.* The acronym stands for Sandia, Los Alamos, Air Force Weapons Laboratory Technical Exchange Committee[11], formed in 1974 to

[11]The Air Force Weapons Laboratory has been renamed the Phillips Laboratory.

"foster the exchange of technical information among the three computing departments". In 1977 a subcommittee undertook the development of a complete, noncommercial Fortran library for numerical supercomputing [**Buz84**]. The primary motivation was that the suppliers of commercial libraries regarded the supercomputing market as too small. The library subcommittee admitted subsequently five additional U. S. Government agencies (the Lawrence Livermore, Oak Ridge and Sandia Livermore National Laboratories, the National Energy Supercomputer Center at Lawrence Livermore, and the National Institute of Standards and Technology). SLATEC Version 1.0 appeared in 1981. Version 4.0, the third major revision and expansion, was released in December 1992. The initial coverage of special functions coincided with FNLIB [**Ful77**], FUNPACK [**Cod75, Cod84a**] and AMOSLIB [**AD79**]. Subroutines from [**ADW77a, ADW77b, Amo80a, Amo83a, Amo83b, Amo86, CN81, LS81, OS83**] were added later. Available from *Energy Science and Technology Center, P. O. Box 1020, Oak Ridge, TN 37831* and from Netlib (§3.1.1 above).

3.4. Interactive Systems. The software packages and libraries considered in the preceding three sections are used in conjunction with standard programming languages. These languages are not fully interactive. A program needs to be written, compiled and linked to libraries before it can be executed, and after the results are examined the cycle may need to be repeated to correct errors or change parameters. An *interactive system* provides a powerful set of commands which the user can enter at the keyboard. The response to each command is displayed immediately. The burden of programming and the compile-link-execute cycle is reduced. Programming in an interactive system serves a new purpose: to extend or customize the command set.

A striking characteristic of interactive systems is their ability to integrate nonnumerical tasks with numerical computation. Graphical and symbolic computing work best in an interactive environment, and one or both are combined powerfully with numerical computing in commercially available interactive systems. The trend toward increased integration of these computational components is being recognized by recent developments of the IMSL, NAG and Scientific Desk libraries (§§3.3.2, 3.3.3 and 3.3.9 above).

A particular type of interactive system with a special capability for the numerical evaluation of special functions is the *computer algebra system*, developed to provide symbolic processing of mathematical formulas and intended, primarily, to assist in mathematical developments. These systems contain basic mathematical information that enables them to manipulate algebraic expressions, make substitutions, differentiate and integrate functions, solve algebraic, transcendental and differential equations, manipulate power series, and the like. Some knowledge of mathematical properties of special functions is built in, and more can be added by programming extensions to the command set. Numerical approximations are to be avoided, in keeping with the primary purpose of supporting exact mathematical developments, but floating-point computation is provided as a secondary capability. This often comes with a bonus when compared to the usual programming languages which simply use the hardware computer arithmetic: the precision can be set arbitrarily.

The rationale for arbitrary precision is not entirely clear. It is clear that exact rational arithmetic is essential in computer algebra applications. Perhaps multiple-precision floating-point, being relatively easy to implement, is considered a worthwhile additional capability. Also, evaluation of symbolic expressions may require high precision because of numerical sensitivity. Whatever the reason, for occasional usage of arbitrary-precision floating-point, computer algebra systems are well worth considering.

3.4.1. *HiQ* (*Apple Macintosh, Sun*). This system [**Bim93**] approaches the goal of reducing the need to write programs by making concentrated use of the graphical user interface (the image displays and controls associated with the computer screen). The system opens with a blank worksheet (in a window on the screen) and an array of icons. Each icon corresponds to a particular kind of task.

For example, one icon is called the "expression evaluator". When activated by the mouse, this icon presents a window with three areas. An algebraic expression is entered into the input area in a conventional programming-language syntax (like Fortran). This expression can contain numbers, symbols representing numbers, and symbols representing built-in HiQ functions. The symbols representing numbers are assigned numerical values in the options area as constants or finite arithmetical sequences. Output icons are generated in the output area when the "run button" (another icon) is "pushed" by clicking the mouse. The output icons, when activated, display tables and graphs of the computed data.

Other tasks that can be performed by similar sequences of manipulations with icons are numerical integration, optimization, data fitting, finding roots of polynomials and nonlinear functions of one variable, and solving nonlinear systems, integral equations, and initial-value and boundary-value problems in ordinary differential equations.

Special functions included in HiQ are Airy functions; beta, gamma, log gamma, psi, incomplete beta, incomplete gamma, and complementary incomplete gamma functions; Kelvin functions; Bessel functions of integer and half-integer order; Struve and Weber parabolic cylinder functions; hypergeometric function and series; confluent hypergeometric function and series; Riemann zeta function. Although HiQ performs complex arithmetic, most if not all the special functions are evaluated for real arguments only.

3.4.2. *Macsyma.* Macsyma [**Sym92**] is a computer algebra system that supports symbolic, graphical and numerical computing on personal computers, scientific workstations and mainframes. Its built-in capabilities can be extended by programming in either Lisp or an Algol-like procedural language.

Macsyma avoids numerical approximations unless floating-point numbers are introduced, either explicitly or as the result of special commands. Floating-point numbers are represented internally in machine single precision, machine double precision, or software arbitrary precision. When arbitrary precision is being used there is a precision specifier. Operand precisions are adjusted, if necessary, to the specified operational precision by truncating or extending with zero digits before arithmetic operations are performed. The precision specifier can be changed at any time.

Macsyma supports the numerical evaluation of elliptic, error, gamma, polygamma, polylogarithmic and zeta functions; Airy, Bessel and Legendre functions; complete elliptic integrals and the exponential integral; classical orthogonal polynomials. Only the gamma, polygamma and Riemann zeta functions are computable in arbitrary precision. Some of the other functions are restricted to single precision. Complex arguments are allowed for the error, gamma and Bessel functions.

3.4.3. ♣ *Maple.* Maple V Release 2 [**CGG**$^+$**91**] is a computer algebra system containing symbolic, numerical and graphical capabilities. It is available for a wide range of computers, from personal computers and Unix workstations to vector supercomputers. The normal mode of operation is interactive. A Pascal-like programming language, called the Maple language, is provided also. Much of the Maple system is programmed in this language. This part, called the Maple Library, can be viewed on the screen or printed, and serves as useful supplementary documentation or as a guide for the preparation of additional library modules. The core of Maple, written in the C programming language, is not normally accessible to users.

Because of its emphasis on symbolic computing, Maple avoids any evaluation which would introduce an inexact result unless the user specifically requests it. Expressions are evaluated symbolically, with numbers rendered as rational fractions with arbitrarily long numerators and denominators or represented as symbols. The user can request floating-point evaluation to arbitrary precision. For mathematical functions, Maple detects certain special values and can make appropriate substitutions. Otherwise the functions are left as symbolic representations until the user explicitly requests their evaluation in floating-point format. These evaluations, if they can be done at all, are to the precision specified by the user.

3.4.4. *Mathcad (PC with Microsoft Windows).* This system [**Mat93a**] is oriented toward the engineering professions but is useful also in educational, mathematical, scientific and statistical applications. Mathcad can be regarded as an editor and calculator that can be used to create complete documents. These documents can include graphics, ordinary text, and mathematical text resulting from input commands and their associated numerical or symbolic output. Commands are selected from an extensive array of icons (similar to HiQ, §3.4.1 above) or they can be activated by appropriate keystrokes. Numerical commands support real and complex computations with scalars, vectors and matrices; numerical differentiation and integration ; solution of algebraic equations; constrained and unconstrained minimization; Fourier transforms; statistical operations. Symbolic commands are supported by a subset of Maple (§3.4.3 above). Coverage of special functions includes error, gamma and polylogarithmic functions; sine, cosine and Fresnel integrals; Bessel functions of integer order. Except for the gamma function, all arguments must be real.

3.4.5. ♣ *Mathematica.* Mathematica [**Wol91**] is a computer algebra system for symbolic, graphical and numerical computing on personal computers, scientific workstations, and larger computers. A highly developed user interface, available on some of this hardware, integrates Mathematica output with ordinary text for the preparation of complete documents entirely within the Mathematica system. A programming language, based on pattern matching, is included and can be used for extending the capabilities of the system.

As with other computer algebra systems, Mathematica uses floating-point numbers only when requested explicitly. If numbers are introduced with no more significant figures than the precision of the machine floating-point system, and if the machine underflow and overflow limits are not exceeded, then computations proceed in hardware floating-point arithmetic. On the other hand, numbers that are not machine-representable are stored in a software floating-point format. Each such number is tagged with its own precision, and computations are performed in software floating-point arithmetic. The precision of nonmachine numbers is arbitrary but the internal representation is set to the highest justifiable precision. This is determined by the number of significant figures in an input number and by the precision of the operands or arguments in arithmetic operations and function evaluations. If numbers in a hardware computation underflow or overflow, then the software arithmetic takes over automatically.

3.4.6. *Matlab.* This system [**Mat92**] uses matrix notation to provide a built-in set of commands for standard algorithms of numerical computation. A graphics capability is included also. Additional commands can be coded in concise procedures using Matlab notation. Symbolic computing is supported through a recently introduced option using Maple (§3.4.3 above). Matlab runs on a broad range of computers from personal computers and scientific workstations to vector supercomputers. One of its strengths is that it treats complex arithmetic as the natural extension of real arithmetic: variables do not have a fixed real or complex type as in Fortran. Nevertheless, Matlab's coverage of complex functions is limited. It supports Bessel functions of real order and complex argument z but warns in the online help system that the functions "may produce inaccurate results" for large order and $|z|$. Built-in special functions for real arguments and parameters include error and inverse error functions; gamma function; incomplete gamma and beta functions; Bessel functions I, J, K and Y; complete elliptic integrals; Jacobi's elliptic functions.

4. Functions of One Variable

In the references that follow an indication is made of the programming language where applicable. Also, special note is made of references that include surveys. Libraries and interactive systems are listed separately.

In the subsections of §4 and §5, a library or interactive system is listed only if it employs an algorithm tailored to the restrictions of the subsection. For example, NAG is listed in §4.1.1 and §4.1.2 because it has separate capabilities for Airy functions of real and complex argument. Maple is listed only in §4.1.1 because it does not have Airy functions of complex argument. Mathematica is listed only in §4.1.2 because it does not use a restricted algorithm for real arguments. Because these distinctions are sometimes difficult to infer from software documentation and even, when available, from source code, they should be regarded only as a guide, both in §4 and §5.

4.1. Airy Functions. This section includes Scorer's functions.

4.1.1. *Real Arguments.* [**Ném71**], [**Pri75**, Fortran], [**RS81**]. Libraries: [**Bak92**], [**Mos89**], [**ULI90**], **IMSL**, **NAG**, **Numerical Recipes**, **Scientific Desk**, **SLATEC**. Systems: **Maple**.

4.1.2. *Complex Arguments.* [**Amo86**, Fortran], [**CJR92**, Fortran]. Libraries: **NAG**, **Scientific Desk**, **SLATEC**. Systems: **Mathematica**.

4.1.3. *Articles.* [**CCF83**], [**Lee80**], [**LO93**], [**Moo81**], [**SAG79**].

4.2. Error Functions, Dawson's Integral, Fresnel Integrals.

4.2.1. *Error Functions of Real Argument.* [**Ada69**, Algol], [**Cle62**], [**CMW63**, Algol], [**Cod69**], [**Cod90a**, Fortran], [**CT85**, Fortran], [**Hil73**, Fortran], [**Luk69b**], [**Luk75**], [**Sch78**], [**SL81**], [**SZ70**, Fortran], [**Tem94**, Pascal]. Libraries: [**Bak92**], [**Mos89**], [**ULI90**], **IMSL**, **NAG**, **Numerical Recipes**, **Scientific Desk**, **SLATEC**.

4.2.2. *Inverse Error Functions of Real Argument.* [**BEJ76**], [**Cun69**, Fortran], [**HD73**, Algol]. Libraries: [**Mos89**], [**ULI90**], **IMSL**, **NAG**, **Scientific Desk**.

4.2.3. *Integrals of the Error Function.* [**Gau77a**, Fortran], [**Woo67**]. Libraries: [**Bak92**].

4.2.4. *Dawson's Integral of Real Argument.* [**CPT70**], [**Hum64**], [**Ryb89**, Fortran]. Libraries: [**Bak92**], [**Mos89**], [**ULI90**], **IMSL**, **NAG**, **Numerical Recipes**, **Scientific Desk**, **SLATEC**.

4.2.5. *Fresnel Integrals of Real Argument.* [**Bul67**, Algol], [**Cod68**], [**Hea85**], [**LG64**, Algol], [**Luk69b**], [**Luk75**], [**Sny93**, Fortran]. Libraries: [**Bak92**], [**Mos89**], [**ULI90**], **IMSL**, **NAG**, **Numerical Recipes**. Systems: **Maple**.

4.2.6. *Complex Arguments.* [**Gau69a**, Algol], [**Luk69b**], [**Lyn93**, Fortran], [**PW90a**, Fortran], [**SZ81**, Fortran]. Libraries: [**Bak92**], **IMSL**, **NAG**. Systems: **Maple**, **Mathematica**.

4.2.7. *Articles.* [**BR71**], [**Cod90b**, includes survey], [**Col87a**], [**Fle68**], [**Gau70**], [**Gau77b**], [**Hen79**], [**HR72**], [**LW90**], [**LW91**], [**McC74**], [**Mor83**], [**MR71**], [**PW90b**], [**Str68**], [**vdLT84**], [**Wei94a**, includes survey], [**Wei94b**].

4.3. Exponential Integrals, Logarithmic Integral, Sine and Cosine Integrals.

4.3.1. *Exponential Integrals of Real Argument.* [**Amo80a**, Fortran], [**Cle62**], [**CMW63**, Algol], [**CT69**], [**Gau73**, Algol], [**Luk69b**], [**Luk76**], [**Pac70**, Fortran], [**SZ76**, Fortran]. Libraries: [**Bak92**], [**ULI90**], **IMSL**, **NAG**, **Numerical Recipes**, **Scientific Desk**, **SLATEC**.

4.3.2. *Logarithmic Integral of Real Argument.* Libraries: [**Bak92**], [**ULI90**], **IMSL**, **Scientific Desk**, **SLATEC**.

4.3.3. *Sine and Cosine Integrals and Hyperbolic Sine and Cosine Integrals of Real Argument.* [**Bul67**, Algol], [**Luk69b**]. Libraries: [**Bak92**], [**Mos89**], [**ULI90**], **IMSL**, **NAG**, **Numerical Recipes**, **Scientific Desk**.

4.3.4. *Complex Arguments.* [**Amo90a**, Fortran], [**Luk69b**]. Libraries: [**Bak92**]. Systems: **Maple**, **Mathematica**.

4.3.5. *Articles.* [**Amo80b**], [**Amo90b**], [**CT68**], [**TM68**], [**vdLT84**].

4.4. Gamma, Psi, and Polygamma Functions.

4.4.1. *Gamma Function of Real Argument.* [**CH67**], [**Cle62**], [**CMW63**, Algol], [**CT85**, Fortran], [**FS67**, Algol], [**Luk69b**], [**Luk75**], [**Mac89**, Fortran], [**Tem94**, Pascal]. Libraries: [**Bak92**], [**Mos89**], [**ULI90**], **IMSL**, **NAG**, **Numerical Recipes**, **Scientific Desk**, **SLATEC**.

4.4.2. *Psi and Polygamma Functions of Real Argument.* [**Amo83b**, Fortran], [**Bow84**, Fortran], [**CST73**], [**Luk69b**], [**Luk75**]. Libraries: [**Bak92**], [**Mos89**], [**ULI90**], **IMSL**, **NAG**, **Scientific Desk**, **SLATEC**.

4.4.3. *Complex Arguments.* [**BD80**, Fortran], [**Köl72a**, Fortran], [**Kuk72a**, Fortran], [**Luk69b**]. Libraries: [**Bak92**], **IMSL**, **Scientific Desk**, **SLATEC**. Systems: **Maple**, **Mathematica**.

4.4.4. *Articles.* [**AB87b**], [**Cha80**], [**Cod91**, includes survey], [**FW80**], [**Kat78**], [**Krä90**], [**Kuk72b**], [**Luk70a**], [**McC81**], [**Ng75**, includes survey], [**vdLT84**].

4.5. Landau Density and Distribution Functions.

4.5.1. *Real Variables.* [**KS84d**, Fortran], [**Sch74**, Fortran].

4.6. Polylogarithms.

4.6.1. *Dilogarithms.* [**GZ75**, Fortran], [**Luk75**]. Libraries: [**Bak92**], [**Mos89**], **IMSL**, **SLATEC**. Systems: **Maple**.

4.6.2. *Higher Polylogarithms.* Libraries: [**Bak92**]. Systems: **Mathematica**.

4.6.3. *Articles.* [**GT81**], [**JL72**], [**Mor79**].

4.7. Zeta Function.

4.7.1. *Real Arguments.* [**CHT71**], [**Luk69b**], [**Mar65**, Algol], [**PB72**]. Libraries: [**Bak92**], [**Mos89**]. Systems: **Maple**.

4.7.2. *Complex Arguments.* [**BD80**, Fortran], [**YKK88**, Fortran]. Systems: **Mathematica**.

4.7.3. *Articles.* [**AB89**], [**EKK85**], [**Ker80**, includes survey].

5. Functions of Two or More Variables

As in §4, an indication is made of the programming language where applicable and special note is made of references that include surveys. Libraries and interactive systems are listed separately, and similar remarks apply about the inclusiveness of the subsections.

5.1. Bessel Functions. All of the following subsections apply to the ordinary Bessel functions (J and Y) and the modified Bessel functions (I and K).

5.1.1. *Orders 0 and 1, Real Arguments.* [**Bla74**], [**BS92**, Fortran], [**Cle62**], [**Hil81**, Fortran], [**Luk69b**], [**Luk75**], [**WBR82**]. Libraries: [**Bak92**], [**Mos89**], **IMSL**, **NAG**, **Numerical Recipes**, **Scientific Desk**, **SLATEC**.

5.1.2. *Integer or Half-Integer Orders, Real Arguments.* This subsection includes spherical Bessel functions. [**AM61**], [**AM78**, Fortran], [**Col80**, Fortran], [**Hil81**, Fortran], [**MM90**], [**PB82**], [**RF93**, Fortran]. Libraries: [**Bak92**], [**Mos89**], [**ULI90**], **IMSL**, **Numerical Recipes**.

5.1.3. *Real Orders, Real Arguments.* [**ADW77a**, Fortran], [**Bar82b**, Fortran], [**Cam79**, Fortran], [**Cod83**, Fortran], [**CP66**], [**Luk69b**], [**Luk71a**], [**Luk71b**], [**Luk72a**], [**Luk75**], [**Mat93b**, Fortran], [**Pie84b**, Fortran], [**Tem75**, Algol], [**Tem76**, Algol]. Libraries: [**Mos89**], [**ULI90**], **IMSL**, **Numerical Recipes**, **Scientific Desk**, **SLATEC**. Systems: **Maple**.

5.1.4. *Integer or Half-Integer Orders, Complex Arguments.* This subsection includes Kelvin functions. [**BKN88a**, Fortran], [**BKN88b**, Fortran], [**Bur63**], [**CM83**], [**Mas83**, Fortran]. Libraries: [**Bak92**], [**ULI90**], **IMSL**, **NAG**.

5.1.5. *Real Orders, Complex Arguments.* [**Amo86**, Fortran], [**Cam81**, Fortran], [**Luk69b**], [**Luk75**], [**TB87**, Fortran]. Libraries: **IMSL**, **NAG**, **Scientific Desk**, **SLATEC**.

5.1.6. *Complex Orders, Complex Arguments.* [**TB85**, Fortran]. Systems: **Mathematica**.

5.1.7. *Integrals of Bessel Functions.* [**Amo83a**, Fortran], [**And82a**, Fortran], [**BEJ78**], [**Cha83**, Fortran], [**Feu91**, Fortran], [**GP64**], [**PB84**, Fortran], [**Pie82**, Fortran], [**SZ79**, Fortran], [**Tal83**, Fortran]. Libraries: [**Bak92**], **SLATEC**.

5.1.8. *Zeros of Bessel Functions.* [**Cam84**, Fortran], [**Pie84a**], [**Pie90**, Fortran], [**Tem79**, Algol]. Libraries: [**Bak92**].

5.1.9. *Articles—Functions.* [**Ach86**], [**ADW77b**], [**Amo74**], [**Bar81a**], [**BGV93**], [**Cam80**], [**CF87**], [**CMF77**], [**Cod80**, includes survey], [**Col87b**], [**CS89**, includes survey], [**Gau91b**], [**GB87**], [**GS78**], [**Hit68**], [**KS84b**], [**Luk72b**], [**Luk77b**], [**Nes84**], [**OS72**], [**Rem73**], [**TB86**], [**VGK+91**], [**Wal84**], [**WC90**], [**YN74**], [**Yos92**].

5.1.10. *Articles—Integrals.* [**Amo83c**], [**And82b**], [**BFST86**], [**BGV93**], [**Can81**], [**Chr90**], [**Cof91**], [**Cor72**], [**DK90**], [**Gab79**], [**Gab80**], [**GM81**], [**Han85**], [**Joh75**], [**Lew91**], [**Lin72**], [**LK73**], [**LPM81**], [**Lun85**], [**MDS92**], [**Moo83**], [**OFM78**], [**PB82**], [**PB83**], [**PB85**, includes survey], [**Puo88**], [**SBK92**], [**Sie77**].

5.1.11. *Articles—Zeros.* [**CH70a**], [**IKF91**], [**KS84a**], [**KS84c**], [**KS85a**], [**KS85b**], [**KS85c**], [**KS87**], [**MF86**], [**Sko85**].

5.2. Coulomb Wave Functions.

5.2.1. *Real Arguments and Parameters.* [**Bar76**, Fortran], [**Bar81b**, Fortran], [**Bar82b**, Fortran], [**BDG+72**, Fortran], [**BS80**, Fortran], [**CH70b**], [**NT84**, Fortran], [**Sea82**, Fortran], [**She74**]. Libraries: [**Bak92**].

5.2.2. *Complex Arguments and Parameters.* [**TB85**, Fortran], [**TR69**, Fortran].

5.2.3. *Articles.* [**AS92**], [**Bar81a**], [**Bar82a**], [**Bar82c**], [**Gau69b**], [**Köl72b**, includes survey], [**Nes84**], [**Pex70**], [**SG72**], [**TB86**].

5.3. Elliptic Integrals and Functions. An important recent change in the old subject of elliptic integrals is a renormalization of the definitions of the integrals. This is due to B. C. Carlson: references will be found in §5.3.5.

5.3.1. *Complete Elliptic Integrals.* [**Bel88**], [**Bul65a**, Algol], [**Bul65b**, Algol], [**Bul69b**, Algol], [**Cod65a**], [**Cod65b**], [**Luk69b**], [**MH73**, Algol]. Libraries: [**Bak92**], [**Mos89**], [**ULI90**], **IMSL**, **Numerical Recipes**. Systems: **Mathematica**.

5.3.2. *Incomplete Elliptic Integrals.* [**Bul65a**, Algol], [**Bul69b**, Algol], [**Car87**, Fortran], [**Car88**, Fortran], [**CN81**, Fortran], [**Luk69b**], [**PT90**, Fortran]. Libraries: [**Bak92**], [**Mos89**], [**ULI90**], **IMSL**, **NAG**, **Numerical Recipes**, **Scientific Desk**, **SLATEC**. Systems: **Mathematica**.

5.3.3. *Jacobi's Elliptic Functions.* This subsection includes the theta functions. [**Bul65a**, Algol]. Libraries: [**Bak92**], [**Mos89**], [**ULI90**], **IMSL**, **NAG**, **Numerical Recipes**. Systems: **Mathematica** (includes inverse functions).

5.3.4. *Weierstrass' Elliptic Functions.* [**Eck76**], [**Eck77**], [**Eck80**, Fortran]. Libraries: [**Bak92**], [**ULI90**], **IMSL**. Systems: **Mathematica**.

5.3.5. *Articles.* [**ACJP85**, includes survey], [**Bul69a**], [**Car65**], [**Car77a**], [**Car77b**], [**Car79**], [**Car87**], [**Car88**], [**Car89**], [**Car91**], [**Car92**], [**CGL90**], [**Cri89**], [**FGG82**], [**FL67**], [**Lee90**], [**Lee92**], [**Luk68**], [**Luk70b**], [**LY88**], [**Mid75**], [**NC66**], [**Sal89**], [**War60**].

5.4. Fermi-Dirac, Bose-Einstein, and Debye Integrals. This section includes the Lerch transcendent.

5.4.1. *Real Parameter and Argument.* [**BDM81**, Fortran], [**CT67**], [**FR86**, Fortran], [**NDT69**]. Libraries: [**Bak92**].

5.4.2. *Complex Argument and/or Parameters.* Systems: **Mathematica**.

5.4.3. *Articles.* [**Bui91**], [**Gau93a**], [**Gau93b**], [**LS91**], [**NM93**], [**Pas88**], [**Pas91**], [**Pic89**], [**Sag91a**], [**Sag91b**].

5.5. Hypergeometric and Confluent Hypergeometric Functions.

5.5.1. *Hypergeometric Functions.* Libraries: [**Bak92**], [**Mos89**], [**ULI90**], **Numerical Recipes**. Systems: **Mathematica**.

5.5.2. *Confluent Hypergeometric Functions.* [**BS80**, Fortran], [**NPB92a**, Fortran], [**NT84**, Fortran], [**Tem83**, Algol]. Libraries: [**Bak92**], [**Mos89**], [**ULI90**], **SLATEC**. Systems: **Mathematica**.

5.5.3. *Other Hypergeometric Functions.* [**CM84**, Pascal], [**PBN93**]. Libraries: [**Bak92**], [**Mos89**]. Systems: **Maple**, **Mathematica**.

5.5.4. *Articles.* [**BMOF92**], [**CG89**], [**Luk75**], [**Luk77a**], [**Ném74**], [**NPB92b**], [**Wim74**].

5.6. Incomplete Bessel Functions, Incomplete Beta Function. This section includes F–, t– and von Mises' distribution functions.

5.6.1. *Incomplete Bessel Functions.* [**Hil77**, Fortran].

5.6.2. *Incomplete Beta Function.* [**DM92**, Fortran], [**Dor68**, Algol], [**Gau64**, Algol], [**Hil70a**, Algol], [**Lev69**, Fortran], [**MB73a**, Fortran], [**Mor69**, Algol], [**Phi90**, Basic]. Libraries: [**Bak92**], [**Mos89**], [**ULI90**], **IMSL**, **NAG**, **Numerical Recipes**, **Scientific Desk**, **SLATEC**. Systems: **Mathematica**.

5.6.3. *Inverse Incomplete Beta Function.* [**AS93a**, Fortran], [**Hil70b**, Algol], [**MB73b**]. Libraries: [**Mos89**], [**ULI90**], **IMSL**, **NAG**, **Scientific Desk**. Systems: **Maple**, **Mathematica**.

5.6.4. *Articles.* [**AS93b**], [**DJ67**], [**OM68**], [**Tem92b**].

5.7. Incomplete Gamma Functions, Generalized Exponential Integrals. These functions are essentially equivalent; thus $E_p(z) = z^{p-1}\Gamma(1-p, z)$. This section includes the chi-square distribution function.

5.7.1. *Real z and Integer or Half-Integer p.* [**Amo80a**, Fortran], [**SP75**, Fortran], [**SZ74**, Fortran]. Libraries: [**Mos89**], [**ULI90**], **SLATEC**. Systems: **Maple**.

5.7.2. *Real z and Real p.* [**CLM90a**, Fortran], [**CLM90b**, Fortran], [**DM87**, Fortran], [**Ful72**, Fortran], [**Gau79a**, Fortran], [**Moo82**, Fortran], [**She88**, Fortran], [**Tem94**, Pascal]. Libraries: [**Bak92**], [**Mos89**], [**ULI90**], **IMSL**, **NAG**, **Numerical Recipes**, **Scientific Desk**, **SLATEC**.

5.7.3. *Complex z and Real or Complex p.* Systems: **Maple**, **Mathematica**.

5.7.4. *Inverse Function.* [**DM87**, Fortran], [**Phi88**, Fortran]. Libraries: [**Mos89**], [**ULI90**], **IMSL**, **NAG**, **Scientific Desk**. Systems: **Maple**, **Mathematica**.

5.7.5. *Articles.* [**AB87a**], [**Amo80b**], [**Bar61**], [**CLM87**], [**CLM88**], [**CLM90c**], [**DM86**], [**Gau79b**], [**JT85**], [**LDP93**], [**Luk75**], [**Mar82**], [**Tem85**], [**Tem87**], [**Tem92a**].

5.8. Legendre Functions and Associated Legendre Functions. This section includes the conical and toroidal functions. See also hypergeometric functions (§5.5) and orthogonal polynomials (§5.10).

5.8.1. *Real Argument and Parameters.* [**Bra73**, Fortran], [**Del79**, Fortran], [**Gau65**, Algol], [**LS81**, Fortran], [**OS83**, Fortran]. Libraries: [**Bak92**], **Numerical Recipes**, **SLATEC**.

5.8.2. *Complex Argument and/or Parameters.* [**Köl81**, conical with real argument]. Libraries: [**Bak92**]. Systems: **Mathematica**.

5.8.3. *Articles.* [**CM78**], [**CM79**], [**EWB84**], [**Fet70**], [**SOL81**].

5.9. Mathieu, Lamé, and Spheroidal Wave Functions.

5.9.1. *Characteristic Values of Mathieu's Equation.* [**Cle69**, Fortran], [**Del73**, Algol], [**Lee79**, Fortran], [**RL80**, Fortran], [**Shi93a**, Fortran]. Libraries: [**Bak92**], **IMSL**.

5.9.2. *Mathieu Functions.* [**Cle69**, Fortran], [**Del73**, Algol], [**RL80**, Fortran], [**Shi93a**, Fortran]. Libraries: [**Bak92**], **IMSL**.

5.9.3. *Spheroidal Wave Functions.* [**BC83a**, Fortran], [**BC83b**, Fortran], [**KBH70**, Fortran], [**KvB70**, Fortran], [**vBBH70**, Fortran]. Libraries: [**Bak92**].

5.9.4. *Articles.* [**ADK$^+$84**], [**ADKL89**], [**ADKL91**], [**ATZ83**], [**Bla46**], [**Cal88**], [**Can71**], [**Egl84**], [**EP69**], [**Hod70**], [**Pal69**], [**Shi93b**], [**SM75**], [**TP83**], [**vBBHK72**].

5.10. Orthogonal Polynomials. See also hypergeometric functions (§5.5), Legendre functions (§5.8), and Weber parabolic cylinder functions (§5.13).

5.10.1. *Classical Polynomials (Chebyshev, Hermite, Jacobi, Laguerre, Legendre etc.), Real Arguments.* [**LPT80**, Fortran], [**Sim64**, Algol], [**Wit68**, Fortran]. Libraries: [**Bak92**], [**ULI90**].

5.10.2. *Classical Polynomials, Complex Arguments.* Systems: **Maple**, **Mathematica**.

5.10.3. *Other Orthogonal Polynomials.* [**Bis91**, Maple], [**Coo68**, Fortran], [**EK92**], [**Gau94**, Fortran], [**Öpi87**, Fortran].

5.10.4. *Articles.* [**BEGG91**], [**BR91**], [**Chi92**], [**FG91**], [**FG92**], [**Gau82**], [**Gau85**], [**Gau90**], [**Gau91a**], [**Luk75**], [**PA92**].

5.11. Polylogarithms (Generalized).

5.11.1. *Real Variables.* [**KMR70**, Algol].

5.11.2. *Articles.* [**Bar74**].

5.12. Struve and Anger-Weber Functions.

5.12.1. *Struve Functions or Integrals of Struve Functions.* [**Luk69b**], [**Luk75**], [**Mac93**], [**New84**]. Libraries: [**Bak92**], [**Mos89**], [**ULI90**].

5.12.2. *Integrals of Anger-Weber Functions.* Libraries: [**Bak92**].

5.12.3. *Articles.* [**Zan75**].

5.13. Weber Parabolic Cylinder Functions. See also confluent hypergeometric functions (§5.5).

5.13.1. *Real Arguments and Parameters.* [**Tau92**, Fortran]. Libraries: [**Bak92**].

5.13.2. *Articles.* [**LR74**], [**MMV81**], [**RL76**], [**SGA81**].

5.14. Zeta Function (Generalized).

5.14.1. *Real arguments.* [**AB89**]. Libraries: [**Bak92**], [**Mos89**]. Systems: **Maple**, **Mathematica**.

5.14.2. *Articles.* [**CHT71**], [**Moi88**].

6. Testing and Library Construction

In this section we list articles and books that provide general observations on the testing of software and/or the construction of software libraries for the special functions. For information on individual libraries see §3.

[**Cod74**], [**Cod76**], [**Cod82**], [**Cod84b**], [**Cod85**], [**CS91**], [**Eva74**, especially pp. 275–301 and 357–435], [**Ful77**], [**Gaf88**], [**Kuk71**], [**LMS73a**], [**LMS73b**], [**Mos89**], [**PTVF92**, example books], [**Ric83**], [**Sch76**], [**SL73**].

7. Future Trends

Great progress has been made in recent years in the construction of software for generating the special functions, yet enormous gaps remain for functions having variable parameters in addition to the argument. This is especially true when the variables are complex. In this concluding section we offer some general suggestions concerning future work in this area.

First, because of the sheer magnitude of the effort required, there should be a perceived physical or other applied need before a decision is made to embark on the construction of extensive new software for functions of two or more variables. At present there are simply too many gaps to fill to be able to indulge in the luxury of arbitrary selection. Moreover, great care should be exercised in the choice of actual functions to be generated. For example, neither the Airy function $\mathrm{Bi}(z)$ nor the Bessel function of the second kind $Y_\nu(z)$ has a useful role when the argument z is not real; compare [**Olv74**, Chapters 7 and 11].

Second, coverage of a chosen region should be dictated by uniform accuracy requirements (in an appropriate measure), not by the limitations of the methods that happen to be used. At the very least it is frustrating for users to discover that the precision yielded by a package varies widely, or worse still disappears altogether, in parts of the claimed regions of coverage.

Third, the potential offered by the ongoing increase in power of computers should be exploited with a view to reducing the number and complexity of algorithms to be used. This includes, for example, the use of parallel or vector methods for summing series [**Kar91**] or solving differential or difference equations [**LO93**].

Fourth—and here we are looking further into the future—the use of systems of computer arithmetic other than floating-point should be considered. The floating-point system has two disadvantages which become especially annoying and time-consuming in the construction of special-function software. One is that the associated error measure, relative precision, is quite inappropriate in the neighborhoods of zeros. The other stems from failure due to overflow or underflow: here the usual remedy of rescaling can be difficult to apply, owing to the extremely varied asymptotic behavior of functions of several variables. A system of computer arithmetic that is capable of overcoming both problems in an elegant manner is the so-called level-index system [**COT89**].

Lastly, any new algorithm or package should be documented fully. It should also be subjected to exhaustive testing procedures, and these, too, need to be documented. Indeed, the proposed testing procedures should be considered at an early stage in the planning of the main algorithms[12]. There are so many pitfalls in the construction of algorithms for the special functions that the use of undocumented or insufficiently tested packages is a risky proposition[13].

Acknowledgments

We are grateful to the following individuals for supplying references and making helpful comments: D. E. Amos, W. L. Anderson, A. R. Barnett, E. Battiste, C. Brezinski, B. C. Carlson, B. Gabutti, P. W. Gaffney, W. Gautschi, K. S. Kölbig, S. D. Leigh, L. C. Maximon, M. A. McClain, B. R. Miller, W. Parke, N. M. Temme, M. Vuorinen.

A Note on the Reference Acronyms

In the references that follow, the acronyms within the identifying square brackets follow the AMS BibTeX scheme. Initial letters pertain to the author(s) or editor(s). These are followed by two digits representing the year of publication; a letter may also be appended to distinguish between publications in the same year, e.g. [**AB87a**], [**AB87b**]. In the case of papers or books with more than four authors, initial letters from the names of the first three authors are used, followed by a $^+$ sign, e.g. [**ADK$^+$84**]. **It is also important to note that the references are listed according to the alphabetical order of the acronyms and not according to the alphabetical order of the authors' names.**

[12]For example, it is better to avoid the use of Wronskian and Casoratian relations in the main computing package, if possible, in order to reserve these identities for consistency checks.

[13]For a striking example see [**Olv91**].

References

[AB87a] G. Allasia and R. Besenghi, *Numerical calculation of incomplete gamma functions by the trapezoidal rule*, Numer. Math. **50** (1987), 419–428.

[AB87b] G. Allasia and R. Besenghi, *Numerical computation of Tricomi's psi function by the trapezoidal rule*, Computing **39** (1987), 271–279.

[AB89] G. Allasia and R. Besenghi, *Numerical calculation of the Riemann zeta function and generalizations by means of the trapezoidal rule*, Numerical and Applied Mathematics, Part 2 (Paris 1988) (C. Brezinski, ed.), IMACS Ann. Comput. Appl. Math., 1.2, Baltzer, Basel, 1989, pp. 467–472.

[Ach86] J.-J. Achenbach, *Numerik. Implementierung von Zylinderfunktionen*, Friedr. Vieweg & Sohn, Braunschweig/Wiesbaden, 1986.

[ACJP85] J. Arazy, T. Claesson, S. Janson, and J. Peetre, *Means and their iterations*, Proceedings of the Nineteenth Nordic Congress of Mathematicians, Reykjavik 1984, Icelandic Mathematical Society, Reykjavik, 1985, pp. 191–212.

[ACM] *Collected algorithms from ACM*, Association for Computing Machinery, 1515 Broadway, New York, New York 10036, periodically updated looseleaf collection.

[ACM64] *Index by subject to algorithms, 1960–63*, Comm. ACM **7** (1964), 146–148.

[AD79] D. E. Amos and S. L. Daniel, *AMOSLIB, a special function library, version 9/77*, Report 77-1390, Sandia Laboratories, August 1979.

[Ada69] A. G. Adams, *Algorithm 39. Areas under the normal curve*, Comput. J. **12** (1969), 197–198.

[ADK+84] A. A. Abramov, A. L. Dyshko, N. B. Konyukhova, T. V. Pak, and B. S. Pariiskii, *Evaluation of prolate spheroidal function by solving the corresponding differential equations*, U.S.S.R. Comput. Math. and Math. Phys. **24** (1984), no. 1, 1–11.

[ADKL89] A. A. Abramov, A. L. Dyshko, N. B. Konyukhova, and T. V. Levitina, *Evaluation of Lamé angular wave functions by solving auxiliary differential equations*, U.S.S.R. Comput. Math. and Math. Phys. **29** (1989), no. 3, 119–131.

[ADKL91] A. A. Abramov, A. L. Dyshko, N. B. Konyukhova, and T. V. Levitina, *Computation of radial wave functions for spheroids and triaxial ellipsoids by the modified phase function method*, Comput. Math. and Math. Phys. **31** (1991), no. 2, 25–42.

[ADW77a] D. E. Amos, S. L. Daniel, and M. K. Weston, *Algorithm 511. CDC 6600 subroutines IBESS and JBESS for Bessel functions $I_\nu(x)$ and $J_\nu(x)$, $x \geq 0$, $\nu \geq 0$*, ACM Trans. Math. Software **3** (1977), 93–95, for erratum see same journal v. 4 (1978), p. 411.

[ADW77b] D. E. Amos, S. L. Daniel, and M. K. Weston, *CDC 6600 subroutines IBESS and JBESS for Bessel functions $I_\nu(x)$ and $J_\nu(x)$, $x \geq 0$, $\nu \geq 0$*, ACM Trans. Math. Software **3** (1977), 76–92.

[Air84] T. J. Aird, *The IMSL library*, Sources and Development of Mathematical Software (W. R. Cowell, ed.), Prentice-Hall, Inc., Englewood Cliffs, New Jersey, 1984, pp. 264–301.

[AM61] A. M. Arthurs and R. McCarroll, *Expansion of spherical Bessel functions in a series of Chebyshev polynomials*, Math. Comp. **15** (1961), 159–162.

[AM78] R. W. B. Ardill and K. J. M. Moriarty, *Spherical Bessel functions j_n and y_n of integer order and real argument*, Comput. Phys. Comm. **14** (1978), 261–265.

[Amo74] D. E. Amos, *Computation of modified Bessel functions and their ratios*, Math. Comp. **28** (1974), 239–251.

[Amo80a] D. E. Amos, *Algorithm 556. Exponential integrals*, ACM Trans. Math. Software **6** (1980), 420–428, for remark see same journal v. 9 (1983), p. 525.

[Amo80b] D. E. Amos, *Computation of exponential integrals*, ACM Trans. Math. Software **6** (1980), 365–377.

[Amo83a] D. E. Amos, *Algorithm 609. A portable Fortran subroutine for the Bickley functions* $\mathrm{Ki}_n(x)$, ACM Trans. Math. Software **9** (1983), 480–493.

[Amo83b] D. E. Amos, *Algorithm 610. A portable Fortran subroutine for derivatives of the psi function*, ACM Trans. Math. Software **9** (1983), 494–502.

[Amo83c] D. E. Amos, *Uniform asymptotic expansions for exponential integrals $E_n(x)$ and*

Bickley functions $\mathrm{Ki}_n(x)$, ACM Trans. Math. Software **9** (1983), 467–479

[Amo86] D. E. Amos, *Algorithm 644. A portable package for Bessel functions of a complex argument and nonnegative order*, ACM Trans. Math. Software **12** (1986), 265–273, for remark see same journal v. 16 (1990), p. 404.

[Amo90a] D. E. Amos, *Algorithm 683. A portable Fortran subroutine for exponential integrals of a complex argument*, ACM Trans. Math. Software **16** (1990), 178–182.

[Amo90b] D. E. Amos, *Computation of exponential integrals of a complex argument*, ACM Trans. Math. Software **16** (1990), 169–177.

[And82a] W. L. Anderson, *Algorithm 588. Fast Hankel transforms using related and lagged convolutions*, ACM Trans. Math. Software **8** (1982), 369–370.

[And82b] W. L. Anderson, *Fast Hankel transforms using related and lagged convolutions*, ACM Trans. Math. Software **8** (1982), 344–368.

[AS64] M. Abramowitz and I. A. Stegun (eds.), *Handbook of mathematical functions with formulas, graphs and mathematical tables*, National Bureau of Standards Applied Mathematics Series, vol. 55, U. S. Government Printing Office, Washington, D. C., 1964.

[AS68] *Statistical algorithms editorial note*, Appl. Statist. **17** (1968), 79–82.

[AS92] J. Abad and J. Sesma, *Computation of Coulomb wave functions at low energies*, Comput. Phys. Comm. **71** (1992), 110–124.

[AS93a] R. W. Abernathy and R. P. Smith, *Algorithm 724. Program to calculate F-percentiles*, ACM Trans. Math. Software **19** (1993), 481–483.

[AS93b] R. W. Abernathy and R. P. Smith, *Applying series expansion to the inverse beta distribution to find percentiles of the F-distribution*, ACM Trans. Math. Software **19** (1993), 474–480.

[Ask89] R. A. Askey, *Handbooks of special functions*, A Century of Mathematics in America, Part III, Hist. Math., vol. 3, American Mathematical Society, Providence, Rhode Island, 1989, pp. 369–391.

[ATZ83] F. M. Arscott, P. J. Taylor, and R. V. M. Zahar, *On the numerical construction of ellipsoidal wave functions*, Math. Comp. **40** (1983), 367–380.

[Bai93] D. H. Bailey, *Algorithm 719. Multiprecision translation and execution of Fortran programs*, ACM Trans. Math. Software **19** (1993), 288–319.

[Bak92] L. Baker, *C mathematical function handbook*, McGraw-Hill, Inc., New York, 1992, includes diskette.

[Bar61] R. Barakat, *Evaluation of the incomplete gamma function of imaginary argument by Chebyshev polynomials*, Math. Comp. **15** (1961), 7–11.

[Bar74] R. H. Barlow, *Convergent continued fraction approximants to generalised polylogarithms*, BIT **14** (1974), 112–116.

[Bar76] A. R. Barnett, *RCWFF—A modification of the real Coulomb wavefunction program RCWFN*, Comput. Phys. Comm. **11** (1976), 141–142.

[Bar81a] A. R. Barnett, *An algorithm for regular and irregular Coulomb and Bessel functions of real order to machine accuracy*, Comput. Phys. Comm. **21** (1981), 297–314.

[Bar81b] A. R. Barnett, *KLEIN: Coulomb functions for real λ and positive energy to high accuracy*, Comput. Phys. Comm. **24** (1981), 141–159.

[Bar82a] A. R. Barnett, *Continued-fraction evaluation of Coulomb functions $F_\lambda(\eta, x)$, $G_\lambda(\eta, x)$ and their derivatives*, J. Comput. Phys. **46** (1982), 171–188.

[Bar82b] A. R. Barnett, *COULFG: Coulomb and Bessel functions and their derivatives, for real arguments, by Steed's method*, Comput. Phys. Comm. **27** (1982), 147–166.

[Bar82c] A. R. Barnett, *High-precision evaluation of the regular and irregular Coulomb wavefunctions*, J. Comput. Appl. Math. **8** (1982), 29–33.

[BC83a] T. A. Beu and R. I. Câmpeanu, *Prolate angular spheroidal wave functions*, Comput. Phys. Comm. **30** (1983), 187–192.

[BC83b] T. A. Beu and R. I. Câmpeanu, *Prolate radial spheroidal wave functions*, Comput. Phys. Comm. **30** (1983), 177–185.

[BD80] A. Bañuelos and R. A. Depine, *A program for computing the Riemann zeta func-*

tion for complex argument, Comput. Phys. Comm. **20** (1980), 441–445.

[BDG+72] C. Bardin, Y. Dandeu, L. Gauthier, J. Guillerman, T. Lena, and J.-M. Pernet, *Coulomb functions in the entire (η, ρ)-plane*, Comput. Phys. Comm. **3** (1972), 73–87.

[BDM81] A. Bañuelos, R. A. Depine, and R. C. Mancini, *A program for computing the Fermi-Dirac functions*, Comput. Phys. Comm. **21** (1981), 315–322.

[BEGG91] D. L. Boley, S. Elhay, G. H. Golub, and M. H. Gutknecht, *Nonsymmetric Lanczos and finding orthogonal polynomials associated with indefinite weights*, Numer. Algorithms **1** (1991), 21–43.

[BEJ76] J. M. Blair, C. A. Edwards, and J. H. Johnson, *Rational Chebyshev approximations for the inverse of the error function*, Math. Comp. **30** (1976), 827–830.

[BEJ78] J. M. Blair, C. A. Edwards, and J. H. Johnson, *Rational Chebyshev approximations for the Bickley functions* $\mathrm{Ki}_n(x)$, Math. Comp. **32** (1978), 876–886.

[Bel88] V. N. Belykh, *Calculation on a computer of the complete elliptic integrals $K(x)$ and $E(x)$*, Boundary value problems for partial differential equations, Akad. Nauk SSSR Sibirsk. Otdel., Inst. Mat., Novosibirsk, 1988, pp. 3–15, 137 (Russian).

[BFST86] V. Bezvoda, R. Farzan, K. Segeth, and G. Takó, *On numerical evaluation of integrals involving Bessel functions*, Apl. Mat. **31** (1986), 396–410.

[BG81a] G. A. Baker, Jr. and P. Graves-Morris, *Padé approximants, part I*, Encyclopedia of Mathematics and its Applications, vol. 13, Addison-Wesley Publishing Company, Reading, Massachusetts, 1981.

[BG81b] G. A. Baker, Jr. and P. Graves-Morris, *Padé approximants, part II*, Encyclopedia of Mathematics and its Applications, vol. 14, Addison-Wesley Publishing Company, Reading, Massachusetts, 1981.

[BGV93] K. Balla, O. S. Guk, and M. Vicsek, *On the computation of Bessel functions of first kind*, Computing **50** (1993), 77–85.

[BH75] N. Bleistein and R. A. Handelsman, *Asymptotic expansions of integrals*, Holt, Rinehart and Winston, New York, 1975.

[BHK91] R. F. Boisvert, S. E. Howe, and D. K. Kahaner, *The guide to available mathematical software problem classification system*, Comm. Statist. B—Simulation Comput. **20** (1991), 811–842.

[BHKS90] R. F. Boisvert, S. E. Howe, D. K. Kahaner, and J. L. Springmann, *Guide to available mathematical software*, Tech. Report NISTIR 90-4237, National Institute of Standards and Technology, Center for Computing and Applied Mathematics, Gaithersburg, Maryland 20899, 1990.

[Bim93] *HiQ reference manual, version 2.0*, Bimillennium Corporation, 16795 Lark Avenue, Suite 200, Los Gatos, California 95030, 1993.

[Bis91] A. K. Bisoi, *A Maple program to generate orthonormal polynomials*, Comput. Math. Appl. **22** (1991), no. 9, 1–5.

[BKN88a] L. V. Babushkina, M. K. Kerimov, and A. I. Nikitin, *Algorithms for computing Bessel functions of half-integer order with complex arguments*, U.S.S.R. Comput. Math. and Math. Phys. **28** (1988), no. 5, 109–117.

[BKN88b] L. V. Babushkina, M. K. Kerimov, and A. I. Nikitin, *Algorithms for evaluating spherical Bessel functions in the complex domain*, U.S.S.R. Comput. Math. and Math. Phys. **28** (1988), no. 6, 122–128.

[Bla46] G. Blanch, *On the computation of Mathieu functions*, J. Math. and Phys. **25** (1946), 1–20.

[Bla74] J. M. Blair, *Rational Chebyshev approximations for the modified Bessel functions $I_0(x)$ and $I_1(x)$*, Math. Comp. **28** (1974), 581–583.

[BMOF92] E. Badralexe, P. Marksteiner, Y. Oh, and A. J. Freeman, *Computation of the Kummer functions and Whittaker functions by using Neumann type series expansions*, Comput. Phys. Comm. **71** (1992), 47–55.

[Bow84] K. O. Bowman, *Computation of the polygamma functions*, Comm. Statist. B—Simulation Comput. **13** (1984), 409–415.

[BR71] R. D. Bardo and K. Ruedenberg, *Numerical analysis and evaluation of normalized repeated integrals of the error function and related functions*, J. Comput. Phys. **8** (1971), 167–174.

[BR91] C. Brezinski and M. Redivo Zaglia, *A new presentation of orthogonal polynomials with applications to their computation*, Numer. Algorithms **1** (1991), 207–221.

[Bra73] W. J. Braithwaite, *Associated Legendre polynomials, ordinary and modified spherical harmonics*, Comput. Phys. Comm. **5** (1973), 390–394.

[Bre76] R. P. Brent, *Fast multiple-precision evaluation of elementary functions*, J. Assoc. Comput. Mach. **23** (1976), 242–251.

[Bre78a] R. P. Brent, *Algorithm 524. MP, a Fortran multiple-precision arithmetic package*, ACM Trans. Math. Software **4** (1978), 71–81, for remark see same journal v. 5 (1979), pp. 518–519.

[Bre78b] R. P. Brent, *A Fortran multiple-precision arithmetic package*, ACM Trans. Math. Software **4** (1978), 57–70.

[Bre91] C. Brezinski, *History of continued fractions and Padé approximants*, Springer Series in Computational Mathematics, vol. 12, Springer-Verlag, Berlin, 1991.

[BS80] K. L. Bell and N. S. Scott, *Coulomb functions (negative energies)*, Comput. Phys. Comm. **20** (1980), 447–458.

[BS92] R. F. Boisvert and B. V. Saunders, *Portable vectorized software for Bessel function evaluation*, ACM Trans. Math. Software **18** (1992), 456–469, for corrigendum see same journal v. 19 (1993), p. 131.

[Bui91] Bui Doan Khanh, *A computation of the Fermi-Dirac integrals by asymptotics and the Hermite corrector formula*, Appl. Math. Comput. **41** (1991), 61–68.

[Bul65a] R. Bulirsch, *Numerical calculation of elliptic integrals and elliptic functions*, Numer. Math. **7** (1965), 78–90.

[Bul65b] R. Bulirsch, *Numerical calculation of elliptic integrals and elliptic functions. II*, Numer. Math. **7** (1965), 353–354.

[Bul67] R. Bulirsch, *Numerical calculation of the sine, cosine and Fresnel integrals*, Numer. Math. **9** (1967), 380–385.

[Bul69a] R. Bulirsch, *An extension of the Bartky-transformation to incomplete elliptic integrals of the third kind*, Numer. Math. **13** (1969), 266–284.

[Bul69b] R. Bulirsch, *Numerical calculation of elliptic integrals and elliptic functions. III*, Numer. Math. **13** (1969), 305–315.

[Bur63] F. D. Burgoyne, *Approximations to Kelvin functions*, Math. Comp. **17** (1963), 295–298.

[Buz84] B. L. Buzbee, *The SLATEC common mathematical library*, Sources and Development of Mathematical Software (W. R. Cowell, ed.), Prentice-Hall, Inc., Englewood Cliffs, New Jersey, 1984, pp. 302–320.

[BvI93] C. Brezinski and J. van Iseghem, *Padé approximations*, Handbook of Numerical Analysis (P. G. Ciarlet and J. L. Lions, eds.), vol. 3, North-Holland, Amsterdam, 1993, in press.

[Cal88] J. Caldwell, *Computation of eigenvalues of spheroidal harmonics using relaxation*, J. Phys. A **21** (1988), 3685–3693.

[Cam79] J. B. Campbell, *Bessel functions $J_\nu(x)$ and $Y_\nu(x)$ of real order and real argument*, Comput. Phys. Comm. **18** (1979), 133–142.

[Cam80] J. B. Campbell, *On Temme's algorithm for the modified Bessel function of the third kind*, ACM Trans. Math. Software **6** (1980), 581–586.

[Cam81] J. B. Campbell, *Bessel functions $I_\nu(z)$ and $K_\nu(z)$ of real order and complex argument*, Comput. Phys. Comm. **24** (1981), 97–105, for erratum see same journal v. 25 (1982), p. 207.

[Cam84] J. B. Campbell, *Determination of ν-zeros of Hankel functions*, Comput. Phys. Comm. **32** (1984), 333–339.

[Can71] J. Canosa, *Numerical solution of Mathieu's equation*, J. Comput. Phys. **7** (1971), 255–272.

[Can81] S. M. Candel, *An algorithm for the Fourier-Bessel transform*, Comput. Phys. Comm. **23** (1981), 343–353.

[Car65] B. C. Carlson, *On computing elliptic integrals and functions*, J. Math. and Phys. **44** (1965), 36–51.

[Car77a] B. C. Carlson, *Elliptic integrals of the first kind*, SIAM J. Math. Anal. **8** (1977), 231–242.

[Car77b] B. C. Carlson, *Special functions of applied mathematics*, Academic Press, New York, 1977.

[Car79] B. C. Carlson, *Computing elliptic integrals by duplication*, Numer. Math. **33** (1979), 1–16.

[Car87] B. C. Carlson, *A table of elliptic integrals of the second kind*, Math. Comp. **49** (1987), 595–606 and S13–S17.

[Car88] B. C. Carlson, *A table of elliptic integrals of the third kind*, Math. Comp. **51** (1988), 267–280 and S1–S5.

[Car89] B. C. Carlson, *A table of elliptic integrals: Cubic cases*, Math. Comp. **53** (1989), 327–333.

[Car91] B. C. Carlson, *A table of elliptic integrals: One quadratic factor*, Math. Comp. **56** (1991), 267–280.

[Car92] B. C. Carlson, *A table of elliptic integrals: Two quadratic factors*, Math. Comp. **59** (1992), 165–180.

[CCF83] J. N. L. Connor, P. R. Curtis, and D. Farrelly, *A differential equation method for the numerical evaluation of the Airy, Pearcey and swallowtail canonical integrals and their derivatives*, Molecular Phys. **48** (1983), 1305–1330.

[CER93] *CERNLIB short writeups*, CERN Program Library Office, CERN–CN Division, CH–1211 Geneva 23, Switzerland, 1993, electronic mail address is cernlib@cernvm.cern.ch.

[CF87] Ll. Closas and J. Fernández Rubio, *Calculo rapido de las funciones de Bessel modificadas $K_{is}(X)$ e $I_{is}(X)$ y sus derivadas*, Stochastica **11** (1987), no. 1, 53–61.

[CG89] G. Chiocchia and B. Gabutti, *A new transformation for computing hypergeometric series and the exact evaluation of the transonic adiabatic flow over a smooth bump*, Comput. & Fluids **17** (1989), no. 1, 13–23.

[CGG$^+$91] B. W. Char, K. O. Geddes, G. H. Gonnet, B. L. Leong, M. B. Monaghan, and S. M. Watt, *Maple V library reference manual*, Springer-Verlag, New York, 1991.

[CGL90] R. Coquereaux, A. Grossmann, and B. E. Lautrup, *Iterative method for calculation of the Weierstrass elliptic function*, IMA J. Numer. Anal. **10** (1990), 119–128.

[CH67] W. J. Cody and K. E. Hillstrom, *Chebyshev approximations for the natural logarithm of the gamma function*, Math. Comp. **21** (1967), 198–203.

[CH70a] J. A. Cochran and J. N. Hoffspiegel, *Numerical techniques for finding ν-zeros of Hankel functions*, Math. Comp. **24** (1970), 413–422.

[CH70b] W. J. Cody and K. E. Hillstrom, *Chebyshev approximations for the Coulomb phase shift*, Math. Comp. **24** (1970), 671–677, for corrigendum, see same journal v. 26 (1972), p. 1031.

[Cha80] B. W. Char, *On Stieltjes' continued fraction for the gamma function*, Math. Comp. **34** (1980), 547–551.

[Cha83] A. D. Chave, *Numerical integration of related Hankel transforms by quadrature and continued fraction expansion*, Geophysics **48** (1983), 1671–1686.

[Chi92] R. C. Y. Chin, *A domain decomposition method for generating orthogonal polynomials for a Gaussian weight on a finite interval*, J. Comput. Phys. **99** (1992), 321–336.

[Chr90] N. B. Christensen, *Optimized fast Hankel transform filters*, Geophysical Prospecting **38** (1990), 545–568, for comment and reply, see same journal v. 39 (1991), pp. 445–447 and 449–450.

[CHT71] W. J. Cody, K. E. Hillstrom, and H. C. Thacher, Jr., *Chebyshev approximations for the Riemann zeta function*, Math. Comp. **25** (1971), 537–547.

[CJR92] R. M. Corless, D. J. Jeffrey, and H. Rasmussen, *Numerical evaluation of Airy functions with complex arguments*, J. Comput. Phys. **99** (1992), 106–114.

[Cle62] C. W. Clenshaw, *Chebyshev series for mathematical functions*, National Physical Laboratory Mathematical Tables, vol. 5, Her Majesty's Stationery Office, London, 1962.

[Cle69] D. S. Clemm, *Algorithm 352. Characteristic values and associated solutions of Mathieu's differential equation*, Comm. ACM **12** (1969), 399–407, for remarks see same journal v. 13 (1970), p. 750 and v. 15 (1972), p. 1074.

[CLM87] C. Chiccoli, S. Lorenzutta, and G. Maino, *A numerical method for generalized exponential integrals*, Comput. Math. Appl. **14** (1987), 261–268.

[CLM88] C. Chiccoli, S. Lorenzutta, and G. Maino, *On the evaluation of generalized exponential integrals* $E_\nu(x)$, J. Comput. Phys. **78** (1988), 278–287.

[CLM90a] C. Chiccoli, S. Lorenzutta, and G. Maino, *An algorithm for exponential integrals of real order*, Computing **45** (1990), 269–276.

[CLM90b] C. Chiccoli, S. Lorenzutta, and G. Maino, *Calculation of exponential integrals of real order*, Internat. J. Comput. Math. **31** (1990), 125–135.

[CLM90c] C. Chiccoli, S. Lorenzutta, and G. Maino, *On a Tricomi series representation for the generalized exponential integral*, Internat. J. Comput. Math. **31** (1990), 257–262.

[CM78] J. N. L. Connor and D. C. Mackay, *Accelerating the convergence of the zonal harmonic series representation in the Schumann resonance problem*, J. Atmospheric Terrestrial Phys. **40** (1978), 977–980.

[CM79] J. N. L. Connor and D. C. Mackay, *Calculation of angular distributions in complex angular momentum theories of elastic scattering*, Molecular Phys. **37** (1979), 1703–1712.

[CM83] J. P. Coleman and A. J. Monaghan, *Chebyshev expansions for the Bessel function* $J_n(x)$ *in the complex plane*, Math. Comp. **40** (1983), 343–366.

[CM84] A. P. Clarke and W. Marwood, *A compact mathematical function package*, Austral. Comput. J. **16** (1984), 107–114.

[CMF77] W. J. Cody, R. M. Motley, and L. W. Fullerton, *The computation of real fractional order Bessel functions of the second kind*, ACM Trans. Math. Software **3** (1977), 232–239.

[CMW63] C. W. Clenshaw, G. F. Miller, and M. Woodger, *Algorithms for special functions I*, Numer. Math. **4** (1963), 403–419.

[CN81] B. C. Carlson and E. M. Notis, *Algorithm 577. Algorithms for incomplete elliptic integrals*, ACM Trans. Math. Software **7** (1981), 398–403.

[Cod65a] W. J. Cody, *Chebyshev approximations for the complete elliptic integrals* K *and* E, Math. Comp. **19** (1965), 105–112, for corrigenda see same journal v. 20 (1966), p. 207.

[Cod65b] W. J. Cody, *Chebyshev polynomial expansions of complete elliptic integrals*, Math. Comp. **19** (1965), 249–259.

[Cod68] W. J. Cody, *Chebyshev approximations for the Fresnel integrals*, Math. Comp. **22** (1968), 450–453, with Microfiche Supplement.

[Cod69] W. J. Cody, *Rational Chebyshev approximations for the error function*, Math. Comp. **23** (1969), 631–637.

[Cod70] W. J. Cody, *A survey of practical rational and polynomial approximation of functions*, SIAM Rev. **12** (1970), 400–423.

[Cod74] W. J. Cody, *The construction of numerical subroutine libraries*, SIAM Rev. **16** (1974), 36–46.

[Cod75] W. J. Cody, *The FUNPACK package of special function routines*, ACM Trans. Math. Software **1** (1975), 13–25.

[Cod76] W. J. Cody, *An overview of software development for special functions*, Lecture Notes in Mathematics 506: Numerical Analysis Dundee, 1975 (G. A. Watson, ed.), Springer-Verlag, Berlin, 1976, pp. 38–48.

[Cod80] W. J. Cody, *Preliminary report on software for the modified Bessel functions of the first kind*, Tech. Memorandum TM–357, Argonne National Laboratory, 9700 South Cass Avenue, Argonne, Illinois 60439–4801, 1980.

[Cod82] W. J. Cody, *Implementation and testing of function software*, Lecture Notes in Computer Science No. 142. Problems and Methodologies in Mathematical Software Production (P. C. Messina and A. Murli, eds.), Springer-Verlag, Berlin, 1982, pp. 24–47.

[Cod83] W. J. Cody, *Algorithm 597. Sequence of modified Bessel functions of the first kind*, ACM Trans. Math. Software **9** (1983), 242–245.

[Cod84a] W. J. Cody, *FUNPACK—A package of special function routines*, Sources and Development of Mathematical Software (W. R. Cowell, ed.), Prentice-Hall, Inc., Englewood Cliffs, New Jersey, 1984, pp. 49–67.

[Cod84b] W. J. Cody, *Observations on the mathematical software effort*, Sources and Development of Mathematical Software (W. R. Cowell, ed.), Prentice-Hall, Inc., Englewood Cliffs, New Jersey, 1984, pp. 1–19.

[Cod85] W. J. Cody, *Software for special functions*, Rend. Sem. Mat. Univ. Politec. Torino **Fascicolo Speciale. Special Functions: Theory and Computation** (1985), 91–116.

[Cod90a] W. J. Cody, *The normal integral*, Tech. Report MCS–89–1090, Mathematics and Computer Science Division, Argonne National Laboratory, 9700 South Cass Avenue, Argonne, Illinois 60439–4801, 1990.

[Cod90b] W. J. Cody, *Performance evaluation of programs for the error and complementary error functions*, ACM Trans. Math. Software **16** (1990), 29–37.

[Cod91] W. J. Cody, *Performance evaluation of programs related to the real gamma function*, ACM Trans. Math. Software **17** (1991), 46–54.

[Cod93a] W. J. Cody, *Algorithm 714. CELEFUNT: A portable test package for complex elementary functions*, ACM Trans. Math. Software **19** (1993), 1–21.

[Cod93b] W. J. Cody, *Algorithm 715. SPECFUN: A portable Fortran package of special function routines and test drivers*, ACM Trans. Math. Software **19** (1993), 22–32.

[Cof91] M. W. Coffey, *Calculation of generalized Lommel integrals for modified Bessel functions*, J. Phys. A **24** (1991), 23–33.

[Col80] J. P. Coleman, *A Fortran subroutine for the Bessel function $J_n(x)$ of order 0 to 10*, Comput. Phys. Comm. **21** (1980), 109–118.

[Col87a] J. P. Coleman, *Complex polynomial approximation by the Lanczos τ–method: Dawson's integral*, J. Comput. Appl. Math. **20** (1987), 137–151.

[Col87b] J. P. Coleman, *Polynomial approximations in the complex plane*, J. Comput. Appl. Math. **18** (1987), 193–211.

[Coo68] B. E. Cooper, *Algorithm AS 10. The use of orthogonal polynomials*, Appl. Statist. **17** (1968), 283–287.

[Cor72] P. Cornille, *Computation of Hankel transforms*, SIAM Rev. **14** (1972), 278–285.

[COT89] C. W. Clenshaw, F. W. J. Olver, and P. R. Turner, *Level-index arithmetic: An introductory survey*, Lecture Notes in Mathematics 1397: Numerical Analysis and Parallel Processing (Lancaster 1987) (P. R. Turner, ed.), Springer-Verlag, Berlin, 1989, pp. 95–168.

[CP66] C. W. Clenshaw and S. M. Picken, *Chebyshev series for Bessel functions of fractional order*, National Physical Laboratory Mathematical Tables, vol. 8, Her Majesty's Stationery Office, London, 1966.

[CPC84] *Master index volumes 1–30, July 1969—December 1983*, Comput. Phys. Comm. **35** (1984), B1–B75, C1–C928.

[CPC87] *Program master index volumes 1–40, July 1969—June 1986*, Comput. Phys. Comm. (1987), 1–75.

[CPC90] *Master index volumes 41–50, July 1986—July 1988*, Comput. Phys. Comm. (1990), 17–30.

[CPT70] W. J. Cody, K. A. Paciorek, and H. C. Thacher, Jr., *Chebyshev approximations for Dawson's integral*, Math. Comp. **24** (1970), 171–178.

[Cri89] C. L. Critchfield, *Computation of elliptic functions*, J. Math. Phys. **30** (1989), 295–297.

[CS89] W. J. Cody and L. Stoltz, *Performance evaluation of programs for certain Bessel functions*, ACM Trans. Math. Software **15** (1989), 41–48.

[CS91] W. J. Cody and L. Stoltz, *The use of Taylor series to test accuracy of function programs*, ACM Trans. Math. Software **17** (1991), 55–63.

[CST73] W. J. Cody, A. J. Strecok, and H. C. Thacher, Jr., *Chebyshev approximations for the psi function*, Math. Comp. **27** (1973), 123–127.

[CT67] W. J. Cody and H. C. Thacher, Jr., *Rational Chebyshev approximations for Fermi-Dirac integrals of orders* $-1/2$, $1/2$ *and* $3/2$, Math. Comp. **21** (1967), 30–40.

[CT68] W. J. Cody and H. C. Thacher, Jr., *Rational Chebyshev approximations for the exponential integral* $E_1(x)$, Math. Comp. **22** (1968), 641–649.

[CT69] W. J. Cody and H. C. Thacher, Jr., *Chebyshev approximations for the exponential integral* $\mathrm{Ei}(x)$, Math. Comp. **23** (1969), 289–303.

[CT85] M. Carmignani and A. Tortorici Macaluso, *Calcolo delle funzioni speciali* $\Gamma(x)$, $\log\Gamma(x)$, $\beta(x,y)$, $\mathrm{erf}(x)$, $\mathrm{erfc}(x)$ *alle alte precisioni*, Atti Accad. Sci. Lett. Arti Palermo (5) **2 (1981–82)** (1985), no. 1, 7–25.

[Cun69] S. W. Cunningham, *Algorithm AS 24. From normal integral to deviate*, Appl. Statist. **18** (1969), 290–293.

[CW80] W. J. Cody and W. Waite, *Software manual for the elementary functions*, Prentice Hall, Englewood Cliffs, New Jersey, 1980.

[Del73] Delft Numerical Analysis Group, *On the computation of Mathieu functions*, J. Engrg. Math. **7** (1973), 39–61.

[Del79] G. Delic, *Chebyshev expansion of the associated Legendre polynomial* $P_L^M(x)$, Comput. Phys. Comm. **18** (1979), 63–71.

[Dev93] K. Devlin (ed.), *Computers and mathematics*, Notices Amer. Math. Soc. **40** (1993), 613–623.

[DG87] J. J. Dongarra and E. Grosse, *Distribution of mathematical software via electronic mail*, Comm. ACM **30** (1987), 403–407.

[DJ67] A. R. DiDonato and M. P. Jarnagin, *The efficient calculation of the incomplete beta-function ratio for half-integer values of the parameters* a, b, Math. Comp. **21** (1967), 652–662.

[DK90] S. L. Dvorak and E. F. Kuester, *Numerical computation of the incomplete Lipschitz-Hankel integral* $\mathrm{Je}_0(a,z)$, J. Comput. Phys. **87** (1990), 301–327.

[DKK81] V. A. Ditkin, K. A. Karpov, and M. K. Kerimov, *The computation of special functions*, U.S.S.R. Comput. Math. and Math. Phys. **20** (1981), no. 5, 3–12.

[DM86] A. R. DiDonato and A. H. Morris, Jr., *Computation of the incomplete gamma function ratios and their inverse*, ACM Trans. Math. Software **12** (1986), 377–393.

[DM87] A. R. DiDonato and A. H. Morris, Jr., *Algorithm 654. Fortran subroutines for computing the incomplete gamma function ratios and their inverse*, ACM Trans. Math. Software **13** (1987), 318–319.

[DM92] A. R. DiDonato and A. H. Morris, Jr., *Algorithm 708. Significant digit computation of the incomplete beta function ratios*, ACM Trans. Math. Software **18** (1992), 360–373.

[Dor68] E. Dorrer, *Algorithm 322. F-distribution*, Comm. ACM **11** (1968), 116–117, for certification see same journal v. 12 (1969), p. 39.

[Eck76] U. Eckhardt, *A rational approximation to Weierstrass' $\mathcal{P}$-function*, Math. Comp. **30** (1976), 818–826.

[Eck77] U. Eckhardt, *A rational approximation to Weierstrass' $\mathcal{P}$-function. II. The lemniscatic case*, Computing **18** (1977), 341–349.

[Eck80] U. Eckhardt, *Algorithm 549. Weierstrass' elliptic functions*, ACM Trans. Math. Software **4** (1980), 112–120.

[Egl84] A. P. Eglaya, *Eigenvalues of wave spheroidal functions with a complex parameter*, Latv. Mat. Ezhegodnik (1984), no. 28, 143–150 (Russian).

[EK92] Ö. Eğecioğlu and Ç. K. Koç, *A parallel algorithm for generating discrete orthogonal polynomials*, Parallel Comput. **18** (1992), 649–659.

[EKK85] A. Yu. Eremin, I. E. Kaporin, and M. K. Kerimov, *The calculation of the Riemann zeta-function in the complex domain*, U.S.S.R. Comput. Math. and Math. Phys. **25** (1985), no. 2, 111–119, see also [YKK88].

[EMOT53a] A. Erdélyi, W. Magnus, F. Oberhettinger, and F. G. Tricomi, *Higher transcendental functions*, vol. 1, McGraw-Hill, New York, 1953, reprinted and published in 1981 by Krieger Publishing Company, Melbourne, Florida.

[EMOT53b] A. Erdélyi, W. Magnus, F. Oberhettinger, and F. G. Tricomi, *Higher transcendental functions*, vol. 2, McGraw-Hill, New York, 1953, reprinted and published in 1981 by Krieger Publishing Company, Melbourne, Florida.

[EMOT55] A. Erdélyi, W. Magnus, F. Oberhettinger, and F. G. Tricomi, *Higher transcendental functions*, vol. 3, McGraw-Hill, New York, 1955, reprinted and published in 1981 by Krieger Publishing Company, Melbourne, Florida.

[EP69] S. P. Eraševskaja and A. A. Pal'cev, *The computation of spheroidal functions and their first derivatives on a computer. II*, Vescī Akad. Navuk BSSR Ser. Fīz.-Mat. Navuk (1969), no. 4, 37–46 (Russian).

[Eva74] D. J. Evans (ed.), *Software for numerical mathematics*, Proceedings of the Loughboro University of Technology Conference of the IMA held in April 1973, Academic Press, London, 1974.

[EWB84] A. S. Elder, J. N. Walbert, and E. C. Benck, *Calculation of Legendre functions on the cut for integral order and complex degree by means of Gauss continued fractions*, Tech. Report ARBRL–MR–03335, U. S. Army Armament Research and Development Center, Ballistic Research Laboratory, Aberdeen Proving Ground, Maryland, 1984, copies obtainable from National Technical Information Service, U. S. Dept. of Commerce, Springfield, VA 22161.

[Fet70] H. E. Fettis, *A new method for computing toroidal harmonics*, Math. Comp. **24** (1970), 667–670.

[Feu91] F. Feuillebois, *Numerical calculation of singular integrals related to Hankel transform*, Comput. Math. Appl. **21** (1991), no. 2-3, 87–94.

[FG91] B. Fischer and G. H. Golub, *On generating polynomials which are orthogonal over several intervals*, Math. Comp. **56** (1991), 711–730.

[FG92] B. Fischer and G. H. Golub, *How to generate unknown orthogonal polynomials out of known orthogonal polynomials*, J. Comput. Appl. Math. **43** (1992), 99–115.

[FGG82] J. D. Fenton and R. S. Gardiner-Garden, *Rapidly-convergent methods for evaluating elliptic integrals and theta and elliptic functions*, J. Austral. Math. Soc. Ser. B **24** (1982), 47–58.

[FHS78a] P. A. Fox, A. D. Hall, and N. L. Schryer, *Algorithm 528. Framework for a portable library*, ACM Trans. Math. Software **4** (1978), 177–188.

[FHS78b] P. A. Fox, A. D. Hall, and N. L. Schryer, *The PORT mathematical subroutine library*, ACM Trans. Math. Software **4** (1978), 104–126.

[Fik68] C. T. Fike, *Computer evaluation of mathematical functions*, Prentice-Hall, Inc., Englewood Cliffs, New Jersey, 1968.

[FL67] W. G. Fair and Y. L. Luke, *Rational approximations to the incomplete elliptic integrals of the first and second kinds*, Math. Comp. **21** (1967), 418–422.

[Fle68] O. L. Fleckner, *A method for the computation of the Fresnel integrals and related functions*, Math. Comp. **22** (1968), 635–640.

[Fox56] L. Fox, *The use and construction of mathematical tables*, National Physical Laboratory Mathematical Tables, vol. 1, Her Majesty's Stationery Office, London, 1956.

[Fox84] P. A. Fox, *The PORT mathematical subroutine library*, Sources and Development of Mathematical Software (W. R. Cowell, ed.), Prentice-Hall, Inc., Englewood Cliffs, New Jersey, 1984, pp. 346–374.

[FP68] L. Fox and I. B. Parker, *Chebyshev polynomials in numerical analysis*, Oxford University Press, London, 1968.

[FP84] B. Ford and J. C. T. Pool, *The evolving NAG library service*, Sources and Development of Mathematical Software (W. R. Cowell, ed.), Prentice-Hall, Inc., Englewood Cliffs, New Jersey, 1984, pp. 375–397.

[FR86] L. W. Fullerton and G. A. Rinker, *Generalized Fermi-Dirac integrals—FD, FDG, FDH*, Comput. Phys. Comm. **39** (1986), 181–185.

[FS67] A. M. S. Filho and G. Schwachheim, *Algorithm 309. Gamma function with arbitrary precision*, Comm. ACM **10** (1967), 511–512.

[Ful72] L. W. Fullerton, *Algorithm 435. Modified incomplete gamma function*, Comm. ACM **15** (1972), 993–995, for remark see ACM Trans. Math. Software v. 4 (1978), pp. 296–304.

[Ful77] L. W. Fullerton, *Portable special function routines*, Lecture Notes in Computer Science 57: Portability of Numerical Software (Oak Brook 1976) (W. R. Cowell, ed.), Springer-Verlag, Berlin, 1977, pp. 452–483.

[Ful80] L. W. Fullerton, *A bibliography on the evaluation of mathematical functions*, Computing Science Tech. Report No. 86, Bell Laboratories, Murray Hill, New Jersey 07974, 1980.

[FW80] A. Fransén and S. Wrigge, *High-precision values of the gamma function and of some related coefficients*, Math. Comp. **34** (1980), 553–566, for addendum and corrigendum see same journal v. 37 (1981), pp. 233–235.

[Gab79] B. Gabutti, *On high precision methods for computing integrals involving Bessel functions*, Math. Comp. **33** (1979), 1049–1057.

[Gab80] B. Gabutti, *On the generalization of a method for computing Bessel function integrals*, J. Comput. Appl. Math. **6** (1980), 167–168.

[Gaf88] P. W. Gaffney, *When things go wrong ...*, Pitman Research Notes in Mathematics Series (D. F. Griffiths and G. A. Watson, eds.), vol. 170, Longman Scientific and Technical, Harlow, Essex, U. K., 1988, pp. 67–114.

[Gau64] W. Gautschi, *Algorithm 222. Incomplete beta function ratios*, Comm. ACM **7** (1964), 143–144, for certification see same journal v. 7 (1964), p. 244.

[Gau65] W. Gautschi, *Algorithm 259. Legendre functions for arguments larger than one*, Comm. ACM **8** (1965), 488–492, for remark see ACM Trans. Math. Software, v. 3 (1977), pp. 204-205.

[Gau69a] W. Gautschi, *Algorithm 363. Complex error function*, Comm. ACM **12** (1969), 635, for certification see same journal v. 15 (1972), pp. 465–466.

[Gau69b] W. Gautschi, *An application of minimal solutions of three-term recurrences to Coulomb wave functions*, Aequationes Math. **2** (1969), 171–176.

[Gau70] W. Gautschi, *Efficient computation of the complex error function*, SIAM J. Numer. Anal. **7** (1970), 187–198.

[Gau73] W. Gautschi, *Algorithm 471. Exponential integrals*, Comm. ACM **16** (1973), 761–763.

[Gau75] W. Gautschi, *Computational methods in special functions—a survey*, Theory and application of special functions, Proc. Advanced Sem., Math. Res. Center, Univ. Wisconsin, Madison, Wis., Academic Press, New York, 1975, pp. 1–98.

[Gau77a] W. Gautschi, *Algorithm 521. Repeated integrals of the coerror function*, ACM Trans. Math. Software **3** (1977), 301–302.

[Gau77b] W. Gautschi, *Evaluation of the repeated integrals of the coerror function*, ACM Trans. Math. Software **3** (1977), 240–252.

[Gau79a] W. Gautschi, *Algorithm 542. Incomplete gamma functions*, ACM Trans. Math. Software **5** (1979), 482–489.

[Gau79b] W. Gautschi, *A computational procedure for incomplete gamma functions*, ACM Trans. Math. Software **5** (1979), 466–481.

[Gau82] W. Gautschi, *On generating orthogonal polynomials*, SIAM J. Sci. Statist. Comput. **3** (1982), 289–317.

[Gau85] W. Gautschi, *Orthogonal polynomials—constructive theory and applications*, J. Comput. Appl. Math. **12/13** (1985), 61–76.

[Gau90] W. Gautschi, *Computational aspects of orthogonal polynomials*, Orthogonal Polynomials: Theory and Practice (P. Nevai, ed.), Kluwer Academic Publishers, Dordrecht, 1990, NATO ASI Series, vol. C294, pp. 181–216.

[Gau91a] W. Gautschi, *Computational problems and applications of orthogonal polynomi-*

als, Orthogonal Polynomials and their Applications (C. Brezinski, L. Gori, and A. Ronveaux, eds.), J. C. Baltzer AG, Scientific Publishing Company, Basel, 1991, pp. 61–71.

[Gau91b] W. Gautschi, *On the paper "A continued fraction approximation of the modified Bessel function* $I_1(t)$*" by P. R. Parthasarathy and N. Balakrishnan*, Appl. Math. Lett. **4** (1991), no. 5, 47–51.

[Gau93a] W. Gautschi, *Gauss-type quadrature rules for rational functions*, International Series of Numerical Mathematics, vol. 112, Birkhäuser Verlag, Basel, 1993, pp. 111–130.

[Gau93b] W. Gautschi, *On the computation of generalized Fermi-Dirac and Bose-Einstein integrals*, Comput. Phys. Comm. **74** (1993), 233–238.

[Gau94] W. Gautschi, *Algorithm. ORTHPOL: A package of routines for generating orthogonal polynomials and Gauss-type quadrature rules*, ACM Trans. Math. Software **20** (1994), in press.

[GB87] A. Ganguli and R. Baskaran, *Generation of Bessel functions with complex arguments and integer orders*, Internat. J. Comput. Math. **21** (1987), 43–64.

[GH85] P. Griffiths and I. D. Hill (eds.), *Applied statistics algorithms*, Ellis Horwood Limited, Chichester, 1985.

[GM81] B. Gabutti and B. Minetti, *A new application of the discrete Laguerre polynomials in the numerical evaluation of the Hankel transform of a strongly decreasing even function*, J. Comput. Phys. **42** (1981), 277–287.

[GP64] I. Gargantini and T. Pomentale, *Rational Chebyshev approximations to the Bessel function integrals* $\mathrm{Ki}_s(x)$, Comm. ACM **7** (1964), 727–730.

[GS78] W. Gautschi and J. Slavik, *On the computation of modified Bessel function ratios*, Math. Comp. **32** (1978), 865–875.

[GT81] R. Gastmans and W. Troost, *On the evaluation of polylogarithmic integrals*, Simon Stevin **55** (1981), 205–219.

[GZ75] E. S. Ginsberg and D. Zaborowski, *Algorithm 490. The dilogarithm function of a real argument*, Comm. ACM **18** (1975), 200–202, for remark see ACM Trans. Math. Software v. 2 (1976), p. 112.

[Ham85] D. E. Hamilton, *Remark on Algorithm 620: References and keywords for collected algorithms from ACM*, ACM Trans. Math. Software **11** (1985), 305–307.

[Han85] E. W. Hansen, *Fast Hankel transform algorithm*, IEEE Trans. Acoust. Speech Signal Process. **33** (1985), 666–671.

[HCL+68] J. F. Hart, E. W. Cheney, C. L. Lawson, H. J. Maehly, C. K. Mesztenyi, J. R. Rice, H. C. Thacher, Jr., and C. Witzgall, *Computer approximations*, John Wiley and Sons, Inc., New York, 1968.

[HD73] G. W. Hill and A. W. Davis, *Algorithm 442. Normal deviate*, Comm. ACM **16** (1973), 51–52.

[Hea85] M. A. Heald, *Rational approximations for the Fresnel integrals*, Math. Comp. **44** (1985), 459–461.

[Hem81] P. W. Hemker (ed.), *NUMAL numerical procedures in Algol 60*, Mathematisch Centrum Syllabus 47.1–7, Centrum voor Wiskunde en Informatica, Kruislaan 413, 1098 SJ, Amsterdam, 1981.

[Hen79] P. Henrici, *Zur numerischen Berechnung der Fresnelschen Integrale*, Z. Angew. Math. Phys. **30** (1979), 209–219.

[Hil70a] G. W. Hill, *Algorithm 395. Student's t-distribution*, Comm. ACM **13** (1970), 617–619, for remarks see ACM Trans. Math. Software v. 5 (1979), pp. 238–239, v. 7 (1981), pp. 247–249.

[Hil70b] G. W. Hill, *Algorithm 396. Student's t-quantiles*, Comm. ACM **13** (1970), 619–620, for remarks see ACM Trans. Math. Software v. 5 (1979), pp. 238–239, v. 7 (1981), pp. 250–251.

[Hil73] I. D. Hill, *Algorithm AS 66. The normal integral*, Appl. Statist. **22** (1973), 424–427.

[Hil77] G. W. Hill, *Algorithm 518. Incomplete Bessel function* I_0*: The von Mises distri-*

bution, ACM Trans. Math. Software **3** (1977), 279–284.

[Hil81] G. W. Hill, *Algorithm 571. Statistics for von Mises' and Fisher's distributions of directions*: $I_1(x)/I_0(x), I_{1.5}(x)/I_{0.5}(x)$, *and their inverses*, ACM Trans. Math. Software **7** (1981), 233–238.

[Hit68] S. Hitotumatu, *On the numerical computation of Bessel functions through continued fraction*, Comment. Math. Univ. St. Paul. **16** (1967/68), 89–113.

[HM90a] T. R. Hopkins and D. R. Morse, *Index of statistical algorithms*, Appl. Statist. **39** (1990), 177–187.

[HM90b] T. R. Hopkins and D. R. Morse, *Remark on Algorithm 620: References and keywords for collected algorithms from ACM*, ACM Trans. Math. Software **16** (1990), 401–403.

[Hod70] D. B. Hodge, *Eigenvalues and eigenfunctions of the spheroidal wave equation*, J. Math. Phys. **11** (1970), 2308–2312.

[HR72] D. B. Hunter and T. Regan, *A note on the evaluation of the complementary error function*, Math. Comp. **26** (1972), 539–541.

[Hum64] D. G. Hummer, *Expansion of Dawson's function in a series of Chebyshev polynomials*, Math. Comp. **18** (1964), 317–319.

[IKF91] Y. Ikebe, Y. Kikuchi, and I. Fujishiro, *Computing zeros and orders of Bessel functions*, J. Comput. Appl. Math. **38** (1991), 169–184.

[IMS91] *MATH/LIBRARY special functions, version 2.0*, IMSL, Inc., 2500 CityWest Boulevard, Houston, Texas 77042, September 1991, current address is Visual Numerics, Inc., 9990 Richmond Ave, Suite 400, Houston, Texas 77042-4548.

[JE45] E. Jahnke and F. Emde, *Tables of functions with formulae and curves*, fourth ed., Dover Publications, Inc., New York, 1945.

[JEL60] E. Jahnke, F. Emde, and F. Lösch, *Tables of higher functions*, sixth ed., McGraw-Hill, New York, 1960.

[JL72] D. Jacobs and F. Lambert, *On the numerical calculation of polylogarithms*, BIT **12** (1972), 581–585.

[Joh75] H. K. Johansen, *An interactive computer/graphic-display-terminal system for interpretation of resistivity soundings*, Geophysical Prospecting **23** (1975), 449–458.

[JT80] W. B. Jones and W. J. Thron, *Continued fractions: analytic theory and applications*, Encyclopedia of Mathematics and its Applications, vol. 11, Addison-Wesley Publishing Company, Reading, Massachusetts, 1980.

[JT85] W. B. Jones and W. J. Thron, *On the computation of incomplete gamma functions in the complex domain*, J. Comput. Appl. Math. **12/13** (1985), 401–417.

[Kar91] E. A. Karatsuba, *Fast evaluation of transcendental functions*, Problems Inform. Transmission **27** (1991), 339–360.

[Kat78] C. R. Katholi, *On the computation of values of the psi function from rapidly converging power series expansions*, J. Statist. Comput. Simulation **8** (1978), 25–42.

[KBH70] B. J. King, R. V. Baier, and S. Hanish, *A Fortran computer program for calculating the prolate spheroidal radial functions of the first and second kind and their first derivatives*, NRL Report No. 7012, Naval Res. Lab., Washington, D. C., 1970.

[Ker80] M. K. Kerimov, *Methods of computing the Riemann zeta-function and some generalizations of it*, U.S.S.R. Comput. Math. and Math. Phys. **20** (1980), no. 6, 212–230.

[KG80] W. J. Kennedy, Jr. and J. E. Gentle, *Statistical computing*, Marcel Dekker, New York, 1980.

[KMR70] K. S. Kölbig, J. A. Mignaco, and E. Remiddi, *On Nielsen's generalized polylogarithms and their numerical calculation*, BIT **10** (1970), 38–73.

[Köl72a] K. S. Kölbig, *Programs for computing the logarithm of the gamma function, and the digamma function, for complex argument*, Comput. Phys. Comm. **4** (1972), 221–226.

[Köl72b] K. S. Kölbig, *Remarks on the computation of Coulomb wavefunctions*, Comput. Phys. Comm. **4** (1972), 214–220.

[Köl81] K. S. Kölbig, *A program for computing the conical functions of the first kind* $P^m_{-\frac{1}{2}+i\tau}(x)$ *for* $m = 0$ *and* $m = 1$, Comput. Phys. Comm. **23** (1981), 51–61.

[Krä90] W. Krämer, *Berechnung der Gammafunktion* $\Gamma(x)$ *für reelle Punkt- und Intervallargumente*, Z. Angew. Math. Mech. **70** (1990), T581–T584.

[Kro91] F. T. Krogh, *ACM algorithms policy*, ACM Trans. Math. Software **17** (1991), 427–430.

[KS84a] M. K. Kerimov and S. L. Skorokhodov, *Calculation of modified Bessel functions in the complex domain*, U.S.S.R. Comput. Math. and Math. Phys. **24** (1984), no. 3, 15–24.

[KS84b] M. K. Kerimov and S. L. Skorokhodov, *Calculation of the complex zeros of the modified Bessel function of the second kind and its derivatives*, U.S.S.R. Comput. Math. and Math. Phys. **24** (1984), no. 4, 115–123.

[KS84c] M. K. Kerimov and S. L. Skorokhodov, *Evaluation of complex zeros of Bessel functions* $J_\nu(z)$ *and* $I_\nu(z)$ *and their derivatives*, U.S.S.R. Comput. Math. and Math. Phys. **24** (1984), no. 5, 131–141.

[KS84d] K. S. Kölbig and B. Schorr, *A program package for the Landau distribution*, Comput. Phys. Comm. **31** (1984), 97–111.

[KS85a] M. K. Kerimov and S. L. Skorokhodov, *Calculation of the complex zeros of a Bessel function of the second kind and its derivatives*, U.S.S.R. Comput. Math. and Math. Phys. **25** (1985), no. 5, 117–128.

[KS85b] M. K. Kerimov and S. L. Skorokhodov, *Calculation of the complex zeros of Hankel functions and their derivatives*, U.S.S.R. Comput. Math. and Math. Phys. **25** (1985), no. 6, 26–36.

[KS85c] M. K. Kerimov and S. L. Skorokhodov, *Calculation of the multiple zeros of the derivatives of the cylindrical Bessel functions* $J_\nu(z)$ *and* $Y_\nu(z)$, U.S.S.R. Comput. Math. and Math. Phys. **25** (1985), no. 6, 101–107.

[KS87] M. K. Kerimov and S. L. Skorokhodov, *On the calculation of the multiple complex roots of the derivatives of cylindrical Bessel functions*, U.S.S.R. Comput. Math. and Math. Phys. **27** (1987), no. 6, 18–25.

[Kuk71] H. Kuki, *Mathematical function subprograms for basic system libraries— objectives, constraints and trade-off*, Mathematical Software (J. R. Rice, ed.), Academic Press, New York, 1971, pp. 187–199.

[Kuk72a] H. Kuki, *Algorithm 421. Complex gamma function with error control*, Comm. ACM **15** (1972), 271–272.

[Kuk72b] H. Kuki, *Complex gamma function with error control*, Comm. ACM **15** (1972), 262–267.

[KvB70] B. J. King and A. L. van Buren, *A Fortran computer program for calculating the prolate and oblate angle functions of the first kind and their first and second derivatives*, NRL Report No. 7161, Naval Res. Lab., Washington, D. C., 1970.

[LCY65] L. A. Lyusternik, O. A. Chervonenkis, and A. R. Yanpol'skii, *Handbook for computing elementary functions*, International Series of Monographs in Pure and Applied Mathematics, vol. 76, Pergamon Press, Oxford, 1965.

[LDP93] Y. Li, X. Dong, and S. Pan, *Computation of auxiliary functions in STO molecular integrals up to arbitrary accuracy. I. Evaluation of incomplete gamma function* $E_n(x)$ *by forward recursion*, Internat. J. Quantum Chem. **45** (1993), 3–14.

[Lee79] W. R. Leeb, *Algorithm 537. Characteristic values of Mathieu's differential equation*, ACM Trans. Math. Software **5** (1979), 112–117.

[Lee80] S.-Y. Lee, *The inhomogeneous Airy functions* $\mathrm{Gi}(z)$ *and* $\mathrm{Hi}(z)$, J. Chem. Phys. **72** (1980), 332–336.

[Lee90] D. K. Lee, *Application of theta functions for numerical evaluation of complete elliptic integrals of the first and second kinds*, Comput. Phys. Comm. **60** (1990), 319–327.

[Lee92] D. K. Lee, *Calculation of coefficients in a power-series expansion of the nome* $q = \exp[-\pi K(k')/K(k)]$, Comput. Phys. Comm. **70** (1992), 292–296.

[Lev69] D. A. Levine, *Algorithm 344. Student's t-distribution*, Comm. ACM **12** (1969),

37–38, for remarks see same journal v. 13 (1970), pp. 124 and 449.

[Lew91] S. Lewanowicz, *Evaluation of Bessel function integrals with algebraic singularities*, J. Comput. Appl. Math. **37** (1991), 101–112.

[LG64] H. Lotsch and M. Gray, *Algorithm 244. Fresnel integrals*, Comm. ACM **7** (1964), 660–661.

[Lin72] P. Linz, *A method for computing Bessel function integrals*, Math. Comp. **26** (1972), 509–513.

[LK73] P. Linz and T. E. Kropp, *A note on the computation of integrals involving products of trigonometric and Bessel functions*, Math. Comp. **27** (1973), 871–872.

[LMS73a] D. W. Lozier, L. C. Maximon, and W. L. Sadowski, *A bit comparison program for algorithm testing*, Comput. J. **16** (1973), 111–117.

[LMS73b] D. W. Lozier, L. C. Maximon, and W. L. Sadowski, *Performance testing of a Fortran library of mathematical function routines—A case study in the application of testing techniques*, J. Res. Nat. Bur. Standards **77B** (1973), 101–110.

[LO93] D. W. Lozier and F. W. J. Olver, *Airy and Bessel functions by parallel integration of ODEs*, Proceedings of the Sixth SIAM Conference on Parallel Processing for Scientific Computing, vol. 2 (R. F. Sincovec, D. E. Keyes, M. R. Leuze, L. R. Petzold, and D. A. Reed, eds.), Society for Industrial and Applied Mathematics, Philadelphia, 1993, pp. 531–538.

[LPM81] A. R. Lehman, W. C. Parke, and L. C. Maximon, *Numerical evaluation of integrals containing a spherical Bessel function by product integration*, J. Math. Phys. **22** (1981), 1399–1413.

[LPT80] M. C. Lorenzini, G. Puleo, and A. Tortorici Macaluso, *Un package di aritmetica in multiprecisione ed applicazione al calcolo dei polinomi di Jacobi, Hermite, Laguerre, Legendre, Chebyshev*, Atti Accad. Sci. Lett. Arti Palermo Ser. 4, Parte 1 **39** (1980), 339–376.

[LR74] W. P. Latham and R. W. Redding, *On the calculation of the parabolic cylinder functions*, J. Comput. Phys. **16** (1974), 66–75.

[LS81] D. W. Lozier and J. M. Smith, *Algorithm 567. Extended-range arithmetic and normalized Legendre polynomials*, ACM Trans. Math. Software **7** (1981), 141–146.

[LS91] A. I. Litvin and S. D. Simonzhenkov, *Computation of Fermi-Dirac functions*, Comput. Math. and Math. Phys. **31** (1991), no. 8, 100–103.

[Luk68] Y. L. Luke, *Approximations for elliptic integrals*, Math. Comp. **22** (1968), 627–634.

[Luk69a] Y. L. Luke, *The special functions and their approximations*, vol. 1, Academic Press, New York, 1969.

[Luk69b] Y. L. Luke, *The special functions and their approximations*, vol. 2, Academic Press, New York, 1969.

[Luk70a] Y. L. Luke, *Evaluation of the gamma function by means of Padé approximations*, SIAM J. Math. Anal. **1** (1970), 266–281.

[Luk70b] Y. L. Luke, *Further approximations for elliptic integrals*, Math. Comp. **24** (1970), 191–198.

[Luk71a] Y. L. Luke, *Miniaturized tables of Bessel functions*, Math. Comp. **25** (1971), 323–330.

[Luk71b] Y. L. Luke, *Miniaturized tables of Bessel functions. II*, Math. Comp. **25** (1971), 789–795, for corrigendum see same journal v. 26 (1972), Microfiche Supplement A1–A7.

[Luk72a] Y. L. Luke, *Miniaturized tables of Bessel functions. III*, Math. Comp. **26** (1972), 237–240, with Microfiche Supplement.

[Luk72b] Y. L. Luke, *On generating Bessel functions by use of the backward recurrence formula*, Tech. Report ARL 72–0030, Aerospace Research Laboratories, Wright-Patterson Air Force Base, Ohio, 1972.

[Luk75] Y. L. Luke, *Mathematical functions and their approximations*, Academic Press, New York, 1975.

[Luk76] Y. L. Luke, *On the expansion of exponential type integrals in series of Chebyshev polynomials*, Theory of Approximation with Applications (A. G. Law and B. N. Sahney, eds.), Academic Press, Inc., New York, 1976, pp. 180–199.

[Luk77a] Y. L. Luke, *Algorithms for rational approximations for a confluent hypergeometric function*, Utilitas Math. **11** (1977), 123–151.

[Luk77b] Y. L. Luke, *Algorithms for the computation of mathematical functions*, Academic Press, New York, 1977.

[Lun85] J. Lund, *Bessel transforms and rational extrapolation*, Numer. Math. **47** (1985), 1–14.

[LW90] F. G. Lether and P. R. Wenston, *An algorithm for the numerical computation of the Voigt function*, Appl. Math. Comput. **35** (1990), 277–289.

[LW91] F. G. Lether and P. R. Wenston, *The numerical computation of the Voigt function by a corrected midpoint quadrature rule for* $(-\infty, \infty)$, J. Comput. Appl. Math. **34** (1991), 75–92.

[LY88] T. Y. Lemczyk and M. M. Yovanovich, *Efficient evaluation of incomplete elliptic integrals and functions*, Comput. Math. Appl. **16** (1988), 747–757.

[Lyn93] A. E. Lynas-Gray, *VOIGTL—A fast subroutine for Voigt function evaluation on vector processors*, Comput. Phys. Comm. **75** (1993), 135–142.

[Mac89] A. J. MacLeod, *Algorithm AS 245. A robust and reliable algorithm for the logarithm of the gamma function*, Appl. Statist. **38** (1989), 397–402.

[Mac93] A. J. MacLeod, *Chebyshev expansions for modified Struve and related functions*, Math. Comp. **60** (1993), 735–747.

[Mar65] B. Markman, *Contribution no. 14. The Riemann zeta function*, BIT **5** (1965), 138–141.

[Mar82] A. L. Marshak, *On the approximation of the exponential integral by quadrature formulae*, U.S.S.R. Comput. Math. and Math. Phys. **22** (1982), no. 5, 20–28.

[Mas83] J. P. Mason, *Cylindrical Bessel functions for a large range of complex arguments*, Comput. Phys. Comm. **30** (1983), 1–11.

[Mat92] *MATLAB high performance numeric computation and visualization software reference guide*, The MathWorks, Inc., Cochituate Place, 24 Prime Park Way, Natick, Massachusetts 01760, August 1992, electronic mail address is info@mathworks.com.

[Mat93a] *Mathcad 4.0 user's guide*, MathSoft Inc., 201 Broadway, Cambridge, Massachusetts 02139, March 1993.

[Mat93b] G. Matviyenko, *On the evaluation of Bessel functions*, Appl. Comput. Harmonic Anal. **1** (1993), 116–135.

[MB73a] K. L. Majumder and G. P. Bhattacharjee, *Algorithm AS 63. The incomplete beta integral*, Appl. Statist. **22** (1973), 409–411, for remark see same journal v. 26 (1977), pp. 111–114.

[MB73b] K. L. Majumder and G. P. Bhattacharjee, *Algorithm AS 64. Inverse of the incomplete beta function ratio*, Appl. Statist. **22** (1973), 411–414, for remarks and corrections see same journal v. 26 (1977), pp. 111–114, v. 39 (1990), pp. 309–310, v. 40 (1991), p. 236.

[McC74] J. H. McCabe, *A continued fraction expansion, with a truncation error estimate, for Dawson's integral*, Math. Comp. **28** (1974), 811–816.

[McC81] P. McCullagh, *A rapidly convergent series for computing* $\psi(z)$ *and its derivatives*, Math. Comp. **36** (1981), 247–248.

[MDS92] R. C. McPhredan, D. H. Dawes, and T. C. Scott, *On a Bessel function integral*, Appl. Algebra Engrg. Comm. Comput. **2** (1992), 207–216.

[MF86] V. P. Modenov and A. V. Filonov, *Calculation of zeros of cylindrical functions and their derivatives*, Vestnik Moskov. Univ. Ser. XV Vychisl. Mat. Kibernet. (1986), no. 2, 63–64, 71 (Russian).

[MH73] T. Morita and T. Horiguchi, *Convergence of arithmetic-geometric mean procedure for the complex variables and the calculation of the complete elliptic integrals with complex modulus*, Numer. Math. **20** (1973), 425–430, for correction see same journal v. 29 (1978), pp. 233–236.

[Mid75] P. Midy, *An improved calculation of the general elliptic integral of the second kind in the neighbourhood of $x = 0$*, Numer. Math. **25** (1975), 99–101.

[MM90] P. Maličký and M. Maličká, *On the computation of Riccati-Bessel functions*, Apl. Mat. **35** (1990), 487–493.

[MMV81] G. Maino, E. Menapace, and A. Ventura, *Computation of parabolic cylinder functions by means of a Tricomi expansion*, J. Comput. Phys. **40** (1981), 294–304.

[Moi88] A. I. Moiseyev, *Computation of certain functions related to the Hurwitz zeta-function*, U.S.S.R. Comput. Math. and Math. Phys. **28** (1988), no. 3, 1–6.

[Moo81] W. Moon, *Airy function with complex arguments*, Comput. Phys. Comm. **22** (1981), 411–417.

[Moo82] R. J. Moore, *Algorithm AS 187. Derivatives of the incomplete gamma integral*, Appl. Statist. **31** (1982), 330–335.

[Moo83] D. R. Mook, *An algorithm for the numerical evaluation of the Hankel and Abel transforms*, IEEE Trans. Acoust. Speech Signal Process. **31** (1983), 979–985.

[Mor69] J. Morris, *Algorithm 346. F-test probabilities*, Comm. ACM **12** (1969), 184–185, for remark see ACM Trans. Math. Software v. 14 (1988), pp. 288–289.

[Mor79] R. Morris, *The dilogarithm function of a real argument*, Math. Comp. **33** (1979), 778–787.

[Mor80] M. Mori, *Analytic representations suitable for numerical computation of some special functions*, Numer. Math. **35** (1980), 163–174.

[Mor83] M. Mori, *A method for evaluation of the error function of real and complex variable with high relative accuracy*, Publ. Res. Inst. Math. Sci. **19** (1983), 1081–1094.

[Mor93] A. H. Morris, Jr., *NSWC library of mathematics subroutines*, NSWCDD/TR–92/425, Naval Surface Warfare Center, Dahlgren Division, Dahlgren, Virginia, 22448, January 1993.

[MOS66] W. Magnus, F. Oberhettinger, and R. P. Soni, *Formulas and theorems for the special functions of mathematical physics*, Die Grundlehren der mathematischen Wissenschaften in Einzeldarstellungen, vol. 52, Springer-Verlag, New York, 1966.

[Mos89] S. L. B. Moshier, *Methods and programs for mathematical functions*, Ellis Horwood Limited, Chichester, 1989, separate diskette.

[MR71] F. Matta and A. Reichel, *Uniform computation of the error function and other related functions*, Math. Comp. **25** (1971), 339–344.

[MY91] P. Midy and Y. Yakovlev, *Computing some elementary functions of a complex variable*, Math. Comput. Simulation **33** (1991), 33–49.

[NAG91] *NAG Fortran library manual, mark 15*, vol. 10, NAG Inc., 1400 Opus Place, Suite 200, Downers Grove, Illinois 60515-5702, June 1991.

[NC66] W. J. Nellis and B. C. Carlson, *Reduction and evaluation of elliptic integrals*, Math. Comp. **20** (1966), 223–231.

[NDT69] E. W. Ng, C. J. Devine, and R. F. Tooper, *Chebyshev polynomial expansion of Bose-Einstein functions of orders 1 to 10*, Math. Comp. **23** (1969), 639–643.

[Ném71] G. Németh, *Chebyshev polynomial expansions of Airy functions, their zeros, derivatives, first and second integrals*, Magyar Tud. Akad. Mat. Fiz. Oszt. Közl. **20** (1971), 13–33 (Hungarian).

[Ném74] G. Németh, *Expansion of generalized hypergeometric functions in Chebyshev polynomials*, Collection of Scientific Papers in Collaboration of Joint Institute for Nuclear Research, Dubna, USSR and Central Research Institute for Physics, Budapest, Hungary, Hungarian Academy of Sciences, Central Research Institute for Physics, Budapest, 1974, reference No. KFKI-74-34, pp. 57–91.

[Nes84] R. K. Nesbet, *Algorithms for regular and irregular Coulomb and Bessel functions*, Comput. Phys. Comm. **32** (1984), 341–347.

[New84] J. N. Newman, *Approximations for the Bessel and Struve functions*, Math. Comp. **43** (1984), 551–556.

[Ng75] E. W. Ng, *A comparison of computational methods and algorithms for the complex gamma function*, ACM Trans. Math. Software **1** (1975), 56–70.

[NM93] A. Natarajan and N. Mohan Kumar, *On the numerical evaluation of the gener-

alised Fermi-Dirac integrals, Comput. Phys. Comm. **76** (1993), 48–50.

[NPB92a] M. Nardin, W. F. Perger, and A. Bhalla, *Algorithm 707. CONHYP: A numerical evaluator of the confluent hypergeometric function for complex arguments of large magnitudes*, ACM Trans. Math. Software **18** (1992), 345–349.

[NPB92b] M. Nardin, W. F. Perger, and A. Bhalla, *Numerical evaluation of the confluent hypergeometric function for complex arguments of large magnitudes*, J. Comput. Appl. Math. **39** (1992), 193–200.

[NT84] C. J. Noble and I. J. Thompson, *COULN, a program for evaluating negative energy Coulomb functions*, Comput. Phys. Comm. **33** (1984), 413–419.

[OFM78] A. V. Oppenheim, G. V. Frisk, and D. R. Martinez, *An algorithm for the numerical evaluation of the Hankel transform*, Proc. Inst. Electr. Electron. Engrg. **66** (1978), 264–265.

[Olv74] F. W. J. Olver, *Asymptotics and special functions*, Academic Press, New York, 1974.

[Olv91] F. W. J. Olver, *Review of United Laboratories, Inc., Mathematical Function Library for Microsoft-Fortran, Wiley, New York, 1989*, Math. Comp. **56** (1991), 879–885.

[OM68] D. Osborn and R. Madey, *The incomplete beta function and its ratio to the complete beta function*, Math. Comp. **22** (1968), 159–162.

[Öpi87] U. Öpik, *A program to set up systems of orthogonal polynomials*, Comput. Phys. Comm. **46** (1987), 263–296.

[OS72] F. W. J. Olver and D. J. Sookne, *Note on backward recurrence algorithms*, Math. Comp. **26** (1972), 941–947.

[OS83] F. W. J. Olver and J. M. Smith, *Associated Legendre functions on the cut*, J. Comput. Phys. **51** (1983), 502–518.

[PA92] C. U. Pabon-Ortiz and M. Artoni, *Laguerre polynomials: Novel properties and numerical generation scheme for analysis of a discrete probability distribution*, Comput. Phys. Comm. **71** (1992), 215–221.

[Pac70] K. A. Paciorek, *Algorithm 385. Exponential integral* $\mathrm{Ei}(x)$, Comm. ACM **13** (1970), 446–447, for certification and remarks see same journal v. 13 (1970), pp. 448–449 and p. 750; v. 15 (1972), p. 1074.

[Pal69] A. A. Pal'cev, *The computation of spheroidal functions and their first derivatives on a computer. I*, Vescī Akad. Navuk BSSR Ser. Fīz.-Mat. Navuk (1969), no. 1, 19–25 (Russian).

[Pas88] S. Paszkowski, *Evaluation of Fermi-Dirac integral*, Nonlinear Numerical Methods and Rational Approximation (A. Cuyt, ed.), D. Reidel Publishing Company, Dordrecht, 1988, pp. 435–444.

[Pas91] S. Paszkowski, *Evaluation of the Fermi-Dirac integral of half-integer order*, Zastos. Mat. **21** (1991), 289–301.

[PB72] R. Piessens and M. Branders, *Chebyshev polynomial expansions of the Riemann zeta function*, Math. Comp. **26** (1972), 1022.

[PB82] R. Piessens and M. Branders, *Approximation for Bessel functions and their application in the computation of Hankel transforms*, Comput. Math. Appl. **8** (1982), 305–311.

[PB83] R. Piessens and M. Branders, *Modified Clenshaw-Curtis method for the computation of Bessel function integrals*, BIT **23** (1983), 370–381.

[PB84] R. Piessens and M. Branders, *Algorithm 28. Algorithm for the computation of Bessel function integrals*, J. Comput. Appl. Math. **11** (1984), 119–137.

[PB85] R. Piessens and M. Branders, *A survey of numerical methods for the computation of Bessel function integrals*, Rend. Sem. Mat. Univ. Politec. Torino **Fascicolo Speciale. Special Functions: Theory and Computation** (1985), 249–265.

[PBN93] W. F. Perger, A. Bhalla, and M. Nardin, *A numerical evaluator for the generalized hypergeometric series*, Comput. Phys. Comm. **77** (1993), 249–254.

[Pea22] K. Pearson, *Tables of the incomplete* Γ*-function*, H. M. Stationery Office, London, 1922, reissued by Biometrika, 1934.

[Pex70] R. L. Pexton, *Computer investigation of Coulomb wave functions*, Math. Comp. **24** (1970), 409–411.

[Phi88] H. N. Phien, *A Fortran routine for the computation of gamma percentiles*, Adv. Engrg. Software **10** (1988), 159–164.

[Phi90] H. N. Phien, *A note on the computation of the incomplete beta function*, Adv. Engrg. Software **12** (1990), 39–44.

[Pic89] B. Pichon, *Numerical calculation of the generalized Fermi-Dirac integrals*, Comput. Phys. Comm. **55** (1989), 127–136.

[Pie82] R. Piessens, *Automatic computation of Bessel function integrals*, Comput. Phys. Comm. **25** (1982), 289–295.

[Pie84a] R. Piessens, *Chebyshev series approximations for the zeros of the Bessel functions*, J. Comput. Phys. **53** (1984), 188–192.

[Pie84b] R. Piessens, *The computation of Bessel functions on a small computer*, Comput. Math. Appl. **10** (1984), 161–166.

[Pie90] R. Piessens, *On the computation of zeros and turning points of Bessel functions*, Bull. Soc. Math. Grèce (N.S.) **31** (1990), 117–122.

[Pri75] P. J. Prince, *Algorithm 498. Airy functions using Chebyshev series approximations*, ACM Trans. Math. Software **1** (1975), 372–379.

[PT84] B. A. Popov and G. S. Tesler, *Computation of functions on electronic computers—handbook*, Naukova Dumka, Kiev, 1984 (Russian), see review by K. S. Kölbig in Math. Comp. v. 55 (1990), pp. 395–397.

[PT90] W. H. Press and S. A. Teukolsky, *Elliptic integrals*, Computers in Physics **4** (1990), Jan/Feb, 92–96.

[PTVF92] W. H. Press, S. A. Teukolsky, W. T. Vetterling, and B. P. Flannery, *Numerical recipes. The art of scientific computing*, second ed., Cambridge University Press, 1992, diskettes and example books available. Editions exist in Basic (1991), C (1992), Fortran (1992), Macintosh Fortran (1988) and Pascal (1989).

[Puo88] M. Puoskari, *A method for computing Bessel function integrals*, J. Comput. Phys. **75** (1988), 334–344.

[PW90a] G. P. M. Poppe and C. M. J. Wijers, *Algorithm 680. Evaluation of the complex error function*, ACM Trans. Math. Software **16** (1990), 47.

[PW90b] G. P. M. Poppe and C. M. J. Wijers, *More efficient computation of the complex error function*, ACM Trans. Math. Software **16** (1990), 38–46.

[Rem73] G. F. Remenets, *Computation of Hankel (Bessel) functions of complex index and argument by numerical integration of a Schläfli contour integral*, U.S.S.R. Comput. Math. and Math. Phys. **13** (1973), no. 6, 58–67.

[RF93] Yu. L. Ratis and P. Fernández de Córdoba, *A code to calculate (high order) Bessel functions based on the continued fractions method*, Comput. Phys. Comm. **76** (1993), 381–388.

[RH84] J. R. Rice and R. J. Hanson, *Algorithm 620. References and keywords for collected algorithms from ACM*, ACM Trans. Math. Software **10** (1984), 359–360.

[Ric83] J. R. Rice, *Numerical methods, software and analysis: IMSL reference edition*, McGraw-Hill Book Company, New York, 1983.

[Riv90] T. J. Rivlin, *Chebyshev polynomials. From approximation theory to algebra and number theory*, second ed., John Wiley and Sons, Inc., New York, 1990.

[RL76] R. W. Redding and W. P. Latham, *On the calculation of the parabolic cylinder functions. II. The function $V(a,x)$*, J. Comput. Phys. **20** (1976), 256–258.

[RL80] S. R. Rengarajan and J. E. Lewis, *Mathieu functions of integral orders and real arguments*, IEEE Trans. Microwave Theory Tech. **28** (1980), 276–277.

[RS81] M. Razaz and J. L. Schonfelder, *Remark on Algorithm 498. Airy functions using Chebyshev series approximations*, ACM Trans. Math. Software **7** (1981), 404–405.

[RWGH87] J. P. Royston, J. B. Webb, P. Griffiths, and I. D. Hill, *The construction and description of algorithms*, Appl. Statist. **36** (1987), 94–103.

[Ryb89] G. B. Rybicki, *Dawson's integral and the sampling theorem*, Computers in Physics **3** (1989), Mar/Apr, 85–87.

[SAG79] Z. Schulten, D. G. M. Anderson, and R. G. Gordon, *An algorithm for the evaluation of the complex Airy functions*, J. Comput. Phys. **31** (1979), 60–75.

[Sag91a] R. P. Sagar, *A Gaussian quadrature for the calculation of generalized Fermi-Dirac integrals*, Comput. Phys. Comm. **66** (1991), 271–275.

[Sag91b] R. P. Sagar, *On the evaluation of the Fermi-Dirac integrals*, Astrophys. J. **376** (1991), 364–366.

[Sal89] K. L. Sala, *Transformations of the Jacobian amplitude function and its calculation via the arithmetic-geometric mean*, SIAM J. Math. Anal. **20** (1989), 1514–1528.

[SBK92] O. A. Sharafeddin, H. F. Bowen, and D. J. Kouri, *Numerical evaluation of spherical Bessel transforms via fast Fourier transforms*, J. Comput. Phys. **100** (1992), 294–296.

[Sch74] B. Schorr, *Programs for the Landau and the Vavilov distributions and the corresponding random numbers*, Comput. Phys. Comm. **7** (1974), 215–224.

[Sch76] J. L. Schonfelder, *The production of special function routines for a multi-machine library*, Software—Practice and Experience **6** (1976), 71–82.

[Sch78] J. L. Schonfelder, *Chebyshev expansions for the error and related functions*, Math. Comp. **32** (1978), 1232–1240.

[Sea82] M. J. Seaton, *Coulomb functions analytic in the energy*, Comput. Phys. Comm. **25** (1982), 87–95.

[SG72] A. J. Strecok and J. A. Gregory, *High precision evaluation of the irregular Coulomb wave functions*, Math. Comp. **26** (1972), 955–961.

[SGA81] Z. Schulten, R. G. Gordon, and D. G. M. Anderson, *A numerical algorithm for the evaluation of Weber parabolic cylinder functions* $U(a,x), V(a,x)$, *and* $W(a,\pm x)$, J. Comput. Phys. **42** (1981), 213–237.

[She74] V. B. Sheorey, *Chebyshev expansions for wave functions*, Comput. Phys. Comm. **7** (1974), 1–12.

[She88] B. L. Shea, *Algorithm AS 239. Chi-squared and incomplete gamma integral*, Appl. Statist. **37** (1988), 466–473.

[Shi93a] R. B. Shirts, *Algorithm 721. MTIEU1 and MTIEU2: Two subroutines to compute eigenvalues and solutions to Mathieu's differential equation for noninteger and integer order*, ACM Trans. Math. Software **19** (1993), 391–406.

[Shi93b] R. B. Shirts, *The computation of eigenvalues and solutions of Mathieu's differential equation for noninteger order*, ACM Trans. Math. Software **19** (1993), 377–390.

[Sie77] A. E. Siegman, *Quasi fast Hankel transform*, Optics Lett. **1** (1977), 13–15.

[Sim64] J. M. S. Simões Pereira, *Algorithm 234. Poisson-Charlier polynomials*, Comm. ACM **7** (1964), 420, for certification see same journal v. 8 (1965), p. 105.

[Sko85] S. L. Skorokhodov, *On the computation of complex zeros of the modified Bessel function of the second kind*, Soviet Math. Dokl. **31** (1985), 78–81.

[SL73] W. L. Sadowski and D. W. Lozier, *A unified standards approach to algorithm testing*, Program Test Methods (W. C. Hetzel, ed.), Prentice-Hall, Inc., Englewood Cliffs, New Jersey, 1973, pp. 277–290.

[SL81] M. M. Shepherd and J. G. Laframboise, *Chebyshev approximation of* $(1+2x)\exp(x^2)\,\mathrm{erfc}\,x$ *in* $0 \le x < \infty$, Math. Comp. **36** (1981), 249–253.

[SM75] B. P. Sinha and R. H. MacPhie, *On the computation of the prolate spheroidal radial functions of the second kind*, J. Math. Phys. **16** (1975), 2378–2381.

[Smi89] D. M. Smith, *Efficient multiple-precision evaluation of elementary functions*, Math. Comp. **52** (1989), 131–134.

[Smi91] D. M. Smith, *Algorithm 693. A Fortran package for floating-point multiple-precision arithmetic*, ACM Trans. Math. Software **17** (1991), 273–283.

[Sny93] W. V. Snyder, *Algorithm 723. Fresnel integrals*, ACM Trans. Math. Software **19** (1993), 452–456.

[SO87] J. Spanier and K. B. Oldham, *An atlas of functions*, Hemisphere Publishing Corporation, Washington, D. C., 1987.

[SOL81] J. M. Smith, F. W. J. Olver, and D. W. Lozier, *Extended-range arithmetic and normalized Legendre polynomials*, ACM Trans. Math. Software **7** (1981), 93–105.

[SP75] P. Spellucci and P. Pulay, *Effective calculation of the incomplete gamma function for parameter values* $\alpha = (2n+1)/2, n = 0, \ldots, 5$, Angew. Informatik **17** (1975), 30–32.

[Spr91] J. C. Sprott, *Numerical recipes: routines and examples in BASIC*, Cambridge University Press, 1991.

[Str68] A. J. Strecok, *On the calculation of the inverse of the error function*, Math. Comp. **22** (1968), 144–158.

[Sym92] *MACSYMA reference manual, version 13*, Symbolics, Inc., 20 Academy St., Arlington, Massachusetts 02174-6436, November 1992.

[SZ70] I. A. Stegun and R. Zucker, *Automatic computing methods for special functions*, J. Res. Nat. Bur. Standards **74B** (1970), 211–224.

[SZ74] I. A. Stegun and R. Zucker, *Automatic computing methods for special functions. Part II. The exponential integral* $E_n(x)$, J. Res. Nat. Bur. Standards **78B** (1974), 199–216.

[SZ76] I. A. Stegun and R. Zucker, *Automatic computing methods for special functions. Part III. The sine, cosine, exponential integrals, and related functions*, J. Res. Nat. Bur. Standards **80B** (1976), 291–311.

[SZ79] B. Sommer and J. G. Zabolitzky, *On numerical Bessel transformation*, Comput. Phys. Comm. **16** (1979), 383–387.

[SZ81] I. A. Stegun and R. Zucker, *Automatic computing methods for special functions. Part IV. Complex error function, Fresnel integrals, and other related functions*, J. Res. Nat. Bur. Standards **86** (1981), 661–686.

[Tal83] J. D. Talman, *LSFBTR: A subroutine for calculating spherical Bessel transforms*, Comput. Phys. Comm. **30** (1983), 93–99.

[Tau92] G. Taubmann, *Parabolic cylinder functions* $U(n, x)$ *for natural* n *and positive* x, Comput. Phys. Comm. **69** (1992), 415–419.

[TB85] I. J. Thompson and A. R. Barnett, *COULCC: A continued-fraction algorithm for Coulomb functions of complex order with complex arguments*, Comput. Phys. Comm. **36** (1985), 363–372.

[TB86] I. J. Thompson and A. R. Barnett, *Coulomb and Bessel functions of complex arguments and order*, J. Comput. Phys. **64** (1986), 490–509.

[TB87] I. J. Thompson and A. R. Barnett, *Modified Bessel functions* $I_\nu(z)$ *and* $K_\nu(z)$ *of real order and complex argument, to selected accuracy*, Comput. Phys. Comm. **47** (1987), 245–257.

[Tem75] N. M. Temme, *On the numerical evaluation of the modified Bessel function of the third kind*, J. Comput. Phys. **19** (1975), 324–337.

[Tem76] N. M. Temme, *On the numerical evaluation of the ordinary Bessel function of the second kind*, J. Comput. Phys. **21** (1976), 343–350.

[Tem77] N. M. Temme, *The numerical computation of special functions by use of quadrature rules for saddle point integrals. I. Trapezoidal integration rules*, Tech. Report TW 164/77, Mathematisch Centrum, Amsterdam, 1977.

[Tem78] N. M. Temme, *Some aspects of applied analysis: asymptotics, special functions and their numerical computation*, Mathematisch Centrum, Amsterdam, 1978.

[Tem79] N. M. Temme, *An algorithm with ALGOL 60 program for the computation of the zeros of ordinary Bessel functions and those of their derivatives*, J. Comput. Phys. **32** (1979), 270–279.

[Tem83] N. M. Temme, *The numerical computation of the confluent hypergeometric function* $U(a, b, z)$, Numer. Math. **41** (1983), 63–82.

[Tem85] N. M. Temme, *Special functions as approximants in uniform asymptotic expansions of integrals; a survey*, Rend. Sem. Mat. Univ. Politec. Torino **Fascicolo Speciale. Special Functions: Theory and Computation** (1985), 289–317.

[Tem87] N. M. Temme, *On the computation of the incomplete gamma functions for large values of the parameters*, Inst. Math. Appl. Conf. Ser. New Ser., vol. 10: Algorithms for Approximation (Shrivenham, 1985), Oxford Univ. Press, New York,

1987, pp. 479–489.

[Tem92a] N. M. Temme, *Asymptotic inversion of incomplete gamma functions*, Math. Comp. **58** (1992), 755–764.

[Tem92b] N. M. Temme, *Asymptotic inversion of the incomplete beta function*, J. Comput. Appl. Math. **41** (1992), 145–157.

[Tem94] N. M. Temme, *A set of algorithms for the incomplete gamma functions*, Probab. Engrg. Inform. Sci. (1994), in press.

[TM68] R. F. Tooper and J. Mark, *Simplified calculation of* $\mathrm{Ei}(x)$ *for positive arguments, and a short table of* $\mathrm{Shi}(x)$, Math. Comp. **22** (1968), 448–449.

[TP83] N. Tuğbay and E. Panayirci, *An efficient algorithm for generation of prolate spheroidal wave functions*, Bull. Tech. Univ. Istanbul **36** (1983), 563–577.

[TR69] T. Tamura and F. Rybicki, *Coulomb functions for complex energies*, Comput. Phys. Comm. **1** (1969), 25–30, for erratum see same journal v. 3 (1972), p. 276.

[ULI90] *Mathematical function library for Microsoft–C*, United Laboratories, Inc., John Wiley & Sons, 1990, includes diskettes. Edition also exists in Fortran (1989).

[vBBH70] A. L. van Buren, R. V. Baier, and S. Hanish, *A Fortran computer program for calculating the oblate spheroidal radial functions of the first and second kind and their first derivatives*, NRL Report No. 6959, Naval Res. Lab., Washington, D. C., 1970.

[vBBHK72] A. L. van Buren, R. V. Baier, S. Hanish, and B. J. King, *Calculation of spheroidal wave functions*, J. Acoust. Soc. Amer. **51** (1972), 414–416.

[vdLT84] C. G. van der Laan and N. M. Temme, *Calculation of special functions: The gamma function, the exponential integrals and error-like functions*, CWI Tract, vol. 10, Centrum voor Wiskunde en Informatica, Amsterdam, 1984.

[VGK+91] V. K. Vlasov, M. N. Glukhova, L. N. Korolev, M. S. Pelina, S. N. Razumovskiĭ, and O. L. Ulasik, *On the calculation of Bessel functions of a complex index and a complex argument*, Vestnik Moskov. Univ. Ser. XV Vychisl. Mat. Kibernet. (1991), no. 3, 46–51, 64 (Russian).

[Wal84] J. L. Walmsley, *On the efficient evaluation of modified Bessel functions of zeroth and first orders for arguments of the form* $x\exp(i\pi/4)$, J. Comput. Phys. **56** (1984), 349–355.

[War60] M. Ward, *The calculation of the complete elliptic integral of the third kind*, Amer. Math. Monthly **67** (1960), 205–213.

[WBR82] C. A. Wills, J. M. Blair, and P. L. Ragde, *Rational Chebyshev approximations for the Bessel functions* $J_0(x), J_1(x), Y_0(x), Y_1(z)$, Math. Comp. **39** (1982), 617–623.

[WC90] E. J. Weniger and J. Cižek, *Rational approximations for the modified Bessel function of the second kind*, Comput. Phys. Comm. **59** (1990), 471–493.

[Wei94a] J. A. C. Weideman, *Computation of the complex error function*, SIAM J. Numer. Anal. (1994), in press.

[Wei94b] J. A. C. Weideman, *Computing integrals of the complex error function*, These proceedings, 1994.

[Wim74] J. Wimp, *On the computation of Tricomi's Ψ function*, Computing **13** (1974), 195–203.

[Wim84] J. Wimp, *Computation with recurrence relations*, Pitman, London, 1984.

[Wit68] B. F. W. Witte, *Algorithm 332. Jacobi polynomials*, Comm. ACM **11** (1968), 436–437, for remarks see same journal v. 13 (1970), p. 449 and v. 18 (1975), pp. 116–117.

[Wol91] S. Wolfram, *Mathematica, a system for doing mathematics by computer*, second ed., Addison-Wesley, Redwood City, California, 1991.

[Won89] R. Wong, *Asymptotic approximations of integrals*, Academic Press, New York, 1989.

[Woo67] V. E. Wood, *Chebyshev expansions for integrals of the error function*, Math. Comp. **21** (1967), 494–496.

[YKK88] A. Yu. Yeremin, I. E. Kaporin, and M. K. Kerimov, *Computation of the derivatives of the Riemann zeta-function in the complex domain*, U.S.S.R. Comput. Math.

and Math. Phys. **28** (1988), no. 4, 115–124, see also [EKK85].

[YN74] T. Yoshida and I. Ninomiya, *Computation of Bessel functions $K_n(z)$ with complex argument by τ-method*, J. Inform. Process. **14** (1974), 32–37.

[Yos92] T. Yoshida, *Error analysis of the recurrence technique for calculation of Bessel functions $I_\nu(x)$*, J. Inform. Process. **15** (1992), 213–221.

[Zan75] R. Zanovello, *Sul calcolo numerico della funzione di Struve $H_\nu(z)$*, Rend. Sem. Mat. Univ. Politec. Torino **32** (1975), 251–269.

[Ziv91] A. Ziv, *Fast evaluation of elementary mathematical functions with correctly rounded last bit*, ACM Trans. Math. Software **17** (1991), 410–423.

Applied and Computational Mathematics Division, National Institute of Standards and Technology, Gaithersburg, Md 20899
E-mail address: dlozier@nist.gov

Institute for Physical Science and Technology, University of Maryland, College Park, MD 20742
E-mail address: olver@bessel.umd.edu

Proceedings of Symposia in Applied Mathematics
Volume 48, 1994

A Survey of Numerical Cubature over Triangles

J. N. LYNESS AND RONALD COOLS

ABSTRACT. This survey collects together theoretical results in the area of numerical cubature over triangles and is a vehicle for a current bibliography. We treat first the theory relating to regular integrands and then the corresponding theory for singular integrands with emphasis on the "full corner singularity." Within these two sections we treat successively approaches based on transforming the triangle into a square, formulas based on polynomial moment fitting, and extrapolation techniques. Within each category we quote key theoretical results without proof, and relate other results and references to these. Nearly all the references we have found may be readily placed in one of these categories. This survey is theoretical in character and does not include recent work in adaptive and automatic integration.

1. Introduction

1.1. A historical perspective. The numerical calculation of areas has a history that precedes the history of the infinitesimal calculus by several millenia. But the arrival of an analytic method in the late eighteenth century led to a surge in both theoretical and numerical methods for quadrature. The development of numerical methods for cubature (multidimensional integration) came perhaps a century and a half later, following closely the development of computing machines capable of handling these more intensive numerical calculations. Not until the early 1970s did textbooks begin to address cubature.

Undoubtedly, in the early days, cubature was treated as an iterative application of one-dimensional quadrature. But it was quickly realized that it was often more cost-effective to treat cubature over standard regions as a single entity. Two factors inhibiting the development of the theory were the large number of

1991 *Mathematics Subject Classification.* Primary 41A55, 41A63; Secondary 65B05, 65B15, 65D32.

This work was supported by the Office of Scientific Computing, U.S. Department of Energy, under Contract W-31-109-Eng-38.

This paper is in final form, and no version of it will be submitted for publication elsewhere.

© 1994 American Mathematical Society
0160-7634/94 $1.00 + $.25 per page

standard regions, each demanding special attention, and the cost of experimentation.

The triangle is simply one of these standard regions; the development of dedicated cubature methods is comparatively recent. Our earliest reference is to Radon's seven-point rule of polynomial degree 5 [**Rad48**]. A version of this rule is applicable to any planar region. The theory of integration over a triangle has lagged behind the theory for the square, but it has recently been spurred by two particular applications. One is the application to adaptive cubature, where triangles seem to be replacing squares as the basic module. The other application is to the finite element; here integrands with specified singularities may be involved.

The century and a half of development of one-dimensional quadrature has left us with a rich legacy in terms of variety of methods. We have rules involving derivative values, rules with all sorts of weighting functions, equally spaced rules, copy rules, minimum norm rules, equal weight rules, and rules of specified trigonometric degree. There are methods for using the same sets of function values to evaluate sets of integral transforms. We have techniques involving subtracting out singularities, and the method of steepest descent in the complex plane. And we have a wealth of expressions for the discretization error. For an overview of all these, we suggest [**DR84**].

Compared with all this, the available theory for the triangle is sparse indeed.

References. [**Rad48**], [**DR84**]

1.2. Outline and notation. Throughout this paper, Ω denotes a specified triangle, and $I(\Omega)[f]$ denotes the integral of an integrand function $f(x, y)$ over this triangle. We refer to

$$Q[f] = \sum_{i=1}^{\nu} w_i f(x_i, y_i) \ , \ w_i \in \mathbb{R} \ , \ (x_i, y_i) \in \mathbb{R}^2 \tag{1.1}$$

as a *cubature rule,* usually some sort of approximation to $I(\Omega)[f]$. This rule involves $\nu = \nu(Q)$ weights and points. It is of *polynomial degree* $d = d(Q)$ *for* Ω if $Q[f] = I(\Omega)[f]$ whenever $f(x, y)$ is a polynomial of degree d or less. We do not discuss cubature rules requiring derivatives.

In §2.1 we introduce the Duffy transformation. This reduces the problem to that of cubature over a square. In §3.1 the same transformation is applied to advantage in situations where the integrand has certain singularities.

Section 2.2 is devoted to cubature rules of the form (1.1). Special attention is paid to different approaches to the construction of such rules and to the number of function values $\nu(Q)$ required to obtain polynomial degree $d(Q)$. Because this theory is an immediate generalization of one-dimensional Gaussian quadrature, we refer to these rules generically as GQ. In §3.2, the few analogous results known to us involving weight functions are given.

In §2.3 we outline the theory of cubature using extrapolation (EQ). It appears that the result of extrapolation is yet another cubature rule, obtained in a somewhat roundabout way. This rule turns out to have nontrivial polynomial degree, but is marginally less cost-effective in terms of polynomial degree than the GQ rules. In §3.3, extrapolation cubature is discussed in the context of integrand functions having certain algebraic or logarithmic singularities on the edge or at the vertex. This approach to these cubatures has significant advantages in terms of generality and convenience over the corresponding GQ.

In much of the theory, the results are independent of the actual triangle used for the calculation. This is especially true of the polynomial degree of a formula. Any nonsingular linear transformation takes one triangle into another and transforms a cubature formula in an obvious manner. It appears that the polynomial degree of the cubature formula is retained. (But this happens only occasionally when nontrivial weight functions are involved.) Different authors have found it convenient to use several different triangles, some of which are listed below. Originally,

$$\text{◺} = \{(x,y) : x \geq 0\ ,\ y \geq 0\ ,\ x+y \leq 1\}, \tag{1.2}$$

which we refer to as the *standard triangle*, was used almost exclusively. Parts of the theory will involve the *unit square*,

$$\square = [0,1]^2, \tag{1.3}$$

and what we term the *complementary triangle*,

$$\text{◹} = \{(x,y) : x \leq 1\ ,\ y \leq 1\ ,\ x+y \geq 1\}. \tag{1.4}$$

Clearly, $\text{◺} \cup \text{◹} = \square$. More recent work has been based on the *equilateral triangle*

$$\lhd = \{(x,y) : x \leq \frac{1}{2}\ ,\ \sqrt{3}y - x \leq 1\ ,\ -\sqrt{3}y - x \leq 1\}, \tag{1.5}$$

which has vertices at $(1/2, \pm\sqrt{3/2})$ and $(-1,0)$. Finally, we have used the triangle

$$\text{◿} = \{(x,y) : 0 \leq y \leq x \leq 1\} \tag{1.6}$$

for the Duffy transformation below, simply because this makes the transformation simpler to follow. A consequence of the fact that different authors prefer different triangles is that the same cubature formula can appear in several forms in the literature.

2. A Regular Integrand Function

2.1. Transformation into a square. A natural approach to integration over a triangle is to transform the triangle $\triangle$ into the square $\square$ as follows:

$$\int_0^1 \left\{ \int_0^x f(x,y)dy \right\} dx = \int_0^1 x \int_0^1 f(x,xt)dtdx.$$

In the days before the formal theory for multidimensional integration, one supposed that a scientist would do this naturally, partly because, in the early days, numerical integration over a square was more familiar. This transformation is now known as the *Duffy transformation* [**Duf82**].

To integrate over the square, the scientist might use the product of two one-dimensional rules, for example,

$$\sum_{i=1}^{n_t} T_i g(t_i) = \int_0^1 g(t)dt + E_g \text{ , where } E_g = 0 \ \ \forall g \in \mathcal{P}_{d_t},$$

and

$$\sum_{i=1}^{n_x} X_j g(x_j) = \int_0^1 xg(x)dx + E_g \text{ , where } E_g = 0 \ \ \forall g \in \mathcal{P}_{d_x}.$$

Thus, he would calculate an approximation of the form

$$Q[f] = \sum_{i=1}^{n_t} \sum_{j=1}^{n_x} T_i X_j f(x_j, t_i x_j).$$

A short calculation shows that this is of polynomial degree $d = \min(d_t, d_x)$. While, in principle, any pair of one-dimensional quadrature rules can be used, an erudite scientist might choose the Gauss-Legendre rule and the appropriate Gauss-Jacobi rule. The weights and abscissas for these Gaussian rules are often available for moderate to high values of n (see, e.g., [**SS66**]). If the scientist also chose $n_t = n_x$, he would recover the well-known *Stroud Conical Product rule* [**Str71**], $Q_{CP}^{(d)}$, for which the number of points is

$$\nu_{CP}^{(d)} = \nu(Q_{CP}^{(d)}) = \frac{(d+1)^2}{4} \text{ for all odd } d. \tag{2.1}$$

We note that this is a somewhat indirect way to obtain a rule for a triangle. The rule is not symmetric, and there are three distinct Stroud Conical Product rules for each degree d, depending on which vertex is the preferred one.

References. [**Duf82**],[**Str71**],[**SS66**]

2.2. Polynomial moment fitting. Nearly all the theory of numerical integration over a triangle is related in one way or another to polynomial approximation. In the cubature rule (1.1), the points (x_i, y_i) and weights w_i are conventionally chosen so that the formula is exact for all polynomials up to a certain degree d. If the polynomials $p_1, p_2, \ldots, p_M$ form a basis for the space $\mathcal{P}_d^2$ of all polynomials in two variables of degree at most d, then a cubature formula of degree d is determined by a system of polynomial equations, known as the *moment equations*,

$$(2.2) \qquad Q[p_i] = I(\Omega)[p_i]\ ,\ i = 1, 2, \ldots, M = \dim \mathcal{P}_d^2 = \frac{1}{2}(d^2 + 3d + 2).$$

Here the *moments*, $I(\Omega)[p_i]$, are known, and the unknowns are $x_i, y_i, w_i; i = 1, 2, \ldots, \nu$, which occur when the left-hand side is replaced by the expression in (1.1). Since a cubature formula of a specified polynomial degree for a specified triangle Ω can be scaled onto any other triangle with preservation of its degree, one may choose any convenient triangle Ω for this calculation.

One can distinguish between two approaches to construct cubature formulas:

1. one may proceed directly to solve this system of nonlinear equations (most of this section describes this); or
2. one can search for polynomials that vanish at the points of the formula.

This second approach has been very successful in (one-dimensional) quadrature. In cubature, it has turned out to be difficult for the square, and significantly more difficult for the triangle. Radon constructed the first cubature formula for a triangle using the common zeros of three orthogonal polynomials. Schmid [**Sch83**] used the theory of real polynomial ideals to construct a minimal formula of degree 4 with positive weights. To our knowledge, only very few rules for the triangle have been obtained by using orthogonal polynomials or polynomial ideals. However, this approach did lead to the following important result.

THEOREM 2.1. *The number of points $\nu(Q)$ in a cubature formula Q of degree d for a triangle satisfies $\nu(Q) \geq \nu_L(d)$, where*

$$(2.3) \qquad \nu_L(d) = \frac{(d+2)(d+4)}{8} \qquad \textit{if } d \textit{ is even},$$

$$(2.4) \qquad \nu_L(d) = \frac{(d+1)(d+3)}{8} + \left\lfloor \frac{d+1}{4} \right\rfloor \quad \textit{if } d \textit{ is odd}.$$

The first bound can be traced back to the same paper by Radon that contains the first cubature formula for a triangle. In its general form it was probably first published by Stroud in [**Str60**]. The second bound was obtained by Möller for $d < 12$ [**Möl76**] in 1976. This result was generalized by Rasputin in 1983 [**Ras83b**]. The same bound had been established for centrally symmetric regions, such as a square, in 1973 [**Möl73**]. It is revealing that it took ten years to modify the theory for a triangle with constant weight function and another ten years for a triangle with a special weight function (see §3.2). It is also known

TABLE 1. Known minimal formulas, their properties, and original discoverers

d	$\nu_L(d)$	Quality	Reference
2	3	PI(3)	**[HS56]**
		PI(3)	**[Hil77]**
3	4	NI	**[HS56]**
		PI(2)	**[Hil77]**
4	6	PI	**[Cow73]**
		PI	**[Sch83]**
5	7	PI	**[Rad48]**
6	10	PO	**[Ras83a]**
7	12	PI	**[Gat88, Bec87]**
8	15	PO	**[CH87]**

that a formula of even degree that attains this lower bound has all weights positive [**Mys68, Str71, CH88b**]. This property is, however, not guaranteed for formulas of odd degree.

In Table 1 we give an overview of the known minimal formulas. In the third column, P indicates that no negative weights occur and I indicates that there are no abscissas strictly outside the triangle. N and O are the negations of P and I, respectively. The number of distinct formulas (if more than one) is indicated in parentheses.

We now return to the direct approach, item 1 above. This has been fruitful, and an arsenal of cubature formulas now exists. At first, however, progress was painfully slow. In retrospect, it turns out that progress has been governed by the willingness of researchers to appreciate the significance of symmetry and structure and to realize that the detailed application of these is quite different for the triangle than it is for the square.

The term *symmetry* will be familiar to readers. By the symmetry group of a triangle, we mean the group of six linear transformations which take that specific triangle into itself. Any such group is isomorphic with the permutation group S_3. For the equilateral triangle $\lhd$, this symmetry group is termed the dihedral group $\mathcal{D}_3$. The term *structure* is less precise. In cubature it refers to conditions applied to the rule. These take the form of constraints on the solutions of the moment equations. For example, if we look ahead to (2.7), once numerical values for K_0, K_1, and K_2 have been assigned, the rule has been structured. One may then try to solve the moment equations. It is common experience among those who attempt to solve systems like (2.2) that exploiting the symmetry of the region and imposing a simple structure to the rule form has a simplifying effect on their task. This simplification is a consequence of Sobolev's theorem [**Sob62**].

THEOREM 2.2. *Let the integral $I(\Omega)$ and a cubature formula Q be invariant under the transformations of a group $\mathcal{G}$. The cubature formula Q has polynomial* *degree d if $Q[f] = I(\Omega)[f]$ whenever $f(x,y)$ is a polynomial of degree d or less that is invariant with respect to $\mathcal{G}$.*

The larger the symmetry group $\mathcal{G}$, the lower is the dimension of the space of all $\mathcal{G}$-invariant polynomials of degree at most d and, consequently, the lower is the number of nonlinear equations that determine a $\mathcal{G}$-invariant cubature formula.

It is convenient and appropriate to describe these formulas in a partly historic perspective.

1a. If no structure is imposed on the cubature formula, one can use the monomials
$$x^i y^j\ ,\ 0 \le i+j \le d\,,$$
as a basis of $\mathcal{P}_d^2$. The number of moment equations in this case is then $\frac{1}{2}(d^2+3d+2)$. We have no evidence that any serious attempt was ever made to solve this set of equations directly.

1b. In the 1960s, significant progress was made in cubature for the square by exploiting its symmetry. In particular, by imposing a rule structure symmetric under coordinate interchange, one could reduce the number of independent moment equations. It appeared that for the standard triangle ◺, a significant reduction was available if one structured the rule to be invariant under coordinate interchange, and fitted only monomials $x^\alpha y^\beta$ with $\alpha \ge \beta \ge 0$. The number of independent moment equations is then $\lfloor \frac{1}{4}(d^2+4d+4) \rfloor$. Formulas of degrees 2, 3, 4, 5, and 6 with this symmetry exist with their number of points equal to the lower bound of Theorem 2.1.

In retrospect, one can see that Sobolev's theorem is being applied, in this case with $\mathcal{G}$ being a group of order 2, whose elements are the unit and a reflection about one median. Later (see 1d below), it was applied using as $\mathcal{G}$ the full dihedral group $\mathcal{D}_3$. Later still (see 1c below), when the overall theory had become more familiar, the same theorem was applied using another subgroup of $\mathcal{D}_3$.

1c. In the late 1980s, Gatermann [**Gat88**] and Cools and Haegemans [**CH87**] searched for minimal formulas for degrees larger than 5, using the subgroup of $\mathcal{D}_3$ generated by the rotations. This approach led to minimal formulas of degree 7 and 8. It was later established that no formula of degree 6 with 10 points exists with this symmetry [**Gat90**]. For this symmetry, a basis in polar coordinates using the equilateral triangle $\triangleleft$ is
$$(r^2)^i (r^3 \cos 3\theta)^j (r^3 \sin 3\theta)^l\ ,\ 0 \le 2i + 3(j+l) \le d\ ,\ l = 0 \text{ or } 1,$$
and the number of independent moment equations is $\lfloor \frac{1}{6}(d^2+3d+6) \rfloor$.

1d. The full symmetry of a triangle was systematically exploited for the first time in 1975 by Lyness and Jespersen [**LJ75**], who introduced the equilateral triangle $\triangleleft$ for this purpose. Since then, this choice has been popular. One of the advantages of $\triangleleft$ is that it is straightforward to write down a basis for the $\mathcal{D}_3$-invariant polynomials of degree d and less. This is

$$(r^2)^i(r^3\cos 3\theta)^j\ ,\ 0 \le 2i+3j \le d,$$

and the number of independent moment equations is

$$E(d) = \left\lfloor \frac{1}{12}(d^2+6d+12) \right\rfloor . \tag{2.5}$$

In the case of the triangle, then, the larger the symmetry group, the smaller the number of independent moment equations $E(d)$. We now describe in detail how this theory is applied. We shall employ the equilateral triangle $\triangleleft$, denote its area by A, and work in polar coordinates. We shall apply the convention that (r,θ) and $(-r,\theta+\pi)$ refer to the same point. Since $Q[f]$ is to have this symmetry, it follows that, if any point (r,α) appears in $Q[f]$, so does any point of the form $(r,\pm\alpha+\frac{2\pi j}{3})$, and all distinct points of this set carry the same weight. Such a set of points is sometimes called an *orbit.* A moment's reflection will convince the reader that no orbit has more than six elements; and a basic cubature rule, one that contains only one orbit, can be expressed in the form

$$Q(r,\alpha)[f] = \frac{A}{6}\sum_{j=1}^{3}\left\{ f\left(r,\alpha+\frac{2\pi j}{3}\right) + f\left(r,-\alpha+\frac{2\pi j}{3}\right)\right\}, \tag{2.6}$$

where negative values of r are allowed, $\alpha \in [0,\pi/6)$, and A is the area of the triangle. Moreover, any rule $Q[f]$ is expressible as a linear combination of these basic rules.

Three geometrically distinct types of orbit occur. The first (type 0), with $r=0$, involves only the centroid. There can be at most one of these in $Q[f]$. The type-1 orbits have $\alpha = 0$ and positive or negative assignments of r. These have precisely three distinct points all lying on a median of the triangle. All other orbits are termed type-2 orbits and include six distinct points, none of which are on any median.

Let K_i be the number of orbits of type i in a $\mathcal{D}_3$-invariant cubature formula. These nonnegative integers specify the *structure* of the rule and are known as *structure parameters.* A $\mathcal{D}_3$-invariant cubature formula must have the following form:

$$Q[f] = K_0 w_0 f(0,0) + \sum_{i=1}^{K_1} w_i Q(r_i,0)[f] + \sum_{i=K_1+1}^{K_1+K_2} w_i Q(r_i,\alpha_i)[f], \tag{2.7}$$

where $Q(r,\alpha)[f]$ is defined in (2.6) above. The number of points required by this rule is

$$\nu(Q) = K_0 + 3K_1 + 6K_2. \tag{2.8}$$

The process of solving sets of nonlinear equations can be very involved. Some regard it as more of an art than a science. There is a set of inequalities, known as the consistency conditions, that are heuristic but extraordinarily useful. They are based on the inaccurate premise that one may obtain a solution to a set of E equations in N unknowns only if $E \geq N$. This premise is applied to the set of $E(d)$ independent moment equations and to any conveniently determined subset of these equations that does not involve some of the unknowns. Some consistency conditions for cubature formula (2.7) have been determined in [**LJ75**]. These are

$$\begin{array}{rcl} K_0 + 2K_1 + 3K_2 & \geq & E(d)\ , \\ 2K_1 + 3K_2 & \geq & E(d) - 1\ , \\ 3K_2 & \geq & E(d-6)\ , \\ K_0 & \leq & 1, \end{array} \tag{2.9}$$

where $E(d)$ is given by (2.5).

A modified problem, much easier than that of finding an unrestricted optimal cubature formula, is that of finding a $\mathcal{D}_3$-invariant optimal cubature formula subject to these consistency conditions. It is straightforward to show that, when Q given by (2.7) has polynomial degree d, then

$$\nu(Q) \geq \left\lfloor \frac{1}{6}(d^2 + 3d + 2) \right\rfloor . \tag{2.10}$$

In principle, one could use an integer programming routine to determine the set of nonnegative K_i that, for a given degree d, satisfies (2.9) and minimizes $\nu(Q)$ given by (2.8). In practice, this particular set is so simple that it can almost be done by inspection. Then one has to solve the set of nonlinear equations. If no solution can be found, one determines the next best solution of the minimization problem, and iterates. In the end, one is brought to a stop when the system becomes too large for one's computing aids. This search was initiated by Lyness and Jespersen [**LJ75**], who went as far as $d = 11$; it was continued up to $d = 20$ by Dunavant [**Dun85**]. To our knowledge, apart from conical product rules, no rules of higher degree have been determined explicitly.

In the foregoing account we have implicitly assumed that the search has been for all real solutions to the moment equations. As in the theory of one-dimensional quadrature, various additional criteria have been applied in the construction of cubature formulas. For example, a *good* formula (also known as a PI formula in [**LJ75**]) is one all of whose weights are positive and all points inside the region of integration. Restricting the search to formulas with all weights positive has largely been discontinued. Instead, all rules are found, and, when of interest, the value of a *condition number*

$$\sigma(Q) = \sum_{i=1}^{\nu} |w_i| / \sum_{i=1}^{\nu} w_i \tag{2.11}$$

is reported. Rules having some points outside the triangle are also reported. A different cost criterion, the *cytolic* point count (conventionally termed $\overline{\nu}$), is occasionally used. Its justification is set in a context in which a large triangular region is subdivided into m^2 small triangular regions of equal area, and the cubature rule is applied to each smaller triangle. This reflects the average number of points per cell by discounting appropriately the corner and edge point count. These variations have added little to the theory.

On the other hand, Silvester [**Sil70**], by means of a somewhat elegant use of homogeneous coordinates, introduced the theory necessary to construct analogues of the open and the closed type of one-dimensional Newton-Cotes rules. The closed type uses all points on ◺ of the form $(j/m, k/m)$ except the three vertices when m is even, is of degree $d = m$, and has $\mathcal{D}_3$ symmetry. Like the one-dimensional Newton-Cotes rules, the weights are rationals, and the value of σ, ((2.11) above), increases indefinitely with increasing m. For these the point count is

$$\nu_{NCC}(d) = \frac{1}{2}(d+1)(d+2) - \frac{3}{2}(1+(-1)^d). \tag{2.12}$$

Embedded formulas are pairs (or longer sequences) of cubature formulas where some or all of the points of a less precise formula are also used by a more precise formula. In contrast with the one-dimensional situation, where the gap between the degree of a Gaussian quadrature formula and its Kronrod extension is large, in two dimensions one can construct embedded formulas of successive degrees.

For constructing such embedded formulas, the approach using orthogonal polynomials has been more successful. Although published results are scarce, the highest-degree embedded sequences available have been constructed by using this approach.

THEOREM 2.3. [**CH90**] *Let a set of ν distinct points support two distinct cubature formulas of respective degrees $2k-1$ and $2k+m, m \in \mathbb{N}$, in such a way that, for every point, the two assigned weights are different from each other. Then $\nu \geq (k+1)^2$.*

The lower bound of Theorem 2.3 is known to be sharp for pairs of degrees (1,3) and (5,7). The confirming rules appeared in references [**GM78, CH88a**]. Simpler, but less effective, embedded formulas are obtained naturally by extrapolation (which is treated in the following section).

Nearly all the cubature formulas for the triangle that were known before 1971 are collected in Stroud's standard work on multiple integration [**Str71**]. A list of more recent references to available rules, together with information about the quality of these rules, has been compiled by Cools and Rabinowitz [**CR93**].

References. [**Bec87**], [**CH87**], [**CH88b**], [**CH88a**], [**CH90**], [**Cow73**], [**CR93**], [**Dun85**], [**Gat88**], [**Gat90**], [**GM78**], [**Hil77**], [**HS56**], [**LJ75**],

[**Möl73**], [**Möl76**], [**Mys68**], [**Rad48**], [**Ras83a**], [**Ras83b**], [**Sch83**], [**Sil70**], [**Sob62**], [**Str60**], [**Str71**]

Additional reference material for this section. [**BE90**], [**dD79**], [**Eng70**], [**Eng80**], [**Fra71**], [**Gat92**], [**Gün75**], [**Hem73**], [**HMS56**], [**Lau55**], [**Lau82**], [**LG78**], [**Moa74**], [**Ras86**], [**SF73**], [**Str61**], [**Str64**], [**Str66**], [**Str69**]

2.3. Extrapolation. In 1927, Richardson [**Ric27**] proposed his deferred approach to the limit. Possibly the first application of this in one-dimensional quadrature became known as Romberg integration [**Rom55**]. In the mid-1960s, extrapolation was applied by several authors to integration over a square. In the 1970s the techniques used to construct this theory were modified to provide corresponding results for the triangle. We describe first the theory for the square and then the significantly more difficult theory for the triangle.

We define a one-point rule

$$Q_{\alpha,\beta}[f] = f(\alpha,\beta). \tag{2.13}$$

The m^2-copy version of this rule for integration over the square $\square$ is

$$Q^{(m)}_{\alpha,\beta}(\square)[f] = \frac{1}{m^2}\sum_{j=0}^{m-1}\sum_{k=0}^{m-1} f\left(\frac{j+\alpha}{m}, \frac{k+\beta}{m}\right). \tag{2.14}$$

Using (2.13), we may express a general cubature formula (1.1) as

$$Q[f] = \sum_{i=1}^{\nu} w_i Q_{\alpha_i,\beta_i}[f]. \tag{2.15}$$

Q is *centrally symmetric* for the square $\square$ if it may be re-expressed in the form

$$Q(\square)[f] = \sum_{j=1}^{\mu} \frac{w_j}{2}\left(Q_{\alpha_j,\beta_j}[f] + Q_{1-\alpha_j,1-\beta_j}[f]\right).$$

The m^2 copy of (2.15) is defined as

$$Q^{(m)}(\square)[f] = \sum_{i=1}^{\nu} w_i Q^{(m)}_{\alpha_i,\beta_i}(\square)[f]. \tag{2.16}$$

This rule is an approximation to $I(\square)[f]$ obtained by subdividing the square $\square$ into m^2 identical squares, each of side $1/m$, and applying a properly scaled version of $Q[f]$ to each. If Q is centrally symmetric, then $Q^{(m)}(\square)[f]$ is also centrally symmetric.

For regular $f(x,y)$, extrapolation is based on the two-dimensional version of the *Euler-Maclaurin expansion.* When $f \in C^p(\square)$, $p \in \mathbb{N}$, the almost self-evident extension of the standard one-dimensional expansion may be expressed in the form

$$Q^{(m)}(\square)[f] - I(\square)[f] = \sum_{q=1}^{p-1} \frac{B_q(\square;Q;f)}{m^q} + O(m^{-p}). \tag{2.17}$$

The coefficients in this expansion have the form

$$B_q(\square;Q;f) = \sum_{\substack{q_1+q_2=q \\ q_i \geq 0}} c_{q_1,q_2}(Q) \int_{\square} f^{(q_1,q_2)}(x,y)dxdy, \tag{2.18}$$

where $c_{q_1,q_2}(Q)$ is obtained by applying the rule Q to

$$\frac{B_{q_1}(x)B_{q_2}(y)}{q_1!q_2!},$$

$B_q(t)$ being the Bernoulli polynomial of degree q in t. Integral representations for the remainder term of the expansion (2.17) have been given in [**LM70**].

These coefficients have several properties that are useful in constructing individual extrapolation procedures and in assessing them.

1. If Q is centrally symmetric, then

$$B_q(\square;Q;f) = 0 \ \forall \text{ odd } q, \tag{2.19}$$

and so the Euler-Maclaurin expansion (2.17) is an even expansion.

2. If Q is of polynomial degree d, then

$$B_q(\square;Q;f) = 0\ ,\ q = 1,2,\ldots,d. \tag{2.20}$$

3. If $f(x,y)$ is a polynomial of degree $d(f)$, then the asymptotic expansion (2.17) reduces to a finite sum, and

$$B_q(\square;Q;f) = 0 \ \forall\, q > d(f). \tag{2.21}$$

There are curiosities in the theory. For example, when $f(x,y)$ is periodic with unit period, one finds that all coefficients B_q are zero, but $Q^{(m)}[f] - I[f]$ need not be zero. The vanishing of the remainder term is a stronger result than, and is not implied by, the vanishing of all individual terms.

Romberg integration over the square is conventionally based on the product trapezoidal rule

$$Q^{(m)}(\square)[f] = \frac{1}{4}\sum_{\alpha=0}^{1}\sum_{\beta=0}^{1} Q^{(m)}_{\alpha,\beta}(\square)[f]$$

and occasionally on the mid-square rule $Q^{(m)}_{1/2,1/2}(\square)[f]$.

The extrapolation process will be familiar to the reader. The numerical result, conventionally denoted by $T_{k,p}$, is a linear combination of $p+1$ distinct approximations to $I[f]$, namely, $Q^{(m_i)}[f]$ with $i = k, k+1, \ldots, k+p$. This combination is constructed so that the first p nonzero terms on the right of (2.17) disappear from the corresponding expansion of $T_{k,p} - I[f]$. When $f(x,y)$ is a polynomial of degree p, the result in item 3 assures us that other terms also disappear. In fact, the remainder term vanishes too, leaving the result that $T_{k,p}$ is exact, and so the implied rule is of polynomial degree p. More familiar is the case of an even expansion when the extrapolation is designed to recognize only even terms, and the result $T_{k,p}$ is of polynomial degree $2p+1$.

We refer to these generalizations of Romberg integration collectively as *extrapolation quadrature* (EQ). The reader should note that there are many parameters to be set. These include the rule Q to be used, the mesh ratio sequence m_i, and, of course, for an individual result, the values of k and p.

The theory given above for the square is extraordinarily simple to establish. Only integral representations for the remainder term pose any serious difficulty at all. In contrast, development of the corresponding theory for the standard triangle ◺ has not been straightforward. When one subdivides the standard triangle ◺ into m^2 equal triangles using lines parallel to $x = 0, y = 0$ and $x + y = 0$, one finds $m(m+1)/2$ identical triangles oriented in the same way as ◺ together with $m(m-1)/2$ identical triangles oriented in the same way as ◹. If one proceeds on a cell-by-cell basis, it is clear how to assign points to the first set of triangles, but not the second set. In [**Lyn78**] this incongruity was resolved by introducing a *rule pair*. The resulting theory is complicated.

DEFINITION 2.1. A rule pair $(Q(◺), Q(◹))$ comprises two cubature rule operators

$$
\begin{aligned}
Q^{(1)}(◺)[f] &= \sum_{i=1}^{\nu} w_i f(\alpha_i, \beta_i), \text{ and} \\
Q^{(1)}(◹)[f] &= \sum_{i=1}^{\overline{\nu}} \overline{w}_i f(\overline{\alpha}_i, \overline{\beta}_i),
\end{aligned} \tag{2.22}
$$

where $\sum_{i=1}^{\nu} w_i + \sum_{i=1}^{\overline{\nu}} \overline{w}_i = 1$.

These are two quite independent rules, pertaining to the two different triangles. To construct the m-copy, one requires both rules. Then we have the following definition.

DEFINITION 2.2. The m-copy rule for the triangle ◺ using the rule pair of Definition 2.1 is

$$
\begin{aligned}
Q^{(m)}(◺)[f] &= \frac{1}{m^2} \sum_{i=1}^{\nu} w_i \sum_{j=0}^{m-1} \sum_{k=0}^{m-1-j} f\left(\frac{\alpha_i + j}{m}, \frac{\beta_i + k}{m}\right) \\
&+ \frac{1}{m^2} \sum_{i=1}^{\overline{\nu}} \overline{w}_i \sum_{j=0}^{m-2} \sum_{k=0}^{m-2-j} f\left(\frac{\overline{\alpha}_i + j}{m}, \frac{\overline{\beta}_i + k}{m}\right).
\end{aligned} \tag{2.23}
$$

Note that when $m = 1$, $Q^{(m)}(◺)[f]$ may be zero, but in an extrapolation technique, this zero value may be used as a meaningful result.

When $f \in C^p(\triangle)$, $p \in \mathbb{N}$, the Euler-Maclaurin expansion for the triangle may be written in the form

$$Q^{(m)}(\triangle)[f] - I(\triangle)[f] = \sum_{q=1}^{p-1} \frac{B_q(\triangle; Q; f)}{m^q} + O(m^{-p}). \tag{2.24}$$

This was established in [**Lyn78**]. The coefficients $B_q(\triangle; Q; f)$ have a more complicated structure than those for the square in (2.18). Clearly,

$$B_q(\triangle; Q; f) = \sum_{i=1}^{\nu} w_i B_q(\triangle; Q_{\alpha_i,\beta_i}; f) + \sum_{i=1}^{\overline{\nu}} \overline{w}_i B_q(\triangle; Q_{\overline{\alpha}_i,\overline{\beta}_i}; f).$$

It may be shown that

$$B_q(\triangle; Q_{\alpha,\beta}; f) = \sum_{\substack{q_1+q_2=q \\ q_i \geq 0}} B_{q_1,q_2}(\triangle; Q_{\alpha,\beta}; f),$$

where, so long as $\alpha, \beta \in (0,1)$ and $\alpha + \beta \neq 1$,

$$\begin{aligned} B_{q_1,q_2}(\triangle; Q_{\alpha,\beta}; f) \quad = \quad & \frac{B_{q_1}(\alpha)B_{q_2}(\{\alpha+\beta\})}{q_1!q_2!} \int_0^1 \frac{\partial^{q_1}}{\partial x^{q_1}} f^{(0,q_2-1)}(x, 1-x)dx \\ & - \frac{B_{q_1}(\alpha)B_{q_2}(\beta)}{q_1!q_2!} \int_0^1 \frac{\partial^{q_1}}{\partial x^{q_1}} f^{(0,q_2-1)}(x, 0)dx. \end{aligned}$$

Here $\{x\}$ means the fractional part of x. Moreover, when Q includes abscissas on the edges and even outside its triangle, the expansion (2.24) remains valid, and the coefficients may be obtained from those given above by using analytic continuation.

Clearly, an associated rule for the square is

$$Q(\square) = Q(\triangle) + Q(\bigtriangledown).$$

This rule $Q(\square)$ is centrally symmetric if $\nu = \overline{\nu}$, $w_i = \overline{w}_i$ and $\alpha_i + \overline{\alpha}_i = \beta_i + \overline{\beta}_i = 1$.

Results analogous to (2.19), (2.20), and (2.21) hold, namely,

1. if $Q(\square)$ is centrally symmetric, then $B_q(\triangle; Q; f) = 0 \ \forall$ odd q;
2. if both $Q^{(1)}(\triangle)$ and $Q^{(1)}(\bigtriangledown)$ have polynomial degree d for their respective triangles, then $B_q(\triangle; Q; f) = 0$, $q = 1, 2, \ldots, d$; and
3. if $f(x, y)$ is a polynomial of degree $d(f)$, then the asymptotic expansion (2.24) reduces to a finite sum, and

$$B_q(\triangle; Q; f) = 0 \ \forall\, q > d(f) + 1. \tag{2.25}$$

These results were obtained earlier using an entirely different approach [**LP73**] involving product integration. This equally complicated theory was generalized in 1979 by de Doncker [**dD79**]. Let μ be half an odd integer, $f \in C^p(\triangle)$, $p \in \mathbb{N}$, and

$$Q^{(\mu)}[f] = \frac{1}{\mu^2} \sum_{j=1}^{\mu+1/2} \sum_{k=1}^{\mu-j+1/2} f\left(\frac{j-1/2}{\mu}, \frac{k-1/2}{\mu}\right).$$

Then

$$Q^{(\mu)}[f] = I[f] + \sum_{\substack{q \text{ even} \\ q<p}} \frac{B_q}{\mu^q} + O(m^{-p}).$$

In any extrapolation procedure based on this expansion, one is allowed to employ only values of μ for which 2μ is an odd integer. De Doncker has shown that when f is a polynomial of degree $d(f)$, then $B_q = 0 \ \forall\ q > d(f)$. This means that an extrapolation procedure that eliminates B_q, $q \leq d(f)$, leaves an exact result. This circumstance may be used to generate rules of specified degree. Grundmann and Möller [**GM78**], in an independent approach, established a set of rules of the form

$$Q_t[f] = \sum_{j=1}^{t} a_{t,j} Q^{(j-1/2)}[f] \tag{2.26}$$

of degree $2t-1$. These rules are identical with rules obtained later by De Doncker using extrapolation. (For larger t, the rule (2.26) has unduly high condition number; see (2.11) above. An alternative is to extrapolate using only a selection of the approximations $Q^{(j-1/2)}[f]$. This reduces the conditioning error at a cost in extra function values.) So far as we are aware, this approach to integration over the triangle has not been developed further, nor has the corresponding theory for squares.

References. [**dD79**], [**GM78**] [**LM70**], [**LP73**], [**Lyn78**] [**Ric27**] [**Rom55**]

3. Specified Singularities

3.1. Transformation into a square. The Duffy transformation, introduced in §2.1, may be used to advantage in the case where the integrand function has a full corner singularity and a line singularity along the opposite edge. In particular, for the triangle $\triangle$, when the integrand is of the form

$$f(x,y) = y^\lambda (x-y)^\mu (1-x)^\nu r_\rho(x,y) g(x,y),$$

where $r_\rho(x,y) = (x^2+y^2)^{\rho/2}$ and $g(x,y)$ is regular in the triangle of integration, one finds readily that

(3.1)

$$\int_0^1 \left\{ \int_0^x f(x,y)dy \right\} dx = \int_0^1 \int_0^1 x^{1+\lambda+\mu+\rho}(1-x)^\nu (1-t)^\mu r_\rho(1,t) g(x,tx) dt dx.$$

Note that, while $r_\rho(x,xt)$ has a singularity at $x=0$, this has been extracted, leaving the innocuous function $r_\rho(1,t)$ which is regular for all t. This integral may be approximated by the product of a pair of Gauss-Jacobi formulas, just as described in §2.1.

This transformation is correct as written when $r_\rho(x,y)$ is replaced by any other function homogeneous of degree ρ in the triangle of integration, as defined in Definition 3.1 below.

3.2. Polynomial moment fitting. In this section we discuss cubature rules with a nonconstant weight function for the triangle. This weight function $w(x,y)$ is given, and its moments are known, usually in analytic form. The rule $Q[f]$ is of the same form as in preceding sections, but it is designed to approximate the integral

$$I(\Omega;w)[f] = \int_\Omega\int w(x,y)f(x,y)dxdy. \tag{3.2}$$

As before, when this approximation is exact for all polynomials of degree d or less, the rule $Q[f]$ is said to be of degree d.

The major difference between this theory and the special case with unit weight function is that, in general, one cannot transform the triangle into another of different shape and retain both the nature of the singularity and the polynomial degree. What happens is that the affine transformation alters the weight function too. When $w(x,y) = 1$, this is no problem. But, in general, one finds a rule of the original polynomial degree for a new triangle and a new weight function.

The shape of the triangle and the weight function are now important, and, for a fixed type of weight function like $1/r$, one needs a new rule for each differently shaped triangle. For reasons unknown to us, only one triangle has been treated, this having sides in the ratio $\sqrt{5}, \sqrt{5}, 2$. One set of rules has a radial weight function $1/r$, r being measured from the mid-point of the shorter side. The other is similar, except that r is now measured from the vertex opposite the shorter side. Only results for low degrees (1, 2, 3, 5, and 7) are published [**CL78, Hae93, PFB81**]. Most of them were constructed by solving the nonlinear equations that define them and imposing symmetry with respect to one median (the only symmetry available). The highest-degree results were obtained using the common zeros of three orthogonal polynomials, a variant of the method used by Radon.

The narrowness of the scope for possible application of these numerical results is manifest. If one has an equilateral triangle, or a right-angled triangle, these weights and abscissas are useless. They are of value for triangles of one specified shape only. Unfortunately, the authors do not make this clear. The casual reader could easily retain the impression that these results are of wider application.

Lyness and Gatteschi [**LG82**] have treated a variant problem. They use a general triangle and an integrand function $w(x,y)f(x,y)$ closely related to that in the preceding subsection. In terms of the triangle $\triangle$, the weight function is

$$w(x,y) = y^\lambda(x-y)^\mu(1-x)^\nu r^\beta, \tag{3.3}$$

and the rest of the integrand function is of the form $f(x, y) = h(r)g(x, y)$. The cubature rule is of *joint polynomial degree* d if it is exact for all functions $h(r)g(x, y)$, where g and h are polynomials of degree γ and δ, respectively, and $\gamma + \delta \leq d$. The authors provide the specifications for a product rule of the same type as in the preceding subsection. However, one of the product rule components is a one-dimensional *quasi-degree* quadrature rule.

The only theoretical results known to us about formulas for triangles with weight function are the following. First, the lower bound for the number of points presented in Theorem 2.1 is valid for even degrees whenever the integral has a positive weight function. Recently, Berens and Schmid [**BS92**] proved that Theorem 2.1 is valid for the integral

$$I[g] = \int_0^1 \int_0^x y^\lambda (x - y)^\mu (1 - x)^\nu g(x, y) dy dx \tag{3.4}$$

for odd degrees also.

References. [**BS92**] [**CL78**], [**Hae93**], [**LG82**] [**PFB81**]

3.3. Extrapolation. A powerful method of handling integration over a triangle of an integrand having certain boundary singularities is by extrapolation. The results resemble closely corresponding results for integration over a square, and much of this theory may be established as a minor corollary to the theory for the square.

The theory is based on homogeneous functions:

DEFINITION 3.1. The function $f(x, y)$ is homogeneous of degree γ about the origin in a region $\mathcal{R}$ if $f(kx, ky) = k^\gamma f(x, y)$ for all $k > 0$ and $(x, y) \in \mathcal{R} \setminus \vec{0}$.

This is in fact an n-dimensional definition. In one dimension the only homogeneous functions of degree γ are of the form kx^γ. In higher dimensions more sophisticated functions may be homogeneous. Two-dimensional examples include $r^\gamma = (x^2 + y^2)^{\gamma/2}$, x^γ, $(x - 3y + r)^\gamma$, and $(x^5 + 6y^5)^{\gamma/5}$ which are homogeneous of degree γ, while any function of y/x, such as $\theta = \arctan(y/x)$, is of degree zero. The singularities that can be handled through extrapolation are closely related to homogeneous functions.

Two early and fundamental papers [**LM80, Sid83**] establish error functional expansions for singular integrals over triangles. The basic result in [**LM80**] appears in Theorem 3.1 below. This is a generalization of (2.24) above to the singular integrand. The cubature rule operator is the one of Definition 2.2

THEOREM 3.1. *Let $f_\gamma(x, y)$ be homogeneous of degree γ about the origin in the first quadrant $x \geq 0$, $y \geq 0$ and be C^p, $p \in \mathbb{N}$, there except possibly at the origin. Then*

$$Q^{(m)}(\triangle)[f_\gamma] - I(\triangle)[f_\gamma] =$$

(3.5)

$$\frac{A_{2+\gamma}(\text{◺};Q;f_\gamma)}{m^{2+\gamma}}+\frac{C_{2+\gamma}(\text{◺};Q;f_\gamma)(\ln m)}{m^{2+\gamma}}+\sum_{s=1}^{p-1}\frac{B_s(\text{◺};Q;f_\gamma)}{m^s}+O(m^{-p}),$$

where $C_{2+\gamma}=0$ *unless* $\gamma\in\mathbb{N}$.

The identical expansion had been previously established for the square. The proof in the case of the square is not easy. However, once established for the square, it can be established for the triangle almost trivially. One notes that $\square=\text{◺}\cup\text{◹}$. Then, since

$$Q^{(m)}(\text{◺})+Q^{(m)}(\text{◹})=Q^{(m)}(\square),$$

any expansion for ◺ is simply the difference of the corresponding expansion for $\square$ and for ◹. However, $f_\gamma(x,y)$ is regular within ◹, and so $Q^{(m)}(\text{◹})[f_\gamma]$ has an expansion of the same type as (2.24). Taking this difference then simply alters the coefficients in some of the existing terms and does not introduce any additional terms.

Detailed expressions for the coefficients in the case of the square are given in [**Lyn76**]. The coefficients in (3.5) have the following properties:

1. if both $Q^{(1)}(\text{◺})$ has polynomial degree d for ◺ and Q has polynomial degree d for $\square$, then

   (3.6) $$C_s=B_s=0\ \text{ for all odd }\ s;$$

2. if Q is of polynomial degree d, then

   (3.7) $$C_s=B_s=0\ \text{ for }\ s=1,2,\dots,d.$$

The result as stated applies only to $f_\gamma(x,y)$, which is homogeneous of degree γ. A natural extension to

$$F(x,y)=f_\gamma(x,y)g(x,y),$$

where g is regular in $\square$, follows by expanding $g(x,y)$ as a Taylor series about the origin and noting that each term $f_\gamma(x,y)g^{(r,s)}(0,0)x^ry^s/r!s!$ is itself homogeneous of degree $\gamma+r+s$, and so the theorem applies also to this term in its own right. Taking care to handle the remainder term properly, we are led to an expansion of the form

$$Q^{(m)}(\text{◺})[F]-I(\text{◺})[F]\simeq\sum_{j=0}\frac{A_{2+\gamma+j}}{m^{2+\gamma+j}}+\sum_{j=0}\frac{C_{2+\gamma+j}\ln m}{m^{2+\gamma+j}}+\sum_{s=1}\frac{B_s}{m^s}.$$

Sidi's result for $F(x,y)=x^{\gamma+1}(\ln x)\,g(x,y)$ [**Sid83**] has the same form as this.

A standard result from classical analysis allows differentiation of any asymptotic expansion with respect to an incidental parameter so long as the coefficients are differentiable. The proof (not given here) of Theorem 3.1 above indicates that the coefficients are analytic functions of γ in any region not including integer γ.

Setting $\Phi(x,y) = \frac{\partial}{\partial\gamma} f_\gamma(x,y) = f_\gamma(x,y) \ln f_\gamma(x,y)$, we find

(3.8)

$$Q^{(m)}(\triangle)[\Phi] - I(\triangle)[\Phi] \simeq \frac{\partial A_{2+\gamma}/\partial\gamma}{m^{2+\gamma}} - \frac{A_{2+\gamma} \ln m}{m^{2+\gamma}} + \sum_{s=1} \frac{\partial B_s/\partial\gamma}{m^s}, \; \gamma \notin \mathbb{Z},$$

where $A_{2+\gamma}$ and B_s are the coefficients in (3.5). In the case that $\gamma \in \mathbb{Z}$, a serious calculation is required. This yields an expansion of the same form as one would have obtained if one had ignored the fact that the coefficients are discontinuous. This effect is to include in (3.8) an additional term in $(\ln m)^2 m^{-(2+\gamma)}$.

Several generalizations of this result extend or alter the singularity structure of the integrand in a way that is quite infeasible for Gaussian quadrature. For example, let

$$\Phi(x,y) = (f_1(x,y) + f_2(x,y) + f_3(x,y))g(x,y),$$

where f_1 and f_2 are homogeneous of degree γ_1 and γ_2 about the origin and f_3 is homogeneous of degree γ_3 about another vertex. Then the expansion for $Q(\triangle)[\Phi] - I(\triangle)[\Phi]$ is simply a concatenation of three expansions, each having the same form. The justification for concatenating the first two is obvious. We have to appeal to the Darboux theorem to establish the theoretical basis for including the third; see [**LM80**].

Another generalization is established in [**Lyn92**]. One may apply the same expansion to any differently shaped triangle so long as the integrand function has the same singular behavior at the vertex. The point here is that an affine transformation, besides transforming the triangle, transforms the singularity. In the case of GQ, this means that one must start anew to calculate a rule. However, a singularity of homogeneous degree γ is transformed into another of the same degree; and if the technique depends only on this degree, then it is invariant under the transformation. We consider this to be perhaps the most important attribute of EQ.

Recently, in the case of the square, an expansion has been developed for the full corner singularity [**LdD93, VH93**]

$$F(x,y) = x^\lambda y^\mu f_\gamma(x,y) g(x,y).$$

It is beyond the scope of this article to pursue this further. The situation seems to be that for many algebraic or logarithmic singularities that occur at a vertex or along a side, an extrapolation expansion exists.

References. [**LdD93**], [**LM80**], [**Lyn76**], [**Lyn92**], [**Sid83**], [**VH93**]

4. Concluding Remarks

The prevailing situation for regular integrands seems to be qualitatively different from that for singular integrands.

For regular integrands, so far as rules of specified polynomial degree are concerned, recent progress has been less than spectacular. Fifty years ago, what are now called conical product rules were available, allowing a result of polynomial degree d based on $(d+1)^2/4$ function values. Present theory reveals that the minimal formulas require more than half this number; these formulas are available only up to $d = 8$. The intermediate $\mathcal{D}_3$ formulas, using approximately $(d+1)^2/6$, are available up to $d = 20$. The general development of extrapolation quadrature (a set of methods based on Richardson's deferred approach to the limit) has been slow. The theory is complicated. Many variant and potentially interesting formulas exist; but so far as polynomial degree is concerned, the number of points required increases at best like $d^3/48$.

For singular integrands, the situation is not the same. Gaussian rules of specified polynomial degree are difficult to construct, and very few are available in the literature. Compounding this scarcity is the circumstance that with the same weighting function, a different rule is required for each differently shaped triangle. On the other hand, progress on extrapolation quadrature for singular integrands has been significant and a wide class of algebraic and logarithmic singularities is now within the reach of this theory. Moreover, for these, while possibly expensive, the integration can be carried out effectively in an iterative manner without the need for tables of weights and abscissas. The theory is to a significant extent independent of the shape of the triangle.

However, different parts of the theory are connected and may be useful in a single application. For example, suppose one requires a Gaussian formula for an integrand function having an extensive but known singularity structure of the type encountered in §3.3. To construct any Gaussian formula, one needs accurate numerical values of the moments. These might be determined by using extrapolation based on expansions of §3.3, using for Q one of the minimal formulas of §2.2. In this context, one might bear in mind that the expansions of §3.3 require, in their proof, the expansions of §2.3. The reader may find the description above somewhat artificial and self-serving. However, calculations of this general nature are habitually used to construct special finite elements. Thus, all four principal parts of the theory may be said to have contributed to this application in a cooperative manner.

The ever-present problem of constructing minimal formulas remains, and we are convinced that the more courageous and idealistic of us will continue to find the time to face this daunting set of successively harder challenges. We are particularly gratified that these formulas have found a role, albeit minor, in the related area of extrapolation quadrature.

The scope of this article has been limited to a review of the theory of integration over the triangle. We refer the reader interested in the important application to adaptive quadrature over the triangle to [**HK85**] and to [**Kea92**]. In [**Lyn83**], he will find a somewhat dated but detailed practical guide to handling applications. In [**Sid79**] a general framework is presented for all sorts of extrapolation, linear and nonlinear. The cogent problems of convergence and stability in extrapolation are treated in [**Sid90**]. A useful list of references about cubature over all standard regions has recently appeared in [**CR93**]. This provides a welcome update to the reference list contained in the standard works on this topic, namely, [**Str71, Mys81**]. Some very recent work in integration over curved surfaces and in the integration of expressions containing derivatives is not treated here. Our hope is that this synopsis of the theory, with its linked bibliography, will be helpful to individuals interested in further research in this and related areas.

References. [**CR93**], [**HK85**] [**Kea92**], [**Lyn83**], [**Mys81**], [**Sid79**], [**Sid90**], [**Str71**]

REFERENCES

[BE90] J. Berntsen and T.O. Espelid, *Degree 13 symmetric quadrature rules for the triangle*, Reports in Informatics 44, Dept. of Informatics, University of Bergen, 1990.

[Bec87] T. Becker, *Konstruktion von interpolatorischen Kubaturformeln mit Anwendungen in der Finit-Element-Methode*, Ph.D. thesis, Technische Hochschule Darmstadt, 1987.

[BS92] H. Berens and H.J. Schmid, *On the number of nodes of odd degree cubature formulae for integrals with Jacobi weights on a simplex*, Numerical Integration – Recent Developments, Software and Applications (Dordrecht) (T.O. Espelid and A. Genz, eds.), NATO ASI Series C: Math. and Phys. Sciences, vol. 357, Kluwer Academic Publishers, 1992, pp. 37–44.

[CH87] R. Cools and A. Haegemans, *Construction of minimal cubature formulae for the square and the triangle using invariant theory*, Report TW 96, Dept. of Computer Science, K.U. Leuven, 1987.

[CH88a] R. Cools and A. Haegemans, *An embedded pair of cubature formulae of degree 5 and 7 for the triangle*, BIT **28** (1988), 357–359.

[CH88b] R. Cools and A. Haegemans, *Why do so many cubature formulae have so many positive weights?*, BIT **28** (1988), 792–802.

[CH90] R. Cools and A. Haegemans, *A lower bound for the number of function evaluations in an error estimate for numerical integration*, Constr. Approx. **6** (1990), 353–361.

[CL78] M. Critescu and G. Loubignac, *Gaussian quadrature formulas for functions with singularities in $1/r$ over triangles and quadrangles*, Recent Advances in Boundary Element Methods (C.A. Brebbia, ed.), Pentech. Press, 1978.

[Cow73] G.R. Cowper, *Gaussian quadrature formulas for triangles*, Internat. J. Numer. Methods Engrg. **7** (1973), 405–408.

[CR93] R. Cools and P. Rabinowitz, *Monomial cubature rules since 'Stroud': A compilation*, J. Comput. Appl. Math. **48** (1993), 309–326.

[dD79] E. de Doncker, *New Euler-Maclaurin expansions and their application to quadrature over the s-dimensional simplex*, Math.Comp. **33** (1979), 1003–1018.

[DR84] P.J. Davis and P. Rabinowitz, *Methods of numerical integration*, Academic Press, London, 1984.

[Duf82] M.G. Duffy, *Quadrature over a pyramid or cube of integrands with a singularity at a vertex*, SIAM J. Numer. Anal. **19** (1982), 1260–1262.

[Dun85] D.A. Dunavant, *High degree efficient symmetrical Gaussian quadrature rules for the triangle*, Internat. J. Numer. Methods Engrg. **21** (1985), 1129–1148.

[Eng70] H. Engels, *Über gleichgewichtete Kubaturformeln für ein Dreiecksgebiet*, Elek. Daten. **12** (1970), 535–539.

[Eng80] H. Engels, *Numerical quadrature and cubature*, Academic Press, London, 1980.

[Fra71] R. Franke, *Obtaining cubatures for rectangles and other planar regions by using orthogonal polynomials*, Math. Comp. **25** (1971), 803–813.

[Gat88] K. Gatermann, *The construction of symmetric cubature formulas for the square and the triangle*, Computing **40** (1988), 229–240.

[Gat90] K. Gatermann, *Gruppentheoretische Konstruktion von symmetrischen Kubaturformeln*, Technical Report TR 90-1, Konrad-Zuse-Zentrum für Informationstechnik Berlin, 1990.

[Gat92] K. Gatermann, *Linear representations of finite groups and the ideal theoretical construction of G-invariant cubature formulas*, Numerical Integration – Recent Developments, Software and Applications (Dordrecht) (T.O. Espelid and A. Genz, eds.), NATO ASI Series C: Math. and Phys. Sciences, vol. 357, Kluwer Academic Publishers, 1992, pp. 25–35.

[GM78] A. Grundmann and H.M. Möller, *Invariant integration formulas for the n-simplex by combinatorial methods*, SIAM J. Numer. Anal. **15** (1978), 282–290.

[Gün75] C. Günther, *Zweidimensionale Quadraturformeln vom Grad 7 mit 14 Punkten*, Numer. Math. **24** (1975), 309–316.

[Hae93] A. Haegemans, *Cubature formulas for triangles and squares with a 1/r singularity*, Report TW 192, Dept. of Compter Science, K.U. Leuven, 1993.

[Hem73] P.W. Hemker, *A sequence of nested cubature rules*, NW 3/73, Stichting Mathematisch Centrum, 1973.

[Hil77] P. Hillion, *Numerical integration on a triangle*, Internat. J. Numer. Methods Engrg. **11** (1977), 797–815.

[HK85] M. Hollosi and P. Keast, *Numerical integration of singular functions over a triangle*, Technical Report 1985CS-3, Computing Science Division, Dalhousie University, 1985.

[HMS56] P.C. Hammer, O.J. Marlowe, and A.H. Stroud, *Numerical integration over simplexes and cones*, Math. Tables Aids Comput. **10** (1956), 130–137.

[HS56] P.C. Hammer and A.H. Stroud, *Numerical integration over simplexes*, Math. Tables Aids Comput. **10** (1956), 137–139.

[Kea92] P. Keast, *Software for integration over triangles and general simplices*, Numerical Integration – Recent Developments, Software and Applications (Dordrecht) (T.O. Espelid and A. Genz, eds.), NATO ASI Series C: Math. and Phys. Sciences, vol. 357, Kluwer Academic Publishers, 1992, pp. 283–294.

[Lau55] R. Lauffer, *Interpolation mehrfacher Integrale*, Arch. Math. **6** (1955), 159–164.

[Lau82] D.P. Laurie, *CUBTRI - automatic cubature over a triangle*, ACM Trans. Math. Software **8** (1982), 210–218.

[LdD93] J.N. Lyness and E. de Doncker, *Quadrature error expansions. II. The full corner singularity*, Numer. Math. **64** (1993), 355–370.

[LG78] M.E. Laursen and M. Gellert, *Some criteria for numerically integrated matrices and quadrature formulas for triangles*, Internat. J. Numer. Methods Engrg. **12** (1978), 67–76.

[LG82] J.N. Lyness and L. Gatteschi, *A note on cubature over a triangle of a function having specified singularities*, Numerical Integration (Basel) (G. Hämmerlin, ed.), Internat. Ser. Appl. Math., vol. 57, Birkhäuser Verlag, 1982, pp. 164–169.

[LJ75] J.N. Lyness and D. Jespersen, *Moderate degree symmetric quadrature rules for the triangle*, J. Inst. Math. Appl. **15** (1975), 19–32.

[LM70] J.N. Lyness and B.J.J. McHugh, *On the remainder term in the N-dimensional Euler-Maclaurin expansion*, Numer. Math. **15** (1970), 333–344.

[LM80] J.N. Lyness and G. Monegato, *Quadrature error functional expansion for the simplex when the integrand has singularities at vertices*, Math. Comp. **34** (1980), 213–225.

[LP73] J.N. Lyness and K.K. Puri, *The Euler-Maclaurin expansion for the simplex*, Math. Comp. **27** (1973), 273–293.

[Lyn76] J.N. Lyness, *An error functional expansion for N-dimensional quadrature with an integrand function singular at a point*, Math. Comp. **30** (1976), 1–23.

[Lyn78] J.N. Lyness, *Quadrature over a simplex*, SIAM J. Numer. Anal. **15** (1978), 122–133 and 870–877.

[Lyn83] J.N. Lyness, *QUG2 - integration over a triangle*, ANL/MCS-TM- 13, Mathematics and Computer Science Division, Argonne National Laboratory, 1983.

[Lyn92] J.N. Lyness, *On handling singularities in finite elements*, Numerical Integration – Recent Developments, Software and Applications (Dordrecht) (T.O. Espelid and A. Genz, eds.), NATO ASI Series C: Math. and Phys. Sciences, vol. 357, Kluwer Academic Publishers, 1992, pp. 219–233.

[Moa74] T. Moan, *Experiences with orthogonal polynomials and "best" numerical integration formulas on a triangle; with particular reference to finite element approximations*, Z. Angew. Math. Mech. **54** (1974), 501–508.

[Möl73] H.M. Möller, *Polynomideale und Kubaturformeln*, Ph.D. thesis, Universität Dortmund, 1973.

[Möl76] H.M. Möller, *Kubaturformeln mit minimaler Knotenzahl*, Numer. Math. **25** (1976), 185–200.

[Mys68] I.P. Mysovskikh, *On the construction of cubature formulas with fewest nodes*, Dokl. Akad. Nauk SSSR **178** (1968), 1252–1254 (Russian). Soviet Math. Dokl. **9** (1968) 277–280 (English).

[Mys81] I.P. Mysovskikh, *Interpolatory cubature formulas*, Izdat. 'Nauka', Moscow-Leningrad, 1981 (Russian).

[PFB81] H.L.G. Pina, J.L.M. Fernandes, and C.A. Brebbia, *Some numerical integration formulae over triangles and squares with a $1/r$ singularity*, Appl. Math. Modelling **5** (1981), 209–211.

[Rad48] J. Radon, *Zur mechanischen Kubatur*, Monatsh. Math. **52** (1948), 286–300.

[Ras83a] G.G. Rasputin, *On the construction of cubature formulas containing preassigned knots*, Metody Vychisl. **13** (1983), 122–128 (Russian).

[Ras83b] G.G. Rasputin, *On the question of numerical characteristics for orthogonal polynomials of two variables*, Metody Vychisl. **13** (1983), 145–154.

[Ras86] G.G. Rasputin, *Construction of cubature formulas containing preassigned nodes*, Izv. Vyssh. Uchebn. Zaved. Mat. **11** (1986), 44–51 (Russian). Soviet Math. (Iz. VUZ) **30** (1986) 58–67 (English).

[Ric27] L.F. Richardson, *The deferred approach to the limit*, Philos. Trans. Royal Soc. London **226** (1927), 261–299.

[Rom55] W. Romberg, *Vereinfachte numerische Integration*, Norske Vid. Selsk. Forh. (Trondheim) **28** (1955), 30–36.

[Sch83] H.J. Schmid, *Interpolatorische Kubaturformeln*, Dissertationes Math. **CCXX** (1983).

[SF73] G. Strang and G.J. Fix, *An Analysis of the Finite Element Method*, Series in Automatic Computation, Prentice-Hall, London, 1973.

[Sid79] A. Sidi, *Some properties of a generalization of the Richardson extrapolation process*, J. Inst. Math. Appl. **24** (1979), 327–346.

[Sid83] A. Sidi, *Euler-Maclaurin expansions for integrals over triangles and squares of functions having algebraic/logarithmic singularities along an edge*, J. Approx. Theory **39** (1983), 39–53.

[Sid90] A. Sidi, *On a generalization of the Richardson extrapolation process*, Numer. Math. **57** (1990), 365–377.

[Sil70] P. Silvester, *Symmetric quadrature formulae for simplexes*, Math. Comp. **24** (1970), 95–100.

[Sob62] S.L. Sobolev, *Cubature formulas on the sphere invariant under finite groups of rotations*, Dokl. Akad. Nauk SSSR **146** (1962), 310–313 (Russian). Soviet Math. Dokl. **3** (1962) 1307–1310 (English).

[SS66] A.H. Stroud and D. Secrest, *Gaussian quadrature formulas*, Prentice-Hall, Englewood Cliffs, N.J., 1966.

[Str60] A.H. Stroud, *Quadrature methods for functions of more than one variable*, New York Acad. Sci. **86** (1960), 776–791.

[Str61] A.H. Stroud, *Numerical integration formulas of degree 3 for product regions and cones*, Math. Comp. **15** (1961), 143–150.

[Str64] A.H. Stroud, *Approximate integration formulas of degree 3 for simplexes*, Math. Comp. **18** (1964), 590–597.

[Str66] A.H. Stroud, *Some approximate integration formulas of degree 3 for an n-dimensional simplex*, Numer. Math. **9** (1966), 38–45.

[Str69] A.H. Stroud, *A fifth degree integration formula for the n-simplex*, SIAM J. Numer. Anal. **6** (1969), 90–98.

[Str71] A.H. Stroud, *Approximate calculation of multiple integrals*, Prentice-Hall, Englewood Cliffs, N.J., 1971.

[VH93] P. Verlinden and A. Haegemans, *An error expansion for cubature with an integrand with homogeneous boundary singularities*, Numer. Math. **65** (1993), 383–406.

MATHEMATICS AND COMPUTER SCIENCE DIVISION, ARGONNE NATIONAL LABORATORY, 9700 SOUTH CASS AVENUE, ARGONNE, IL 60439, U.S.A.

E-mail address: lyness@mcs.anl.gov

DEPARTMENT OF COMPUTER SCIENCE, KATHOLIEKE UNIVERSITEIT LEUVEN, CELESTIJNENLAAN 200A, B-3001 HEVERLEE, BELGIUM

E-mail address: ronald.cools@cs.kuleuven.ac.be

Proceedings of Symposia in Applied Mathematics
Volume 48, 1994

New trends in the use and analysis of integral equations

J.C. NÉDÉLEC

ABSTRACT. This work is a survey paper on the use and numerical analysis of integral equations. Our emphasis is on hypersingular operators. We also present some new developments, especially those related to the delayed potential for the wave equation.

1. Introduction

The study and use of integral equations for solving partial differential equations is a very old subject that began during the last century. We will not try to describe any part of this classical theory. We can refer to the classical treatise of Kellogg [23] . The essential tools were the theory of compact operators and the alternative of Fredholm.

A turn in the subject appears with the study of singular integrals, which began with the work of Giraud [18] . This subject has subsequently been developed by many authors, among them Stein, Calderon and Zygmund in the U.S.A., and Mikhlin, Muskhelishvili in the U.S.S.R.. Most integral equations related to elliptic problems lead to the study of singular integrals on compact manifolds. In that sense the works of Seeley [34, 35] are also seminal studies.

The main concern in the numerical analysis of any problem is the stability of the scheme. For linear elliptic problems, the stability is easily obtained as soon as we use the Galerkin method in a variational formulation of the problem in a Hilbert space, satisfying a coercivity property. This framework can be used for most elliptic integral equations.

The variational formulation for integral equations is always related to the variational formulation of the original problem, and also to the order of the operator in the theory of elliptic singular integrals.

1991 *Mathematics Subject Classification.* Primary 45A05,45Exx; Secondary 65N38.
This paper is in final form and no version of it will be submitted for publication elsewhere.

We will emphasize the derivation of these variational formulations in the case of hypersingular operators (where some geometrical tools are needed).

When studying Helmholtz's equation for exterior problems, the associated integral equations turn out to be compact perturbations of the ones associated with the Laplace equation. This property and the use of a variational formulation lead to stability and error analysis. By means of this technique, the result depends heavily on the wave number k.

This method can be extended with some care to the case of Maxwell's equation. New developments then occur in the use of integral equations in the time-dependent case.

Some work has been done for the heat equation, and a satisfactory mathematical framework exists in that case.

The wave equation leads to the classical delayed potential. The associated integral equations involve no longer pseudodifferential operators of classical type. In that case the mathematical framework available is rather weak. Nevertheless, the use of variational techniques leads to stable numerical schemes and nice results.

This can be extended with some difficulty to Maxwell's equations. There is an urgent need of any type of analysis of the stability criterion for such schemes.

2. Equations Related to the Laplace Operator

We will introduce here the integral representations of the solutions of the Helmholtz equation. When the wave number k is zero, we obtain the Laplace operator.

2.1. Integral representation of solutions. For the sake of simplicity, we will consider only the case of bounded domains in $\mathbb{R}^3$. The interior Ω_i is delimited by its boundary Γ, which is a regular closed bounded surface in $\mathbb{R}^3$.

The unit normal n is exterior to the interior domain. The exterior domain is denoted Ω_e.

The variable will be x or y, and the modulus is $|x| = r = \sqrt{x_1^2 + x_2^2 + x_3^2}$.

The following theorem is the basic tool for establishing all integral equations related to the Helmholtz equation.

The fundamental solution of this equation in $\mathbb{R}^3$ is

$$E(x) = \frac{1}{4\pi}\frac{e^{ikr}}{r}, \quad \text{where } k \text{ is the wave number.} \tag{2.1}$$

THEOREM 1. *Let u, regular in $\bar{\Omega}_i$ and $\bar{\Omega}_e$, be such that*

$$\Delta u + k^2 u = 0 \quad \text{in } \Omega_i\,, \tag{2.2}$$

$$\Delta u + k^2 u = 0 \quad \text{in } \Omega_e\,. \tag{2.3}$$

Let $[\,\cdot\,]$ denote the jump accros Γ in the direction of the normal,

$$[u] = u_{\text{int}} - u_{\text{ext}}, \tag{2.4}$$

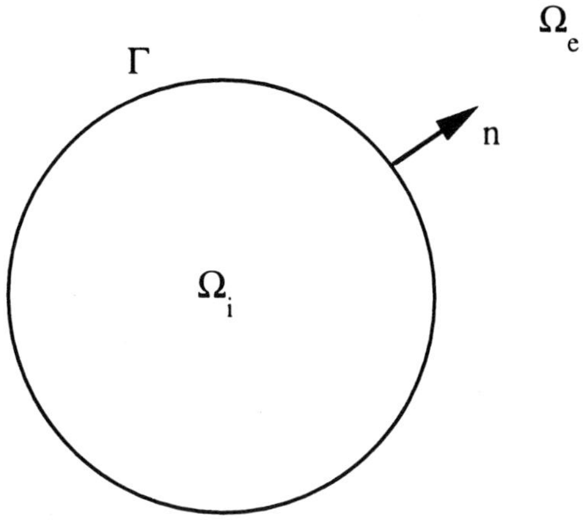

Figure 1

$$\left[\frac{\partial u}{\partial n}\right] = \left.\frac{\partial u}{\partial n}\right|_{\text{int}} - \left.\frac{\partial u}{\partial n}\right|_{\text{ext}} . \tag{2.5}$$

Let u satisfy the radiation condition in the exterior domain or, equivalently,

$$u \in H;\ H = \left\{v;\ \frac{v}{r} \in L^2(\Omega_e);\ \frac{\nabla v}{r} \in L^2(\Omega_e);\ \frac{\partial u}{\partial r} - iku \in L^2(\Omega_e)\right\} . \tag{2.6}$$

Then for $y \notin \Gamma$ we have

$$u(y) = \int_\Gamma E(x-y)\left[\frac{\partial u}{\partial n}(x)\right] d\gamma(x) - \int_\Gamma \frac{\partial}{\partial n_x}(E(x-y))[u(x)]d\gamma(x) , \tag{2.7}$$

where $d\gamma(x)$ is the surface element, and for $y \in \Gamma$,

$$\left\{\begin{aligned} \frac{u_{\text{int}}(y) + u_{\text{ext}}(y)}{2} &= \int_\Gamma E(x-y)\left[\frac{\partial u}{\partial n}(x)\right] d\gamma(x) \\ &\quad - \int_\Gamma \frac{\partial}{\partial n_x}(E(x-y))[u(x)]d\gamma(x). \end{aligned}\right. \tag{2.8}$$ □

The ***single***-*layer potential* associated with a function q is

$$u(y) = Sq = \int_\Gamma E(x-y)q(x)d\gamma(x). \tag{2.9}$$

The *double-layer potential* associated with a function φ is

$$u(y) = D\varphi = \int_\Gamma \frac{\partial}{\partial n_x}(E(x-y))\varphi(x)d\gamma(x). \tag{2.10}$$

The following operators are also relevant:

$$u(y) = D^*q = \int_\Gamma \frac{\partial}{\partial n_y}(E(x-y))q(x)d\gamma(x), \tag{2.11}$$

$$u(y) = N\varphi = \oint_\Gamma \frac{\partial^2}{\partial n_x \partial n_y}(E(x-y))\varphi(x)d\gamma(x). \tag{2.12}$$

The operator N is not a usual Lebesgue integral but a finite part to be defined later in §2.2.2. It corresponds to a hypersingular kernel in the language of pseudodifferential operators.

The four operators give rise to the so-called Calderon Projectors C_{int} and C_{ext} defined by

$$H = \begin{pmatrix} -D & S \\ -N & D^* \end{pmatrix}, \tag{2.13}$$

$$C_{\text{int}} = \frac{I}{2} + H, \tag{2.14}$$

$$C_{\text{ext}} = \frac{I}{2} - H, \tag{2.15}$$

and we have

$$C_{\text{ext}}^2 = C_{\text{ext}}, \quad C_{\text{int}}^2 = C_{\text{int}}, \quad C_{\text{int}} + C_{\text{ext}} = I. \tag{2.16}$$

These last properties are also related to the expression of the normal derivatives of a single-layer potential and to the value on Γ of the discontinuous double-layer potential given by

THEOREM 2. *The normal derivatives at the surface Γ of a single-layer potential* (2.9) *are*

$$\left.\frac{\partial u}{\partial n}\right|_{\text{int}}(y) = \frac{q(y)}{2} + \int_\Gamma \frac{\partial}{\partial n_y}(E(x-y))q(x)d\gamma(x), \tag{2.17}$$

$$\left.\frac{\partial u}{\partial n}\right|_{\text{ext}}(y) = -\frac{q(y)}{2} + \int_\Gamma \frac{\partial}{\partial n_y}(E(x-y))q(x)d\gamma(x). \tag{2.18}$$

The values of a double-layer potential (2.10) *on the surface Γ are*

$$u_{\text{int}}(y) = -\frac{\varphi(y)}{2} + \int_\Gamma \frac{\partial}{\partial n_x}(E(x-y))\varphi(x)d\gamma(x), \tag{2.19}$$

$$u_{\text{ext}}(y) = \frac{\varphi(y)}{2} + \int_\Gamma \frac{\partial}{\partial n_x}(E(x-y))\varphi(x)d\gamma(x). \tag{2.20}$$

2.2. Integral equations for the Laplace operator. When the wave number k is zero, the fundamental solution E is just associated with the Laplace operator,

$$E(x) = \frac{1}{4\pi r}. \tag{2.21}$$

2.2.1. *The operator S.* When solving the *Dirichlet interior or exterior problem* using a **single**-layer potential, we obtain the following basic integral equation:

$$
(2.22)\qquad \begin{cases} \dfrac{1}{4\pi}\displaystyle\int_\Gamma \frac{1}{|x-y|} q(x)d\gamma(x) = u_d(y)\ ,\ y\in\Gamma\ ,\ \text{or} \\ Sq = u_d. \end{cases}
$$

Using the results on pseudodifferential operators on compact manifolds, we can see that the operator S is well defined and continuous on any Sobolev space on the surface Γ. Moreover, it takes $H^s(r)$ into $H^{s+1}(r)$. Its principal symbol is the one for a flat surface, which is $\frac{1}{|\xi|}$. So it is an elliptic pseudodifferential operator. We need more to obtain numerical stability results. The space $H^{-1/2}(\Gamma)$ plays a distinct role in the sense that S operates from this space to $H^{1/2}(\Gamma)$, which is the dual space. So it is quite natural to replace equation (2.22) by the following variational formulation:

$$
(2.23)\qquad \begin{cases} \dfrac{1}{4\pi}\displaystyle\int_\Gamma\int_\Gamma \frac{1}{|x-y|} q(x)q^t(y)d\gamma(x)d\gamma(y) \\ \qquad = \displaystyle\int_\Gamma u_d(y)q^t(y)d\gamma(y)\ \ \forall q^t \in H^{-1/2}(\Gamma). \end{cases}
$$

It was first proved by Nédélec and Planchard [30] that the above bilinear form is coercive in $H^{-1/2}(\Gamma)$, i.e.,

$$
(2.24)\qquad \begin{cases} \dfrac{1}{4\pi}\displaystyle\int_\Gamma\int_\Gamma \frac{q(x)q(y)}{|x-y|} d\gamma(x)d\gamma(y) \ \ \geq \alpha||q||^2_{H^{-1/2}(\Gamma)} \\ \qquad\qquad \forall q \in H^{-1/2}(\Gamma)\ , \text{where } \alpha > 0. \end{cases}
$$

So the operator S is an isomorphism from $H^{-1/2}(\Gamma)$ onto $H^{1/2}(\Gamma)$.

The corresponding result in the case of two dimensions is associated with the potential $-\frac{1}{2\pi}\text{Log}r$. Le Roux [26] proved that in the 2D case we have

$$
(2.25)\qquad \begin{cases} \dfrac{1}{2\pi}\displaystyle\int_\Gamma \int_\Gamma \text{Log}|x-y| q(x)q(y)d\gamma(y) \geq \alpha||q||^2_{H^{-1/2}(\Gamma)} \\ \qquad \forall\ q\ \in H^{-1/2}(\Gamma) \text{ such that } \displaystyle\int_\Gamma q d\gamma = 0, \text{ where } \alpha > 0. \end{cases}
$$

A slightly different form of this result was obtained independently a little later by Hsiao and Wendland [21] .

With the variational formulation (2.23) at hand, it is quite natural to approximate the problem using a Galerkin method.

Strictly speaking, we must take into account the approximation of the surface Γ by an approximate one. This can be done by classical finite elements. For instance, if we use $\mathbb{P}_1$ finite elements, the surface will be a union of connected flat triangles. In that case, the simplest Galerkin space for approximating $H^{-1/2}(\Gamma_h)$ consists of having one constant unknown for each triangle. Denote this space by

V_h. Then the approximate problem is

$$(2.26)\qquad \begin{cases} \dfrac{1}{4\pi}\displaystyle\int_{\Gamma_h}\int_{\Gamma_h}\frac{1}{|x-y|}q_h(x)q_h^t(y)d\gamma_h(x)d\gamma_h(y) = \\ \qquad\qquad\displaystyle\int_{\Gamma_h} u_{dh}q_h^t d\gamma_h \quad \forall q_h^t \in V_h\,. \end{cases}$$

In Nédélec [29], we were able to prove the following error estimate, using the key property (2.24):

$$(2.27)\qquad \begin{cases} \|q-\tilde{q}_h\|_{L^2(\Gamma)} \le ch, & \text{where } h \text{ is the small parameter} \\ & \text{associated with the mesh.} \end{cases}$$

The $\tilde{q}_h$ is defined on the surface Γ by a kind of local projection of the surface Γ_h onto the surface Γ.

If we associate with the approximate density q_h the approximate potential u_h given by

$$(2.28)\qquad u_h(y) = \frac{1}{4\pi}\int_{\Gamma_h}\frac{q_h(x)}{|x-y|}d\gamma(x)\ ,$$

we gain an extra power of convergence, for any y not close to the surface Γ_h, i.e.,

$$(2.29)\qquad |u(y) - u_h(y)| \le ch^2\ .$$

So the Galerkin method seems very attractive. In spite of that, most *engineers continue* to use the *collocation method* for this problem. In the case under consideration here, the latter can be interpreted as follows: we need to compute the exact coefficients of the matrix arising from (2.26), which are

$$(2.30)\qquad \begin{cases} a_{ij} = \dfrac{1}{4\pi}\displaystyle\int_{T_i}\int_{T_j}\frac{1}{|x-y|}d\gamma(x)d\gamma(y), \\ T_i \text{ and } T_j \text{ are triangles of the approximate surface.} \end{cases}$$

If we decide to approximate these coefficients by

$$(2.31)\qquad \tilde{a}_{ij} = \frac{1}{4\pi\ \text{meas}(T_i)}\int_{T_j}\frac{1}{|x-y_i|}d\gamma(x)\ ,$$

this is the collocation method.

It is quite clear that the difference between the two coefficients is not large when T_i and T_j are not close. It becomes important only for $T_i = T_j$ or T_j adjacent to T_i. Nevertheless, there exists no proof of stability (and convergence) of this collocation method in the $3D$ case. In the $2D$ case, the situation is much better, and there are some results due to Arnold and Wendland [2,3] and several other authors, including Prössdorf and Rathsfeld [32] and more recently, **von Petersdorff**.

2.2.2. *The operators* $\frac{I}{2} \pm D$ *and* $\frac{I}{2} \pm D^*$. When solving a Dirichlet problem using a double-layer potential, or a Neumann problem using a **single**-layer potential, the operators $\frac{I}{2} \pm D$ and $\frac{I}{2} \pm D^*$ arise. The operators D and D^* are adjoint to each other. They are pseudodifferential operators of order -1; that means they are continuous from $H^s(\Gamma)$ into $H^{s+1}(\Gamma)$, and so compact from $L^2(\Gamma)$ in $L^2(\Gamma)$. This proves that the Fredholm alternative is applicable to the above operators. The eigenfunctions and eigenvalues can be related to those of the interior problem, leading to the isomorphism results. We will not further develop this theory, which is very classical and well known.

Again, concerning the approximation, we can use the Galerkin method. The error analysis is then based on the continuity of the operator D (or D^*) from $L^2(\Gamma)$ into $H^1(\Gamma)$, which allows an approximation in the norm of the operator from $L^2(\Gamma)$ into $L^2(\Gamma)$.

Most engineers use the collocation method in that case too. Although the situation is a bit better than for the operator S, there are no clear stability results for this method.

These formulations are nonsymmetric.

2.2.3. *The operator N.* When solving the interior or exterior Neumann problem using a double-layer potential, we obtain formally the equation

$$-\frac{1}{4\pi}\oint_\Gamma \frac{\partial^2}{\partial n_x \partial n_y}\left(\frac{1}{|x-y|}\right)\varphi(x)d\gamma(x) = u_n(y)\ ,\ y\in\Gamma. \tag{2.32}$$

The kernel is

$$\frac{\partial^2}{\partial n_x \partial n_y}\frac{1}{|x-y|} = \frac{(n_x\cdot n_y)}{|x-y|^3} - 3\frac{(n_x\cdot x-y)(n_y\cdot x-y)}{|x-y|^5}. \tag{2.33}$$

It is odd and nonintegrable. So the operator is a hypersingular integral operator.

The pseudodifferential theory shows that it is an elliptic operator of order 1 and of principal symbol $|\xi|$. So it maps $H^s(\Gamma)$ into $H^{s-1}(\Gamma)$.

From these results, it is clear that $H^{1/2}(\Gamma)$ plays a special role since $-N$ operates from $H^{1/2}(\Gamma)$ into its dual space.

A new difficulty consists in evaluating the principal value expression. This is curiously known, in the engineering literature, under the name of "regularization" . We need some differential geometry tools in order to accomplish this regularization.

Let our surface Γ be regular. In a neighborhood of Γ any point y has a unique projection on Γ denoted by $\wp(y)$.

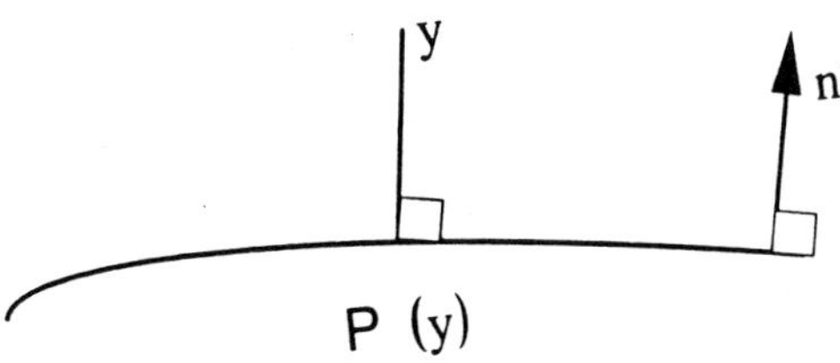

Figure 2

With any function defined on Γ, we associate the extension

$$\tilde{\varphi}(y) = \varphi(\wp(y)). \tag{2.34}$$

Then the tangential gradient of φ is

$$\nabla_\Gamma \varphi = \nabla \tilde{\varphi}|_\Gamma . \tag{2.35}$$

The $\overrightarrow{\text{curl}}$ of φ is

$$\overrightarrow{\text{curl}}_\Gamma\, \varphi = \text{curl}\, (\varphi \vec{n}) = \nabla_\Gamma \varphi \wedge n. \tag{2.36}$$

Let now v be a vector defined in the tangent plane to the surface Γ. We extend it, using the curvature operator:

$$y = x + sn(x)\ , \quad x \in \Gamma, \tag{2.37}$$

$$\Re_s = \nabla n(y), \tag{2.38}$$

$$\tilde{v}(y) = v(x) - s\Re_s v(x). \tag{2.39}$$

Then we define the curl_Γ of the vector as

$$\text{curl}_\Gamma\, v = (\text{curl}\, \tilde{v} \cdot n)|_\Gamma \tag{2.40}$$

and the divergence as

$$\text{div}_\Gamma\, v = \text{div}\, \tilde{v}|_\Gamma . \tag{2.41}$$

The Laplace-Beltrami operator is

$$\Delta_\Gamma u = \text{div}_\Gamma \nabla_\Gamma u = -\text{curl}_\Gamma\, \overrightarrow{\text{curl}}_\Gamma u. \tag{2.42}$$

The operators $\overrightarrow{\text{curl}}_\Gamma$ and curl_Γ are adjoint, that is,

$$\int_\Gamma (\overrightarrow{\text{curl}}_\Gamma\, u \cdot v)\, d\gamma - \int_\Gamma u\, \text{curl}_\Gamma\, v\, d\gamma = 0. \tag{2.43}$$

We are now in a position to give the result of regularization.

THEOREM 3. *The operator $-N$ is an isomorphism from $H^{1/2}(\Gamma)/R$ onto the subspace of $H^{-1/2}(\Gamma)$ such that $(q, 1) = 0$.*

The problem (2.32) *admits the following variational formulation*:

$$\frac{1}{4\pi}\int_\Gamma\int_\Gamma \frac{(\overrightarrow{\text{curl}}_\Gamma \varphi(x) \cdot \overrightarrow{\text{curl}}_\Gamma \psi(y))}{|x-y|} d\gamma(x) d\gamma(y) = \int_\Gamma u_n \psi\, d\gamma \qquad \forall \psi \in H^{1/2}(\Gamma)/R. \tag{2.44}$$

We also have

$$\left\{ \begin{array}{l} \dfrac{1}{4\pi}\displaystyle\int_\Gamma\int_\Gamma \frac{(\overrightarrow{\text{curl}}_\Gamma \varphi(x) \cdot \overrightarrow{\text{curl}}_\Gamma \varphi(y))}{|x-y|} d\gamma(x) d\gamma(y) \geq \alpha ||\varphi||^2_{H^{1/2}(\Gamma)/R} \\ \qquad \forall \varphi \in H^{1/2}(\Gamma),\ \textit{where}\ \alpha > 0\ . \end{array} \right. \tag{2.45}$$

An expression for $-N$, where only integrable kernels appear, is

$$
(2.46)\qquad \left\{\begin{aligned}
-N\varphi = \frac{1}{4\pi}\Bigg[&-\int_\Gamma \frac{\Delta_\Gamma\varphi(x)}{|x-y|}d\gamma(x)\\
&+\int_\Gamma\left(\nabla_\Gamma\varphi(x)\cdot\nabla_\Gamma\frac{1}{|x-y|}\right)(1-(n_x\cdot n_y))d\gamma(x)\\
&+\int_\Gamma\frac{\partial}{\partial n_x}\left(\frac{1}{|x-y|}\right)(\nabla_\Gamma\varphi(x)\cdot n_y)d\gamma(x)\Bigg].\quad\square
\end{aligned}\right.
$$

The variational approximation for this problem is now straightforward. We introduce an approximate surface Γ_h, say a union of flat triangles ($\mathbb{P}_1$ elements). The "natural" approximate subspace of $H^{1/2}(\Gamma_h)$ is the space of functions of degree 1 on each triangle and continuous ($\mathbb{P}_1$ element). The approximate problem is

$$
(2.47)\qquad \left\{\begin{aligned}
&\frac{1}{4\pi}\int_{\Gamma_h}\int_{\Gamma_h}\frac{(\overrightarrow{\mathrm{curl}}_\Gamma\varphi_h(x)\cdot\overrightarrow{\mathrm{curl}}_\Gamma\psi_h(y))}{|x-y|}d\gamma_h(x)d\gamma_h(y)\\
&\qquad=\int_{\Gamma_h}u_{n_h}\psi_h d\gamma_h \qquad \forall\psi_h\in V_h\,/\,R.
\end{aligned}\right.
$$

We are able to prove the following error estimates:

$$
(2.48)\qquad |\varphi-\tilde{\varphi}_h|_{L^2(\Gamma)/R}\le c_1h^2+c_2h,
$$

and for the potential

$$
(2.49)\qquad |u(y)-u_h(y)|\le ch^2.
$$

The linear system is full, symmetric and positive definite.

2.3. Coupling procedure. Suppose we have to solve in an exterior domain a variable-coefficient problem such as

$$
(2.50)\qquad \left\{\begin{aligned}
-\mathrm{div}(a(x)\nabla u) &= f(x), \quad x\in\Omega,\\
u|_{\Gamma_0} &= u_d,
\end{aligned}\right.
$$

and let us also suppose that, at the exterior of a bounded surface Γ, the coefficient a is identically 1, and that $f\equiv 0$ in this domain. Let Ω_2 be the exterior domain, and let $\Omega_1=\Omega/\Omega_2$.

It is then possible to represent the solution in the exterior of Γ by an integral formula and thus obtain an equivalent problem. There are several ways of doing this, each related to an integral representation.

For instance, in Johnson and Nédélec [22], we chose to represent the exterior solution as a **single** layer,

$$
(2.51)\qquad u(y)=\frac{1}{4\pi}\int_\Gamma\frac{q(x)}{|x-y|}d\gamma(x),
$$

$$
(2.52)\qquad \frac{\partial u}{\partial n}(y)\bigg|_{\mathrm{ext}}=-\frac{q(y)}{2}+\frac{1}{4\pi}\int_\Gamma\frac{\partial}{\partial n_y}\left(\frac{1}{|x-y|}\right)q(x)d\gamma(x).
$$

The equivalence between the original problem and the new one consists in saying that u and $\frac{\partial u}{\partial n}$ are the same on Γ.

The variational formulation gives rise to a system

$$(2.53)\quad \begin{cases} \displaystyle\int_{\Omega_1}(a(x)\nabla u\cdot\nabla v)dx+\frac{1}{2}\int_\Gamma qv d\gamma \\ \displaystyle\qquad -\frac{1}{4\pi}\int_\Gamma\int_\Gamma \frac{\partial}{\partial n_y}\frac{1}{|x-y|}q(x)v(y)d\gamma(x)d\gamma(y) \\ \displaystyle\qquad =\int_{\Omega_1} fv dx \qquad \forall v\in H^1(\Omega_1)\,,\ v|_{\Gamma_0}=0\,; \\ \displaystyle\frac{1}{4\pi}\int_\Gamma\int_\Gamma \frac{q(x)q^t(y)}{|x-y|}d\gamma(x)d\gamma(y)-\int_\Gamma uq^t d\gamma=0 \\ \displaystyle\qquad\qquad \forall q^t\in H^{-1/2}(\Gamma)\,. \end{cases}$$

This system in (v,q) is a compact perturbation of a symmetric coercive system. So, if we introduce the usual variational formulation consisting of $\mathbb{P}_1$ finite elements for the first part, and the above approximation for the integral part, we are able to prove stability results and obtain some error analysis.

This way of proceeding has the disadvantage of being nonsymmetric.

Costabel [12], and subsequently Costabel and Stephan [13], have introduced in this case, and also for other applications, a symmetric system.

They start from the Green representation formula, which is

$$(2.54)\quad u(y)=-\frac{1}{4\pi}\int_\Gamma\frac{1}{|x-y|}\frac{\partial u}{\partial n}(x)d\gamma(x)+\frac{1}{4\pi}\int_\Gamma\frac{\partial}{\partial n_x}\frac{1}{|x-y|}u(x)d\gamma(x).$$

We have, on the surface Γ,

$$(2.55)\quad \begin{cases} \displaystyle u(y)=\frac{1}{2}u(y)-\frac{1}{4\pi}\int_\Gamma\frac{1}{|x-y|}\frac{\partial u}{\partial n}(x)d\gamma(x) \\ \displaystyle\qquad +\frac{1}{4\pi}\int_\Gamma\frac{\partial}{\partial n_x}\left(\frac{1}{|x-y|}\right)u(x)d\gamma(x), \end{cases}$$

$$(2.56)\quad \begin{cases} \displaystyle\frac{\partial u}{\partial n}(y)=\frac{1}{2}\frac{\partial u}{\partial n}(y) \\ \displaystyle -\frac{1}{4\pi}\int_\Gamma\frac{\partial}{\partial n_x}\left(\frac{1}{|x-y|}\right)\frac{\partial u}{\partial n}(x)d\gamma(x) \\ \displaystyle\qquad +\frac{1}{4\pi}\oint_\Gamma\frac{\partial^2}{\partial n_x\partial n_y}\left(\frac{1}{|x-y|}\right)u(x)d\gamma(x). \end{cases}$$

The variational formulation is

$$(2.57)\quad \begin{cases} \displaystyle\int_{\Omega_1}(a(x)\nabla u\cdot\nabla v)dx-\frac{1}{2}\int_\Gamma\frac{\partial u}{\partial n}v d\gamma \\ \displaystyle +\frac{1}{4\pi}\int_\Gamma\int_\Gamma\frac{\partial}{\partial n_y}\left(\frac{1}{|x-y|}\right)\frac{\partial u}{\partial n}(x)v(y)d\gamma(x)d\gamma(y) \\ \displaystyle \qquad -(Nu,v)=\int_\Omega fvdx \qquad \forall v\in H^1(\Omega_1), v|_\Gamma=0; \\ \displaystyle -\frac{1}{2}\int_\Gamma u\psi d\gamma+\frac{1}{4\pi}\int_\Gamma\int_\Gamma\frac{\partial}{\partial n_x}\left(\frac{1}{|x-y|}\right)u(x)\psi(y)d\gamma(x)d\gamma(y) \\ \displaystyle -\frac{1}{4\pi}\int_\Gamma\int_\Gamma\frac{\frac{\partial u}{\partial n}(x)\psi(y)}{|x-y|}d\gamma(x)d\gamma(y)=0 \qquad \forall\psi\in H^{-1/2}(\Gamma). \end{cases}$$

This system is now fully symmetric in the two unknowns u and $\frac{\partial u}{\partial n}$ and coercive. So it leads to stable numerical approximations.

Levillain in his thesis [25] has implemented several versions of this coupling procedure for a system of Maxwell equations. It appears clearly that the symmetric ones are more stable and are much less sensitive to the problem of interior eigenvalues. There is yet no analysis of this observed qualitative property.

3. Equations Related to the Helmholtz Equation

The main structure of the integral equations associated with the Helmholtz equations are very similar to those associated with the Laplace equation. But the physics of these problems, at least for exterior problems, is very different.

For instance, there are now several fundamental solutions. The one we have chosen is

$$(3.1)\qquad E(x)=\frac{1}{4\pi}\frac{e^{ikr}}{r}.$$

It satisfies the so-called Sommerfeld condition, that is,

$$(3.2)\qquad \left|\frac{\partial E}{\partial r}-ikE\right|\le\frac{c}{r^2}.$$

We could have chosen

$$\bar{E}(x)=\frac{1}{4\pi}\frac{e^{-ikr}}{r},$$

which satisfies the Sommerfeld condition

$$(3.3)\qquad \left|\frac{\partial\bar{E}}{\partial r}+ik\bar{E}\right|\le\frac{c}{r^2}.$$

It is also possible to choose

$$(3.4)\qquad \frac{E+\bar{E}}{2}=\frac{1}{4\pi}\frac{\cos kr}{r},$$

which is real.

When dealing with interior problems whose solutions are real, this last choice is indicated.

We will now focus on the exterior problem satisfying the Sommerfeld condition, and in that case the only possible choice is (3.1).

3.1. The operator S. When solving the interior and exterior problem using a single-layer potential, we arrive at the equation

$$\frac{1}{4\pi}\int_\Gamma \frac{e^{ik|x-y|}}{|x-y|}q(x)d\gamma(x) = u_d(y), \quad y\in\Gamma\,. \tag{3.5}$$

The structure seems very similar to that of the Laplace case, except that this equation solves also the interior Dirichlet problem, which happens to have eigenfrequencies. *So the equation* (3.5) *cannot be invertible when k^2 is an eigenfrequency of the interior Dirichlet problem.*

We can also write

$$\frac{e^{ikr}}{r} = \frac{1}{r} + ik + \frac{e^{ikr}-1-ikr}{r}. \tag{3.6}$$

Equivalently (S_0 is associated with the Laplace case $k=0$),

$$S = S_0 + R. \tag{3.7}$$

The theory of pseudodifferential operators shows that R is an operator of order –3 with principal symbol of the form $\frac{ck^2}{|\xi|^3}$. It maps H^s into $H^{s+3}(\Gamma)$.

And so we can rewrite the operator S as

$$S = S_0(I + S_0^{-1}R). \tag{3.8}$$

It is now clear that the Fredholm alternative is applicable and that S is invertible except for the values of k^2 mentioned above.

The numerical approximation follows the same approach as for the Laplace operator and gives rise to the same type of error analysis.

But it is very important to notice that the *physics* is *completely different as soon as k is large.* Our analysis fails in that case, since the compact term contains k^2 as a factor. We also observe in the computations that we need to respect a kind of CFL condition. We need a certain number of triangles of the mesh per wavelength. For the classical $\mathbb{P}_0$ approximation, 5 to 20 points per wavelength are necessary.

Experimentally, the numerical results, the stability, and the condition number of the matrix are good, even for large objects of 100 wavelengths. The perturbation due to the interior eigenfrequencies is quite sharp.

There is no analysis for this phenomenon, even in the case of the sphere. This is especially observed with the symmetric integral formulation. In the other cases, the perturbation due to the interior eigenfrequencies is larger

The matrix obtained is full, complex, and symmetric (non-Hermitian).

There is quite a lot of activity on the linear algebra of such systems. One first objective is to compute the coefficients cheaply. Rokhlin and Greengard have done some work in this direction.

The method of biconjugate gradients is quite efficient for solving the linear system. But there is a lack of good preconditioners.

3.2. The operator N. When solving the interior and exterior Neumann problem using a double-layer potential, we are led to the equation

$$-\frac{1}{4\pi}\oint_\Gamma \frac{\partial^2}{\partial n_x \partial n_y}\left(\frac{e^{ik|x-y|}}{|x-y|}\right)\varphi(x)d\gamma(x) = u_n(y), \quad y \in \Gamma. \tag{3.9}$$

Exactly as in the case of the Laplace equation, this kernel is nonintegrable and it corresponds to a pseudodifferential operator of order 1. Its principal symbol is again $|\xi|$, as can be seen from the expansion (3.6) . This equation solves the interior Neumann problem, and so it *cannot be invertible when k^2 is an eigenfrequency of this interior Neumann problem.*

As in the case of the Laplace equation, there exists a variational formulation which uses only integrable kernels. This is a result of Hamdi [20].

THEOREM 4. *The operator $-N$ is an isomorphism from $H^{1/2}(\Gamma)$ onto the space $H^{-1/2}(\Gamma)$, when k^2 is not an eigenvalue of the interior Neumann problem. The problem* (3.9) *has the following variational formulation*:

$$\left\{\begin{aligned} &\frac{1}{4\pi}\int_\Gamma\int_\Gamma \frac{e^{ik|x-y|}}{|x-y|}(\overrightarrow{\text{curl}}_\Gamma\varphi(x)\cdot\overrightarrow{\text{curl}}_\Gamma\psi(y))d\gamma(x)d\gamma(y)\\ &\qquad -\frac{k^2}{4\pi}\int_\Gamma\int_\Gamma \frac{e^{ik|x-y|}}{|x-y|}\varphi(x)\psi(y)(n_x\cdot n_y)d\gamma(x)d\gamma(y)\\ &\qquad = \int_\Gamma u_n\psi d\gamma \qquad \forall\psi\in H^{1/2}(\Gamma)\,. \end{aligned}\right. \tag{3.10}$$

An expression for $-N$ is

$$\left\{\begin{aligned} -N\varphi = \frac{1}{4\pi}\Big[&-\int_\Gamma \frac{e^{ik|x-y|}}{|x-y|}\Delta_\Gamma\varphi(x)d\gamma(x)\\ &+\int_\Gamma(\nabla_\Gamma\varphi(x)\cdot\nabla_\Gamma\frac{e^{ik|x-y|}}{|x-y|})(1-(n_x\cdot n_y))d\gamma(x)\\ &+\int_\Gamma\frac{\partial}{\partial n_x}\left(\frac{e^{ik|x-y|}}{|x-y|}\right)(\nabla_\Gamma\varphi(x)\cdot n_y)d\gamma(x)\\ &-k^2\int_\Gamma\frac{e^{ik|x-y|}}{|x-y|}\varphi(x)(n_x\cdot n_y)d\gamma(x)\Big]\,. \end{aligned}\right. \tag{3.11}$$

This variational formulation remains valid in the case where the interior domain is reduced to a crack, i.e., when the domain is just the exterior of a flat bounded surface. In that case of great practical importance, the double-layer potential is the only usable representation for solving a Neumann problem.

There are nice results on this problem by Èskin [17] for the elastic case. There are also some results by Stephan [36] and Ervin and Stephan [16].

Comments.

– When k is very large, the phenomenon observed for the operator S can also be seen in the case of the operator N, when using a $\mathbb{P}_1$ continuous finite element approximation. There are studies on this subject by De La Bourdonnaye [14], which present a new approximation to this equation based on some high-frequency analysis.

– The coupling techniques described for the Laplacian can be extended to the Helmholtz case without modifications.

– *The trick of Brakhage and Werner.*

The problem of noninvertibility for the interior eigenvalues can be avoided by means of the following trick. Instead of using a single-layer potential or a double-layer potential, we can use a combination of both. So, u is written as

$$u(y) = \frac{1}{4\pi}\int_\Gamma \left[\frac{e^{ik|x-y|}}{|x-y|} + i\alpha \frac{\partial}{\partial n_x}\frac{e^{ik|x-y|}}{|x-y|}\right] q(x)d\gamma(x)\,. \tag{3.12}$$

Using Theorem 2, we can establish that such a potential satisfies

$$\left[\frac{\partial u}{\partial n} - i\alpha u\right]_\Gamma = 0\,. \tag{3.13}$$

The interior problem associated with this potential is

$$\begin{cases} \Delta u + k^2 u = 0\,, \\ \dfrac{\partial u}{\partial n} - i\alpha u|_\Gamma = 0\,, \end{cases} \tag{3.14}$$

which has no eigenvalues when α is real. This potential therefore leads to invertible integral equations for any value of k.

A good choice of α seems to be $\alpha = k$.

4. The Maxwell Equation

4.1. The time-harmonic case. The unknowns in the time-harmonic Maxwell equations are the electric field E and the magnetic field H. They are related, in a homogeneous medium, by the relations (μ and ε are assumed real positive constant coefficients and ω is the wave number)

$$\begin{cases} \text{curl } E - i\omega\mu H = 0, \\ \text{curl } H + i\omega\varepsilon E = 0. \end{cases} \tag{4.1}$$

Moreover, we have the resulting equations

$$\text{div } E = \text{div } H = 0. \tag{4.2}$$

The fundamental solution of the system (4.1) is a matrix.

If we denote by

$$G(x) = \frac{1}{4\pi}\frac{e^{ikr}}{r}, \quad k^2 = \omega^2 \varepsilon\mu, \tag{4.3}$$

the Green's function of the Helmholtz equation, the "E part" of this fundamental solution is

$$E_i^j = i\omega\mu\delta_i^j G - \frac{1}{i\omega\varepsilon}\frac{\partial^2}{\partial x_i \partial x_j} G. \tag{4.4}$$

Let us consider the problem of an incoming plane wave E^{inc} arriving on a perfectly conducting object. Then the unknown is the diffracted field E, which satisfies the usual radiation condition

$$\left|\frac{\partial E}{\partial n} - ikE\right| \leq \frac{c}{r^2}, \tag{4.5}$$

or equivalently, the Sylver-Muller radiation condition

$$|E - H \wedge n| \leq \frac{c}{r^2}. \tag{4.6}$$

A well-posed exterior problem can be formulated with the only unknown E, and has the form

$$\begin{cases} E \text{ satisfies (4.6)}, \\ -\text{curl curl } E + k^2 E = 0, \\ E \wedge n|_\Gamma = -E^{\text{inc}} \wedge n. \end{cases} \tag{4.7}$$

An integral representation of the electric field is obtained by expressing the fact that the total field $E + E^{\text{inc}}$ is zero inside the object, and leads to

$$\begin{cases} E(y) = E^{\text{inc}}(y) - \dfrac{1}{i\omega\varepsilon} \text{grad} \displaystyle\int_\Gamma G(x,y)\, \text{div}_\Gamma \vec{j}(x) d\gamma(x) \\ \qquad\qquad + i\omega\mu \displaystyle\int_\Gamma G(x,y) \vec{j}(x) d\gamma(x). \end{cases} \tag{4.8}$$

The new variable $\vec{j}$ that appears is

$$\vec{j} = H \wedge n = \frac{1}{i\omega\mu} \text{curl } E \wedge n \quad \text{on the surface } \Gamma. \tag{4.9}$$

In this case $\vec{j}$ is the tangential current which is developed at the skin of the conductor by the incident field.

The integral equation is obtained by writing down that the tangential part of the electric field is prescribed.

A variational formulation for this equation can be obtained in the following Hilbert space:

$$\mathcal{J} = \left\{ \vec{j} \in (H^{-1/2}(\Gamma))^2 \quad ; \quad \vec{j} \text{ tgt to } \Gamma \quad ; \quad \text{div}_\Gamma \vec{j} \in H^{-1/2}(\Gamma) \right\}.$$

It is

$$
(4.10)\qquad \begin{cases} \dfrac{1}{i\omega\varepsilon}\displaystyle\int_\Gamma\int_\Gamma G(x,y)\,\mathrm{div}_\Gamma\vec{j}(x)\,\mathrm{div}_\Gamma\vec{j}^t(y)d\gamma(x)d\gamma(y) \\ \quad +i\omega\mu\displaystyle\int_\Gamma\int_\Gamma G(x,y)(\vec{j}(x)\cdot\vec{j}^t(y))d\gamma(x)d\gamma(y) = \\ \qquad -\displaystyle\int_\Gamma (E^{\mathrm{inc}}\cdot\vec{j}^t)d\gamma \qquad \forall j^t\in\mathcal{J}\,. \end{cases}
$$

Bendali [8] has proved that this equation is well posed and has a unique solution. He also introduced the following approximation:

The surface Γ_h is approximated as usual by a closed union of flat triangles ($\mathbb{P}_1$ finite elements). The space $\mathcal{J}$ is approximated using the basic mixed finite element

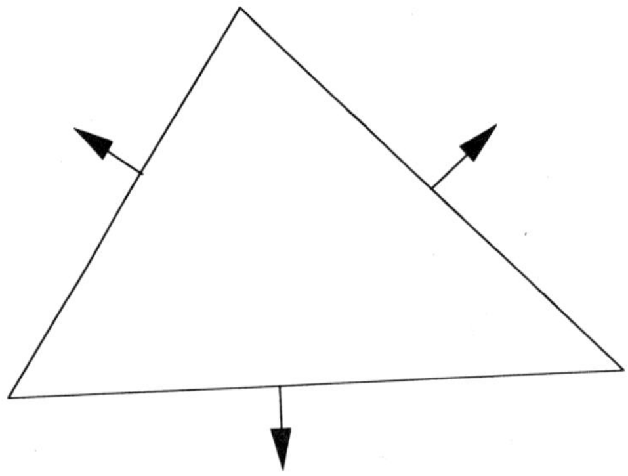

$$
\mathbb{P} = \begin{cases} \alpha_1 + \beta x_1 \\ \alpha_2 + \beta x_2 \end{cases}
$$

unknowns $\displaystyle\int_{a_i}(j\cdot\nu)ds$

for each edge a_i with normal ν .

Figure 3

$$
\mathcal{J}_h = \left\{\vec{j}_h\,;\ \vec{j}_h\Big|_{T_i} \in \mathbb{P}\,;\ \text{the vector } \vec{j}_h \text{ is in the triangle } T_i \text{ for every } T_i\in\Gamma_h\right\}.
$$

The unknowns are the flux of the current going from a triangle to the next through the edges. The number of unknowns is the number of edges.

Bendali [9] has obtained some error estimates for this approximation of the type

$$
(4.11)\qquad ||\vec{j}-\vec{j}_h||_{H^{1/2}(\Gamma)} \le c\sqrt{h}.
$$

The extension of this type of study to the more general problem of electromagnetics, where we have also dielectric and metallic objects, is a very active subject. We must mention the book of Colton and Kress [11], and also Knauff and Kress [24], and more recently the paper by Abboud and Nédélec [1].

The extension of the above approximation to the general case is relatively easy. On each interface between two dielectric media, we introduce both the electric current $\vec{j}$ and the magnetic current $\vec{m}$, which are respectively the trace $H \wedge n$ and $E \wedge n$. Both currents are approximated using the same finite elements.

An industrial code has been developed using this approximation by the society Thomson C.S.F.

4.1. The time-dependent case. Let us introduce this subject by the study of the Dirichlet exterior problem for the wave equation:

$$\left\{\begin{array}{r} \dfrac{\partial^2 u}{\partial t^2} - c^2 \Delta u = 0 \text{ in } \Omega_{\text{ext}} , \\ c \text{ is the speed of sound or the speed of light,} \\ u|_\Gamma = u_d . \end{array}\right. \tag{4.12}$$

The causal fundamental solution of the wave equation is

$$G(x,t) = \frac{1}{4\pi|x|}\delta\left(t - \frac{|x|}{c}\right) \qquad (\delta \text{ is the space-time Dirac mass}). \tag{4.13}$$

We can introduce the Fourier transform in time. We have

$$\hat{G}(\omega) = \frac{1}{\sqrt{2\pi}} \int_R e^{i\omega t} G(t) dt , \tag{4.14}$$

which then satisfies

$$-\Delta \hat{G}(x,\omega) - \frac{\omega^2}{c^2}\hat{G}(x,\omega) = \frac{1}{\sqrt{2\pi}}\delta(x), \tag{4.15}$$

and so the causal fundamental solution G is the Fourier transform of the outgoing fundamental solution of the Helmholtz equation:

$$\hat{G} = \frac{1}{\sqrt{2\pi}} \frac{1}{4\pi r} e^{i\frac{\omega}{c}r}. \tag{4.16}$$

Using the Fourier transform, we obtain the representation of the solution of (4.12) by a single-layer delayed potential:

$$\left\{\begin{array}{l} u(y,t) = \displaystyle\int_\Gamma \frac{q(x, t - \frac{|x-y|}{c})}{4\pi|x-y|} d\gamma(x), \\ q(x,t) = [u(x,t)]_\Gamma . \end{array}\right. \tag{4.17}$$

The associated integral equation is

$$u_d(y,t) = \int_\Gamma \frac{q(x, t - \frac{|x-y|}{c})}{4\pi|x-y|} d\gamma(x). \tag{4.18}$$

This equation has been used numerically by some pioneers with some success in the 1960s via a collocation method. Bamberger and Ha Duong [5] first introduced the idea of using a variational method in space-time variables. They were able to prove some coercivity, but not in the Hilbert space, where we have continuity. Ding [15], in his thesis, has implemented this method, which consists in solving

$$(4.19)\qquad \begin{cases} \displaystyle\int_R\int_\Gamma\int_\Gamma \frac{q_h(x,t-\frac{|x-y|}{c})}{4\pi|x-y|} q_h^t(x,t)d\gamma(x)d\gamma(y)dt = \\ \displaystyle\qquad\qquad \int_R\int_\Gamma u_d q_h^t d\gamma dt \quad \forall q_h^t \in V_h. \end{cases}$$

The choice of approximation was constant in space and $\mathbb{P}_1$ continuous in time.

The experiments showed that this scheme is relatively stable, at least for low CFL number $\frac{\Delta t}{\Delta x}$ and only for a finite time. There are no direct ways to analyze such stability criteria, the operator being nonlocal.

Later experiments showed that it is much better to have different approximations in time for q and q^t. The choice of q being $\mathbb{P}_1$ continuous and q^t constant at each time step leads to a scheme which seems unconditionally stable and gives much better results.

Let us return to the time-dependent Maxwell equation. The equations in the case of a perfectly conducting problem are

$$(4.20)\qquad \begin{cases} \text{curl } E + \mu \dfrac{\partial B}{\partial t} = 0\,, \\ \text{curl } H - \varepsilon \dfrac{\partial E}{\partial t} = 0\,, \\ E \wedge n|_\Gamma = -(E^{\text{inc}} \wedge n)\,. \end{cases}$$

The speed of light is $c = \frac{1}{\sqrt{\varepsilon\mu}}$.

We can also introduce the scalar and vector potential V and $\vec{A}$ such that

$$(4.21)\qquad E = -\text{ grad } V - \frac{\partial \vec{A}}{\partial t}.$$

If we introduce the auxiliary unknowns

$$(4.22)\qquad \begin{cases} \vec{j}|_\Gamma = [H \wedge n]_\Gamma\,, \\ \rho|_\Gamma = -[E \cdot n]_\Gamma\,, \end{cases}$$

we can verify, using the representation formula (4.17), that

$$(4.23)\qquad \left\{ V(y,t) = \int_\Gamma \frac{\rho(x,t-\frac{|x-y|}{c})}{4\pi|x-y|} d\gamma(x)\,, \right.$$

$$(4.24)\qquad \left\{ \vec{A}(y,t) = \mu \int_\Gamma \frac{\vec{j}(x,t-\frac{|x-y|}{c})}{4\pi|x-y|} d\gamma(x)\,. \right.$$

We have obtained the integral representation of the electric and the magnetic field:

$$(4.25)\qquad \begin{cases} E(y,t) = -\operatorname{grad}_y \displaystyle\int_\Gamma \frac{\rho(x,t-\frac{|x-y|}{c})}{4\pi|x-y|}d\gamma(x) \\ \qquad -\dfrac{\partial}{\partial t}\displaystyle\int_\Gamma \mu \frac{\vec{j}(x,t-\frac{|x-y|}{c})}{4\pi|x-y|}d\gamma(x)\,, \end{cases}$$

$$(4.26)\qquad H(y,t) = \operatorname{curl}_y \int_\Gamma \frac{\vec{j}(x,t-\frac{|x-y|}{c})}{4\pi|x-y|}d\gamma(x)\,.$$

Moreover, $\vec{j}$ and ρ are linked by the conservation of the electric charge, i.e.,

$$(4.27)\qquad \operatorname{div}_\Gamma \vec{j} + \varepsilon\frac{\partial\rho}{\partial t} = 0\,.$$

It is convenient to introduce another auxiliary unknown,

$$(4.28)\qquad \vec{\varphi} = [\operatorname{curl} E \wedge n]_\Gamma .$$

With this unknown, the integral equation becomes

$$(4.29)\qquad \begin{cases} -n\wedge(E^{\text{inc}}\wedge n)(y,t) \\ \quad = -\operatorname{grad}_{\Gamma_y}\displaystyle\int_\Gamma \frac{c^2}{4\pi|x-y|}\int_0^{t-\frac{|x-y|}{c}} ds \int_0^s \operatorname{div}_\Gamma\vec{\varphi}(x,\sigma)d\sigma\, d\gamma(x) \\ \qquad + \pi_{\Gamma_y}\displaystyle\int_\Gamma \frac{1}{4\pi|x-y|}\vec{\varphi}(x,t-\frac{|x-y|}{c})d\gamma(x)\,, \end{cases}$$

where π_{Γ_y} denotes projection onto the tangent plane. Terrasse [37] has introduced in his thesis a rather technical mathematical framework with some exotic Hilbert spaces (time-space dependent) such that this equation has a unique solution and admits a variational formulation which is coercive. A basic feature is again that continuity and coercivity are not in the same Hilbert space.

The "natural" variational formulation is obtained from the one associated with the harmonic problem by using a Fourier transform and a Parseval equality in order to be close to the physical energy of the problem.

It is

$$(4.30)\qquad \begin{cases} b(\vec{\varphi},\vec{\psi}) = \displaystyle\int_R\int_\Gamma\int_\Gamma \frac{\vec{\varphi}(x,t-\frac{|x-y|}{c})}{4\pi|x-y|}\frac{\partial\vec{\psi}}{\partial t}(x,t)d\gamma(x)d\gamma(y)dt \\ + \displaystyle\int_R\int_\Gamma\int_\Gamma \frac{c^2}{4\pi|x-y|}\left(\int_0^{t-\frac{|x-y|}{c}} ds\int_0^s \operatorname{div}_\Gamma\vec{\varphi}(x,\sigma)\right)d\sigma \\ \qquad\qquad \cdot \operatorname{div}_\Gamma \dfrac{\partial\vec{\psi}}{\partial t}(x,t)d\gamma(x)d\gamma(y)dt. \end{cases}$$

Discretization.

It was natural for this problem to have a space approximation similar to that of the harmonic case, as indicated in Figure 3 for $\vec{j}_h$. If we then choose $\mathbb{P}_1$ finite elements in time, the corresponding scheme is always unstable.

The change of having $\vec{j}^t \in \mathbb{P}_0$ is not allowed, as $\vec{j}^t$ appears through its second derivative $\left(\frac{\partial\psi}{\partial t} = \frac{\partial^2 j}{\partial t^2}\right)$.

So it is necessary to adopt a new variational formulation.

Terrasse chose one with three unknowns, $(\vec{j}, \rho, \mu)$, which is

$$\tag{4.31} \begin{cases} b((\rho,\vec{j}),(\rho^t,\vec{j}^t)) + < c(\rho^t,\vec{j}^t),\mu >= L(\rho^t,\vec{j}^t), \\ < c(\rho,\vec{j}),\mu^t >= 0, \end{cases}$$

$$\tag{4.32} \begin{cases} b((\rho,\vec{j}),(\rho^t,\vec{j}^t)) = \\ \quad -\mu \displaystyle\int_R\int_\Gamma\int_\Gamma \frac{\frac{\partial\vec{j}}{\partial t}(x,t-\frac{|x-y|}{c})}{4\pi|x-y|}\frac{\partial^2\vec{j}^t}{\partial t^2}(x,t)d\gamma(x)d\gamma(y)dt \\ \quad -\varepsilon \displaystyle\int_R\int_\Gamma\int_\Gamma \frac{\rho(x,t-\frac{|x-y|}{c})}{4\pi|x-y|}\frac{\partial^3\rho}{\partial t^3}(x,t)d\gamma(x)d\gamma(y)dt, \end{cases}$$

$$\tag{4.33} L(\rho^t,\vec{j}^t) = \int_R\int_\Gamma E^{\rm inc}(x,t)\frac{\partial^2\vec{j}^t}{\partial t^2}(x,t)d\gamma(x)dt,$$

$$\tag{4.34} c(\rho,\vec{j}^t) = {\rm div}_\Gamma\vec{j} + \varepsilon\frac{\partial\rho}{\partial t}.$$

One choice which led to a conditionally stable scheme is

$$\begin{aligned} &\vec{j} \in \mathbb{P}_0({\rm div})\ (x) \times \mathbb{P}_1\ (t), \\ &\rho \in \mathbb{P}_0(x) \times \mathbb{P}_1\ (t), \\ &\mu \in \mathbb{P}_0(x) \times \mathbb{P}_0\ (t). \end{aligned}$$

Many other choices are unstable or accumulate rounding errors systematically. With this choice, the conservation of the charge is not exact. It is approximated to second order in time.

Nevertheless, this technique of delayed potential seems to be very promising and leads to relatively inexpensive and precise computational techniques. In that direction, we can also mention the work of Bachelot and Pujols [4], and the work of Rao and Wilton [33]. This subject clearly needs much more investigation, especially to understand the stability properties.

5. Elasticity

The system of classical isotropic elasticity is another important example leading to wave propagation for both interior and exterior domains. The unknowns are $u = (u_1, u_2, u_3)$, the local displacement of the medium. We introduce the following tensors:

$$\varepsilon_{ij} = \frac{1}{2}\left(\frac{\partial u_i}{\partial x_j} + \frac{\partial u_j}{\partial x_i}\right) \qquad \text{(strain tensor)}, \tag{5.1}$$

$$\sigma_{ij} = 2\mu\,\varepsilon_{ij} + \lambda\,\mathrm{trace}(\varepsilon)\,\delta_i^j \qquad \text{(stress tensor)}, \tag{5.2}$$

(5.3) λ, μ are real positive coefficients (the Lamé constants).

Then the elastic wave problem involves the equation

$$\rho\frac{\partial^2 u_i}{\partial t^2} - \sum \frac{\partial}{\partial x_j}\sigma_{ij}(\vec{u}) = 0. \tag{5.4}$$

The boundary conditions are of Dirichlet type

$$u_i|_\Gamma = u_d \tag{5.5}$$

or of Neumann type

$$\sigma_{ij}n_j|_\Gamma = g_i. \tag{5.6}$$

The elastic system has two wave speeds, one related to div u, the second related to curl u. This can be seen, for instance, when one computes the fundamental solution. Let us introduce

$$G_1 = \frac{e^{ik_s r}}{4\pi r}, \qquad k_s = \frac{\omega}{c_s}, \qquad c_s = \sqrt{\frac{\mu}{\rho}}; \tag{5.7}$$

$$G_2 = \frac{e^{ik_p r}}{4\pi r}, \qquad k_p = \frac{\omega}{c_p}, \qquad c_p = \sqrt{\frac{\lambda + 2\mu}{\rho}}. \tag{5.8}$$

Then the fundamental solution corresponding to the harmonic problem

$$\rho\omega^2\vec{u} - \mathrm{div}(\sigma(\vec{u})) = 0 \tag{5.9}$$

is given by

$$U_i^j(x) = \frac{1}{4\pi\mu}\left[\frac{e^{ik_s r}}{r}\delta_i^j + \frac{1}{k_s^2}\frac{\partial^2}{\partial x_i \partial x_j}\left(\frac{e^{ik_s r}}{r} - \frac{e^{ik_p r}}{r}\right)\right]. \tag{5.10}$$

The **single**-layer potential associated with this problem is (using the summation convention on repeated indices)

$$u_i(y) = \int_\Gamma U_i^j(x-y)q_j(x)d\gamma(x)\,. \tag{5.11}$$

The integral equation of the first kind which solves the interior and exterior Dirichlet problem is

$$u_{d,i}(y) = \int_\Gamma U_i^j(x-y)q_j(x)d\gamma(x)\,. \tag{5.12}$$

The kernel is integrable. Its principal symbol is obtained from the expression (5.10). When ω tends to zero, this expression has a limit which corresponds to the elastostatic problem

$$U^{\circ j}_{i}(x) = \frac{1}{4\pi\mu}\left(\frac{1}{r}\delta_i^j + \left(1 - \frac{\mu}{\lambda+2\mu}\right)\frac{\partial^2}{\partial x_i \partial x_j} r\right). \tag{5.13}$$

This kernel is now a pseudodifferential operator of order −1. Moreover, it has been proved by Giroire [19] that the natural variational formulation for this problem is coercive, i.e.,

$$\int_\Gamma\int_\Gamma U^{\circ j}_{i}(x-y)q_j(y)q_i(x)d\gamma(x)d\gamma(y) \geq \alpha||q||^2_{(H^{1/2}(\Gamma))^3}\ , \text{where } \alpha > 0\ . \tag{5.14}$$

So, concerning this first-kind equation, we are in a situation very similar to that of the Helmholtz case. It is a compact perturbation of the static case, which is coercive.

The pseudodifferential theory shows also that this operator is continuous from $(H^s(\Gamma))^3$ into $(H^{s+1}(\Gamma))^3$. We will be in trouble for the values of ω which are eigenvalues of the interior Dirichlet problem.

The resulting system is symmetric (non-Hermitian).

The case of the hypersingular kernel. The situation of the hypersingular kernel is quite different from what happens for the Helmholtz equation.

A double-layer potential is associated with the variable σ, i.e., with the kernel

$$\Sigma_{ik}^{jl} = {''\sigma''}(U). \tag{5.15}$$

This kernel Σ is the Σ-part of the fundamental solution of the full system, i.e.,

$$\begin{cases} \operatorname{div}\Sigma + \rho\,\omega^2\,U = 0, \\ \varepsilon(U) - A\Sigma = \delta\ I, \\ I_{ij}^{kl} = \frac{1}{2}(\delta_{ki}\delta_{ij} + \delta_{kj}\delta_{ji}), \\ A = \frac{1}{2\mu}\left[\sigma - \frac{\lambda}{3\lambda+2\mu}\operatorname{trace}\sigma I\right]. \end{cases} \tag{5.16}$$

Although the algebra seems quite tricky, we have proved in Bécache, Nédélec, Nishimura [7] that the following decomposition of Σ holds.

LEMMA. *There exist a stress function $\phi_{::}$ and a kernel R such that*

$$\Sigma_{ij}^{kl} = \operatorname{curl}_i \operatorname{curl}_j \operatorname{curl}_k \operatorname{curl}_l \phi_{::} + \rho\omega^2 R_{ij}^{kl}, \tag{5.17}$$

$$\Lambda = \hat{\Gamma} - \rho\omega^2 I, \tag{5.18}$$

$$\hat{\Gamma}_{ik} = C_{ijkl}\xi_j\xi_l, \tag{5.19}$$

$$C = A^{-1}, \quad i.e., \quad C = 2\mu\,\varepsilon + \lambda\operatorname{trace}(\varepsilon)I. \tag{5.20}$$

Then we have

$$\hat{\phi} = \frac{P(\xi)}{\det \Lambda}, \tag{5.21}$$

$$\hat{R} = \frac{Q(\omega,\xi)}{\det \Lambda}, \tag{5.22}$$

where P is a homogeneous polynomial of degre 2 and Q a homogeneous polynomial of degree 4 in all variables.

The hypersingular equation is

$$g_i(y) = \oint_\Gamma \Sigma_{ij}^{kl}(x-y)n_j(x)n_l(y)\varphi_k(x)d\gamma(x). \tag{5.23}$$

Using the above lemma, we can prove that the problem admits a variational formulation with integrable kernels. It is

$$b(\varphi,\psi) = \int_\Gamma g_i\psi_i d\gamma \qquad \forall\psi \in (H^{1/2}(\Gamma))^3\,, \tag{5.24}$$

$$\left\{\begin{aligned} b(\varphi,\psi) = {} & \int_\Gamma\int_\Gamma F_{ij}^{kl}(x-y)(\overrightarrow{\mathrm{curl}}_\Gamma\varphi_k(x))_l(\overrightarrow{\mathrm{curl}}_\Gamma\psi_i(y))_j d\gamma(x)d\gamma(y) \\ & +\rho\omega^2\int_\Gamma\int_\Gamma R_{ij}^{kl}\varphi_k(x)n_l(x)\psi_i(y)n_j(y)d\gamma(x)d\gamma(y), \end{aligned}\right. \tag{5.25}$$

$$F_{ij}^{kl} = -e_{ima}\, e_{knb}\, \frac{\partial^2}{\partial m\partial n}\phi_{aj}^{bl}, \tag{5.26}$$

$$e_{imk} = \begin{cases} +1 & \text{if } (i,m,k) \text{ is a circular permutation,} \\ 0 & \text{if two indices are equal,} \\ -1 & \text{otherwise.} \end{cases}$$

The expressions of F and R are explicit:

$$\left\{\begin{aligned} R_{ij}^{kl} = {} & \frac{\mu}{\rho\omega^2(\lambda+\mu)}\Big[[(\lambda+3\mu)(\delta_{ki}\delta_{lj}+\delta_{kj}\delta_{li})+3\lambda\delta_{kl}\delta_{ij}]\Delta \\ & - 2\lambda(\delta_{kl}\partial_{ij}^2+\delta_{ij}\partial_{kl}^2) \\ & - \mu(\delta_{ik}\partial_{jl}^2+\delta_{il}\partial_{jk}^2+\delta_{jk}\partial_{il}^2+\delta_{jl}\partial_{ik}^2)\Big]N(r,\omega) \\ & + \frac{1}{\lambda+\mu}\,[\mu(\delta_{ki}\delta_{lj}+\delta_{kj}\delta_{li})+\lambda\delta_{kl}\delta_{ij}]\,N(r,\omega), \end{aligned}\right. \tag{5.27}$$

$$N(r,\omega) = \frac{e^{i\omega\frac{r}{c_s}} - e^{i\omega\frac{r}{c_p}}}{4\pi r}, \tag{5.28}$$

$$\left\{\begin{aligned} F_{ij}^{kl} = {} & \frac{-\mu^2}{(\lambda+\mu)\rho\omega^2}\,[2\lambda e_{inj}e_{kql} \\ & + (\lambda+2\mu)(e_{inp}e_{kqp}\delta_{jl}+\varepsilon_{inl}\varepsilon_{kqj})]\,\partial_{np}^2N(r,\omega). \end{aligned}\right. \tag{5.29}$$

The above regularization technique has the advantage of keeping the causality of the initial equation, which is obvious from the fact that it involves only some derivatives of the two outgoing fundamental solutions of the scalar wave equation (with k_s and k_p). This is quite important when one wants to solve the time-dependent problem. This was the work of Bécache [6], in her thesis, for the $2D$ case. She has observed that the $P_1\,P_0$ scheme seems to be the best, and it is unconditionally stable. She has computed with this method the scattering of an incident wave by a crack.

There is some work on this subject by Nishimura and Kobayashi [31], and also Nishimura with applications to inverse scattering problems.

6. Conclusion and Final Remarks

We have presented in this paper some aspects of the use and development of integral equations techniques. This method appears to be more complex to use than finite difference or finite element methods. In particular, the problem of stability is always difficult.

This presentation is far from exhausting the subject and reflects only the personal experience of the author. We have not mentioned, for example, the work on the heat equation by Costabel and also Arnold, who obtain some complete mathematical results. We have not mentioned the problem of wire antennas, which is a singular perturbation problem (the small parameter is the thickness of the wire). Mazari and Rogier have some results on this problem, which was already studied by Jones in his book

Neither have we mentioned the problems involving corner singularities. Costabel and Stephan obtained one of the first results of this type. The group at St. Petersburg is also very active on this subject (Maz'ya , Kozlov).

References

1. Abboud, T. and Nédélec, J.C., *Electromagnetic waves in inhomogeneous medium*, J. Math. Anal. Appl. **164** (1992), 40–58.
2. Arnold, D.N. and Wendland, W.L., *On the asymptotic convergence of collocation methods*, Math. Comp. **41** (1983), 349–381.
3. Arnold, D.N. and Wendland, W.L., *The convergence of spline collocation for strongly elliptic equations on curves*, Numer. Math. **47** (1985), 313–341.
4. Bachelot, A. and Pujols, A., *Equations intégrales espace-temps pour le système de Maxwell*, C.R. Acad. Sci. Paris Sér. I **314** (1992), 639–644.
5. Bamberger, A. and Ha Duong, T., *Formulation variationnelle pour le calcul de la diffraction d'une onde acoustique par une surface rigide*, Math. Methods Appl. Sci. **8** (1986), 598–608.
6. Bécache, E., *Résolution par une méthode d'équations intégrales d'un problème de diffraction d'ondes élastiques transitoires par une fissure*, Thesis, Ecole Polytechnique, 1991.
7. Bécache, E., Nédélec, J.C. and Nishimura, N., *Regularization in 3 D for anisotropic elastodynamic crack and obstacle problems*, J. Elasticity **31** (1993), 25–46.
8. Bendali, A., *Numerical analysis of the exterior boundary value problem for the time-harmonic Maxwell equations by a boundary finite element method. Part 1: The continuous problem*, Math. Comp. **43** (1984), 29–46.

9. Bendali, A., *Numerical analysis of the exterior boundary value problem for the time-harmonic Maxwell equations by a boundary finite element method. Part 2: The discrete problem*, Math. Comp. **43** (1984), 47–68.
10. Brakhage, H. and Werner, P., *Über das Dirichletsche Ausssenraum Problem für die Helmholtzsche Schwingungsgleichung*, Arch. Math. **16** (1965), 325–329.
11. Colton, P. and Kress, R., *Integral equation methods in scattering theory*, John Wiley, New York, 1983.
12. Costabel, M., *Principles of boundary element methods*, Comput. Phys. Rep. **6** (1987), 243–274.
13. Costabel, M.and Stephan, E. P., *Coupling of finite and boundary element methods for an elastoplastic interface problem*, SIAM J. Numer. Anal. **27** (1990), 1212–1226.
14. De La Bourdonnaye, A., Thesis, Ecole Polytechnique, 1992.
15. Ding, Y., *Méthodes numériques sur l'équation intégrale aux bords pour le problème des ondes acoustiques diffractées par une surface rigide en 3 D*, Thesis, University Paris 11, 1989.
16. Ervin, V.J. and Stephan, E.P., *A boundary element Galerkin method for a hypersingular integral equation on open surfaces*, Math. Methods Appl. Sci. **13** (1990), 281–289.
17. Èskin, G.I., *Boundary value problems for elliptic pseudodifferential equations*, Translations of Mathematical Monographs, Vol. 52, Amer. Math. Soc., Providence, R.I., 1981.
18. Giraud, G., *Equations à intégrales principales*, Ann. Ec. Norm. **51**, 3 and 4 (1934), 251–272.
19. Giroire, J., Thesis, Paris, 1988.
20. Hamdi, M.A., *Une formulation variationnelle par équations intégrales pour la résolution de l'équation de Helmholtz avec des conditions aux limites mixtes*, C.R. Acad. Sci. Paris Sér. II **292** (1981), 17–20.
21. Hsiao, G. and Wendland, W.L., *A finite element method for some integral equations of the first kind*, J. Math. Anal. Appl. **58** (1977), 449–481.
22. Johnson, C. and Nédélec, J.C., *On the coupling of boundary integral and finite element methods*, Math. Comp. **35** (1980), 1063–1079.
23. Kellogg, O.D., *Foundations of potential theory*, Springer, Berlin, 1929.
24. Knauff, W. and Kress, R., *On the exterior boundary value problem for the time harmonic equations*, J. Math. Anal. Appl. **72** (1979), 215–235.
25. Levillain, V., *Couplage éléments finis-équations intégrales pour la résolution des équations de Maxwell en milieu hétérogène*, Thesis, Ecole Polytechnique, 1991.
26. Le Roux, M.N., *Méthode d'éléments finis par la résolution numérique de problèmes extérieurs en dimension 2*, RAIRO Anal. Numér. **11** (1977), 27–60.
27. Mikhlin, S.G. and Prössdorf, S., *Singular integral operators*, Springer, Berlin, 1986.
28. Muskhelishvili, N.I., *Singular integral equations*, Noordhoff, Groningen, 1953.
29. Nédélec, J.C., *Curved finite element methods for the solution of singular integral equations on surfaces in* $\mathbb{R}^3$, Comp. Methods Appl. Mech. Engrg. **8** (1976), 61–80.
30. Nédélec, J.C. and Planchard, J., *Une méthode variationnelle d'éléments finis pour la résolution numérique d'un problème extérieur dans* $\mathbb{R}^3$, RAIRO R3 **7** (1973), 105–129.
31. Nishimura, N. and Kobayashi, S., *A regularized boundary integral equation method for elastodynamic crack problems*, Comput. Mech. **4** (1989), 319–328.
32. Prössdorf, S. and Rathsfeld, A., *A finite element collocation method for singular integral equations*, Math. Nachr. **100** (1981), 33–60.
33. Rao, S.M. and Wilton, D.R., *Transient scattering by conducting surfaces of arbitary shape*, IEEE Trans. Antennas and Propagation **39** (1991), 56–61.
34. Seeley, R.T., *Singular integrals on compact manifolds*, Amer. J. Math. **81** (1959), 658–690.
35. Seeley, R.T., *Topics in pseudo-differential operators,* in: Pseudo-Differential Operators (ed. L. Nirenberg), Edizioni Cremonese, Roma, 1969, pp. 169–305.
36. Stephan, E.P., *Boundary integral equations for screen problems in* $\mathbb{R}^3$, Integral Equations Operator Theory **10** (1987), 236–257.

37. Terrasse, I., *Résolution mathématique et numérique des équations de Maxwell instationnaires par une méthode de potentiels retardés*, Thesis, Ecole Polytechnique, 1993.

CENTRE DE MATHÉMATIQUES APPLIQUÉES, ECOLE POLYTECHNIQUE, 91128 PALAISEAU CEDEX, FRANCE

Proceedings of Symposia in Applied Mathematics
Volume 48, 1994

Applications of Multivariate Splines

Larry L. Schumaker

ABSTRACT. An extensive (although still incomplete) theory of multivariate splines has been developed over the past twenty years. Much of this development was driven by a desire to produce tools for various classical applications such as data fitting, surface approximation, and the numerical solution of partial differential equations and other operator equations. More recently, whole new fields of applications have emerged, including computer-aided design, computer vision, numerical machine control, image compression, etc. Our aim in this paper is to give an overview of some of these applications of multivariate splines.

1. Introduction

Univariate splines in their simplest and most useful form are nothing more than pieces of polynomials joined together smoothly at certain knots. They were studied intensely in the 60s, and by the mid 70s were sufficiently well understood to permit fairly comprehensive treatment in book form (see [8,66]). Univariate splines remain very important tools in a multitude of applications involving curve fitting and design.

There are, of course, many applications which involve functions of more than one variable, and so starting in the early 70s, a number of researchers began to develop an analogous multivariate spline theory. The expectation was that this could be completed in 10 to 15 years. But some 20 years later, there are still huge gaps. We realize now that the multivariate case is far richer and far more complex than originally thought.

The development of multivariate splines proceeded along three rather different lines: 1) the study of certain variational problems analogous to the one defining univariate polynomial splines, 2) construction of analogs of the B-spline, leading to a theory of so-called *simplex splines*, and the development of *box splines* which prove to be far more tractable and far more applicable, and 3) the study of multivariate splines as linear spaces of *smooth piecewise polynomials* defined over appropriately tessellated domains. These three approaches overlap significantly, and all three are

---

1991 *Mathematics Subject Classification*. Primary 65D07,41A05; Secondary 41A15.

Supported by NSF DMS-9208413.

This paper is in final form, and no version of it will be submitted for publication elsewhere.

© 1994 American Mathematical Society
0160-7634/94 $1.00 + $.25 per page

still being actively pursued. To get an idea of what is known, we recommend that the reader consult

1) the surveys [11,23,37] for simplex splines,
2) the surveys [21,42], and the very recent book [12] on box splines,
3) the lecture notes [16], the surveys of [10,11,17], and the forthcoming book [46] for splines on triangulations,
4) the books [30,43] for a computer-aided design approach,
5) the bibliography [33] for help in locating some of the original literature.

It is not our purpose here to go into the *theory* of multivariate splines in any detail. Instead, we want to discuss some of the many interesting *applications*. These include data fitting (§3), computer-aided design (§4), the finite element method (§5), and image analysis (§6). For ease of exposition, we focus primarily on the *bivariate case*; see §7 for some remarks on the case of more than two variables.

The reader should be warned that this survey is severely limited by space constraints, and that it represents a highly biased personal view. Many interesting applications get short shrift here, and still others are left out altogether, see the remarks in §7. Also, because of space limitations, our list of references consists mostly of surveys and books. There is, of course, a huge body of original literature on multivariate splines and their applications (for a list prepared in 1987, see [33]).

2. Bivariate Spline Spaces

We begin with a general description of what we mean by a space of *bivariate splines*. Let Ω be a closed simply connected subset of the plane, and let $\Delta = \{\Omega_i\}_{i=1}^N$ be a collection of closed subsets of Ω. We say that Δ is a *partition of* Ω provided that

1) $\cup_{i=1}^N \Omega_i = \Omega$
2) any pair Ω_i and Ω_j intersect only at points on their common boundary.

Given a finite-dimensional space $\mathcal{P}$ of real-valued functions which are at least r-times differentiable on Ω, we call

$$\mathcal{S}^r(\mathcal{P};\Delta) := \{s \in C^r(\Omega) \; : \; s|_{\Omega_i} \in \mathcal{P}, \quad \text{all } i = 1,\dots,N\} \tag{2.1}$$

the *space of (bivariate) splines of smoothness r with respect to the partition* Δ.

2.1. Tensor-product splines. The simplest example of a space of bivariate splines is the space of *tensor-product splines*, where Ω is a rectangle, the partition Δ consists of rectangles corresponding to a rectangular grid on Ω (see Fig. 1a), and where $\mathcal{P}$ is chosen to be the space of *tensor-product polynomials of degree* (d_x, d_y):

$$\mathcal{P}_{d_x,d_y} := \{p(x,y) = \sum_{i=0}^{d_x}\sum_{j=0}^{d_y} a_{ij}x^i y^j\}.$$

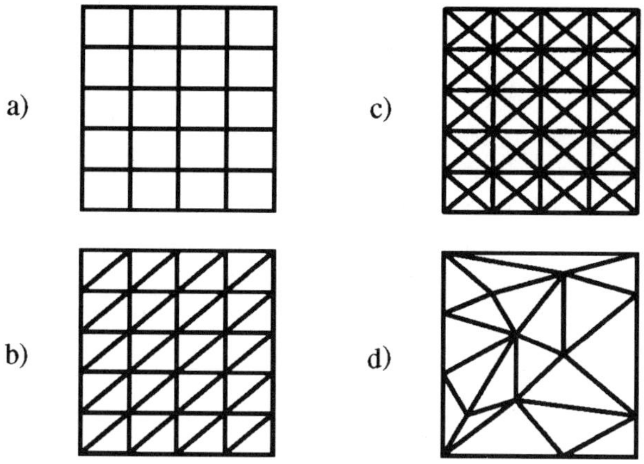

Fig. 1. Partitions.

Since both the theory and applications of tensor-product splines are well established (see [8,27,47,66]), we do not discuss them further here.

2.2. Polynomial splines on triangulations. We get a particularly useful bivariate spline space if we choose Δ to be a triangulation (see Fig. 1d) and take $\mathcal{P}$ to be the *space of polynomials of total degree d*:

$$\mathcal{P}_d := \{p(x,y) = \sum_{i+j\le d} a_{ij}x^i y^j\}. \tag{2.2}$$

Given a triangulation Δ and integers $0 \le r < d$, we define the space of *polynomial splines of degree d and smoothness r on* Δ to be

$$\mathcal{S}_d^r(\Delta) := \{s \in C^r(\Omega) \ : \ s|_{\Omega_i} \in \mathcal{P}_d, \quad \text{all } i = 1,\dots,N\}. \tag{2.3}$$

Clearly, $\mathcal{S}_d^r(\Delta)$ is a linear space of smooth piecewise polynomials. It is a natural extension to two variables of the well-known univariate splines (associated with a partition of an interval). In analogy with the univariate case, a complete constructive theory of these spline spaces should include

1) explicit formulae for the dimension of $\mathcal{S}_d^r(\Delta)$;
2) a convenient construction of locally supported basis elements;
3) efficient algorithms for storing and evaluating the splines and their derivatives, integrals, moments, etc.;
4) precise estimates of how well smooth functions can be approximated by splines.

Despite efforts by spline researchers over the past twenty years, there are still substantial gaps in this theory. For surveys of some of these topics, see [16,68,71] for dimension, and [10,11,16,17,19] for local bases and approximation power.

Although the theory is still incomplete, polynomial splines on triangulations have nevertheless become very useful tools for the numerical solution of a wide variety of problems involving surfaces. Our aim is to survey some of these applications. We devote the remainder of this section to a brief discussion of some useful properties of $\mathcal{S}_d^r(\Delta)$.

2.2.1. The Bernstein-Bézier form. To be useful for numerical applications, we have to be able to store and manipulate our splines on a digital computer. Unless we are on a very special partition (see §2.3 below), the most convenient way to store a polynomial spline is to store each of its polynomial pieces in *Bernstein-Bézier* (BB-) *form*. Given a piecewise polynomial s, let s_l be its restriction to the triangle Ω_l. Since this is a polynomial of total degree d, we can write it as

$$s_l(x,y) := \sum_{i+j+k=d} c_{ijk}^l B_{ijk}^d(\beta_1,\beta_2,\beta_3), \tag{2.4}$$

where $(\beta_1, \beta_2, \beta_3)$ are the *barycentric coordinates* of (x, y) with respect to the triangle Ω_l, and where

$$B_{ijk}^d(\beta_1,\beta_2,\beta_3) := \frac{d!}{i!j!k!}\beta_1^i\beta_2^j\beta_3^k, \quad i+j+k=d, \tag{2.5}$$

are the classical *Bernstein basis polynomials*. These polynomials are linearly independent and form a basis for the space $\mathcal{P}_d$ of polynomials of total degree d. For details, see [9,30,43].

It is common to assign each coefficient a physical *location* in the triangulation: the coefficient c_{ijk}^l can be associated with the point $\xi_{ijk}^l = (iv_1^l + jv_2^l + kv_3^l)/d$, where v_1^l, v_2^l, v_3^l are the vertices of Ω_l. One of the major advantages of this notation is that it leads immediately to a simple way to describe when two neighboring polynomial patches join together smoothly.

To insure C^0 continuity between adjoining patches, we identify the coefficients on the common edges between neighboring triangles. If we relabel the remaining coefficients, we get a vector $\alpha = \{\alpha_i\}_{i=1}^{n_c}$ with

$$n_c := V + (d-1)E + \frac{(d-2)(d-1)}{2}N, \tag{2.6}$$

where V is the number of vertices, E is the number of edges, and N is the number of triangles in the triangulation.

In view of the above discussion, each spline in the space $\mathcal{S}_d^0(\Delta)$ is uniquely defined by its coefficient vector α, and the dimension of $\mathcal{S}_d^0(\Delta)$ is precisely n_c. Given a spline $s \in \mathcal{S}_d^0(\Delta)$, we introduce the notation $\boldsymbol{c}_{ijk}^l = (\xi_{ijk}^l, c_{ijk}^l)$. These are points lying in $\mathbb{R}^3$, and are referred to as *control points*. If we connect each of these points to its neighbors with straight lines, we get a wire-frame figure which is called the *control net*. Filling in the triangles of this wire-frame object with linear facets, we get a *control surface*.

To store a spline $s \in \mathcal{S}_d^0(\Delta)$ in the computer, we need only store the triangulation Δ (see [72] for references to algorithms for creating triangulations and for a discussion of various storage schemes), along with the collection $\{c_{ijk}^l\}$ of its coefficients.

2.2.2. Evaluation. The BB-form is very convenient for numerical manipulation of a spline. To evaluate a spline s at a given point in a triangle Ω_l, we may use the following well-known algorithm for computing the value of s at a point in a triangle T with barycentric coordinates $(\beta_1, \beta_2, \beta_3)$:

(DE CASTELJAU ALGORITHM)
Let $a_{ijk}^{(0)} = c_{ijk}^{l}$ for $i + j + k = d$
For $m = 1$ step 1 until d
For all $i + j + k = d - m$
$a_{ijk}^{(m)} = \beta_1 a_{i+1,j,k}^{(m-1)} + \beta_2 a_{i,j+1,k}^{(m-1)} + \beta_3 a_{i,j,k+1}^{(m-1)}$

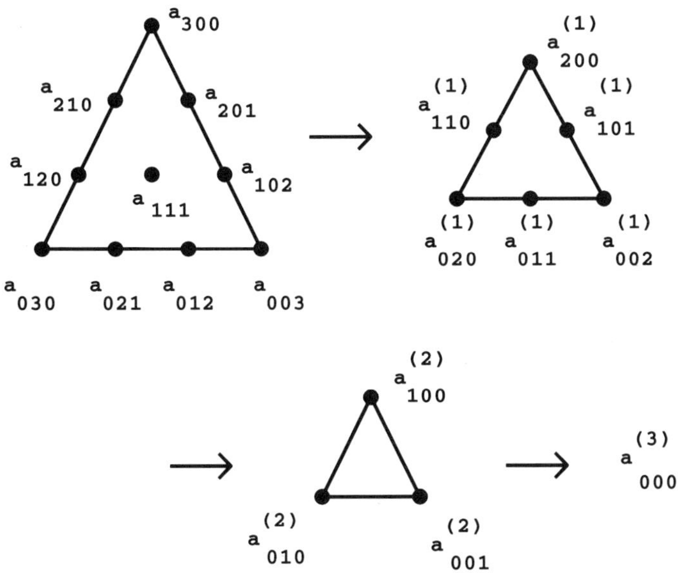

Fig. 2. The de Casteljau algorithm.

Fig. 2 illustrates the progress of the algorithm. It produces a sequence of control nets, where the mth net involves $\binom{d-m+2}{2}$ points, and is obtained from the previous one by combining triples of neighboring points using the weights $(\beta_1, \beta_2, \beta_3)$. This algorithm is numerically very stable since the barycentric coordinates β_i will be nonnegative and sum to one, and so each step of the algorithm involves taking a convex combination of previously computed quantities.

2.2.3. Derivatives. We now give a formula for the *directional derivative* of a polynomial in BB-form. Suppose (x, y) and $(\tilde{x}, \tilde{y})$ are two points in the plane whose barycentric coordinates relative to T are given by $(\beta_1, \beta_2, \beta_3)$ and $(\tilde{\beta}_1, \tilde{\beta}_2, \tilde{\beta}_3)$. Then the vector δ from the first point to the second defines a direction. Let $\delta_i = \beta_i - \tilde{\beta}_i$, for $i = 1, 2, 3$.

Now suppose p is a bivariate polynomial on T as in (2.4). Then the mth-order *directional derivative* of p in the direction δ is given by

$$D_\delta^m p(x, y) = \frac{d!}{(d-m)!} \sum_{i+j+k=d-m} a_{ijk}^{(m)}(\delta) B_{ijk}^{d-m}(\beta_1, \beta_2, \beta_3), \tag{2.7}$$

where $a_{ijk}^{(m)}(\delta)$ are the coefficients arising in the mth step of the de Casteljau algorithm using $(\delta_1, \delta_2, \delta_3)$.

It is clear from (2.7) that the mth-order derivative restricted to the side ϵ_1 opposite v_1 depends only on the coefficients c_{ijk} of p with $i \le m$. These are the coefficients in the Bézier net which are in the first $m+1$ rows along the side ϵ_1. In particular, the first derivative depends only on the coefficients in the first two rows.

2.2.4. Integrals and inner products. The integral of a BB-polynomial over a triangle T is given by

$$\int_T p(x,y)dxdy = \frac{|T|}{\binom{d+2}{2}} \sum_{i+j+k=d} c_{ijk}, \tag{2.8}$$

where $|T|$ denotes the area of the triangle T. To compute inner products of BB-polynomials, we may use the fact that

$$\int_T B_{ijk}^d(x,y)\, B_{lmn}^d(x,y)dxdy = \frac{\binom{i+l}{i}\binom{j+m}{j}\binom{k+n}{k}|T|}{\binom{i+j+k+l+m+n}{i+j+k}\binom{2d+2}{2}}. \tag{2.9}$$

2.2.5. Subdivision of BB-polynomials. In carrying out the de Casteljau algorithm discussed in §2.2.2 using given $(\beta_1, \beta_2, \beta_3)$, it turns out that the intermediate values $a_{ijk}^{(m)}$ produced by the algorithm can be used to rewrite a given BB-polynomial p on a triangle T in terms of new BB-polynomials defined on three subtriangles of T. In particular, suppose P is the point in T with barycentric coordinates $(\beta_1, \beta_2, \beta_3)$. Connecting P to each of the vertices of T leads to three subtriangles. Then the Bézier net of p considered as a spline on the triangulation Δ defined by these three triangles is obtained by combining the nets $\{a_{ijk}^{(m)}\}_{i+j+k=m}$ for $m = 1, \ldots, d$ in a simple way [30,43].

Subdivision is useful for a variety of purposes. One application is in *rendering* a surface, since it can be shown [30,43] that if we continually subdivide a given BB-polynomial, the corresponding sequence of control surfaces converges to the surface defined by p over T.

2.2.6. Degree-raising of BB-polynomials. It is clear that a BB-polynomial of degree d can also be regarded as a BB-polynomial of any degree $\tilde{d} > d$. In particular, given

$$p(x,y) := \sum_{i+j+k=d} c_{ijk} B_{ijk}^d(\beta_1, \beta_2, \beta_3), \tag{2.10}$$

multiplying both sides by $1 = \beta_1 + \beta_2 + \beta_3$, we can also write it as

$$p(x,y) = \sum_{i+j+k=d+1} \tilde{c}_{ijk} B_{ijk}^{d+1}(\beta_1, \beta_2, \beta_3), \tag{2.11}$$

where

$$\tilde{c}_{ijk} := (ic_{i-1,j,k} + jc_{i,j-1,k} + kc_{i,j,k-1})/(d+1), \quad i+j+k = d+1. \tag{2.12}$$

This *degree-raising formula* follows easily from properties of Bernstein polynomials; see [30,43]. It is also known (cf. [30,43]) that the sequence of control surfaces obtained by repeated degree-raising converges to the surface defined by p.

2.2.7. Higher-order smoothness. Since $\mathcal{S}_d^r(\Delta)$ is a subspace of $\mathcal{S}_d^0(\Delta)$, we can use the BB-form to store and evaluate splines in the spaces $\mathcal{S}_d^r(\Delta)$. To single out the C^r splines in $\mathcal{S}_d^0(\Delta)$, we must impose smoothness conditions across the edges of the triangulation. It is well known [9,30,43] that if two Bézier polynomials of degree d join with C^{r-1} continuity along a common edge, then they also join with C^r continuity if and only if the BB-coefficients in the first r rows on both sides of the edge satisfy a set of $d+1-r$ linear conditions.

It follows that a spline s in $\mathcal{S}_d^0(\Delta)$ belongs to $C^r(\Omega)$ if and only if its coefficients satisfy a homogeneous linear system of equations of the form

$$H\alpha = 0. \tag{2.13}$$

Here α is the n_c-vector of Bernstein-Bézier coefficients of s, and H is a matrix of size $n_{cont} \times n_c$, where

$$n_{cont} := [d + (d-1) + \cdots + (d-r+1)]E_I = \frac{2rd + r - r^2}{2} E_I, \tag{2.14}$$

and E_I is the number of interior edges in Δ. Each row of H corresponding to a C^j continuity condition contains at most $(j+1)(j+2)/2 + 1$ nonzero elements.

2.2.8. Locally supported basis elements. In analogy with the univariate case, it would be useful if we could build locally supported splines (which would hopefully form a basis for $\mathcal{S}_d^r(\Delta)$).

The most natural support set for such a spline would be the *star of a vertex* v which consists of the union of all triangles which have one vertex at v. A spline which vanishes outside of such a set is called a *vertex spline*. It is known that for $d \geq 3r+2$, it is possible to construct a basis for $\mathcal{S}_d^r(\Delta)$ consisting only of vertex splines. In general, it is not known when such bases exist for smaller values of d. On certain special triangulations (see the next section), one can get bases which are locally supported (but whose supports are typically somewhat larger than a single star of a vertex) by using box splines (to be discussed in §2.5 below).

For a detailed treatment of vertex splines, and their use in defining certain quasi-interpolation operators (see §3.1 below), see [17,19].

2.3. Special triangulations. Much of the theory of §2.2 greatly simplifies if the triangulation Δ has special properties. Two of the most commonly used special triangulations are

1) the *type-1* triangulation shown in Fig. 1b which arises from a rectangular grid by drawing in north-easterly diagonals in each subrectangle (when the grid lines are equally spaced, this is also called a *three-direction mesh*),
2) the *type-2* triangulation shown in Fig. 1c which arises from a rectangular grid by drawing in both diagonals in each subrectangle (when the grid lines are equally spaced, this is also called a *four-direction mesh*).

For these types of triangulations one can say much more about dimension, the existence of locally supported bases, approximation power, and so forth [16,46].

2.4. Macro elements. We have seen in §2.2.8 that building locally supported bases for spline spaces on arbitrary triangulations is difficult. One way to overcome this difficulty is to work with triangulations which are formed by starting with a given triangulation Δ and subdividing each triangle in some regular way. In this section we illustrate this with two simple examples.

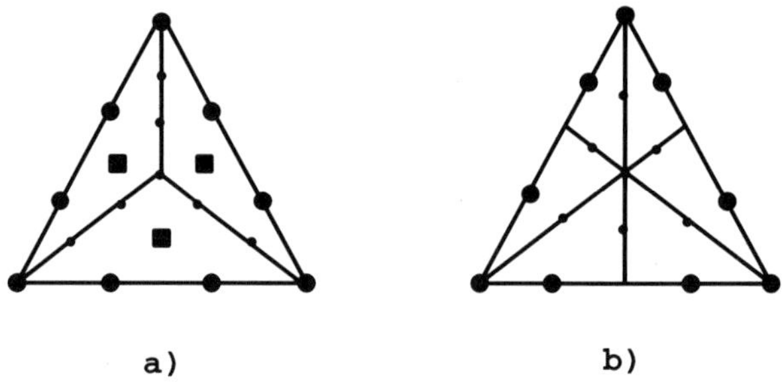

Fig. 3. Macro elements.

In our first example, we divide each triangle T in Δ into three subtriangles as shown in Fig. 3a. This is called the *Clough-Tocher split* of T, and we denote the corresponding (refined) triangulation by Δ_{CT}. Now suppose we want to construct C^1 cubic splines on Δ_{CT}. It is easy to see [11,27,39] that a spline in $\mathcal{S}_3^1(\Delta_{CT})$ is uniquely determined if we specify the gradient at each vertex and the cross derivative at some point (say the midpoint) of each edge. This corresponds to assigning values to the Bézier coefficients marked with heavy dots and squares in Fig. 3a.

Still simpler macro elements can be obtained if we look at the subspace $\mathcal{S}$ of $\mathcal{S}_3^1(\Delta_{CT})$ whose normal derivatives across edges of Δ are linear rather than quadratic. In this case each spline in $\mathcal{S}$ is uniquely determined by the gradient values at the vertices. This is an example of a *condensed* macro element.

A similar construction is possible using C^1 quadratic splines. Now we subdivide each triangle of the original triangulation Δ into six subtriangles as shown in Fig. 3b. If this is done carefully [11,14,76], we can build a C^1 quadratic spline space which depends only on the gradients at each vertex of the original triangulation Δ. These macro elements are called *Powell-Sabin elements*.

2.5. Box splines. As discussed above, for general triangulations, it is difficult (and often impossible) to build locally supported bases for the spline spaces $\mathcal{S}_d^r(\Delta)$. In some cases, there may be a sufficient number of vertex splines (cf. §2.2.8) for applications, although usually this is not the case.

In such cases we may need to work with locally supported splines whose supports are still local (but not as small as for the vertex splines). On certain regular triangulations including the type-1 and type-2 triangulations shown in Fig. 1, there is an elegant way to construct such splines. To describe this approach, we first recall that the classical univariate B-spline (unnormalized) can be defined recursively by

(2.15) $$M_k(x) := \int_0^1 M_{k-1}(x-t)dt,$$

where

(2.16) $$M_1(x) = \begin{cases} 1 & \text{if } 0 \le x < 1, \\ 0 & \text{otherwise.} \end{cases}$$

To generalize this construction to the bivariate case, suppose we are given a $2 \times m$ matrix $A_m = [a^1 \cdots a^m]$ where we may assume that a^1 and a^2 are linearly independent. Let

(2.17) $$M(x,y \mid A_m) := \int_0^1 M(x - ta_x^m, y - ta_y^m \mid A_{m-1})dt,$$

where a^i has components (a_x^i, a_y^i), and

(2.18) $$M(x,y \mid A_2) := \begin{cases} 1/|\det\,(A_2)| & \text{if } (x,y) \in [A_2), \\ 0 & \text{otherwise,} \end{cases}$$

with

(2.19) $$[A_j) := \{t_1a^1 + \cdots + t_ja^j \ : \ 0 \le t_i < 1, \quad 1 \le i \le j\}$$

for $j = 2, \ldots, m$. Each of the functions $M(x,y \mid A_j)$ is referred to as a *box spline*.

Clearly, $M(x,y \mid A_2)$ is just a constant for $(x,y) \in [A_2)$, and is zero outside. In general, one can show (cf. [12,21,42]) the following:

1) $M(x,y \mid A_m)$ is positive on the interior of the set $[A_m)$, and vanishes outside of its closure $[A_m]$ (called the *support* of M),
2) $\int_{[A_m]} M(x,y \mid A_m)\, dxdy = 1$,
3) $M(x,y \mid A_m)$ is a piecewise polynomial of degree at most $m-2$ defined on the partition of $[A_m]$ obtained by connecting each of its vertices with all others,
4) $M(x,y \mid A_m) \in C^{r-1}$, where r is the maximum of all integers k such that all subsets of $m-k$ vectors chosen from A_m still span $\mathbb{R}^2$,
5) the set of integer translates $\{M(x-i, y-j \mid A_m)$ are linearly independent on $\mathbb{R}^2$ if and only if all sets of pairs of vectors b_1, b_2 from A_m are such that det $[b_1 b_2]$ has values -1, 0, or 1.

Fig. 4a shows the support of the box spline corresponding to the set A_3 consisting of the three vectors $(1,0)^T$, $(0,1)^T$, and $(1,1)^T$. This spline is a C^0 piecewise linear function. Its translates live on a three-direction mesh as in Fig. 1b. This box spline is an example of a *Courant hat function*.

Fig. 4b shows the support of the box spline corresponding to the set A_4 consisting of the four vectors $(1,0)^T$, $(0,1)^T$, $(1,1)^T$, and $(1,-1)^T$. This spline is a C^1 piecewise quadratic function. Its translates live on a four-direction mesh as in Fig. 1c. This box spline is called the *Zwart-Powell element*.

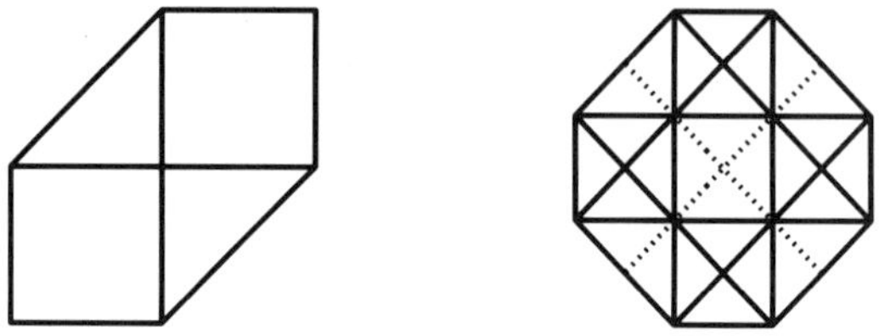

a) b)

Fig. 4. Supports of two simple box splines.

A lot of work has gone into deriving properties of box splines (such as approximation power, formulae for derivatives and integrals, local linear independence, subdivision algorithms, etc.). See the surveys [21,42] and the book [12]. Clearly, if we are working with spline spaces of the form $\mathcal{S}_d^r(\Delta)$ on either a three-direction or four-direction mesh, then box splines provide a convenient way of constructing locally supported elements which lie in the spline space. Usually, they do not actually form a basis, but often can be used anyway to construct interpolating and approximating surfaces.

The box splines described above can be generalized in various ways. For example, repeating the same vector in A_m is allowed (this leads to box splines with reduced smoothness). It is also straightforward to define box splines in any number of variables ([12,21]).

2.6. Simplex splines. Our overview of multivariate splines would not be complete without at least mentioning simplex splines. Suppose B is a convex set in $\mathbb{R}^m$, and that P is an affine map of $\mathbb{R}^m$ into $\mathbb{R}^2$. Then the corresponding (analog of the univariate) B-spline M_B is defined to be the distribution which represents the linear functional $f \to \int_B f \circ P$. The most interesting case is when B is a *polytope* (the convex hull of a finite set). In this case, it is known (cf. [11,23]) that

1) M_B is positive on the interior of the set Ω_B which is the convex hull of $\{Px : x \in B\}$, and vanishes outside of Ω_B,
2) $\int_{\Omega_B} M_B dxdy = 1$,
3) if B is in general position and P is onto $\mathbb{R}^2$, then M_B is a piecewise polynomial of degree at most $m-2$ defined on the partition of Ω_B obtained by connecting each of its vertices with all others,
4) $M_B \subset C^{r-1}$, where r is the smallest integer k such that P maps all k-dimensional faces of B into a set in $\mathbb{R}^2$ with interior.

These so-called *polyhedral* splines include box splines as a special case (we simply choose B to be a box in $\mathbb{R}^m$). If B is chosen to be a simplex in $\mathbb{R}^m$, we get what are called *simplex splines*. For surveys of the theory of simplex splines, see [11,23].

So far, attempts to use simplex splines in applications has been limited (see [75] for a survey and references). The two main problems with using them in applications are:

1) it seems to be difficult to design a convenient stable evaluation algorithm,

2) it seems to be difficult to build useful finite-dimensional linear spaces which can be used in applications.

Both problems are currently under study, and (see the remarks in [11]) there is new hope that both are tractable.

3. Fitting Surfaces

In this section we discuss a classical application of multivariate splines: the fitting of surfaces.

3.1. Quasi-interpolation. Suppose we are given a surface defined by a function f defined on a set $\Omega \subset \mathbb{R}^2$, and suppose that $\{B_1, \ldots, B_n\}$ is a set of bivariate splines defined on Ω. One way to construct an approximation to f is to define

$$Sf(x,y) := \sum_{i=1}^{n} (\lambda_i f) B_i(x,y), \tag{3.1}$$

where $\lambda_1, \ldots, \lambda_n$ are prescribed linear functionals defined on a space containing f. Linear operators of the form (3.1) are called *quasi-interpolants*.

For example, to get a quasi-interpolant defined on $C(\Omega)$, we might simply take $\lambda_i f = f(x_i, y_i)$, where (x_i, y_i) are given points in the supports of B_i, $i = 1, \ldots, n$. This may not lead to a useful approximation method, however. To get useful quasi-interpolants, one usually has to work with more complicated linear functionals, carefully chosen so that the operator S at least reproduces total-degree polynomials of some degree. Although it sounds complicated, such quasi-interpolants can be constructed (for example using vertex splines [16,17,19]), and can be used to construct surfaces. They are also useful for establishing how well classes of smooth functions can be approximated by multivariate splines.

3.2. Interpolation. The following problem is one of the fundamental problems of numerical analysis.

PROBLEM 1. *Let $(x_1, y_1), \ldots, (x_n, y_n)$ be points in a domain $\Omega \in \mathbb{R}^2$, and suppose that $z_1, \ldots, z_n$ are given real numbers. The scattered data interpolation problem is to find s defined on Ω such that*

$$s(x_i, y_i) = z_i, \qquad i = 1, \ldots, n. \tag{3.2}$$

If the data points lie on a regular grid, we call this gridded data; otherwise we call it scattered data.

This problem has been heavily studied. For a survey up to 1976, see [65]. A comparison of methods [32] was carried out in 1979. For further references, see also [33]. Gridded data interpolation is usually attacked using tensor-product splines. Our discussion here will concentrate on the scattered case.

3.2.1. Combinations of basis functions. Given a collection of basis splines $B_1, \ldots, B_n$ defined on Ω, the usual approach to solving an interpolation problem is to look for a spline of the form

$$s(x,y) := \sum_{i=1}^{n} b_i B_i(x,y). \tag{3.3}$$

To force s to interpolate as in (3.2), we must choose the coefficients in (3.3) to satisfy

$$Ab = z, \tag{3.4}$$

where A is the $n \times n$ matrix $A = (B_i(x_j, y_j))_{i,j=1}^n$, $b = (b_1, \ldots, b_n)^T$, and $z = (z_1, \ldots, z_n)^T$.

The main difficulty with this approach is that we have to show that the matrix A is nonsingular. In the case where the B_i are univariate B-splines, the classical *Schoenberg-Whitney* conditions [8,66] completely characterize when A is nonsingular. Finding analogous conditions in the bivariate case seems to be extremely difficult [16].

One situation where we can sometimes guarantee the nonsingularity of A is the case where the B_i's are taken to be box splines, and the interpolation points lie on a regular grid. Indeed, there is a substantial theory of *cardinal interpolation* by box splines [44,45,61].

3.2.2. Minimal energy methods. Another approach to Problem 1 proceeds as follows:

1) construct a triangulation Δ,
2) choose a spline space $\mathcal{S}_d^r(\Delta)$,
3) find Bézier coefficients so that the corresponding spline $s \in \mathcal{S}_d^r(\Delta)$ interpolates.

We say more about choosing Δ below. Typically, we choose it to be the well-known *Delaunay* triangulation, which is defined as a triangulation in the set of all triangulations (with vertices at the data points) whose smallest angle is maximal [67,72].

We now discuss Step 3). In view of the fact that on each triangle, only one Bernstein polynomial is nonzero at a given vertex, to get an interpolating spline, all we have to do is choose its Bézier net α such that at each vertex, the associated BB-coefficient is the desired z-value at that vertex. This specifies exactly V of the $n_c = V + (d-1)E + \frac{(d-2)(d-1)}{2}N$ coefficients of a spline in $\mathcal{S}_d^0(\Delta)$. To make s belong to $C^r(\Omega)$, we must enforce smoothness conditions as described in §2.2.7, i.e., we must enforce (2.13). This is a set of n_{cont} conditions, and even after forcing interpolation at the vertices, we still have at least $n_c - n_{cont} - V$ free parameters. One way to use these free parameters is to minimize some quantity of interest.

With the case of univariate splines in mind, the natural thing to minimize would be a measure of the energy of the surface. The standard choice is the *thin plate energy functional*

$$\mathcal{E}(s) := \sum_{i=1}^{N} \int_{\Omega_i} [s_{xx}^2 + 2s_x s_y + s_{yy}^2]\, dxdy. \tag{3.5}$$

It turns out [59] that $\mathcal{E}(s)$ can be expressed in terms of the Bézier net α of s as a quadratic form

$$\mathcal{E}(s) = \alpha^T E \alpha. \tag{3.6}$$

Now we are in a position to construct a *smoothest interpolating spline* from the space. To do so, we simply have to minimize the quadratic form (3.6) subject to interpolation and smoothness conditions on α. Using Lagrange multipliers, we are led to a linear system of the form

$$\begin{pmatrix} E & C^T \\ C & 0 \end{pmatrix} \begin{pmatrix} \alpha \\ \lambda \end{pmatrix} = \begin{pmatrix} z \\ 0 \end{pmatrix}. \tag{3.7}$$

Since the matrix C is generally sparse, sparse-matrix techniques can be used to solve (3.7).

We should point out that in order for the system (3.7) to have a solution, the space $\mathcal{S}_d^r(\Delta)$ must be big enough to include interpolants of arbitrary data at the vertices. In practice, a common choice for $\mathcal{S}_d^r(\Delta)$ is the space $\mathcal{S}_3^1(\Delta)$ of cubic C^1 splines, but as of this writing, it is still an open question as to whether interpolation with this space is always possible.

3.2.3. Macro-element techniques. The interpolation methods in §§ 3.2.1 and 3.2.2 are global in nature. Thus, if the number of data is large, they require solving a large linear system of equations. This disadvantage can be avoided by working with *local interpolation schemes*.

Local interpolation is possible in the spaces $\mathcal{S}_d^r(\Delta)$ if one chooses d sufficiently large compared to r. For example, by examining the BB-net, it is easy to see that specifying derivatives up to order 2 at each vertex of a triangulation Δ plus the normal derivative at the center of each edge uniquely determines a spline in $\mathcal{S}_5^1(\Delta)$. Each polynomial piece is locally determined by the data. Of course, if we are not given the needed derivatives, we would have to estimate them from the scattered data (there exist a variety of methods for this [47,65]).

To build local schemes using polynomials whose degrees are lower than $4r+1$, we have to split the triangles. For example, if we use the Clough-Tocher split discussed in §2.4, then there exists a unique spline $s \in \mathcal{S}_3^1(\Delta_{CT})$ which takes on given values and gradients at each of the vertices of the original triangulation Δ, along with given cross-derivatives at each of the mid-points of edges of Δ (see Fig. 3a).

The cross-derivatives can be estimated by various local approximation schemes. Alternatively, it is also possible to get by without these derivatives by using a *condensed macro element* as discussed in §2.4. Explicit BB-representations of the condensed Clough-Tocher element can be found in [39].

There are a variety of similar macro elements using polynomials of other degrees, and also with higher orders of smoothness. For example, if we want to use C^1 quadratic splines, we can use the Powell-Sabin split [11,14,76].

3.2.4. Data-dependent triangulations. Recently, it has been observed that the quality of surface fits obtained using polynomial splines on triangulations can be vastly improved if the triangulation itself is adjusted to the data.

To give an example of how this works, suppose Δ is a given triangulation of the data. Then clearly there exists a unique spline $s \in \mathcal{S}_1^0(\Delta)$ which interpolates given values at the vertices (in this case there are no free parameters). But we are free to choose different triangulations, which suggests that we should look for a best triangulation, depending on the data. This idea was first proposed in [29], where several different criteria were discussed for distinguishing a *best* triangulation. A

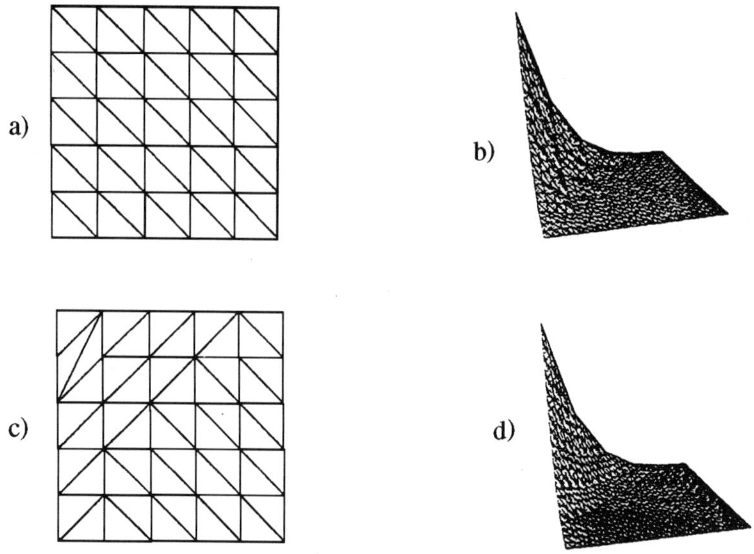

Fig. 5. Data-dependent interpolation.

typical criterion would be to choose Δ to minimize the sizes of the angles between the normal vectors to adjacent triangular patches.

Suppose we want to interpolate the surface associated with the function

$$f(x,y) := (y-x^2)_+ = \begin{cases} (y-x^2) & \text{if } y > x^2, \\ 0 & \text{otherwise} \end{cases} \tag{3.8}$$

at the points on a 6×6 grid on the unit square. Fig. 5b shows the interpolating surface associated with the Delaunay triangulation in Fig. 5a, while Fig. 5d shows the surface associated with the alternate triangulation shown in Fig. 5c. Clearly the surface based on the second triangulation is a much better fit of f.

To construct a data-dependent triangulation, we may start with an arbitrary triangulation, and then swap edges. The idea is to look at *swappable edges* (those whose adjoining triangles form a convex quadrilateral), and to swap them when the alternate triangulation results in an improvement in the swap criterion.

(Local Swap Algorithm)
do *until no swap is possible*
  choose *a swappable edge e*
  if *swapping edge e improves the triangulation, make the swap.*

In practice, it is convenient to sort the edges after each swap according to the criterion of interest, and then examine the edges in order (the idea being to find the swappable edge which leads to the most improvement in our criterion). Unfortunately, for most criteria, this local swap algorithm is not guaranteed to lead to a globally optimal triangulation. See [72] for a discussion and examples. It was shown in [73] that *simulated annealing* can be a useful tool for searching for globally optimal triangulations.

This same approach can be used to construct data-dependent C^1 interpolating splines based on Clough-Tocher macro elements [58].

3.3. Noisy data. Frequently one is asked to solve Problem 1 even when the values $z_1, \ldots, z_n$ are contaminated by noise. In this case it usually doesn't make sense to interpolate; instead we should smooth out the noise.

3.3.1. Discrete least squares. The classical approach to noisy data is to perform a least squares fit, i.e., to find a function s from some space $\mathcal{S}$ to minimize

$$\mathcal{F}(s) := \sum_{i=1}^{n} [s(x_i, y_i) - z_i]^2. \tag{3.9}$$

One can of course seek s in a spline space $\mathcal{S}_d^r(\Delta)$. If we have a convenient set of basis functions which are linearly independent over the data, the least squares fit can be determined from the associated set of *normal equations*. If not, we can still work with $\mathcal{S}_d^r(\Delta)$ in Bézier form. Since $\mathcal{F}(s)$ is a quadratic form, there is a matrix F such that

$$\mathcal{F}(s) = \alpha^T F \alpha, \tag{3.10}$$

where α is the vector of Bézier coefficients. To get a spline in $\mathcal{S}_d^r(\Delta)$ which minimizes $\mathcal{F}(s)$, we enforce the smoothness conditions (2.13). Using Lagrange multipliers, we see that this linearly constrained minimization problem reduces to solving a (sparse) linear system of equations.

3.3.2. Penalized least squares. In constructing least squares fits of surfaces using splines, it can happen that the resulting surfaces oscillate too much. This can be controlled by replacing the fit criterion $\mathcal{F}(s)$ in (3.9) by a combination of goodness of fit and energy, say

$$\mathcal{G}(s) := \lambda \mathcal{E}(s) + \mathcal{F}(s), \tag{3.11}$$

where $\mathcal{E}$ is the thin plate energy functional of (3.5), and λ is a smoothing parameter controlling the trade-off between fit and smoothness.

For results on minimizing $\mathcal{G}(s)$ over a general finite-dimensional linear space $\mathcal{S}$, see [36]. This problem again reduces to solving a linear system of equations. The nature of the fit depends heavily on the choice of λ. For $\lambda = 0$, we clearly get the least squares fit from $\mathcal{S}$. As λ goes to infinity, the solution approaches the least squares fit from the subspace of $\mathcal{S}$ with zero energy. The smoothing parameter can be selected automatically by *generalized cross validation* [36].

3.3.3. Data-dependent least squares. In performing least squares (or penalized least squares) using a space of splines, it clearly may be advantageous to allow the triangulation to depend on the data. For example, we can adjust the triangulation

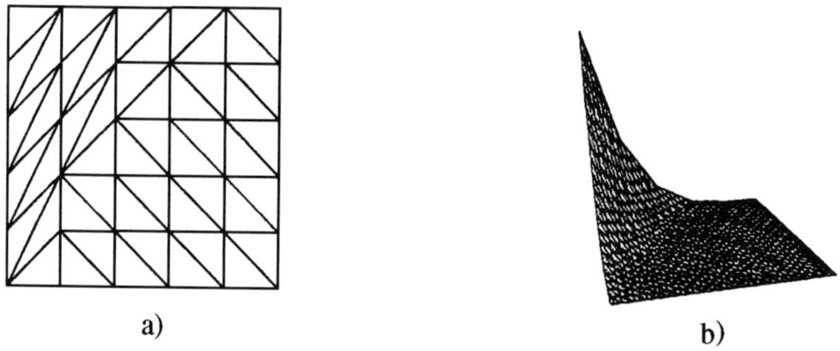

Fig. 6. Data-dependent least squares.

using the swap algorithm of §3.2.4. The idea is to swap an edge in the triangulation whenever the swap leads to a reduction in $\mathcal{E}(s)$.

This idea has been explored for linear C^0 splines in [60,63]. Fig. 6 shows the effect of adjusting the triangulation in fitting the function $f(x,y) = (y - x^2)_+$ defined in (3.8) on $\Omega = [0,1] \times [0,1]$ using C^0 linear splines with vertices at the corners of a 6×6 uniform grid, based on a set of measurements at 100 scattered points in Ω. In particular, Fig. 6b shows the surface associated with the triangulation in Fig. 6a. Clearly, this is a much better fit than the surface based on the Delaunay triangulation (see Figs. 5a,b).

3.3.4. Knot insertion and removal. The techniques of adding knots and removing knots is well established in the univariate and tensor-product spline literature [30,43,52]. Both techniques can also be employed when working with spline spaces $\mathcal{S}_d^r(\Delta)$. In this setting, *adding a knot* corresponds to inserting a new vertex, adjusting the triangulation, and recomputing an associated spline. Similiarly, *removing a knot* corresponds to deleting some vertex, adjusting the triangulation, and recomputing an associated spline. Both methods are most effective when the spline fit is accomplished by a local procedure. For a specific algorithm, see [50].

3.4. Shape control. In many applications it is important to construct a surface defined on a domain Ω which has certain shape properties. For example, we may want the surface to be positive, monotone, or convex in some sense. We discuss each of these briefly in the following subsections.

3.4.1. Positivity. There are a number of situations where we would like a surface to be nonnegative. One such case arises in approximating density functions. Suppose we are working with linear combinations of basis functions as in (3.3), and that each of the basis functions is itself nonnegative (which is always the case if we use box splines or simplex splines). Then to insure that the fitting surface s is nonnegative, we need only require that all coefficients in (3.3) be nonnegative. Vertex splines usually do not have this nonnegativity property.

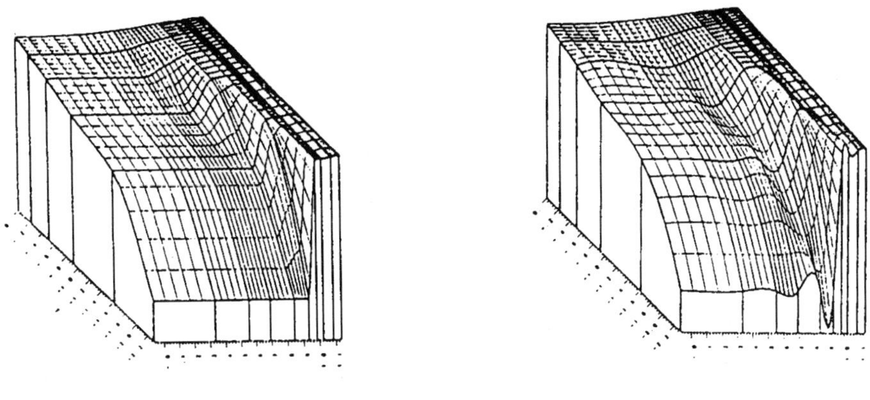

Fig. 7. Monotone and nonmonotone fits.

If we work with splines in $\mathcal{S}_d^r(\Delta)$ written in BB-form, then to insure that our fitting surface is nonnegative, we will have to enforce some conditions on the vector α of Bézier coefficients. For the space $\mathcal{S}_1^0(\Delta)$, a spline is nonnegative if and only if all of its coefficients are nonnegative. For higher-degree splines, weaker conditions on the Bézier coefficients α for nonnegativity can be found, see [56,57].

3.4.2. Monotonicity. We first have to agree on what we mean by monotonicity of a bivariate function. The following seems to be the most reasonable definition.

DEFINITION 2. *Let K be a cone in the plane $\mathbb{R}^2$ corresponding to two lines passing through the origin such that the angle at the origin does not exceed $\pi/2$. We say that a bivariate function s is monotone increasing relative to K provided that*

$$s(P_2) - s(P_1) \geq 0 \quad \textit{whenever the vector } P_2 - P_1 \textit{ lies in } K.$$

The most natural choice for the cone K is the quadrant defined by the x- and y-axes, and we restrict the remainder of our discussion to this choice. Given data on a rectangular grid which is consistent with the construction of a monotone increasing surface, there are four natural local methods for constructing a corresponding monotone increasing surface. They are based on

1) tensor-product C^1 bicubic splines [13],
2) C^1 cubic splines defined over the *Sibson split* of each rectangle (obtained by drawing in both diagonals of each rectangle) [39],
3) C^1 biquadratic splines based on a rectangular partition obtained by inserting one extra knot line between each of the original grid lines [4],
4) C^1 quadratic splines on the type-2 triangulation obtained by drawing in the diagonals in the subrectangle of the grid in 3), see [6].

A common feature of all of these methods is that appropriate gradient values must be provided at all vertices of the grid. In all cases, setting all of the gradients to zero leads to a monotone surface, but in general it will exhibit flat spots in regions near the grid points. These can be eliminated by choosing more realistic values for the gradients (by estimating them with local approximation methods), but then to assure monotonicity, these gradients typically have to be adjusted (downward). In all of these methods this involves multiple sweeps (usually three) through the grid points. For still other methods, see the survey [31]. Fig. 7 shows a typical example of what happens when we try to fit monotone data without shape control as compared to using one of these monotonicity-preserving methods.

If we are given data at points which do not lie on a grid, it is possible [39] to first convert this scattered data to gridded data which can then be fit with one of the above methods.

3.4.3. Convexity. Much work has gone into finding conditions on a BB-polynomial of degree d on a triangle T which guarantee that it is convex there, see [30,43]. An exact set of necessary and sufficient conditions must include nonlinear inequalities, although there exist sufficient conditions which involve only linear inequalities on the Bézier coefficients.

Suppose we are given data on a grid which is consistent with building a convex surface, and we want to build an interpolating surface which is also convex. For general triangulations, it does not seem easy to use the above results to define an algorithm for constructing a convex spline interpolant. One approach is to use Powell-Sabin macro elements, see [14].

It may also be of interest to construct a convex least squares fit. This has been done in [76] again using Powell-Sabin elements. This problem could also be approached using an arbitrary space $\mathcal{S}_d^r(\Delta)$ by enforcing the (sufficient) linear constraints for convexity on the Bézier coefficients. This leads to a linearly constrained quadratic minimization problem (which may not have any solution).

3.5. Surfaces on surfaces. So far, we have been discussing surfaces which are defined by a bivariate function defined on a set $\Omega \in \mathbb{R}^2$. For many applications, one needs to approximate real-valued functions which are defined on a domain Ω which itself lies on a surface in $\mathbb{R}^3$. The most common situation is when Ω lies on the surface of a sphere (such problems arise frequently in geophysics and meteorology). But other more complicated surfaces also arise naturally — the surface of a wing, for example [5].

One approach to dealing with surfaces on surfaces is to convert them to bivariate problems. For example, to work on the surface of a sphere, we can map it to a rectangle in $\mathbb{R}^2$, and then use bivariate methods [27]. This approach has one major problem: it is difficult to handle the north and south poles. This defect can be overcome by using tensor-products of polynomial and trigonometric splines [74].

A more natural spline approach on the sphere would be to use analogs of the polynomial spline space $\mathcal{S}_d^r(\Delta)$. To build such spaces, one needs a form of *spherical barycentric coordinates*. Until recently, it was generally believed that such coordinates did not exist, but in fact a complete theory of *spherical BB forms* can be constructed [2] which also works on general *sphere-like surfaces*.

3.6. 3D-reconstruction. Multivariate splines have also proved to be useful in an important problem which arises in medical imaging, nondestructive testing, and elsewhere. Given a sequence of contour lines obtained by slicing through a 3D object with a series of parallel planes, the problem is to construct a 3D-surface which passes through these contours. For a survey of this problem and how multivariate splines can be used to help solve it, see [70].

4. Computer-Aided Geometric Design

The idea in computer-aided geometric design (CAGD) is to find representations of curves and surfaces which are easy to store and manipulate on a digital computer, and which are easy to render on a graphical device (such as a computer screen). To be of the most use, these representations should have convenient *handles* consisting of a set of parameters which can be varied by the user to make well-defined changes in the curve or surface. In many cases one would also like to use the representation to drive some manufacturing device (such as a lathe or other cutting tool). This is usually referred to as *computer-aided manufacture* (CAM).

While many of the surfaces in CAGD applications can be described by bivariate functions on some appropriate domain (in which case standard spline fitting methods can be applied), more often the surfaces of interest have to be represented *parametrically*. The standard way to do this is to choose a domain $\Omega \in \mathbb{R}^2$ and three bivariate functions $f_i(x,y)$ defined on Ω. Then as (x,y) runs over Ω, the set of points

$$\boldsymbol{f}(x,y) = \begin{pmatrix} f_1(x,y) \\ f_2(x,y) \\ f_3(x,y) \end{pmatrix} \tag{4.1}$$

defines a surface lying in $\mathbb{R}^3$. For a survey of fitting scattered data using splines in parametric form, see [55].

A discussion of issues relating to CAGD (and extensive bibliographies) can be found in the textbooks [30,43]. In addition to designing and modelling surfaces, some of the questions of interest include

1) constructing *offset surfaces* to a given surface, i.e., given a surface s, find a second surface s_δ which (measured perpendicularly) lies at a fixed distance δ from s everywhere,
2) finding the *intersection curves* of two given surfaces,
3) filling in a given wire-frame with a smooth surface,
4) *crowning* a given surface, i.e., raising or lowering the surface in some locally defined area while leaving the rest of it unchanged,
5) *conversion* of one representation of a surface to some other.

Polynomial splines (on both rectangular and triangular partitions) are key tools in CAGD. Indeed, the whole idea of the Bernstein-Bézier representation of splines comes out of the CAGD literature.

There is one topic in applications of splines in CAGD which is unique to the parametric setting. One way to make a parametric surface appear smooth (say with continuously varying *tangent plane* or curvature) is to make its components smooth (say C^1 or C^2 continuous). However, it is also possible to achieve smooth-looking surfaces without this stringent condition on its components. This weaker type of smoothness is referred to as *visual* or *geometric* continuity. For details and references, see the survey [38] or the books [30,43].

5. The Finite Element Method

Engineers have been using spaces of piecewise polynomials defined on triangulations to solve boundary value problems involving partial differential equations (and other operator equations) for many years. The idea is to choose a finite-dimensional linear space of suitable functions (possibly satisfying the boundary conditions), and then to try to find an element in the space which approximately solves the operator equation. Given a basis for the approximating space, the problem usually reduces to finding the associated coefficients by solving an appropriate linear system of equations.

For problems involving bivariate functions, a common choice for the approximating space is the span of some finite collection of locally supported piecewise polynomials, usually defined on a triangulation. Typically, the sizes of triangles varies considerably in such meshes, with more triangles being used in regions where the function to be approximated is expected to change more rapidly. In practice, the most commonly used basis functions are macro elements as discussed in §2.4, or certain more general *supersplines* [19,49,68,69,71].

6. Image Analysis

Given a black and white image on a rectangle H, we can create a digital approximation to it by dividing H into a collection of rectangles, and then associating with each subrectangle (called a *pixel*) an integer (say between 0 and 255) which describes the brightness of that pixel (with 0 representing black and 255 representing white, for example). Thus, an image is just an integer-valued function on H. Such digital images can be easily stored, transmitted, and processed with computers.

To get high-resolution images, one has to use a large number of pixels. A typical size is $512 \times 512 = 262{,}144$ pixels. Since it takes 8 bits to represent an integer in the range $[0, 255]$, a monochrome digital image requires 2,097,152 bits to store. Color images are typically made up of three components (for example, an RGB image consists of combinations of red, green, and blue images). Thus, the same size color image requires 6,291,456 bits.

These large file sizes can become a problem when images have to be stored over a period of time (as happens with medical images). They are also a problem when one wants to transmit images using a fixed bandwidth channel. For example, to get smooth motion in video, one needs at least 24 frames per second. Thus transmitting color television pictures at 512×512 resolution would require transmitting 150,994,944 bits per second. The typical Ethernet network is able to handle about 10 million bits per second, while even fiber optic networks are limited to about 140 million bits per second. The requirements for high-definition television are even higher.

In view of this discussion, it is obvious that there is a significant need for *image compression methods*. We need to be able to represent (or code) the image in some way which requires many fewer bits, but which still allows 1) the exact reconstruction of the image (*lossless compression*, important in medical applications and elsewhere), or 2) an approximate reconstruction (*lossy compression*, acceptable for television and other applications).

Finding ways to mathematically describe an image is not only important for compression purposes, but is also useful if we want to *process the image* in some way. A typical application would be to sharpen the image by removing noise. Multivariate splines are finding more and more application in image processing, and there is room here to describe only a few of the main ideas.

6.1. Segmentation methods. One approach to approximating an image f defined on the square $H = [0, 511] \times [0, 511]$ is to use piecewise constants on a partition of H into subsquares. It can be computed by a *segmentation algorithm* which first approximates f by a single constant on all of H. If that approximation is not good enough, we split H into four subsquares and approximate f on each by a constant. Then we examine each subsquare, and if the approximation there is not good enough, we subdivide again. This process is repeated until an acceptable approximation is obtained. Any of the discrete ℓ_p-norms can be used as a measure of approximation error (although there is some evidence [25] that the ℓ_1-norm is preferable).

The result of this process is a piecewise constant surface involving pieces defined on squares of varying sizes (but always in dyadic position), see Fig. 8. It is not a tensor-product type surface. To store and reconstruct this surface, we have to know where each square is located, what size it is, and what constant value is assigned to it. This sounds complicated, but in fact all of the needed information can be encoded in a compact tree structure.

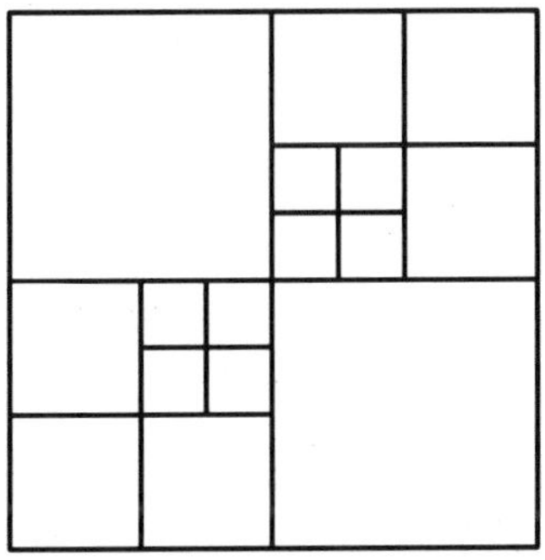

Fig. 8. A dyadic segmentation of an image.

6.2. Spline-based methods. Since an image is just an integer-valued surface defined on a rectangle H, we can also approximate it by a spline defined on a triangulation Δ based on some set of vertices lying in H (including the four corners). Since images are typically not very smooth, it seems natural to work with $\mathcal{S}_1^{-1}(\Delta)$ or $\mathcal{S}_1^0(\Delta)$. Clearly, we can use edge swapping to help improve the approximation (cf. §§3.2.4 and 3.3.3), and can add and remove vertices (cf. §3.3.4). This approach is currently under study, and preliminary results seem promising.

6.3. Wavelets. We begin with a quick review of basic wavelet theory as can be found in any of the recent books, e.g. [18,24]. To explain what a wavelet is, it is easiest to talk about univariate functions first. Suppose $\phi(x)$ is a function in $L_2(\mathbb{R})$, and define

$$(6.1) \qquad \mathcal{S}_k := \text{span } \{\phi_{jk}\}_{j=-\infty}^{\infty},$$

where

$$(6.2) \qquad \phi_{jk}(x) := \phi(2^k x - j), \quad \text{all integers } j, k.$$

Under appropriate conditions on ϕ, it is known [18,24] that

$$(6.3) \qquad \cdots \subset \mathcal{S}_{k-1} \subset \mathcal{S}_k \subset \mathcal{S}_{k+1} \subset \cdots,$$

$$(6.4) \qquad \cap\, \mathcal{S}_k = \{0\}, \qquad \cup\, \mathcal{S}_k \text{ is dense in } L_2(\mathbb{R}),$$

and there exist finite positive constants c and C such that for all $\{c_{jk}\}$,

$$(6.5) \qquad c\Big(\sum c_{jk}^2\Big)^{1/2} \le \|\sum c_{jk}\phi_{jk}\|_2^2 \le C\Big(\sum c_{jk}^2\Big)^{1/2},$$

where $\|\cdot\|_2$ denotes the usual L_2 norm.

This setup is usually referred to as a *multiresolution analysis*. The basic function ϕ is usually called a *scaling function*. By the nestedness (6.3) of the spaces, we can write

$$(6.6) \qquad \mathcal{S}_{k+1} = \mathcal{S}_k \oplus \mathcal{W}_k,$$

where $\mathcal{W}_k$ is the *orthogonal complement* of $\mathcal{S}_k$ in $\mathcal{S}_{k+1}$.

Under further conditions on ϕ, there exists a function ψ (usually called the *mother wavelet*) such that

$$(6.7) \qquad \mathcal{W}_k := \text{span } \{\psi_{jk}\},$$

where the ψ_{jk} are defined by *dilation* and *translation* just as in (6.2).

The simplest example of a multiresolution analysis is provided by the choice

$$(6.8) \qquad \phi(x) := \begin{cases} 1 & x \in [0,1), \\ 0 & \text{otherwise.} \end{cases}$$

This is just a classical B-spline of degree 0. In this case the corresponding mother wavelet (called the *Haar wavelet*) is given by

$$(6.9) \qquad \psi(x) := \begin{cases} 1 & x \in [0,.5), \\ -1 & x \in [.5,1), \\ 0 & \text{otherwise.} \end{cases}$$

Given a positive integer m, any $s \in \mathcal{S}_n$ has a decomposition of the form

$$(6.10) \qquad s := s_{n-m} + w_{n-m} + \cdots + w_n,$$

where $s_{n-m} \in \mathcal{S}_{n-m}$, and $w_i \in \mathcal{W}_i$, for $i = n-m, \ldots, n$. Each of these functions can be stored in terms of its vector of coefficients. These can be found using a convenient *decomposition algorithm*. A related *reconstruction algorithm* can then be used to reconstruct the original function s from its pieces.

What is important about the *wavelet decomposition* is that the function s is expanded in terms of a coarse function s_{n-m} which models the basic behavior of s, plus a sequence of wavelet expansions which give finer and finer details. This is of great use in signal processing since it allows compressing and processing signals in various ways by operating on the wavelet coefficients.

Scaling functions and associated mother wavelets can be constructed in many ways [18,24]. One natural approach is to use a polynomial B-spline for the scaling function, in which case the corresponding mother wavelet is also a spline [18].

The simplest way to get a multiresolution analysis for $L_2(\mathbb{R}^2)$ is to take tensor-products of two univariate scaling functions. In the bivariate case, three mother wavelets are required to define bases for the complementary spaces W_k. For example, *tensor-product splines* can be used [18].

To get bivariate wavelets without relying on tensor-products, we can use a *box spline* for the scaling function. Again the corresponding three mother wavelets can be found explicitly. For examples of the use of box splines on the three-direction mesh to do image compression, see [22,62].

7. Remarks

1) For functions of three variables, the natural analog of a triangulation is a *tetrahedral partition*. Given a set of points in $\mathbb{R}^3$ which are to serve as the vertices, it turns out that in contrast to the bivariate case, the number of tetrahedra is not uniquely determined [1]. Given a tetrahedral partition, the natural analog of $\mathcal{S}_d^r(\Delta)$ is the space of piecewise polynomials in three variables of total degree d, joined together with C^r smoothness. For theoretical results on such spaces, see [3]. For a survey of some interpolation methods using these trivariate splines, see [1]. They are also heavily used as finite elements.

2) Both simplex and box splines can be easily defined in any number of variables, and much of the theory carries over directly [11,12,21]. I am not aware of many applications in more than two variables.

3) In the univariate case a lot of work has gone into developing generalized spline spaces where the pieces are no longer polynomials, but are instead drawn from more general finite-dimensional linear spaces (such as trigonometric or exponential functions). It is also possible to build multivariate splines whose pieces are other than polynomials. For example, there is a well-developed theory of *exponential box splines* [64].

4) It is worth emphasizing the role of subdivision in applications of multivariate splines, especially in CAGD and in the theory of wavelets. The reader who wants to learn more about subdivision should consult the survey [28] and the monograph [15].

5) Multivariate splines are also useful in robotics and computer vision. A typical application would involve constructing a mathematical model of the 3D objects which a robot sees, and then using it to control the robot's movement.

6) A discussion of some aspects of *numerical machine control* can be found in [40,43].

7) Multivariate splines are also of use in various statistical settings. For example, in [34] they are used to do adaptive multivariate regression for data in many variables.

8) For the reader who wants to learn more about the topics discussed here, I recommend the monographs [8,12,15,16,24,27,30,43,47,66]. There is also a lot of related material in the conference proceedings [48,52,53,54]. The 1987 bibliography [33] contains over 1100 references on multivariate splines and applications.

9) I would like to thank Peter Alfeld, Carl de Boor, and Tom Lyche for useful comments on an earlier draft of this paper.

References

1. Alfeld, P., Scattered data interpolation in three or more variables, in *Mathematical Methods in Computer Aided Geometric Design*, T. Lyche and L. Schumaker (eds.), Academic Press (New York), 1989, 1–33.
2. Alfeld, P., M. Neamtu, and L. L. Schumaker, Bernstein-Bézier polynomials on the circle, sphere, and sphere-like surfaces, manuscript, 1993.
3. Alfeld, P., L. L. Schumaker, and W. Whiteley, The generic dimension of the space of C^1 splines of degree $d \geq 8$ on tetrahedral decompositions, SIAM J. Numer. Anal. **30** (1993), 889–920.
4. Asaturyan, S., and K. Unsworth, A C^1 monotonicity preserving surface interpolation scheme, in *Mathematics of Surfaces III*, D. C. Handscomb (ed.), Clarendon Press (Oxford), 1989, 243–266.
5. Barnhill, R., and H. Ou, Surfaces defined on surfaces, Comput. Aided Geom. Design **7** (1990), 323–336.
6. Beatson, R., and Z. Ziegler, Monotonicity preserving surface interpolation, SIAM J. Numer. Anal. **22** (1985), 401–411.
7. de Boor, C., Splines as linear combinations of B–splines, a survey, in *Approximation Theory II*, G. G. Lorentz, C. K. Chui, and L. L. Schumaker (eds.), Academic Press (New York), 1976, 1–47.
8. de Boor, C., *A Practical Guide to Splines*, Springer Verlag, New York, 1978.
9. de Boor, C., B–form basics, in *Geometric Modeling: Algorithms and New Trends*, G. E. Farin (ed.), SIAM Publications (Philadelphia), 1987, 131–148.
10. de Boor, C., Quasiinterpolants and approximation power of multivariate splines, in *Computation of Curves and Surfaces*, W. Dahmen, M. Gasca and C. Micchelli (eds.), Kluwer (Dordrecht), 1990, 313–345.
11. de Boor, C., Multivariate piecewise polynomials, Acta Numerica (1993), 65–109.
12. de Boor, C., K. Höllig, and S. D. Riemenschneider, *Box Splines*, Springer-Verlag, Berlin, 1993.
13. Carlson, R. E., and F. N. Fritsch, Monotone piecewise bicubic interpolation, SIAM J. Numer. Anal. **22** (1985), 386–400.
14. Carnicer, J. M., and W. Dahmen, Convexity preserving interpolation and Powell-Sabin elements, Comput. Aided Geom. Design **9** (1992), 279–289.
15. Cavaretta, A. S., W. Dahmen, and C. A. Micchelli, *Stationary Subdivision*, Mem. Amer. Math. Soc., Providence, 1993.
16. Chui, C. K., *Multivariate Splines*, CBMS-NSF Reg. Conf. Series in Appl. Math., vol. 54, SIAM, Philadelphia, 1988.
17. Chui, C. K., Vertex splines and their applications to interpolation of discrete data, in *Computation of Curves and Surfaces*, W. Dahmen, M. Gasca and C. Micchelli (eds.), Kluwer (Dordrecht), 1990, 137–181.
18. Chui, C. K., *An Introduction to Wavelets*, Academic Press, Boston, 1992.

19. Chui, C. K., and M. J. Lai, Multivariate vertex splines and finite elements, J. Approx. Theory **60** (1993), 245–343.

20. Daehlen, M., Modelling with box spline surfaces, in *Curve and Surface Design*, H. Hagen (ed.), SIAM Publications, SIAM (Philadelphia), 1992, 191–205.

21. Dæhlen, M., and T. Lyche, Box splines and applications, in *Geometric Modeling, Methods and Applications*, H. Hagen and D. Roller (eds.), Springer Verlag (Berlin), 1991, 35–93.

22. Dæhlen, M., and T. Lyche, Decomposition of splines, in *Mathematical Methods in Computer Aided Geometric Design II*, T. Lyche and L. Schumaker (eds.), Academic Press (New York), 1992, 135–160.

23. Dahmen, W., and C. A. Micchelli, Recent progress in multivariate splines, in *Approximation Theory IV*, C. Chui, L. Schumaker, and J. Ward (eds.), Academic Press (New York), 1983, 27–121.

24. Daubechies, I., *Ten Lectures on Wavelets*, SIAM, Philadelphia, 1992.

25. DeVore, R.A., Jawerth, B., and Lucier, B., Surface compression, Comput. Aided Geom. Design **9** (1992), 219–239.

26. DeVore, R.A., Jawerth, B., and Lucier, B., Image compression through transform coding, IEEE Trans. Inf. Theory **IT-38** (1992), 719–746.

27. Dierckx, P., *Curve and Surface Fitting with Splines*, Clarendon Press, Oxford, 1993.

28. Dyn, N., Subdivision schemes in CAGD, in *Advances in Numerical Analysis Vol. II: Wavelets, Subdivision Algorithms and Radial Basis Functions*, W.A. Light (ed.), Oxford University Press (Oxford), 1992, 36–104.

29. Dyn, N., D. Levin, and S. Rippa, Algorithms for the construction of data dependent triangulations, in *Algorithms for Approximation II*, J. C. Mason and M. G. Cox (eds.), Chapman & Hall (London), 1990, 185–192.

30. Farin, G., *Curves and Surfaces for Computer Aided Geometric Design*, Academic Press, New York, 1988.

31. Fontanella, F., Shape preserving interpolation, in *Computation of Curves and Surfaces*, W. Dahmen, M. Gasca and C. Micchelli (eds.), Kluwer (Dordrecht), 1990, 183–214.

32. Franke, R., Scattered data interpolation: tests of some methods, Math. Comp. **38** (1982), 181–200.

33. Franke, R., and L. L. Schumaker, A bibliography of multivariate approximation, in *Topics in Multivariate Approximation*, C. K. Chui, L. L. Schumaker, and F. Utreras (eds.), Academic Press (New York), 1987, 275–335.

34. Friedman, J., Multivariate adaptive regression splines, Ann. Statist. **19** (1991), 1–141.

35. Goldman, R., and T. Lyche (eds.), *Knot Insertion and Deletion Algorithms for B-spline Curves and Surfaces*, SIAM (Philadelphia), 1993.

36. von Golitschek, M., and L. L. Schumaker, Data fitting by penalized least squares, in *Algorithms for Approximation II*, J. C. Mason and M. G. Cox (eds.), Chapman & Hall (London), 1990, 210–227.

37. Goodman, T. N. T., Polyhedral splines, in *Computation of Curves and Surfaces*, W. Dahmen, M. Gasca and C. Micchelli (eds.), Kluwer (Dordrecht), 1990, 347–382.

38. Gregory, J. A., Geometric continuity, in *Mathematical Methods in Computer Aided Geometric Design*, T. Lyche and L. Schumaker (eds.), Academic Press (New York), 1989, 353–371.

39. Han, Lu, and L. L. Schumaker, Monotone fitting of scattered data using C^1 cubic splines, preprint, 1991.

40. Held, M, *On the Computational Geometry of Pocket Machining*, Lecture Notes in Computer Science, vol. 500, Springer Verlag, Berlin, 1991.

41. Höllig, K., Multivariate splines, in *Approximation Theory, Proc. Symp. Appl. Math.*, C. de Boor (ed.), Amer. Math. Soc. (Providence), 1986, 103–127.

42. Höllig, K., Box-spline surfaces, in *Mathematical Methods in Computer Aided Geometric Design*, T. Lyche and L. Schumaker (eds.), Academic Press (New York), 1989, 385–402.

43. Hoschek, J., and D. Lasser, *Fundamentals of Computer Aided Design*, AKPeters, Boston, 1993.

44. Jetter, K., A short survey on cardinal interpolation by box splines, in *Topics in Multivariate Approximation*, C. K. Chui, L. L. Schumaker, and F. Utreras (eds.), Academic Press (New York), 1987, 125–139.

45. Jetter, K., Multivariate approximation: A view from cardinal interpolation, in *Approximation Theory VII*, E. W. Cheney, C. Chui, and L. Schumaker (eds.), Academic Press (New York), 1992, 131–161.

46. Lai, Ming-Jun, and L. L. Schumaker, *Splines on Triangulations*, book, in progress.

47. Lancaster, P., and K. Salkauskas, *Curve and Surface Fitting, An Introduction*, Academic Press, New York, 1986.

48. Laurent, P.-J., A. Le Méhauté, and L. L. Schumaker (eds.), *Curves and Surfaces*, Academic Press (New York), 1991.

49. Le Méhauté, A., A finite element approach to surface reconstruction, in *Computation of Curves and Surfaces*, W. Dahmen, M. Gasca and C. Micchelli (eds.), Kluwer (Dordrecht), 1990, 237–274.

50. Le Méhauté, A., and Y. Lafranche, A knot removal strategy for scattered data in $\mathbb{R}^2$, in *Mathematical Methods in Computer Aided Geometric Design*, T. Lyche and L. Schumaker (eds.), Academic Press (New York), 1989, 419–426.

51. Lucier, B., Wavelets and Image Compression, in *Mathematical Methods in Computer Aided Geometric Design II*, T. Lyche and L. Schumaker (eds.), Academic Press (New York), 1992, 391–400.

52. Lyche, T., and R. Goldman (eds.), *Knot Insertion and Deletion Algorithms for B-spline Modeling*, SIAM (Philadelphia), 1992.

53. Lyche, T., and L. L. Schumaker (eds.), *Mathematical Methods in Computer Aided Geometric Design*, Academic Press, Boston, 1989.

54. Lyche, T., and L. L. Schumaker (eds.), *Mathematical Methods in CAGD and Image Processing*, Academic Press, Boston, 1992.

55. Mann, S., C. Loop, M. Lounsbery, D. Meyers, J. Painter, T. DeRose, and K. Sloan, A survey of parametric scattered data fitting using triangular interpolants, in *Curve and Surface Design*, H. Hagen (ed.), SIAM Publications, SIAM (Philadelphia), 1992, 145–172.

56. Micchelli, C. A., and A. Pinkus, Some remarks on nonnegative polynomials on polyhedra, in *Probability, Statistics and Mathematics: Papers in Honor of Samuel Karlin*, T. Anderson, K. Athreya and D. Iglehart (eds.), Academic Press (San Diego), 1989, 163–186.

57. Nadler, E., Nonnegativity of bivariate quadratic functions on a triangle, Comput. Aided Geom. Design **9** (1992), 195–205.

58. Quak, E., and L. L. Schumaker, Cubic spline fitting using data dependent triangulations, Comput. Aided Geom. Design **7** (1990), 293–301.

59. Quak, E., and L. L. Schumaker, Calculation of the energy of a piecewise polynomial surface, in *Algorithms for Approximation II*, J. C. Mason and M. G. Cox (eds.), Chapman & Hall (London), 1990, 134–143.

60. Quak, E., and L. L. Schumaker, Least squares fitting by linear splines on data dependent triangulations, in *Curves and Surfaces*, P.-J. Laurent, A. Le Méhauté, and L. L. Schumaker (eds.), Academic Press (New York), 1991, 387–390.

61. Riemenschneider, S. D., Multivariate cardinal interpolation, in *Approximation Theory VI*, C. Chui, L. Schumaker, and J. Ward (eds.), Academic Press (New York), 1989, 561–580.

62. Riemenschneider, S. D., and Zuowei Shen, Box splines, cardinal series, and wavelets, in *Approximation Theory and Functional Analysis*, C.K. Chui (ed.), Academic Press (New York), 1991, 133–149.

63. Rippa, S., Long and thin triangles can be good for linear interpolation, SIAM J. Numer. Anal. **29** (1992), 257–270.

64. Ron, A., Linear independence of the translates of an exponential box spline, Rocky Mountain J. Math. **22** (1992), 331–351.

65. Schumaker, L. L., Fitting surfaces to scattered data, in *Approximation Theory II*, G. G. Lorentz, C. K. Chui, and L. L. Schumaker (eds.), Academic Press (New York), 1976, 203–268.

66. Schumaker, L.L., *Spline Functions: Basic Theory*, Wiley, New York, 1981 (reprinted Krieger, Malabar, Florida, 1993).

67. Schumaker, L. L., Triangulation methods, in *Topics in Multivariate Approximation*, C. K. Chui, L. L. Schumaker, and F. Utreras (eds.), Academic Press (New York), 1987, 219–232.

68. Schumaker, L. L., Constructive aspects of bivariate piecewise polynomials, in *Mathematics of Finite Elements and Applications VI*, J. Whiteman (ed.), Academic Press (London), 1988, 513–520.

69. Schumaker, L. L., On super splines and finite elements, SIAM J. Numer. Anal. **26** (1989), 997–1005.

70. Schumaker, L. L., Reconstructing 3D objects fom cross-sections, in *Computation of Curves and Surfaces*, W. Dahmen, M. Gasca and C. Micchelli (eds.), Kluwer (Dordrecht), 1990, 275–309.

71. Schumaker, L. L., Recent progress on multivariate splines, in *Mathematics of Finite Elements and Applications VII*, J. Whiteman (ed.), Academic Press (London), 1991, 535–562.

72. Schumaker, L. L., Triangulation methods in CAGD, IEEE Computer Graphics and Applications **13** (1993), 47–52.

73. Schumaker, L. L., Computing optimal triangulations using simulated annealing, Comput. Aided Geom. Design **10** (1993), 329–345.

74. Schumaker, L. L, and C. Traas, Fitting scattered data on spherelike surfaces using tensor products of trigonometric and polynomial splines, Numer. Math. **60** (1991), 133–144.

75. Traas, C. R., Practice of bivariate quadratic simplicial splines, in *Computation of Curves and Surfaces*, W. Dahmen, M. Gasca and C. Micchelli (eds.), Kluwer (Dordrecht), 1990, 383–422.

76. Willemans, K., and P. Dierckx, Surface fitting using convex Powell-Sabin splines, preprint, 1992.

DEPARTMENT OF MATHEMATICS, VANDERBILT UNIVERSITY, NASHVILLE, TN 37240

E-mail address: s@mars.cas.vanderbilt.edu

Proceedings of Symposia in Applied Mathematics
Volume 48, 1994

INITIAL VALUE PROBLEMS FOR ORDINARY DIFFERENTIAL EQUATIONS: DEVELOPMENT OF IDEAS, TECHNIQUES, AND IMPLEMENTATION

HANS J. STETTER

ABSTRACT. Some aspects of the development of computational techniques for solving ODEs since 1950 are surveyed. Necessarily, the selection has a very personal flavor; many important facets had to be omitted for shortage of space. Note that this is *not* a survey of the present status of the numerical solution of ODEs. Likewise, the bibliography had to be restricted to a few entries per subject area typical for its development; it should not be considered representative for the period after 1960.

1. EARLY DEVELOPMENT

Numerical solution of ordinary differential equations is an old subject; this is best shown by the names associated with the fundamental approaches:

Leonhard Euler (1707-1783), one of the great mathematicians of all times, Swiss-born, worked in St. Petersburg and Berlin.
John C. Adams (1819-1892), an English astronomer whose computations led to the discovery of Neptune, eventually professor in London.
Carl Runge (1856-1927), one of the most renowned applied mathematicians in Germany, eventually professor in Göttingen.
Martin W. Kutta (1867-1944), also known for the Kutta-Joukowsky formula in aerodynamic wing theory, eventually professor in Stuttgart.

An excellent presentation of the state-of-the-art of 50 years ago is to be found in Collatz's comprehensive volume *Numerische Behandlung von Differentialgleichungen* ([10]) of 1951 whose first chapter deals with initital value problems for ordinary differential equations. It describes the well-known classical methods; naturally, the treatment is entirely based on "hand computation", i.e., on the use of desk calculators, which necessitates elaborate discussions about suitable controls. (In the preface, Collatz mentions the development of new automatic computing machinery "whose impact cannot yet be reliably judged"!) The derivation of the formulas follows classical lines, there is no presentation of a more comprehensive underlying theory. On the other hand, there are rather elaborate a priori error estimates. The

1991 *Mathematics Subject Classification.* Primary 01-08, 65-03, 65L05.
Key words and phrases. Ordinary differential equations, initial value problems, numerical methods, historical development.
This paper is in final form and no version of it will be submitted for publication elsewhere.

© 1994 American Mathematical Society
0160-7634/94 $1.00 + $.25 per page

other textbook from that time, Milne's *Numerical Solution of Differential Equations* ([37]), supplements this picture of the situation around 1950.

In this connection, it is exciting to find the first use of *one-sided Lipschitz constants* for perturbation error estimates in systems of ODEs in a Supplement of Collatz's book [10]. The definition is as in (4.4) and the potential for a negative value is emphasized. Collatz reports about a lecture given by H. Eltermann in January 1951; Eltermann's paper [18] did not appear until 1955.

In comparison, the mid-century state-of-the-art of the numerical treatment of initial value problems for (hyperbolic and parabolic) *partial differential equations* appears more advanced, as a result of the fundamental work of the Göttingen-New York school of R. Courant and his students. Also, in PDEs, a proper analysis of the stability situation was required before any useful computational work could be undertaken while the analogous requirements for classical ODE methods (Runge-Kutta and Adams) were trivial.

The first paper in *MTAC* on computational ODEs is found in volume 4 of 1950: F.J. Murray ([39]) reports on "Planning and Error Considerations for the Numerical Solution of a System of Differential Equations on a Sequence Calculator". He uses an Adams predictor-corrector method of order 4, and his analysis contains a number of modern aspects of the subject:

> classification of local errors (discretization, truncation after one corrector step, round-off),
> appraisal of the local errors by comparison with a local solution of the differential system,
> analysis of the error propagation where, however, only the variational equation is considered while a potential instability (weak or strong) of the discretization is ignored,
> procedures for stepchanging and other implementation questions for a system with as many as 14 equations.

The same volume of *MTAC* also contains J. Todd's "Solution of differential equations by recurrence relations" ([54]) which points out the phenomenon of strong instability which may occur in multipstep methods based on *several* preceding y-values. H. Rutishauser's *ZAMP* paper [44] of 1952 analyzes the true nature of the weak instability of Simpson's method which had been described as a numerical effect ("Aufrauhungserscheinung") in Collatz's book; similar analyses were presented by a few others at about the same time. (Actually, Rutishauser *and* Dahlquist had presented papers on weak instability at the GAMM Meeting in 1951, cf. [12]). Thus, by the mid-50s, the time was ripe for a firm theoretical foundation of the stepwise computation of approximate solution values of ordinary initial value problems. This foundation was laid by Germund Dahlquist, a (then) young Swedish mathematician, in his *Math. Scand.* paper [13] of 1956 and his Ph.D. thesis published as a KTH report [14] in 1959.

The 1956 paper introduced *stable convergence* (= convergence under perturbation) of the sequence of approximations generated by a difference method, and it identified the *stability* of the consistent difference operator as the necessary and sufficient condition for stable convergence. Stability was shown to be equivalent to the *root criterion* for the "ρ-part" of a multistep difference operator. (The ρ, σ-notation for the polynomials associated with a multistep operator in this paper has remained in use until this day.) The paper also contained the so-called *Dahlquist barrier* for the consistency order of a stable multistep operator, and the "Roman-Greek transformation" $\zeta = \frac{z+1}{z-1}$ for its proof. Thus, it also led the way in the use of complex function theory for the analysis of ODE discretizations (cf. § 3.4 for a much more refined example). Also the asymptotic approach ($h \to 0$) was widely used in the analysis.

In his thesis, Dahlquist extended the theory through an analysis of *weak instability*, introducing *growth factors* and proving related results. He also introduced the matrix function $\mu[A] := \lim_{h \to 0^+}(\| I + hA \| - 1)/h$, which was later called *logarithmic norm*, and used it for more realistic a priori error bounds.

The fundamental platform for the great subsequent expansion of ODE methods was completed with the appearance of P. Henrici's *Discrete Variable Methods in Ordinary Differential Equations* ([29]) and its supplement *Error Propagation for Difference Methods* ([30]) in 1962 and 1963 resp. This didactically excellent text organized the available knowledge (for an important exception, see § 4.2) about discretization methods for ordinary initial value problems in a systematic and transparent way; it introduced classifications, terminology and notations, which have stood the test of time. In particular, it introduced the $h \to 0$ analysis in a rigorous form. In the preface of my monograph [48] of 1973, I have called Henrici's text a "lighthouse", a distinction which still appears justified 20 years later.

Meanwhile, *electronic digital computers* had become generally available, and procedural, machine-independent programming languages made their use easier. Systems of ODEs of all varieties and sizes waited for their numerical solution. The relatively low speed and small memory of the computers of that time led to a search for more powerful methods and for clever implementations, so that larger and more complicated systems could be solved in a shorter time. A tremendous growth of the subject began, which continued for about 20 years.

2. Fundamentals of Computational ODE Solving

For the historical sketch of § 1, we have assumed that the reader has a vague knowledge of computational ODE solving. Now we will shortly survey its fundamentals to provide some background and notation for later technicalities.

The initial value problem for a general *explicit* system of s ordinary differential equations of the first order has the form

$$y'(t) = f(t, y(t)), \quad y(t_0) = y_0 , \tag{2.1}$$

where – for some suitable domain D in $\mathbb{R}^s$ and $T > t_0$ – we have $y_0 \in D$, the solution y maps $[t_0, T]$ into D and $f : [t_0, T] \times D \to \mathbb{R}^s$. We will denote the value of the solution of (2.1) at t by $y(t; t_0, y_0)$ where appropriate.

For the sake of existence and uniqueness, f is generally assumed to be continuous in t and Lipschitz continuous in y. Almost all discretization methods for (2.1) make sense only with further smoothness assumptions on f. However, finite jumps in f or its derivatives do occur in real world models and must be handled by general-purpose codes.

Without further special assumptions on f, the problem (2.1) has very little mathematical structure, which is even more obvious from the equivalent integral equation reformulation:

$$(2.2) \qquad y(t;t_0,y_0) = y_0 + \int_{t_0}^{t} f(\tau, y(\tau;t_0,y_0))d\tau \;, \qquad t \in [t_0,T] \;.$$

Naturally, we may split the integral into finitely many parts:

(2.3)

$$y(t;t_0,y_0) = y_0 + \int_{t_0}^{t_1} f(\tau,y(\tau))d\tau + \int_{t_1}^{t_2} f(\tau,y(\tau))d\tau + \cdots + \int_{t_{n-1}}^{t} f(\tau,y(\tau))d\tau \;,$$

and we may consider the solution value at some $\hat{t}$ as a new initial value:

$$(2.4) \qquad y(t;t_0,y_0) = y(t;\hat{t},y(\hat{t};t_0,y_0)) \;, \quad t_0 < \hat{t} < t \;.$$

The relations (2.3) and (2.4) suggest that a computational solution of (2.1) should proceed through a *recursive solution* of *local initial value problems*

$$(2.5) \qquad y'(t) = f(t,y(t)), \qquad y(t_{\nu-1}) = y_{\nu-1},$$

on a finite sequence of *steps* $[t_{\nu-1},t_\nu]$. This has led to the *discretization methods* or *discrete variable methods* which generate a sequence

$$(2.6) \qquad \{y_0,y_1,\ldots,y_\nu,\ldots\} \;, \qquad y_\nu \in \mathbb{R}^s \;,$$

of approximate solution values on a *grid*

$$(2.7) \qquad G := \{t_0,t_1,\ldots,t_\nu,\ldots\} \subset [t_0,T] \;,$$

with

$$(2.8) \qquad y_\nu \approx y(t_\nu;t_0,y_0) \;.$$

The grid G may be prespecified, e.g. equidistant ($t_\nu := t_0 + \nu \cdot h$); more often it will be determined in the course of the computation.

Actually, in a *one-pass* forward-stepping sequence of local problem solutions, the goal of achieving a good agreement in (2.8) can be approached only indirectly. The deviation in (2.8) – the *global discretization error* – reflects the complete computational history; it depends on the accuracy with which the local problems have been solved – the *local discretization error* – *and* on the sensitivity of y_ν on these errors in all previous steps – the *stability* of our discretization scheme.

This sensitivity should – qualitatively and quantitatively – *reflect the sensitivity of the analytic solution* of (2.1) to perturbations along its path, and this sensitivity should not depend significantly on the chosen grid. The exclusion of an unlimited growth of the sensitivity with an arbitrary *refinement* of the grid represents the original 0-*stability* requirement of Dahlquist and others, which is necessary for convergence. The requirement of a uniformly moderate sensitivity in the integration of *stable systems* of ODEs leads to considerable restrictions in the structure of the numerical method. We will return to the subject of "stiff problems" in § 4.2.

The natural request for the computation of an approximate solution *function* defined on the whole interval of integration has received little attention for a long time. We will shortly consider this wider view of a computational solution of (2.1) in § 5.1; until then, we will concentrate on the customary discretized view of (2.5) - (2.8).

In the approximate solution of the local problem (2.5), we have to determine y_ν such that

$$y_\nu \approx y_{\nu-1} + \int_{t_{\nu-1}}^{t_\nu} f(\tau, y(\tau; t_{\nu-1}, y_{\nu-1}))d\tau \; ; \tag{2.9}$$

obviously, the principal difficulty is the implicit occurrence of the unknown local solution in the integrand of (2.9). Therefore, a crude *classification* of methods for the computational solution of initial value problems for ODE systems (2.1) may be based on the approach used to overcome this implicitness of what would otherwise be a simple quadrature.

Such classification aspects are:

1) Is the local problem (2.5) solved
   - in a detached fashion,
   - with information from the preceding computation?

   The terms "one-step" and "multistep" have been associated with this distinction.
2) In the computational solution of (2.5), does the right-hand side f enter
   - only through its values at a number of argument $(s+1)$-tuples,
   - also through values of its derivatives along the trajectory of (2.5)?

   The second approach (e.g. Taylor series methods) has remained largely theoretical so far; it may gain significance with the availability of efficient automatic differentiation codes (cf., e.g., the implementations of Chang and Corliss, [9]).
3) Is the computation of y_ν from (2.9) effected
   - by a *recursive* evaluation of successive values of f (and derivatives),
   - through the (approximate) solution of a *system of equations*?

   The terms "explicit" and "implicit" have been associated with this distinction.
4) In the computational solution of the local or the global problem, do we pass over the respective interval of integration
   - once or several times?

   While the Gragg-Bulirsch-Stoer (GBS) extrapolation scheme ([24], [5]) is the only multipass method in practical use today, differently conceived multipass methods (e.g., based on defect correction) could become attractive for the use on concurrent computer architectures.

Many more options are available beyond these principal ones, and the mathematical simplicity of the problem permits almost unlimited variations. The count of publications dealing with yet another variant of yet another approach easily reaches three-digit numbers, even without those turned down by reluctant referees. The more sobering is a look at the main numerical software collections [32] and [40]: The codes which have accounted for 99.9 % of the computational ODE solving over the past 20 years have been based on four different discretization methods, with only very few variants.

For non-stiff systems, we find

- pairs of explicit Runge-Kutta (RK) methods, with either a fixed pair or automatic switching between a few pairs of different orders;

- the Adams predictor-corrector method, with automatic order selection;
- the GBS extrapolation method, with automatic control of the extrapolation depth;

for stiff systems, all standard codes employ

- some variant of the BDF method, with automatic order selection.

The harsh requirement that a useful numerical method must permit an efficient, robust, and reliable implementation has withered the beautiful flowers which bloomed on thousands of journal pages. A few valuable suggestions may have been overlooked through unfortunate coincidences, but only changes in the *paradigms* of scientific computing – like widespread parallel computation or efficient automatic differentiation – are likely to change the composition of the ODE chapters of numerical software libraries, at least as far as general purpose codes are concerned.

In the area of special purpose codes for restricted but important classes of problems, there is still a demand for good implementations of proposed methods, or even for new approaches. We will return to this subject in § 4.

3. Some Mathematical Concepts and Tools

The design of new methods, their analysis and understanding has been facilitated by the development of a number of mathematical techniques mainly in the 60s and 70s. In the following, we will sketch a few of them.

3.1. Asymptotic analysis for $h \to 0$. As in many areas of mathematics, limit considerations have proved to be a powerful tool in discretization analysis. Although the process of "letting the stepsize h go to zero" cannot be realized computationally, it has led to important concepts and to results which are qualitatively correct for realistic stepsizes.

The technique is straightforward for equidistant grids where all quantities depend on the step h in a well-defined fashion. Neglection of all but the lowest powers of h leads to explicit expressions for the lowest order terms of quantities like the local and global discretization errors and to very transparent asymptotic representations of many discretization mechanisms, e.g. in predictor-corrector methods. The *order* of a quantity is the exponent of the lowest h-power associated with it. This approach has been introduced as a general tool by Henrici ([29]).

For non-equidistant grids, one has to introduce sequences of *coherent* grids (which arise through proportional subdivision of the mother grid) to retain a clean analysis, cf., e.g., [48]. But a good number of important qualitative results are derivable under the more realistic assumptions

$$\frac{h_{max}}{h_{min}} \le c < \infty \quad \text{and} \quad h_{max} \to 0 \,. \tag{3.1}$$

Stronger smoothness assumptions on the problem (2.1) permit the establishment of more terms in an asymptotic expansion, notably of the global discretization error, which forms a basis for the use of the *Richardson extrapolation* technique which had become well-known through Romberg's quadrature algorithm, cf., e.g., [47]. Since the implicitness of (2.9) prohibits the reuse of integrand values, the local integration routine for (2.5) must be very cheap to justify the multipass approach. The two-step midpoint rule turned out to be the appropriate scheme for an analog to Romberg quadrature, the Gragg-Bulirsch-Stoer method of [24] and [5].

Recently, E. Hairer ([27]) has announced a rigorous asymptotic formulation for the *backward error* of discretization methods: The approximate solution values y_n generated by a 0-stable method of order p for the system (2.1) are exact values of the solution of a nearby system whose perturbation permits an asymptotic expansion which may be specified.

3.2. Symmetry. Some integration schemes for ODEs remain invariant under a reversal of time, e.g. the two-step midpoint rule

$$y_{\nu+1} = y_{\nu-1} + 2hf(t_\nu, y_\nu) \ . \tag{3.2}$$

The asymptotic expansions for such *symmetric schemes* can contain only even powers of h, which makes them ideal, e.g., for Richardson extrapolation as in the GBS-method. In the case of multistep schemes, like (3.2), the starting and stepsize change mechanisms must be kept consistent with the time invariance of the forward-stepping procedure, otherwise the symmetry is lost, cf., e.g., [48].

The design of pairs of schemes such that their alternate use constitutes a symmetric method, makes this concept more widely applicable.

Symmetry is also of importance in the integration of Hamiltonian systems; see § 4.5.

3.3. Trees and Butcher series. In the analysis of the local discretization error of higher order methods, it becomes more and more complicated to follow the ramifications of successive substitutions into, or differentiations of, the right-hand side f of (2.5), even if one restricts consideration (without loss of generality) to autonomous systems with $f : \mathbb{R}^s \to \mathbb{R}^s$.

J. C. Butcher ([6]) has established a formal one-to-one relation between the *elementary differentials* of f, which appear through repeated differentiation of $y'(t) = f(y(t))$ with respect to t, and the set $\mathbf{T}$ of rooted trees. This makes it possible to index the formal expansion of $y_{\nu+1}$ about y_ν in powers of h by the set $\mathbf{T}$; such expansions have been named *Butcher series.* Various relations for their coefficients may then be characterized through tree operations. Thus, e.g., the rapidly growing number of algebraic relations which must be satisfied by the coefficients of RK-methods of increasing order, become manageable.

For a number of other related problems, similar tree models have been established by the research group of E. Hairer and G. Wanner in Geneva, e.g. for 2nd order equations $y'' = f(y, y')$ and for differential-algebraic equations of indices 1, 2 and 3, and for Hamiltonian systems by Sanz-Serna. Such models and the associated Butcher series permit not only the derivation of the order conditions for discretization methods for these problems but also a deeper understanding of the structure of such methods. For a survey, see, e.g., [27].

3.4. Stability regions and order stars. While the implications of 0-stability (the "root condition" for multistep methods) have been known since the late 50s, the general requirement of a qualitative and quantitative *agreement* between the perturbation sensitivities of the analytic problem and its discretization may be interpreted in various ways.

It was soon commonly accepted that a perturbation of the discretization of a *stable* system of ODEs should "die out" as the integration continues. Necessary conditions for this situation may obviously be obtained from the discretization of the "test problem"

$$y'(t) = \lambda \cdot y(t) \ , \qquad y(0) = 1 \ , \tag{3.3}$$

with a constant step h: The solution sequence $\{y_n\}$ of the discretization should tend to zero for $\lambda \in \mathbb{C}_- = \{z \in \mathbb{C}, \text{Re } z < 0\}$.

For explicit methods, it was found that this requirement could be satisfied only for small values of $h|\lambda|$; therefore, one determined those sets $H_0 \subset \mathbb{C}$ such that $h\lambda \in H_0$ implied $y_n \to 0$ for $n \to \infty$ for a particular discretization method and called them *stability regions*. For weakly (un)stable methods in the sense of Dahlquist's thesis, H_0 is empty. A large H_0 was considered desirable; the natural *A-stability* requirement $H_0 \supseteq \mathbb{C}_-$, which was introduced in [15], can only be achieved for (special) *implicit* methods.

Various authors considered *relative stability regions* $H_\mu, \mu > 0$, which contain the values of $h\lambda$ such that $e^{\mu h n} y_n$ tends to zero for the respective discretization of (3.3), which corresponds to the decay of the analytic solution for Re $\lambda < -\mu$. The method is called *L-stable* (cf. [17]) if every $H_\mu, \mu > 0$, contains the "far negative end" of the complex plane.

In 1978, Wanner, Hairer and Nørsett ([55]) discovered that the regions obtained by a direct comparison of the growth rates of the solutions of equation (3.3) and of its discretization for each $\lambda \in \mathbb{C}$ are not only esthetically more pleasing but also convey far more insight into the mathematical structure of the discretization. For an implicit one-step method, where $y_n = [R(h\lambda)]^n$ with a rational growth function R, they called the domain

$$\{z \in \mathbb{C} : \ |R(z)| > |e^z|\} \tag{3.4}$$

the *order star* of the method; later they generalized the concept to multistep methods. For more details, cf. the monographs [28] and [33].

4. Diversification

Not all ODEs problems are equal!

The concentration on explicit first-order systems, without further special assumptions, was helpful in the early stages of the development because it assisted the introduction of commonly accepted concepts and approaches. However, practically from the beginning of numerical ODE solving, it has been realized that different methods are not equally effective for different classes of problems, so that special methods for special problem classes were considered.

4.1. Second order systems $y''(t) = f(t, y(t), y'(t))$. Such systems, even without an explicit occurrence of y', appear in a great number of physical models, notably in astronomy. Although they may be reduced to first order systems, the development of methods for their direct numerical integration appeared more natural. The multistep formulas of Störmer (1907), Cowell (1910) and Numerov (1924), and the one-step methods of Nyström (1925) are well-known early methods which have been in general use ever since. They have been described in the early texts on numerical ODEs, e.g. [10] and [37].

Theoretical foundations for these and also more general methods for 2nd order systems were laid in Dahlquist's thesis [14] and in Henrici's book [29].

4.2. Stiff problems. For the purposes of this historical survey, let us denote a problem (2.1) as *stiff* when it admits strongly damped solution components and possesses an ultimately smooth solution. Due to the ubiquity of such problems, it must have been a well-known phenomenon around 1950 that the use of a reasonably large stepsize may lead to violent error growth, with all methods in use at that time. However, the phenomenon is not mentioned in Collatz's or Milne's books, nor is it treated in Henrici's book [29] of 1962, a fact which has definitely had a retarding effect on the numerical community's awareness of this important problem.

The term "stiff" was apparently coined in the 1952 paper [11] of Curtiss and Hirschfelder who also proposed a valid cure, viz. BDF (backward differentiation formulas). In 1960, the first volume of the then famous "Ralston-Wilf" – an early predecessor of *Numerical Recipes* – contains a chapter by J. Certaine on "The solution of ordinary differential equations with large time constants" ([7]), which are "systems of 1st order coupled differential equations where the derivatives take on a form which makes the system unsuitable for direct use of substitution processes such as that of Runge-Kutta".

Serious mathematical analysis began again in Sweden, with H.O. Kreiss's ([34]) and G. Dahlquist's ([15]) *BIT* papers of 1962 and 1963, respectively. But it took another decade until it became standard to exclude stiffness in investigations which used the "classical" approach of assuming that the coefficients and remainder terms in asymptotic representations are of moderate size. At the same time, scientific activities in the area of stiff problems began to explode. (From the point of view of this survey, it is noteworthy how the additional structure in the original problem (2.1) implied by stiffness has made both its efficient computational solution and the mathematical analysis of its discretizations so much more difficult and interesting.)

I cannot even begin to enumerate the many results which have been accumulated over the last 20 years. Major centers in this effort have been (in alphabetical order)

Amsterdam, Auckland, Geneva, Leiden, Stockholm, Vienna.

The volume [3], compiled and edited by R.C. Aiken, gives a good overview of the development of the research on stiff problems.

A major tool for progress has been the consideration of *model problems*:

– linear problems:

$$y'(t) = Ay(t) \quad \text{and} \quad y'(t) = A(t)y(t) \;; \tag{4.1}$$

– weakly nonlinear problems:

$$y'(t) = Ay(t) + g(t, y(t)) \;; \tag{4.2}$$

– singular perturbation problems:

$$\begin{aligned} y'(t) &= f_1(y(t), z(t)) \;, \\ \epsilon z'(t) &= f_2(y(t), z(t)) \;, \; \epsilon > 0 \text{ small}; \end{aligned} \tag{4.3}$$

– one-sided Lipschitz bounds:

$$
(4.4) \qquad (2.1) \text{ with } \quad \begin{array}{rcl} \langle f(t,y) - f(t,z), y - z \rangle & \leq & m \parallel y - z \parallel^2 , \\ \text{or} \quad \mu \left[\frac{\partial f(t,y(t))}{\partial y} \right] & \leq & m ; \end{array}
$$

– Prothero-Robinson model ([43]):

$$
(4.5) \qquad y'(t) = \lambda(y(t) - \varphi(t)) + \varphi'(t) ;
$$

and others. Each of these models has led to elaborate analyses and to new concepts and techniques and given insight into the mechanisms of error generation and propagation in discretizations of stiff systems.

The essential difficulty consists in establishing that the terms which arise in the analysis are not adversely affected by the presence of extremely fast decay rates, even after the corresponding solution components have died out, cf., e.g., [21]. Within suitably restricted problem classes, convergence for $h \to 0$ and even asymptotic expansion results can be proved. The convergence orders may be lower than in the classical analysis of the same method (*order reduction*), cf., e.g., [22].

In spite of the worldwide research in this area, some important aspects, notably for nonlinear systems with a wide range of potential decay rates, have not been fully understood. And – while a host of methods has been proposed – various implementations of BDF methods continue to serve as workhorses for the computational solution of real-life stiff problems.

4.3. Differential-algebraic equations. Differential systems on manifolds or "differential-algebraic" equations may be regarded as the limiting case of the singular perturbation model (4.3) for $\epsilon \to 0$; thus, the behavior of discretizations for DAEs sheds light on the situation with stiff problems and vice versa. DAEs are considered in detail in the paper by L. Petzold in this volume.

4.4. Highly oscillatory problems. These are problems (2.1) whose solutions are composed of a smoothly varying and of a "nearly periodic" vector-valued function, i.e., they are oscillations whose waveform and period varies slowly with time (relative to the period), and where the solution is sought over a very large number of cycles. For such problems, one cannot and does not want to follow the trajectories; instead one resorts to the computation of *quasi-envelopes*. A quasi-envelope is a smooth function z which interpolates the values taken by the highly oscillatory function y at points progressing by a quantity T which is approximately equal to the momentary period of the oscillation. Thus, for some time interval beginning at $\bar{t}$, a quasi-envelope z will satisfy

$$
(4.6) \qquad z(\bar{t} + mT) = y(\bar{t} + mT) , \quad m = 0(1)M ,
$$

where y is the exact solution of the given problem and M is compatible with the variation in the period.

If a variable s is introduced which counts the cycles, we have

$$
(4.7) \qquad \begin{array}{rcl} t(s+1) - t(s) & = & T(t(s)) \\ z(s+1) - z(s) & = & g(s, z(s)) , \end{array}
$$

with

$$g(s,z) \quad := \quad y(t(s+1);t(s),z) - z \; . \tag{4.8}$$

Since s is a fast variable for the problem, (4.7) may be approximated by

$$\begin{aligned} t'(s) &= T(t(s)) \; , \\ z'(s) &= g(s,z(s)) \; . \end{aligned} \tag{4.9}$$

Computational methods for such problems have been developed in L. Petzold's thesis [42]; cf. also [23]. The *outer integration* solves (4.9) with large steps; but an evaluation of the right-hand side requires a (re)determination of the current period and the change in waveform. For this, several cycles have to be computed in detail from the original system (2.1) with sufficiently small steps (*inner integration*).

The period determination – or detection when the highly oscillatory behavior begins after an initial transient behavior – is based on finding recurrent values of the derivative y' through an investigation of successive zeros of some $c^T y'$, where $c^T y' = 0$, at the start of the inner iteration.

A good number of theoretical and implementational questions for discretizations of problems of this type are still in a preliminary form.

4.5. Hamiltonian systems. Time-independent Hamiltonian systems with s degrees of freedom are of the form

$$p'(t) = -\frac{\partial H}{\partial q}(p(t),q(t)), \qquad q'(t) = \frac{\partial H}{\partial p}(p(t),q(t)) \; , \tag{4.10}$$

with the Hamiltonian function $H : \mathbb{R}^s \times \mathbb{R}^s \to \mathbb{R}$, $p, q : \mathbb{R} \to \mathbb{R}^s$. A great number of models, e.g. in astronomy, molecular dynamics etc., are of this type. The *flow* of such systems $(p(0), q(0)) \to (p(t), q(t))$ preserves the value of the exterior quadratic form $dp \wedge dq$, i.e. the sum of the areas of the projections of a surface element in phase space onto the p_i, q_i-planes. A numerical method for which the generated discrete flow $(p(t), q(t)) \to (p(t+h), q(t+h))$ has the same property is called *canonical* or *symplectic*.

The investigation of symplectic methods began about 10 years ago, e.g. by Feng K. in China ([20]); it has been flourishing within the last 5 years. Symplectic methods of consistency order p generate the solution of a perturbed Hamiltonian system, with an $O(h^p)$ perturbation. Therefore, in the computational solution of systems (4.10) over long time intervals, symplectic methods are superior, both in their error vs. time and in their error vs. cost behavior.

While early designs of symplectic methods used higher derivatives of H explicitly, it has been found that standard methods may also be symplectic if they satisfy particular *symmetry constraints*, e.g.

$$b_i b_j = b_i a_{ij} + b_j a_{ji} \; , \quad i,j = 1(1)m \; , \tag{4.11}$$

for an m-stage RK-method with the coefficient scheme (a_{ij}, b_j). Thus, the m-stage Gauss methods are symplectic. The explicit 2-step midpoint method is also symplectic, which is of importance for the use of the *leap-frog method* in PDEs. *Partitioned* RK-methods which use different RK-schemes for the integration of p and q may even be explicit and yet be symplectic.

For Hamiltonians of the form $H(p,q) = \frac{1}{2}p^Tp + V(q)$, one has the equivalent 2nd order system

$$q''(t) = -\frac{\partial V}{\partial q}(q(t)) \,. \tag{4.12}$$

There exist explicit RK-Nyström methods for such systems which are symplectic.

For an overview of the development and the current status of the research on this subject, cf. Sanz-Serna's review [45].

4.6. Delay-differential equations. Systems of *retarded* functional differential equations

$$y'(t) = f(t, y(t), y(t-\alpha)) \tag{4.13}$$

where $\alpha \geq 0$ may depend on t and/or $y(t)$, and where there may be several retarded arguments, are natural extensions of standard ODE systems (2.1), since many real-life phenomena have "memory". Except when $\alpha(t) \leq t$, initial values for (4.13) must be specified on a suitable initial inteval.

For the computational solution of (4.13), standard discretization methods may be used which possess a procedure for generating a *continuous approximate solution* in each step, after the approximation at the next gridpoint has been computed, cf., e.g., [56]. The f-evaluations can use these solution values for the retarded arguments, except when $t_n - \alpha > t_{n-1}$. Naturally, the accuracy of the continuous approximation procedure must be related to that of the main method if no reduction in its order is to take place.

Since delay-differential systems (4.13) exhibit a number of phenomena which do not occur in standard ODE systems (cf., e.g., [16]), some special properties of their discretizations may be needed for special types of such problems. In particular, the computational solution should remain bounded for *stable* delay differential systems. This leads to restrictions for discretization and interpolation.

5. Implementation

5.1. Computational models. Unfortunately, a bad implementation of a good method will produce a poor code just as a good implementation of a bad method. Implementation is the engineering part of numerical mathematics, and it is just as demanding as the analysis and design of methods. It is an illusion to believe that a good engineer merely follows his experience and intuition. A good engineer casts his thorough theoretical and practical knowledge of a situation into a *model* and fits his design to that model. The better the model, the better the design.

In the design of computational modules, we have to model the *computational situations* with which we deal (cf., e.g., [53] and [26]). Once we have analyzed and understood the computational situation and its potential degeneracies sufficiently well, we know at least what our design has to achieve. The realization is largely a matter of craftsmanship.

Since we are dealing with an analytic problem about which we can only collect finite information, we can never completly determine the model parameters or the model structure. In this situation, the assumed validity of the model is equivalent to "silent assumptions" about the admissible problems, which can rarely be rigorously formulated.

These difficulties could be significantly reduced if some analytic information were available during the computation, like bounds for derivatives etc. Actually, such information could often be obtained through *automatic differentiation* of the arithmetic expressions specifying the differential equations. So far, none of the standard software for ODEs makes use of automatic differentiation, although this tool has matured in recent years; cf. the excellent volume [25] compiled by Griewank and Corliss. An example of what may be achieved in this fashion is the estimation of the location and the character of the *nearest singularity* from the decay rates of the Taylor coefficients of the local solution at some (t_n, y_n). Chang and Corliss ([8]) have shown that this is feasible in an automated fashion.

Typical modules which are needed in codes for initial value problems of ODEs must deal with the following tasks:

acceptance test for the currently computed step,
computation of a stepsize for the next step,
computation of a first step (at start or restart),
order or method control if applicable,
event detection and localization (of a specified condition or of an extraordinary situation),
stiffness detection (also of onset and/or disappearance of stiffness),
input-output control,
communication with calling program or user (in interactive mode),
graphical representation of the computed approximate solution.

Each one of these and other tasks has had its historical development and may still find better realizations. A recent interesting approach ([26]) to stepsize control applies principles of *feedback control* to achieve a smoother and more efficient stepsize sequence in critical situations.

The last task in the above list requires the generation of an approximate solution $\tilde{y}_n(t)$ in each step $[t_{n-1}, t_n]$ whose accuracy order should be consistent with that of the values y_ν on the computational grid. The $\tilde{y}_n$ may be formed either by a posteriori interpolation of the y_ν, f_ν, or during the computation of the y_ν, e.g. in "continuous RK-methods" ([56]) and multistep predictor-corrector methods ([52]).

5.2. Testing. The performance of a code for a reasonably wide class of analytic problems like (2.1) cannot be rigorously established by deduction; hence testing is mandatory. In the early 70s, a more systematic approach to the testing of ODEs software was proposed by F. Krogh ([35]) and by T. E. Hull and his group in Toronto ([31]). The latter project eventually led to the DETEST and STIFF DETEST packages [19] which are universally used for a comparative testing of software for ordinary initial value problems.

DETEST has achieved a *standardization* of the testing procedure which is a prerequisite for comparative testing in three ways:

(i) the performance criteria and their measurement have been rigorously defined in a code-independent fashion;
(ii) the cost of a computation has been defined in a standardized way;
(iii) a set of test problems has been collected and standardized.

The package has been tailored for the evaluation of general-purpose software; but its philosophy and approach could be extended to special-purpose situations.

Objective (i) has been met by the generation of a smoothed continuous error vs. cost performance function for each code from the measured data at a discrete set of tolerances. Thus all codes may be compared on the same error levels while the test computations yield different errors for each code, with a potentially non-smooth dependence on the tolerance values. (The achievement of a reasonably smooth *tolerance proportionality* is also a quality mark of a code.)

The two sets (nonstiff and stiff) of test problems have been the source of some controversy. They reflect the state-of-the-art of the late 70s and have all kinds of bias. But a code which performs uniformly worse on the test set than some established code is very unlikely to qualify as a superior general-purpose code on an enlarged test set. Many additional test problems have been suggested; but no supplementary set has gained universal acceptance so far.

Naturally, the selection of a production code for a narrow problem area requires a customized testbed. Today, the design of standardized testbeds appears desirable for codes which are to be used on concurrent architectures. This might encourage the design of ODE codes which run uniformly well on a relatively wide range of concurrent architectures, a research and development topic of growing importance.

6. Error Estimates and Bounds

In the early times of numerical analysis, rigorous a priori error bounds for the result of a numerical computation were considered indispensible. A number of early papers on computational ODEs were largely or exlusively devoted to error bounds. Although these bounds were often unrealistically pessimistic, they were accepted as proof for the "correctness" of the numerical method. In his thesis [14], Dahlquist introduced *logarithmic norms* (not under this name) to obtain more realistic a priori error bounds for multistep methods without weak instability; Henrici's books [29,30] devote considerable space to bounds for the discretization error and probabilistic bounds for the round-off errors.

With the general availability of computers and with their greater speed and wordlength, the interest in rigorous a priori error bounds for the computational solution of ODEs disappeared, like in other areas of numerical mathematics. They were replaced by asymptotic results; it became customary to run problems with different stepsizes (or tolerances) and to believe in the coinciding digits.

6.1. Error estimates. An *error estimate* is an *approximate computational solution* of the following mathematical problem: Given the actual computation for the approximate value of some quantity, find the error of this approximation. A 10% relative accuracy (i.e., one digit) is usually satisfactory for an error estimate, sometimes the correct order of magnitude and sign.

Estimates of the *local error* (however defined) are needed in each ODEs code for the control of the stepsize (and possibly order) and for various diagnostic purposes. The fact that a sound error estimate permits a correction of the computed approximate value poses an *efficiency dilemma*. Many codes choose to continue with the corrected value ("local extrapolation") which may make the previous error estimate meaningless even in a qualitative sense.

An estimate of the *global error* (deviation between computed and true trajectory) requires (cf., e.g., [50]):

– either parallel computation on two coherent grids with local steps $2h$ and h, so that *Richardson extrapolation* yields an error estimate

$$e_h(t) = \frac{1}{2^p - 1}[\tilde{y}_{2h}(t) - \tilde{y}_h(t)] \ , \qquad t \in G_{2h} \ , \tag{6.1}$$

for a method of order p,

– or a reliable estimate of the *local defect* $\bar{d}(t)$ of the computational solution, e.g. via the defect of a continuous approximate local solution $\bar{y}(t)$, and an approximate integration of the *error equation:*

$$e'(t) = f(t, \bar{y}(t)) - f(t, \bar{y}(t) - e(t)) + \bar{d}(t), \tag{6.2}$$

with a vanishing initial value.

For *nonstiff* systems (2.1), the transition to the variational equation

$$e'(t) = \frac{\partial f}{\partial y}(t, \bar{y}(t)) \cdot e(t) + \bar{d}(t) \tag{6.3}$$

requires Jacobian values which are normally not available from the primary computation. A *defect correction* approach, on the orginal grid but with a less expensive integrator, gives a very cheap procedure with a reasonable accuracy, cf. [51].

For *stiff* systems, the situation is less trivial. Even if the Jacobian values needed in (6.3) are now often available, it appears unreasonable to follow local defect components which are completely damped out after a few steps. What we want to estimate is the difference between the *smooth* solutions of the discretized and of the analytic problem, at least in cases with a clear distinction between the transient and the smooth phase of the computation. This may require projections on "smooth subspaces", which is easy only for suitably partitioned systems.

In the NAG library [40], a global error estimate may be optionally obtained for their RK code, with the extrapolation approach.

It is often claimed that global error estimates are superfluous because the effects of modelling errors will exceed the computational errors considerably. Even then, a user should be able to know how accurate a code solves his problem so that he can adjust this accuracy to the error level of his model. Besides, the concurrent computation of a decent estimate of the global error may also permit a better control of the primary computation. When we put so much ingenuity into the computation of (hopefully!) accurate trajectory values, why should we neglect the less demanding task of error estimation? Reruns with different tolerance values are both more costly and less reliable. Maybe, the spread of concurrent architectures where the global error estimation may be delegated to a seperate processor will make error estimation more a matter of course.

6.2. A posteriori error bounds. The current interest in dynamical systems has shed more light on the fact that even in ODE systems without chaotic behavior there may exist segments along a trajectory where tiny pertubations lead to a qualitative change in the trajectory's further course. E. Adams et al. ([2]) have analyzed computations of trajectories which pass closely by a singularity of the differential system; they have exhibited cases where an insufficient containment of the local error in just one step of the computation leads to an excessive qualitative deviation of the subsequently computed trajectory.

It is clear that a discrete local error control, however refined, can be mistaken about the current situation, so that - for a variety of locally sensitive differential systems - it cannot be guaranteed that the computation is not following a completely wrong trajectory. In such case, a global error estimation may similarly be fooled about this transition to a wrong trajectory.

From this and other points of view it appears desirable to have a computational tool which permits a rigorous *checking of the validity* of a computed ODE solution. Such a validation tool must generate rigorous and realistic a posteriori error bounds; this is equivalent to the generation of a *set-valued* trajectory $Y(t)$ such that

$$y(t) \in Y(t) \quad \text{for} \quad t \in [t_0, T] \tag{6.4}$$

is a correct mathematical assertion. The *function tube* Y will be called an *inclusion* of the exact solution y of (2.1).

In principle, the computation of such set-valued inclusions appears straightforward since (see (2.2))

$$y(t) = y_0 + \int_{t_0}^{t} f(\tau, y(\tau))d\tau, \qquad t \in [t_0, T], \tag{6.5}$$

is an equivalent formulation of (2.1) in *fixed point form*:

$$y = Fy. \tag{6.6}$$

As is well known from the study of Volterra integral equations, F is contractive on the Banach space of continuous functions on $[t_0, T]$, within the domain of f, if this Banach space if furnished with a suitably weighted norm. Hence, if we have succeeded in generating a function tube Y_0 such that

$$Y_1(t) := y_0 + \int_{t_0}^{t} f(\tau, Y_0(\tau))d\tau \subseteq Y_0(t), \qquad t \in [t_0, T], \tag{6.7}$$

we have mathematically established Y_1 as an inclusion for y, by Banach's fixed point theorem. Also, an iterative repetition of (6.7) should lead to successively narrower inclusions. In (6.7), $f(t, Z)$ denotes the range set of the f-values on some set Z in $\mathbb{R}^s$.

Naturally, one can execute this procedure in a stepwise fashion:

$$Y_{1,n}(t) := Y_{n-1}(t_{n-1}) + \int_{t_{n-1}}^{t} f(\tau, Y_{0,n}(\tau))d\tau \subseteq Y_{0,n}(t), \quad t \in [t_{n-1}, t_n], \tag{6.8}$$

with $y(t_{n-1}) \in Y_{n-1}(t_{n-1})$, where Y_{n-1} has been the final inclusion on the previous step. This approach has been introduced by R. E. Moore in his book [38].

The computational realization of (6.8) such that satisfactorily narrow inclusions are obtained over sufficiently long time intervals is a delicate problem. R. Lohner ([36]) has elaborated and successfully implemented an approach which works well for nonstiff systems. In his code, the inclusion sets $Y_n(t)$ are affine images $A_n \cdot \left[\underline{y}_n(t), \bar{y}_n(t)\right]$ of s-dimensional intervals whose bounds are polynomials in t: Upper and lower bounds for the integral in (6.8) are computed by Taylor expansion, with automatic differentiation, and a rigorous inclusion of the remainder term which requires a coarse a priori inclusion of y on $[t_{n-1}, t_n]$. The A_n have to be updated such that the widening effect of a rotation of the coordinate system along the trajectory on the interval inclusion is minimized. All inaccuracies, including all rounding errors, must be accounted for.

For stiff systems, an efficient inclusion method can naturally not be based on Taylor polynomials. So far, no efficiently implementable procedure which would permit reasonably large steps has been proposed.

6.3. Validated solution of ODEs as a mathematical tool. Since all steps (6.8) except the first one must use a set-valued initial value $Y_{n-1}(t_{n-1})$, a variation of the orginal initial values y_0 of (6.5) (i.e. of (2.1)) in a set Y_0 may also be allowed. Furthermore, f may depend on parameters c_j which are only known to be within some interval C_j; this is formally handled by treating the c_j as solution components, with differential equations $c_j' = 0$, but interval initial values $c_j(0) \in C_j$. Thus a code for the validated solution of an ODE system permits also the rigorous bounding of the *effects of uncertain data* in ODE systems.

Furthermore, such a code may be used to *prove* mathematical assertions about such systems *computationally*, e.g. the *existence of periodic solutions.* If the flow issuing from some surface element in phase space may be shown to return into the interior of that element, the associated Poincaré mapping is contractive on the element and hence has a fixed point. In [1], this approach has been successfully realized for a mathematical model of the famous Oregonator, an apparently periodic chemical process:

The question whether the dynamical system

$$\begin{aligned} y'(t) &= \tfrac{1}{s}\cdot\left(-y(t)-\alpha\left(y(t)\right)y(t)+z(t)\right) , \\ z'(t) &= w\cdot\left(\alpha\left(y(t)\right)-z(t)\right) , \end{aligned} \tag{6.9}$$

where

$$\alpha(y) := \frac{1}{2q}(1-y+\sqrt{4qy+(1-y)^2})$$

and

$$q = 8.375\cdot 10^{-6} , \quad s = 77.27 , \quad w = 0.161 ,$$

admits a periodic solution cannot be answered analytically. By the computation of an inclusion tube for all solutions emanating from a suitable segment in phase space, the group of E. Adams has found that the tube crosses the *interior* of the initial segment after one cycle; he has thus provided a *computational proof* that the periodicity has been retained in the simple model (6.9) of the Oregonator. For this delicate computation, the stepsizes in Lohner's code had to be adjusted interactively, with a variation between $5\cdot 10^{-5}$ and 1, over a time intervall of 300.

Homoclinic and heteroclinic orbits of the Lorenz equation have been enclosed (and thus proven to exist) in the study [46].

Similary, validated solution of ODE systems may permit a *computational proof* that a given dynamical system admits *chaotic motion*:

The existence of a *transversal homoclinic point* implies the potential for chaos. Thus, if one has followed the unstable and stable manifolds M_u and M_s from an equilibrium to an intersection and then computationally carried the positions of the tangential manifolds along the trajectories in M_u and M_s to this intersection point, one may "see" the transversality. This computation is turned into a mathematical proof if the *computed inclusions* of the tangential manifolds at the homoclinic point contain no common linear manifold. In this spirit, A. Neumaier [41] has been able to establish the potential for chaos for a discrete dynamical system.

References

1. E. Adams, A. Holzmüller, D. Straub: The Periodic Solutions of the Oregonator and Verification of Results, in: Scientific Computation with Automatic Result Verification, U. Kulisch, H.J. Stetter (eds.), Computing Suppl. **6** (1988), 111-121.
2. E. Adams, W.F. Ames, W. Kühn, W. Rufeger, H. Spreuer: Computational Chaos May Be Due to a Single Local Error, J. Comput. Phys. **104** (1993), 241-250.
3. R.C. Aiken (Ed.), Stiff Computation, Oxford University Press, New York–Oxford, 1985.
4. A. Bellen, M. Zennaro: Stability Properties of Interpolants for Runge Kutta Methods, SIAM J. Numer. Anal. **25** (1988), 411-432.
5. R. Bulirsch, J. Stoer: Numerical Treatment of Ordinary Differential Equations by Extrapolation Methods, Numer. Math. **8** (1966), 1-13.
6. J.C. Butcher: Coefficients for the Study of Runge-Kutta Integration Processes, J. Austral. Math. Soc. **3** (1963), 185-201.
7. J. Certaine: The Solution of Ordinary Differential Equations with Large Time Constants; in: Mathematical Methods for Digital Computers, A. Ralston, H. Wilf (eds.), Wiley, New York, 1960, pp. 128-132.
8. Y.F. Chang, G.F. Corliss: Ratio-like and Recurrence Relation Tests for Convergence of Series, J. Inst. Math. Appl. **25** (1980), 349-359.
9. Y.F. Chang, G.F. Corliss: Solving Ordinary Differential Equations Using Taylor Series, ACM Trans. Math. Software **8** (1982), 114-144.
10. L. Collatz, Numerische Behandlung von Differentialgleichungen, Springer, Berlin–Göttingen–Heidelberg, 1951.
11. C.F. Curtiss, J.O. Hirschfelder: Integration of Stiff Problems, Proc. Nat. Acad. Sci. USA **38** (1952), 235-243.
12. G. Dahlquist: Fehlerabschätzungen bei Differenzenmethoden zur numerischen Integration gewöhnlicher Differentialgleichungen, Z. Angew. Math. Mech. **31** (1951), T 239-240.
13. G. Dahlquist: Convergence and Stability in the Numerical Integration of Ordinary Differential Equations, Math. Scand. **4** (1956), 33-53.
14. G. Dahlquist: Stability and Error Bounds in the Numerical Integration of Ordinary Differential Equations, Trans. Roy. Inst. Technol. Stockholm, No. 130 (1959).
15. G. Dahlquist: A Special Stability Problem for Linear Multistep Methods, BIT **3** (1963), 27-43.
16. R.D. Driver, Ordinary and Delay Differential Equations, Appl. Math. Sci. 20, Springer, 1977.
17. B.L. Ehle: On Padé Approximations to the Exponential Function and A-stable Methods for the Numerical Solution of Initial Value Problems, Thesis, Univ. Waterloo, 1969.
18. H. Eltermann: Fehlerabschätzungen bei näherungsweiser Lösung von Systemen von Differentialgleichungen erster Ordnung, Math. Z. **62** (1955), 469-501.
19. W.H. Enright, J.D. Pryce: Two FORTRAN Packages for Assessing Initial Value Methods, ACM Trans. Math. Software **13** (1987), 1-27.
20. Feng K.: Difference Schemes for Hamiltonian Formalism and Symplectic Geometry, J. Comput. Math. **4** (1986), 279-289.
21. R. Frank, J. Schneid, C.W. Überhuber: The Concept of B-Convergence, SIAM J. Numer. Anal. **18** (1981), 753-780.
22. R. Frank, J. Schneid, C.W. Überhuber: Order Results for Implicit Runge-Kutta Methods Applied to Stiff Systems, SIAM J. Numer. Anal. **22** (1985), 515-534.
23. C.W. Gear: Automatic Detection and Treatment of Oscillatory and/or Stiff Ordinary Differential Equations, in: Numerical Integration of Differential Equations and Large Linear Systems, Lecture Notes in Math. **968** (1982), 190-206.
24. W.B. Gragg: On Extrapolation Algorithms for Ordinary Initial Value Problems, SIAM J. Numer. Anal. **2** (1965), 384-403.
25. A. Griewank, G.F. Corliss (Eds.), Automatic Differentiation: Theory, Implementation, and Application, SIAM, Philadelphia, 1991.
26. K. Gustafsson: Control of Error and Convergence in ODE Solvers; Ph.D. Thesis, Dept. Autom. Control, Lund Inst. Tech., 1992. Partly published in "Control Theoretic Techniques for

Stepsize Selection in Explicit Runge-Kutta Methods, ACM Trans. Math. Software **17** (1991) 533-554.
27. E. Hairer: Trees – a Survey, presented at Intern. Conf. Sci. Comput. and Differential Equations, Auckland NZ, January 1993, to appear in Annals of Numer. Math, vol. 1: SCADE'93, J.C. Baltzer, Basel, 1994.
28. E. Hairer, G. Wanner, Solving Ordinary Differential Equations II - Stiff and Differential-Algebraic Problems, Springer, Berlin–Göttingen–Heidelberg, 1991.
29. P. Henrici, Discrete Variable Methods in Ordinary Differential Equations, J. Wiley & Sons, New York–London, 1962.
30. P. Henrici, Error Propagation for Difference Methods, J. Wiley & Sons, New York–London, 1963.
31. T.E. Hull, W.H. Enright, B.M. Fellen, A.E. Sedgwick: Comparing Numerical Methods for Ordinary Differential Equations, SIAM J. Numer. Anal. **9** (1972), 603-637.
32. IMSL, MATH/LIBRARY, User's Manual, FORTRAN Subroutines for Mathematical Applications, Version 1.0, April 1987.
33. A. Iserles, S.P. Nørsett, Order Stars, Chapman & Hall, London–New York, 1991.
34. H.O. Kreiss: Über die Stabilitätsdefinition für Differenzengleichungen, die partielle Differentialgleichungen approximieren, BIT **2** (1962), 153-181.
35. F. Krogh: On Testing a Subroutine for the Numerical Integration of Ordinary Differential Equations, Calif. Inst. Tech., JPL TM 217 (1970).
36. R. Lohner: Enclosing the Solutions of Ordinary Initial and Boundary Value Problems, in: Computer Arithmetic, E. Kaucher, U. Kulisch, Ch. Ullrich (eds.), Teubner, Stuttgart, 1987, pp. 255-286.
37. W.E. Milne, Numerical Solution of Differential Equations, J. Wiley & Sons, New York–London, 1953.
38. R.E. Moore, Interval Analysis, Prentice Hall, Englewood Cliffs, N.J., 1966.
39. F.J. Murray: Planning and Error Considerations for the Numerical Solution of a System of Differential Equations on a Sequence Calculator, Math. Tables Aids Comput. **4** (1950), 133-144.
40. NAG, FORTRAN Library, Library Manual, Mark 12, March 1987.
41. A. Neumaier, Th. Rage: Rigorous Chaos Verification in Discrete Dynamical Systems, to appear in Physica D.
42. L.R. Petzold: An Efficient Numerical Method for Highly Oscillatory Ordinary Differential Equations, Ph. D. Thesis, Dept. Comput. Sci., Univ. Illinois, 1978.
43. A. Prothero, A. Robinson: On the Stability and Accuracy of One-Step Methods for Solving Stiff Systems of Ordinary Differential Equations, Math. Comp. **28** (1974), 145-162.
44. H. Rutishauser: Über die Instabilität von Methoden zur Integration gewöhnlicher Differentialgleichungen, Z. Angew. Math. Phys. **3** (1952), 65-74.
45. J.M. Sanz-Serna: Symplectic Integrators for Hamiltonian Problems: an Overview, Acta Numerica **1** (1991), 243-286.
46. H. Spreuer, E. Adams: On the Existence and the Verified Determination of Homoclinic and Heteroclinic Orbits of the Origin for the Lorenz Equations, in: Validation Numerics, R. Albrecht, G. Alefeld, H.J. Stetter (eds.), Computing Suppl. **9** (1993), 233-246.
47. H.J. Stetter: Asymptotic Expansions for the Error of Discretization Algorithms for Nonlinear Functional Equations, Numer. Math. **7** (1965), 18-31.
48. H.J. Stetter, Analysis of Discretization Methods for Ordinary Differential Equations, Springer, Berlin–Heidelberg–New York, 1973.
49. H.J. Stetter: The Defect Correction Principle and Discretization Methods, Numer. Math. **29** (1978), 425-443.
50. H.J. Stetter: Global Error Estimation in ODE-Solvers, in: Numerical Analysis, Lecture Notes in Math. **630** (1978), 179-189.
51. H.J. Stetter: Global Error Estimation in Adams PC-Codes, ACM Trans. Math. Software **5** (1979), 415-430.
52. H.J. Stetter: Interpolation and Error Estimation in Adams PC-Codes, SIAM J. Numer. Anal. **16** (1979), 311-323.
53. H.J. Stetter: Performance Evaluation of O.D.E. Software through Modelling, in: Performance Evaluation of Numerical Software, Ll. Fosdick (ed.), North-Holland, 1979, pp. 175-184.

54. J. Todd: Solution of Differential Equations by Recurrence Relations, Math. Tables Aids Comput. **4** (1950), 39-44.
55. G. Wanner, E. Hairer, S.P. Nørsett: Order Stars and Stability Theorems, BIT **18** (1978), 475-489.
56. M. Zennaro: Natural Continuous Extensions of Runge-Kutta Methods, Math. Comp. **46** (1986), 119-133.

INSTITUTE FOR APPLIED AND NUMERICAL MATHEMATICS, TECHNICAL UNIVERSITY OF VIENNA, WIEDNER HAUPTSTR. 8–10, A–1040 VIENNA, AUSTRIA

E-mail address: stetter@uranus.tuwien.ac.at

Proceedings of Symposia in Applied Mathematics
Volume **48**, 1994

Multiresolution Methods for Partial Differential Equations

ROGER TEMAM

ABSTRACT. Decomposition methods have played, and keep playing, an important role in numerical analysis: examples include the decomposition of operators, as in the splitting or alternating direction methods, and domain decomposition methods currently much used in parallel computing.

Multiresolution methods are another type of decomposition methods: they consist in decomposing the set of unknowns into subsets playing different roles. This is the case, for instance, with multigrid methods. For dissipative evolution equations (such as Kuramato-Sivashinsky equations, Navier-Stokes equations, thermohydraulics, etc.), the recently introduced concept of inertial manifolds suggests a differentiated treatment of high and low frequencies, or small-scale and large-scale components of the solution. In this lecture we will present some algorithms based on such decompositions of unknowns, including the nonlinear Galerkin method for spectral and finite element approximations, and the incremental unknowns method for finite differences.

Introduction

Computational Mathematics has undergone profound changes during the past half century since the time where a significant part of computing was devoted to the compilation of tables and their judicious utilization. Nevertheless, all those involved in computational mathematics know that this field will undergo again profound changes in coming years with the advent of parallel computing.

The introduction of parallel computing into scientific computing will affect programming as well as algorithms.

The utilization of parallel architectures will range from an automatic but clever redistribution of elementary operations among processors to the development of adapted algorithms and of appropriate procedures allowing a complete application problem to be split into subproblems treated by the processors. The emphasis at this stage is on programming and the development of suitable algorithms such as domain decomposition and splitting methods (decomposition of operators).

Another impact of massively parallel computing is perhaps less known. Because of the considerable increase in computing power that it will generate, bringing

---

1991 *Mathematics Subject Classification.* Primary 35A40, 35Q30, 65M20.

Parts of the work presented in this lecture have been supported by NSF grant DMS9024769.

This paper is in final form, and no version of it will be submitted for publication elsewhere.

© 1994 American Mathematical Society
0160-7634/94 $1.00 + $.25 per page

us closer to the teraflops, parallel computing will lead, for continuous problems, to the utilization of very fine discretizations involving a very large number of unknowns. Hence, and however efficient and friendly the computers may be, the specialist of scientific computing will have to learn how to manage, to compress and to store efficiently the huge amount of data that computing will involve. In that respect, it seems to us that multiresolution methods will play an important role, and our aim in this lecture is to survey and describe some aspects of the multiresolution methods, in particular when they are applied to nonlinear evolution equations.

Multiresolution methods are based on the decomposition of the set of unknowns into a hierarchy of subsets and on a systematically differentiated treatment of these subsets. Such ideas have been already successfully used in related areas of mathematics and computing. Let us mention, for example:

(1) Multigrid methods, particularly oriented towards the solution of elliptic linear and nonlinear problems.

(2) Wavelets, which play already a key role in vision and data compression.

(3) The concept of inertial manifolds for nonlinear dissipative evolution equations, such as the Navier-Stokes equations or reaction-diffusion equations.

The multiresolution algorithms that we study here are related (either closely or loosely) to the concept of inertial manifolds. When an inertial manifold exists, the corresponding system tends very rapidly towards a state where the high frequencies are slaved to the low frequencies. It is tempting, of course, to try to use in computations such a fundamental property which exists by itself, "naturally". Beside the concept of (exact) inertial manifold, the concept of approximate inertial manifold produces similarly an approximate slaving law of the high frequencies to the low frequencies. These concepts of exact and approximate inertial manifolds have been introduced respectively in [11] and [10], and much work has been devoted to the utilization and the study of the effectiveness of these manifolds in computing. We will return to this question in the concluding remarks, but most of this article will be devoted to simpler and more basic aspects of multiresolution methods.

We now describe briefly the content of this article. In §1 we describe a few types of multiresolution decompositions of the unknowns into a hierarchy of unknowns. Then, in §§2 and 3, we present and study some typical algorithms based on such decompositions. Section 2 is devoted to spectral or finite element discretizations and algorithms of the type of the nonlinear Galerkin method, while §3 is devoted to finite difference schemes using the incremental unknowns.

1. Multiresolution decompositions

Assume that we are interested in the numerical solution of an equation of Navier-Stokes type, either in the evolution case, i.e.,

$$\frac{du}{dt} + Au + B(u) = f, \tag{1.1}$$

or in the stationary case,

$$Au + B(u) = f. \tag{1.2}$$

The framework, which we describe in more detail in §2, is that of abstract evolution (or stationary) equations; A is a linear unbounded operator in a Hilbert space H, and $B = B(u)$ is a nonlinear operator. The unknown function $u = u(x,t)$ depends on x and t for (1.1), and on x only for (1.2).

The introduction of multiresolution discretizations of such equations depends on the type of discretization considered. We will address successively the case of spectral and finite difference discretizations; some comments on finite element discretizations are made in Remark 1.1.

1.1. Spectral and pseudospectral discretizations. When spectral discretizations are used, the solution u of either (1.1) or (1.2) is represented by a series

$$u = \sum_{j=1}^{\infty} \hat{u}_j w_j, \tag{1.3}$$

where $w_j = w_j(x)$, while the coefficients $\hat{u}_j$ depend on time in case (1.1), and are independent of time (constant) in case (1.2).

When such an expansion is used, the decomposition of u into its large-scale component y and its small-scale component z is done in a very natural way; choosing a cutoff number m, we write

$$y = \sum_{j=1}^{m} \hat{u}_j w_j, \quad z = \sum_{j=m+1}^{\infty} \hat{u}_j w_j. \tag{1.4}$$

More generally, choosing r cutoff numbers $m_1 < \cdots < m_r$, we can decompose u as

$$\begin{gathered} u = y + z_1 + \cdots + z_r, \\ y = \sum_{j=1}^{m} \hat{u}_j w_j, \; z_1 = \sum_{j=m_1+1}^{m_2} \hat{u}_j w_j, \ldots, z_r = \sum_{j=m_r+1}^{\infty} \hat{u}_j w_j. \end{gathered} \tag{1.5}$$

Expansions of type (1.3) occur in many different situations. First, in the space-periodic case, the w_j would be sine and cosine functions, and (1.3) the Fourier series of u. More generally, for spectral discretizations, the w_j could be Chebyshev or Legendre polynomials, or other similar orthogonal polynomials (cf. [3, 12, 13]). In the case of proper orthogonal decompositions or Karhunen-Loève expansions, the w_j would be the so-called empirical eigenfunctions [20]. Finally, without forgetting the wavelets, the w_j could be the eigenvectors of A or of a related linear operator.

In all physically relevant cases, y is associated with the large wavelengths and z with the small wavelengths. Hence, when series expansions are considered, the decomposition $u = y + z$ corresponds to a decomposition into *low and high frequencies*. From the numerical viewpoint, the $\hat{u}_j$ tend to 0 as $j \to \infty$. Hence, z in (1.4) and the z_i in (1.5) are much smaller than y (in some appropriate norms). Therefore, the coupling of y and z in the discretization of (1.1) (or (1.2)) induces *a stiffness problem*, which is new in numerical analysis and simply due to the fact that we can, nowadays and in the future, afford using very fine discretizations. Not addressing this stiffness question could lead to a waste of computing power and even, in some cases, to incorrect results. It is one of our long-term objectives to try to address this stiffness problem by proposing differentiated treatments of the low- and high-frequency components.

1.2. Finite differences. While the decomposition of a function u into its small- and large-scale components is obvious when spectral representations are used, this is less obvious when finite difference approximations are used. In this case such a decomposition can be obtained with the concept of Incremental Unknowns (cf. [5,6,7,22]) but, as we shall see, several decompositions of this type can be introduced, each possessing its own advantages.

In space dimension one we assume that (1.2) is a Dirichlet-type problem on the interval $(0,1)$. Then we consider a discretization mesh of the form $h = 1/(2N)$, N an integer, and we denote the approximate nodal values of u as

$$u_\alpha \simeq u(\alpha h), \quad \alpha = 0, \dots, 2N.$$

Restricting ourselves, for the moment, to two-level decompositions, we would consider also a coarse mesh $2h = 1/N$, so that we have, like in multigrid methods, two nested meshes, the fine one with mesh h, and the coarse one with mesh $2h$.

Several types of incremental unknowns can be associated with these two grids. Let us first define the second-order incremental unknowns (cf. [22]): for $\alpha = 2i$, $i = 0, \dots, N$, we set $y_{2i} = u_{2i}$, and then for $\alpha = 2i+1, i = 0, \dots, N-1$, we set

$$z_{2i+1} = u_{2i+1} - \frac{1}{2}(u_{2i+2} + u_{2i}). \tag{1.6}$$

Thus, z_{2i+1} is an increment of u to the average of u at the neighboring points, $2ih$ and $(2i+2)h$, hence the name. Obviously, by Taylor's formula, for h small, $z_{2i+1} = O(h^2)$. Of course, this decomposition can be repeated several times; the details of the procedure will be described in §3.

Concerning data compression, let us assume that, say, $h = 10^{-3}$ and that, working in simple precision, we retain 16 digits for each u_α; then the whole set of u_α's necessitates 10^{19} digits of storage. On the other hand, for the incremental unknowns we will only need $(10^{19} + 10^{13})/2$ digits, essentially half as many. As we shall see, the compression of data is even more significant in higher dimensions and can reach several orders of magnitude when several levels are used.

Other incremental unknowns which can be associated with the two levels mentioned before include the following: the first-order incremental unknowns [22], where $y_{2i} = u_{2i}$ for $i = 0, \dots, N$ while, for $i = 0, \dots, N-1$,

$$z_{2i+1} = u_{2i+2} - u_{2i};$$

finally, the wavelets-like incremental unknowns [5, 6] are defined by setting $h = 1/(2N+1), N \in \mathbb{N}$, and

$$
(1.7)\qquad
\begin{aligned}
y_{2i} &= \frac{1}{2}(u_{2i-1} + u_{2i}), \quad i = 1, \dots, N,\\
z_{2i-1} &= \frac{1}{2}(u_{2i-1} - u_{2i}) = u_{2i-1} - y_{2i}, \quad i = 1, \dots, N.
\end{aligned}
$$

By Taylor's formula the corresponding z's are of the order of h for h small, and obviously these decompositions can be repeated on several levels.

Let us define now the second-order incremental unknowns in space dimension two. We are interested, for instance, in a boundary value problem in the square $\Omega = (0,1)^2$. We define the mesh in both directions as $h = 1/(2N), N \in \mathbb{N}$, and denote by $u_{\alpha,\beta}$ the approximate value of the nodal values of u,

$$u_{\alpha,\beta} \simeq u(\alpha h, \beta h), \quad \alpha, \beta = 0, \dots, 2N.$$

Beside the fine grid of mesh h, we consider the coarse grid of mesh $2h = 1/N$ (see Figure 1). Associated with these two grids are the following incremental unknowns

$\times$ $2i, 2j+2$	$\circ$ $2i+1, 2j+2$	$\times$ $2i+2, 2j+2$
$\circ$ $2i, 2j+1$	$\circ$ $2i+1, 2j+1$	$\circ$ $2i+2, 2j+1$
$\times$ $2i, 2j$	$\circ$ $2i+1, 2j$	$\times$ $2i+2, 2j$

Figure 1. The coarse grid ($\times$) and the fine grid ($\circ$)

(IU): $y_{2i,2j} = u_{2i,2j}$ for $i, j = 0, \dots, N$. Then we observe that the points of the fine grid which do not belong to the coarse grid are of three different types:

- the points of type $(2ih, (2j+1)h)$ for which we write

$$
\begin{aligned}
z_{2i,2j+1} = u_{2i,2j+1} &- \frac{1}{2}(u_{2i,2j} + u_{2i,2j+2}),\\
i = 1, \dots, N-1,\ j &= 0, \dots, N-1;
\end{aligned}
$$

- the points of type $((2i+1)h, 2jh)$ for which we write

$$
\begin{aligned}
z_{2i+1,2j} = u_{2i+1,2j} &- \frac{1}{2}(u_{2i,2j} + u_{2i+2,2j}),\\
i = 0, \dots, N-1,\ j &= 1, \dots, N-1;
\end{aligned}
$$

– the points of type $((2i+1)h,(2j+1)h)$ for which

$$z_{2i+1,2j+1} = u_{2i+1,2j+1} - \frac{1}{4}(u_{2i,2j} + u_{2i+2,2j}, + u_{2i,2j+2} + u_{2i+2,2j+2}),\ i,j = 0,\dots,N-1.$$

The role of the IU as a preconditioner for the linear systems associated with linear elliptic equations is studied in [5]; in the present article we emphasize the utilization of incremental unknowns for evolution equations.

REMARK 1.1. When finite element discretizations are used, a multiresolution decomposition of the unknowns occurs automatically when hierarchical bases are implemented; see [2, 24]. In fact, the second-order incremental unknowns considered before are the same as the components of u for $\mathcal{Q}_1$ finite elements, using hierarchical bases; however, as usual, the corresponding discretized linear systems are not the same with finite elements and finite differences.

2. The nonlinear Galerkin method

In the case of evolution equations, decompositions of the set of unknowns into different arrays of unknowns pave the way to the introduction of new classes of numerical schemes leading to improved computational efficiency and perhaps to a better description of the physics of the phenomenon for equations related to multiscale turbulence. Hereafter, in §§2 and 3, we give examples of such schemes and refer the reader to other articles for more details.

Section 2 is devoted to spectral (Galerkin) discretizations. Section 2.1 describes the framework for equations (1.1). Then in §2.2 we describe the *new schemes*: we decompose the unknown u in the form $y+z$ as in (1.4) and, this is the key point, the schemes that we present treat y and z differently. Section 2.3 gives some indications on the stability analysis of the scheme, exhibiting at no or little cost an improved CFL stability condition.

2.1. The framework for the equation. We make the hypotheses on equation (1.1) more precise but refrain, in this article, from giving too detailed a mathematical description.

Equation (1.1) is set in a Hilbert space H, so that u is a function from $\mathbb{R}_+ = [0,\infty)$ into H. The scalar product and the norm in H are denoted $(\cdot,\cdot), |\cdot|$. The operator A is an abstract operator associated with the linear elliptic part of the evolution equation (remember that (1.1) is dissipative). As usual, A is a positive linear unbounded operator in H; we assume also that A is self-adjoint and that A^{-1} is compact. Therefore, there exists an orthonormal basis of H consisting of eigenvectors w_j of A, namely

$$(2.1)\qquad \begin{gathered} Aw_j = \lambda_j w_j,\\ 0 < \lambda_1 \le \lambda_2 \le \cdots,\ \ \lambda_j \to \infty \text{ as } j \to \infty. \end{gathered}$$

We can define the powers A^s of A, for $s \in \mathbb{R}$. For $s > 0$, A^s is an unbounded operator in H with domain $D(A^s)$. In particular, we usually write $V = D(A^{1/2})$;

the scalar product and the norm in V are written $((\cdot,\cdot)), \|\cdot\|)$. By (2.1),

$$\langle Au, u\rangle_{V',V} = \|u\|^2 \geq \lambda_1 |u|^2, \quad \forall\, u \in V.$$

For several physically relevant equations, the term B in (1.1) is a quadratic function of u while $f \in H$ is a forcing term. We assume that $B(u) = B(u,u)$ and define a trilinear form $b(u,v,w)$ on V by setting

$$b(u,v,w) = \langle B(u,v), w\rangle_{V',V}, \quad \forall\, u, v, w \in V.$$

We assume (see examples in [21]):

$$b(u,v,w) = -b(u,w,v), \quad \forall\, u,v,w \in V, \tag{2.2}$$

$$|b(u,v,w)| \leq c_1 |u|^{1/2}\|u\|^{1/2}\|v\|\; |w|^{1/2}\|w\|^{1/2}. \tag{2.3}$$

It is well known that under these hypotheses, for $u_0 \in H$ and $f \in L^2(0,T;H)$ given, equation (1.1), supplemented with the initial condition

$$u(0) = u_0, \tag{2.4}$$

possesses a unique solution:

$$u \in \mathcal{C}([0,T]; H) \cap L^2(0,T;V). \tag{2.5}$$

By assuming that $f \in L^\infty(\mathbb{R}_+; H)$, we see that u is defined for all $t > 0$ and satisfies (2.5), for every $T > 0$.

2.2. The multiresolution schemes. We consider a space and time discretization of equation (1.1) with initial condition (2.4). The spatial discretization is a Galerkin (spectral) discretization based on the w_j; hence, $u = u(t)$ is approximated by a function $u_N = u_N(t)$, where N is an integer to be determined:

$$u_N(t) = \sum_{j=1}^{N} g_{j,N}(t) w_j. \tag{2.6}$$

Furthermore, we choose an integer $N_1, 0 < N_1 < N$, and write

$$\begin{gathered} u_N = y_N + z_N, \\ y_N = \sum_{j=1}^{N_1} g_{j,N} w_j, \quad z_N = \sum_{j=N_1+1}^{N} g_{j,N} w_j. \end{gathered} \tag{2.7}$$

We denote by V_N (resp. V_{N_1}) the space spanned by $w_1, \ldots, w_N$ (resp. $w_1, \ldots, w_{N_1}$), and by W_N the space orthogonal to V_{N_1} in V_N. Hence, $u_N = u_N(t)$ is V_N-valued, $y_N = y_N(t)$ is V_{N_1}-valued, while $z_N = z_N(t)$ is W_N-valued.

Let $T > 0$ be fixed, let $M \in \mathbb{N}$ and $k = \Delta t = T/M$ be the time step. For $m = 0, \dots, M$ we define by induction elements $y_N^m \in V_{N_1}, z_N^m \in W_N$. The two schemes described hereafter are semi-implicit; the first is implicit in y and z for the linear part, explicit in z for the nonlinear part. The second is totally explicit in y and totally implicit in z.

For both schemes we start with

$$u_N^0 = y_N^0 + z_N^0, \tag{2.8}$$

where y_N^0 (resp. z_N^0) is the projection of u_0 onto V_{N_1} (resp. W_N).

Scheme I

When y_N^{m-1} and z_N^{m-1} are known, $y_N^m \in V_{N_1}$ and $z_N^m \in W_N$ are defined by

$$\begin{aligned} &\frac{1}{\Delta t}(y_N^m - y_N^{m-1}, \hat{y}) + ((y_N^m + z_N^m, \hat{y})) + b(y_N^{m-1}, y_N^{m-1}, \hat{y}) \\ &+b(y_N^{m-1}, z_N^m, \hat{y}) + b(z_N^m, y_N^{m-1}, \hat{y}) = (f^m, \hat{y}), \quad \forall \hat{y} \in V_{N_1}, \end{aligned} \tag{2.9}$$

$$\begin{aligned} &\frac{1}{\Delta t}(z_N^m - z_N^{m-1}, \hat{z}) + ((y_N^m + z_N^m, \hat{z})) + b(y_N^{m-1}, y_N^{m-1}, \hat{z}) \\ &= (f^m, \hat{z}), \quad \forall \hat{z} \in W_N, \end{aligned} \tag{2.10}$$

where

$$f^m = \frac{1}{\Delta t} \int_{(m-1)\Delta t}^{m\Delta t} f(t)dt. \tag{2.11}$$

Equations (2.9) and (2.10) are linear in y_N^m and z_N^m, and the existence and uniqueness of y_N^m and z_N^m is easy.

Scheme II

When y_N^{m-1} and z_N^{m-1} are known, $y_N^m \in V_{N_1}$ and $z_N^m \in W_N$ are defined by

$$\begin{aligned} &\frac{1}{\Delta t}(y_N^m - y_N^{m-1}, \hat{y}) + ((y_N^{m-1} + z_N^m, \hat{y})) + b(y_N^{m-1}, y_N^{m-1}, \hat{y}) \\ &+b(y_N^{m-1}, z_N^m, \hat{y}) + b(z_N^m, y_N^{m-1}, \hat{y}) \\ &= (f^m, \hat{y}), \quad \forall \hat{y} \in V_{N_1}, \end{aligned} \tag{2.12}$$

$$\begin{aligned} &\frac{1}{\Delta t}(z_N^m - z_N^{m-1}, \hat{z}) + ((y_N^{m-1} + z_N^m, \hat{z})) + b(y_N^{m-1}, y_N^{m-1}, \hat{z}) \\ &= (f^m, \hat{z}), \quad \forall \hat{z} \in W_N. \end{aligned} \tag{2.13}$$

For this scheme, we first determine z_N^m by solving (2.13), then we determine y_N^m through (2.12). The existence and uniqueness of y_N^m and z_N^m are easy to show.

2.3. A stability result. One of the first requirements when approximating an evolution equation is stability, i.e., the requirement that the approximations of

$u(t)$ remain bounded in some norm as the mesh Δt tends to 0. Unless the scheme is unconditionally stable, this property is only guaranteed if the time step Δt and the space step Δx satisfy a condition of CFL (Courant-Friedrichs-Levy) type. For spectral discretizations like (2.6), Δx is replaced by N.

Assume as before that we are given $u_0 \in H$ and $f \in L^\infty(\mathbb{R}_+; H)$, and assume the following:

$$\Delta t \leq \frac{4}{\lambda_1}, \tag{2.14}$$

$$\Delta t \, \lambda_{N_1} \leq \frac{1}{16 c_1^2 K_1}, \tag{2.15}$$

where

$$K_1 = \frac{2}{\lambda_1} \|u_0\|^2 + \frac{8}{\lambda_1^2} |f|^2_{L^\infty(\mathbb{R}_+; H)}. \tag{2.16}$$

We then have, for Scheme I, the following result, which can be proved using the methods of [23]:

Theorem 2.1. *Under the above hypotheses, and in particular* (2.14) – (2.16), *Scheme I is stable in the following sense:*

(2.17) *the* $y_N^m, z_N^m, u_N^m = y_N^m + z_N^m$ *are bounded in* H *independently of* N *and* Δt.

(2.18) *the RMS (root mean square) value of the* y_N^m, z_N^m, u_N^m *is similarly bounded in* V, *i.e.,*

$$\Delta t \sum_{m=1}^{M} \|y_N^m\|^2 \leq K, \quad \Delta t \sum_{m=1}^{M} \|z_N^m\|^2 \leq K,$$

where K *is independent of* N *and* Δt.

REMARK 2.1. The essential point in Theorem 2.1 is that the CFL condition (2.15) involves λ_{N_1} and not λ_N. Hence, the limitation on the time step is based on the long-wave component and not on the short-wave component of the flow. This is more advantageous and reflects an improved stability.

REMARK 2.2. A similar stability result is valid for Scheme II.

REMARK 2.3. We do not address here implementation issues and the question of computational efficiency. For the incompressible two- and three-dimensional Navier-Stokes equations, the reader is referred to [8,9], where extensive computations involving similar multilevel schemes are presented.

3. Incremental unknowns

We turn in this section to the study of multiresolution schemes when finite difference discretizations are used; these schemes are based on the utilization of the incremental unknowns (IU).

For the sake of simplicity we restrict ourselves to the case where equation (1.1) is a reaction-diffusion equation in the square $\Omega = (0,1)^2$, and we consider only two-level incremental unknowns.

The reaction-diffusion equation, which is a generalization of the Chaffee-Infante equation, has a polynomial nonlinearity and reads

$$\frac{\partial u}{\partial t} - \nu \Delta u + g(u) = 0, \quad x \in \Omega, \quad t > 0, \tag{3.1}$$

where $\nu > 0$, and g is a polynomial of odd degree with a positive leading coefficient:

$$g(s) = \sum_{j=0}^{2p-1} b_j s^j, \quad b_{2p-1} > 0. \tag{3.2}$$

The boundary condition is

$$u = 0 \text{ on } \partial\Omega, \tag{3.3}$$

and the initial condition is written

$$u(x,0) = u_0(x), \quad x \in \Omega. \tag{3.4}$$

The existence and uniqueness of solution of (3.1) – (3.4) is standard. The study of the large-time behavior of the solutions has been the object of much study, including attractors and inertial manifolds; see e.g. [1, 2, 17, 18, 21].

3.1. The incremental unknown discretization. Let $\Delta t = k$ be the time step, and let $h = 1/(2N), N \in I\!N$, be the space step. We denote by $u_{i,j}^m$ the approximation of u at time $m\Delta t$ and at the point ih, jh,

$$u_{i,j}^m \simeq u(ih, jh, m\Delta t).$$

The discretization of (3.1) – (3.4) with the implicit Euler scheme and a five-point discretization of the Laplace operator gives the relations

$$\begin{aligned} u_{i,j}^m - u_{i,j}^{m-1} + \frac{\nu \Delta t}{h^2}(4u_{i,j}^m - u_{i+1,j}^m - u_{i-1,j}^m \\ - u_{i,j+1}^m - u_{i,j-1}^m) + \Delta t\, g(u_{i,j}^m) = 0, \\ i,j = 1,\ldots,2N-1, \quad m \geq 1, \end{aligned} \tag{3.5}$$

which we supplement with the discrete analog of the boundary and initial conditions:

$$u_{i,j}^m = 0 \text{ if } i \text{ or } j = 0 \text{ or } 2N, \tag{3.6}$$

$$u_{i,j}^0 = \frac{1}{h^2} \int_{(i-\frac{1}{2})h}^{(i+\frac{1}{2})h} \int_{(j-\frac{1}{2})h}^{(j+\frac{1}{2})h} u_0(x)dx. \tag{3.7}$$

With this type of schemes, at each time step m all the $u_{i,j}^m$ play the same role. As indicated before, our aim here is to consider new numerical schemes where all the $u_{i,j}^m$ do not play the same role, the motivation being to treat differently the large eddies and the small eddies, which behave differently.

For each m, the $u_{i,j}^m$ belong to a space of dimension $(2N+1)^2$, $\mathbb{R}^{(2N+1)^2}$. We can associate with $u_{i,j}^m$ the $y_{i,j}^m$ and $z_{i,j}^m$ defined as in §1.2 (second-order incremental unknowns)[1]. Now it is easily seen that the mapping

$$u = \{u_{i,j}^m\}_{i,j} \xrightarrow{S_h} \{y_h, z_h\} = \{y_{i,j}^m z_{i,j}^m\}_{i,j}$$

is invertible (m fixed). Hence, conversely, a family $u_{i,j}$ can be defined for each given family of $y_{i,j}, z_{i,j}$. For instance, mixing now the time levels, we associate with the family

$$\{y_{i,j}^{m-1} z_{i,j}^m\}_{i,j}$$

a vector u denoted $\tilde{u}_h^m = \{\tilde{u}_{i,j}^m\}_{i,j}$. Or similarly, with the family

$$\{y_{i,j}^m,\ 0\}_{i,j}$$

we associate a vector u denoted $\hat{u}_h^m = \{\hat{u}_{i,j}^m\}_{i,j}$.

Now we give some examples of multilevel schemes involving the incremental unknowns. They are justified by the fact that they have potential for producing approximations with the same level of accuracy as usual schemes, but with less computing effort; note, however, that implementation issues and the question of computational efficiency will not be addressed here; see [5, 6] for linear systems, [7] for evolution equations, and [4, 14] for implementation issues concerning nonlinear stationary equations and bifurcation problems.

Scheme A

$$\begin{aligned} u_{i,j}^m - u_{i,j}^{m-1} + \frac{\nu\Delta t}{h^2}(4u_{i,j}^m - u_{i+1,j}^m - u_{i-1,j}^m \\ - u_{i,j+1}^m - u_{i,j-1}^m) + \Delta t\ g(\tilde{u}_{i,j}^m) = 0, \\ i,j = 1,\dots,2N-1, \quad m \geq 0; \end{aligned} \tag{3.8}$$

Scheme B

$$\begin{aligned} u_{i,j}^m - u_{i,j}^{m-1} + \frac{\nu\Delta t}{h^2}(4\tilde{u}_{i,j}^m - \tilde{u}_{i+1,j}^m \\ - \tilde{u}_{i-1,j}^m - \tilde{u}_{i,j+1}^m - \tilde{u}_{i,j-1}^m) + \Delta t\ g(\tilde{u}_{i,j}^m) = 0, \\ i,j = 1,\dots,2N-1, \quad m \geq 1. \end{aligned} \tag{3.9}$$

Of course, both schemes are supplemented by the "boundary conditions" (3.6) and by the same initial condition for $m = 0$ as (3.7).

[1] $y_{i,j}^m = 0$, $z_{i,j}^m = 0$ for the values of i, j for which this number has not been defined.

It is clear that Scheme A like Scheme I is implicit in the z-incremental component, while for the y-component, it is implicit for the linear part and explicit for the nonlinear part. Scheme B like Scheme II is fully implicit in the z-component and fully explicit in its y-component.

Both schemes can be conveniently written in a matrix-vector form, which is more transparent. We denote by U^m the vector consisting of the $\{u^m_{i,j}\}_{i,j}$ properly ordered, and we denote by $G(U^m)$ the vector consisting of the $g(u^m_{i,j})$ ordered similarly. Then, for instance, with an obvious notation ($A = A_h,\ G = G_h$), we can write the standard implicit Euler scheme (3.5) in the form

$$U^m + \nu\Delta t \quad AU^m + \Delta t \quad G(\tilde{U}^m) = U^{m-1}. \tag{3.10}$$

Similarly, let Y^m (resp. Z^m) be the vector consisting of the $\{y^m_{i,j}\}_{i,j}$ (resp. $\{z^m_{i,j}\}_{i,j}$) properly ordered. We write

$$\bar{U}^m = \begin{pmatrix} Y^m \\ Z^m \end{pmatrix}$$

and we denote by S the transfer matrix

$$U^m = S\bar{U}^m. \tag{3.11}$$

Then Scheme A can be rewritten as

$$(I + \nu\Delta t A)U^m + \Delta t G\ (\tilde{U}^m) = U^{m-1}, \tag{3.12}$$

where, of course,

$$\tilde{U}^m = S\begin{pmatrix} Y^{m-1} \\ Z^m \end{pmatrix}.$$

Hence also

$$S\bar{U}^m + \nu\Delta t A\bar{U}^m + \Delta t\ G\left(S\begin{pmatrix} Y^{m-1} \\ Z^m \end{pmatrix}\right) = S\bar{U}^{m-1}, \tag{3.13}$$

or, multiplying from the left by ${}^tS$,

$$({}^tSS + \nu\Delta t\bar{A})\begin{pmatrix} Y^m \\ Z^m \end{pmatrix} + \Delta t\,{}^tSG\left(S\begin{pmatrix} Y^{m-1} \\ Z^m \end{pmatrix}\right) = {}^tSS\begin{pmatrix} Y^{m-1} \\ Z^{m-1} \end{pmatrix}, \tag{3.14}$$

with $\bar{A} = {}^tSAS$. Similarly, for Scheme B,

$$U^m + \nu\Delta t A\tilde{U}^m + \Delta t\ G(\tilde{U}^m) = U^{m-1}, \tag{3.15}$$

$$S\bar{U}^m + \nu\Delta t AS\begin{pmatrix} Y^{m-1} \\ Z^m \end{pmatrix} + \Delta t\ G\left(S\begin{pmatrix} Y^{m-1} \\ Z^m \end{pmatrix}\right) = S\bar{U}^{m-1}, \tag{3.16}$$

$${}^tSS\begin{pmatrix} Y^m \\ Z^m \end{pmatrix} + \nu\Delta t\bar{A}\begin{pmatrix} Y^{m-1} \\ Z^m \end{pmatrix} + \Delta t\,{}^tSG\left(S\begin{pmatrix} Y^{m-1} \\ Z^m \end{pmatrix}\right) = {}^tSS\begin{pmatrix} Y^{m-1} \\ Z^m \end{pmatrix}. \tag{3.17}$$

REMARK 3.1. Schemes A and B are nonlinear with respect to U^m; the proof of existence and uniqueness of U^m follows standard methods.

REMARK 3.2. Other related schemes can be obtained by replacing $g(\tilde{u}^m_{i,j})$ by $g(\hat{u}^m_{i,j})$ in (3.8) and (3.9). The issue then, which will be addressed elsewhere, is that of accuracy.

3.2. A stability result. We describe now a stability result for Scheme A. A similar result is valid for Scheme B; the stability for both schemes can be proved using the methods of [7] and [23].

From the properties of g (see (3.2)) we see that there exist $c_2, c_3 > 0$ such that ($\beta = b_{2p-1} > 0$):

$$g(s)s \geq \frac{1}{2}\beta s^{2p} - c_2,$$
$$|g(s)| \leq 2\beta|s|^{2p-1} + c_3, \quad \forall s \in \mathbb{R}.$$

We then assume that $\Delta t \leq k_0, k_0 = 1/\nu$, and set

$$K_1 = |u_0|^2 + \frac{1}{\nu}(c_2 + h_0 c_3),$$
$$K_1' = |u_0|^2 + \frac{3}{\nu}(c_2 + k_0 c_3),$$

where $|u_0|$ is the norm of u_0 in $L^2(\Omega)$,

$$|u_0| = \left\{\iint_\Omega |u_0(x)|^2 dx\right\}^{1/2}.$$

The stability result concerning Scheme A can then be stated as follows:

Theorem 3.1. *We assume that $\Delta t \leq k_0 = 1/\nu$ and*

$$\frac{\Delta t}{h^{2p-2}} \leq \frac{1}{32c\beta(K_1 + K_1')^{p-1}}, \tag{3.18}$$

where c is an appropriate constant[2]*. Then the RMS norm of the u^m is uniformly bounded, independently of Δt and h:*

$$h^2 \sum_{i,j=1}^{2N-1} |u^m_{i,j}|^2 \leq K_1.$$

---

[2] There exist absolute constants c, c', such that

$$c'|U|^2 \leq \left|S\begin{pmatrix}Y\\0\end{pmatrix}\right|^2 + \left|S\begin{pmatrix}0\\Z\end{pmatrix}\right|^2 \leq c|U|^2 .$$

Furthermore, for $T > 0$ fixed and $T = M\Delta t,\ M \in I\!N$,

$$2\nu\Delta t \sum_{m=1}^{M} \sum_{i,j=0}^{2N-1} \{|u_{i+1,j}^m - u_{i,j}^m|^2 + |u_{i,j+1}^m - u_{i,j}^m|^2\}$$
$$\leq |u_0|^2 + 2(c_2 + k_0 c_3)T, \ \textit{etc.}.$$

Concluding remarks. The increase of computing power expected from the forthcoming massively parallel computers does not prevent us from looking for more efficient algorithms and, possibly, for algorithms closer to physics, which could take into account the complexity of the interaction of scales for nonlinear phenomena related to turbulence.

In this lecture we have shown that the multiresolution discretizations permit the development of algorithms of a new type. Some significant gain in computing times have been observed using such algorithms; the corresponding numerical experiments are reported elsewhere (see e.g. [5, 6, 8, 9]).

It is hoped that, in the long range, the multiresolution methods will help build better and better approximate inertial manifolds providing algorithms which are computationally efficient for parallel computers, and which give a relevant description of the complex phenomena related to chaotic (turbulent) behaviors.

References

1. Babin, A.V. and Vishik, M.I., *Attractors of partial differential equations and estimate of their dimension*, Russian Math. Surveys **38** (1983), 151–213.

2. Babin, A.V. and Vishik, M.I., *Maximal attractors of semigroups corresponding to evolution differential equations*, Math. USSR-Sbornik **54** (1986), 387–408.

3. Canuto, C., Hussaini, M.Y., Quarteroni, A., and Zang, T.A., *Spectral Methods in Fluid Dynamics*, Springer Series in Computational Physics, Springer-Verlag, New York (1988).

4. Chehab, J.P., Thesis, University Paris-Orsay (1993).

5. Chen, M. and Temam, R., *Incremental unknowns in finite differences: condition number of the matrix*, SIAM J. Matrix Anal. Appl. **14** (1993), 432–455.

6. Chen, M. and Temam, R. *Nonlinear Galerkin method in the finite difference case and wavelet-like incremental unknowns*, Numer. Math. **64** (1993), 271–294.

7. Chen, M. and Temam, R., *Nonlinear Galerkin Method with multilevel incremental unknows*, in *Contributions in Numerical Mathematics*, R.P. Agarwal, ed., World Scientific Series in Applicable Analysis, Singapore (1993).

8. Debussche, A., Dubois, T. and Temam, R., *The nonlinear Galerkin method: A multi-scale method applied to the simulation of homogeneous turbulent flows*, ICASE preprint 1993; to appear.

9. Dubois, T., Jauberteau, F. and Temam, R., *Solutions of the incompressible Navier-Stokes equations by the nonlinear Galerkin method*, to appear in J. Scientific Computing.

10. Foias, C., Manley, O. and Temam, R., *Sur l'interaction des petits et grands tourbillons dans les écoulements turbulents*, C.R. Acad. Sci. Paris, Série I, **305** (1987), 497–500.
Modelling of the interaction of small and large eddies in two-dimensional turbulent flows, Modél. Math. Anal. Numér. (M^2AN) **22** (1988), 93–118.

11. Foias, C., Sell, G. and Temam, R., *Variétés Inertielles des équations différentielles dissipatives*, C.R. Acad. Sci. Paris, Série I, **301** (1985), 139–142.
Inertial Manifolds for nonlinear evolutionary equations, J. Differential Equations **73** (1988), 309–353.

12. Gottlieb, D., Hussaini, Y.M. and Voigt, R.G., eds. *Spectral Methods for Partial Differential Equations*, SIAM, Philadelphia (1984).

13. Gottlieb, D. and Orszag, S.A., *Numerical Analysis of Spectral Methods: Theory and Application*, CBMS Regional Conference Series in Applied Mathematics **26**, SIAM, Philadelphia, 1977.

14. Goyon, O., Thesis, University Paris-Orsay, to appear.

15. Lions, J.L., *Quelques Méthodes de Résolution des Problèmes aux Limites non Linéaires*, Dunod, Paris (1969).

16. Lions, J.L., Temam, R. and Wang, S., *Models of the coupled atmosphere and ocean*, Computational Mechanics Advances, J.T. Oden, ed., Elsevier, Amsterdam, Vol. **1** (1993), 1–112.

17. Marion, M., *Attractors for reaction-diffusion equations: existence and estimate of their dimension*, Appl. Anal. **25** (1987), 101–147.

18. Marion, M., *Approximate inertial manifolds for reaction-diffusion equations in high space dimension*, J. Dynamics Differential Equations **1** (1989), 245–267.

19. Poullet, P., Thesis, University Paris-Orsay, to appear.

20. Sirovich, L., *Turbulence and the dynamics of coherent structures I, II*, Quart. Appl. Math. **45** (1987), 561–573; 573-590.

21. Temam, R., *Infinite Dimensional Dynamical Systems in Mechanics and Physics*, Springer-Verlag, New York, Applied Mathematical Sciences Series, Vol. **68** (1988).

22. Temam, R., *Inertial manifolds and multigrid methods*, SIAM J. Math. Anal. **21** (1990), 154–178.

23. Temam, R., *Stability analysis of the nonlinear Galerkin method*, Math. Comp. **57** (1991), 477–505.

24. Yserentant, H., *Two preconditioners based on the multilevel splitting of finite element spaces*, Numer. Math. **58** (1990), 163–184.

25. Zienckiewicz, A.C., Kelly, D.W., Gago, J., and Babuška, I., *Hierarchical finite element approaches, error estimates and adaptative refinement*, in *The Mathematics of Finite Elements and Applications IV*, J.R. Whiteman, ed., Mafelap 1981, London, 1982.

Laboratoire d'Analyse Numérique, Université Paris Sud, Bâtiment 425, Orsay, France 91405 and Institute for Scientific Computing & Applied Mathematics, Indiana University, 618 East Third Street, Bloomington, Indiana 47405

Contributed Papers

Proceedings of Symposia in Applied Mathematics
Volume **48**, 1994

A Comparison of Techniques for Solving Ill-Conditioned Problems Arising from the Immersed Boundary Method

LOYCE ADAMS AND ZHIYUN YANG

1. Introduction

One approach to solving incompressible fluid problems on complicated geometries is the immersed boundary method developed by Peskin [**8**]. The idea is to use a larger underlying Cartesian grid and modify the right-hand side of the partial differential equation so that the resulting solution still satisfies the boundary conditions. The advantage is that fast techniques for solving equations on Cartesian grids can be employed, including parallelization strategies, and the problems associated with grid generation for a body-fitted grid can be avoided. Mathematically, if the original problem is

$$(1.1)\qquad \begin{aligned} Au &= f(x,y) \quad \text{in}\,\Omega', \\ u &= \bar{u}(x,y) \quad \text{on}\,\partial\Omega', \end{aligned}$$

where A is a differential operator and $\Omega' \subset \mathrm{R}^2$, then the modified problem is

$$(1.2)\qquad \begin{aligned} Au &= f(x,y) + \textstyle\int_{\partial\Omega'} \phi(X(s))\delta(X - X(s))\,ds \quad \text{in}\ \Omega, \\ u &= 0 \text{ on } \partial\Omega \quad \text{or} \quad u \text{ is periodic on } \partial\Omega, \end{aligned}$$

$$(1.3)\qquad Ru = \bar{u}(x,y) \quad \text{on}\,\partial\Omega',$$

where ϕ is the necessary boundary force to produce the correct value $\bar{u}$ on $\partial\Omega'$. The last equation ensures that u restricted to Ω' is $\bar{u}$. Unlike Peskin's original method, we consider ϕ to be an unknown function that can be determined implicitly by solving (1.2) and (1.3). Mayo [**5, 6**] also employed Cartesian grids to solve (1.1) by using a Green's function formulation to find ϕ followed by a different modification of the right hand to find u. In this paper, we give a particular discretization of (1.2) and (1.3) for a model problem. We show that a conjugate

1991 *Mathematics Subject Classification.* Primary 65F10; Secondary 65N06.

The second author was supported in part by NASA grant NAG 5-2213 while a student in the Applied Mathematics Department, University of Washington.

The final version of this paper will be submitted for publication elsewhere.

© 1994 American Mathematical Society
0160-7634/94 \$1.00 + \$.25 per page

gradient approach to solving these equations gives $O(h)$ accuracy in u with less work than either a truncated SVD or the Tikhonov regularization method.

2. Discrete equations for a model problem

As a model problem, we consider the operator A to be the two-dimensional Laplacian, the boundary $\partial\Omega'$ to be the unit circle, $\bar{u} = 1$, $f = 0$, Ω to be the square $-2 < x < 2$, $-2 < y < 2$, and $u = \log r + 1$ on $\partial\Omega$. The exact solution is $u = 1$ in Ω', $u = \log r + 1$ outside Ω' and $\phi = 1$. To discretize (1.2), we use N^2 points on the square with uniform mesh spacing h in each coordinate direction and K evenly distributed points along the boundary of the circle. We approximate the delta function by D, a product of two one-dimensional discrete delta functions, and R corresponding to (1.3) is taken to be $h^2 D^T$. Then the equations become

$$\begin{pmatrix} A & -D\Delta s \\ R & 0 \end{pmatrix} \begin{pmatrix} u \\ \phi \end{pmatrix} = \begin{pmatrix} f \\ \bar{u} \end{pmatrix}, \tag{2.1}$$

where Δs is the arc length on the circle, A is $N^2 \times N^2$, D is $N^2 \times K$, u and f are $N^2 \times 1$, and ϕ is $K \times 1$. We solve (2.1) by forming the Schur complement for ϕ. That is,

$$B\phi = g, \tag{2.2}$$

where $B = \Delta s R A^{-1} D$ and $g = \bar{u} - R A^{-1} f$. Once (2.2) is solved for ϕ, we recover u from the equation $Au = \Delta s D\phi + f$.

The advantage of solving (2.2) instead of (2.1) directly is that (2.2) is only a $K \times K$ system and $K = O(N)$. However, problems do arise in solving this equation since it can be very poorly conditioned or nearly singular, owing to the approximation of the delta function in (1.2) and the value used for Δs. We have found experimentally that choosing Δs to be close to h is the best choice.

3. Numerical methods

We now discuss methods for the solution of (2.2). If direct methods are to be considered, then the matrix B must be explicitly formed. This costs essentially K Laplacian solves since D has K columns. Beyer [**1, 2**] observed that B had singular vectors associated with small singular values that should be ignored in the computation of ϕ. This TSVD approach in total costs $O(KN^2 \log_2 N, K^3)$, assuming the Laplacian solves are done by a fast solver. When $K = O(N)$, the procedure is dominated by the formation of B. Another approach that also involves the calculation of an SVD or generalized SVD is Tikhonov regularization, [**3**], which again requires that B be formed.

We consider using a conjugate gradient iterative approach to solve (2.2). This strategy would allow B to be left in its factored form since only the multiplication of B by a vector is required. Each iteration would require one such multiplication, which involves only one Laplacian solve. We would thereby reduce the cost tremendously if only a few iterations were required to compute a good regularized ϕ, and a good u. Now, if we apply the standard conjugate gradient (CG) method to the normal equations for (2.2), with initial conditions $\phi^{(0)} = 0$, the solution will converge to B^+g. Also, preconditioned versions can be developed that solve the normal equations and minimize $||L\phi||_2$ ([**3**] and the references therein, [**9**], and [**7**]). It is also known that if one stops the CG or PCG iteration before convergence, the resulting iterate can be viewed as a regularized solution. This is because the effects of the singular values in iterate $\phi^{(k)}$ that are smaller than the smallest of the k Ritz values associated with the CG procedure are damped out. For our problem, we choose $\phi^{(0)} = 0$, $L = I$ and apply CG to the normal equations.

4. Numerical results and conclusions

In this section we give numerical results for the CG, TSVD, and Tikhonov methods described in the last section for the solution of ϕ and discuss the quality of the resulting solutions for u.

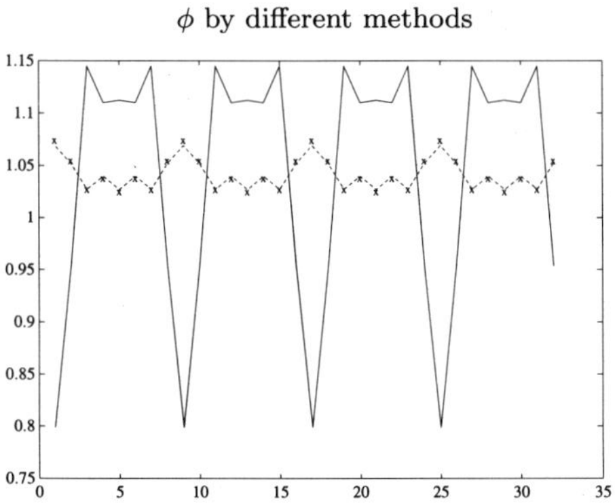

FIG. 1. ϕ computed by $CG_{k=1}$ (dashed), $TSVD_{k=1}$ (x-values) and Tikhonov (solid).

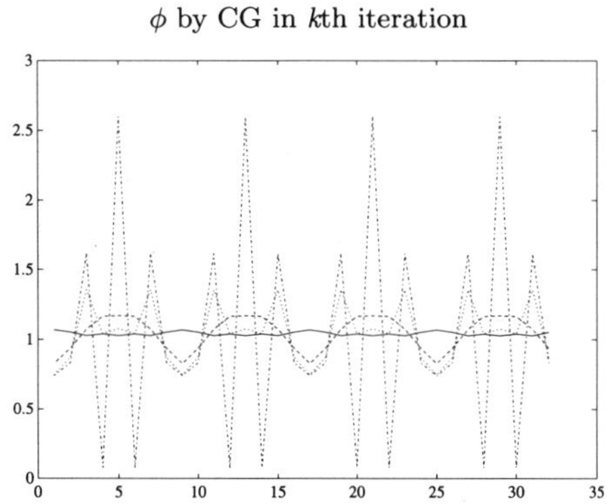

FIG. 2. ϕ for CG, $k = 1$ (solid), $k = 2$ (dashed), $k = 3$ (dotted), $k = 8$ (dashed-dotted). $N = 19$.

Figure 1 shows that one step of CG and TSVD, keeping the largest singular value, produced the best results for ϕ. Figure 2 shows that $\phi^{(k)}$ becomes more oscillatory as k increases. This indicates that we must look for a regularized solution to (2.1). It also shows that even though CG converges in 8 iterations, the best value for ϕ is obtained when $k = 1$.

Table 1 summarizes the 2-norm error in ϕ and u. Note that even though it is crucial to know what k-value to use to determine a good regularized ϕ, all possible k values gave a relative error in u of about two percent in the 2-norm. This means that just a few iterations are sufficient, and the cost can be greatly reduced over direct approaches.

	k	$\parallel \phi_e \parallel_2$	$\parallel u_e \parallel_2$
	1	0.0440	0.0196
C	2	0.1287	0.0212
	3	0.1748	0.0216
	4	0.2119	0.0219
G	5	0.8000	0.0238
	8	0.8009	0.0238
T	1	0.0448	0.0196
S	9	0.1267	0.0211
V	24	0.2015	0.0218
D	*all*	0.8009	0.0238
Tik		0.1243	0.0199

Table 1. $\parallel \phi_e \parallel$ and $\parallel u_e \parallel$ for $N = 19$.

	$N = 19$		$N = 39$	
	$\parallel \cdot \parallel_\infty$	$\parallel \cdot \parallel_2$	$\parallel \cdot \parallel_\infty$	$\parallel \cdot \parallel_2$
T_e	2.3302	0.4269	5.3706	0.4550
$\tilde{T}_e$	0.2164	0.0605	0.4260	0.0295
u_e	0.0343	0.0196	0.0228	0.0102
$\tilde{u}_e$	0.0350	0.0194	0.0190	0.0101
$u_e - \tilde{u}_e$	0.0169	0.0031	0.0128	0.0012
$\frac{u_e - \tilde{u}_e}{u_e}$	0.4931	0.1574	0.5646	0.1154

Table 2. T_e, $\tilde{T}_e$ and u_e, $\tilde{u}_e$ for $N = 19$, $N = 39$.

As the mesh size decreases, we have observed that the error in u, u_e,behaves like $O(h)$ despite the fact that we are spreading values of magnitude $O(h^{-1})$ to nearby grid points with the discrete delta function. To explain this, we note that the first equation in (2.1) is $Au = S\phi + f$, where $S = \Delta s D$. The true u, u_{tr}, and the true ϕ, ϕ_{tr}, satisfy $Au_{tr} = S\phi_{tr} + f + t$. Therefore, we get $Au_e = S\phi_e + t \equiv T_e$. The first row in Table 2 shows that $||T_e||_\infty$ and $||T_e||_2$ both grow as the mesh is refined. This is also shown in Figures 3.1 and 3.2, where the horizontal axis represents the grid points of the Cartesian square grid numbered bottom to top, left to right. Nevertheless, Table 2 and Figures 4.1 and 4.2 show that $||u_e||_\infty$ and $||u_e||_2$ behave as $O(h)$ as the mesh is refined. (This is also confirmed for smaller h-values as well.) This is because A^{-1} acts as a smoother to reduce the

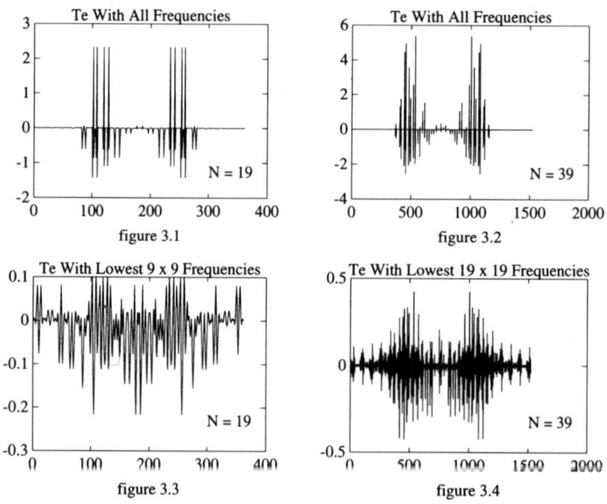

FIG. 3. Te for $N = 19$, $N = 39$ and $\tilde{T}_e$ for $N = 19$, $N = 39$.

high-frequency components of T_e. The eigenvalues of A^{-1} range from $O(h^2)$ to $O(1)$ and multiply the highest to lowest frequency components of T_e, respectively. Figures 3.3 and 3.4 show that if we consider only 1/4 of the components of T_e which correspond to the lowest frequencies, then the modified T_e does not look

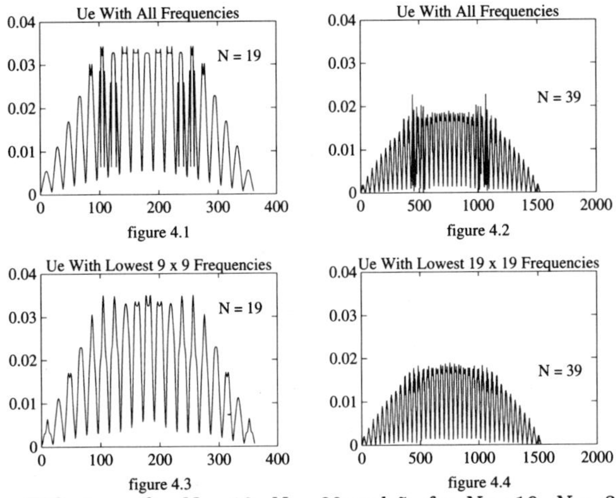

FIG. 4. u_e for $N = 19$, $N = 39$ and $\tilde{u}_e$ for $N = 19$, $N = 39$.

like the real T_e since the high-frequency components cannot be ignored. On the contrary, Figures 4.3 and 4.4 show that the u_e computed from the modified T_e has the same behavior as the true u_e. Therefore, the discrete delta function contributed to high-frequency components in T_e that were taken care of by the smoothing properties of A^{-1}, and the resulting method is $O(h)$.

References

1. R.P. Beyer, *A computational model of the cochlea using the immersed boundary method*, Ph.D. thesis, University of Washington, Department of Applied Mathematics, 1989.
2. R.P. Beyer, *A computational model of the cochlea using the immersed boundary method*, J. Comput. Phys. **98**(1992), 145-162.
3. Per Christian Hansen, *Regularization Tools: A MATLAB package for analysis and solution of discrete ill-posed problems*, Linear Algebra Appl. **130**(1992), 257-267.
4. Zhilin Li, R.J. LeVeque, *The immersed interface method for elliptic equations with discontinuous coefficients and singular sources*, Department of Applied Mathematics Technical Report No. 92-12, University of Washington, Seattle, WA, December 1992.
5. A. Mayo, *The fast solution of Poisson's and the biharmonic equations on irregular regions*, SIAM J. Numer. Anal. **21**(1984), 285-299.
6. A. Mayo, A. Greenbaum, *Fast parallel iterative solution of Poisson's and the biharmonic equations on irregular regions*, SIAM J. Sci. Statist. Comput. **13**(1992), 101-118.
7. C.C. Paige and M.A. Saunders, *LSQR: An algorithm for sparse linear equations and sparse least squares*, ACM Trans. Math. Software **8**(1982), 43-71.
8. C.S. Peskin, *Numerical analysis of blood flow in the heart*, J. Comput. Phys. **25**(1977), 220-252.
9. A. van der Sluis, H.A. van der Vorst, *SIRT and CG type methods for iterative solution of sparse linear least squares problems*, Linear Algebra Appl. **130**(1990), 257-302.

Department of Applied Mathematics, University of Washington, Seattle, WA 98195

E-mail address: adams@amath.washington.edu

Department of Applied Mathematics, University of Washington, Seattle, WA 98195

E-mail address: dazhi@amath.washington.edu

Proceedings of Symposia in Applied Mathematics
Volume 48, 1994

A Mixed Spectral-Collocation and Operator Splitting Method for the Wigner-Poisson Equation

ANTON ARNOLD

ABSTRACT. The Wigner-Poisson equation is a kinetic pseudodifferential equation to model quantum transport. A combined spectral-collocation and operator splitting method is presented and analyzed.

1. Introduction

In this paper we present and analyze an operator splitting method for the Wigner-Poisson (WP) equation. The Wigner formalism ([10,6]), which represents a phase-space description of quantum mechanics, has in recent years received a great deal of attention as a tool to simulate transport phenomena in ultra-integrated quantum effect semiconductor devices.

The real-valued Wigner distribution function $w(x, v, t)$ describes the motion of an electron ensemble in position-velocity (x, v)-phase space under the action of the electrostatic potential $V(x, t)$. The time evolution of w is governed by the Wigner equation, which reads in scaled form

$$\partial_t w + v\partial_x w + \Theta[V]w = 0, \quad x \in (-\pi, \pi),\ v \in I\!R, \tag{1.1}$$

with the pseudodifferential operator (PDO)

$$\Theta[V]w = \frac{i}{2\pi} \int\int \delta V(x, \eta, t) w(x, v', t) e^{i(v-v')\eta}\, dv' d\eta, \tag{1.2}$$

$$\delta V(x, \eta, t) = V(x + \frac{\eta}{2}, t) - V(x - \frac{\eta}{2}, t).$$

1991 *Mathematics Subject Classification.* Primary 65M70, 81S30; Secondary 65J15.
The author was partially supported by the grants ERBCHRXCT930413 from the EC, "Acciones Integradas, 1994 " from the DAAD, and the DFG ("Analysis und Numerik von kin. Quantentransportmodellen").
The detailed version of this paper has been submitted for publication elsewhere.

© 1994 American Mathematical Society
0160-7634/94 $1.00 + $.25 per page

To incorporate many-body effects in a simple mean-field approximation, the potential V is selfconsistently coupled to the Poisson equation

$$\partial_x^2 V(x,t) = \rho(x,t) - D(x), \tag{1.3}$$

with the particle density $\rho = \int w dv$ and some fixed ion density D.

Mainly for simplicity of the presentation we will here only consider the $1D$-problem, which we supplement with periodic boundary conditions (BC) in x. For realistic simulations of resonant tunneling diodes ([4]), absorbing BCs ([8]) have given very satisfactory results, but the analysis of the resulting initial-boundary value problem is still under investigation ([1]). In order for the Poisson equation to be solvable with periodic BCs, charge neutrality of the initial state, i.e.,

$$\int_{I\!R}\int_{-\pi}^{\pi} w^I(x,v)\,dx\,dv = \int_{-\pi}^{\pi} D(x)\,dx, \tag{1.4}$$

has to hold, which will then persist under temporal evolution. Also we will require the potential V to satisfy the normalization $\int_{-\pi}^{\pi} V\,dx = 0$ for all time t.

In §2 we will present the numerical scheme for the solution of the coupled system (1.1) – (1.3), and we outline its stability and convergence proof in §3 (for mathematical details we refer to [2,3]).

2. Numerical scheme

From the numerical point of view, the major problem in the discretization of the WP system stems from the PDO which is nonlocal in v. In [2] an operator splitting method has been analyzed (first used in [9]), which consists in separating the transport operator $A = -v\partial_x$ and the nonlinear PDO $Bw = -\Theta[V[w]]$ ($V[w]$ is the solution of (1.3) with periodic BCs) for each time step:

$$\partial_t f = Af\,,\quad t_s \le t \le t_{s+1} = t_s + h;\quad f(t_s) = w_s\,,\quad w_{s+\frac{1}{2}} = f(t_{s+1}), \tag{2.1}$$

$$\partial_t f = Bf\,,\quad t_s \le t \le t_{s+1};\quad f(t_s) = w_{s+\frac{1}{2}}\,,\quad w_{s+1} = f(t_{s+1}). \tag{2.2}$$

The advantage of this operator splitting method lies in the fact that, at least theoretically, both steps can be carried out explicitly, since the density $\rho[f]$, and thus also $V[f]$, stay constant during the second evolution step (2.2):

$$w_{s+\frac{1}{2}}(x,v) = w_s(x-hv,v),\quad \hat{w}_{s+1}(x,\eta) = \hat{w}_{s+\frac{1}{2}}(x,\eta)e^{-ih\delta V_{s+1}(x,\eta)}, \tag{2.3}$$

where $\hat{w}$ denotes the Fourier transform with respect to v, and $\delta V_{s+1} = V[w_{s+\frac{1}{2}}](x+\frac{\eta}{2}) - V[w_{s+\frac{1}{2}}](x-\frac{\eta}{2})$.

For the linear Wigner equation with a given potential $V \in L^\infty(I\!R)$ the L^2-convergence (this is the natural framework for the Wigner function [5]) of this splitting scheme as $h \to 0$ is an immediate consequence of Trotter's product formula for linear semigroups, yielding a first-order method (second-order for 'Strang splitting', [2]).

We will now specify the finite-dimensional representation of w_s. Because of the definition of the PDO via Fourier transforms it is natural to expand the Wigner function in the trigonometric functions $\Phi(n,v) = \sqrt{\frac{\alpha}{2\pi}}\exp(i\alpha n v)$, which are eigenfunctions of $\Theta[V]$. Since we discuss here the case of periodic BCs, we will expand the Wigner function in trigonometric functions in the spatial direction as well:

$$(2.4)\quad w_s(x,v) \sim u_s(x,v) = \sum_{n=-N+1}^{N} \hat{u}_s(x,n)\Phi(n,v) = \sum_{m=-M+1}^{M} \overline{u}_s(m,v)e^{imx}$$

$$= \sum_{m=-M+1}^{M}\sum_{n=-N+1}^{N} \tilde{u}(m,n)e^{imx}\Phi(n,v),$$

where $\hat{u}_s(x,-n) = \hat{u}_s(x,n)^*$ for $|n| < N$ and $\hat{u}_s(x,N) \in I\!R$, and similarly for $\overline{u}_s$. As the shift step in (2.3) will in general not preserve the structure of $\overline{u}_s$, it will have to be carried out by collocation. We define the collocation points x_k and v_j by

$$(2.5)\quad x_k = \frac{k\pi}{M}\ ,\quad k = -M+1,...,M,\quad v_j = \frac{j\pi}{\alpha N}\ ,\quad j = -N+1,...,N.$$

For the approximation of (1.3) the integral over w has to be truncated at $v = \pm\frac{\pi}{\alpha}$:

$$(2.6)\qquad \rho_s(x) \sim \int_{-\frac{\pi}{\alpha}}^{\frac{\pi}{\alpha}} u_s(x,v)dv = \sqrt{\frac{2\pi}{\alpha}}\hat{u}_s(x,0).$$

After also expanding ρ_s, V_s, and D into trigonometric polynomials in x, the fully discretized fractional step (2.3) takes the form $u_{s+\frac{1}{2}} = G(h)u_s$, $u_{s+1} = F_M(h, u_{s+\frac{1}{2}})$, with

$$(2.7)\qquad [G(h)u](x,v_j) = \sum_{m=-M+1}^{M} \overline{u}(m,v_j)e^{-imhv_j}e^{imx},$$

$$(2.8)\qquad V(x) = \sum_{m\neq 0} \frac{[\overline{D}(m) - \sqrt{\frac{2\pi}{\alpha}}\tilde{u}(m,0)]}{m^2}e^{imx},$$

$$[F_M(h,u)](x_k,v_j) = \sum_{n=-N+1}^{N} \hat{u}(x_k,n)e^{-ih\delta V(x_k,\alpha n)}\Phi(n,v_j).$$

Here, $\overline{V}(0) = 0$ reflects the chosen normalization of the potential.

When taking $\frac{\alpha N}{2\pi} \in I\!N$, then $\delta V(x,\alpha N) = 0$ follows, and the scheme (2.7), (2.8) can easily be seen to preserve the total charge $\sqrt{\frac{2\pi}{\alpha}}\frac{\pi}{M}\tilde{u}(0,0)$ and the discrete L^2-norm of u_s. This will be the basis for the stability obtained in §3.

When the WP system is discretized in the v-direction only, we obtain the hyperbolic system

$$\partial_t u(x,v_j,t) + v_j \partial_x u(x,v_j,t) + (\Theta[V]u)(x,v_j,t) = 0, \tag{2.9}$$

$$\partial_x^2 V(x,t) = \frac{\pi}{\alpha N} \sum_{j=-N+1}^{N} u(x,v_j,t) - D(x). \tag{2.10}$$

In [7] it has been shown that the solution u of (2.9), (2.10) converges to the solution of the WP system with spectral accuracy for $\alpha \to 0$ and $\alpha N \to \infty$. Therefore, we will here only focus on the convergence of the splitting method (2.7), (2.8) to the solution of the hyperbolic system (2.9), (2.10).

3. Stability and convergence

In the mathematical analysis of the WP system one has to require w and $vw \in L^2$ in order to allow for a proper definition of ρ. In the stability analysis of (2.7), (2.8) we have to establish a discrete analogue of this framework. The potential in (2.8) satisfies the estimate

$$\|V\|_{1,\infty}^2 \le \text{const}\,[\frac{2\pi}{\alpha} \sum_{m=-M+1}^{M} |\tilde{u}(m,0)|^2 + 1], \tag{3.1}$$

with a constant independent of α, M and N. The approximation for the density $\rho[u]$ can be estimated by

$$\frac{2\pi}{\alpha} \sum_{m=-M+1}^{M} |\tilde{u}(m,0)|^2 \le \text{const}\|u\|_{X_0}^2, \quad \|u\|_{X_0}^2 = \|u\|_2^2 + \|Hu\|_2^2, \tag{3.2}$$

for all $u \in B_{MN}$, which is the space of functions represented by (2.4). Here the operator $H : B_{MN} \to B_{MN}$, defined by $Hu = \frac{\exp(i\alpha v_j)-1}{\alpha} u$, is the discrete approximation for ivu.

The scheme (2.7), (2.8) conserves $\|u_s\|_2$, and the shift step also conserves $\|Hu_s\|_2$. A straightforward estimate of (2.8) gives

$$\|HF_M(h,u)\|_2^2 \le \text{const}(1+h)(1+\|V\|_{1,\infty}^2)\|u\|_{X_0}^2, \tag{3.3}$$

which then yields a uniform bound for $\|u_s\|_{X_0} + \|V_s\|_{1,\infty}$, $sh \le T$. In order to obtain nonlinear stability of the numerical scheme, one can show that (2.8) is locally Lipschitz in the X_0-norm.

To discuss the consistency of the method, we introduce the projection (by collocation at x_k) $P_M : B_N \to B_{MN}$, with $B_N = \{\sum_n \hat{u}(x,n)\Phi(n,v), \hat{u}(\cdot,n) \in L^2(-\pi,\pi)\}$. Now we extend the fractional step (2.7), (2.8) to B_N : G is defined like in (2.7), but the nonlinear operator F now involves the exact solution of the Poisson equation with periodic BCs. For the formulation of the main result of this section we need the periodic (in x) Sobolev space $X_r = \{u \in B_N | u(\cdot,v),\, Hu(\cdot,v) \in H_p^r(-\pi,\pi)\}$.

Theorem 3.1. *Let* $u \in C^1([0,T], X_r)$, $\partial_x u$, $v_j u \in C([0,T], X_r)$, $\int_{-\pi}^{\pi} \int_{-\frac{\pi}{\alpha}}^{\frac{\pi}{\alpha}} u^I \, dx dv = \int_{-\pi}^{\pi} D \, dx$, *and* $D \in B_{MN} \cap H_p^{r+1}(-\pi, \pi)$. *Then*

$$\|P_M u(t+h) - F_M(h, G(h) P_M u(t))\|_{X_0} \leq \text{ const} \, h(h + M^{-r}). \tag{3.4}$$

The left-hand side of (3.4) can be split as $P_M[u(t+h) - F(h, G(h)u(t))] + P_M[\{F(h, G(h)\cdot) - I\} \circ (I - P_M)u(t)]$, where the first term represents the error of the operator splitting in B_N, and the second term is due to the collocation projection of $u(t)$.

Combining this theorem with the above nonlinear stability result shows the first-order convergence of the scheme with respect to h, and the spectral accuracy of the spacial discretization.

References

1. Arnold A., *On absorbing boundary conditions for quantum transport equations*, to appear in Math. Model. Numer. Anal. (1994).

2. Arnold A., Ringhofer C., *An operator splitting method for the Wigner–Poisson problem*, submitted to SIAM J. Numer. Anal. (1994).

3. Arnold A., Ringhofer C., *Operator splitting methods applied to spectral discretizations of quantum transport equations*, to appear in SIAM J. Numer. Anal. (1994).

4. Kluksdahl N., Kriman A. M., Ferry D. K., Ringhofer C., *Selfconsistent study of the resonant tunneling diode*, Phys. Rev. B **39** (1989), 7720–7735.

5. Markowich P. A., Ringhofer C., *An analysis of the quantum Liouville equation*, Z. Angew. Math. Mech. **69** (1989), 121–127.

6. Markowich P. A., Ringhofer C., Schmeiser C., *Semiconductor Equations*, Springer-Verlag, Wien and New York, 1990.

7. Ringhofer C., *A spectral collocation technique for the solution of the Wigner-Poisson problem*, SIAM J. Numer. Anal. **29** (1992), 679-700.

8. Ringhofer C., Ferry D., Kluksdahl N., *Absorbing boundary conditions for the simulation of quantum transport phenomena*, Transport Theory Statist. Phys. **18** (1989), 331–346.

9. Suh N. D., Feix M. R., Bertrand P., *Numerical simulation of the quantum Liouville–Poisson system*, J. Comput. Phys. **94** (1991), 403–418.

10. Tatarskii V. I., *The Wigner representation of quantum mechanics*, J. Soviet Phys. Usp. **26** (1983), 311–327.

FB Mathematik, MA 6-2, TU-Berlin, D-10623 Berlin, Germany
E-mail address : arnold@math.tu-berlin.de

Proceedings of Symposia in Applied Mathematics
Volume 48, 1994

Finite Volume Methods for Irregular One-Dimensional Grids

M. J. BERGER, R. J. LEVEQUE AND L. G. STERN

ABSTRACT. We consider an approach for hyperbolic conservation laws on irregular grids in which the flux is determined based on an average value of the approximate solution over an interval of fixed size h to the left and right of the interface, where h is some measure of the average or maximum grid spacing. This interval may overlap several grid cells and hence the stencil is enlarged so that the method remains stable with a time step chosen relative to h.

1. Introduction

We consider finite volume methods for the hyperbolic system of conservation laws

$$u_t + f(u)_x = 0, \tag{1.1}$$

in one space dimension on an irregular grid. For simplicity we discuss scalar equations although the methods generalize easily to systems. Let U_i^n be the finite volume approximation to the cell average

$$U_i^n \approx \frac{1}{h_i}\int_{x_{i-1/2}}^{x_{i+1/2}} u(x,t_n)\,dx,$$

where $h_i = x_{i+1/2} - x_{i-1/2}$. The finite volume method takes the form

$$U_i^{n+1} = U_i^n - \frac{k}{h_i}\left(F_{i+1/2}^n - F_{i-1/2}^n\right), \tag{1.2}$$

where k is the time step and $F_{i-1/2}^n$ is the numerical flux at the grid interface between cells $i-1$ and i (we will often omit the superscript n when it is clear

1991 *Mathematics Subject Classification*. Primary 65M06; Secondary 65M12, 76M20.

This work was supported in part by NSF Grants DMS-8657319 and ACS-8858101, AFOSR Contract 91-0063, and DOE Contract DEFG02ER25053.

The final version of this paper will be submitted for publication elsewhere.

© 1994 American Mathematical Society
0160-7634/94 \$1.00 + \$.25 per page

from context). The simplest methods such as upwind or Lax-Wendroff are 3-point methods in which $F_{i-1/2} = \bar{F}(U_{i-1}, U_i)$ for some fixed function $\bar{F}$.

Our goal is to develop robust methods that compute smooth and accurate solutions with reasonable time steps even if the grid is highly nonuniform. This cannot be achieved with 3-point methods since the CFL condition requires that k satisfy

$$k \max_u |f'(u)| \leq \min_i h_i, \tag{1.3}$$

and so it is the *smallest* cell that limits the time step, which may then be unreasonable relative to the majority of cells.

2. The h-box Method

The basic idea of the class of methods we are now studying is to define the flux $F_{i-1/2}$ based on a standard numerical flux $\bar{F}$ but applied to modified data $U^L_{i-1/2}$ and $U^R_{i-1/2}$ rather than to the cell values U_{i-1} and U_i. The modified data is obtained by the following steps:

(i) Reconstruct some function $\tilde{u}(x)$ from the cell-average data $\{U_i\}$ (*e.g.*, piecewise constant, piecewise linear, or higher order).

(ii) Compute the average of $\tilde{u}(x)$ over virtual cells of fixed length h to the left and right of the interface $x_{i-1/2}$:

$$U^L_{i-1/2} = \frac{1}{h}\int_{x_{i-1/2}-h}^{x_{i-1/2}} \tilde{u}(x)\,dx \qquad \text{and} \qquad U^R_{i-1/2} = \frac{1}{h}\int_{x_{i-1/2}}^{x_{i-1/2}+h} \tilde{u}(x)\,dx.$$

Note that this is easily done in practice by using an interpolation of the primitive function $W_i = \sum_{j=0}^{i} U_j\ h_j \approx \int_0^{x_{i+1/2}} u(x)\,dx$.

Here, h is some measure of the average or maximum cell size on the grid. We refer to the interval $[x_{i-1/2} - h,\ x_{i-1/2}]$ as the *h-box* to the left of the interface $i - 1/2$. After computing $U^L_{i-1/2}$ and $U^R_{i-1/2}$, we then set

$$F_{i-1/2} = \bar{F}(U^L_{i-1/2}, U^R_{i-1/2})$$

and use the flux-differencing expression (1.2) to update U_i. On regular grids, with $h_i = h$, this is identical to the regular scheme.

Note that each h-box may overlap several grid cells, so that the method is no longer a 3-point method in general. Because the flux at $x_{i-1/2}$ is now computed based on information a distance h away on either side, the CFL condition now only requires that

$$k \max_u |f'(u)| \leq h, \tag{2.1}$$

and the methods are typically stable with this relaxed time step restriction. Note, however, that we still divide by h_i in (1.2), which could be much smaller than h. Stability is still maintained since for very small cells the h-boxes used to compute $F_{i-1/2}$ and $F_{i+1/2}$ overlap nearly completely, so that these fluxes are nearly equal and in fact $F_{i+1/2} = F_{i-1/2} + O(h_i)$ as $h_i \to 0$. (See [**1**].)

Besides increasing the time step, this approach yields much smoother solutions than other methods on highly nonuniform grids. The use of fixed-size h-boxes to compute the fluxes introduces a projection onto a uniform grid.

3. Numerical Examples

The upwind method in the case $f'(u) > 0$ has the flux

$$\bar{F}(U^L, U^R) = f(U^L).$$

The standard upwind method on a nonuniform grid takes the form

$$U_i^{n+1} = U_i - \frac{k}{h_i}\,(f(U_i) - f(U_{i-1}))$$

and is stable only if (1.3) is satisfied. Moreover, a truncation error analysis shows that this method is formally not even consistent unless the grid is nearly uniform in the sense that $h_i/h_{i-1} = 1+O(h)$ as $h \to 0$. In fact it can be shown[**4**] (see also [**3**], [**5**]) that the method is convergent and even first-order accurate in spite of this, on arbitrary grids, but the error is typically not smooth. Figure 1 shows an example calculation for Burgers' equation, $f(u) = u^2/2$, with data $u(x,0) = \frac{1}{2}(1 - \sin 2\pi x)$ at time $t = 0.2$, which gives a smooth solution. The grid has random cell sizes h_i uniformly distributed between $h/10$ and h, with $h = 1/25$. There are 48 grid cells. In Figure 1a, the standard upwind method is used with time step $k = h/10$, as required for stability.

Figures 1b and 1c show the result obtained on the same problem when h-boxes are used to define the values $U_{i-1/2}^L$ and then

$$U_i^{n+1} = U_i - \frac{k}{h_i}\,(f(U_{i+1/2}^L) - f(U_{i-1/2}^L)).$$

In this case, $k = h$ is used on the same random grid as before. Figure 1b shows results when $\tilde{u}(x)$ is taken to be the piecewise constant function with value U_i in the ith cell. Figure 1c shows the results obtained if $\tilde{u}(x)$ is the piecewise linear function

$$\tilde{u}(x) = U_i + (x - x_i)(U_i - U_{i-1})/h_{i-1/2} \qquad \text{on cell } [x_{i-1/2},\ x_{i+1/2}].$$

Here, $h_{i-1/2} = (h_{i-1} + h_i)/2$ and $x_i = (x_{i-1/2} + x_{i+1/2})/2$, the cell midpoint.

Figure 1d shows results when the Lax-Wendroff flux is used together with piecewise linear reconstruction for $\tilde{u}(x)$. The Lax-Wendroff flux for $f'(u) \geq 0$ is given by

$$\bar{F}(U^L, U^R) = \frac{1}{2}(f(U^L) + f(U^R)) - \frac{k}{2h}\left(\frac{(f(U^R) - f(U^L))^2}{U^R - U^L}\right).$$

Grid refinement studies show that this method is second-order accurate (for smooth solutions) in h, while all of the previous methods are first-order accurate.

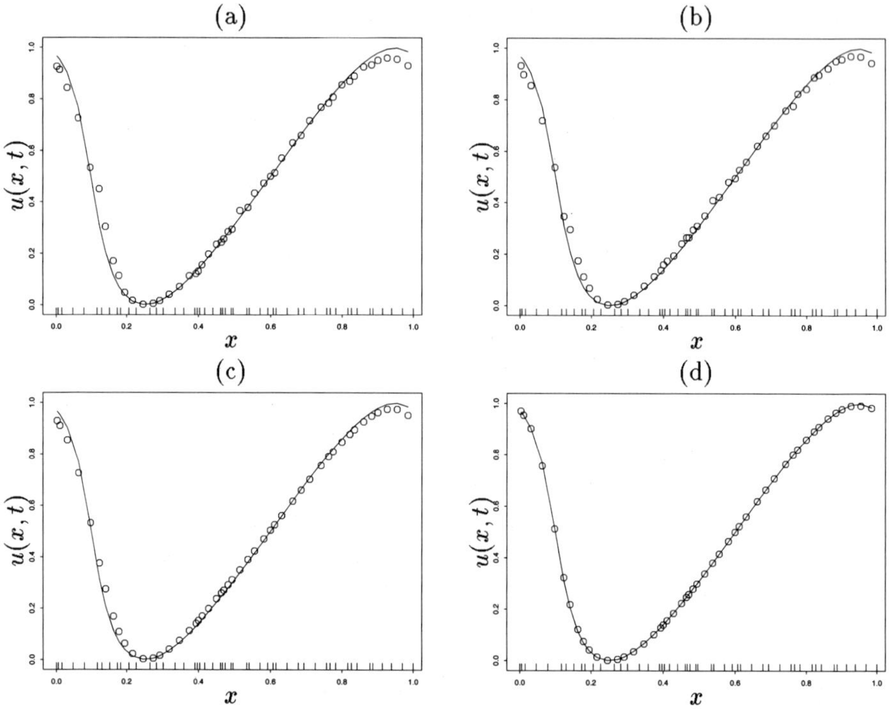

FIGURE 1. Results on a random grid for Burgers' equation with a smooth solution. (a) The standard upwind method with $k = h/10$. (b) The h-box upwind method with piecewise constant $\tilde{u}(x)$. (c) The h-box upwind method with piecewise linear $\tilde{u}(x)$. (d) The h-box Lax-Wendroff method with piecewise linear $\tilde{u}(x)$.

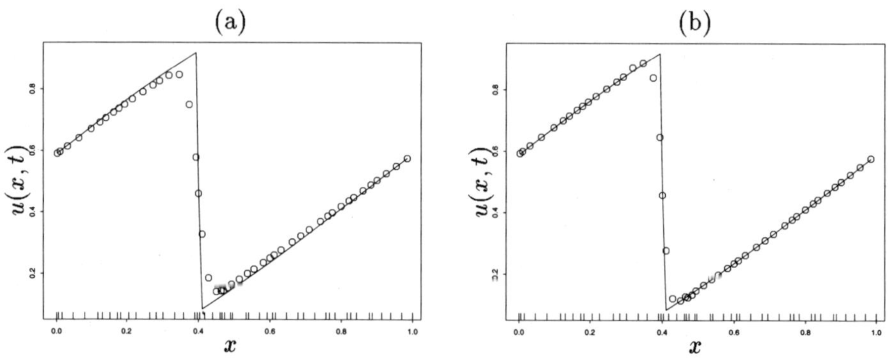

FIGURE 2. Results on a random grid for Burgers' equation with a discontinuous solution. (a) The h-box upwind method with piecewise linear $\tilde{u}(x)$. (b) The h-box high resolution flux limiter method with piecewise linear $\tilde{u}(x)$.

4. Extensions

The ideas presented here can be extended to situations where the basic flux $\bar{F}$ depends on more than two points, and in particular to flux-limiter methods for the high-resolution computation of shock waves. This will be presented in more detail elsewhere. As an example, Figure 2 shows the same techniques as Figure 1c and 1d at time $t = 0.8$, after a shock has developed.

There remain some problems, particularly when small cells are located near a transonic rarefaction. Several interfaces may appear to be transonic points because their h-boxes overlap the transonic region, causing large inaccuracies and oscillations. This can be corrected by reducing the size of the h-boxes near transonic points, where the propagation speed is smaller.

Although we consider only one-dimensional grids here, the original motivation for this problem comes from problems in more than one dimension, where nonuniform grids arise more commonly. The idea of creating h-boxes that overlap several grid cells as a way to compute stable fluxes was introduced by Berger and LeVeque[**1**], [**2**] as a way to deal with small cells created near the boundary when a Cartesian grid is used on an irregular region. The same idea could be used at interfaces between different grids with a composite grid method or perhaps at all cell interfaces on an unstructured grid. A more complete study of such methods is underway. Even in one space dimension these ideas could find application in conjunction with moving-mesh or front-tracking algorithms where nonuniformities in the grid are created.

References

1. M. Berger and R. J. LeVeque, *Cartesian meshes and adaptive mesh refinement for hyperbolic partial differential equations*, Proc. Third Int'l Conf. Hyperbolic Problems, Uppsala, 1990.
2. ———, *Stable boundary conditions for Cartesian grid calculations*, Computing Systems in Engineering **1** (1990), 305–311.
3. H.-O. Kreiss, T. A. Manteuffel, B. Swartz, B. Wendroff, and A. B. White, Jr., *Supraconvergent schemes on irregular grids*, Math. Comp. **47** (1986), 537-554.
4. B. Wendroff and A. B. White, Jr., *Some supraconvergent schemes for hyperbolic equations on irregular grids*, Second International Conference on Hyperbolic Problems, Aachen, Vieweg, 1988, pp. 671–677.
5. ———, *A supraconvergent scheme for nonlinear hyperbolic systems*, Comput. Math. Appl. **18** (1989), 761–767.

Courant Institute of Mathematical Sciences, 251 Mercer St., New York, NY 10012

E-mail address: berger@cims.nyu.edu

Department of Applied Mathematics, FS-20, University of Washington, Seattle, WA 98195

E-mail address: rjl@amath.washington.edu

Department of Applied Mathematics, FS-20, University of Washington, Seattle, WA 98195

E-mail address: stern@amath.washington.edu

Proceedings of Symposia in Applied Mathematics
Volume 48, 1994

LINEAR RATIONAL INTERPOLATION OF CONTINUOUS FUNCTIONS OVER AN INTERVAL

JEAN–PAUL BERRUT

ABSTRACT. To avoid the drawbacks of polynomial interpolation between arbitrary points, we suggest using rational interpolation with a fixed denominator depending only on the points and not on the interpolated function. We determine here the "optimal" denominator by minimizing the Lebesgue constant of the corresponding linear projection.

1. POLYNOMIAL INTERPOLATION

Let f be a complex–valued function defined on some interval $[a, b]$ of the real line, let $x_0, x_1, \ldots, x_n$ be $n+1$ distinct points of $[a, b]$ and let $f_k := f(x_k)$, $k = 0(1)n$. Then $p_n(x) = \sum_{k=0}^{n} f_k \ell_k(x)$, $\ell_k(x) = \prod_{\substack{i=0\\ i\neq k}}^{n} (x - x_i) / \prod_{\substack{i=0\\ i\neq k}}^{n} (x_k - x_i)$, is the Lagrangian representation of the unique polynomial of degree n which interpolates f between the points x_k: $p_n(x_k) = f_k$. The polynomial p_n can also be written in its *barycentric form* [5, p. 238]

$$p_n(x) = \sum_{k=0}^{n} \frac{w_k}{x - x_k} f_k \bigg/ \sum_{k=0}^{n} \frac{w_k}{x - x_k} \tag{1.1}$$

in which the weights $w_k := 1 / \prod_{\substack{i=0\\ i\neq k}}^{n} (x_k - x_i)$ depend only on the interpolation points and not on the interpolated function.

The quality of the approximation of f by p_n is usually measured in the maximum norm $\|\cdot\|_\infty$ in the class $C[a, b]$ of continuous functions. For given interpolation points, the norm of the interpolation operator is its Lebesgue constant $\Lambda_n = \max_{a\leq x\leq b} \sum_{k=0}^{n} |\ell_k(x)|$ [6, p. 41]. For arbitrary points the latter can be very large (see the table at the end), and it is well known that the corresponding approximation can be disastrous, e.g. when f has poles close to the interval $[a, b]$ as in Runge's famous example $f(x) = \frac{1}{1+x^2}$ with a large number of equidistant points in $[-5, 5]$ [7]. On the other hand, if one can choose the x_k, e.g. as Čebyšev or Legendre points, then polynomial interpolation can be very accurate. However, the user often cannot choose the interpolation points or does not want to take them clustered at the extremities of the interval.

Classical rational interpolation is known to give excellent results in some cases [4, p. 288], but unattainable points and/or poles in the interval of interpolation

1991 *Mathematics Subject Classification.* Primary 65D05, 41A05; Secondary 41A20.
Key words and phrases. Interpolation, rational interpolation, linear interpolation.
The final version of this paper will be submitted for publication elsewhere.

© 1994 American Mathematical Society
0160-7634/94 $1.00 + $.25 per page

often render it useless. Moreover, it cannot be used as "Ansatz" in the numerical solution of equations like those arising in boundary value problems.

We suggest here to replace polynomial interpolation by *rational interpolation with a denominator that depends on the interpolation points but not on the interpolated function.* In this way we preserve the linearity of the interpolation process with respect to the interpolated functions, and this leads us immediately to a Lagrange interpolation formula and the corresponding Lebesgue constant, which we can write as function of the barycentric weights. We can then choose the weights according to some appropriate criterion. Here we will minimize the Lebesgue constant. Once we have the corresponding (numerical) "optimal" weights as "function" of the points, we can, as in polynomial interpolation, determine in a further stage the optimal points as those with minimal optimal Lebesgue constant.

2. Rational interpolation in barycentric form

One can replace the w_k in (1.1) by any set $\{u_k\}$ of numbers, all different from 0; the quotient remains a (rational) interpolant of f [10] (if one of the u_k is 0, the interpolant will take the value of the quotient without the corresponding terms, so that at the corresponding point all values of f but one will lead to an "unattainable" point [9, p. 56]). And, conversely, one sees immediately [3] that every interpolant in $\mathcal{R}_n$, the set of real rational functions with numerator and denominator degrees less than or equal to n, can be written in barycentric form

$$r_n(x) = \sum_{k=0}^{n} \frac{u_k}{x - x_k} f_k \Bigg/ \sum_{k=0}^{n} \frac{u_k}{x - x_k} \tag{2.1}$$

for some $\{u_k\}_{k=0}^n$, $u_k \in \mathbb{R}$.

Schneider and Werner [8] have (implicitly) used the above observation to restate in terms of barycentric formulae the classical rational interpolation problem in which the number of unknown coefficients in the canonical representation of numerator and denominator add up to the number of interpolation conditions.

Here we will pursue our investigations of [1]: noting that "in most applications the degrees of the numerator and the denominator give no decisive criteria for judging the quality of the interpolation", we leave the degree condition out of consideration and try to choose the weights u_k in (2.1) in such a way that

— no unattainable point occurs for any f, which implies $u_k \neq 0$ for all k;

— no pole lies in the interval of interpolation. A necessary, but not sufficient condition (counterexample: $n = 2$, $x_0 = -1$, $x_1 = 0$, $x_2 = 1$, $u_0 = 1$, $u_1 = -1$, $u_2 = 6$) for the absence of poles is the alternating of the signs of the u_k ([8, Prop. 8], [2, p. 710]).

Each choice of *fixed* weights $\{u_k\}$ leads to a rational interpolation process which is a projection and is *linear* in the interpolated functions. And it is obvious that, conversely, fixed u_k are a necessary condition for linearity. With such weights, r_n can be written as

$$r_n(x) = \sum_{k=0}^{n} f_k \ell_k^{(u)}(x) \quad \text{where} \quad \ell_k^{(u)}(x) = \frac{u_k}{x - x_k} \Bigg/ \sum_{j=0}^{n} \frac{u_j}{x - x_j}. \tag{2.2}$$

Here, $\ell_k^{(u)}$ is the Lagrange fundamental rational function with denominator $q_u(x) = (x-x_0)(x-x_1)\cdots(x-x_n)\sum_{j=0}^{n}\frac{u_j}{x-x_j}$: it interpolates the values $f_j := \delta_{jk}$, $j = 0(1)n$. The polynomial q_u is independent of k. The following theorem can be proven in the same way as in the polynomial case [6, p. 41].

2.1. Theorem. *Let $x_0, \ldots, x_n$ be $n+1$ distinct points in $[a,b]$. To each set of weights $\{u_k\}_{k=0}^{n}$ such that $u_k \neq 0$ for all k and q_u does not vanish on $[a,b]$, there corresponds a linear projection $\mathrm{R}_n^{(u)}$ in $C[a,b]$ which associates to any function f the unique rational function $\mathrm{R}_n^{(u)} f := r \in \mathcal{R}_n$ from (2.2) that interpolates f between the x_k's. The norm (Lebesgue constant) of $\mathrm{R}_n^{(u)}$ is given by*

$$\|\mathrm{R}_n^{(u)}\| = \Lambda_n^{(u)} := \max_{a\le x\le b}\sum_{k=0}^{n}\left|\ell_k^{(u)}(x)\right|.$$

Moreover, the error of the approximation of f by $\mathrm{R}_n^{(u)} f$ satisfies the bound

$$\|f - \mathrm{R}_n^{(u)} f\| \le \left[1 + \Lambda_n^{(u)}\right] d_u^*,$$

where $d_u^* := \|f - r^{(u)^*}\|_\infty$ is the error of the best approximation $r^{(u)^*}$ of f by any function in $\mathcal{R}_n$ with denominator q_u.

The unicity of $\mathrm{R}_n^{(u)} f$ follows obviously from the unicity of polynomial interpolation for the numerator. The rational functions that can be written as quotients (2.1) are interpolated exactly; this is the case for all constants, and more generally for all polynomials of degree less than or equal to $n - \deg q_u$. A theorem comparing d_u^* and the error of best polynomial approximation will be given in [3].

In [1] we determined the u_k so that the resulting interpolation would be (experimentally) well conditioned. For that purpose, we preliminarily transformed the problem to an equivalent one on the circle where we used trigonometric interpolation with weights $u_k = (-1)^k$, independent of the given points; transforming back to the interval, we obtained for $[a,b] := [-1,1]$ the interpolant

$$R(x) = \sum_{k=0}^{n}\frac{(-1)^k\delta_k\eta_k}{x-x_k}f_k \Bigg/ \sum_{k=0}^{n}\frac{(-1)^k\delta_k\eta_k}{x-x_k} \tag{2.3a}$$

with the "Čebyšev" weights

$$\eta_k = \begin{cases} \sqrt{1-x_k^2}, & -1 \text{ and } 1 \text{ are not interpolation points,}\\ \sqrt{\frac{1}{2}(1+x_k)}, & 1 \text{ is interpolation point, } -1 \text{ is not,}\\ \sqrt{\frac{1}{2}(1-x_k)}, & -1 \text{ is interpolation point, } 1 \text{ is not,}\\ 1, & -1 \text{ and } 1 \text{ are both interpolation points,}\end{cases} \tag{2.3b}$$

and

$$\delta_k = \begin{cases} 1/2, & x_k = -1 \text{ or } x_k = 1,\\ 1, & \text{otherwise.}\end{cases} \tag{2.3c}$$

(To avoid possible cancellation in the η_k, one should use their trigonometric representation, see [1].)

3. Linear rational interpolation with minimal Lebesgue constant

We can now take advantage of the independence of the weights from the points resulting from substituting linear rational interpolation for polynomial interpolation, and we can choose the u_k's so as to minimize the Lebesgue constant $\Lambda_n^{(u)}$ of the corresponding projection.

Details of the minimization process will be given elsewhere [3]. Here we only display numerical results which demonstrate that this kind of linear rational interpolation can work in practice. The following table contains, for $n+1=13$ equidistant points, in the second column the Lebesgue constants Λ_n, $\Lambda_n^{(R)}$ and $\Lambda_n^{(u^*)}$ for interpolation respectively by the polynomial (1.1), by R from (2.3), and by the (approximate) linear rational interpolant with minimal Lebesgue constant, and in the third column the corresponding maximal interpolation errors for Runge's function $\frac{1}{1+25x^2}$ on the interval $[-1,1]$.

	Λ_n	Error
p_n	89.32	3.66
R	2.38	$2.18 \cdot 10^{-2}$
$R_n^{(u)} f$	2.29	$1.98 \cdot 10^{-2}$

These values demonstrate that at least in some examples the linear rational interpolant with minimal Lebesgue constant is much better than the interpolating polynomial and also better than the rational interpolant R. However, we note that in many examples it is only marginally — if at all — better than the much easier-to-compute R.

References

1. Berrut, J.–P., *Rational functions for guaranteed and experimentally well–conditioned global interpolation*, Comput. Math. Appl. **15** (1988), 1–16.
2. Berrut, J.–P., *Barycentric formulae for cardinal (SINC–)interpolants*, Numer. Math. **54** (1989), 703–718.
3. Berrut, J.–P., *Lebesgue constant minimizing linear rational interpolation of continuous functions over the interval* (1994), Report Institut de Mathématiques, Université de Fribourg (Suisse).
4. Bulirsch, R., Rutishauser, H., *Interpolation und genäherte Quadratur*, Mathematische Hilfsmittel des Ingenieurs, Grundlehren der Math. Wissenschaften (Sauer, R., Szabó, I., eds.), vol. 141, Springer-Verlag, Berlin–Heidelberg, 1968, pp. 232–319.
5. Henrici, P., *Essentials of Numerical Analysis*, Wiley, New York, 1982.
6. Powell, M. J. D., *Approximation theory and methods*, Cambridge University Press, 1981.
7. Runge, C., *Über empirische Funktionen und die Interpolation zwischen äquidistanten Ordinaten*, Z. Math. Phys. **46** (1901), 224–243.
8. Schneider, C., Werner, W., *Some new aspects of rational interpolation*, Math. Comp. **47** (1986), 285–299.
9. Stoer, J., *Einführung in die Numerische Mathematik I*, 4th ed., Springer, Berlin–Heidelberg–New York, 1983.
10. Werner, W., *Polynomial interpolation: Lagrange versus Newton*, Math. Comp. **43** (1984), 205–217.

Institut de Mathématiques, Université de Fribourg, CH–1700 Fribourg, Switzerland

E-mail address: berrut@cfruni51.bitnet

Proceedings of Symposia in Applied Mathematics
Volume **48**, 1994

A von Neumann reflection for the 2-D Burgers equation

M. BRIO AND J. K. HUNTER

ABSTRACT. We present a numerical solution of a Riemann problem for the 2-D Burgers equation which contains a von Neumann reflection.

The 2-D Burgers equation,

$$\begin{aligned} &u_t + \left(\frac{1}{2}u^2\right)_x + v_y = 0, \\ &u_y - v_x = 0, \end{aligned} \tag{1}$$

arises as an asymptotic limit of the Euler equation of gas dynamics in various contexts [**4**]. In [**1,8**], this equation is used to study the reflection of weak shocks by thin wedges, and the von Neumann paradox in particular. See also [**5**].

Here, following [**2**], we use the term "von Neumann paradox" to refer to the fact that earlier numerical and experimental studies of the irregular reflection of weak shocks seemed to show a single Mach reflection, with three shocks and a contact discontinuity meeting at a point (the triple point) — a solution which violates the jump relations when the shocks are sufficiently weak [**1,3,5,8**].

Various alternatives to Mach reflection have been considered over the years: in [**6**], different local expansions near the triple point were considered; in [**8**], a singular local expansion for triple points in solutions of (1) was proposed; in [**3**], von Neumann reflection (NR) was suggested. In NR, there is no triple point; the incident shock bends smoothly into the boundary, and a smooth diffracted compression wave comes off the bend, steepening into the reflected shock a short distance away.

1991 *Mathematics Subject Classification.* Primary 35L67, 35M10, 76H05; Secondary 35L65, 76L05.

The first author was supported in part by AFOSR Grant #F49620-92-J-0054.

The second author was supported in part by NSF Grant #DMS-9011548.

The final version of this paper will be submitted for publication elsewhere.

© 1994 American Mathematical Society
0160-7634/94 $1.00 + $.25 per page

Colella and Henderson [**2**] gave strong evidence for a NR in numerical solutions of the 2-D Euler equations. Experimental observations of a NR and further computations are shown in [**7**].

In this paper, we present a numerical solution of a Riemann problem for the 2-D Burgers equation which contains a NR. Since solutions of the 2-D Burgers equation give a (formal) asymptotic approximation of solutions of the Euler equations, this is further evidence for the occurrence of NR in the irregular reflection of weak shocks.

We solve (1) with the Riemann initial data,

$$(u,v) = \begin{cases} (0,0), & x > ay, \\ (1,-a), & -by < x < ay, \\ (1+a/b, 0), & x < -by, \end{cases} \tag{2}$$

where a and b are given positive parameters. This is the simplest initial data exhibiting a NR. The problem is self-similar, so the numerical solution at one time provides a solution at all times after rescaling.

The main numerical problem is to resolve the diffracted wave, since an unresolved NR can look like a perfect triple point. This is the source of the von Neumann paradox. For example, fig. 8 in [**1**] shows an apparent triple point, but computations on a more refined grid show a NR similar to fig. 1 below.

The Riemann initial data (2), with b finite, does not correspond to the physical problem of the reflection of a weak shock by a wedge. The initial data for the wedge problem is obtained by matching the solution of (1) with the linearized solution for the reflection of a discontinuity off a wedge. The details of this matching procedure and the study of the resulting initial value problem will be presented in a forthcoming paper.

The main difficulty in solving the 2-D Burgers equation numerically is that time is a characteristic direction, so that its domains of influence and dependence are unbounded. In [**8**], the $x-t$ coordinate frame was rotated to overcome this difficulty. Here, we use the same numerical method as in [**1**]. We write (1) in potential form and use an upwind (ENO) treatment of the nonlinear term. We also add an artificial viscosity proportional to the x-grid size to suppress numerical oscillations and use absorbing boundary conditions at the top boundary.

In fig. 1, we show the v-contours from a numerical solution of (1)–(2) with the parameter values $a = 0.2$ and $b = 1.0$. The u-contours are similar, although the variations in u in the diffracted wave behind the incident shock are much smaller than the variations in v. This solution was computed on a 2000×500 grid; for comparison, see fig. 4 in [**1**], which shows the same problem computed on a 200×50 grid.

The numerical solution in fig. 1 is a von Neumann reflection. There is no corner in the incident shock, and u and v change continuously behind the bend in the incident shock. The sonic line [**4**], $x/t + y^2/(4t^2) = u(x,y,t)$, across which

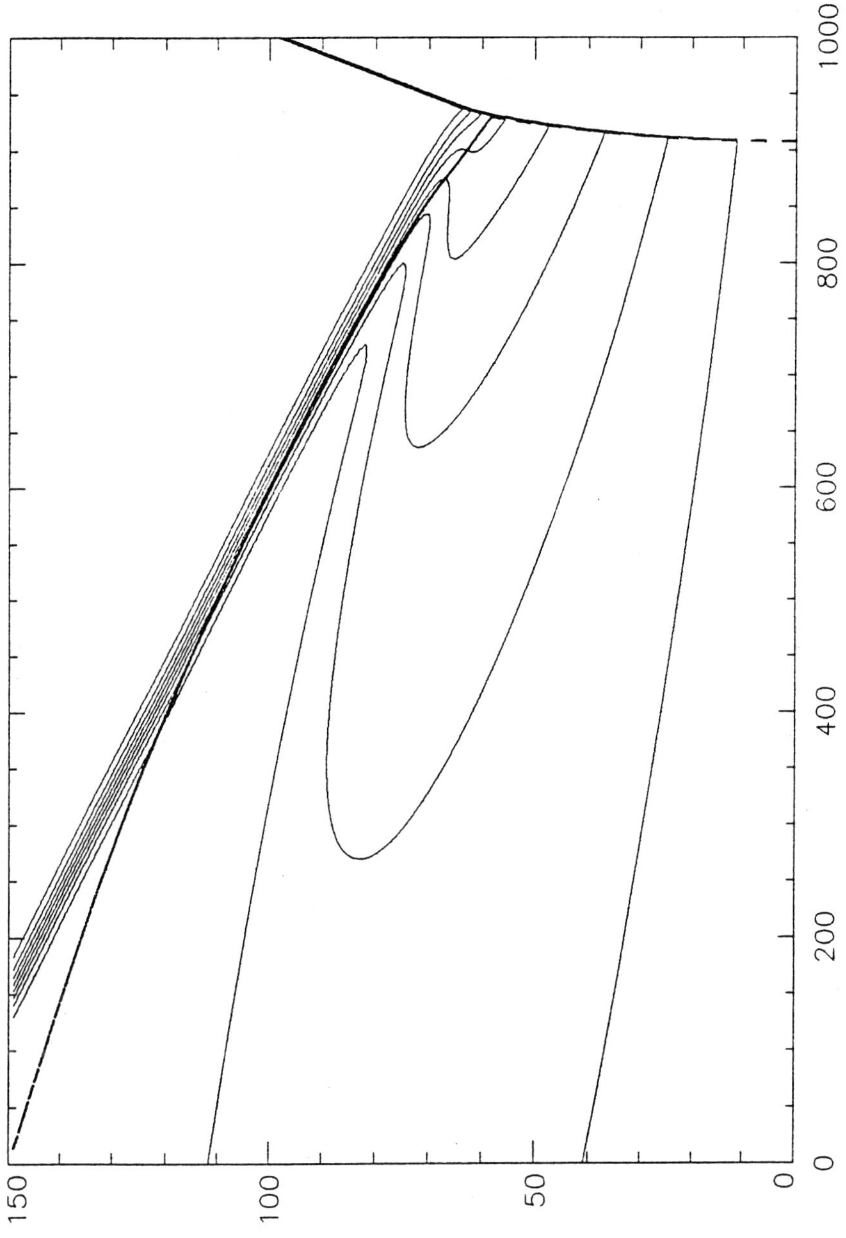

figure 1. Contour plot of v showing the von Neumann reflection.

the type of the equation in self-similar coordinates changes from hyperbolic to elliptic, is shown in fig. 1 as a dashed line.

Above the sonic line is a small hyperbolic region with nonuniform flow. Inside this region, a compression wave comes off the bend in the incident shock and steepens into the reflected shock a short distance away. Below the sonic line and the reflected shock, and to the left of the incident shock, is an elliptic region with nonuniform flow.

We are currently investigating how the structure of the NR depends on the parameter values (a, b). In particular, the analysis of Tabak [**8**] suggests that for small enough a it may be possible to have a NR in which the sonic line moves to the edge of the region of nonuniform flow, so that the flow is piecewise constant in the hyperbolic region.

References

1. M. Brio and J. K. Hunter, *Mach reflection for the two-dimensional Burgers equation*, Physica D **60** (1992), 194–207.
2. P. Colella and L. F. Henderson, *The von Neumann paradox for the diffraction of weak shock waves*, J. Fluid Mech. **213** (1990), 71–94.
3. L. F. Henderson, *Regions and boundaries for diffracting shock wave systems*, Z. Angew. Math. Mech. **67** (1987), 73–86.
4. J. K. Hunter, *Nonlinear geometrical optics*, Multidimensional Hyperbolic Problems and Computations, Vol. 29, IMA Volumes in Mathematics and its Applications, eds. A. J. Majda and J. Glimm, Springer-Verlag, New York (1991), 179–197.
5. C. S. Morawetz, *Potential theory for regular and Mach reflection of a shock at a wedge*, Comm. Pure Appl. Math. **45** (1993), 1–32.
6. R. D. Richtmyer, *Principles of Advanced Mathematical Physics, Vol. I*, Springer-Verlag, New York, 1981.
7. A. Sasoh, K. Takayama, and T. Saito, *A weak shock wave reflection over wedges*, Shock Waves **2** (1992), 277–281.
8. E. G. Tabak, *Focusing of weak shock waves and the von Neumann paradox of oblique shock reflection*, Ph.D. thesis, MIT, 1992.

Department of Mathematics, University of Arizona, Tucson, AZ 85721

E-mail address: brio@math.arizona.edu

Department of Mathematics and Institute of Theoretical Dynamics, University of California, Davis, CA 95616

Proceedings of Symposia in Applied Mathematics
Volume 48, 1994

Slow evolution from the boundary: a new stabilizing constraint in ill-posed continuation problems

ALFRED S. CARASSO

ABSTRACT. A new a-priori constraint sharply reduces noise contamination in ill-posed continuation problems. One application concerns a significant class of image deblurring problems that can be formulated as ill-posed time-reversed parabolic initial value problems.

1. Introduction

Many ill-posed continuation problems in partial differential equations obey a logarithmic convexity inequality and can be stabilized in an appropriate Banach space by imposing an a-priori bound on the solutions. Together with [**8**] and [**9**], the surveys [**1, 5, 6**] and [**10**] discuss many topics and contain a wealth of references. In the simplest cases, logarithmic convexity leads to the sharp stability estimate $\| u_1(t) - u_2(t) \| \leq 2M^{1-t} \epsilon^t$, $0 \leq t \leq 1$, for the difference of any two continuations, where t is the continuation variable, M is an a-priori bound on $\| u(0) \|$, and ϵ is an upper bound on the norm of the error in the continuation data at $t = 1$. For small $t > 0$, such Hölder-continuous dependence on the data is not useful at the levels of data error ϵ typically found in practice, and noise contamination as $t \downarrow 0$ is a characteristic feature of many stabilized ill-posed computations. The present summary outlines the main results of [**3**] and [**4**], concerning a new stabilizing constraint for the numerical computation of such problems from noisy data, together with the application of this theory to a significant class of image deblurring problems that can be reformulated as ill-posed time-reversed parabolic initial value problems.

1991 *Mathematics Subject Classification.* Primary 35R25, 35B60, 35B35, 65M30, 68U10.

Key words and phrases. Ill-posed continuation, Hölder continuity, SECB constraint, noise suppression, image deblurring, Tikhonov-Miller regularization.

The detailed version of this paper has been submitted for publication elsewhere.

A United States patent is pending on part of the work described herein.

© 1994 American Mathematical Society
0160-7634/94 \$1.00 + \$.25 per page

2. Slow evolution from the boundary

The use of supplementary information in the form of a *slow evolution from the continuation boundary* constraint, (SECB), is proposed and analyzed in [**3**]. Attention is focused on problems involving continuation in a single variable t on $0 \leq t \leq 1$, with continuation boundary at $t = 0$. In spatial continuation problems with radial symmetry, [**7**], t is a radial coordinate, while t is related to time in applications involving continuation in the time variable. In an appropriate Banach space X, the continuation $u(t)$ is an X-valued function with the norm $\| u(t) \|$ for fixed t. With $\| u(0) \| \leq M$, and $\| u(1) - f \| \leq \epsilon$, let $u_1(t)$, $u_2(t)$, be any two possible continuations from the given data f at $t = 1$, and let $w(t) = u_1(t) - u_2(t)$. It is assumed that the following 'log-convex' estimate holds for $w(t)$:

$$\|w(t)\| \leq \|w(0)\|^{1-t} \|w(1)\|^{t}, \qquad 0 \leq t \leq 1. \tag{1}$$

As in the case of prescribed bounds, the SECB constraint presupposes knowledge of the particular physical processes underlying the mathematical continuation problem. Let ϵ, M be as above, with $\epsilon/M \ll 1$. For given K with $0 < K \ll M/\epsilon$, let s^* be such that $1 - (\epsilon/M)^{s^*} = K\epsilon/M$, i.e.,

$$s^*(\epsilon, M, K) = \log\{M/(M - K\epsilon)\}/\log(M/\epsilon). \tag{2}$$

The SECB constraint is expressed as follows:

There exists a known small constant $K > 0$, and a known fixed $s > 0$, with $s > s^*$, such that $\|u(s) - u(0)\| \leq K\epsilon$.

By continuity as $t \downarrow 0$, given any $\epsilon > 0$, there always exists a sufficiently small s such that the last inequality holds with a small K. However, the requirement that s be known and greater than s^* represents a further limitation on the class of admissible solutions. It is desirable that K be as small as possible, with $s < 1$ as large as possible ($s/s^* \gg 1$). In the case of the backward diffusion equation, SECB expresses the *a-priori* knowledge that the relevant solution on $0 \leq t \leq 1$ is such that the values at time s are a close approximation to the initial values. This is typically the case when, for example, the diffusion coefficient α is small, and the high-frequency components in $u(x, 0)$ do not play an overwhelming role. However, the SECB constraint is distinctly different from prescribing bounds on the x-derivatives of $u(x, 0)$, inasmuch as SECB does not imply, and is not implied by, differentiability of $u(x, 0)$ with respect to the x-variables, [**3**]. When SECB is applicable, it has a decisive effect in suppressing noise contamination as $t \downarrow 0$, leading to an improved stability estimate for a large class of ill-posed continuation problems. The following result is proved in [**3**].

THEOREM 1. *Let ϵ, M, K be given positive constants, with $\epsilon \ll M$ and $K \ll M/\epsilon$. Let X be a Banach space with norm $\| \cdot \|$, and let $f \in X$. Let $\mathcal{C}$ be a continuation problem from the data f for the X-valued function $u(t)$, $0 \leq t \leq 1$,* *where* $\|u(0)\| \le M$ *and* $\|u(1) - f\| \le \epsilon$. *Let* $\mathcal{C}$ *be such that the difference* $w(t)$ *of any two possible continuations satisfies*

$$\|w(t)\| \le \|w(0)\|^{1-t}\, \|w(1)\|^t, \qquad 0 \le t \le 1. \tag{3}$$

If the solutions of $\mathcal{C}$ *also satisfy* $\|u(s) - u(0)\| \le K\epsilon$ *for some fixed* $s > s^*$, *where* s^* *is defined in* (2), *then*

$$\|w(t)\| \le 2\Gamma^{1-t}\, \epsilon, \qquad 0 \le t \le 1, \tag{4}$$

where $\Gamma(K, s)$ *is the unique root of the transcendental equation* $x = K + x^{1-s}$, *and* $\Gamma \ll M/\epsilon$ *provided* $s/s^* \gg 1$.

3. Image deblurring

In [**3**] and [**4**], the usefulness of the SECB constraint is demonstrated by applying it to a class of image deblurring problems whereby the original image is reconstructed from a noisy blurred version, [**2, 11**]. Attention is restricted to shift-invariant blurring kernels $p(x, y)$ with Fourier transform given by

$$\begin{aligned} \hat{p}(\xi, \eta) &\equiv \int_{R^2} p(x, y) e^{-2\pi i(\xi x + \eta y)}\, dx dy \\ &= e^{-\sum_{m=1}^{J} \lambda_m (\xi^2+\eta^2)^{\beta_m}}, \qquad \lambda_m \ge 0,\ 0 < \beta_m \le 1. \end{aligned} \tag{5}$$

Kernels satisfying (5) occur in numerous applications, [**4**]. For such kernels, the blurred noiseless image $f_e(x, y)$ may be identified with $u(x, y, 1)$, where $u(x, y, t)$ is the unique bounded solution of the well-posed direct problem

$$\begin{aligned} &u_t = -\textstyle\sum_{m=1}^{J} \gamma_m (-\Delta)^{\beta_m} u, \qquad t > 0,\ x, y \in R^2,\ \gamma_m = \lambda_m (4\pi^2)^{-\beta_m}, \\ &u(x, y, 0) = g(x, y), \end{aligned} \tag{6}$$

and $g(x, y)$ is the original unblurred image. This follows by recognizing (5) as the Fourier transform of the causal fundamental solution for the initial value problem (6). If all $\beta_m = 1$, (6) is the classical heat conduction equation. For $0 < \beta_m < 1$, (6) represents a generalized diffusion process. The image restoration problem is equivalent to backwards in time continuation of the solution of (6), from noisy data $f(x, y)$ at $t = 1$, where $f(x, y)$ is the blurred recorded image. The unblurred image $g(x, y)$ is the continuation at $t = 0$. For given fixed t with $0 < t < 1$, $u(x, y, t)$ represents a *partial restoration.*

3.1. SECB restoration. Write the deblurring problem in the form $Pg = f$, where P is the integral operator in $L^2(R^2)$ with kernel $p(x - u, y - v)$. For fixed t with $0 \le t \le 1$, define P^t to be the convolution integral operator in $L^2(R^2)$, with kernel $p(t; x - u, y - v)$, where $p(t; x, y)$ is the inverse Fourier transform of (5) with λ_m replaced by $t\lambda_m$. Then, $P^t P^s = P^{t+s}$, for all $s, t \ge 0$, and $P^0 = I$. The solution of (6) may be written as $u(t) = P^t g = P^{t-1} f_e$, $0 \le t \le 1$.

SECB restoration is based on the following *a-priori* information:

$$(7) \qquad \|Pg - f\| \leq \epsilon, \quad \|g\| \leq M, \quad \|g - P^s g\| \leq K\epsilon, \qquad \text{fixed } s > s^*,$$

for given ϵ, $M > 0$, $\epsilon/M \ll 1$, $0 < K \ll M/\epsilon$, where $\|\cdot\|$ denotes the L^2 norm and s^* is defined in (2). It is desirable that the blurring operator P and the unblurred image $g(x,y)$ both be such that (7) holds with a small K and a relatively large s, so that $s/s^* \gg 1$. The SECB restoration is defined as the unique function $g^C(x,y)$ which minimizes the functional

$$(8) \qquad \|Ph - f\|^2 + \omega^2 \|h\|^2 + K^{-2} \|h - P^s h\|^2, \qquad \omega = \epsilon/M,$$

over all $h \in L^2(R^2)$. We can write $g^C = P^{-1} f^C$, where $f^C = [I + \omega^2 (P^*P)^{-1} + K^{-2}(P^*P)^{-1}(I - P^s)^*(I - P^s)]^{-1} f$. If $s = 0$ in (8), SECB restoration reduces to the canonical Tikhonov-Miller restoration, [**8**]. The *partial SECB restoration at time* t is defined to be $P^{t-1} f^C$, which is an approximation to $u(t) = P^{t-1} f_e$. The following result is proved in [**3**].

THEOREM 2. *Let the unblurred image $g(x,y)$ satisfy the constraints (7). Let $g^C(x,y)$ be the function which minimizes (8). Then, g^C is the unique solution of*

$$(9) \qquad \left[P^*P + \omega^2 I + K^{-2}(I - P^s)^*(I - P^s)\right] g^C = P^* f, \qquad \omega = \epsilon/M,$$

and g^C satisfies

$$(10) \qquad \|Pg^C - f\|^2 + \omega^2 \|g^C\|^2 + K^{-2} \|g^C - P^s g^C\|^2 \leq 3\epsilon^2.$$

Moreover, if $\Gamma(K,s)$ is the constant defined in Theorem 1, the partial SECB restorations $P^{t-1} f^C$ satisfy

$$(11) \qquad \|P^{t-1}(f^C - f_e)\| \leq \sqrt{3}\, \Gamma^{1-t}\, \epsilon, \qquad 0 \leq t \leq 1.$$

In contrast to (11), Tikhonov-Miller restoration results in an $O(\epsilon^t)$ estimate. Both SECB and Tikhonov-Miller restorations can be efficiently implemented in the Fourier transform domain, using FFT algorithms. For 512 × 512 images, SECB restoration requires about 30 seconds of cpu time on currently available desktop workstations. Documented experiments are described in [**3**] and [**4**], comparing SECB and Tikhonov-Miller deblurring of the same image, using optimal values of the regularization parameters. These experiments confirm the noise-suppressing property of the SECB constraint.

REFERENCES

1. K. A. Ames, *Improperly posed problems for nonlinear partial differential equations*, in *Nonlinear Equations in the Applied Sciences*, W. F. Ames and C. Rogers, eds., Academic Press, Boston, MA, 1992, 1–29.
2. H. C. Andrews and B. R. Hunt, *Digital Image Restoration*, Prentice-Hall, Englewood Cliffs, NJ, 1977.
3. A. S. Carasso, *Overcoming Hölder continuity in ill-posed continuation problems*, (1993). To appear in SIAM J. Numer. Anal.
4. A. S. Carasso, *Image restoration and diffusion processes*, SPIE Proceedings Series, 2035 (1993), Mathematical Methods in Medical Imaging II, 255–266.
5. R. J. Knops, ed., *Symposium on Non-Well-Posed Problems and Logarithmic Convexity*, Lecture Notes in Math. 316, Springer-Verlag, New York, 1973.
6. M. M. Lavrent'ev, V. G. Romanov, and S. P. Shishatskii, *Ill-Posed Problems of Mathematical Physics and Analysis*, Translation of Mathematical Monographs, Vol. 64, American Mathematical Society, Providence, RI, 1986.
7. H. A. Levine, *Continuous data dependence, regularization, and a three lines theorem for the heat equation with data in a space like direction*, Ann. Mat. Pura Appl. 134 (1983), 267–286.
8. K. Miller, *Least squares methods for ill-posed problems with a prescribed bound*, SIAM J. Math. Anal. 1 (1970), 52–74.
9. K. Miller, *Stabilized numerical analytic prolongation with poles*, SIAM J. Appl. Math. 18, (1970), 346–363.
10. L. E. Payne, *Improperly Posed Problems in Partial Differential Equations*, CBMS Regional Conference Series, 22, Society for Industrial and Applied Mathematics, Philadelphia, PA, 1975.
11. W. K. Pratt, *Digital Image Processing*, Second ed., John Wiley, New York, 1991.

NATIONAL INSTITUTE OF STANDARDS AND TECHNOLOGY, GAITHERSBURG, MD 20899.
E-mail address: alfred@cam.nist.gov

Proceedings of Symposia in Applied Mathematics
Volume 48, 1994

A Finite Element Method for the 2D Drift-Diffusion Semiconductor Model

ZHANGXIN CHEN

Abstract. Stability and convergence properties are stated for a finite element method for numerically solving the 2D drift-diffusion semiconductor device equations. The method combines a mixed finite element method for the approximation of the electric field and a discontinuous upwinding finite element method for the approximation of the electron and hole concentrations. The mixed method gives an approximate electric field in the precise form needed by the discontinuous method, which is conservative and fully parallelizable.

1. Introduction

In this paper we consider a finite element method for the 2D drift-diffusion semiconductor model

$$u_t - \operatorname{div} J_u = -\lambda^2 \alpha_u \mathcal{R}(u,p), \quad (x,y) \in \Omega, \tag{1.1a}$$

$$p_t + \operatorname{div} J_p = -\lambda^2 \alpha_p \mathcal{R}(u,p), \quad (x,y) \in \Omega, \tag{1.1b}$$

$$u = u_D, \quad (x,y) \in \partial\Omega_D, \tag{1.1c}$$

$$p = p_D, \quad (x,y) \in \partial\Omega_D, \tag{1.1d}$$

$$\partial u/\partial \nu = 0, \quad (x,y) \in \partial\Omega_N, \tag{1.1e}$$

$$\partial p/\partial \nu = 0, \quad (x,y) \in \partial\Omega_N, \tag{1.1f}$$

$$u(0,x,y) = u_{\text{init}}(x,y), \quad (x,y) \in \Omega, \tag{1.1g}$$

$$p(0,x,y) = p_{\text{init}}(x,y), \quad (x,y) \in \Omega, \tag{1.1h}$$

where $t > 0$, $\Omega = (0,1)^2$, $\partial\Omega = \partial\Omega_D \cup \partial\Omega_N$ with $\partial\Omega_D \cap \partial\Omega_N = \emptyset$ and Ω_N containing the endpoints of its segments, u and p are the (scaled) electron and hole concentrations, $\mathcal{R}$ is the carrier recombination-generation rate, λ is the normed Debye length, α_u and α_p are the lifetime-dependent constants, and J_u and J_p are the current densities

$$J_u = \mu_u(\beta)(\lambda^2 \nabla u - u\beta), \quad t > 0, \quad (x,y) \in \Omega, \tag{1.2a}$$

$$J_p = -\mu_p(\beta)(\lambda^2 \nabla p + p\beta), \quad t > 0, \quad (x,y) \in \Omega, \tag{1.2b}$$

1991 *Mathematics Subject Classification.* Primary 65N30, 65N12; Secondary 35L60, 35L65.

Partly supported by the Army Research Office under contract DAAL03-89-C-0038 with the University of Minnesota Army High Performance Computing Research Center.

This paper is in the final form, and no version of it will be submitted for publication elsewhere.

© 1994 American Mathematical Society
0160-7634/94 $1.00 + $.25 per page

where β is the (scaled) negative electric field given by

$$\text{(1.3a)} \qquad -\operatorname{div}\beta = C - u + p, \quad t > 0, \quad (x,y) \in \Omega,$$

$$\text{(1.3b)} \qquad \beta = \nabla\phi, \quad t > 0, \quad (x,y) \in \Omega,$$

$$\text{(1.3c)} \qquad \phi = \phi_D, \quad t \geq 0, \quad (x,y) \in \partial\Omega_D,$$

$$\text{(1.3d)} \qquad \partial\phi/\partial\nu = 0, \quad t \geq 0, \quad (x,y) \in \partial\Omega_N,$$

with C being the (scaled) doping profile and ϕ being the (scaled) potential, and μ_u and μ_p are the field-dependent electron and hole mobilities. The numerical method considered in this paper is an extension of the finite element method introduced in [5, 6, 2, 3] for the 1D case, and combines a mixed method for a piecewise linear approximation of the electric field, $-\beta$, with an explicit upwinding method for piecewise constant approximations of the electron and hole concentrations, u and p. The reason for using the mixed finite element approximation of the electric field is that the electron and hole concentration equations (1.1a–b) depend on the potential only through this field, and the mixed method provides a better approximation of it than more standard Galerkin approaches would give [1, 7]. The motivation for including the discontinuous upwinding finite element method in approximating u and p is that, since the normed Debye length ranges from 10^{-3} to 10^{-5}, the concentration equations, while formally parabolic, are in fact more nearly hyperbolic. Thus, the upwinding method is applied to follow the transport more accurately than the standard finite difference or finite element method does. In particular, this method can capture discontinuities in the solution without producing spurious oscillations [4]. Another computational advantage of the method is its full parallelizability.

2. The Finite Element Method

Let $\{x_{i+1/2}\}_{i=0}^{n_x} \times \{y_{j+1/2}\}_{j=0}^{n_y}$ be a partition of Ω with $x_{1/2} = y_{1/2} = 0$ and $x_{n_x+1/2} = y_{n_y+1/2} = 1$, and let $\{t^n\}_{n=0}^{n_T}$ be a partition of $[0,T]$ with $t^0 = 0$ and $t^{n_T} = T$. Then, set $I_i^x = (x_{i-1/2}, x_{i+1/2})$, $I_j^y = (y_{j-1/2}, y_{j+1/2})$, $\Delta x_i = x_{i+1/2} - x_{i-1/2}$, $\Delta y_j = y_{j+1/2} - y_{j-1/2}$, $J^n = [t^n, t^{n+1})$, and $\Delta t^n = t^{n+1} - t^n$. Associated with these partitions, we introduce the spaces

$$\begin{aligned}
V_h &= \{v \in H(\operatorname{div};\Omega) : \; v|_{I_i^x \times I_j^y} = (a_{i,j}^1 + a_{i,j}^2 x, a_{i,j}^3 + a_{i,j}^4 y), \quad a_{i,j}^k \in \mathbb{R}, \\
&\qquad i = 1, \ldots, n_x, \; j = 1, \ldots, n_y, \quad v \cdot \nu|_{\partial\Omega_N} = 0\}, \\
W_h^\phi &= \{w \in L^2(\Omega) : \; w|_{I_i^x \times I_j^y} \in P^0(I_i^x \times I_j^y), \quad i = 1, \ldots, n_x, \; j = 1, \ldots, n_y\}, \\
W_h &= \{w \in L^\infty(\Omega) : \; w|_{I_i^x \times I_j^y} \in P^0(I_i^x \times I_j^y), \quad i = 1, \ldots, n_x, \; j = 1, \ldots, n_y\}, \\
W_{\Delta t} &= \{w \text{ right continuous} ; \; w|_{J^n} \in P^0(J^n), \; n = 0, \ldots, n_T - 1\}.
\end{aligned}$$

If $v \in V_h$, then $v_{i+1/2,j}$ and $v_{i,j+1/2}$ will denote $v(x_{i+1/2}, y_j)$ and $v(x_i, y_{j+1/2})$, respectively. If $w \in W_h$, then $w_{i,j}$ represents the constant value $w(x,y)$, $(x,y) \in I_i^x \times I_j^y$. Finally, w^n will indicate the constant $w(t)$, $t \in J^n$, if $w \in W_{\Delta t}$.

Let P_h and $P_{\Delta t}$ denote the L^2-projections into W_h and $W_{\Delta t}$, respectively, and set

$$\begin{aligned}
u_{\text{init},h} &= P_h u_{\text{init}}, \\
p_{\text{init},h} &= P_h p_{\text{init}}, \\
C_h &= P_h C.
\end{aligned}$$

Then our numerical method is formulated as follows.

i) The electric potential and field: Find $(\beta_h, \phi_h) \in W_{\Delta t} \otimes V_h \times W_{\Delta t} \otimes W_h^\phi$ for $n = 0, \dots, n_T$ such that

$$
\begin{aligned}
-(\operatorname{div} \beta_h^n, w) &= (C_h - u_h^n + p_h^n, w), \quad && \forall w \in W_h^\phi, \\
(\beta_h^n, v) + (\phi_h^n, \operatorname{div} v) &= \langle \phi_D^n, v \cdot \nu \rangle_{\partial\Omega_D}, && \forall v \in V_h.
\end{aligned}
$$

ii) The electron concentration: Find $u_h \in W_{\Delta t} \otimes W_h$ for $n = 0, \dots, n_T - 1$, $i = 1, \dots, n_x$, and $j = 1, \dots, n_y$ such that

$$
\begin{aligned}
&\frac{u_{i,j}^{n+1} - u_{i,j}^n}{\Delta t^n} + \frac{f_{1,i+1/2,j}^n - f_{1,i-1/2,j}^n}{\Delta x_i} + \frac{f_{2,i,j+1/2}^n - f_{2,i,j-1/2}^n}{\Delta y_j} \\
&\quad - \frac{2\lambda^2}{\Delta x_i + \Delta x_{i+1}} \left(\mu_{u,1,i+1,j}^n q_{1,i+1/2,j}^n - \mu_{u,1,i,j}^n q_{1,i-1/2,j}^n \right) \\
&\quad - \frac{2\lambda^2}{\Delta y_j + \Delta y_{j+1}} \left(\mu_{u,2,i,j+1}^n q_{2,i,j+1/2}^n - \mu_{u,2,i,j}^n q_{2,i,j-1/2}^n \right) = -\lambda^2 \alpha_u \mathcal{R}(u_{i,j}^n, p_{i,j}^n),
\end{aligned}
$$

where

$$
\begin{aligned}
f_{1,i-1/2,j}^n &= \mu_{u,1,i,j}^n \left(u_{i-1,j}^n \beta_{1,i-1/2,j}^{n^+} + u_{i,j}^n \beta_{1,i-1/2,j}^{n^-} \right), \\
f_{2,i,j-1/2}^n &= \mu_{u,2,i,j} \left(u_{i,j-1}^n \beta_{2,i,j-1/2}^{n^+} + u_{i,j}^n \beta_{2,i,j-1/2}^{n^-} \right), \\
\mu_{u,1,i,j}^n &= \mu_u \left(\beta_{i-1/2,j}^n \right), \\
\mu_{u,2,i,j}^n &= \mu_u \left(\beta_{i,j-1/2}^n \right), \\
\beta_{i-1/2,j}^n &= \left(\beta_{1,i-1/2,j}^n, \left(\beta_{2,i-1,j-1/2}^n + \beta_{2,i-1,j+1/2}^n + \beta_{2,i,j-1/2}^n + \beta_{2,i,j+1/2}^n \right) /4 \right), \\
\beta_{i,j-1/2}^n &= \left(\left(\beta_{1,i-1/2,j-1}^n + \beta_{1,i+1/2,j-1}^n + \beta_{1,i-1/2,j}^n + \beta_{1,i+1/2,j}^n \right) /4, \beta_{2,i,j-1/2} \right),
\end{aligned}
$$

and the function $q_h = (q_1, q_2) \in W_{\Delta t} \otimes V_{\Delta x}$ is the solution of

$$
(q_h(t^n), v_h) = (u_h(t^n), \operatorname{div} v_h) + \langle u_{D,\Delta t}, v_h \cdot \nu \rangle_{\partial\Omega_D}, \quad \forall v_h \in V_h,
$$

after the mass matrix has been mass-lumped. In our case, the expression for the degrees of freedom of q_h is taken as follows:

$$
\begin{aligned}
q_{1,i-1/2,j}^n &= (u_{i,j}^n - u_{i-1,j}^n)/\Delta x_i, \\
q_{2,i,j-1/2}^n &= (u_{i,j}^n - u_{i,j-1}^n)/\Delta y_j.
\end{aligned}
$$

iii) The hole concentration: Find $p_h \in W_{\Delta t} \otimes W_h$ for $n = 0, \dots, n_T - 1$, $i = 1, \dots, n_x$, and $j = 1, \dots, n_y$ such that

$$
\begin{aligned}
&\frac{p_{i,j}^{n+1} - p_{i,j}^n}{\Delta t^n} - \frac{g_{1,i+1/2,j}^n - g_{1,i-1/2,j}^n}{\Delta x_i} - \frac{g_{2,i,j+1/2}^n - g_{2,i,j-1/2}^n}{\Delta y_j} \\
&\quad - \frac{2\lambda^2}{\Delta x_i + \Delta x_{i+1}} \left(\mu_{p,1,i+1,j}^n r_{1,i+1/2,j}^n - \mu_{p,1,i,j}^n r_{1,i-1/2,j}^n \right) \\
&\quad - \frac{2\lambda^2}{\Delta y_j + \Delta y_{j+1}} \left(\mu_{p,2,i,j+1}^n r_{2,i,j+1/2}^n - \mu_{p,2,i,j}^n r_{2,i,j-1/2}^n \right) = -\lambda^2 \alpha_p \mathcal{R}(u_{i,j}^n, p_{i,j}^n),
\end{aligned}
$$

where

$$g^n_{1,i-1/2,j} = \mu_{p,1,i,j}\left(p^n_{i-1,j}\beta^{n^-}_{1,i-1/2,j} + p^n_{i,j}\beta^{n^+}_{1,i-1/2,j}\right),$$
$$g^n_{2,i,j-1/2} = \mu_{p,2,i,j}\left(p^n_{i,j-1}\beta^{n^-}_{2,i,j-1/2} + p^n_{i,j}\beta^{n^+}_{2,i,j-1/2}\right).$$

Similar expressions hold for $\mu_{p,1,i,j}$ and $\mu_{p,2,i,j}$, and the function $r_h = (r_1, r_2) \in W_{\Delta t} \otimes V_{\Delta x}$ is similarly computed (after mass-lumped) in the form

$$r^n_{1,i-1/2,j} = (p^n_{i,j} - p^n_{i-1,j})/\Delta x_i,$$
$$r^n_{2,i,j-1/2} = (p^n_{i,j} - p^n_{i,j-1})/\Delta y_j.$$

iv) The Neumann boundary conditions in (1.1e–f) are imposed by the reflection principle:

$$\pi_{i,j-1} = \pi_{i,j} \quad \text{if } (x_i, y_{j-1}) \text{ or } (x_i, y_j) \in \partial\Omega_N, \ \pi = u, p,$$
$$\pi_{i-1,j} = \pi_{i,j} \quad \text{if } (x_{i-1}, y_j) \text{ or } (x_i, y_j) \in \partial\Omega_N, \ \pi = u, p.$$

On $\partial\Omega_D$, we have

$$\pi_{D,\Delta t,i,j} = P_{\Delta t}\pi_D(x_i, y_j) \quad \text{if } (x_i, y_j) \in \partial\Omega_D, \ \pi = u, p.$$

We remark that it follows from the definition of V_h that the elements in V_h have continuous normal components on interelements edges. Thus, the numerical fluxes $f^n_{1,i-1/2,j}$, $f^n_{2,i,j-1/2}$, $g^n_{1,i-1/2,j}$, and $g^n_{2,i,j-1/2}$ are all well defined.

3. Stability and Convergence

Stability and convergence properties for the numerical method of the previous section will be stated only in case of $\partial\Omega_D = \emptyset$; i.e., we only consider the Neumann boundary conditions. It turns out that the standard approach used in the 1D case [5, 6, 2, 3] cannot be extended to the 2D case if the Dirichlet boundary conditions are included. The reason for this is that the numerical electric field, on which the flux operators in (1.1a–b) depend globally, does not have the required smoothness in the proof of the stability and convergence properties. Thus, new techniques will be needed in the case of Dirichlet boundary conditions. This will be for future work.

It is assumed that C is constant and that

(3.1a) $$u_{\text{init}}(x,y) \in [0, u^\star], \ p_{\text{init}}(x,y) \in [0, p^\star], \qquad (x,y) \in \Omega,$$

(3.1b) $$\nabla u_{\text{init}}, \ \nabla p_{\text{init}} \in (L^\infty(\Omega))^2.$$

THEOREM 3.1 (STABILITY). *Assume that (3.1) and the following hypothesis are satisfied:*

(3.2) $$u^\star = C + p^\star.$$

Then there is a constant $Q_1 = Q_1(u^\star, p^\star, C, \alpha_u, \alpha_p)$ such that if for $n = 0, \dots, n_T - 1$ the following CFL *condition is satisfied:*

(3.3) $$\Delta t^n \le Q_1 \left(\frac{1}{\Delta x_i} + \frac{1}{\Delta y_j} + \frac{\lambda^2}{\Delta x_i \Delta y_j}\right)^{-1},$$

we have

$$u_h(t,x,y) \in [0,u^\star], \qquad (t,x,y) \in [0,T]\times\Omega, \tag{3.4a}$$

$$p_h(t,x,y) \in [0,p^\star], \qquad (t,x,y) \in [0,T]\times\Omega. \tag{3.4b}$$

Moreover, there exists a constant Q_2 *independent of* h *such that*

$$||\beta_h||_{L^\infty(0,T;L^\infty(\Omega))} \le Q_2, \tag{3.5a}$$

$$||\operatorname{div}\beta_h||_{L^\infty(0,T;L^1(\Omega))} \le Q_2. \tag{3.5b}$$

We denote below by $(v_h, q_h, \sigma_h, \psi_h)$ the approximate solution $(u_h, p_h, \beta_h, \phi_h)$ of (1.1)–(1.3) with u_{init} and p_{init} replaced by v_{init} and q_{init}, respectively.

THEOREM 3.2 (CONTINUITY WITH RESPECT TO THE INITIAL DATA). *Suppose that the hypotheses of Theorem* 3.1 *are satisfied for both sets of data. Then,*

$$\begin{aligned} ||u_h(t) - v_h(t)||_{L^\infty(\Omega)} &+ ||p_h(t) - q_h(t)||_{L^\infty(\Omega)} \\ &\le Q_3 \left(||u_{\text{init}} - v_{\text{init}}||_{L^\infty(\Omega)} + ||p_{\text{init}} - q_{\text{init}}||_{L^\infty(\Omega)} \right), \end{aligned}$$

with Q_3 *independent of* h.

THEOREM 3.3 (CONVERGENCE). *Assume that the hypotheses of Theorem* 3.1 *are satisfied. Then, the numerical sequence* $\{(u_h, p_h, \beta_h, \phi_h)\}_{h>0}$ *converges in* $L^\infty(0,T;L^\infty(\Omega)) \times L^\infty(0,T;L^\infty(\Omega)) \times L^\infty(0,T;H(\text{div};\Omega)) \times L^\infty(0,T;BV(\Omega))$ *to the unique solution of the equations* (1.1)–(1.3).

References

[1] Z. CHEN, *Analysis of mixed methods using conforming and nonconforming finite element methods*, RAIRO Modél. Math. Anal. Numér. **37** (1993), pp. 9–34.

[2] Z. CHEN AND B. COCKBURN, *Error estimates for a finite element method for the drift-diffusion semiconductor device equations*, to appear in SIAM J. Numer. Anal.

[3] Z. CHEN, *Finite element analysis of the 1D full drift-diffusion semiconductor model*, to appear in SIAM J. Numer. Anal.

[4] B. COCKBURN AND C.W. SHU, *TVB Runge-Kutta local projection discontinuous Galerkin finite element method for scalar conservation laws II: general framework*, Math. Comp. **52** (1989), pp. 411–435.

[5] B. COCKBURN AND I. TRIANDAF, *Convergence of a finite element method for the drift-diffusion semiconductor device equations*, Math. Comp. **59** (1992), pp. 383–401.

[6] B. COCKBURN AND I. TRIANDAF, *Error estimates for a finite element method for the drift-diffusion semiconductor device equations: the zero diffusion case*, to appear in Math. Comp.

[7] P.A. RAVIART AND J.M. THOMAS, *A mixed finite element method for 2nd order elliptic problems*, Lecture Notes in Math. **606**, Springer, Berlin, 1977, pp. 292–315.

DEPARTMENT OF MATHEMATICS AND THE INSTITUTE FOR SCIENTIFIC. COMPUTATION, TEXAS A&M UNIVERSITY, COLLEGE STATION, TX 77843.

E-mail address: zchen@isc.tamu.edu

Proceedings of Symposia in Applied Mathematics
Volume **48**, 1994

Splitting Functions and Numerical Analysis of WR-type Methods for Evolutionary and Stationary Problems

S. DE MARCHI, M. VIANELLO AND R. ZANOVELLO

ABSTRACT. We show that, by resorting to the fundamental concept of *splitting function*, we can consider *Waveform Relaxation* (WR)-*type iterative methods* for both *evolutionary* and *stationary* problems in the framework of Hilbert and Banach spaces. The main purpose of such an abstract setting is the *application to nonlinear* PDEs and *systems of* PDEs. Yet, the performed analysis and the possible applications do present some original features also in finite-dimensional instances. We wish to emphasize two important computational advantages: *linearization*, for semilinear problems and *decoupling* of systems into independent or weakly coupled subsystems, useful for parallel implementation.

1. The AWR (Abstract Waveform Relaxation) method

The "classical" WR method, originally introduced for the time domain simulation of electrical networks, is nowadays also applied in the framework of evolutionary PDEs. Indeed, it is used to solve the high-dimensional ordinary initial value problems (IVPs) arising from some semidiscretization in space of the original problem (cf., e.g., [**7**] and references therein).

Here we want to show that an analogous method can be *directly* applied to *infinite-dimensional instances*, and hence to various partial differential equations and systems. To the authors' knowledge, so far the only steps in extending the WR method have been made by Nevanlinna and Odeh [**4**], for abstract evolution equations in the framework of Hilbert spaces, and more recently by Bjørhus [**2**], for hyperbolic PDEs (via a domain decomposition approach).

1991 *Mathematics Subject Classification.* Primary 65J15, 65M99; Secondary 65H10.
This work has been supported by the MURST (40%-Funds "Analisi Numerica"), the GNIM-CNR and the Istituto Nazionale di Alta Matematica "F. Severi".
This paper is a survey, in its final form, of some papers in preparation.

© 1994 American Mathematical Society
0160-7634/94 $1.00 + $.25 per page

We consider an abstract IVP of the form

$$\dot{u}(t) = F(t, u(t)) = f(t, u(t), u(t)) \, , \quad t \in (0, T] \, ; \quad u(0) = u_0 \, , \tag{1.1}$$

where the *splitting function* $f(t, u, v)$ is such that $f(t, \cdot, v) : (D_t \subseteq X) \to X$ with X a real (or complex) Banach space, and $f(t, u, \cdot) : D \to X$, $\bigcup_{t \in (0,T]} D_t \subseteq D \subseteq X$. Although the theory could be developed in a general complex Banach space, for the sake of conciseness and simplicity we restrict our analysis to real *Hilbert spaces*, $X = \mathcal{H}$. For a full treatment of the AWR method the reader is referred to [**3**].

We assume that $f(t, \cdot, v)$ satisfies a *one-sided Lipschitz condition* like

$$\langle f(t, u_1, v) - f(t, u_2, v), u_1 - u_2 \rangle_{\mathcal{H}} \leq K_1 \|u_1 - u_2\|^2 \, , \quad \forall \, u_1, u_2 \in D_t \, , \tag{1.2}$$

for each fixed $t \in (0, T]$ and $v \in D$, with $K_1 \in \mathbf{R}$; on the other hand, $f(t, u, \cdot)$ satisfies a *classical Lipschitz condition* in D with constant $K_2 > 0$, uniformly with respect to $t \in (0, T]$ and $u \in D_t$. Note that $f(t, \cdot, v)$ can be *discontinuous*. This is exactly the case when (1.1) represents an evolutionary PDE written in abstract form, since $f(t, \cdot, v)$ is a differential operator in the space variables. We recall that when $K_1 = 0$, the operator $f(t, \cdot, v)$ is usually termed *dissipative*, and *strongly dissipative* when $K_1 < 0$.

By a *solution* to (1.1) we mean a strongly continuous function $u \in C([0, T]; \mathcal{H}) \cap AC_{\text{loc}}((0, T]; \mathcal{H})$ such that $u(t) \in D_t$ for all $t \in (0, T]$, which is *strongly differentiable a.e.* in $(0, T]$ and satisfies (1.1) *a.e.* in $(0, T]$. We assume also that suitable additional hypotheses are fulfilled (when necessary) that guarantee existence of solutions (cf., e.g., [**5**] for existence results in the semilinear case and for connections with semigroup theory). As for uniqueness, we only observe that it can be proved by means of differential inequalities.

Finally, we assume that any abstract IVP like

$$\dot{z}(t) = f(t, z(t), \gamma(t)) \, , \quad t \in (0, T] \, ; \quad z(0) = u_0 \, , \tag{1.3}$$

has a solution (in the sense above) for each fixed $\gamma \in S \subseteq C([0, T]; D)$; such a solution is also required to belong to S (again uniqueness holds).

The last hypothesis ensures that the iterative scheme

$$\dot{u}^{n+1}(t) = f(t, u^{n+1}(t), u^n(t)) \, , \quad t \in (0, T] \, ; \quad u^{n+1}(0) = u_0 \, , \quad n = 0, 1, 2, \ldots \, , \tag{1.4}$$

is well defined for any initial approximation $u^0 \in S$. Together with (1.4), we consider the *perturbed scheme*

$$\dot{z}^{n+1}(t) = f(t, z^{n+1}(t), \tilde{u}^n(t)), t \in (0, T]; z^{n+1}(0) = u_0, S \ni \tilde{u}^{n+1} = z^{n+1} + \delta^{n+1}, \tag{1.5}$$

where the *functional perturbation* δ^{n+1} takes into account *all* errors and approximations affecting the $(n + 1)$st step of the continuous-time scheme (1.4).

We give now some convergence results and estimates. The analysis is based on nonstandard *differential inequalities*, and is particularly cumbersome in the case of general Banach spaces (cf. [**3**]). Denoting by u^* the (unique) solution of (1.1), we can prove that

$$\|u^n(t) - u^*(t)\| \leq a_n(t) \sup_{s\in[0,t]} \|u^0(s) - u^*(s)\| \,, \quad \forall t \in [0,T] \,, \quad n = 1,2,\dots \,, \tag{1.6}$$

where $a_n(t) = (\exp\{K_1 t\})(K_2 t)^n/n!$ for $K_1 \geq 0$, and $a_n(t) = (\exp\{K_1 t\})\, {}_1F_1(1; n+1; -K_1 t)(K_2 t)^n/n!$ for $K_1 < 0$, where ${}_1F_1(a;b;c)$ is the classical *confluent hypergeometric function*. The latter estimate represents an *a priori* pointwise *convergence estimate*, which immediately yields the *uniform convergence* of the AWR iterative scheme on $[0,T]$ and the *superlinear* order of convergence (as for the finite-dimensional WR method, cf. [**1**]). It is also possible to derive an *a posteriori convergence estimate*, which becomes very useful in numerical implementations to achieve realistic stopping criteria. For the sake of conciseness we present only the case $K_1 = 0$, which is important for applications to PDEs:

$$\|u^n(t) - u^*(t)\| \leq (\exp\{K_2 t\} - 1) \sup_{s\in[0,t]} \|u^n(s) - u^{n-1}(s)\|, \quad \forall t \in [0,T] \,, \tag{1.7}$$

for $n = 1,2,\dots$. As for the stability of the AWR method, again for $K_1 = 0$,

$$\|\tilde{u}^n(t) - u^n(t)\| \leq \exp\{K_2 t\} \max_{1\leq j\leq n} \sup_{s\in[0,t]} \|\delta^j(s)\| \,, \quad \forall t \in [0,T] \,, \quad n = 1,2,\dots \,. \tag{1.8}$$

To conclude this section, we present an application of the AWR method to semilinear autonomous problems. Suppose that $f(t,u,v) = Au + B(v)$, where A is a *maximal dissipative* linear operator with dense domain $D(A) \equiv D_t$, and B is a Lipschitz continuous operator defined on the whole space $\mathcal{H}$. For such f, existence results for (1.1), (1.3) can be found, e.g., in [**5**]; here, $S = \mathrm{Lip}([0,T];\mathcal{H})$. This example shows that the AWR method is applicable to *reaction-diffusion systems* like

$$\begin{aligned} \frac{\partial w}{\partial t} &= D\Delta w + g(w) \,, \quad (x,t) \in \Omega \times (0,T] \,, \\ w(x,0) &= w_0(x) \,, \quad w(x,t) \equiv 0 \ \text{ for } x \in \partial\Omega \,, \end{aligned} \tag{1.9}$$

where Ω is a bounded domain of $\mathbf{R}^d$, $w(x,t) \in \mathbf{R}^m$, D is an m-dimensional diagonal diffusion matrix, and g is an m-dimensional Lipschitz continuous vector field, with Lipschitz constant K_g. Such a system can be easily translated, as is known, into an abstract evolution equation as above, where $u(t) = w(\cdot,t)$, $\mathcal{H} = [L^2(\Omega)]^m$, $D(A) = [H^2(\Omega)\cap H_0^1(\Omega)]^m$. Here, $A = D\Delta$, and B is the "substitution operator" associated with the vector field g. Hence, we have $K_1 = 0$, $K_2 = K_g$.

The AWR iteration (1.4) applied to (1.9) *reduces the problem* to the solution of a *sequence of approximating linear systems* (linearization), each given by m

independent inhomogeneous *linear heat equations* (decoupling). The decoupling allows parallel implementation and individual numerical treatments of different components, which can be crucial in the presence of very different scales.

2. The SWR (Stationary Waveform Relaxation) method

In this section we consider the following class of *fixed point equations*:

$$u = \phi(u) = \psi(u, u) \ , \tag{2.1}$$

where the *splitting function* $\psi(u, v)$ is such that $\psi : D_1 \times D_2 \to X$ and X is a real (or complex) Banach space, $D_1 \times D_2 \subseteq X^2$. We assume that $\psi(\cdot, v)$ satisfies a *one-sided Lipschitz condition* with constant $C_1 \in \mathbf{R}$, while $\psi(u, \cdot)$ satisfies a *classical Lipschitz condition* with constant $C_2 > 0$. Again, we allow discontinuities (in the first argument), so that various partial differential equations and systems are included in the formulation (2.1).

We assume that $C_1+C_2 < 1$, and that any fixed point equation like $u = \psi(u, \eta)$ has a solution in $D_1 \cap D_2$ for each fixed $\eta \in D_2$ (uniqueness is almost trivial by the first hypothesis).

Consider now the following iterative method (SWR method):

$$u^{n+1} = \psi(u^{n+1}, u^n) \ , \quad n = 0, 1, 2 ... \ ; \quad u^0 \in D_2 \ , \tag{2.2}$$

which is well defined in view of the previous assumptions. It can be proved that the sequence in (2.2) is a Cauchy sequence. Denoting its limit by u^*, we can show that, when D_2 is *closed*, u^* is the *unique fixed point* of ϕ in (2.1) (cf. [**6**]). Moreover, even assuming existence for (2.1) with an arbitrary subset D_2, we obtain the *a priori convergence estimate*

$$\|u^n - u^*\| \le \frac{C^n}{1-C} \|u^0 - u^*\| \ , \quad C = \frac{C_2}{1-C_1} < 1 \ , \quad n = 1, 2, ... \ , \tag{2.3}$$

and the *a posteriori convergence estimate*

$$\|u^n - u^*\| \le \frac{C}{1-C} \|u^n - u^{n-1}\| \ , \quad n = 1, 2, ... \ . \tag{2.4}$$

Other typical results of fixed point approximation (cf. [**8**]), like *local convergence* and *stability*, can also be proved for the recursion (2.2).

As an infinite-dimensional application, consider the reaction-diffusion system (1.9), which has been *semidiscretized* in time by the well-known abstract *backward Euler method*. In this case, the SWR method reduces the problem to the solution of a sequence of *linear elliptic systems*, each given by m *independent Helmoltz equations*. The computational advantages pointed out at the end of §1 still hold in this case.

Actually, the SWR method can be applied at any discretization level. In order to exhibit a finite-dimensional application, we consider the following scalar one-dimensional reaction-diffusion IBVP: $\partial w/\partial t = d\Delta w + g(w)$, $(x, t) \in (a, b) \times (0, T]$;

$w(x,0) = w_0(x)$, $w(a,t) = h_1(t)$, $w(b,t) = h_2(t)$; $d > 0$ is the diffusion coefficient, and $g : \mathbf{R} \to \mathbf{R}$ is a C^1 and globally Lipschitz continuous function. Suppose that the problem has been *semidiscretized in space* by central differences, with N internal nodes. Moreover, suppose that the resulting N-dimensional system of ODEs be approximated by an *implicit* θ-method, $0 < \theta \leq 1$. *At each time step* we have then to solve a *semilinear* system in $\mathbf{R}^N$ like

$$u = \frac{\theta h d}{k^2} \mathcal{T} u + \theta h G(u) + \beta \, , \quad \beta \in \mathbf{R}^N \, , \tag{2.5}$$

where $\mathcal{T} = (\tau_{i,j})$ is the *tridiagonal negative definite* matrix with $\tau_{i,i} = -2$, $\tau_{i,i+1} = 1$, $\tau_{i-1,i} = 1$. Moreover, G is the "diagonal" vector field associated with g, $G_j(u) = g(u_j)$, and h, k are the discretization steps in time and space, respectively. The SWR method is applicable to (2.5), by choosing $\psi(u,v) = (\theta hd/k^2)\mathcal{T}u + \theta h G(v) + \beta$, and h sufficiently small; here (using the Euclidean norm), $C_1 = -4(\theta hd/k^2)\sin^2[\pi/2(N+1)]$, $C_2 = \theta h K_g$. On the other hand, the "local solver" for the θ-method usually adopted in the literature is Newton's method, or one of its variants. It is worth noting that the SWR method, whose convergence is only *linear* in general, possesses an *important computational advantage* in this case: The *tridiagonal* (positive definite) *matrix* of the linear system appearing at each iterative step in (2.2) is *fixed* for a fixed mesh, so that its LU *decomposition* can be computed *once and for all* at the beginning. Numerical examples have shown a global saving of computing time varying from 25% up to 40%.

References

1. A. Bellen and M. Zennaro, guest eds., *Parallel methods for ordinary differential equations*, Appl. Numer. Math., Special Issue, Vol. 11, Numbers 1–3, 1993.
2. M. Bjørhus, *Dynamic iteration for time-dependent partial differential equations: A first approach*, Tech. Report Numerics no. 1/1992, Department of Mathematical Sciences, University of Trondheim, Norway.
3. S. De Marchi, M. Vianello and R. Zanovello, *Numerical analysis of the WR method for infinite-dimensional evolution equations*, in preparation, 1993.
4. O. Nevanlinna and F. Odeh, *Remarks on the convergence of waveform relaxation method*, Numer. Funct. Anal. and Optim. **9** (1987), 435–445.
5. A. Pazy, *Semigroups of linear operators and applications to partial differential equations*, Springer-Verlag, New York, 1983.
6. A. Sommariva, *A numerical method for a class of fixed point equations*, Degree Thesis, Univ. of Padua, 1993 (Italian).
7. S. Vandewalle and R. Piessens, *On dynamic iteration methods for solving time-periodic differential equations*, SIAM J. Numer. Anal. **30** (1993), 286–303.
8. E. Zeidler, *Nonlinear functional analysis and its applications. I. Fixed-point theorems*, Springer-Verlag, New York, 1986.

Dipartimento di Matematica Pura e Applicata, Università di Padova, Via Belzoni 7, 35131 Padova - ITALY

E-mail address: {demarchi,marcov,zanov}@pdmat1.math.unipd.it

Proceedings of Symposia in Applied Mathematics
Volume 48, 1994

Error Estimates for a Quadrature Rule for Cauchy Principal Value Integrals

Kai Diethelm

ABSTRACT. The Cauchy principal value integral $⨍_{-1}^{1} \frac{f(x)}{x-\lambda}\,dx$ with Hölder continuous f and $\lambda \in (-1,1)$ is approximated using the method of "subtracting the singularity" and application of the classical trapezoidal formula with n nodes to the resulting integral. Using Peano kernel methods, we prove that, for such f, the error is $O(n^{-1}\ln n)$ uniformly for $\lambda \in (-1,1)$. Explicit upper and lower bounds for the Peano constants are given. Similar results hold if the trapezoidal rule is replaced by the midpoint rule.

1. Introduction and statement of the main results

We consider quadrature formulae Q for the numerical evaluation of the Cauchy principal value integral

$$(1)\qquad I[f;\lambda] := ⨍_{-1}^{1} \frac{f(x)}{x-\lambda}\,dx := \lim_{\varepsilon\to 0+}\left(\int_{-1}^{\lambda-\varepsilon} \frac{f(x)}{x-\lambda}\,dx + \int_{\lambda+\varepsilon}^{1} \frac{f(x)}{x-\lambda}\,dx\right),$$

where $\lambda \in (-1,1)$ and $f \in H_1$, the class of all functions which fulfill a Hölder condition of order 1. Under these assumptions, $I[f;\lambda]$ exists.

Our approach is based on the representation (cf. Davis and Rabinowitz [1, p. 184])

$$(2)\qquad I[f;\lambda] = \int_{-1}^{1} \frac{f(x)-f(\lambda)}{x-\lambda}\,dx + f(\lambda)\ln\frac{1-\lambda}{1+\lambda}$$

and approximation of the proper Riemann integral on the right-hand side of (2) by a classical quadrature formula $\tilde{Q}$. Writing $f_\lambda(x) := \frac{f(x)-f(\lambda)}{x-\lambda}$ for $x \neq \lambda$ and $f_\lambda(\lambda) := f'(\lambda)$, we obtain a Cauchy principal value quadrature formula Q defined by

$$(3)\qquad Q[f;\lambda] := \tilde{Q}[f_\lambda] + f(\lambda)\ln\frac{1-\lambda}{1+\lambda}.$$

We denote the error by $R[f;\lambda] := I[f;\lambda] - Q[f;\lambda]$. This method has been considered frequently; see, e.g., [4, 5] and the literature cited therein.

Here, we will choose the classical quadrature formula to be either the midpoint

1991 *Mathematics Subject Classification.* Primary 65D30; Secondary 41A55.

This paper is in final form, and no version of it will be submitted for publication elsewhere.

© 1994 American Mathematical Society
0160-7634/94 $1.00 + $.25 per page

formula Q_n^{Mi} or the trapezoidal formula Q_{n+1}^{Tr}. The resulting Cauchy principal value quadrature formulae will be denoted by $Q_{n+1}^{\mathrm{Mi\,SS}}$ and $Q_{n+2}^{\mathrm{Tr\,SS}}$, respectively. The additional superscript "SS" indicates the method of "Subtraction of the Singularity".

In particular, we are interested in estimates of the type

$$|R[f;\lambda]| \leq c(\lambda)\|f'\|_\infty \tag{4}$$

with best possible constant $c(\lambda)$. The best possible constant $c(\lambda)$ in inequality (4) is known as the first Peano constant of $R[\,\cdot\,;\lambda]$. In §2, we will give an exact expression for the Peano constants, which we use to obtain explicit upper and lower bounds for $c(\lambda)$ differing essentially, for $n \to \infty$, only by a constant factor. The main results are stated in the following theorems and a straightforward corollary.

THEOREM 1. *Let $x_\nu = -1 + 2\frac{\nu}{n}$. For the formula $Q_{n+2}^{\mathrm{Tr\,SS}}$, we have, if $|\lambda| \in [x_\mu, x_{\mu+1})$ $(\mu < n)$,*

$$\frac{1}{2}n^{-1}\ln\left(\mu\left(n-\mu\right)\right) \leq c(\lambda) \leq \frac{3}{4}n^{-1}\ln\left(\left(\mu-\frac{1}{2}\right)\left(n-\mu-\frac{1}{2}\right)\right)+\frac{35}{2}n^{-1}.$$

THEOREM 2. *Let $x_\nu = -1 + \frac{2\nu-1}{n}$. For the formula $Q_{n+1}^{\mathrm{Mi\,SS}}$, we have, if $|\lambda| \in [x_\mu, x_{\mu+1})$ $(\mu < n)$,*

$$\frac{1}{2}n^{-1}\left(\frac{1}{2}-\ln 4+\ln\left(\mu\left(n-\mu\right)\right)\right) \leq c(\lambda) \leq \frac{3}{4}\,n^{-1}\ln\left(\left(\mu-\frac{3}{2}\right)\left(n-\mu-\frac{1}{2}\right)\right)+20\,n^{-1}.$$

If $|\lambda| \geq x_n$, then

$$\frac{1}{2}n^{-1}\left(\frac{1}{4}-\ln 2+\ln n\right) \leq c(\lambda) \leq \frac{3}{4}n^{-1}\ln n+\frac{61}{6}n^{-1}.$$

We will prove Theorem 1 in §3; Theorem 2 can be shown analogously.

COROLLARY 3. *For the formula $Q_{n+2}^{\mathrm{Tr\,SS}}$, we have, for every $\lambda \in (-1,1)$,*

$$\frac{1}{2}n^{-1}\ln(n-1) \leq c(\lambda) \leq \frac{3}{2}n^{-1}\ln n+\left(\frac{35}{2}-\ln 2\right)n^{-1}.$$

For the formula $Q_{n+1}^{\mathrm{Mi\,SS}}$, we have, for every $\lambda \in (-1,1)$,

$$\frac{1}{2}n^{-1}\left(\frac{1}{4}-\ln 2+\ln(n-1)\right) \leq c(\lambda) \leq \frac{3}{2}n^{-1}\ln n+20\,n^{-1}.$$

Thus, we obtain, for both these quadrature formulae and for every $f \in H_1$, that

$$R[f;\lambda] = O(n^{-1}\ln n) \tag{5}$$

uniformly for $\lambda \in (-1,1)$. This substantially improves a result of Rabinowitz [5, Theorem 4]. Uniform convergence is an important feature when these formulae are used for the numerical solution of integral equations with Cauchy principal value integrals, which arise in fields such as aerodynamics or fluid mechanics [3, 5].

REMARK. In [2], we have shown that there does not exist a sequence of Cauchy principal value quadrature formulae for which $c(\lambda) = o(n^{-1} \ln n)$ holds uniformly for $|\lambda| < 1$. Hence, the error of the rules considered here is of an optimal order.

2. Peano's theorem and the Peano kernel

The main tool which we will use to prove upper bounds for the Peano constants is the following theorem, which is simply a generalization of Peano's Theorem for classical quadrature formulae (cf. Davis and Rabinowitz [1, p. 286]):

THEOREM 4 (Peano). *Let $Q[\cdot\,;\lambda]$ be a Cauchy principal value quadrature formula given by $Q[f;\lambda] = \sum_{\nu=1}^{N} a_\nu f(x_\nu) + bf'(\lambda)$ which is exact for all polynomials of degree zero. Then, we have, for every $f \in H_1$,*

$$R[f;\lambda] = \int_{-1}^{1} K_\lambda(x) f'(x)\, dx, \tag{6}$$

where K_λ is the so-called (first) Peano kernel of $R[\cdot\,;\lambda]$ given by

$$K_\lambda(x) = \ln\left|\frac{1-\lambda}{x-\lambda}\right| - \sum_{\nu=1}^{N} a_\nu (x_\nu - x)_+^0 - b\,\delta(x-\lambda) = \ln\left|\frac{1+\lambda}{x-\lambda}\right| + \sum_{\nu=1}^{N} a_\nu (x - x_\nu)_+^0 - b\delta(x-\lambda).$$

(δ denotes Dirac's distribution, and $(\cdot)_+^0$ is the truncated power function [1, p. 286].)

This theorem can be shown simply by an explicit evaluation of the integral in equation (6) using either of the representations of K_λ.

As in the case of classical quadrature formulae [1, (4.3.11)], we can see that the Peano constant $c(\lambda)$ is now given by

$$c(\lambda) = \int_{-1}^{1} |K_\lambda(x)|\, dx. \tag{7}$$

Using the notation of Theorem 4, we can immediately deduce

COROLLARY 5. *Assuming that $-1 \le x_1 < x_2 < \cdots < x_N \le 1$, we have:*

1. *If $\lambda \ge x_{\nu+1}$, then K_λ is strictly monotone increasing and convex on the interval $(x_\nu, x_{\nu+1})$.*
2. *If $\lambda \le x_\nu$, then K_λ is strictly monotone decreasing and convex on $(x_\nu, x_{\nu+1})$.*
3. *If $x_\nu < \lambda < x_{\nu+1}$, then K_λ is strictly monotone increasing and convex on (x_ν, λ) and strictly monotone decreasing and convex on $(\lambda, x_{\nu+1})$.*
4. *$\lim_{x\to\lambda} K_\lambda(x) = \infty$.*

3. Proof of Theorem 1

In this section, we are going to give a proof for the explicit upper bounds for $c(\lambda)$ based on the representation from Theorem 4. This theorem is applicable to the quadrature formulae under consideration here because we have that, for all

$\lambda \in (-1,1)$, $R^{\mathrm{Tr\,SS}}_{n+2}[f;\lambda] = R^{\mathrm{Mi\,SS}}_{n+1}[f;\lambda] = 0$ whenever f is a polynomial of degree less than or equal to 2 (cf. Monegato [4]).

We will also prove the explicit lower bounds for $c(\lambda)$.

In both cases, we will only give proofs for the case $R = R^{\mathrm{Tr\,SS}}_{n+2}$; the case of the midpoint formulae can be treated in a very similar fashion.

Owing to symmetry, we can restrict our analysis to the case $\lambda \geq 0$.

3.1. The lower bound for the Peano constant. First, we show the validity of the lower bound in Theorem 1. The proof is based on the relation

$$c(\lambda) = \sup\{|R[f;\lambda]|\,;\ f \in H_1 \text{ and } \|f'\|_\infty \leq 1\}. \tag{8}$$

Defining

$$g_\lambda(x) := \begin{cases} x - x_\nu & \text{if } x_\nu \leq x < x_\nu + \frac{1}{n} \text{ and } \lambda \leq x_\nu, \\ x_{\nu+1} - x & \text{if } x_\nu + \frac{1}{n} \leq x < x_{\nu+1} \text{ and } \lambda \leq x_\nu, \\ x_\nu - x & \text{if } x_\nu \leq x < x_\nu + \frac{1}{n} \text{ and } \lambda \geq x_{\nu+1}, \\ x - x_{\nu+1} & \text{if } x_\nu + \frac{1}{n} \leq x < x_{\nu+1} \text{ and } \lambda \geq x_{\nu+1}, \\ 0 & \text{else,} \end{cases}$$

we can deduce, for $\lambda \in [x_\mu, x_{\mu+1})$,

$$\begin{aligned} c(\lambda) &\geq \left|R^{\mathrm{Tr\,SS}}_{n+2}[g_\lambda;\lambda]\right| = |I[g_\lambda;\lambda]| = \sum_{\nu=0}^{n-1} \left| \int_{x_\nu}^{x_{\nu+1}} \frac{g_\lambda(x)}{x-\lambda}\,dx \right| \\ &\geq \frac{1}{2n}\left(\sum_{\nu=1}^{\mu} \frac{1}{\nu} + \sum_{\nu=2}^{n-\mu} \frac{1}{\nu} \right) \geq \frac{1}{2} n^{-1} \ln\left(\mu(n-\mu)\right). \end{aligned}$$

3.2. The upper bound for the Peano constant. For the proof of the upper bound, we will use the representation of $c(\lambda)$ via the Peano kernel as given by equation (7). Let K_λ denote the Peano kernel of $R^{\mathrm{Tr\,SS}}_{n+2}[\,\cdot\,;\lambda]$. We will start by giving some elementary properties of K_λ:

$$K_\lambda(-1+) = \frac{1}{n(-1-\lambda)} < 0. \tag{9}$$

If $x_\nu < \lambda$, then

$$\begin{aligned} K_\lambda(x_\nu+) &= \ln\left|\frac{1+\lambda}{x_\nu - \lambda}\right| + \frac{2}{n}\sum_{\kappa=1}^{\nu} \frac{1}{x_\kappa - \lambda} + \frac{1}{n(-1-\lambda)} \\ &\leq \ln\left|\frac{1+\lambda}{x_\nu - \lambda}\right| + \frac{1}{n(-1-\lambda)} + \ln\left|\frac{\nu - \frac{n}{2} - \frac{n\lambda}{2}}{1 - \frac{n}{2} - \frac{n\lambda}{2}}\right| \\ &\quad + \frac{1}{2}\left(\frac{1}{1 - \frac{n}{2} - \frac{n\lambda}{2}} + \frac{1}{\nu - \frac{n}{2} - \frac{n\lambda}{2}} \right) \\ &\leq \frac{1}{n}\left(\frac{1}{\lambda - x_1} - \frac{1}{\lambda - x_\nu} \right) + \frac{1}{n(-1-\lambda)} < 0 \end{aligned} \tag{10}$$

and

$$|K_\lambda(x_\nu+)| = -K_\lambda(x_\nu+) \leq \frac{2}{n(\lambda - x_\nu)} + \frac{1}{n\left(x_\nu - \lambda - \frac{1}{n}\right)}. \tag{11}$$

If $x_{\nu+1} < \lambda$, then

$$0 \le K_\lambda(x_{\nu+1}-) \le \frac{1}{n}\left(\frac{1}{1+\lambda-\frac{2}{n}} - \frac{1}{1+\lambda} + \frac{1}{x_\nu - \lambda} - \frac{2}{x_{\nu+1}-\lambda}\right). \tag{12}$$

We note that the upper bound for $|K_\lambda|$ given by inequality (12) is larger than the bound given by inequality (11). Now, the convexity of K_λ (Corollary 5) in connection with $\frac{d}{dx}|K_\lambda(x)|_{x=x_\nu+} = \frac{1}{x_\nu-\lambda}$ for $x_\nu < \lambda$, yields for $x_{\nu+1} < \lambda$:

$$\int_{x_\nu}^{x_{\nu+1}} |K_\lambda(x)|\,dx \le \frac{3}{2n^2}\left(\frac{1}{1+\lambda-\frac{2}{n}} - \frac{1}{1+\lambda} + \frac{1}{x_\nu - \lambda} - \frac{2}{x_{\nu+1}-\lambda}\right). \tag{13}$$

Analogous statements hold for the cases $\lambda < x_\nu$ and $\lambda < x_{\nu+1}$, respectively.

If $\lambda = x_\nu$ or $\lambda = x_{\nu+1}$, then, using the inequalities stated above, one can show

$$\int_{x_\nu+}^{x_{\nu+1}-} |K_\lambda(x)|\,dx \le \frac{10}{3n}, \tag{14}$$

and an additional summand $\frac{2}{n}$ resulting from the Dirac distribution has to be taken into consideration. (The Dirac distribution only occurs if λ is equal to one of the nodes of Q_{n+1}^{Tr}.)

If λ is in an adjacent interval of $[x_\nu, x_{\nu+1}]$, then

$$\int_{x_\nu}^{x_{\nu+1}} |K_\lambda(x)|\,dx \le \frac{18}{5n}. \tag{15}$$

Finally, if $\lambda \in (x_\nu, x_{\nu+1})$,

$$\int_{x_\nu}^{x_{\nu+1}} |K_\lambda(x)|\,dx \le \frac{16}{3n}. \tag{16}$$

Adding up these bounds for all the intervals and estimating carefully, the statement of the theorem follows.

ACKNOWLEDGEMENT. I would like to thank the referee for his helpful comments.

References

1. Davis, P.J.; Rabinowitz, P.: *Methods of Numerical Integration.* 2nd ed., Academic Press, Orlando, 1984.

2. Diethelm, K.: *Uniform convergence of optimal order quadrature rules for Cauchy principal value integrals.* To appear in J. Comp. Appl. Math. (1994).

3. Gerasoulis, A.: *Piecewise polynomial quadratures for Cauchy singular integrals.* SIAM J. Numer. Anal. **23** (1986), 891–902.

4. Monegato, G.: *The numerical evaluation of one-dimensional Cauchy principal value integrals.* Computing **29** (1982), 337–354.

5. Rabinowitz, P.: *Convergence results for piecewise linear quadratures for Cauchy principal value integrals.* Math. Comp. **51** (1988), 741–747.

INSTITUT FÜR MATHEMATIK, UNIVERSITÄT HILDESHEIM, MARIENBURGER PLATZ 22, 31141 HILDESHEIM, GERMANY.

E-mail address: diethelm@informatik.uni-hildesheim.de

Proceedings of Symposia in Applied Mathematics
Volume 48, 1994

A Numerical Radius Approach to Stable Difference Schemes for Parabolic Systems

Moshe Goldberg

ABSTRACT. In this note we take a numerical radius approach in order to discuss sufficient stability conditions for a wide family of implicit finite difference schemes for the initial value problem associated with a general, well-posed, multi-space-dimensional parabolic system with constant coefficients.

Consider the initial value problem

$$\text{(1a)}\qquad \frac{\partial \mathbf{u}(\mathbf{x},t)}{\partial t} = \sum_{1\le p\le q\le s} A_{pq}\frac{\partial^2 \mathbf{u}(\mathbf{x},t)}{\partial x_p \partial x_q} + \sum_{1\le p\le s} B_p \frac{\partial \mathbf{u}(\mathbf{x},t)}{\partial x_p} + C\mathbf{u}(\mathbf{x},t),$$

$$\text{(1b)}\qquad \mathbf{u}(\mathbf{x},0)=\mathbf{f}(\mathbf{x}),\quad \mathbf{x}=(x_1,\ldots,x_s),\quad -\infty < x_p < \infty,\quad 0\le t\le T,$$

where (1a) is a parabolic system of partial differential equations and (1b) a given initial value. Here, $\mathbf{u}(\mathbf{x},t)=(u^{(1)}(\mathbf{x},t),\ldots,u^{(m)}(\mathbf{x},t))'$ (prime denoting the transpose) is the unknown m-vector; and A_{pq}, B_p and C are fixed $m\times m$ matrices, A_{pq} being Hermitian.

We shall always assume that (1) is well posed in the sense of Petrowski (e.g. [1]). Since the A_{pq} are Hermitian, it is well known that this is equivalent to the existence of a fixed $\delta > 0$ satisfying

$$\text{(2)}\qquad \sum_{1\le p\le q\le s} \sigma_p\sigma_q A_{pq} \ge \delta \sum_{p=1}^{s} \sigma_p^2 I \quad \text{for all real } \sigma_1,\ldots,\sigma_s,$$

where for Hermitian matrices M and N we write $M \ge N$ if $M-N$ is positive semi-definite. Letting all the σ_p but one vanish, we immediately obtain from (2) that the A_{pp} are not just Hermitian but in fact positive definite.

The purpose of this note is to discuss the L_2-stability of a family of finite difference schemes for (1), where $\partial\mathbf{u}/\partial t$ is approximated by a forward difference in time, and the right-hand side of (1a) is approximated by a convex combination of centered space-differences taken at two adjacent time levels.

More precisely, setting a mesh size $\Delta x_p > 0$, $p=1,\ldots,s$, $\Delta t > 0$, so that the ratios

---

1991 *Mathematics Subject Classification*. Primary 65M12; Secondary 65M06.
The final version of this paper will be submitted for publication elsewhere.

© 1994 American Mathematical Society
0160-7634/94 $1.00 + $.25 per page

$$\lambda_p \equiv \Delta t / \Delta x_p^2, \quad p=1,\dots,s,$$

are kept fixed, and using the standard notation

$$\mathbf{u}_\mathbf{k}^n = \mathbf{u}(k_1\Delta x_1,\dots,k_s\Delta x_s,n\Delta t), \quad \mathbf{k}=(k_1,\dots,k_s),$$

we introduce the translations

$$E_p\mathbf{u}_\mathbf{k}^n \equiv \mathbf{u}(k_1\Delta x_1,\dots,k_{p-1}\Delta x_{p-1},(k_p+1)\Delta x_p,k_{p+1}\Delta x_{p+1},\dots,k_s\Delta x_s,t), \quad p=1,\dots,s,$$

and define

$$\begin{aligned}(\Phi\mathbf{u})_\mathbf{k}^n \equiv & \sum_{p=1}^{s}\frac{A_{pp}}{\Delta x_p^2}(E_p-2I+E_p^{-1})\mathbf{u}_\mathbf{k}^n \\ & + \sum_{1\le p<q\le s}\frac{A_{pq}}{4\Delta x_p\Delta x_q}(E_p-E_p^{-1})(E_q-E_q^{-1})\mathbf{u}_\mathbf{k}^n \\ & + \sum_{p=1}^{s}\frac{B_p}{2\Delta x_p}(E_p-E_p^{-1})\mathbf{u}_\mathbf{k}^n + C\mathbf{u}_\mathbf{k}^n.\end{aligned}$$

As $(\Phi\mathbf{u})_\mathbf{k}^n$ provides a centered approximation for the right-hand side of (1a) taken at time level $n\Delta t$, we select θ, $0\le\theta\le 1$, and let our scheme be

$$\frac{\mathbf{v}_\mathbf{k}^{n+1}-\mathbf{v}_\mathbf{k}^n}{\Delta t} = \theta(\Phi\mathbf{v}_\mathbf{k}^{n+1}) + (1-\theta)(\Phi\mathbf{v}_\mathbf{k}^n),$$

where $\mathbf{v}_\mathbf{k}^n$ approximates the unknown vector $\mathbf{u}_\mathbf{k}^n$. Employing the ratios λ_p, we can now put the scheme in its final form,

$$\mathbf{v}_\mathbf{k}^{n+1} - \theta Q\mathbf{v}_\mathbf{k}^{n+1} = \mathbf{v}_\mathbf{k}^n + (1-\theta)Q\mathbf{v}_\mathbf{k}^n, \tag{3}$$

where the finite difference operator Q is given by

$$\begin{aligned}Q\mathbf{v}_\mathbf{k}^n \equiv \Delta t(\Phi\mathbf{v})_\mathbf{k}^n = & \sum_{p=1}^{s}\lambda_p A_{pp}(E_p-2I+E_p^{-1})\mathbf{v}_\mathbf{k}^n \\ & + \sum_{1\le p<q\le s}\tfrac{1}{4}\sqrt{\lambda_p\lambda_q}\,A_{pq}(E_p-E_p^{-1})(E_q-E_q^{-1})\mathbf{v}_\mathbf{k}^n \\ & + \sum_{p=1}^{s}\tfrac{1}{2}\sqrt{\lambda_p\Delta t}\,B_p(E_p-E_p^{-1})\mathbf{v}_\mathbf{k}^n + \Delta t C\mathbf{v}_\mathbf{k}^n.\end{aligned}$$

Clearly, $\theta=0$ gives the standard explicit scheme for (1), whereas all other values of θ yield implicit schemes. While $\theta=1/2$ provides the well-known, second-order Crank-Nicolson scheme, all other schemes are merely first-order accurate.

Our main observation concerning the above family of schemes is the following:

THEOREM. *The scheme* (3) *is stable if either* $1/2\le\theta\le 1$, *or if*

$$0\le\theta<1/2 \quad \textit{and} \quad \sum_{p=1}^{s}\lambda_p\rho(A_{pp}) \le \frac{1}{2(1-2\theta)},$$

where $\rho(A_{pp})$ *denotes the spectral radius of* A_{pp}.

In view of the Lax Equivalence Theorem [5], the theorem tells us that for $1/2 \le \theta \le 1$ (which includes the Crank-Nicolson case) our scheme converges unconditionally. For $0 \le \theta < 1/2$, however, convergence is guaranteed if Δt is small enough to satisfy

$$\Delta t \le \left[2(1-2\theta)\sum_{p=1}^{s} \frac{\rho(A_{pp})}{\Delta x_p^2} \right]^{-1}.$$

Our proof – detailed in a forthcoming paper [3] – is obtained by analyzing the amplification matrix associated with (3):

$$G(\Delta t,\xi) = [I-\theta H(\Delta t,\xi)]^{-1}[I+(1-\theta)H(\Delta t,\xi)], \quad \xi \equiv (\xi_1,\dots,\xi_s) \in \mathbf{R}^s,$$

where

$$\begin{aligned} H(\Delta t,\xi) \equiv & \sum_{p=1}^{s} \lambda_p (e^{i\xi_p} - 2 + e^{-i\xi_p}) A_{pp} \\ & + \sum_{1\le p<q\le s} \tfrac{1}{4}\sqrt{\lambda_p\lambda_q}\,(e^{i\xi_p} - e^{-i\xi_p})(e^{i\xi_q} - e^{-i\xi_q})\, A_{pq} \\ & + \sum_{p=1}^{s} \tfrac{1}{2}\sqrt{\lambda_p \Delta t}\,(e^{i\xi_p} - e^{-i\xi_p}) B_p + \Delta t C \\ = & \sum_{p=1}^{s} 2\lambda_p(\cos\xi_p - 1)A_{pp} \\ & - \sum_{1\le p<q\le s} \sqrt{\lambda_p\lambda_q}\, \sin\xi_p \sin\xi_q\, A_{pq} \\ & + i\sum_{p=1}^{s} \sqrt{\lambda_p \Delta t}\, \sin\xi_p\, B_p + \Delta t C. \end{aligned}$$

The analysis rests on the well-known fact (e.g. [5, 6]) that the L_2-stability of (3) is equivalent to the power boundedness of $G(\Delta t,\xi)$, which amounts to exhibiting a norm $\|\cdot\|$ on $\mathbf{C}^{m\times m}$, the algebra of complex $m\times m$ matrices, and constants $\tau > 0$, $K > 0$, so that under the hypotheses of the theorem,

$$\left\| G(\Delta t,\xi)^n \right\| \le K \quad \text{for all } \Delta t \le \tau,\ n\Delta t \le T \text{ and } \xi \in \mathbf{R}^s. \tag{4}$$

To provide an outline of the proof, let $H_0(\xi)$ be the principal part of $H(\Delta t,\xi)$, i.e.,

$$H_0(\xi) = \sum_{p=1}^{s} 2\lambda_p(\cos\xi_p - 1)A_{pp} - \sum_{1\le p<q\le s} \sqrt{\lambda_p\lambda_q}\, \sin\xi_p \sin\xi_q\, A_{pq}.$$

Using the well-posedness condition in (2), one can show that the matrix $I-\theta H_0(\xi)$ is positive definite; hence, for any choice of ξ,

$$(y,z)_\xi \equiv z^*[I-\theta H_0(\xi)]y, \quad y,z \in \mathbf{C}^m,$$

constitutes an inner product on $\mathbf{C}^m$. Consequently,

$$r_\xi(M) \equiv \max\{|(My,y)_\xi| : y \in \mathbf{C}^m, (y,y)_\xi = 1\}, \quad M \in \mathbf{C}^{m\times m},$$

is the numerical radius associated with $(y, z)_\xi$ (e.g. [2]).

Having r_ξ, we can use the theorem's hypotheses to obtain constants $\alpha > 0,\ \tau > 0$, so that

$$r_\xi(G(\Delta t,\xi)) \le 1 + \alpha\,\Delta t \quad \text{for all } \Delta t \le \tau \text{ and } \xi \in \mathbf{R}^s. \tag{5}$$

With this crucial step at hand, we go on to exhibit a constant $\beta > 0$ such that

$$\|M\|_2 \le \beta\, r_\xi(M) \quad \text{for all } \xi \in \mathbf{R}^s \text{ and } M \in \mathbf{C}^{m\times m}, \tag{6}$$

where

$$\|M\|_2 \equiv \max\{(y^* M^* My)^{1/2} : y \in \mathbf{C}^m,\ y^* y = 1\}$$

is the spectral norm of M.

Now, by (5), (6) and Berger's Inequality [4], it follows that for all $\Delta t \le \tau$, $n\Delta t \le T$, and $\xi \in \mathbf{R}^s$:

$$\begin{aligned}\left\|G(\Delta t,\xi)^n\right\|_2 &\le \beta\, r_\xi(G(\Delta t,\xi)^n) \le \beta\, r_\xi(G(\Delta t,\xi))^n \\ &\le \beta(1+\alpha\Delta t)^n \le \beta e^{\alpha n\Delta t} \le \beta e^{\alpha T} \equiv K.\end{aligned}$$

So by (4), our proof is complete.

Let us point out that a scalar, two-space-dimensional version of our scheme was studied in [5]. There, however, $G(\Delta t,\xi)$ was a scalar function of Δt and ξ that yielded to standard methods of ordinary calculus.

We further remark that it would be interesting to obtain our results in other ways; for example from the priciples of Widlund's Theory in [7].

References

1. A. Friedman, *Generalized Functions and Partial Differential Equations*, Prentice-Hall, Englewood Cliffs, N.J., 1993.

2. M. Goldberg, *On the numerical radius and its applications*, Linear Algebra Appl. **42** (1992), 263—284.

3. M. Goldberg, *Numerical radii and stable difference schemes for parabolic systems*, to be submitted.

4. C. Pearcy, *An elementary proof of the power inequality for the numerical radius*, Michigan Math. J. **13** (1966), 289–291.

5. D. Richtmyer and K. W. Morton, *Difference Methods for Initial-Value Problems*, 2nd edition, Interscience Pub., New York, 1967.

6. V. Thomée, *Finite difference methods for linear parabolic equations*, in: P. G. Ciarlet and J. L. Lions, eds., *Handbook of Numerical Analysis*, Vol. I, North-Holland, Amsterdam, pp. 7–184.

7. O. B. Widlund, *On the stability of parabolic difference schemes*, Math. Comp. **19** (1965), 1–13.

DEPARTMENT OF MATHEMATICS, TECHNION–ISRAEL INSTITUTE OF TECHNOLOGY, HAIFA 32000, ISRAEL
E-mail address: goldberg@gauss.technion.ac.il

Proceedings of Symposia in Applied Mathematics
Volume 48, 1994

AN EXTENSION OF THE OLVER-SOOKNE METHOD FOR THE SOLUTION OF SECOND-ORDER LINEAR DIFFERENCE EQUATIONS

TAKEMITSU HASEGAWA AND TATSUO TORII

ABSTRACT. We extend the stable numerical method due to Olver and Sookne for computing minimal solutions of a homogeneous second-order linear difference equation to the computation of nondominant solutions of an inhomogeneous difference equation. The present algorithm computes a weighted sum of a nondominant solution to a prescribed accuracy by estimating the truncation error. Results of numerical experiments are briefly described.

1. Introduction

Consider the inhomogeneous second-order linear difference equation

$$a_n\, y_{n-1} + b_n\, y_n + c_n\, y_{n+1} = d_n, \qquad n = 1, 2, \ldots, \tag{1.1}$$

where a_n, b_n, c_n, d_n are given sequences of real or complex numbers. Suppose that the homogeneous equation associated with (1.1) (i.e., $d_n = 0$ for all n), where a_n and c_n do not vanish, has a pair of complementary solutions f_n and g_n such that $\lim_{n\to\infty} f_n/g_n = 0$. Then f_n and g_n are referred to as minimal and dominant, respectively [4]. Let the inhomogeneous equation (1.1) have a particular solution h_n for which $\lim_{n\to\infty} h_n/g_n = 0$, so that the general solution of (1.1) may be written as $y_n = \alpha\, f_n + \beta\, g_n + h_n$, where α and β are arbitrary constants. Nondominant solutions of (1.1) in the form $\alpha\, f_n + h_n$ cannot be found simply by forward and backward recurrence of the difference equation, owing to strong instability; see Gautschi [4] and Wimp [12] for surveys of nondominant solutions. See also Levrie and Bultheel [6].

Olver [9] introduced a stable method having a built-in error estimation technique [8] for the computation of nondominant solutions of (1.1). In Olver's method the original initial value problem is replaced by an equivalent boundary value problem, which is solved by Gaussian elimination without pivoting; beginning with $p_0 = 0$

1991 *Mathematics Subject Classification.* Primary 65Q05; Secondary 65F05.
A detailed version of this paper will be submitted for publication elsewhere.

© 1994 American Mathematical Society
0160-7634/94 \$1.00 + \$.25 per page

and $p_1 = 1$, one computes the solution p_n of the homogeneous form of (1.1) for $n = 2, 3, \ldots$ and a sequence $\{q_n\}$ defined by $q_n = (a_n q_{n-1} - d_n p_n)/c_n$ $(n = 1, 2, \ldots)$ with starting value $q_0 = y_0$. Let ϵ be a given tolerance to the approximation $y_m^{(N)}$ $(m < N)$ of the wanted solution y_m. Then the computations of p_n and q_n are carried out until $|y_m^{(N+1)} - y_m^{(N)}| = |p_m q_N/(p_N p_{N+1})| \leq \epsilon$. The approximation $y_m^{(N)}$ can then be obtained by backward substitution.

Van der Cruyssen [11] proposes the use of LU decomposition instead of Gaussian elimination to reduce both the amount of computational effort and the possible occurrence of overflow. Cash [1, 2] extends Olver's method to high-order recurrence relations. See Lozier [7] for a more detailed discussion.

In either method, whether the one based on Gaussian elimination or on LU decomposition, significant loss in accuracy may occasionally occur in the course of computations of, say, p_n when the value of f_0 is very small [12, p.92], [11]. Olver and Sookne [10] showed that in the case of a homogeneous difference equation it is possible to avoid the loss of accuracy in the elimination stage by combining Miller's algorithm and Olver's algorithm and by using a normalizing condition of the type $\sum_{m=0}^{\infty} \lambda_m y_m = s$. Their modification applies when the dominant solution g_n tends to infinity with n in such a way that ultimately $|g_n|$ is monotonic. In this connection, the following lemma due to Olver and Sookne plays an important role.

Lemma 1.1. *For a positive integer M, let the coefficients a_n, b_n and c_n in* (1.1) *satisfy the condition $|b_n| \geq |a_n| + |c_n|$ when $n \geq M + 1$, and $|b_n| < |a_n| + |c_n|$ otherwise. Let p_n be the solution of the homogeneous version of* (1.1) *satisfying $p_M = 0$ and $p_{M+1} = 1$. Then $|p_n| \geq |p_{n-1}|$ for all n greater than M.*

The aim of this paper is to modify the Olver-Sookne method and a summation technique due to Deuflhard [3] for minimal solutions of the homogeneous form of (1.1) to make them applicable in the more general case of inhomogeneous equations. Indeed, for a positive integer K and given constants ξ_m $(0 \leq m \leq K)$, we develop in §2 a numerically stable method for computing the approximation to the weighted sum $S_K = \sum_{m=0}^{K} \xi_m y_m$ of a nondominant solution of (1.1) to a prescribed accuracy by estimating the truncation error.

Further, our method is also an extension of methods due to Olver [9] and Van der Cruyssen [11] for initial value problems to the case using a normalizing condition. In fact, our method uses the LU decomposition scheme [2, 11] along with a rank-one updating technique [5, p.592] to solve a system of linear equations which is obtained from the original difference equation (1.1) and the normalizing condition. The coefficient matrix is a band matrix except for having a row representing the normalizing condition on an appropriate place in the matrix determined so that the loss of accuracy in the elimination stage may be avoided. The linear system is successively solved by increasing the size of the system until an estimated truncation error of the required solution is within the prescribed accuracy. Results of numerical experiments are briefly described in §3.

2. Statement of the algorithm

Similarly to Olver and Sookne, here and henceforth we assume that the coefficients in (1.1) satisfy the condition in Lemma 1.1. Select an integer $N(> K, M)$ and replace the problem of computing the weighted sum $S_K = \xi_0 y_0 + \cdots + \xi_K y_K$

of y_n satisfying (1.1) and the normalizing condition $\lambda_0 y_0 + \lambda_1 y_1 + \cdots = s$, where $\{\lambda_n\}$ is a sequence of real or complex numbers and s is a constant, by the one of solving the linear system of equations $\begin{bmatrix} A^{(N)} & \mathbf{0} \\ \boldsymbol{\xi}^T & -1 \end{bmatrix} \begin{bmatrix} \mathbf{y}^{(N)} \\ S_K^{(N)} \end{bmatrix} = \begin{bmatrix} \mathbf{d}^{(N)} \\ 0 \end{bmatrix}$, where the $(N+1)$-vectors $\mathbf{y}^{(N)}$, $\mathbf{d}^{(N)}$, $\boldsymbol{\xi}$, and the $(N+1) \times (N+1)$ matrix $A^{(N)}$ are defined by $\mathbf{y}^{(N)} = \left[y_0^{(N)}, y_1^{(N)}, \ldots, y_N^{(N)}\right]^T$, $\mathbf{d}^{(N)} = [d_1, \ldots, d_M, s, d_{M+1}, \ldots, d_N]^T$, $\boldsymbol{\xi} = [\xi_0, \xi_1, \ldots, \xi_K, 0, \ldots, 0]^T$, and

$$A^{(N)} = \begin{bmatrix} a_1 & b_1 & c_1 & & & & \cdots & 0 \\ \vdots & \ddots & \ddots & \ddots & & & & \vdots \\ 0 & \cdots & a_M & b_M & c_M & & & \\ \lambda_0 & \cdots & \cdots & \lambda_M & \lambda_{M+1} & \cdots & \cdots & \lambda_N \\ & & & a_{M+1} & b_{M+1} & c_{M+1} & \cdots & 0 \\ & & & & \ddots & \ddots & \ddots & \vdots \\ \vdots & & & & & a_{N-1} & b_{N-1} & c_{N-1} \\ 0 & \cdots & & & & & a_N & b_N \end{bmatrix}, \tag{2.1}$$

respectively; we have omitted the dependence of $\boldsymbol{\xi}$ on N. Then $S_K^{(N)}$ $(= \boldsymbol{\xi}^T \mathbf{y}^{(N)})$ and $y_n^{(N)}$ are approximations to S_K and y_n, respectively.

Now we solve the linear system $A^{(N)}\mathbf{y}^{(N)} = \mathbf{d}^{(N)}$ by using an LU decomposition [2, 11] along with a rank-one updating technique. Let B denote the same $(N+1) \times (N+1)$ matrix as $A^{(N)}$, but having the elements λ_k $(k = M+2, \ldots, N)$ replaced with zeros. Further, let $\mathbf{e}$ be a unit vector, $\mathbf{e} = [e_0, \ldots, e_N]^T$, where $e_k = 0$ $(k = 0, \ldots, N,\ k \neq M)$ and $e_M = 1$, and let $\mathbf{r}$ be a vector defined by $\mathbf{r} = [0, \ldots, 0, \lambda_{M+2}, \ldots, \lambda_N]^T$. Suppose that we have the LU factorization $LU = B$; then we have $A^{(N)} = B + \mathbf{e}\mathbf{r}^T = LU + \mathbf{e}\mathbf{r}^T$, where L and U are lower and upper triangular matrices.

Theorem 2.1. *Let* $\mathbf{z}$, $\mathbf{h}$, $\mathbf{v}$, *and* $\mathbf{w}$ *be* $(N+1)$*-vectors obtained by computing* $L\mathbf{z} = \mathbf{d}^{(N)}$, $L\mathbf{h} = \mathbf{e}$, $U^T\mathbf{v} = \mathbf{r}$, *and* $U^T\mathbf{w} = \boldsymbol{\xi}$ *by the process of forward substitution, respectively. Further, for the solution* $\mathbf{y}^{(N)}$ *of* $A^{(N)}\mathbf{y}^{(N)} = \mathbf{d}^{(N)}$, *let* $\mathbf{x}$ *be a vector defined by* $\mathbf{x} = U\mathbf{y}^{(N)}$. *Then we have*

$$\mathbf{x} = \mathbf{z} - \mathbf{h}\left\{\mathbf{v}^T\mathbf{z}/\left(1 + \mathbf{v}^T\mathbf{h}\right)\right\} \quad \text{and} \quad S_K^{(N)} = \mathbf{w}^T\mathbf{x}. \tag{2.2}$$

Proof. We see that $A^{(N)}\mathbf{y}^{(N)} = \mathbf{d}^{(N)}$ can be rewritten as $\left(LU + L\mathbf{h}\mathbf{v}^T U\right)U^{-1}\mathbf{x} = L\mathbf{z}$, which yields $\mathbf{x} + \mathbf{h}\mathbf{v}^T\mathbf{x} = \mathbf{z}$. The first equation in (2.2) follows easily by solving the above equation for $\mathbf{x}$. The second equation in (2.2) is easily verified from the fact that $S_K^{(N)} = \boldsymbol{\xi}^T\mathbf{y}^{(N)} = \boldsymbol{\xi}^T U^{-1}\mathbf{x} = \mathbf{w}^T\mathbf{x}$. $\square$

Let z_n, h_n, v_n, and w_n denote the nth elements of the vectors $\mathbf{z}$, $\mathbf{h}$, $\mathbf{v}$, and $\mathbf{w}$, respectively. Then, it follows from (2.2) that $S_K^{(N+1)} - S_K^{(N)} = w_{N+1}z_{N+1} - \phi(N+1)\sum_{k=M}^{N+1} w_k h_k + \phi(N)\sum_{k=M}^{N} w_k h_k$, where $\phi(N)$ is defined by $\phi(N) = \sum_{k=M+2}^{N} v_k z_k / (1 + \sum_{k=M+2}^{N} v_k h_k)$. This expression enables us to estimate in practice the effect of changing from N to $N+1$ before any backward substitution has been carried out, provided $S_K^{(N)}$ converges rapidly as N increases, which we assume here. Indeed the value of N is determined so that $|S_K^{(N+1)} - S_K^{(N)}| \leq \varepsilon$.

It should be noted that for the matrix $A^{(N)}$ (2.1) in the special case where $M=0$ and $\lambda_1 = 0$ the computation in the LU factorization of $A^{(N)}$ might occasionally suffer from loss of accuracy similar to [10] when $|f_0| \ll |g_0|$.

3. Numerical example

For the required tolerance $\varepsilon = 10^{-10}$ we compute the approximation $S_{14}^{(N)}$ to the finite sum $S_{14} = \sum_{n=0}^{14} y_n$ of the nondominant solution $y_n = 2^{-n}$ $(n = 0, 1, \ldots)$ of the second-order linear difference equation $y_{n-1} - (2n/x)\, y_n + y_{n+1} = \{2.5 - (2n/x)\}\, 2^{-n}$ $(n \geq 1)$. Here, the minimal and dominant solutions are Bessel functions of the first and second kinds $J_n(x)$ and $Y_n(x)$, respectively. The condition in Lemma 1.1 is satisfied if we choose $M = [x]$.

As mentioned in the previous section, when x is a zero of the minimal solution $J_0(x)$, numerical instability may occur for an inappropriate choice of M in some particular case. To see this, we choose $x = 8.653727912911014$, namely the third positive zero of $J_0(x)$, and a normalizing condition of the type $y_0 + 2(y_2 + y_3 + \ldots) = 2$. Then the estimated N for which $|S_{14}^{(N+1)} - S_{14}^{(N)}| \leq 10^{-10}$ is $N = 35$ when $M = [x] = 8$ in (2.1), but $N = 40$ when we incorrectly set $M = 0$. The approximation $S_{14}^{(35)} = 1.9999389648\mathit{847}$ for $M = 8$ indeed satisfies the tolerance, while $S_{14}^{(40)} = 2.0000\mathit{0000}$ for $M = 0$ fails to achieve the accuracy 10^{-10}, owing to numerical instability, as expected.

References

1. J. R. Cash, *An extension of Olver's method for the numerical solution of linear recurrence relations*, Math. Comp. **32** (1978), 497–510.
2. J. R. Cash, *A note on the numerical solution of linear recurrence relations*, Numer. Math. **34** (1980), 371–386.
3. P. Deuflhard, *A summation technique for minimal solutions of linear homogeneous difference equations*, Computing **18** (1977), 1–13.
4. W. Gautschi, *Computational aspects of three-term recurrence relations*, SIAM Rev. **9** (1967), 24–82.
5. G. H. Golub and C. F. Van Loan, *Matrix Computations*, 2nd ed., The Johns Hopkins University Press, Baltimore, 1989.
6. P. Levrie and A. Bultheel, *Convergence acceleration for the numerical solution of second-order linear recurrence relations*, SIAM J. Numer. Anal. **27** (1990), 166–177.
7. D. W. Lozier, *Numerical solution of linear difference equations*, report NBSIR 80–1976. NBS, Washington, 1980.
8. J. Oliver, *An extension of Olver's error estimation technique for linear recurrence relations*, Numer. Math. **12** (1968), 459–467.
9. F. W. J. Olver, *Numerical solution of second-order linear difference equations*, J. Res. NBS **71 (B)** (1967), 111–129.
10. F. W. J. Olver and D. J. Sookne, *Note on backward recurrence algorithms*, Math. Comp. **26** (1972), 941–947.
11. P. Van der Cruyssen, *A reformulation of Olver's algorithm for the numerical solution of second-order linear difference equations*, Numer. Math. **32** (1979), 159–166.
12. J. Wimp, *Computation with Recurrence Relations*, Pitman, Boston, 1984.

Department of Information Science, Faculty of Engineering, Fukui University, Fukui, 910, Japan

E-mail: hasegawa@fuis.fuis.fukui-u.ac.jp

Department of Information Engineering, Faculty of Engineering, Nagoya University, Nagoya, 464-01, Japan

Proceedings of Symposia in Applied Mathematics
Volume 48, 1994

THE FABER POLYNOMIALS FOR CIRCULAR ARCS

MATTHEW HE

ABSTRACT. In recent years there has been a growing interest in studying the Faber polynomials for specific regions. In this paper we study the Faber polynomials $F_n(z)$ generated by a circular arc with end points at $ae^{\pm 2i\alpha}$ ($a > 1$, $\cos\alpha = \frac{1}{a}$) containing the point $z = -a$. An explicit formula for $F_n(z)$ is derived by using its generating equation. In addition, we illustrate our computation of the zeros of these polynomials. The limiting distribution of the zeros is also determined explicitly.

1. INTRODUCTION

Let E be a compact set (not a single point) whose complement $\overline{C} \setminus E$ with respect to the extended plane is simply connected. The Riemann mapping theorem asserts that there exists a conformal mapping $w = \Phi(z)$ of $\overline{C} \setminus E$ onto the exterior of a circle $|w| = \rho_E$ in the w-plane. For a unique choice of ρ_E, we can insist that $\Phi(\infty) = \infty$, $\Phi'(\infty) = 1$, so that, in a neighborhood of infinity,

$$\Phi(z) = z + a_0 + \frac{a_1}{z} + \frac{a_2}{z^2} + \cdots . \tag{1.1}$$

The polynomial part of $\{\Phi(z)\}^n$, denoted by $F_n(z) := F_n(z; E) = z^n + \cdots$, is called the *Faber polynomial* of degree n generated by the set E. Let

$$\Psi(w) = w + b_0 + \frac{b_1}{w} + \frac{b_2}{w^2} + \cdots \tag{1.2}$$

be the inverse function of $w = \Phi(z)$. Faber [1] proved that

$$\frac{\Psi'(w)}{\Psi(w) - z} = \sum_{n=0}^{\infty} \frac{F_n(z)}{w^{n+1}}, \quad |w| > \rho_E, \; z \in E. \tag{1.3}$$

For the unit disk, the Faber polynomial of degree n is z^n. In this paper we study the Faber polynomials associated with *a circular arc* K with end points at $ae^{\pm 2i\alpha}$ ($a > 1$, $\cos\alpha = \frac{1}{a}$) containing the point $z = -a$. An explicit formula for $F_n(z; K)$ is derived. In addition, we illustrate our computation of the zeros of these polynomials and determine the limiting distribution.

1991 *Mathematics Subject Classification.* Primary 30C15, 41A58.
Key words and phrases. Faber polynomial, circular arc.
This paper is in final form and no version of it will be submitted for publication elsewhere.

© 1994 American Mathematical Society
0160-7634/94 \$1.00 + \$.25 per page

2. Explicit formula for $F_n(z;K)$

Here and throughout the remainder of this paper, $F_n(z) = F_n(z;K)$ denotes the Faber polynomial of degree n associated with the circular arc K. In this section we shall use the generating equation (1.3) of Faber polynomials to derive an explicit formula for these polynomials for the set K.

Theorem 2.0.1. *Let $F_n(z)$ be the Faber polynomial of degree n for K. Then for $n \geq 1$,*

$$F_n(z) = \sum_{k=0}^{[\frac{n}{2}]} \frac{(-1)^k n}{a^k(n-k)} C_{n-k}^k z^k (z+a)^{n-2k} - \frac{1}{a^n}, \tag{2.1}$$

where $[\frac{n}{2}] = \begin{cases} \frac{n}{2} & \text{if } n \text{ is even,} \\ \frac{n-1}{2} & \text{if } n \text{ is odd,} \end{cases}$ *and* $C_n^k = \frac{n!}{(n-k)!k!}$. *In particular, if $a = 1$, then K becomes the unit circle and $F_n(z) = z^n$.*

Proof. Integrate (1.3) with respect to w, and let one of the limits approach ∞; we obtain

$$\log \frac{\Psi(w) - z}{w} = -\sum_{n=1}^{\infty} \frac{F_n(z)}{nw^n}. \tag{2.2}$$

Now let

$$\Psi(w) = w \frac{1 - \frac{a}{w}}{1 - \frac{1}{aw}}, \quad a > 1. \tag{2.3}$$

It is known that $\Psi(w)$ maps $|w| > 1$ onto the exterior of K. For this particular function (2.3), equation (2.2) yields

$$\begin{aligned}
\sum_{n=1}^{\infty} \frac{F_n(z)}{nw^n} &= \log \left[\left(1 - \frac{1}{aw}\right) \left(1 - \frac{a+z}{w} + \frac{z}{aw^2}\right)^{-1} \right] \\
&= \sum_{n=1}^{\infty} \frac{1}{nw^n} \left(a + z - \frac{z}{aw}\right)^n - \sum_{n=1}^{\infty} \frac{1}{na^n w^n} \\
&= \sum_{n=1}^{\infty} \frac{1}{nw^n} \sum_{k=0}^{n} (-1)^k C_n^k (a+z)^{n-k} \left(\frac{z}{aw}\right)^k - \sum_{n=1}^{\infty} \frac{1}{na^n w^n} \\
&= \sum_{n=1}^{\infty} \left(\sum_{k=0}^{[\frac{n}{2}]} \frac{(-1)^k}{a^k(n-k)} C_{n-k}^k z^k (z+a)^{n-2k} \right) \frac{1}{w^n} - \sum_{n=1}^{\infty} \frac{1}{na^n w^n} \\
&= \sum_{n=1}^{\infty} \left[\sum_{k=0}^{[\frac{n}{2}]} \frac{(-1)^k}{a^k(n-k)} C_{n-k}^k z^k (z+a)^{n-2k} - \frac{1}{na^n} \right] \frac{1}{w^n}.
\end{aligned}$$

The formula (2.1) follows by comparing the coefficients of w^{-n} on both sides of the equation. $\square$

3. Zeros of $F_n(z)$

We note first that K becomes the unit circle when $a = 1$, and $F_n(z) = z^n$. Clearly, $z = 0$ is the zero of $F_n(z)$ with multiplicity n. As we shall see from our computation of the zeros of $F_n(z)$, when $a > 1$ the zeros of these polynomials move away from the origin $z = 0$ and approach the arc K.

3.1. Computation of the zeros of $F_n(z; K)$. The Faber polynomial $F_n(z)$ satisfies the following determinant representation [2]

$$F_n(z) = \det(zI_n - H_n), \tag{3.1}$$

where

$$H_n = \begin{bmatrix} b_0 & 1 & 0 & \cdots & 0 & 2b_1 \\ b_1 & b_0 & 1 & \cdots & 0 & 3b_2 \\ b_2 & b_1 & b_0 & \cdots & 0 & 4b_3 \\ \vdots & \vdots & \vdots & & \vdots & \vdots \\ b_{n-2} & b_{n-3} & b_{n-4} & \cdots & b_0 & nb_{n-1} \\ 1 & 0 & 0 & \cdots & 0 & b_0 \end{bmatrix}$$

is an $n \times n$ matrix and I_n the $n \times n$ identity matrix. Consequently, the zeros of $F_n(z)$ are the eigenvalues of H_n. For the set K, we expand (2.3) near ∞,

$$\Psi(w) = w + \frac{1-a^2}{a} + \sum_{n=0}^{\infty} \frac{1-a^2}{a^{n+1}} \frac{1}{w^n}, \quad |w| > 1, \tag{3.2}$$

which provides the coefficients for relation (1.2), hence the elements of H_n. We now illustrate in Fiqures 1 and 2 the location of the zeros of $F_n(z)$.

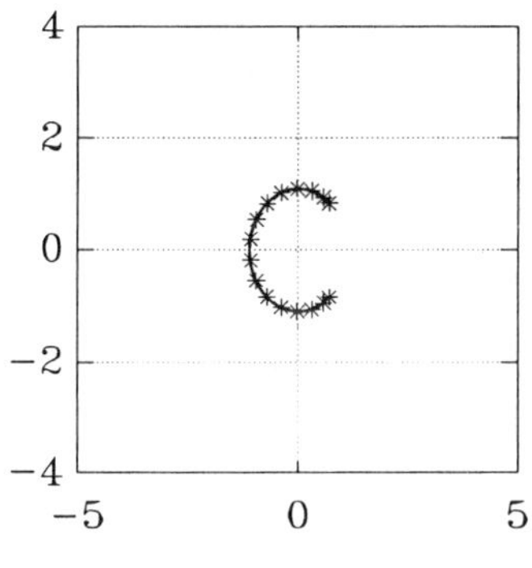

Fig. 1. a=1.1, n=16

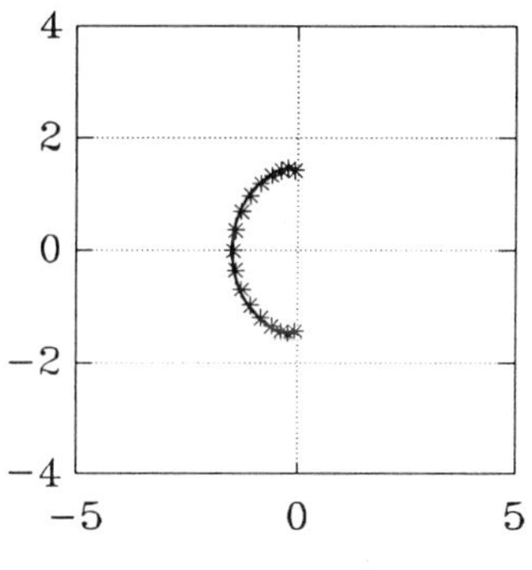

Fig. 2. a=1.5, n=17

3.2. Distribution of the zeros. From our numerical computation of the zeros of $F_n(z)$ in §3.1, we see that the zeros are approaching the arc as $n \to \infty$. The limiting distribution of these zeros is determined by the following theorem.

Theorem 3.2.1. *Let $z_{nj}, j=1,2,\ldots,n$, be the zeros of $F_n(z)$ and $\mu_n(t) = \frac{1}{n}\sum_{j=1}^{n} \delta_{z_{nj}}$ (the Dirac delta δ_z denotes a unit mass at z) be the zero distribution measure. Then*

$$\lim_{n\to\infty} \int_K f(t)d\mu_n(t) = \int_K f(t)d\mu(t),$$

for every continuous function $f(t)$ on K, where

$$(3.3)\qquad d\mu(t) = \frac{1}{\pi}\frac{t-a+\sqrt{t^2+2(a-\frac{2}{a})t+a^2}}{2it\sqrt{t^2+2(a-\frac{2}{a})t+a^2}}dt,\ t\in K.$$

Proof. The first part of the theorem follows from Theorem 16.3.1 in [3]. To prove (3.3), we define

$$(3.4)\qquad M_k := \int_K t^k d\mu(t),\quad k=0,1,...,$$

where M_k is the so-called kth moment of $d\mu(t)$ on K. It can be expressed as (cf. [3, p. 345])

$$(3.5)\qquad M_k = \frac{1}{2\pi i}\int_{|s|=r}[\Psi(s)]^k\frac{ds}{s},$$

for $r>1$, where

$$t=\Psi(s)=s\frac{1-\frac{a}{s}}{1-\frac{1}{as}}$$

maps the exterior of $|s|=1$ onto the exterior of K. We now compute the kth moment M_k from (3.5):

$$dt=\Psi(s)\left[\frac{2s-a-t}{s-t}\right]\frac{ds}{s}=\frac{t\sqrt{t^2+2(a-\frac{2}{a})t+a^2}}{t-a+\sqrt{t^2+2(a-\frac{2}{a})t+a^2}}\frac{ds}{s}.$$

It follows from (3.4) and (3.5) that

$$M_k=\int_{K_r}t^k d\mu(t)=\frac{1}{2\pi i}\int_{K_r}t^k\frac{t-a+\sqrt{t^2+2(a-\frac{2}{a})t+a^2}}{t\sqrt{t^2+2(a-\frac{2}{a})t+a^2}}dt,$$

where K_r is the image of $|w|=r$ under the mapping $\Psi(w)$. Letting $r\to 1$, we get

$$M_k=\int_{K}t^k d\mu(t)=\frac{1}{2\pi i}\int_{K}t^k\frac{t-a+\sqrt{t^2+2(a-\frac{2}{a})t+a^2}}{t\sqrt{t^2+2(a-\frac{2}{a})t+a^2}}dt.$$

We now have (3.3). □

REFERENCES

1. G. Faber, *Über polynomische Entwickelungen,* Math. Ann. **57** (1903), 389–408.
2. M. He, *Numerical results on the zeros of Faber polynomials for m-fold symmetric domains, Exploiting symmetry in applied and numerical analysis,* Lectures in Appl. Math. (E. L. Allgower, K. Georg and R. Miranda, eds.), Vol. 29, Amer. Math. Soc., 1994, pp. 229–240.
3. E. Hille, *Analytic function theory* (Introduction to higher mathematics), Vol. II, Ginn and Co., Boston, 1962.

DEPARTMENT OF MATHEMATICS, NOVA UNIVERSITY, FT. LAUDERDALE, FL 33314, U.S.A.
E-mail address: hem@polaris.nova.edu

Proceedings of Symposia in Applied Mathematics
Volume 48, 1994

Finite Element Approximation for Optimal Control of Electrically Conducting Fluid Flows

L.S. HOU AND S.S. RAVINDRAN

ABSTRACT. We study optimal control problems for an electrically conducting fluid. An optimality system of equations for the optimal solutions is derived using Lagrange multiplier rules. Finite element algorithms for solving the optimality system of equations are defined and optimal error estimates are derived.

1. Introduction

We denote by Ω the flow domain for an electrically conducting fluid. We suppose that Ω is a bounded open set in $\mathbb{R}^2$ or $\mathbb{R}^3$ with a boundary Γ. We assume the flow motion is governed by the following dimensionless equations for the steady incompressible, electrically conducting fluid flow in the presence of a magnetic field (see, e.g., [4]):

$$-\tfrac{1}{M^2}\Delta \mathbf{u} + \tfrac{1}{N}(\mathbf{u}\cdot\nabla)\mathbf{u} + \nabla p - (\mathbf{B}\times\nabla\phi) - (\mathbf{u}\times\mathbf{B})\times\mathbf{B} = \mathbf{f} \quad \text{in } \Omega\,, \tag{1.1}$$

$$\nabla\cdot\mathbf{u} = 0 \quad \text{in } \Omega \tag{1.2}$$

and

$$-\Delta\phi + \nabla\cdot(\mathbf{u}\times\mathbf{B}) = 0 \quad \text{in } \Omega\,, \tag{1.3}$$

where $\mathbf{u}$ denotes the velocity field, p the pressure field, $\mathbf{B}$ the magnetic field, ϕ the electric potential and $\mathbf{f}\in\mathbf{L}^2(\Omega)$ a given body force. N is the interaction number and M is the Hartmann number. We assume that the externally applied magnetic field $\mathbf{B}$ is undisturbed by the flow, i.e., we assume that $\mathbf{B}$ is fixed and given. Such an assumption can be met in a variety of physical applications, e.g.,

---

1991 *Mathematics Subject Classification.* Primary 49J20, 65N30, 76W05; Secondary 49K20.

The first author was supported by the Natural Science and Engineering Research Council of Canada Grant OGP-0137436 and by a Simon Fraser University President's Research grant.

The detailed version of this paper will be submitted for publication elsewhere.

© 1994 American Mathematical Society
0160-7634/94 $1.00 + $.25 per page

in the modelling of electromagnetic pumps and the flow of liquid lithium for fusion reactor cooling blankets ([6, 7]).

The system (1.1)–(1.3) is supplemented with boundary conditions

$$\mathbf{u} = \mathbf{0} \quad \text{on } \Gamma \qquad \text{and} \qquad \frac{\partial \phi}{\partial n} = g \quad \text{on } \Gamma\,, \tag{1.4}$$

where g denotes the normal current through Γ.

In this paper we study optimal control problems for the electrically conducting fluid flows modelled by (1.1)–(1.4). The control is the normal current g through the flow boundary. To fix our ideas, we choose matching the velocity field as our control objective. Mathematically, this task can be described by the following optimal control setting: minimize the cost functional

$$\mathcal{J}(\mathbf{u}, \phi, p, g) = \frac{\gamma}{2} \int_\Omega |\mathbf{u} - \mathbf{u}_d|^2 \, d\Omega + \frac{1}{2} \int_\Gamma |g|^2 \, d\Gamma\,, \tag{1.5}$$

subject to the constraints (1.1)–(1.4). Here γ is a fixed, positive parameter and $\mathbf{u}_d$ is a desired flow field. The appearance of the second term in the functional is customary and it effectively limits the size of the control.

We will freely make use of standard notations for Sobolev spaces and norms, inner products on those spaces (cf. [1] and [3]). We will need the function spaces $L^2_0(\Omega) = \{q \in L^2(\Omega) : \int_\Omega q \, d\Omega = 0\}$ and $H^1_z(\Omega) = H^1(\Omega) \cap L^2_0(\Omega)$.

The weak form of (1.1)–(1.4) is given as follows: seek $\mathbf{u} \in \mathbf{H}^1_0(\Omega)$, $p \in L^2_0(\Omega)$ and $\phi \in H^1_z(\Omega)$ such that

$$\begin{aligned} \frac{1}{M^2} \int_\Omega \nabla \mathbf{u} : \nabla \mathbf{v} \, d\Omega &+ \int_\Omega [\nabla \phi - (\mathbf{u} \times \mathbf{B})] \cdot [\nabla \psi - (\mathbf{v} \times \mathbf{B})] \, d\Omega \\ &+ \frac{1}{N} \int_\Omega (\mathbf{u} \cdot \nabla) \mathbf{u} \cdot \mathbf{v} \, d\Omega - \int_\Omega p \nabla \cdot \mathbf{v} \, d\Omega \\ = \int_\Omega \mathbf{f} \cdot \mathbf{v} \, d\Omega &+ \int_\Gamma g \psi \, d\Gamma \qquad \forall\, (\mathbf{v}, \psi) \in \mathbf{H}^1_0(\Omega) \times H^1_z(\Omega) \end{aligned} \tag{1.6}$$

and

$$- \int_\Omega q \nabla \cdot \mathbf{u} \, d\Omega = 0 \qquad \forall\, q \in L^2_0(\Omega)\,. \tag{1.7}$$

Extending the result of [4] for the Dirichlet case to the present Neumann case, we have

LEMMA 1.1. *For every $g \in H^{-1/2}(\Gamma)$ and $\mathbf{f} \in \mathbf{H}^1_0(\Omega)^*$, (1.6)–(1.7) has a solution $(\mathbf{u}, p, \phi) \in \mathbf{H}^1_0(\Omega) \times L^2_0(\Omega) \times H^1_z(\Omega)$ satisfying the estimate*

$$\|\mathbf{u}\|_1 + \|p\|_0 + \|\phi\|_1 \le C \left(\|\mathbf{f}\|_* + \|g\|_{-1/2,\Gamma} + 1 \right)^2 . \quad \square \tag{1.8}$$

An outline of the remaining two sections is as follows. In §2, we derive an optimality system of equations that the optimum must satisfy. In §3, we define finite element algorithms and derive optimal error estimates. Owing to page limits we will only be able to summarize the main results in this paper. Details can be found elsewhere.

2. An optimality system of equations

For simplicity we assume the set of admissible controls is the entire $L^2(\Gamma)$. The mathematical statement of the optimal control problem is as follows:

$$\text{seek a } (\mathbf{u},\phi,p,g)\in \mathbf{H}_0^1(\Omega)\times H_z^1(\Omega)\times L_0^2(\Omega)\times L^2(\Gamma) \text{ such that}$$

$$\mathcal{J}(\mathbf{u},\phi,p,g)=\frac{\gamma}{2}\int_\Omega |\mathbf{u}-\mathbf{u}_d|^2\,d\Omega+\frac{1}{2}\int_\Gamma |g|^2\,d\Gamma \tag{2.1}$$

$$\text{is minimized subject to the constraints (1.6)–(1.7).}$$

The existence of an optimal solution can be proved based on the a priori estimate (1.8) and standard techniques.

THEOREM 2.1. *There exists an optimal solution* $(\mathbf{u},\phi,p,g)$ *for* (2.1). □

Now we use the method of Lagrange multipliers to derive a system of equations that optimal solutions satisfy. Such a system of equations will be termed *an optimality system of equations.* [5] presents a framework for Lagrange multiplier rules on Banach spaces. It can be verified that, for almost all values of the parameter N, (2.1) fits into that framework. Thus we have

THEOREM 2.2. *Assume* $(\mathbf{u},\phi,p,g)\in \mathbf{H}_0^1(\Omega)\times H_z^1(\Omega)\times L_0^2(\Omega)\times L^2(\Gamma)$ *is an optimal solution to the minimization problem* (2.1). *Then there exists a set* $\Sigma\subset\mathbf{R}$ *of measure zero such that for each* $N\notin\Sigma$ *there exists a* $(\boldsymbol{\mu},\pi,\sigma)\in \mathbf{H}_0^1(\Omega)\times L_0^2(\Omega)\times H_z^1(\Omega)$ *satisfying*

$$\begin{aligned}
&\frac{1}{M^2}\int_\Omega \nabla\boldsymbol{\mu}:\nabla\mathbf{w}\,d\Omega+\int_\Omega[\nabla\sigma-(\boldsymbol{\mu}\times\mathbf{B})]\cdot[\nabla\zeta-(\mathbf{w}\times\mathbf{B})]\,d\Omega\\
&\quad+\frac{1}{N}\int_\Omega(\mathbf{u}\cdot\nabla)\mathbf{w}\cdot\boldsymbol{\mu}\,d\Omega+\frac{1}{N}\int_\Omega(\mathbf{w}\cdot\nabla)\mathbf{u}\cdot\boldsymbol{\mu}\,d\Omega-\int_\Omega\pi\nabla\cdot\mathbf{w}\,d\Omega\\
&\quad=\gamma\int_\Omega(\mathbf{u}-\mathbf{u}_d)\cdot\mathbf{w}\,d\Omega\qquad\forall\,(\mathbf{w},\zeta)\in\mathbf{H}_0^1(\Omega)\times H_z^1(\Omega),
\end{aligned}\tag{2.2}$$

$$-\int_\Omega r\nabla\cdot\boldsymbol{\mu}\,d\Omega=0\qquad\forall\, r\in L_0^2(\Omega)\tag{2.3}$$

and

$$\int_\Gamma gz\,d\Gamma+\int_\Gamma\sigma z\,d\Gamma=0\qquad\forall\,z\in L^2(\Gamma).\ \Box\tag{2.4}$$

The last relation allows us to eliminate g in (1.6) to obtain

$$\begin{aligned}
&\frac{1}{M^2}\int_\Omega\nabla\mathbf{u}:\nabla\mathbf{v}\,d\Omega+\int_\Omega[\nabla\phi-(\mathbf{u}\times\mathbf{B})]\cdot[\nabla\psi-(\mathbf{v}\times\mathbf{B})]\,d\Omega\\
&+\frac{1}{N}\int_\Omega(\mathbf{u}\cdot\nabla)\mathbf{u}\cdot\mathbf{v}\,d\Omega-\int_\Omega p\nabla\cdot\mathbf{v}\,d\Omega=\int_\Omega\mathbf{f}\cdot\mathbf{v}\,d\Omega-\int_\Gamma\sigma\psi\,d\Gamma
\end{aligned}\tag{2.5}$$

for all $(\mathbf{v},\psi)\in\mathbf{H}_0^1(\Omega)\times H_z^1(\Omega)$. Hence, the equations (2.5), (1.7) and (2.2)–(2.3) form an optimality system of equations, which $(\mathbf{u},p,\phi,\boldsymbol{\mu},\pi,\sigma)$ (optimal solutions along with Lagrange multipliers) must satisfy.

3. Finite element approximations

A finite element discretization of the optimality system (2.5), (1.7) and (2.2)–(2.3) is defined as follows. First, one chooses families of finite element subspaces $\mathbf{V}^h \subset \mathbf{H}^1_0(\Omega)$, $S^h \subset L^2_0(\Omega)$ and $Q^h \subset H^1_z(\Omega)$. We assume that these finite element spaces satisfy the following approximation properties (see [2, 3]): there exist an integer k and a constant C, independent of h, $\mathbf{v}$, q and ψ, such that

$$\inf_{\mathbf{v}^h \in \mathbf{V}^h} \|\mathbf{v} - \mathbf{v}^h\|_1 \leq Ch^m \|\mathbf{v}\|_{m+1} \quad \forall\, \mathbf{v} \in \mathbf{H}^{m+1}(\Omega) \cap \mathbf{H}^1_0(\Omega)\,,\ 1 \leq m \leq k\,, \tag{3.1}$$

$$\inf_{q^h \in S^h} \|q - q^h\|_0 \leq Ch^m \|q\|_m \quad \forall\, q \in H^m(\Omega) \cap L^2_0(\Omega)\,,\ 1 \leq m \leq k\,, \tag{3.2}$$

and

$$\inf_{\psi^h \in Q^h} \|\psi - \psi^h\|_1 \leq Ch^m \|\psi\|_{m+1} \quad \forall\, \psi \in H^{m+1}(\Omega) \cap L^2_0(\Omega)\,,\ 1 \leq m \leq k\,. \tag{3.3}$$

Next, we assume the *inf-sup condition*, or *Ladyzhenskaya-Babuška-Brezzi condition* (see [3]): there exists a constant C, independent of h, such that

$$\inf_{0 \neq q^h \in S^h} \ \sup_{0 \neq \mathbf{v}^h \in \mathbf{V}^h} \frac{\int_\Omega q^h \nabla\cdot \mathbf{v}^h \, d\Omega}{\|\mathbf{v}^h\|_1 \, \|q^h\|_0} \geq C\,. \tag{3.4}$$

We define a finite element approximation of the optimality system as follows: seek $\mathbf{u}^h \in \mathbf{V}^h$, $p^h \in S^h$, $\phi^h \in Q^h$, $\boldsymbol{\mu}^h \in \mathbf{V}^h$, $\pi^h \in S^h$ and $\sigma^h \in Q^h$ such that

$$\begin{aligned} \frac{1}{M^2}\int_\Omega \nabla \mathbf{u}^h : \nabla \mathbf{v}^h \, d\Omega + \frac{1}{N}\int_\Omega (\mathbf{u}^h \cdot \nabla)\mathbf{u}^h \cdot \mathbf{v}^h \, d\Omega - \int_\Omega p^h \nabla \cdot \mathbf{v}^h \, d\Omega \\ - \int_\Omega (\mathbf{B} \times \nabla \phi^h) \cdot \mathbf{v}^h \, d\Omega - \int_\Omega (\mathbf{u}^h \times \mathbf{B}) \times \mathbf{B} \cdot \mathbf{v}^h \, d\Omega \\ = \int_\Omega \mathbf{f} \cdot \mathbf{v}^h \, d\Omega \qquad \forall\, \mathbf{v}^h \in \mathbf{V}^h\,, \end{aligned} \tag{3.5}$$

$$\int_\Omega q^h \nabla \cdot \mathbf{u}^h \, d\Omega = 0 \qquad \forall\, q^h \in S^h\,, \tag{3.6}$$

$$\int_\Omega \nabla \phi^h \cdot \nabla \psi^h \, d\Omega + \int_\Omega \nabla \cdot (\mathbf{u}^h \times \mathbf{B}) \psi^h \, d\Omega + \int_\Gamma \sigma^h \psi^h \, d\Gamma = 0 \quad \forall\, \psi^h \in Q^h\,, \tag{3.7}$$

$$\begin{aligned} \frac{1}{M^2}\int_\Omega \nabla \boldsymbol{\mu}^h : \nabla \mathbf{w}^h \, d\Omega - \int_\Omega \pi^h \nabla \cdot \mathbf{w}^h \, d\Omega - \int_\Omega (\mathbf{B} \times \nabla \sigma^h) \cdot \mathbf{w}^h \, d\Omega \\ + \frac{1}{N}\int_\Omega \boldsymbol{\mu}^h \cdot (\nabla \mathbf{u}^h)^t \cdot \mathbf{w}^h \, d\Omega - \frac{1}{N}\int_\Omega (\mathbf{u}^h \cdot \nabla)\boldsymbol{\mu}^h \cdot \mathbf{w}^h \, d\Omega \\ - \int_\Omega (\boldsymbol{\mu}^h \times \mathbf{B}) \times \mathbf{B} \cdot \mathbf{w}^h \, d\Omega = \gamma \int_\Omega (\mathbf{u} - \mathbf{u}_0) \cdot \mathbf{w} \, d\Omega \quad \forall \mathbf{w}^h \in \mathbf{V}^h\,, \end{aligned} \tag{3.8}$$

$$\int_\Omega r^h \nabla \cdot \boldsymbol{\mu}^h \, d\Omega = 0 \qquad \forall\, r^h \in S^h \tag{3.9}$$

and

$$\int_\Omega \nabla\sigma^h \cdot \nabla\tau^h \, d\Omega + \int_\Omega \nabla\cdot(\boldsymbol{\mu}^h \times \mathbf{B})\tau^h \, d\Omega = 0 \qquad \forall\, \tau^h \in Q^h\,. \tag{3.10}$$

We very briefly sketch the proof of optimal error estimates. We first prove optimal error estimates for the approximations of the linearized optimality system of equations, i.e., the system obtained by dropping all nonlinear terms from the optimality system. Then through a careful choice of spaces and operators we may fit the optimality system into the framework of Brezzi-Rappaz-Raviart theory (see [3]). By verifying all the requirements of that theory, we obtain optimal error estimates for the approximation of the optimality system of equations.

THEOREM 3.1. *Assume* $(\mathbf{u},p,\phi,\boldsymbol{\mu},\pi,\sigma)\in \mathbf{H}^{m+1}(\Omega)\times H^m(\Omega)\times H^{m+1}(\Omega)\times \mathbf{H}^{m+1}(\Omega)\times H^m(\Omega)\times H^{m+1}(\Omega)$ *is a nonsingular solution of the optimality system* (2.5), (1.7) *and* (2.2)–(2.3). *Then for each sufficiently small* h, *the approximate optimality system* (3.5)–(3.10) *has a unique solution* $(\mathbf{u}^h,p^h,\phi^h,\boldsymbol{\mu}^h,\pi^h,\sigma^h)\in \mathbf{V}^h\times S^h\times Q^h\times \mathbf{V}^h\times S^h\times Q^h$, *in a neighborhood of* $(\mathbf{u},p,\phi,\boldsymbol{\mu},\pi,\sigma)$, *such that*

$$\begin{aligned}&\|\mathbf{u}-\mathbf{u}^h\|_1+\|p-p^h\|_0+\|\phi-\phi^h\|_1+\|\boldsymbol{\mu}-\boldsymbol{\mu}^h\|_1+\|\pi-\pi^h\|_0+\|\sigma-\sigma^h\|_1\\ &\le Ch^m\{\|\mathbf{u}\|_{m+1}+\|p\|_m+\|\phi\|_{m+1}+\|\boldsymbol{\mu}\|_{m+1}+\|\pi\|_m+\|\sigma\|_{m+1}\}\,.\quad\square\end{aligned}$$

Our numerical results (to be reported elsewhere) indicate that the optimal control approach, when properly implemented, does a very good job in matching velocity and other fields.

REFERENCES

1. R. A. Adams, *Sobolev Spaces*, Academic Press, New York, 1975.
2. Ph. Ciarlet, *The Finite Element Method for Elliptic Problems*, North-Holland, Amsterdam, 1978.
3. V. Girault and P.-A. Raviart, *Finite Element Methods for Navier-Stokes Equations*, Springer-Verlag, Berlin, 1986.
4. J. Peterson, *On the finite element approximation of incompressible flows of an electrically conducting fluid*, Numer. Methods Partial Differential Equations **4** (1988), 57–68.
5. V. Tikhomorov, *Fundamental Principles of the Theory of Extremal Problems*, Wiley, Chichester, 1982.
6. J.S. Walker, *Large interaction parameter magnetohydrodynamics and applications in fusion reactor technology*, Fluid Mechanics in Energy Conversion (J. Buchmaster, ed.), SIAM, Philadelphia, 1980.
7. N. Winowich and W. Hughes, *A finite element analysis of two dimensional MHD flows*, Liquid-Metal Flows and Magnetohydrodynamics (H. Branover, P.S. Lykoudis and A. Yakhot, eds.), AIAA, New York, 1983.

DEPARTMENT OF MATHEMATICS AND STATISTICS, SIMON FRASER UNIVERSITY, BURNABY, B.C. V5A 1S6, CANADA

Current address: Department of Mathematics and Statistics, York University, North York, Ontario M3J 1P3, Canada

E-mail address: hou@mathstat.yorku.ca

DEPARTMENT OF MATHEMATICS AND STATISTICS, SIMON FRASER UNIVERSITY, BURNABY, B.C. V5A 1S6, CANADA

E-mail address: ravindra@cs.sfu.ca

Proceedings of Symposia in Applied Mathematics
Volume **48**, 1994

ADI Methods for Heat Equations With Discontinuities Along an Arbitrary Interface

ZHILIN LI AND ANITA MAYO

ABSTRACT. We develop a second-order accurate ADI (alternating direction implicit) method for parabolic equations with singular sources along an arbitrary but fixed interface. The solution may be nonsmooth or even discontinuous. Theoretical analysis and numerical examples are given to show the efficiency of our ADI method.

1. Introduction

ADI methods are very efficient for time-dependent parabolic PDEs. They are second-order accurate and stable with reasonable time steps, yet in each time step they only require solving tridiagonal systems. However, the classical ADI method introduced by Peaceman and Rachford [**4**] is successful only if the solution is sufficiently smooth. Here we consider a problem in which

$$u_t = (\beta u_x)_x + (\beta u_y)_y - f, \qquad (x,y) \in \Omega, \tag{1}$$

holds on either side of an interface Γ but with given jump conditions in the solution and its normal derivative across Γ. As an example, suppose $f(x,y,t)$ has a delta function singularity along Γ parametrized by $(X(s), Y(s))$,

$$f(x,y,t) = \int_\Gamma C(s,t)\,\delta(x - X(s))\,\delta(y - Y(s))\,ds,$$

i.e., $f(x,y,t)$ is a distribution with the property that

$$\int\int f(x,y,t)\phi(x,y)dxdy = \int_\Gamma C(s,t)\phi(X(s),Y(s))ds$$

for any smooth test function $\phi(x,y)$. Then u is continuous but the normal derivative has a jump of magnitude $[u_n] = C(s,t)$. Such problems arise in

1991 *Mathematics Subject Classification.* Primary 65M06.

The authors were supported in part by Grants DMS-8657319 and DMS-9204329.

This is a preliminary version.

© 1994 American Mathematical Society
0160-7634/94 $1.00 + $.25 per page

various applications (see [**1**]). Problems in which β is discontinuous are also interesting, but are more complicated and will be our future topic.

We assume we know the jump conditions in $[u]$ and $[\partial u/\partial n]$ along the interface which can often be derived from the PDEs as in the above example. They are used to derive appropriate correction terms so that the standard five-point difference operator yields second-order accuracy as in the work of Mayo [**2, 3**] (see also [**1**]). The domain Ω is assumed to be rectangular, say $[a,b]\times[c,d]$. We use a uniform $M\times N$ grid with $x_i = a+ih$, $y_j = a+jh$, where $h=(b-a)/M = (d-c)/N$.

2. The difference scheme

Our ADI method for $\beta = 1$ can be written as

$$\frac{u_{ij}^{n+\frac{1}{2}} - u_{ij}^{n}}{\tau/2} = \delta_x u_{ij}^{n+\frac{1}{2}} - (C_x)_{ij}^{n+\frac{1}{2}} - Q_{ij}^{n} - R_{ij}^{n} + \delta_y u_{ij}^{n} - (C_y)_{ij}^{n} - f_{ij}^{n+\frac{1}{2}}, \tag{2}$$

$$\begin{aligned}\frac{u_{ij}^{n+1} - u_{ij}^{n+\frac{1}{2}}}{\tau/2} &= \delta_x u_{ij}^{n+\frac{1}{2}} - (C_x)_{ij}^{n+\frac{1}{2}} - Q_{ij}^{n} - R_{ij}^{n} \\ &\quad + \delta_y u_{ij}^{n+1} - (C_y)_{ij}^{n+1} - f_{ij}^{n+\frac{1}{2}},\end{aligned} \tag{3}$$

where τ is the time step and

$$\delta_x u_{ij}^n = \left(u_{i-1,j}^n - 2u_{ij}^n + u_{i+1,j}^n\right)/h^2, \quad \delta_y u_{ij}^n = \left(u_{i,j-1}^n - 2u_{ij}^n + u_{i,j+1}^n\right)/h^2.$$

At regular grid points the standard ADI method is used, in which

$$(C_x)_{ij}^{n+\frac{1}{2}} = Q_{ij}^n = R_{ij}^n = (C_y)_{ij}^n = Q_{ij}^n = (C_y)_{ij}^{n+1} = 0.$$

At each irregular grid point we need to determine these correction terms.

Splitting correction terms. Take Figure 1 as an example, and suppose we want to approximate u_{xx}. Using Taylor expansions about some point $(x^*,y^*)\in\Gamma$ and letting $-$ and $+$ superscripts indicate limiting value as we approach (x^*,y^*) from one side or the other, omitting the superscripts n, we find that

$$\begin{aligned}\frac{u_{i-1,j} - 2u_{ij} + u_{i+1,j}}{h^2} &= u_{xx}^- + \frac{[u]}{h^2} + [u_x]\frac{(x_{i+1}-x^*)}{h^2} \\ &\quad + [u_{xx}]\frac{(x_{i+1}-x^*)^2}{2h^2} + O(h) \\ &= u_{xx}^- + (C_x)_{ij} + O(h).\end{aligned} \tag{4}$$

So the correction term C_x is given by

$$(C_x)_{ij} = \frac{[u]}{h^2} + [u_x]\frac{(x_{i+1}-x^*)}{h^2} + [u_{xx}]\frac{(x_{i+1}-x^*)^2}{2h^2}. \tag{5}$$

Similarly, we can get the expression for $(C_y)_{ij}$.

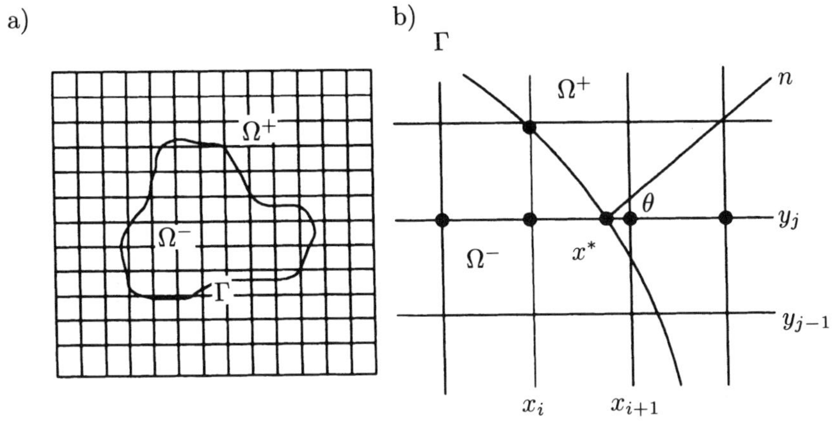

FIGURE 1. a) An arbitrary interface immersed in the uniform grid. b) The geometry at some irregular grid point near the interface.

Now we need to determine the jumps $[u_x]$, $[u_y]$, $[u_{xx}]$, $[u_{yy}]$ from the given jump conditions in $[u]$ and $[\partial u/\partial n]$. As in [**1**], we use a local coordinate transformation at (x^*, y^*) into normal and tangential direction ξ and η,

$$\xi = (x - x^*)\cos\theta + (y - y^*)\sin\theta, \tag{6}$$

$$\eta = -(x - x^*)\sin\theta + (y - y^*)\cos\theta, \tag{7}$$

where θ is the angle between the x-axis and the normal direction at (x^*, y^*). In a neighborhood of this point, the interface lies roughly in the η-direction, so we can parameterize Γ locally by $\xi = \chi(\eta)$, $\eta = \eta$, and we write the jumps as $[u] = w(\eta, t)$, $[u_n] = v(\eta, t)$. For convenience we use the same notation for the solution u both in the original coordinates and the new local coordinates.

After the local coordinate transformation we get

$$[u_\xi] = [u_n] = v(\eta, t), \tag{8}$$

$$[u_\eta] = [u]_\eta = w_\eta(\eta, t), \tag{9}$$

which are known already. For the remaining second-derivative jumps we can use the interface relations from [**1**] and [**2**] with minor modifications to get

$$[u_{\xi\xi}] = \chi''[u_\xi] - w_{\eta\eta} + [f] + w_t, \tag{10}$$

$$[u_{\eta\eta}] = -\chi''[u_\xi] + w_{\eta\eta}, \tag{11}$$

$$[u_{\xi\eta}] = \chi''[u_\eta] + v_\eta. \tag{12}$$

Now we have expressed all the jumps in the local coordinates in terms of the known quantities $[u]$ and $[u_n]$. Mayo [**3**] has also derived these interface relations for a similar problem. From (8)–(12) we can get the jumps in the x- and y-

direction, respectively, using the following formulas:

$$[u_x] = [u_\xi]\cos\theta - [u_\eta]\sin\theta, \tag{13}$$

$$[u_y] = [u_\xi]\sin\theta + [u_\eta]\cos\theta, \tag{14}$$

$$[u_{xx}] = [u_{\xi\xi}]\cos^2\theta - 2[u_{\xi\eta}]\cos\theta\sin\theta + [u_{\eta\eta}]\sin^2\theta, \tag{15}$$

$$[u_{yy}] = [u_{\xi\xi}]\sin^2\theta + 2[u_{\xi\eta}]\cos\theta\sin\theta + [u_{\eta\eta}]\cos^2\theta. \tag{16}$$

The derivations of Q_{ij}^n and R_{ij}^n are too subtle to discuss here, owing to space limitations, and will be presented elsewhere. We simply give the results:

$$Q_{ij}^n = \frac{1}{2}\left((C_x)_{ij}^n + (C_x)_{ij}^{n+1}\right) - (C_x)_{ij}^{n+\frac{1}{2}}, \tag{17}$$

$$R_{ij}^n = \pm\,\frac{[u_{yy}]^{n+1} - [u_{yy}]^n}{4\,\tau}. \tag{18}$$

The sign is determined by the relative position of (x_i, y_j) and Γ.

3. Numerical example

The partial differential equation is

$$u_t = u_{xx} + u_{yy} - f(x,y,t). \tag{19}$$

The function $f(x,y,t)$ is defined as

$$f(x,y,t) = \begin{cases} 0 & \text{if } r \le \frac{1}{2}, \\ \left(\sin t - \frac{\pi^2}{16}\cos t\right)\sin(\frac{\pi}{4}(x+1))\sin(\frac{\pi}{4}(y+1)) & \text{if } r > \frac{1}{2}. \end{cases}$$

where $r = \sqrt{x^2+y^2}$. The boundary conditions and jumps in $u(x,y,t)$ and $\frac{\partial u}{\partial n}$ are specified so that the exact solution is

$$u(x,y,t) = \begin{cases} 1 & \text{if } r \le \frac{1}{2}, \\ \cos t\sin(\frac{\pi}{4}(x+1))\sin(\frac{\pi}{4}(y+1)) & \text{if } r > \frac{1}{2}. \end{cases} \tag{20}$$

Figure 2 shows the global error in the infinity norm and the solution plot. Second-order convergence is clearly seen.

We have tested a number of examples and all of them give second-order accurate results. With a little change we can modify the method for the problems when β is not a constant but a continuous function.

a)

n	$\| E_n \|_\infty$ at $t = 5$	ratio
20	9.28331×10^{-4}	
40	2.46366×10^{-4}	3.6943
80	6.47184×10^{-5}	3.8067
160	1.66843×10^{-5}	3.8790
320	4.25187×10^{-6}	3.9240

b)

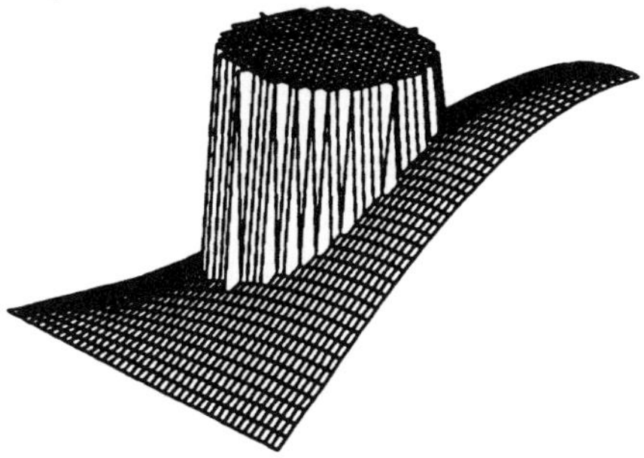

FIGURE 2. a) Error of computed solution at all grid points in infinity norm at $t = 5$ with $n = 20, 40, \ldots, 320$. b) Computed solution at time $t = 2\pi$ with $n = 40$.

REFERENCES

1. R.J. LeVeque, Z. Li, *The immersed interface method for elliptic equations with discontinous coefficients and singular sources.* SIAM J. Numer. Anal., in press.
2. A. Mayo, *The fast solution of Poisson's and the biharmonic equations on irregular regions.* SIAM J. Numer. Anal. **21**(1984), pp. 285-299.
3. A. Mayo, *On the rapid evaluation of heat potential on general regions.* IBM Technical Report 14305.
4. D.W. Peaceman, H.H. Rachford, Jr., *The numerical solution of parabolic and elliptic differential equations.* J. Soc. Indust. Appl. Math. **3**(1955), pp. 28-41.

DEPARTMENT OF APPLIED MATHEMATICS, UNIVERSITY OF WASHINGTON, SEATTLE, WA, 98195

E-mail address: zhilin@amath.washington.edu

IBM RESEARCH DIVISON, T.J. WATSON RESEARCH CENTER, YORKTOWN HEIGHTS, NY 10598

E-mail address: amayo@watson.ibm.com

Proceedings of Symposia in Applied Mathematics
Volume **48**, 1994

Eigenvalue Approximation of Fredholm Integral Operators

E. B. LIN

ABSTRACT. We discuss the degree of accuracy of the numerical solution of eigenvalue problems of Fredholm integral operators by collocation methods and wavelet methods.

1. Introduction

We consider the following integral equation :

$$Ku = \lambda u \ , \tag{1.1}$$

where

$$Ku(x) = \int_0^1 \ k(x, y) \ u(y) \ dy \tag{1.2}$$

with a smooth kernel k and smooth eigenfunction u on [0, 1].

A summary of methods for the approximate solution of (1.1) can be found in [4]. Some of these methods exhibit "superconvergence" in the sense that the order of convergence is higher than one would expect from the degree of the piecewise polynomials employed [9]. Comparisons of different methods for accuracy of eigensolutions are shown in [7,8]. Several convergence results on approximate eigensolutions are obtained in [6]. In this paper we summarize some results on eigenvalue approximations by collocation methods and mention a wavelet basis method for eigenvalue approximation problems [4,5].

In §2 we give an example to illustrate the main results shown in this paper. A fundamental theorem to be used is shown in the third section. The main theorems are presented in the fourth section. A new method will be mentioned in the fifth section, namely, a wavelet basis method. We conclude with some remarks in the last section.

2. An example

Let $k(s,t) = e^{s+t}$. Consider the eigenvalue problem $Ku = \lambda u$, i.e.,

$$\int_0^1 e^{s+t} u(t) dt = \lambda u(s), \quad 0 \leq s \leq 1, \quad (u, u) = 1. \tag{2.1}$$

It is easy to see that $\lambda = \frac{1}{2}(e^2 - 1)$ is an eigenvalue of (2.1). Associated with the eigenvalue λ, the approximate problem is

$$\sum_{i=1}^{n} \int_{(i-1)h}^{ih} \quad e^{\frac{2j-1}{2}h} e^y u_h(\frac{2i-1}{2}h) dy = \lambda_h u_h(\frac{2j-1}{2}h), \tag{2.2}$$

1991 *Mathematics Subject Classification.* 65R20, 65D99.

The final version of this paper will be submitted for publication elsewhere.

© 1994 American Mathematical Society
0160-7634/94 $1.00 + $.25 per page

which is equivalent to the following matrix eigenvalue problem:

$$(\lambda I - A)U = 0, \tag{2.3}$$

where $A = (a_{ij})_{n\times n},\ a_{ij} = (e^h - 1)e^{-\frac{3}{2}h}e^{(i+j)h},\ \ U = (u_h(\frac{h}{2}), \dots, u_h(\frac{2n-1}{2}h))^T$. Solving (2.3), we get

$$\lambda_h = (e^2 - 1)\frac{e^h - 1}{e^{2h} - 1}e^{\frac{h}{2}} = \frac{1}{2}(e^2 - 1)(\cosh\frac{h}{2})^{-1}.$$

On the other hand, $\lim_{h\to 0}\lambda_h = \frac{1}{2}(e^2 - 1) = \lambda$, which shows that λ_h converges to the exact eigenvalue λ. We observe that

$$\lambda - \lambda_h = \frac{1}{16}(e^2 - 1)h^2 + O(h^4),$$

from which, by Richardson extrapolation, an improved eigenvalue approximation, $\lambda_h^{(1)} = (4\lambda_{\frac{h}{2}} - \lambda_h)/3$ is found satisfying

$$\lambda - \lambda_h^{(1)} = O(h^4).$$

3. Bramble-Hilbert theorem

Let Ω be a bounded domain in Euclidean N-space with boundary $\partial\Omega$. Let ρ be the diameter of Ω. We shall assume that Ω satisfies the following property: there exist a finite open covering $\{O_i\},\ i = 1, \dots, n$, of $\partial\Omega$ and corresponding cones $\{C_i\}$ with vertices at the origin such that $x + C_i$ is contained in Ω for any $x \in \Omega \cap O_i$. We shall need the following seminorms:

$$|u|_{p,k,\Omega} = \sum_{|\alpha|=k} \|D^\alpha u\|_{p,\Omega}$$

and

$$|u|_{k,\Omega} = \sum_{|\alpha|=k} |D^\alpha u|_\Omega, \quad \text{where} \quad |u|_\Omega = \sup_{x\in\Omega}|u(x)|.$$

Let $W^{m,p}(\Omega) = \{f \in L^p(\Omega); \forall\alpha \in Z^N_{\geq 0}, |\alpha| \leq m \Rightarrow D^\alpha f \in L^p(\Omega)\}$. We shall denote by P_k the set of polynomials of degree less than or equal to k. Extending the result of Bramble-Hilbert [2,3], we have the following theorem.

Theorem 1. (i) *Suppose that F is a bounded linear functional on $W^{k+2,p}(\Omega)$ and satisfies $F(p) = 0$ for any $p \in P_k$. Then one can expand*

$F(u) = \sum_{|\alpha|=k+1} \frac{F(x^\alpha)}{\alpha!\rho} \int_\Omega D^\alpha u(x)dx + R$, where $|R| \leq C|u|_{k+2,p,\Omega}$.

(ii) *Suppose that F is a bounded linear functional on $W^{k+3,p}(\Omega)$ and $F(p) = 0$ for all $p \in P_k$, $F(x^\alpha) = 0$ for $|\alpha| = k+2$, and the domain Ω is symmetric with respect to the origin. Then $F(u) = \sum_{|\alpha|=k+1} \frac{F(x^\alpha)}{\alpha!\rho} \int_\Omega D^\alpha u(x)dx + R$, where $|R| \leq C|u|_{k+3,p,\Omega}$.*

4. Main theorems

Let the kernel of (1.1) satisfy the following conditions:

(1) $\sup_{0\le s\le 1}\int_0^1 |k(s,t)|dt < \infty$;

(2) $\lim_{\delta\to 0}\int_0^1 |k(s+\delta,t)-k(s,t)|dt = 0, \quad 0\le s\le 1$;

(3) $k(s,t)$ is symmetric, i.e., $k(s,t)=k(t,s)$.

We use the notations of the previous section, $\pi_h K u_h = \lambda_h u_h, \quad u_h \in \Pi_h$, where π_h is a projection operator onto a finite dimensional subspace Π_h of $C[0,1]$.

Let $\hat{u}_h$ be obtained by an iteration $Ku_h = \lambda_h \hat{u}_h$, which gives rise to $\pi_h \hat{u}_h = u_h$. From condition (1) above and $\lambda \neq 0$, we have

$$\lambda_h(u,\hat{u}_h) = (u,\lambda_h\hat{u}_h) = (u,Ku_h). \tag{4.1}$$

Thus,

$$\lambda_h = \frac{(u,Ku_h)}{(u,\hat{u}_h)} = (u,K\pi_h\tilde{u}_h), \tag{4.2}$$

where

$$\tilde{u}_h = \frac{\hat{u}_h}{(u,\hat{u}_h)}.$$

On the other hand,

$$\lambda(u,\hat{u}_h) = (\lambda u,\hat{u}_h) = (Ku,\hat{u}_h), \tag{4.3}$$

$$\lambda = \frac{(Ku,\hat{u}_h)}{(u,\hat{u}_h)} = (Ku,\tilde{u}_h). \tag{4.4}$$

Hence,

$$\begin{aligned}\lambda-\lambda_h &= (Ku,\tilde{u}_h)-(u,K\pi_h\tilde{u}_h)\\ &= (u,K\tilde{u}_h)-(u,K\pi_h\tilde{u}_h)\\ &= (u,K(\tilde{u}_h-\pi_h\tilde{u}_h)),\end{aligned} \tag{4.5}$$

where we use the symmetry property of the kernel (condition (3)).

Based on the above argument, and using the theorem in §3 and other estimates [4], we obtain the following main theorems.

Theorem 2. *If $k(s,t)\in C^4$, then the eigenvalue approximation by piecewise constant collocation satisfies*

$$\lambda-\lambda_h^{(1)} = O(h^4),$$

where $\lambda_h^{(1)} = \frac{4}{3}\lambda_{\frac{h}{2}} - \frac{1}{3}\lambda_h$.

Theorem 3. *If $k(s,t) \in C^6$, then the eigenvalue approximation by piecewise linear collocation at Gauss points satisfies*

$$\lambda - \lambda_h^{(1)} = O(h^6),$$

where

$$\lambda_h^{(1)} = \frac{4}{3}\lambda_{\frac{h}{2}} - \frac{1}{3}\lambda_h.$$

5. Wavelet methods

In terms of Haar functions $h_{j,k}$ defined by

$$h_{j,k}(x) = \begin{cases} 2^{-\frac{j}{2}} & \text{for } 2^j(k-1) < x < 2^j(k-\frac{1}{2}), \\ -2^{-\frac{j}{2}} & \text{for } 2^j(k-\frac{1}{2}) \le x < 2^j k, \\ 0 & \text{elsewhere,} \end{cases} \tag{5.1}$$

which is supported in the dyadic interval $I_{j,k} = [2^j(k-1), 2^j k]$, we can expand $k(x,y)$ in (1.2) as follows:

$$k(x,y) = \sum_{I,I'} \alpha_{II'} h_I(x) h_{I'}(y) + \sum_{I,I'} \beta_{II'} h_I(x)\chi_{I'}(y) + \sum_{I,I'} \gamma_{II'}\chi_I(x) h_{I'}(y), \tag{5.2}$$

where the sum extends over all dyadic squares $I \times I'$ with $|I| = |I'|$, and where

$$\alpha_{II'} = \int\int k(x,y) h_I(x) h_{I'}(y) dx dy, \tag{5.3}$$

$$\beta_{II'} = \int\int k(x,y) h_I(x)\chi_{I'}(y) dx dy, \tag{5.4}$$

$$\gamma_{II'} = \int\int k(x,y)\chi_I(x) h_{I'}(y) dx dy, \tag{5.5}$$

$$\chi_{I_{j,k}}(x) = \begin{cases} |I_{j,k}|^{-\frac{1}{2}} & \text{for } x \in I_{j,k}, \\ 0 & \text{elsewhere.} \end{cases} \tag{5.6}$$

$|I_{j,k}|$ denotes the length of $I_{j,k}$.

Given $u \in L^2(R)$ and an interval $I \subset R$, we define

$$d_I = \int_{-\infty}^{\infty} u(x) h_I(x) dx, \tag{5.7}$$

$$s_I = \int_{-\infty}^{\infty} u(x)\chi_I(x) dx. \tag{5.8}$$

Substituting (5.2) into (1.1), we obtain

$$Ku = \sum_I h_I \sum_{I'} \alpha_{II'} d_{I'} + \sum_I h_I \sum_{I'} \beta_{II'} s_{I'} + \sum_I \chi_I \sum_{I'} \gamma_{II'} d_{I'}. \tag{5.9}$$

Using the Haar system and substituting (5.9) into (4.5), one can obtain error estimates of eigenvalue approximants [5].

6. Remarks

1. If the kernel is smooth enough, we have

$$\lambda - \lambda_h^{(k)} = O(h^{2(k+1)}),$$

where

$$\lambda_h^{(0)} = \lambda_h$$

and

$$\lambda_h^{(k)} = \frac{2^{2k}\lambda_{\frac{h}{2}}^{(k-1)} - \lambda_h^{(k-1)}}{2^{2k}-1} \quad (k = 1, 2, \dots)$$

is the kth extrapolation solution.

2. Under certain conditions on $k(x, y)$, we will obtain error estimates of eigenvalue approximants by applying the results in [1].

References

1. Beylkin, G., Coifman, R., and Rokhlin, V., *Fast wavelet transforms and numerical algorithms I*, Comm. Pure Appl. Math. **44** (1991), 141-183.
2. Bramble, J.H. and Hilbert, S.R., *Estimation of linear functionals on Sobolev spaces with application to Fourier transforms and spline interpolation*, SIAM J. Numer. Anal. **7** (1970), 112-124.
3. Bramble, J.H. and Hilbert, S.R., *Bounds for a class of linear functionals with applications to Hermite interpolation*, Numer. Math. **16** (1971), 362-369.
4. Lin, E.B. and Shen, X., *Approximation methods for eigenvalue problems of Fredholm integral equations*, to appear in Proceedings of First World Congress of Nonlinear Analysis.
5. Lin, E.B., *On collocation and wavelet methods of eigenvalue approximations of Fredholm integral operators*, in preparation.
6. Lin, Qun and Liu Jiaquan, *Extrapolation method for Fredholm integral equations with non-smooth kernels*, Numer. Math. **35** (1980), 459-464.
7. Lin, Q., Sloan, I.H. and Xie, R., *Extrapolation of the iterated-collocation method for integral equations of the second kind*, SIAM J. Numer. Anal. **27** (1990), 1535-1541.
8. Sloan, I.H., *Iterated Galerkin method for eigenvalue problems*, SIAM J. Numer. Anal. **13** (1976), 753-760.
9. Sloan, I.H., *Superconvergence*, "Mathematical Concepts and Methods in Science and Engineering," Plenum Press, New York, 1990, pp. 35-70.

Department of Mathematics, University of Toledo, Toledo, OH 43606

E-mail address:elin@uoft02.utoledo.edu

Proceedings of Symposia in Applied Mathematics
Volume **48**, 1994

Spectral Methods for Singular Perturbation Problems

WENBIN LIU AND TAO TANG

ABSTRACT. Galerkin spectral methods are investigated for singularly perturbed boundary value problems for ordinary differential equations possessing boundary layers. It is known numerically that if the boundary layer is very thin, then boundary layer resolving spectral methods (BLRSMs), which are based on coordinate stretching and spectral methods, are more efficient than the conventional spectral approximations. Theoretical analysis for the BLRSMs is very difficult, owing to the degenerate characters of the transformed equations. Among essential questions to be answered is the following: what are the gains over the conventional spectral methods by applying the BLRSMs? It is the purpose of this work to examine this problem for the Galerkin spectral approximations. We introduce a new class of trial functions, which are not necessarily polynomials, to establish a variational formulation. Our error analysis results show that the BLRSMs are indeed more efficient than the conventional spectral methods.

1. Introduction

We consider the Galerkin spectral method for the following singular perturbation boundary value problem:

$$(1.1)\qquad -\epsilon u''(x)+q(x)u(x)=f(x,\epsilon),\quad x\in(-1,1),\quad u(\pm 1)=0,$$

where $\epsilon>0$ denotes a fixed constant, q and f are smooth functions on $[-1,1]$, $q\geq 0$ and $\|f(\cdot,\epsilon)\|_{L^\infty(-1,1)}$ is independent of ϵ. In many applications, (1.1) possesses boundary layers at the end points with widths of order $O(\epsilon^\gamma)$, where $\gamma>0$. It has been found that spectral methods are attractive in solving this problem (see, e.g., [**1**]).

It has also been observed that if $\epsilon \ll 1$ (e.g. $\epsilon<10^{-6}$) and the problem possesses a boundary layer with width $O(\epsilon)$, then a high accuracy cannot be

1991 *Mathematics Subject Classification.* Primary 65N35; Secondary 65L10.
This work was supported in part by NSERC Canada Grant OGP0105545.
The final version of this paper will be submitted for publication elsewhere.

© 1994 American Mathematical Society
0160-7634/94 $1.00 + $.25 per page

expected no matter how stable the spectral method is ([**2**]). However, by using a suitable coordinate transformation it is possible to achieve a high accuracy for very small ϵ. In fact, it is demonstrated in [**3**] that the Chebyshev spectral methods together with coordinate stretching can resolve a boundary layer as small as $O(10^{-10})$.

In this work we shall present a new BLRSM and give a theoretical interpretation of the high-accuracy behavior of the spectral methods with coordinate transformations. To this end, let $x = g(y)$ with $g \in C^\infty[-1,1]$ be such that $g(\pm 1) = \pm 1$ and $J(y) = g'(y) > 0$ in $(-1,1)$. With this change of variable applied to (1.1), the new equation reads as follows:

$$-\epsilon(a(y)v(y))' + Q(y)v(y) = F(y,\epsilon), \quad y \in (-1,1), \quad v(\pm 1) = 0, \tag{1.2}$$

where $Q(y) = J(y)q(g(y)), F(y,\epsilon) = J(y)f(g(y),\epsilon)$ and $a(y) = 1/J(y)$. To obtain a finer boundary resolution by the transformation, it is necessary that $J(-1) = g'(-1) = 0$, or $J(1) = g'(1) = 0$, or both. Hence, $a(y)$ is not even bounded near the boundary. This causes serious difficulties to the error analysis for the spectral approximation to the solutions of (1.2). It is the purpose of this paper to find suitable trial function spaces for the Galerkin spectral approximations to the solutions of (1.2) and then carry out some error analysis. We compare the error bounds for such approximations with those obtained using the conventional Galerkin spectral approximations on (1.1). For ease of exposition we only examine the case of a constant weight, but our approach is also applicable to the case of the Chebyshev weight.

2. Galerkin spectral method

Note that $a(y) = 1/J(y)$. It is easy to see that there exists an $\alpha > 0$ such that $a(y) \geq \alpha$ for any $y \in (-1,1)$, and that Q and F are smooth with $Q > 0$. The weak formulation of (1.2) cannot be given either in $H_0^1(-1,1)$ or $H_{a,0}^m(-1,1)$ since the integrals $\int_{-1}^1 av^2dy$ or $\int_{-1}^1 a(v')^2dy$ may not exist in these spaces. It is clear that the weak formulation can be established in the space

$$\tilde{H}_0^1(-1,1) = \left\{ v \in H_0^1(-1,1) : \|v\|_{L_J^2(-1,1)} + \|v'\|_{L_a^2(-1,1)} < \infty \right\},$$

which is nothing but the image space of $H_0^1(-1,1)$ under the transformation $(Gu)(y) = u(g(y))$. The weak formulation is stated as follows: Find $v \in \tilde{H}_0^1(-1,1)$ such that for any $w \in \tilde{H}_0^1(-1,1)$,

$$\epsilon \int_{-1}^1 av'w'dy + \int_{-1}^1 Qvwdy = \int_{-1}^1 Fwdy. \tag{2.1}$$

It can be shown by using the Lax–Milgram Theorem that (2.1) is well posed in $\tilde{H}_0^1(-1,1)$.

It is essential to find suitable trial function spaces in order to approximate the solution of (2.1) properly in $\tilde{H}_0^1(-1,1)$. Let $X_N = \{u_N \in \pi_N(-1,1) : u_N(\pm 1) =$

$0\}$, where π_N is the set of polynomials of degree not exceeding N. It is improper to consider the Galerkin approximation for (2.1) in X_N for the same reason as stated before. Let

$$\tilde{X}_N = \{v \in C[-1,1] : v' = JP \text{ with } P \in \pi_N(-1,1), v(\pm 1) = 0\}.$$

It is clear that $\tilde{X}_N$ is a N-dimensional subspace of $\tilde{H}_0^1(-1,1)$. We consider now the following Galerkin approximation to (1.2): Find $v_N \in \tilde{X}_N$ such that for any $w \in \tilde{X}_N$,

$$\epsilon \int_{-1}^{1} a v_N' w' dy + \int_{-1}^{1} Q v_N w dy = \int_{-1}^{1} F w dy. \tag{2.2}$$

It follows from the Lax–Milgram Theorem that (2.2) is well posed in $\tilde{X}_N$. The main results of the present work are given by the following theorem.

THEOREM 1. *Let $u(x)$ be the unique solution of (1.1) and $v_N(y)$ be the solution of (2.2). Assume that there is a $\beta > 0$ such that $C_1 \le J(y)(1-y^2)^{-\beta} \le C_2$ for any $y \in (-1,1)$. Then the following estimates hold:*

$$\begin{aligned} &\epsilon \|u' - \underline{v}_N'\|^2_{L^2(-1,1)} + \|u - \underline{v}_N\|^2_{L^2(-1,1)} \\ \le\ & C\left(N^{-2}\epsilon + N^{-4}\right)\left(\int_{-1}^{1} (u_x')^2 dx + \int_{-1}^{1} (u_x'')^2 \underline{J}^2 dx\right), \end{aligned} \tag{2.3}$$

$$\begin{aligned} &\epsilon \|u' - \underline{v}_N'\|^2_{L^2(-1,1)} + \|u - \underline{v}_N\|^2_{L^2(-1,1)} \\ \le\ & C\left(N^{-4}\epsilon + N^{-6}\right)\Bigg(\int_{-1}^{1} (u_x')^2 dx \\ & + \int_{-1}^{1} (u_x'')^2 \underline{J}^2 dx + \int_{-1}^{1} (u_x''')^2 \underline{J}^4 dx + \int_{-1}^{1} (u_x'')^2 (\underline{J}')^2 dx\Bigg), \end{aligned} \tag{2.4}$$

where $\underline{v}_N(x) = v_N(g^{-1}(x)), \underline{J}(x) = J(g^{-1}(x))$ and $\underline{J}'(x) = J_y'(g^{-1}(x))$.

The most useful feature of the above results is that as $\epsilon \to 0$ the dominating terms on the right-hand sides of (2.3) and (2.4) can be controlled by choosing a suitable J. This is the essential difference between the BLRSMs and the conventional spectral methods. This seems the key gain of the BLRSMs for singular perturbation problems and can be demonstrated by the following example.

EXAMPLE 1. Assume that there are positive constants α, ν, C such that for $x \in [-1,1]$

$$|u^{(i)}(x)| \le C + C\epsilon^{-i/2}\left(e^{-\alpha(1-x)/\sqrt{\epsilon}} + e^{-\nu(1+x)/\sqrt{\epsilon}}\right), \quad i = 1, 2, \dots,$$

where u is the solution of (1.1).

The above assumption is indeed satisfied for many equations of the Helmholtz type. Let us consider the transformation $x = g(y) = \sin\left(\frac{\pi}{2}y\right)$, which was introduced in [**3**]. Straightforward calculation yields that $J(y) = \frac{\pi}{2}\cos\left(\frac{\pi}{2}y\right)$ and $\underline{J}(x) = J(g^{-1}(x)) = \frac{\pi}{2}\sqrt{1-x^2}$. Therefore,

$$\begin{aligned}
&\int_{-1}^{1}(u'_x)^2dx + \int_{-1}^{1}(u''_x)^2\underline{J}^2dx \\
\leq\ & C + C\epsilon^{-1}\int_{-1}^{1}\left(e^{-2\alpha(1-x)/\sqrt{\epsilon}} + e^{-2\nu(1+x)/\sqrt{\epsilon}}\right)dx \\
&+C\epsilon^{-2}\int_{-1}^{1}\left(e^{-2\alpha(1-x)/\sqrt{\epsilon}} + e^{-2\nu(1+x)/\sqrt{\epsilon}}\right)(1-x^2)dx \\
\leq\ & C + C\epsilon^{-1/2} + C\epsilon^{-1} \leq C\epsilon^{-1}.
\end{aligned}$$

It follows from (2.3) of Theorem 1 that

$$\epsilon\|u' - \underline{v}'_N\|^2_{L^2(-1,1)} + \|u - \underline{v}_N\|^2_{L^2(-1,1)} \leq C\left(N^{-2} + N^{-4}/\epsilon\right).$$

The above upper bound can be further improved by using the transformation $x = g(y) = \sin\left(\frac{\pi}{2}\sin\left(\frac{\pi}{2}y\right)\right)$, which gives an upper bound of the order $O(N^{-2}\epsilon^{1/4} + N^{-4}/\epsilon^{3/4})$. When using conventional Galerkin spectral methods to (1.1), one can only expect an upper bound of $O(N^{-2}/\sqrt{\epsilon} + N^{-4}/\epsilon^{3/2})$ (cf. [**1**]). Hence, there is a good improvement in using the BLRSMs when ϵ is sufficiently small.

The error analysis methods introduced in this work are applicable to Chebyshev Galerkin spectral or pseudospectral methods as well. The related work will be reported elsewhere.

References

1. C. Canuto, *Spectral methods and maximum principle*, Math. Comp. **51** (1988), 615–629.
2. H. Eisen and W. Heinrichs, *A new method of stabilization for singular perturbation problems with spectral methods*, SIAM J. Numer. Anal. **29** (1992), 107–122.
3. T. Tang and M. R. Trummer, *Boundary layer resolving pseudospectral methods for singular perturbation problems*, Research Report 93–06, Dept. of Math. and Stat., Simon Fraser University, B.C., Canada, 1993.

Department of Mathematics, Imperial College, London SW7 2BZ, England
E-mail address: w.liu@imperial.ac.uk

Department of Mathematics and Statistics, Simon Fraser University, Burnaby, B.C., Canada V5A 1S6
E-mail address: tangtao@cs.sfu.ca

Proceedings of Symposia in Applied Mathematics
Volume **48**, 1994

A Quaternion-Jacobi Method for Symmetric Matrices

NILOUFER MACKEY

ABSTRACT. The algebra isomorphism between $\mathcal{M}_4(\mathcal{R})$ and $\mathcal{H}\otimes\mathcal{H}$, where $\mathcal{H}$ is the algebra of quaternions, has unexpected computational payoff: it helps construct an orthogonal similarity that diagonalizes a 4×4 symmetric matrix. By replacing plane rotations with these more powerful 4×4 rotations, we obtain a quaternion-Jacobi method in which the 'weight' of 6 elements (instead of one) is transferred all at once onto the diagonal. When the initial matrix has distinct eigenvalues, the new method is proved to converge under any quasi-cyclic ordering.

1. Introduction

Recently, Derek Hacon [**11**] used quaternions to construct an orthogonal similarity transformation that directly reduces a given 4×4 skew-symmetric matrix to its real Schur (2×2 block diagonal) form, thereby obtaining a Jacobi-type algorithm for $n \times n$ skew-symmetric matrices. The method is intriguing, and it is natural to ask whether it can be extended to symmetric matrices.

We show that the algebra isomorphism between $\mathcal{M}_4(\mathcal{R})$ and $\mathcal{H}\otimes\mathcal{H}$, where $\mathcal{H}$ is the algebra of quaternions, has direct and unexpected computational payoff: it leads to the construction of an orthogonal matrix that diagonalizes a 4×4 symmetric matrix. Replacing the plane rotations used in Jacobi's method [**15**] by these more powerful 4-dimensional rotations leads to a quaternion-Jacobi method in which six times as many zeros are created at each iteration. The price we pay for this abundance of zeros is the cost of computing one right-left singular vector pair of a 3×3 matrix, whose entries are easily specifiable in terms of those of the 4×4 symmetric matrix being diagonalized. When the

1991 *Mathematics Subject Classification.* Primary 65F15, 15A18, 15A21; Secondary 15A69.

Key words and phrases. Eigenvalues, symmetric matrix, Jacobi method, quaternion, tensor product.

An earlier version of this paper won the 1993 SIAM Student Paper Competition. This brief note is a partial summary of [**16**] and [**17**].

© 1994 American Mathematical Society
0160-7634/94 $1.00 + $.25 per page

initial matrix has distinct eigenvalues, the new method is proved to converge under any quasi-cyclic ordering.

There is renewed interest in Jacobi-type methods today, because they are easily parallelizable (for example, see [**2, 8, 9**]), and compute small eigenvalues with greater accuracy than the QR method [**6, 19**].

2. The isomorphisms

We refer the reader to [**1, 5, 7, 20**] for the basic facts concerning quaternions and tensor products of finite-dimensional algebras. We will denote the algebra of quaternions by $\mathcal{H}$, and the subspace of pure quaternions by $\mathcal{P}$.

2.1. $\mathcal{P}\otimes\mathcal{P}$ and $\mathcal{M}_3(\mathcal{R})$. The unique linear extension $\psi : \mathcal{P}\otimes\mathcal{P} \longrightarrow \mathcal{M}_3(\mathcal{R})$ of the bilinear map $\mathcal{P}\times\mathcal{P} \longrightarrow \mathcal{M}_3(\mathcal{R})$, $(p,q) \mapsto pq^t$ gives a vector space isomorphism between the 9-dimensional spaces $\mathcal{P}\otimes\mathcal{P}$ and $\mathcal{M}_3(\mathcal{R})$. Note that $\psi(p\otimes q) = pq^t$.

2.2. $\mathcal{H}\otimes\mathcal{H}$ and $\mathcal{M}_4(\mathcal{R})$. To every ordered pair (p,q) in $\mathcal{H}\times\mathcal{H}$, associate the real 4×4 matrix that represents the real linear transformation $\mathcal{H}\to\mathcal{H}$, $v\mapsto pv\bar{q}$ with respect to the standard basis $\{1,i,j,k\}$. Denote this matrix by $\mu(p,q)$. This defines a bilinear map $\mu : \mathcal{H}\times\mathcal{H} \longrightarrow \mathcal{M}_4(\mathcal{R})$, which from the basic properties of tensor product induces a unique *linear* map

$$\phi : \mathcal{H}\otimes\mathcal{H} \to \mathcal{M}_4(\mathcal{R})$$

with the property that $\phi(p\otimes q) = \mu(p,q)$. The map ϕ is an *algebra* isomorphism [**20**], with the additional property that it takes *conjugation in $\mathcal{H}\otimes\mathcal{H}$ to transpose in $\mathcal{M}_4(\mathcal{R})$.*

3. Quaternion representations

We prepare the ground before transplanting the problem of orthogonally diagonalizing a 4×4 symmetric matrix into a corresponding problem in $\mathcal{H}\otimes\mathcal{H}$.

3.1. Canonical Forms. Symmetric and diagonal matrices in $\mathcal{M}_4(\mathcal{R})$ have easily specifiable quaternion representations in $\mathcal{H}\otimes\mathcal{H}$:

- S is symmetric if and only if there exists $c\in\mathcal{R}$ and pure quaternions p,q and r such that $S = c1\otimes 1 + p\otimes i + q\otimes j + r\otimes k$.
- D is diagonal if and only if there exists $c_0,c_1,c_2,c_3 \in \mathcal{R}$ such that $D = c_0 1\otimes 1 + c_1 i\otimes i + c_2 j\otimes j + c_3 k\otimes k$.

Since $\phi(c1\otimes 1) = cI$, we can restrict our attention to symmetric matrices of the form $p\otimes i + q\otimes j + r\otimes k$, i.e., to elements of $\mathcal{P}\otimes\mathcal{P}$.

3.2. Rotations and Similarities. Within a year of the discovery of the quaternions, Hamilton and Cayley had found a connection between rotations of $\mathcal{R}^3$ and quaternions [**3, 7, 12**]. About a decade later, in 1855, Cayley observed that every rotation of $\mathcal{R}^4$ can be represented by pairs of quaternions [**4, 7**]. In particular, every rotation of 3-space can be expressed as $\phi(x\otimes x)$ and every rotation of 4-space as $\phi(x\otimes y)$, where x and y are unit quaternions. It then follows that *the effect of a* 4×4 *special orthogonal similarity on a* 4×4 *matrix can be reduced to the* independent *action of two 3-dimensional rotations.*

4. The symmetric algorithm

The results developed so far immediately lead to the construction of a 4×4 orthogonal matrix that directly reduces a given 4×4 skew-symmetric matrix to its real Schur form, recovering Hacon's method [**11**]. By contrast, the symmetric case is more complicated and requires a new idea.

The problem is that the tensor representation coming directly from the quaternion basis (see §3.1) is the "wrong" one – it reduces the problem of diagonalizing S to that of rotating the triple $\{p, q, r\}$ onto the orthogonal triple $\{i, j, \pm k\}$. Since the first triple is in general oblique, this is impossible.

We manoeuvre out of this apparent impasse by seeking a different quaternion representation of S. This is where the vector space isomorphism ψ introduced in §2.1 and the singular value decomposition (SVD) [**10**] fit beautifully together to produce the "right" quaternion representation of S in $\mathcal{H}\otimes\mathcal{H}$.

THEOREM 4.1. *Given a* 4×4 *symmetric matrix* $S \in \mathcal{P}\otimes\mathcal{P}$*, let* u, v *be left and right singular vectors corresponding to a singular value of the* 3×3 *matrix* $\psi(S)$*. Then* $\phi(x \otimes y) \cdot S \cdot \phi(\overline{x} \otimes \overline{y})$ *is* 2×2 *block diagonal, where*

$$x = \frac{1 - iu}{\mid 1 - iu \mid}, \quad y = \frac{1 - iv}{\mid 1 - iv \mid}.$$

For the convenience of the reader we translate this theorem into matrix form. Let $S = [s_{lm}]$ be a real 4×4 symmetric matrix. Let u, v be left and right singular vectors corresponding to a singular value of the 3×3 matrix

$$\begin{pmatrix} \frac{1}{4}(s_{11} + s_{22} - s_{33} - s_{44}) & \frac{1}{2}(s_{23} + s_{14}) & \frac{1}{2}(s_{24} - s_{13}) \\ \frac{1}{2}(s_{23} - s_{14}) & \frac{1}{4}(s_{11} - s_{22} + s_{33} - s_{44}) & \frac{1}{2}(s_{34} + s_{12}) \\ \frac{1}{2}(s_{24} + s_{13}) & \frac{1}{2}(s_{34} - s_{12}) & \frac{1}{4}(s_{11} - s_{22} - s_{33} + s_{44}) \end{pmatrix}$$

Then RSR^t is 2×2 block-diagonal, where R is the product of the two (commuting) orthogonal matrices

$$\frac{1}{\sqrt{(1+u_1)^2 + u_2^2 + u_3^2}} \begin{pmatrix} 1+u_1 & 0 & -u_3 & u_2 \\ 0 & 1+u_1 & u_2 & u_3 \\ u_3 & -u_2 & 1+u_1 & 0 \\ -u_2 & -u_3 & 0 & 1+u_1 \end{pmatrix}$$

and

$$\frac{1}{\sqrt{(1+v_1)^2 + v_2^2 + v_3^2}} \begin{pmatrix} 1+v_1 & 0 & v_3 & -v_2 \\ 0 & 1+v_1 & v_2 & v_3 \\ -v_3 & -v_2 & 1+v_1 & 0 \\ v_2 & -v_3 & 0 & 1+v_1 \end{pmatrix}.$$

Preliminary investigations indicate that the cost of computing the (partial) SVD of a 3×3 matrix is well worthwhile. Complete diagonalization can be achieved in a number of ways, e.g. by performing two Jacobi rotations[1].

5. Sorting similarities and convergence

If the singular vectors u, v of Theorem 4.1 correspond to the *largest* singular value, then the eigenvalues of the block-diagonalized 4×4 matrix are sorted, i.e., the eigenvalues of the upper block are larger than those of the lower block. This choice is analogous to picking the sorting angle in the traditional Jacobi method, and is relevant for two reasons. First, Mascarenhas [**18**] shows that the sorting angle can lead to significant improvement in performance of the Jacobi method. Second, sorting similarities lead to a globally convergent quaternion-Jacobi method [**16**]:

THEOREM 5.1. *Let A_0 be an $n \times n$ symmetric matrix with distinct eigenvalues $\lambda_1 > \lambda_2 > \cdots > \lambda_n$, and let $A_{i+1} = R_i A_i R_i^t$ be the sequence of iterates of the quaternion-Jacobi method, under any quasi-cyclic ordering. If each R_i effects a sorting, diagonalizing similarity on its target 4×4 submatrix, then the sequence $\{A_i\}$ converges to* $\mathrm{diag}[\lambda_1, \lambda_2, \ldots, \lambda_n]$. *Further, the rate of convergence is asymptotically quadratic.*

Quasi-cyclic orderings were introduced in [**13**]. The proof of the theorem uses a majorization result due to Schur [**14**, pp. 192-193], and an SVD perturbation theorem due to Wedin [**21, 22**].

6. Preliminary experimental results

We present below the results of a MATLAB implementation of Jacobi and quaternion-Jacobi methods on a random 64×64 symmetric matrix with entries uniformly distributed in $[-1, 1]$. This behavior seems to be typical, but more experimentation is needed.

$n = 64$	*Odd Even Ordering; Sorting Similarity*			
	Jacobi		*quaternion-Jacobi*	
Sweep	Number of 2×2 Rotations	BlockOff	Number of 4×4 Rotations	BlockOff
0		$1.800e+01$		$1.800e+01$
1	2016	$1.067e+01$	496	$1.032e+01$
2	2016	$2.883e+00$	496	$2.609e+00$
3	2016	$4.633e-01$	496	$3.301e-01$
4	2016	$1.840e-02$	496	$5.025e-03$
5	2016	$2.877e-05$	496	$1.261e-06$
6	2016	$8.111e-11$	496	$6.702e-14$
7	541	$1.218e-13$		

7. Acknowledgements

Special thanks to Professor P. J. Eberlein, for introducing me to Hacon's work and for suggesting the problem of extending it to the symmetric case. It is also a pleasure to thank Professor Eberlein, Nick Higham, Steve Mackey, Don Schack and Pete Stewart for stimulating discussions, and Cleve Moler and the

[1]In practice, when working on a large matrix, block-diagonalizing at each iteration suffices.

MATLAB team for providing such a convenient environment in which to test matrix algorithms.

REFERENCES

1. Garrett Birkhoff and Saunders MacLane, *A survey of modern algebra*, Macmillan, New York, 1977.
2. R. P. Brent and F. T. Luk, *The solution of singular-value and symmetric eigenvalue problems on multiprocessor arrays*, SIAM J. Sci. Statist. Comput. **6** (1985), 69–84.
3. Arthur Cayley, *On certain results relating to quaternions*, The Collected Mathematical Papers (New York), vol. 1, Johnson Reprint Company, New York, 1963, pp. 123–126. Originally published in The Philos. Mag. **26** (1845), 141–145.
4. ———, *Recherches ultérieures sur les déterminants gauches*, The Collected Mathematical Papers (New York), vol. 2, Johnson Reprint Company, New York, 1963, pp. 202–215. Originally published in J. Reine Angew. Math. **50** (1855), 299–313.
5. H. S. M. Coxeter, *Quaternions and reflections*, Amer. Math. Monthly **53** (1946), 136–146.
6. James Demmel and K. Veselić, *Jacobi's method is more accurate than QR*, SIAM J. Matrix Anal. Appl. **13** (1992), 1204–1245.
7. H. D. Ebbinghaus, H. Hermes, F. Hirzebruch, M. Keocher, K. Mainzer, J. Neukirch, A. Prestel, and R. Remmert, *Numbers*, Springer, New York, 1991.
8. P. J. Eberlein, *On the Schur decomposition of a matrix for parallel computation*, IEEE Trans. Computers **C-36** (1987), 167–174.
9. P. J. Eberlein and Haesun Park, *Efficient implementation of Jacobi algorithms and Jacobi sets on distributed memory architectures*, J. Parallel Distrib. Comput. **8** (1990), 358–366.
10. G. H. Golub and C. Van Loan, *Matrix computations*, 2nd ed., Johns Hopkins University Press, Baltimore, 1989.
11. Derek Hacon, *Jacobi's method for skew-symmetric matrices*, SIAM J. Matrix Anal. Appl. **14** (1993), 619–628.
12. William R. Hamilton, *On quaternions, or a new system of imaginaries in algebra; with some geometrical illustrations*, The Collected Mathematical Papers (Cambridge) (H. Halberstam and R. E. Ingram, eds.), vol. 3, Cambridge University Press, Cambridge, 1967, pp. 355–362. Originally published in *Proc. Royal Irish Academy* **3** (1847), 1–16, communicated November 1844.
13. P. Henrici, *On the speed of convergence of cyclic and quasicyclic Jacobi methods for computing eigenvalues of Hermitian matrices*, J. Soc. Indust. Appl. Math **6** (1958), 144–162.
14. R. A. Horn and C. R. Johnson, *Matrix analysis*, Cambridge University Press, New York, 1990.
15. C. G. J. Jacobi, *Über ein leichtes Verfahren die in der Theorie der Säcularstörungen vorkommenden Gleichungen numerisch aufzulösen*, J. Reine Angew. Math. **30** (1846), 51–94.
16. Niloufer Mackey, *Convergence of the quaternion-Jacobi method*, In preparation.
17. ———, *Hamilton and Jacobi meet again: Quaternions and the eigenvalue problem*, SIAM J. Matrix Anal. Appl., to appear.
18. Walter F. Mascarenhas, *On the convergence of the Jacobi method for arbitrary orderings*, Ph.D. thesis, Massachusetts Institute of Technology, 1991.
19. Roy Mathias, *Accurate eigensystem computations by Jacobi methods*, Preprint, June 1993.
20. Richard S. Pierce, *Associative algebras*, Springer, New York, 1982.
21. G. W. Stewart and Ji-guang Sun, *Matrix perturbation theory*, Academic Press, Boston, 1990.
22. P.-Å. Wedin, *Perturbation bounds in connection with singular value decomposition*, BIT **12** (1972), 99–111.

DEPARTMENT OF COMPUTER SCIENCE, STATE UNIVERSITY OF NEW YORK AT BUFFALO, AMHERST, NY 14260

E-mail address: mackey@cs.buffalo.edu

Proceedings of Symposia in Applied Mathematics
Volume 48, 1994

On Constructing Chebyshev Series Solutions of Differential Equations

Allan J. MacLeod

ABSTRACT. We describe a simple method for generating the system of equations for the Chebyshev series coefficients of the solution to a linear differential equation.

1. Introduction

Consider the equation

$$\sum_{i=0}^{n} p_i(x) \frac{d^i f(x)}{dx^i} = g(x), \quad -1 \le x \le 1, \tag{1.1}$$

where $p_i(x)$, $i = 0, \ldots, n$, are known polynomials, and $g(x)$ is a polynomial or a function with easily-found Chebyshev expansion.

We try to find the coefficients $c_0, c_1, \ldots$ in the expansion

$$f(x) = \frac{1}{2} c_0 T_0(x) + c_1 T_1(x) + c_2 T_2(x) + \ldots, \tag{1.2}$$

where $T_0\, T_1, T_2, \ldots$ are standard Chebyshev polynomials.

There are formulae available for the coefficients of $x^i f^{[j]}$ (see Gottlieb and Orszag [2] for example), but these are only for i and j small (usually ≤ 2). The derivation of results for larger combinations involves a lot of boring algebra.

Papers such as Horner [3] and Evans and Murphy [1] give extensive tables of algebraic identities satisfied by the coefficients, but these are restricted to specific equations of specified degree. Extension to general equations of general degree is not obvious.

1991 *Mathematics Subject Classification*. Primary 65D20, 65L60.

The detailed version of this paper will be submitted for publication elsewhere.

© 1994 American Mathematical Society
0160-7634/94 $1.00 + $.25 per page

2. Method

Let $\mathbf{c}$ and $\mathbf{t}$ be infinite vectors defined as

$$\mathbf{c}^T = (c_0, c_1, c_2, \ldots), \tag{2.1}$$

$$\mathbf{t} = (\tfrac{1}{2}T_0(x), T_1(x), T_2(x), \ldots)^T;$$

then $f(x) = \mathbf{c}^T\,\mathbf{t}$.

Chebyshev polynomials satisfy various well-known relationships. The first of these is

$$\begin{aligned} x\,\tfrac{1}{2}T_0 &= \tfrac{1}{2}T_1, \\ x\,T_1 &= \tfrac{1}{2}T_0 + \tfrac{1}{2}T_2, \\ x\,T_i &= \tfrac{1}{2}(T_{i-1} + T_{i+1}), \quad i \geq 2 . \end{aligned} \tag{2.2}$$

Define the infinite tridiagonal matrix X by $X_{ii} = 0$, $X_{01} = 1$, $X_{i(i-1)} = X_{i(i+1)} = \frac{1}{2}$, $i \geq 1$, assuming subscripts of matrices and vectors run from 0. Using this matrix, we easily see

$$\begin{aligned} x\,f(x) &= \mathbf{c}^T X^T\,\mathbf{t} \\ &= (X\mathbf{c})^T\,\mathbf{t} \end{aligned} \tag{2.3}$$

and, in general,

$$x^m f(x) = (X^m\,\mathbf{c})^T\,\mathbf{t} . \tag{2.4}$$

Similarly, the derivatives of T_i satisfy

$$\begin{aligned} (\tfrac{1}{2}T_0)' &= 0, \\ T_i' &= 2i\,(T_{i-1} + T_{i-3} + \ldots + T_1)\,,\ i \text{ even}, \\ T_i' &= 2i\,(T_{i-1} + T_{i-3} + \ldots + \tfrac{1}{2}T_0)\,,\ i \text{ odd}, \end{aligned} \tag{2.5}$$

and we can define a matrix D to be upper triangular with $D_{ii} = 0$, $D_{ij} = 0$, $j\text{-}i$ even, $D_{ij} = 2j$ otherwise.

Thus,

$$\frac{d^m f}{dx^m} = (D^m\mathbf{c})^T\,\mathbf{t} . \tag{2.6}$$

Applying (2.4) and (2.6) to (1.1), we derive the infinite system of equations

$$\sum_{i=0}^{n} p_i(X)\,D^i\,\mathbf{c} = \mathbf{g}, \tag{2.7}$$

where $\mathbf{g}$ is the vector of the Chebyshev coefficients of $g(x)$.

3. Initial Conditions

Allied to (1.1) will be n initial or boundary conditions, which can also be formulated in this vector form. As an example, consider the condition

$$af'(\alpha) + bf''(\beta) = \gamma \tag{3.1}$$

with α, β arbitrary points in $[-1, 1]$.

$$\text{Let} \quad \mathbf{t}_\alpha = \left(\tfrac{1}{2}T_0(\alpha), T_1(\alpha), T_2(\alpha), \ldots\right)^T,$$
$$\mathbf{t}_\beta = \left(\tfrac{1}{2}T_0(\beta), T_1(\beta), T_2(\beta), \ldots\right)^T;$$

then

$$f'(\alpha) = (D\mathbf{c})^T \mathbf{t}_\alpha = \mathbf{t}_\alpha{}^T D\mathbf{c},$$
$$f''(\beta) = (D^2\mathbf{c})^T \mathbf{t}_\beta = \mathbf{t}_\beta{}^T D^2\mathbf{c},$$

so

$$af'(\alpha) + bf''(\beta) = (a\mathbf{t}_\alpha{}^T D + b\mathbf{t}_\beta{}^T D^2)\,\mathbf{c}$$
$$= \mathbf{h}^T\mathbf{c} = \gamma,$$

where $\mathbf{h}^T = a\,\mathbf{t}_\alpha{}^T D + b\mathbf{t}_\beta{}^T D^2$.

For the n associated conditions let H be the matrix of n rows formed from the specific $\mathbf{h}^T$ vectors. So the initial/boundary conditions can be put in the form

$$H\,\mathbf{c} = \mathbf{r}, \tag{3.2}$$

where $\mathbf{r}$ is the nx1 vector of right-hand sides.

Let $A = \sum_{i=0}^{n} p_i(X)\, D^i$, so that (3.2) combined with (2.7) can be written

$$\begin{pmatrix} H \\ A \end{pmatrix} \mathbf{c} = \begin{pmatrix} \mathbf{r} \\ \mathbf{g} \end{pmatrix}, \tag{3.3}$$

which is the complete system for c.

4. Practical Implementation

In practice, we cannot work with infinite sets of coefficients and infinite matrices. Suppose we wish to estimate $c_0, c_1, \ldots, c_M$ say. What size of matrices (N) must we use? Let X_N, D_N be the upper left NxN segments of X and D, respectively. Let $A_N = \sum_{i=0}^{n} p_i(X_N)\, D_N{}^i$. We would like A_N to be identical to the upper left NxN segment of A.

For each i, the matrix $p_i(X_N)$ will be calculated by nested multiplication. Since X_N is tridiagonal, it is clear that the upper left (N + 1 - deg

(p_i)) square segment is exact. The nature of D means that this segment of $p_i\,(X_n)\,D_n$ is also exact.

Thus, we require

$$M + 1 \le N + 1 - \max\,(\deg\,(p_i))\,, \tag{4.1}$$

so that $N \ge M + \max\,(\deg\,(p_i))$.

With this restriction, we can easily write a program in MATLAB to construct the equations automatically.

To illustrate the power of the method, we consider the eigenvalue problem for the Orr-Sommerfeld equation, as discussed by Orszag [4]. The equation is

$$f^{(iv)} + (Rix^2 - (2 + Ri))\,f'' + (1 - Ri - Rix^2)\,f \;=\; \lambda Ri\,[f - f''] \tag{4.2}$$

where $R = 10000$ is the Reynolds number. Orszag considers even functions f, so there are no odd Chebyshev coefficients. We thus transform the equation using $t = 2x^2 - 1$, since $T_{2n}(x) = T_n\,(2x^2 - 1)$, $n = 0, 1, 2, \ldots$ The equation is now more complex, but still with polynomial coefficients. Applying the ideas of 2 and 3 enables us to calculate the eigenvalues quoted in Table 2 of Orszag in a total elapsed time of 57.3 seconds, using MATLAB 4.0 on a SUN SPARCstation IPC.

References

1. D. J. Evans and C. P. Murphy, *An automatic procedure for the solution of fourth order differential equations by Chebyshev polynomials*, Internat. J. Comput. Math. **38** (1991), 177-191.

2. D. Gottlieb and S. A. Orszag, *Numerical Analysis of Spectral Methods: Theory and Applications*, SIAM, Philadelphia, 1977.

3. T. S. Horner, *Recurrence relations for the coefficients in Chebyshev series solutions of ordinary differential equations*, Math. Comp. **35** (1980), 893-905.

4. S. A. Orszag, *Accurate solution of the Orr-Sommerfeld stability equation*, J. Fluid Mech. **50** (1971), 689-703.

Department of Mathematics and Statistics, University of Paisley, Paisley, Scotland, PA1 2BE

amcl@cs.paisley.ac.uk

Proceedings of Symposia in Applied Mathematics
Volume 48, 1994

Multiquadric collocation methods in the numerical solution of Volterra integral and integro-differential equations

ATHENA MAKROGLOU

ABSTRACT. Volterra integral equations and integro-differential equations with continuous kernel are considered and are solved by applying discrete collocation methods based on multiquadric approximations and quadrature rules. Applications to Hammerstein-type equations are also included with the methods applied in the light of the new collocation methods of Kumar and Sloan.

1. Introduction

This paper considers Volterra integral equations (VIEs) of the form

$$y(x) = f(x) + \int_0^x P(x,s)H(s,y(s))ds, \qquad 0 \le x \le X, \tag{1.1}$$

and Volterra integro-differential equations (VIDEs) of the form

$$y'(x) = G(x, y(x), z(x)), \qquad y(0) = y_0, \qquad 0 \le x \le X, \tag{1.2}$$

where

$$z(x) = \int_0^x K(x,\ s,\ y(s))ds. \tag{1.3}$$

1991 *Mathematics Subject Classification.* Primary 65R20 ; Secondary 65D05.

Key words and phrases. Numerical solution, Volterra integral and integro-differential equations, multiquadrics.

I wish to thank Dr. Edward Kansa for useful communications on the applications of multiquadrics. This paper was completed during a visiting appointment at the Department of Mathematics of the Oregon State University. I wish to express my thanks to the Department.

Detailed versions of parts of this paper will be submitted elsewhere.

© 1994 American Mathematical Society
0160-7634/94 $1.00 + $.25 per page

We apply discrete collocation methods based on multiquadrics (MQ). For equations of the form (1.1) the methods will employ the approach used in the so-called new collocation methods (Kumar and Sloan [**9**], also Brunner [**3**], Makroglou [**12**]).

The plan of this paper is as follows. Section 2 gives a description of the application of the MQ approximations to equations (1.1), (1.2). In §3 are given numerical results, and in §4 conclusions and comments on remaining problems.

2. Description of the methods

First we consider the integral equation (1.1). Setting $z(x) = H(x, y(x))$, we obtain the following integral equation in z:

$$z(x) = H(x, f(x) + \int_0^x P(x,s)z(s)ds). \tag{2.1}$$

Then $y(x)$ can be obtained from (1.1) by replacing $H(s, y(s))$ by $z(s)$. In a piecewise collocation method we consider mesh points $0 = x_0 < x_1 < \cdots < x_N = X$ and seek approximate values of the solution of (2.1) for $0 \le x_{nj} \le X$, where $x_{nj} = x_n + c_j h_n, h_n = x_{n+1} - x_n$ $(n = 0, 1, \ldots, N-1)$, and the collocation parameters are such that $0 \le c_1 < \cdots < c_m \le 1$ (for more details, see for example Brunner and van der Houwen [**2**], Brunner [**4**], Baker [**1**], Makroglou [**11**]).

The form of a multiquadric augmented by a constant term which interpolates a set of points $(c_i, y_i), i = 1, 2, \ldots, m$, is

$$S(x) = \sum_{j=1}^{m} \alpha_j \tilde{u}_j(x), \tag{2.2}$$

where

$$\tilde{u}_j(x) = \begin{cases} 1 & \text{if } j = 1, \\ u_j(x) - u_1(x) & \text{if } j > 1, \end{cases} \tag{2.3}$$

$$u_j(x) = \sqrt{(x - c_j)^2 + R^2} \tag{2.4}$$

and the coefficients α_j are obtained from the solution of the linear system of equations

$$S(c_i) = y_i, \qquad i = 1, 2, \ldots, m. \tag{2.5}$$

For details on multiquadric interpolation, we refer for example to Hardy [**6, 7**], Powell [**13, 14**], Madych and Nelson [**10**], Buhmann and Dyn [**5**], Kansa and Carlson [**8**] and the references therein.

Quadrature rules based on multiquadric approximations can be derived for the approximation of the integrals . They are of the form

$$\int_0^1 \phi(s)ds = \sum_{\mu=1}^{m} w_\mu \phi(c_\mu), \tag{2.6}$$

where

$$w_\mu = \int_0^1 l_\mu(s)ds = \sum_{\lambda=1}^{m} Q_{\mu\lambda} \int_0^1 \tilde{u}_\lambda(s)ds, \tag{2.7}$$

the $Q_{\mu\lambda}$ being the coefficients of $\phi(c_\mu)$ in the expression of α_λ obtained from solving the linear system

$$\sum_{\lambda=1}^{m} \alpha_\lambda \tilde{u}_\lambda(c_j) = \phi(c_j), \qquad j = 1, 2, ..., m,$$

$$\int u_j(s)ds = \frac{R^2}{4}[0.5D + \ln|D| - \frac{1}{2D}], \tag{2.8}$$

$$D = \frac{2(s-c_j)^2}{R^2} + 1 + \frac{2(s-c_j)\sqrt{(s-c_j)^2 + R^2}}{R^2} \tag{2.9}$$

and

$$\int \tilde{u}_j(s)ds = \begin{cases} \int u_j(s)ds - \int u_1(s)ds & \text{if } j > 1, \\ s & \text{if } j = 1. \end{cases} \tag{2.10}$$

Using the above type of approximations for $z(x)$ in $[x_n, x_{n+1}], n = 0, 1, \ldots, N-1$ and the quadrature rules (2.6)–(2.10), we obtain the following systems in the a_{nk}:

$$\sum_{k=1}^{m} a_{nk}\tilde{u}_k(c_j) = H(x_{nj}, f(x_{nj}) + \sum_{i=0}^{n-1} h_i \sum_{\lambda=1}^{m} w_\lambda P(x_{nj}, x_{i\lambda}) z_{i\lambda} \tag{2.11}$$

$$+h_n c_j \sum_{k=1}^{m} a_{nk} \sum_{\lambda=1}^{m} w_\lambda P(x_{nj}, x_n + h_n c_j c_\lambda)\tilde{u}_k(c_j c_\lambda)), j$$

$$= 1, 2, .., m; n = 0, 1, ..., N-1.$$

Consider now the Volterra integro-differential equations (1.2)–(1.3). Proceeding similarly with approximations of the type (2.2)–(2.4) for both $y(x)$ and $z(x)$, and using the quadrature rules (2.6)–(2.10), we obtain the following systems of equations in the a_{nk}, b_{nk}:

$$\sum_{k=1}^{m} a_{nk}\tilde{u}_k(c_j) = \sum_{i=0}^{n-1} h_i \sum_{l=1}^{m} w_l G(x_{il}, y_{il}, z_{il}) + y_0 \tag{2.12}$$

$$+h_n \sum_{l=1}^{m} w_l^j G(x_n + h_n c_l, \sum_{k=1}^{m} a_{nk}\tilde{u}_k(c_l), \sum_{k=1}^{m} b_{nk}\tilde{u}_k(c_l))$$

and

$$\sum_{k=1}^{m} b_{nk}\tilde{u}_k(c_j) = \sum_{i=0}^{n-1} h_i \sum_{l=1}^{m} w_l K(x_{nj}, x_{il}, y_{il}) \tag{2.13}$$

$$+h_n c_j \sum_{l=1}^{m} w_l K(x_{nj}, x_n + h_n c_j c_l, \sum_{k=1}^{m} a_{nk}\tilde{u}_k(c_j c_l)),$$

$$j = 1, 2, ..., m; \quad n = 0, 1, ..., N-1,$$

where

$$w_l^j = \int_0^{c_j} l_l(s)ds.$$

3. Numerical Results

Numerical results are presented for the following examples:

1) $y(x) = 1+\sin^2(x) - \int_0^x 3\sin(x-s)y^2(s)ds$ with true solution $y(x) = \cos x$ (cf. Brunner and van der Houwen [**2,** p. 507]);
2) $y'(x) = -\sin x + \frac{x}{3}[\cos^3(x)-1] + \int_0^x x\sin(s)y^2(s)ds, \quad y(0) = 1$ (constructed), with true solution $y(x) = \cos x$.

All computations were done in double precision.

The entries in Tables 1–3 represent maximum absolute errors for a number of stepsizes. The resulting nonlinear equations were solved by Newton's method. A tolerance equal to 1.D-10 was used. The results with heading "MQ" were obtained with the use of multiquadric quadrature rules (Eqs. 2.6–2.10). Example 1 was solved with both the Radau II points ($c_1 = 1/3, c_2 = 1$) and the Gauss points ($c_1 = \dfrac{3-\sqrt{3}}{6}, c_2 = \dfrac{3+\sqrt{3}}{6}$).

Table 1. Example 1, collocation methods, Radau II points
max. abs. error in $0 \le x \le 2$

h	R	Polynom. Spline	MQ
0.1	6	0.94 D − 4	0.32 D − 4
0.05	12	0.12 D − 4	0.39 D − 5
0.025	24	0.14 D − 5	0.48 D − 6
0.0125	48	0.18 D − 6	0.60 D − 7
0.00625	96	0.22 D − 7	0.75 D − 8
3.125 D − 3	192	0.28 D − 8	0.93 D − 9

Table 2. Example 1, collocation methods, Gauss points
max. abs. error in $0 \le x \le 5$

h	R	Iter.Polyn. Spline	MQ
0.1	6	0.63 D − 4	0.16 D − 4
0.05	12	0.40 D − 5	0.22 D − 6
0.025	24	0.25 D − 6	0.11 D − 7
0.0125	48	0.16 D − 7	0.13 D − 8
6.25 D − 3	96	0.99 D − 9	0.55 D − 9

Table 3. Example 2, collocation methods, Radau II points
max. abs. error in $0 \le x \le 5$

h	R	Polynom. Spline	MQ
0.1	19	0.37 D − 3	0.47 D − 5
0.05	38	0.47 D − 4	0.61 D − 6
0.025	76	0.60 D − 5	0.79 D − 7
0.0125	152	0.75 D − 6	0.99 D − 8
6.25 D − 3	304	0.95 D − 7	0.12 D − 8

4. Concluding Remarks

In this paper the application of collocation methods based on approximations with multiquadrics to Volterra integral and integro-differential equations was considered. The results obtained show that the new methods can produce very accurate results when used in combination with multiquadric quadrature rules. Orders of convergence $O(h^3)$ and $O(h^4)$ were observed with $m = 2$ for the Radau II and Gauss points, respectively. The accuracy of the results depends on the successful choice of the multiquadric parameter R. We are continuing research on the derivation of a method for choosing R, on an error analysis for obtaining the order of convergence, and on applications of the methods to equations with weakly-singular kernels.

References

1. Baker, C.T.H., *The Numerical Solution of Integral Equations*, Clarendon Press, Oxford, 1977.
2. Brunner, H. and Van der Houwen, P.J., *The numerical solution of Volterra equations*, CWI Monographs **3** (1986), North Holland, Amsterdam.
3. Brunner, H., *Implicitly linear collocation methods for nonlinear Volterra equations*, Appl. Numer. Math. **9** (1992), 235–247.
4. Brunner, H., *Iterated collocation methods and their discretizations for Volterra integral equations*, SIAM J. Numer. Anal. **20** (1984), 1132–1145.
5. Buhmann, M.D. and Dyn, N., *Error estimates for multiquadric interpolation*, pp. 51–58 in *Curves and Surfaces* , Laurent, P.J., LeMéhauté, A., Schumaker, L.L. (eds.), Academic Press, 1991.
6. Hardy, R.L., *Multiquadric equations of topography and other irregular surfaces*, J.Geophys. Res. **76** (1971), 1905–1915.
7. Hardy, R.L., *Theory and applications of the multiquadric-biharmonic method*, Comput. Math. Appl. **19** (1990), 163–208.
8. Kansa, E.J. , Carlson, R.E., *Improved accuracy of multiquadric interpolation using variable shape parameters*, Comput. Math. Appl. **24** (1992), 99–120.
9. Kumar, S. and Sloan, I.H., *A new collocation-type method for Hammerstein integral equations*, Math. Comp. **48** (1987), 585–593.
10. Madych, W.R. and Nelson, S.A., *Multivariate interpolation and conditionally positive definite functions. II*, Math. Comp. **54** (1990), 211–230.
11. Makroglou, A., *Convergence of a block-by-block method for nonlinear Volterra integro-differential equations*, Math. Comp. **35** (1980), 783–796.
12. Makroglou, A., *Radial basis functions in the numerical solution of Fredholm integral and integro-differential equations*, pp. 478–484 in *Advances in Computer Methods for Partial Differential Equations - VII* , Vichnevetsky, R., Knight, D., Richter, G. (eds.), Proceedings of 7th IMACS Conf. on Computer Methods for PDEs, IMACS, N.Brunswick, NJ, 1992.
13. Powell, M.J.D., *Univariate multiquadric interpolation : some recent results*, DAMTP 1990/NA9, Dept. of Applied Mathematics and Theoretical Physics, Univ. of Cambridge, England, 1990.
14. Powell, M.J.D., *Radial basis function approximations to polynomials*, pp. 223–241 in *Numerical Analysis* , Griffiths, D.F. and Watson, G.A. (eds.), Longman Scientific and Technical, Harlow, 1988.

DEPARTMENT OF MATHEMATICS, OREGON STATE UNIVERSITY, CORVALLIS, OR 97331, U.S.A.

Current address: School of Mathematical Sciences, University of Portsmouth, Mercantile House, Portsmouth PO1 2EG, England

E-mail address: makrogloua@cv.port.ac.uk

Proceedings of Symposia in Applied Mathematics
Volume 48, 1994

Methods for Solving Large Eigenvalue Problems Associated with Configuration Interaction Electronic Structure Calculations

KRISTYN J. MASCHHOFF

ABSTRACT. Two extensions of the Inverse-Iteration Generalized Davidson method are proposed. First, a method is given to simultaneously solve for multiple eigenpairs. Second, a hybrid of this method and the original Generalized Davidson method is explored.

1. Introduction

Large-Scale Configuration Interaction calculations require the solution to very large ($N = 10^6$ to $N = 10^9$) sparse symmetric eigenvalue problems. A standard method for finding the smallest few eigenvalues and corresponding eigenvectors for CI matrices is the Davidson method [**1**]. This method involves generating a sequence of vectors which are then used within a Ritz procedure to obtain eigenpair estimates. One concern which has become more important with the advent of parallel machines is the balance between I/O time and computational time. With the increase in computational speed provided by parallel machines it has become necessary to limit the maximum number of expansion vectors used to prevent I/O bottlenecks. Many space-saving methods have recently been proposed to address this issue [**3, 4, 5**]. One such method is the Newton-like Inverse-Iteration Generalized Davidson method proposed by Jeppe Olsen [**4**]. This method avoids storing a sequence of correction vectors by directly adding the correction to the current eigenvector estimate. This correction is obtained by a perturbation expansion of the CI eigenvalue equation. Two extensions of the Inverse-Iteration Generalized Davidson method are proposed here. First a method is given to simultaneously solve for several eigenpairs. Second, a hybrid

---

1991 *Mathematics Subject Classification.* Primary 65F15; Secondary 65F50, 81-08.
This research was supported by the U.S. Department of Energy through the Pacific Northwest Laboratory, operated by Battelle Memorial Institute under contract DE-AC-76RLO 1830.
The final version of this paper will be submitted for publication elsewhere.

© 1994 American Mathematical Society
0160-7634/94 $1.00 + $.25 per page

of this method and the original Davidson method is explored. Numerical results for these extensions look encouraging.

2. Olsen method

The Olsen algorithm is a Newton-like method for finding the smallest eigenvalue, E, and corresponding eigenvector, C, of the configuration interaction (CI) system

$$HC = EC, \tag{1}$$

where H is the $N \times N$ Hamiltonian matrix. A first-order perturbation expansion of equation (1) gives

$$(H^0 + H^1)(C^0 + C^1) = (E^0 + E^1)(C^0 + C^1),$$

where H^1, C^1, and E^1 are "small" with respect to H^0, C^0, and E^0, respectively. Neglecting quadratic terms, and solving for the correction vector C^1, yields

$$C^1 = -(H^0 - E^0 I)^{-1}[R - E^1 C^0], \tag{2}$$

where R is the residual vector $R = (H - E^0 I)C^0$. Note that both C^1 and E^1 are unknown quantities. By adding the restriction that C^1 be orthogonal to C^0, C^1 and E^1 can be solved for explicitly: multiplying equation (2) by C^{0T}, we can express E^1 in terms of known quantities,

$$E^1 = \frac{C^{0T}(H^0 - E^0 I)^{-1} R}{C^{0T}(H^0 - E^0 I)^{-1} C^0}, \tag{3}$$

and this value is then used in equation (2) to obtain C^1.

INVERSE-ITERATION GENERALIZED DAVIDSON (OLSEN METHOD)

Generate initial guess vector $C^0 = C^{0(0)}$ (normalized)
For $i = 0, 1, 2, \ldots$

- Compute $\sigma = HC^{0(i)}$
- Calculate the energy estimate using the Rayleigh quotient
  $$E^{0(i)} = C^{0T(i)} H C^{0(i)} = C^{0T(i)} \sigma$$
- Compute the residual vector
  $$R = (H - E^{0(i)} I)C^{0(i)} = \sigma - E^{0(i)} C^{0(i)}$$
- Compute the residual norm — Exit if solution is within tolerance
- Compute the correction energy E^1 using equation (3)
- Construct the correction vector C^1 using equation (2)
- Form the new approximation $C^{0(i+1)} = C^{0(i)} + C^1$ and normalize.

3. Multi-root extension

Suppose the solution of interest is now the smallest K eigenvalues and their corresponding eigenvectors. The CI system to be solved now looks like

$$HC = CE, \tag{4}$$

where C is the $N \times K$ matrix of eigenvectors, and E is a $K \times K$ diagonal matrix of eigenvalues. The approach taken for developing a multi-root method was to combine the original Olsen method with a Ritz method. The size of the subspace from which the eigenpair estimates are obtained equals the number of roots desired and remains fixed. The subspace is then iteratively updated by applying the Olsen correction scheme to each eigenpair approximation individually. The previous subspace is overwritten each iteration. The basis vectors in the subspace are not typically orthogonal to each other after applying the Olsen correction. This complicates the Ritz procedure in that either the solution to a generalized eigenvalue problem is required or the basis set must be orthogonalized after each iteration. For most electronic structure calculations, the overlap matrix for the basis vectors remains positive definite after correction. Solution of the generalized eigenvalue problem is therefore preferred over direct reorthogonalization. The advantages of the multi-root algorithm with respect to the single-root algorithm should be similar to the advantages of subspace iteration with respect to the power method. Multi-root methods are especially important in case of degenerate energy states. A more thorough study of these methods is ongoing.

MULTI-ROOT INVERSE-ITERATION GENERALIZED DAVIDSON

Generate set of initial guess vectors $C^0 = C^{0(0)}$
For $i = 0, 1, 2, \ldots$

- Form the overlap matrix $S = C^{0(i)T} C^{0(i)}$
- Form the subspace matrix $G = C^{0(i)T} H C^{0(i)}$
- Solve the generalized eigenvalue problem $G\alpha = E^{(i)} S\alpha$
- Form the eigenvectors corresponding to these energies and overwrite the current basis set $(C^{0(i)} := C^{0(i)}\alpha)$
- Form the K residual vectors and check each norm for convergence
- Update: If the jth root is not converged, update the jth basis vector, $c_j^{0(i)}$, using the Olsen correction $c_j^{0(i+1)} = c_j^{0(i)} + c_j^1$ and normalize.

4. Numerical results

The matrix used for initial experimentation is a variation of the Modified Nesbet Matrix attributed to Raffenetti (1979):

$$H_{ij} = H_{ji} = 0.1, \qquad i \neq j,$$

$$H_{ii} = \begin{cases} 10 + 0.1(i-1), & 1 \leq i \leq 5, \\ 2i - 1, & 6 \leq i \leq 50. \end{cases}$$

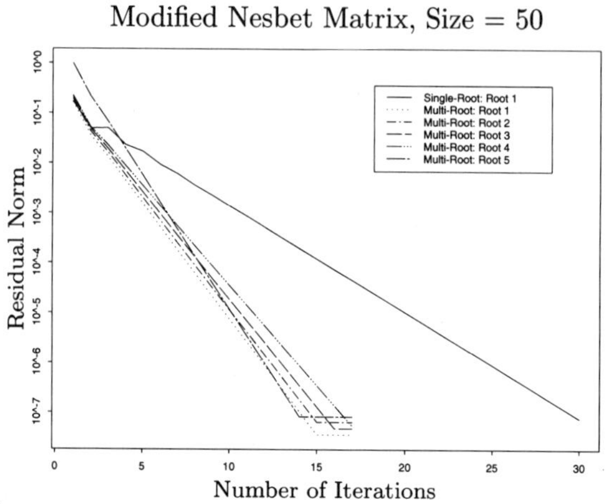

FIGURE 1. Comparison of multi-root and single-root Olsen methods for finding the five lowest eigenvalues. Six basis vectors are included for the multi-root with the sixth root updated only to a tolerance of 10^{-2}

This matrix provides a fairly difficult test because of the clustering of the smallest eigenvalues. This matrix is also representative of the diagonally dominant structure of large CI matrices. The structure of H^0 used here includes the leading 5×5 subblock plus the diagonal. See Figure 1.

5. Observed difficulties

The derivation of the Olsen algorithm does not place any restriction on the form of H^0 except that its perturbation from the full H matrix be considered small. In practice, however, the form of H^0 must be chosen such that the algorithm is practical to implement and efficient computationally. A simple form of H^0 which can be easily implemented is to use a small upper block plus the remaining diagonal of the Hamiltonian matrix H. The dimension of the upper block is chosen such that the eigenstructure of the small block approximates the eigenstructure of the full H matrix for the lower roots. Including the small upper block only requires the solution to two small linear systems in addition to the work required if only the diagonal was used.

For the Olsen algorithm the choice of H^0 is critical. A bad choice not only can slow convergence, but may result in a divergent solution. In addition, an appropriate starting vector must also be chosen carefully. If the matrix H^0 is a good representation of H, then using the eigenvectors of H^0 corresponding to the K smallest eigenvalues as starting vectors should be adequate. The difficulty in using the Olsen method is determining an appropriate form of H^0.

The multi-root algorithm does perform better than the single-root method, especially if the root desired is the lowest root. However, if the energies for higher states are desired, a more reliable method may be required. Numerical studies using moderately large CI matrices have indicated that the multi-root algorithm may not be very reliable for adequately describing the excited state energies. Clustering of the excited state energies makes the usage of perturbation expansions tricky. The observed difficulties with the higher energy states prompted the exploration of a hybrid method.

6. Hybrid method

Combining the Newton-like idea of the Olsen method with the subspace expansion of the Davidson method may provide a viable space-saving method for determining the excited state energies under severe memory and I/O restrictions. The hybrid algorithm is similar to the current Davidson algorithm, but uses the Olsen correction vector

$$(H^0 - EI)^{-1}(R - E^1C^{(i)})$$

instead of the generalized Davidson expansion vector

$$(H^0 - EI)^{-1}R,$$

to expand the basis set. Forming the Olsen correction vector requires additional work per iteration, but if the total number of iterations required to reach the solution is reduced even moderately, this additional expense may be worthwhile. The benefit of using the Olsen correction vector instead of the Davidson expansion vector is most apparent when the maximum size of the expansion space is severely limited.

Acknowledgments. The author would like to thank Rick Kendall, Robert Harrison, and Rik Littlefield of PNL, and Ken Bube of UW, for many useful discussions.

References

1. E. R. Davidson, *The iterative calculation of a few of the lowest eigenvalues and corresponding eigenvectors of large real-symmetric matrices,* J. Comput. Phys. **17** (1975), 87-94.
2. R. B. Morgan and D. S. Scott, *Generalizations of Davidson's method for computing eigenvalues of sparse symmetric matrices,* SIAM J. Sci. Statist. Comput. **7** (1986), 817-825.
3. C. W. Murray, S.C. Racine, and E.R. Davidson, *Improved algorithms for the lowest few eigenvalues and associated eigenvectors of large matrices,* J. Comput. Phys. **103** (1992), 382-389.
4. J. Olsen, P. Jorgensen, and J. Simons, *Passing the one-billion limit in full configuration-interaction (FCI) calculations,* Chem. Phys. Lett. **169** (1990), 463-472.
5. J.H. van Lenthe and P. Pulay, *A space-saving modification of Davidson's eigenvector algorithm,* Comput. Chem. **11** (1990), 1164-1168.

DEPT. OF APPLIED MATHEMATICS, UNIVERSITY OF WASHINGTON, SEATTLE, WA 98195
E-mail address: kristi@amath.washington.edu

Proceedings of Symposia in Applied Mathematics
Volume 48, 1994

Computing Limiting Normals to Real Surfaces

Donal O'Shea and Les Wilson

ABSTRACT. Effective procedures for computing limiting normals to real surfaces are discussed. Examples are given to show that the behavior in the real case is sharply different than what one might expect from the complex case.

1. Introduction

A surface which is given parametrically can be sketched rapidly and accurately. There are even some reasonably effective programs for sketching nonsingular surfaces which are given implicitly. However, the problem of developing good algorithms to systematically sketch an implicitly-given surface near a singularity has been largely untouched.

In this paper, we examine the behavior of limits of normals at a singular point of a real algebraic surface in $\mathbf{R}^3$. There is a nice theory for complex surfaces, which singles out computable algebraic objects furnishing detailed qualitative information on the geometric behavior of a surface near a singular point (see [1,2,3]). Were this theory to carry over to the reals, it would provide a valuable means to algorithmically derive qualitative features of the surface. This in turn would likely be an indispensable first step in devising a program to sketch the surface in a neighborhood of the singular point. In this paper, we consider the extent to which the complex theory carries over to the real setting.

We give a number of examples showing that the complex results require nontrivial reformulation before they can be expected to hold over the reals. We provide suitably formulated versions of some of the complex results which do, in fact, hold over the reals, and make some remarks about how these reformulations affect computability.

2. The Complex Theory

For the convenience of the reader, and to motivate the development that follows, we provide a very quick account of the complex theory with careful references. This theory holds very generally for complex analytic spaces of any dimension and codimension. For our purposes, we restrict ourselves to complex surfaces $V \subset \mathbf{C}^3$ and to real surfaces $W \subset \mathbf{R}^3$.

1991 *Mathematics Subject Classification*. Primary 14J17, 14Q10; Secondary 32C05.
This paper is in final form and no version of it will be submitted for publication elsewhere.

© 1994 American Mathematical Society
0160-7634/94 $1.00 + $.25 per page

Let $V \subset \mathbf{C}^3$ be a surface and $0 \in V$ a singular point. The *tangent cone* $C(V,0)$ to V at 0 is defined to be the set of points on limits of (secant) lines $\overline{0x}$ through 0 and x as x on V tends to 0. We think of this as follows. Let ϕ map $V - \{0\}$ to the set $\mathbf{P}^2$ of lines through 0:

$$\phi : x \mapsto \overline{0x}, \quad x \in V - \{0\},$$

and consider its graph $\operatorname{gr}\phi \subset (V - \{0\}) \times \mathbf{P}^2$. The closure $\operatorname{cl}(\operatorname{gr}\phi) \subset V \times \mathbf{P}^2$ projects onto V. Call this projection π. Then, $\pi^{-1}(0) = \{0\} \times C(V,0)$. It is known classically that if $f = 0$ is a local equation for V near 0, then $f_h = 0$ is a local equation for $C(V,0)$, where f_h is the nonzero homogeneous part of f of lowest degree (in the expansion of f about 0). Thus the tangent cone is a geometric object which is easy to compute algebraically.

Moreover, the tangent cone provides an approximation to a surface near a singular point in the sense that, given any conical neighborhood of the tangent cone, a sufficiently small neighborhood of the surface lies entirely within the conical neighborhood. See [6] for a precise statement. The difficulty with the tangent cone is that it is too coarse – it gives no indication of a number of features of a surface about which one would want to know, and entirely dissimilar surfaces may have the same tangent cone. For example the surfaces $\{y^2 + x^3 - z^2x^2 = 0\}$ and $\{y - x^2 - z^2 = 0\}$ have the same tangent cone, namely the plane $y = 0$.

A more subtle invariant is the space $K(V,0)$ of limiting tangent spaces to V at 0. Since $V \subset \mathbf{C}^3$ (and later $\mathbf{R}^3$), if $x \in V$ is a nonsingular point of V, then the tangent space $T(V,x)$ to V at x is a plane, which we label by its "conormal vector", $df(x)$, whose components are just the partials of f evaluated at x (its complex conjugate is normal to $T(V,x)$ with respect to the usual Hermitian metric on $\mathbf{C}^3$). Let Σ denote the singular set of V and let ψ map $V - \Sigma$ to the set $\check{\mathbf{P}}^2$ of planes in $\mathbf{C}^3$ through 0 by

$$\psi : x \mapsto [df(x)], \quad x \in V - \Sigma,$$

where $[df(x)]$ denotes the canonical image of the vector $df(x)$ in projective space (that is, the set of all nonzero scalar multiples of $df(x)$). Again, we consider the closure of the graph of ψ and let π be the map onto V induced by projection. We define $K(V,0)$ by demanding that

$$\pi^{-1}(0) = \{0\} \times K(V,0).$$

We call an element of $K(V,0)$ a *limiting tangent space* to V at 0. A moment's reflection will convince one that a limiting tangent space to the tangent cone of a variety "ought" to be a limiting tangent space to the variety itself: that is, $K(C(V,0)) \subset K(V,0)$. This inclusion was established by Whitney [7]. However, as Whitney points out, this inclusion can be, and typically is, strict. The question of characterizing the "extra" limiting tangent spaces is answered by a structure theorem due to Lê and Henry [1] for isolated surface singularities, and to Lê [2] for arbitrary surface singularities.

Theorem. *There exists a finite (possibly empty) set of lines* $l_1, \ldots, l_r \subset C(V,0)$, $0 \in l_i$ *for all* $1 \leq i \leq r$ *(called* **exceptional lines**), *such that*

$$K(V,0) = K(C(V,0),0) \cup \cup_{i=1}^{r} K(l_i),$$

where $K(l_i)$ *denotes the set of all planes in* $\mathbf{C}^3$ *containing* l_i.

Thus, one knows $K(V,0)$, given $C(V,0)$ and the exceptional lines. The data {tangent cone, exceptional lines} are called the *aureole* of V at 0 and signal important geometric features of the surface near 0. For example, the z-axis is an exceptional line for $\{y^2 + x^3 - z^2x^2 = 0\}$, and this signals the "vanishing fold". The fact that the z-axis is a line of singular points is incidental: one can easily concoct similar examples in which the surface has an isolated singularity at the origin.

This being the case, one would like a procedure to compute the exceptional lines (the tangent cone being trivial to compute). There are several procedures which allow one to do this. For instance, one can use Lê and Teissier's result that the exceptional lines are the lines in the tangent cones to polar curves of $(V,0)$ which do not vary as the projection varies (the so-called "non-moving" lines) coupled with an elimintation-theoretic procedure (see [3,5]). In practice, it often turns out to be easier to compute equations for $K(V,0)$ and to find the exceptional lines by primary decomposition. If $f = 0$ is a local equation for V about 0 and if f_x, f_y, f_z are the partial derivatives of f, then one finds equations for $K(V,0)$ by eliminating the variable s in the ideal $\langle f, u - sf_x, v - sf_y, w - sf_z \rangle$ (in $\mathbf{C}[x,y,z,s,u,v,w]$) and setting $x = y = z = 0$. One can do the elimination and the primary decomposition by Gröbner basis arguments (see [5]).

3. Real Surfaces

How much of the theory outlined above for the complex case goes over to real case? It turns out that most of the strictly geometric results go over with some modification. However, the more algebraic results (and, hence, the correspondence between algebra and geometry) which render the complex theory so powerful break down.

To see this, let $W = \{x \in \mathbf{R}^3 : f(x) = 0\}$, where f is a polynomial in three real variables which vanishes at the origin. On the one hand, it is fairly natural to take the real tangent cone of W to be the set of real zeros of f_h (here, as before, f_h is the lowest-degree nonvanishing homogeneous piece of f). Call the set $\{x \in \mathbf{R}^3 : f_h(x) = 0\}$ defined in this manner to be the *real algebraic tangent cone* and denote it as $C_a(W,0)$. Likewise, we define the set of *algebraic real exceptional lines* to be the lines in the intersection of the exceptional lines of the complexification $W_{\mathbf{C}} = \{x \in \mathbf{C}^3 : f(x) = 0\}$ with the real tangent cone.

On the other hand, one wants to define the real tangent cone so that it coincides as a set with the set, call it the *geometric tangent cone* $C_g(W,0)$, of points on limiting secant lines. Similarly, we would like to define the set of limiting tangent spaces to be the set of all limits of real tangent spaces to the real nonsingular points of W.

Unfortunately, the two definitions are not the same. For example, if we set $W = \{z^8 = x^4 + y^2\}$, then $C_a(W,0) = \{y = 0\}$ is the xz-plane. However, W

can be parameterized by setting $(x,y,z) = (u, v^2, \sqrt[8]{u^4+v^4})$. So, given any set of points on W tending to the origin, there corresponds a sequence of points (u_k, v_k) tending towards $(0,0)$ and

$$\frac{(u_k, v_k^2, \sqrt[8]{u_k^4+v_k^4})}{\|(u_k, v_k^2, \sqrt[8]{u_k^4+v_k^4})\|} \longrightarrow (0,0,\pm 1).$$

Thus, any limiting secant lies in the direction of the z-axis, and $C_g(W,0)$ is the z-axis.

Not only do the algebraic and geometric definitions of tangent cone differ in this case, but they also have different dimensions: $\dim_0 W = \dim C_a(W,0) = 2$ while $\dim C_g(W,0) = 1$. In particular, the dimension of the geometric tangent cone can be strictly less than the dimension of the algebraic tangent cone. It is not difficult to show that the geometric tangent cone is always contained in the algebraic tangent cone, and that both are semialgebraic sets.

Experience with the behavior of the tangent cone suggests that one must be cautious when considering limits of tangent spaces. Namely, one might expect that the intersection of the limits of tangent spaces to the complexification of a real surface with $\mathbf{R}^3$ could be strictly greater than the limit of real tangent spaces to the real surface. This does, indeed, happen – even in the case of coherent irreducible real surfaces.

Before giving an example, let us record the definitions we will use. The *space of geometric limiting tangent spaces* to a real surface W at 0 will be just the limit of tangent spaces as we have earlier defined it, $K_g(W,0) = K(W,0)$, where we only consider real tangent planes (at real points) of W. We can express this slightly differently as follows. Let TW^0 denote the tangent bundle to the smooth part W^0 of W. Then we can write

$$K_g(W,0) = \pi_{\check{\mathbf{RP}}^3}\left(\overline{\{(x, T_x W) \in TW^0\}} \cap \pi_W^{-1}(0)\right),$$

where the overbar denotes closure. This representation of K_g as the projection of an algebraic set makes it clear that K_g is semialgebraic. We define the *space of algebraic limiting tangent spaces* $K_a(W,0)$ of W to be the set of real planes whose complexifications can be obtained as limits of tangent spaces to the complexification of W:

$$K_a(W,0) = K(W_{\mathbf{C}}, 0) \cap (\mathbf{RP}^3).$$

The above discussion implies that if $W \subset \mathbf{R}^3$ is a real surface with $0 \in W$, then $K_g(W,0) \subset K_a(W,0)$ and $K_a(W,0)$ is a real algebraic set while $K_g(W,0)$ is semialgebraic (see [6]).

Now for the example we promised. Let $W = \{z^2 = y^2(x^4+y^2)\}$. We can parameterize W by setting $x = u, y = v^2, z = \pm v^2\sqrt{u^4+v^4}$. Setting $X(u,v) = (x,y,z)$, and computing partials with respect to u and v, we have

$$X_u \times X_v = 2v\left(\frac{\mp 2v^2u^3}{\sqrt{u^4+v^4}}, \frac{\mp u^4 + 2v^4}{\sqrt{u^4+v^4}}, 1\right).$$

Since $\sqrt{u^4+v^4}$ is essentially order 2, it is clear that

$$\frac{X_u \times X_v}{\|X_u \times X_v\|} \longrightarrow (0,0,1)$$

along any path in which $(u,v) \longrightarrow (0,0)$. So there is a unique limiting tangent plane at $(0,0)$ and $K_g(W,0) = \{(0,0,1)\}$.

On the other hand, it is clear from the above parameterization that the y-axis is an exceptional line for $W^{\mathbf{C}}$ (choose limits along paths for which $\sqrt{u^4+v^4}$ has order greater than 2). So $K_a(W,0) \supset \{\text{all planes containing the } x\text{-axis}\}$ (and a closer analysis establishes that the containment relation is actually an equality). Finally, it is easy to check that W is coherent and irreducible.

Here is an even more illuminating example. Let $W = \{z^4y^2 + y^4 - x^4 = 0\}$. Then a computation using a parameterization of W shows that $K_g(W,0)$ consists of those planes whose normals lie in the xy-plane and which meet the z-axis in an angle with cosine less than or equal to $2/3^{3/4}$ (see [6] for details).

Here, the z-axis certainly deserves to be called an exceptional line (and is an exceptional line for the complexification of W), however, $K_g(W,0)$ does not contain the entire pencil of planes containing the line.

In a forthcoming paper, we will show that an analogue of the theorem cited above holds for the $K_g(W,0)$ where W is real. Namely, we show that there exists a finite (possibly empty) set of lines $l_1,\dots,l_r \subset C_g(W,0), 0 \in l_i$ for all $1 \le i \le r$ (called **exceptional lines**), such that $K_g(W,0)$ is the union of $K_g(C_a(W_{\text{smooth}},0),0)$ and a finite number of sets $K_i(l_i)$, where $K_i(l_i)$ is some real semialgebraic subset of the set of all planes in $\mathbf{R}^3$ containing l_i.

In that paper, we also present some methods for computing the exceptional lines, based on the methods in [5]. Unfortunately, these methods are slow. We also do not, as yet, have a reliable procedure for calculating $K_i(l_i)$, the subset of the pencil of planes containing l_i that are actually realized as limits of real tangent planes.

References

1. J.-P. Henry, Lê Dung Trang, *Limites d'espaces tangents*, in *Séminaire Norguet*, Lecture Notes in Math. **482**, Springer, 1975.

2. Lê Dung Trang, *Limites d'espaces tangents sur les surfaces*, Nova Acta Leopoldina, N.F. **52**, Nr. 240 (1981), 119-137.

3. Lê Dung Trang, B. Teissier, *Limites d'espaces tangents en géometrie analytique*, Comment. Math. Helv. **63** (1988), 540-578.

4. D. Mumford, *Algebraic Geometry I: Complex Projective Varieties*, Springer-Verlag, Berlin, New York, 1976.

5. D. O'Shea, *Computing limits of tangent spaces*, to appear.

6. D. O'Shea, L. Wilson, *Limits of tangent spaces to real surfaces*, preprint, 1993.

7. H. Whitney, *Tangents to an analytic variety*, Ann. of Math. **81** (1965), 496-549.

Mathematics Department, Mount Holyoke College, South Hadley, MA 01075
E-mail address: doshea@mtholyoke.edu

Mathematics Department, University of Hawaii, Honolulu, HI 96822
E-mail address: les@math.hawaii.edu

Proceedings of Symposia in Applied Mathematics
Volume 48, 1994

Orthogonal Spline Collocation Solution of Nonlinear Schrödinger Equations

MARK P. ROBINSON

ABSTRACT. Many phenomena in physics are modeled as nonlinear Schrödinger equations. An orthogonal spline collocation semidiscretization, previously applied in the numerical solution of the cubic Schrödinger equation, is extended to the solution of more complex nonlinear Schrödinger equations. These equations contain general power nonlinearities and nonlinear terms that include derivatives. The accuracy of the method is examined, as well as its ability to conserve certain invariants associated with the theoretical solution. The results of numerical experiments are presented in which the time stepping is carried out using a routine from a software library.

1. Introduction

Spline collocation methods have been extensively employed in the numerical solution of differential equations. One such method proved to be very effective in solving the cubic Schrödinger equation (CSE), an equation which has been examined in numerous numerical studies, many listed in [**5**]. We consider the extension of the method in [**5**] to two equations, of which the CSE is a special case.

The first equation is the Schrödinger equation with general power nonlinearity (GPSE) [**2**],

$$iu_t + u_{xx} + q|u|^{p-1}u = 0, \quad (x,t) \in (-\infty, \infty) \times (0, T], \tag{1.1}$$

where $i^2 = -1$ and, in the cases we consider, q is a positive constant and $p > 1$ is an integer. The second is the generalized nonlinear Schrödinger equation (GNLSE),

$$iu_t + u_{xx} + q_c|u|^2u + q_q|u|^4u + iq_m\left(|u|^2\right)_x u + iq_u|u|^2u_x = 0, \tag{1.2}$$

1991 *Mathematics Subject Classification.* Primary 65M70, 65M20; Secondary 65M15.

A portion of this work was supported by a Summer Research Fellowship from Western Kentucky University's Faculty Research Committee.

The final version of this paper will be submitted for publication elsewhere.

© 1994 American Mathematical Society
0160-7634/94 $1.00 + $.25 per page

$(x,t) \in (-\infty,\infty) \times (0,T]$, where q_c, q_q, q_m and q_u are real constants. Certain cases of this equation are important in applications (cf. [**4**] and references therein). The finite difference solution of (1.1) was examined in [**2**], and four pseudospectral methods for solving (1.2) were studied in [**4**]. In each case the initial condition $u(x,0)$ is given by some function $u_0(x)$, and the solution has the property that $|u(x,t)| \to 0$ as $|x| \to \infty$. For the initial value problem (IVP) associated with (1.1),

$$A_1 = \int_{-\infty}^{\infty} |u|^2 dx \quad \text{and} \quad A_2 = \int_{-\infty}^{\infty} \left(|u_x|^2 - \frac{2q}{p+1}|u|^{p+1} \right) dx \tag{1.3}$$

are conserved in time (cf. [**2**]). For (1.2), conserved quantities include A_1 and

$$\hat{A}_2 = \int_{-\infty}^{\infty} \left(|u_x|^2 - \frac{1}{2}Q|u|^2 \mathrm{Im}(u\bar{u}_x) - \frac{1}{2}q_c|u|^4 + \frac{1}{6}[q_m Q - 2q_q]|u|^6 \right) dx,$$

where $Q = 2q_m + q_u$ (cf. [**4**]). The IVP is replaced by an initial-boundary value problem (IBVP) by restricting x to an interval $[a,b]$, chosen so that $|u(x,t)|$ is negligible for x outside $[a,b]$, and imposing homogeneous Dirichlet boundary conditions at $x = a$ and $x = b$. The IBVP for (1.1) is

$$iu_t + u_{xx} + q|u|^{p-1}u = 0, \quad (x,t) \in (a,b) \times (0,T], \tag{1.4a}$$

$$u(x,0) = u_0(x), \quad x \in (a,b), \tag{1.4b}$$

$$u(a,t) = u(b,t) = 0, \quad t \in [0,T], \tag{1.4c}$$

and that for (1.2) is constructed similarly.

2. Solution of the GPSE

Let $I = (a,b)$, let $\Delta : a = x_1 < x_2 < \cdots < x_{N+1} = b$ be a partition of $\overline{I}$, assumed to belong to a quasi-uniform family, define $I_j = (x_j, x_{j+1})$ and $h_j = x_{j+1} - x_j$ for $j = 1,2,\ldots,N$, and set $h = \max_{1 \le j \le N} h_j$. For $r \ge 3$, let $M_r(\Delta) = \{v \mid v \in C^1(I),\ v \in P_r(\overline{I}_j),\ j = 1,2,\ldots,N\} \cap \{v|v(a) = v(b) = 0\}$, where $P_r(\overline{I}_j)$ is the set of all polynomials of degree at most r on $\overline{I}_j$. Let $\{\lambda_k\}_{k=1}^{r-1}$ be the nodes for the $(r-1)$-point Gaussian quadrature rule on the interval $(0,1)$ with corresponding weights $\{w_k\}_{k=1}^{r-1}$, $w_k > 0$, and set $\lambda_{jk} = x_j + h_j\lambda_k$ for $j = 1,2,\ldots,N$ and $k = 1,2,\ldots,r-1$. We define $\{\cdot,\cdot\}$ by $\{f,g\} = \sum_{j=1}^{N}\{f,g\}_j$, where $\{f,g\}_j = h_j \sum_{k=1}^{r-1} w_k f(\lambda_{jk})\overline{g}(\lambda_{jk})$, and set $|f|_D^2 = \{f,f\}$.

The continuous-time orthogonal spline collocation approximation to the solution u of (1.4) is a differentiable map $U : [0,T] \to M_r(\Delta)$ such that

$$iU_t(\lambda_{jk},t) + U_{xx}(\lambda_{jk},t) + q|U(\lambda_{jk},t)|^{p-1}U(\lambda_{jk},t) = 0, \tag{2.1}$$

for $j = 1,2,\ldots,N$ and $k = 1,2,\ldots,r-1$, and for $t \in (0,T]$, with an appropriate initial condition. An estimate of the error in this approximation is given in the following theorem.

THEOREM 2.1. *Let u be the solution to (1.4) such that $u \in L^\infty(0,T;H^{r+3}(I))$ and $u_t \in L^2(0,T;H^{r+3}(I))$, and let U be the solution to (2.1). Define the differentiable map $W : [0,T] \to M_r(\Delta)$ by $-W_{xx}(\lambda_{jk},t) + W(\lambda_{jk},t) = -u_{xx}(\lambda_{jk},t) + u(\lambda_{jk},t)$ for $j = 1,2,...,N$ and $k = 1,2,...,r-1$. If $\|U - W\|_{L^\infty(I)}(0) \le Ch^{r+1}$, then, for h sufficiently small, $\|u - U\|_{L^\infty(0,T;L^2(I))} \le C(p)\{\|U - W\|_{L^2(I)}(0) + h^{r+1}[\|u\|_{L^\infty(0,T;H^{r+3}(I))} + \|u_t\|_{L^2(0,T;H^{r+3}(I))}]\}$.*

The proof employs the same technique as that of Theorem 3.1 in [**5**] and makes use of the equivalence of (2.1) and the discrete Galerkin method

$$i\{U_t, v\} + \{U_{xx}, v\} + q\{|U|^{p-1}U, v\} = 0, \quad v \in M_r(\Delta). \tag{2.2}$$

Furthermore, $\frac{d}{dt}|U|_D^2 = 0$ and $\frac{d}{dt}\left[\|U_x\|_{L^2(I)}^2 - \frac{2q}{p+1}\left||U|^{(p+1)/2}\right|_D^2 + \psi\right] = 0$, where $\psi = C(r)\sum_{j=1}^N h_j^{2r-1}[\partial^r(\mathrm{Re}U + \mathrm{Im}U)/\partial x^r]|_{I_j}$. The first result is obtained by setting $v = U$ in (2.2) and taking the imaginary part, and the second by setting $v = U_t$ in (2.2) and taking the real part. Note the similarity to the two quantities in (1.3).

We implement the method with $r = 3$ and express the approximate solution using the local representation [**1**]

$$U(x,t) = \sum_{l=1}^{4} \frac{U_{jl}(t)(x - x_j)^{l-1}}{(l-1)!}, \quad x \in \overline{I}_j, \tag{2.3}$$

where $U_{jl}(t) = (\partial^{l-1}U/\partial x^{l-1})(x_j^+, t)$ for $j = 1,2,...,N$ and $l = 1,2,3,4$. We also define $U_{N+1,1}(t) = U(x_{N+1},t)$ and $U_{N+1,2}(t) = U_x(x_{N+1},t)$. We take the Hermite piecewise cubic interpolant of $u_0(x)$ as the initial approximation $U(x,0)$. It is shown in [**5**] that in this case $\|U - W\|_{L^\infty(I)}(0) \le Ch^4\|u_0\|_{H^6(I)}$, so that the hypotheses of Theorem 2.1 are satisfied. The collocation equations (2.1) with U given by (2.3), combined with the algebraic equations from the continuity and boundary conditions, lead to a differential-algebraic system $D\dot{\mathbf{U}}(t) = E\mathbf{U}(t) + q\mathbf{F}(\mathbf{U}(t))$, where a dot denotes differentiation with respect to t, $\mathbf{U}$ contains the real and imaginary parts of the coefficients $U_{jl}(t)$, interlaced, D and E are almost block diagonal matrices, and $\mathbf{F}$ contains the contribution of the nonlinear term. The system is solved using the code D02NNF from the NAG library, employed successfully in [**5**] and [**6**], in which the time stepping is done using backward differentiation formula integrators. All computations were performed in double precision on the IBM 3090-600J at the University of Kentucky.

We consider (1.1) with $q = 2$ and an exact solution $u(x,t) = \Phi(x - 4t)\exp(2i(x - \frac{7}{8}t))$, where $\Phi(x) = \left\{\frac{9(p+1)}{16}\mathrm{sech}^2\left[\frac{3(p-1)}{4}(x+5)\right]\right\}^{1/(p-1)}$ (cf. [**2**]) for $p = 3$ and $p = 5$. In each case the solution is a solitary wave, initially centered at $x = -5$, traveling to the right with velocity 4. When $p = 3$ the amplitude is 1.5; $A_1 = 3$ and $A_2 = 9.75$. In the quintic problem, the amplitude is $(27/8)^{1/4} \approx 1.355$, $A_1 = \pi\sqrt{6}/4 \approx 1.924$, and $A_2 = \pi\sqrt{6} \approx 7.695$.

TABLE 1. Results from solution of GPSE test problem with $p = 3$

t	A_1'	A_2'	L^2 error
0.0	2.997617	9.735139	0.001794
6.0	2.997567	9.734905	0.076685

TABLE 2. Results from solution of GPSE test problem with $p = 5$

t	A_1'	A_2'	L^2 error
0.0	1.921723	7.678105	0.002949
6.0	1.921903	7.683737	1.442422

We solve the problems on $[-30, 30] \times [0, 6]$. Certain output at $t = 0$ and $t = 6$ is presented: A_1' and A_2', approximations to A_1 and A_2, computed using Gauss quadrature to evaluate the integrals $\int_{-30}^{30} |U|^2 dx$ and $\int_{-30}^{30} \left(|U_x|^2 - \frac{2q}{p+1}|U|^{p+1}\right) dx$, respectively; an approximation to the L^2 error, computed using Gauss quadrature; and the graph of $|U|$. Highly satisfactory results are obtained when the cubic problem is solved using $N = 200$, as shown in Table 1 and Figure 1a. The results when the quintic problem is solved using the same value of N are shown in Table 2 and Figure 1b. The solitary wave behavior of the solution deteriorates somewhat; some dispersion is evident at $t = 6$, as the maximum modulus of the solution declines and the wave becomes broader. The growth of the L^2 error is far greater than in the cubic case. Since A_1' and A_2' remain nearly constant, this shows that close adherence to conservation laws is not sufficient to guarantee accuracy of the solution.

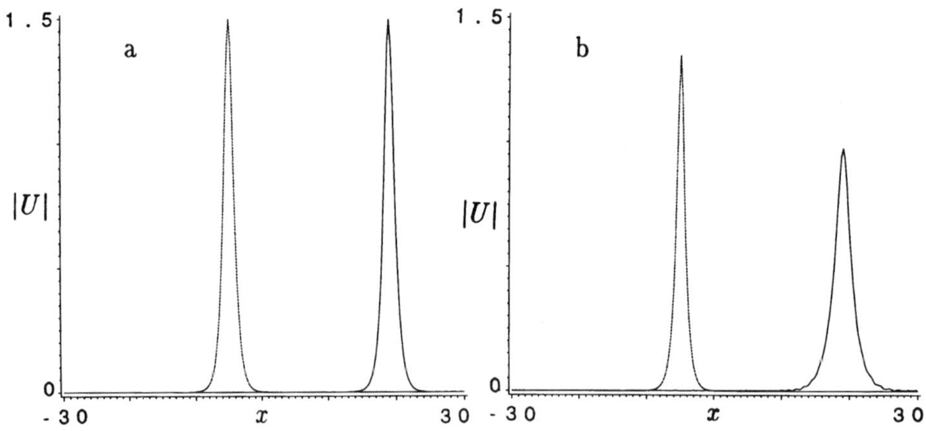

FIGURE 1. GPSE test problems with (a) $p = 3$ and (b) $p = 5$: Graph of $|U|$ at $t = 0$ (dotted) and $t = 6$ (solid)

The solution of the linear Schrödinger equation $iu_t + u_{xx} = 0$, with soliton initial data, is a dispersive wave [**3**]. The dispersive tendency is balanced by the

TABLE 3. Results from solution of solitary wave problem for the GNLSE

t	A_1'	$\hat{A}_2'$	L^2 error
0.0	2.190546	2.861766	0.014512
3.0	2.193853	3.242514	0.033904

addition of the nonlinear term, allowing the formation of solitary wave solutions. Despite its close approximation of conservation laws, the method performs less well in the quintic case than in the cubic case in maintaining this balance, although the advent of dispersive effects may be postponed by refining the spatial mesh.

3. Solution of the GNLSE

We present a numerical experiment in which the method is applied, using $r = 3$, to solve a GNLSE test problem [**4**] in which $q_c = 1/2$, $q_q = -7/4$, $q_m = -1$, $q_u = -2$, and the exact solution is $u(x,t) = \left[4/(4+3\sinh^2(x-2t-15))\right]^{1/2} \exp(i\phi(x,t))$, where $\phi(x,t) = 2\tanh^{-1}(\frac{1}{2}\tanh(x-2t-15)) + x - 15$. This represents a solitary wave, initially centered at $x = 15$, propagating to the right with speed 2. We take the Hermite piecewise cubic interpolant of $u(x,0)$ as the initial approximation $U(x,0)$, and use (2.3) for the approximate solution. We obtain a differential-algebraic system $D\dot{\mathbf{U}}(t) = E\mathbf{U}(t) + \mathbf{G}(\mathbf{U}(t))$, where the matrices D and E are the same as in the case of the GPSE, and $\mathbf{G}$ contains the contribution of the nonlinear terms.

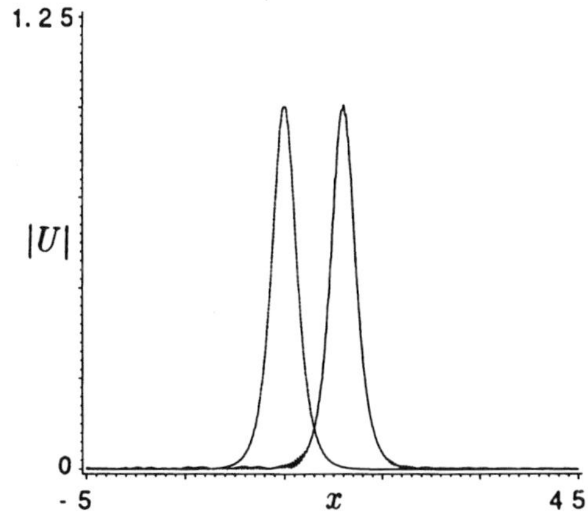

FIGURE 2. GNLSE solitary wave test problem: Graph of $|U|$ at $t = 0$ (dotted) and $t = 3$ (solid)

Results obtained when the problem is solved numerically on $[-5, 45] \times [0, 3]$ using $N = 200$ are shown in Table 3 and Figure 2. The approximations A_1' and $\hat{A}_2'$ are obtained using Gauss quadrature to evaluate the appropriate integrals over the spatial domain. The solitary wave behavior of the solution is preserved, and the L^2 error remains small. While A_1' changes little, and remains close to $A_1 = 2\ln 3 \approx 2.197$, $\hat{A}_2'$ experiences slow growth, which is accompanied by the development of small ripples at the edge of, and away from, the traveling wave.

References

1. U. Ascher, S. Pruess and R. D. Russell, *On spline basis selection for solving differential equations*, SIAM J. Numer. Anal. **20** (1983), 121–142.
2. M. Delfour, M. Fortin and G. Payre, *Finite-difference solutions of a non-linear Schrödinger equation*, J. Comput. Phys. **44** (1981), 277–288.
3. B. M. Herbst, J. Ll. Morris and A. R. Mitchell, *Numerical experiments with the nonlinear Schrödinger equation*, J. Comput. Phys. **60** (1985), 282–305.
4. D. Pathria and J. Ll. Morris, *Pseudo-spectral solution of nonlinear Schrödinger equations*, J. Comput. Phys. **87** (1990), 108–125.
5. M. P. Robinson and G. Fairweather, *Orthogonal spline collocation methods for Schrödinger-type equations in one space variable*, Numer. Math. (to appear).
6. M. P. Robinson, G. Fairweather and B. M. Herbst, *On the numerical solution of the cubic Schrödinger equation in one space variable*, J. Comput. Phys. **104** (1993), 277–284.

Department of Mathematics, Western Kentucky University, Bowling Green, KY 42101

E-mail address: robinson@wkuvx1.wku.edu

Proceedings of Symposia in Applied Mathematics
Volume 48, 1994

Who invented the computer? The debate from the viewpoint of computer architecture

Raúl Rojas

Abstract. Several countries and researchers claim for themselves the distinction of having invented the computer. In this paper we discuss the problem from the viewpoint of computer architecture: a computer should be understood as a universal calculating device. We show which is the minimal structure and the minimal instruction set computing automata should have in order to achieve universality. An analysis and comparison of the first calculating machines provides us with a surprising answer to the question stated in the title of this paper.

1. The problem of universal computation

The discussion about who should be called the true inventor of the computer has never been satisfactorily settled. The debate has been additionally obscured, because priority is claimed only over some part of what we understand today under the name "computer". For anyone interested in this debate, the first point to be clarified is certainly: what do we understand under the name "computer"? The only possible way to arrive at a consensus is to define a computer as a device capable of *universal computation*, that is, capable of implementing all computable functions.

1.1. Minimal structure of a universal computing machine. In the theory of recursive functions, we start from a small number of primitive functions and composition rules in order to define the general recursive functions. These can be implemented in a computing machine with a single accumulator by providing the following primitive instructions: CLR accumulator, INC accumulator, LOAD and STORE. Additionally, function composition and FOR and WHILE loops are required.

It is also necessary to provide some kind of indirect addressing in the machine. Otherwise, it is not possible to access numerical tables of variable length. The FOR and WHILE loops can be implemented in a simpler manner by providing some kind of conditional branch in the machine. An instruction "BR address" which tests the accumulator and branches to "address" if the accumulator is zero can be used to that effect. An unconditional branch can be implemented by executing a CLR instruction followed by a conditional branch.

1991 *Mathematics Subject Classification.* Primary 68M05, 01A65; Secondary 68Q05.
This paper is in final form and no version of it will be submitted for publication elsewhere.

© 1994 American Mathematical Society
0160-7634/94 $1.00 + $.25 per page

1.2. Self-modifying programs. General-purpose programming languages implement iterative calculations over tables of numbers by providing indexed arrays. At the machine level some kind of indirect addressing is required to retrieve the elements of an array from memory. There are two possible solutions to this problem. The first is to make the LOAD instruction introduced above retrieve its parameter from a register in the processor. The second is cheaper and is in fact the one which was used in early computers: just store the program in memory and let the LOAD instruction work only with a constant argument. Since the program is stored in the memory of the machine, all instructions can be modified by the program being executed. A self-modifying program can store the code for the instruction "LOAD 100" or "LOAD 200" in the required address, and when this address is executed, the effect is the same as if we had provided a register for indirect addressing.

Summarizing: we will be looking for the instructions CLR, INC, LOAD, STORE, conditional branches and indirect addressing (or equivalently self-modifying programs), when trying to decide if a particular machine can be called a universal computer.

2. Early computing machines

2.1. The architecture of the Z1. The first machine whose architecture we consider is the Z1 developed by the German engineer Konrad Zuse from 1936 to 1938 in Berlin. The Z1 was a mechanical computer, but it was not built with rotating parts like the calculators used at the time. It used instead sheets of metal capable of controlling binary switches used to store binary digits and to perform logical operations.

The Z1 was capable of computing the four elementary arithmetic operations as well as the square root of numbers. The program was punched on a film tape and was read sequentially by a tape reader. Besides the arithmetic operations, there was an instruction for reading a number from a keyboard and storing it in any of 64 memory cells, as well as an instruction for showing the contents of the accumulator of the machine in a field of electric lamps. Another interesting feature of the Z1 was its use of a floating-point representation. Each number was stored as a mantissa, an exponent and a sign. The processor computed all arithmetic operations using this coding (called by Zuse *logarithmic representation*) [7]. It would be more than fifteen years before someone built another machine with floating-point registers.

2.2. Atanasoff's machine. John Atanasoff built his electronic computer at the University of Iowa from 1938 to 1942. Whereas all other computing machines developed at the time worked with parallel arithmetic hardware, Atanasoff decided to use a serial approach. Two rotating drums stored up to thirty fixed-point numbers. Each number was stored in one sector of the drum using 50 capacitors, that is, 50 bits. The machine could thus hold at any time just two different vectors in its $30 + 30$ memory cells. The vectors were read in decimal form from punched cards and were transformed to binary numbers before being stored. Atanasoff designed his machine in order to solve systems of linear equations with up to 29 variables. Since the machine worked with Gauss reduction, one of the vectors was used repetitively to reduce the other. The leading coefficient of the other vector was reduced to zero through a combination of additions, subtractions and shifts [5].

Since the drums rotated once every second, all bits of corresponding vector elements could be read sequentially and could be fed into a serial adder whose result could be written back to one of the drums. The adder could be implemented with a minimum of hardware by recycling the carriage bit. The adder could also subtract one number from an other, if this was desired. A logical shift of a binary number was nothing but a read and write operation with a specific delay. There was an adder for each vector element.

2.3. The architecture of Mark I. Howard Aiken built the Mark I at Harvard University from 1939 to 1944. Like Atanasoff, Aiken was a physicist anxious to create a machine capable of doing the extensive calculations that he needed. The Mark I was an electromechanical machine, a kind of hybrid between the all-mechanical nature of previous computing devices and the electronics available at the time.

Aiken used a decimal representation for storage. Rotating gears were used to store and transmit the numbers in the memory. There was no sharp distinction between processor and memory, since each of the 72 memory cells was an accumulator capable of adding or subtracting a given fixed-point number. Each accumulator could transmit its contents through what we would today call a bus to other accumulators. There were also special units which stored tables of numbers and a multiplier. Sixty storage words could be set manually by the operator and constituted the constants in a program. The program itself was stored on punched paper tape, and two reading units made possible the execution in parallel of two programs or of the same one (for checking purposes). The Mark I used what we now call a three-address format for its instructions. All elementary arithmetic operations were possible, as well as sequencing of instructions [1].

Aiken's Mark I was not a universal computer since it lacked conditional branches. It was possible to repeat a section of code a given number of times, but no conditional execution of loops was possible. Since the program was stored externally, self-modifying codes could not be implemented.

2.4. The ENIAC. The *Electronic Numerical Integrator and Computer* was built at the Moore School of Electrical Engineering of the University of Pennsylvania from May 1943 to 1945. It solved its first problem in December 1945 and was officially presented in February 1946 [3].

From the viewpoint of computer architecture the ENIAC was a *parallel dataflow machine*. There was no separation between memory and the computing sections. Decimal fixed-point numbers were stored in any of twenty accumulators, each one capable of transmitting its contents to any of the other accumulators, or of receiving a number and adding it to its stored content. There was a multiplying and a dividing unit. Since the machine operated with a basic clock rate of 100,000 pulses per second, it was felt by the designers that an external program (in punched tape, for example) would not fully exploit the calculating speed of the ENIAC. Programs were implemented by physically connecting the different modules, that is, by effectively hardwiring the program into the machine.

Sequencing of arithmetic operations was achieved by connecting the arithmetic units in the desired order, even in parallel. The units worked asynchronously and in a self-clocking manner. A unit which was done with its calculation signaled this with a pulse to the next unit in the computation stream. The ENIAC was thus capable of implementing the CLR operation, addition, load and store of numerical data. It was also capable of implementing primitive recursion since a special unit, called the master controller, could restart a given thread of computation a fixed number of times. This is equivalent to a FOR loop in a high-level language. More important was the ability of the master controller to stop a given computational thread when a given quantity changed its sign. In this way it was also possible to implement WHILE loops and conditional branches in the machine.

2.5. The Analytical Engine. Finally, we must consider the design of the Analytical Engine conceived by Charles Babbage in the nineteenth century. The architecture of Babbage's machine is surprisingly modern. It includes such features as a clean logical separation of numerical storage and processor, a microprogrammed computing unit and even a pipelined design! [2]. Babbage worked on the Analytical Engine from 1834 until his death in 1871.

Although only some fragments of the Analytical Engine were built, and the complete design has only been studied recently, this machine should be considered a serious contender for the distinction of having been the first computer. Each number was stored in a vertical stack of rotating wheels using a decimal representation. The contents of each storage unit could be transported to the processor of the machine, called by Babbage "the mill". The four basic arithmetic operations were implemented in the mill. The program was read from punched cards, similar to the ones used in the Jacquard looms. Another set of cards contained the data for the program. This separation of code and data was a clever way of eluding the problem of indexed addressing. Since Babbage's machine could implement loops, it was possible to repetitively read the code cards at the same time as new data cards were sequentially fed into the machine. Babbage also included the possibility of conditional branching. He was aware of the restrictions imposed by an external program. Because of this, he speculated with the idea of including a card puncher as an output option. The machine could in this way produce its own programs!

3. Architectural comparison

3.1. Summary of characteristics. The following two tables show the most relevant information about the machines discussed in the previous sections. As should be clear from the descriptions given above, no single machine fulfills all the necessary requirements for a universal computer. The Analytical Engine is surprisingly the one which comes closest. Table 1 shows the main features of the machines we analyzed. We also include the Mark 1 machine built at Manchester from 1946 to 1948, because as far as we know this was the first machine to fit our definition of a universal computer. This machine stored its program in random-access digital memory implemented with CRT tubes. All necessary instruction primitives were available (in modified form), and although it lacked indirect addressing, self-modifying programs could be written [4].

3.2. Conclusions. There is no unambiguous answer to the question stated in the title of this paper. Among the earliest machines considered, Babbage's comes closest to a universal computer. Yet this machine was not built and we know about it only through some recent historical research. The ENIAC was the fastest machine of its time and could implement loops, but lacked any kind of soft programming. Atanasoff's machine was the first *electronic* calculating engine. Although the Mark 1 from Manchester was indeed a universal computer, it would be hard to credit its designers with having invented the computer single-handedly, since some of the design ideas were taken from the EDVAC project conceived by the ENIAC creators

Table 1. Comparison of architectural features

Machine	memory and CPU separated?	conditional branching?	soft or hard programming	self-modifying programs?	indirect addressing?
Babbage's	√	√	soft	planned	×
Zuse's Z1	√	×	soft	×	×
Atanasoff's	√	×	hard	×	×
H-Mark I	×	×	soft	×	×
ENIAC	×	partially	hard	×	×
M-Mark 1	√	√	soft	√	×

Table 2. Some additional architectural features

Machine	internal coding	fixed-point or floating-point?	sequential logic?	architecture	technology
Babbage's	decimal	fixed-point	no	pipelined	mechanical
Zuse's Z1	binary	floating-point	no	sequential	mechanical
Atanasoff's	binary	fixed-point	yes	vectorized	electronic
H-Mark I	decimal	fixed-point	no	parallel	electromechanical
ENIAC	decimal	fixed-point	no	data flow	electronic
M-Mark 1	binary	fixed-point	yes	sequential	electronic

and John von Neumann. In fairness, it should be concluded that the invention of the computer spans a period of more than hundred years, in which scientists from at least three different countries provided the theoretical and practical foundations for this enterprise. If there is a technological achievement in which only international cross-fertilization could have led to success, then the invention of the computer is indeed one.

References

[1] H. Aiken and G. Hopper, *The Automatic Sequence Controlled Calculator*, reprinted in [5], 203–222.

[2] A. Bromley, *The evolution of Babbage's calculating engines*, Annals of the History of Computing **9**, no. 2 (1987), 113–136.

[3] A. W. Burks and A. R. Burks, *The ENIAC: First general-purpose electronic computer*, Annals of the History of Computing **3**, no. 4 (1981), 310–399.

[4] S.H. Lavington, *A history of Manchester computers*, NCC Publications, Manchester, 1975.

[5] B. Randell, *The Origins of Digital Computers*, Springer-Verlag, Berlin, 1982.

[6] N. Stern, *From ENIAC to UNIVAC*, Digital Press, Bedford, 1981.

[7] K. Zuse, *Der Computer — Mein Lebenswerk*, Springer-Verlag, Berlin, 1970.

[8] H. H. Goldstine, *The Computer: From Pascal to von Neumann*, Princeton University Press, Princeton, 1993.

Mathematics Department, Freie Universität Berlin, Berlin 14195
E-mail address: rojas@inf.fu-berlin.de

Proceedings of Symposia in Applied Mathematics
Volume **48**, 1994

LOCKING AND BOUNDARY LAYER EFFECTS IN THE FINITE ELEMENT APPROXIMATION OF THE REISSNER-MINDLIN PLATE MODEL

CHRISTOPH SCHWAB AND MANIL SURI

ABSTRACT. We consider the uniform finite element approximation (with respect to the thickness) of the Reissner-Mindlin plate model. We obtain sufficient conditions on the finite element subspaces to ensure the uniform resolution of locking and boundary layer effects.

1. INTRODUCTION

It is well known that the accuracy of finite element (FE) approximations to the Reissner-Mindlin (R-M) plate model can deteriorate as the plate thickness d tends to zero. This deterioration may be due to two phenomena: that of *locking* and the presence of *boundary layers*. Locking occurs because as $d \to 0$, the R-M plate model tends to the Kirchhoff plate model, with the "Kirchhoff Constraint"

$$C_0 u_0 := \vec{\phi}_0 - \text{ grad } w_0 = 0 \tag{1}$$

being imposed in the limit $d \to 0$. (Here, $u_d := (\vec{\phi}_d, w_d)$ represents the rotations and transverse displacements, respectively.) If the FE subspace does not have enough functions satisfying (1), then the approximate solution obtained will be quite poor. Boundary layers, on the other hand, are rapidly varying solution components that are localized in an $O(d)$ neighborhood of the edge of the plate. For their resolution, uniform meshes must be inordinately refined. Our purpose in this note is to give sufficient conditions on FE subspaces that ensure *uniform approximations* (in d) of the R-M plate model. This generalizes the results in [3], where only locking effects, and not boundary layers, were considered.

2. THE R-M PLATE MODEL AND PROPERTIES OF ITS SOLUTION

Let the open, bounded and simply connected region $\Omega \subset \mathbb{R}^2$ represent the mid-plane of the homogeneous, isotropic plate of thickness d, occupying the region

---

1991 *Mathematics Subject Classification.* Primary 65N30, 35B30, 65N15.

The authors were supported in part by the Air Force Office of Scientific Research, Air Force Systems Command, USAF, under Grant F49620-92-J-0100.

This paper is in final form, and no version of it will be submitted for publication elsewhere.

© 1994 American Mathematical Society
0160-7634/94 $1.00 + $.25 per page

$R = \Omega \times (-d/2, d/2)$. We assume the boundary $\Gamma = \partial\Omega$ is smooth. With $\vec{\phi}_d$, w_d as before, the R-M model is given by the second-order elliptic system

$$-\frac{D}{2}\{(1-\nu)\Delta\vec{\phi}_d + (1+\nu)\nabla\nabla\cdot\vec{\phi}_d\} - \kappa\mu d^{-2}(\nabla w_d - \vec{\phi}_d) = 0, \tag{2}$$

$$-\kappa\mu d^{-2}\nabla\cdot(\nabla w_d - \vec{\phi}_d) = g,$$

which gives the bending of the plate in equilibrium under the normal load density per unit area given by gd^3. Here, ν is the Poisson ratio, E the Young's modulus, $D = E/12(1-\nu^2)$, $\mu = E/2(1+\nu)$ and κ is the shear correction factor. We consider the case of *natural* boundary conditions (though our main theorem applies for other conditions as well). Defining $\mathcal{L}$ to be the space of linear functions on Ω, we assume that g is L^2 orthogonal to $\mathcal{L}$. Then this problem will have a solution unique up to functions in $(\text{grad}\,\mathcal{L})\times\mathcal{L}$.

By the results of [1], the case of free boundary conditions considered above has the strongest boundary layer. We summarize some results from this reference. Denote by $\rho_0 > 0$ the minimum radius of curvature of Γ and define $\Omega_0 = \{z - \rho n_z | z \in \Gamma,\ 0 < \rho < \rho_0\} \subset \Omega$, where n_z is the outward unit normal to Ω at z. We parameterize $\partial\Omega$ in terms of the arc length $\theta \in [0, L)$ (which is extended L-periodically to $\theta \in \mathbb{R}$) by writing $z(\theta) = (X(\theta), Y(\theta))$. We will use the boundary-fitted coordinates (ρ, θ) in Ω_0 defined by the correspondence

$$(\rho,\theta) \to z - \rho n_z = (X(\theta) - \rho Y'(\theta), Y(\theta) + \rho X'(\theta)).$$

We assume $g \in C^\infty(\bar{\Omega})$, so that by standard elliptic regularity, $u_d \in [C^\infty(\bar{\Omega})]^3$. However, this does not reflect the nonuniform behavior of the solution as $d \to 0$. For the latter, we use the results in [1], according to which

$$w_d \sim \Sigma_{i=0}^{\infty} d^i w_i,\ \vec{\phi}_d \sim \Sigma_{i=0}^{\infty} d^i \vec{\phi}_i^s + \chi \Sigma_{i=1}^{\infty} d^i \vec{\phi}_i^b, \tag{3}$$

where w_i and $\vec{\phi}_i^s, \vec{\phi}_i^b$ are smooth and independent of d, $\chi = \chi(\rho/d)$ with $0 \le \chi \in C^\infty[0,\infty)$, $\chi(r) = 1$ for $0 \le r \le 1/3$, $= 0$ for $r \ge 2/3$, and $\vec{\phi}_i^b = \vec{\phi}_i^b(\rho/d, \theta)$ are the boundary layers, given by

$$\vec{\phi}_i^b(r,\theta) = \vec{\Pi}_i(r,\theta)e^{-\sqrt{12\kappa}r}. \tag{4}$$

Here, $\vec{\Pi}_i$ are polynomials of degree i in r with smooth, L-periodic (in θ) coefficients. As shown in [1], for any integer $k \ge 1$, the solution of (2) may be decomposed as

$$u_d = u_0 + du_{d,k}^s + du_{d,k}^b + u_{d,k}^r, \tag{5}$$

where, with the notation of (3),

(i) $u_0 = (\vec{\phi}_0, w_0) = (\text{grad} w_0, w_0)$ solves the limiting Kirchhoff problem ,
(ii) $u_{d,k}^s := (\vec{\phi}_{d,k}^s, w_{d,k}^s) = \Sigma_{i=1}^k d^{i-1}(\vec{\phi}_i^s, w_i)$ with all derivatives of $\vec{\phi}_{d,k}^s, w_{d,k}^s$ uniformly bounded in d,
(iii) $u_{d,k}^b := (\chi\vec{\phi}_{d,k}^b, 0) = \chi\Sigma_{i=1}^k d^{i-1}(\vec{\phi}_i^b, 0)$ is the boundary layer,
(iv) $u_{d,k}^r = (\vec{\phi}_{d,k}^r, w_{d,k}^r)$ is the remainder after $k+1$ terms in (3) and satisfies

$$\|\vec{\phi}_{d,k}^r\|_s \le C(g)d^{\min(k+1,k+\frac{3}{2}-s)},\quad \|w_{d,k}^r\|_s \le C(g)d^{k+1}. \tag{6}$$

3. The finite element approximation

The variational form of (2) is given by:
Find $u_d = (\vec{\phi}_d, w_d) \in H = ([H^1(\Omega)]^2/\text{ grad } \mathcal{L}) \times (H^1(\Omega)/\mathcal{L})$ such that for all $v = (\vec{\theta}, \xi) \in H$,

$$a_d(u_d, v) = \int_\Omega g\xi =: F(v), \tag{7}$$

where, with C_0 as in (1), and $\lambda = \kappa\mu$,

(8)
$$a_d(u_d, v) = \frac{D}{2}\int_\Omega \{(1-\nu)\nabla\vec{\phi}_d \cdot \nabla\vec{\theta} + (1+\nu)(\nabla\cdot\vec{\phi}_d)(\nabla\cdot\vec{\theta})\}dx + \lambda d^{-2}(C_0 u_d, C_0 v).$$

Let $H^N = (Y^N \times Z^N) \subset H$ be a sequence of finite-dimensional subspaces (which may depend upon d). The FE approximations $u_d^N = (\vec{\phi}_d^N, w_d^N) \in Y^N \times Z^N$ satisfy (7) for all $v \in H^N$. Let us define the energy norm $\|u\|_{E,d} := (a_d(u,u))^{1/2}$; then we obviously have

$$\|u_d - u_d^N\|_{E,d} = \inf_{v\in H^N} \|u_d - v\|_{E,d}. \tag{9}$$

Note that for any $u = (\vec{\phi}, w) \in H$,

$$K_1\|u\|_{1,1} \le \|u\|_{E,d} \le K_2(\|\vec{\phi}\|_1 + d^{-1}(\|\vec{\phi}\|_0 + \|w\|_1)) \le K_3 d^{-1}\|u\|_{1,1}, \tag{10}$$

where K_i are independent of d, and with $\vec{\phi} = (\phi_1, \phi_2)$,

$$\|u\|_{k,\ell}^2 = \|(\vec{\phi}, w)\|_{k,\ell}^2 = \|\phi_1\|_k^2 + \|\phi_2\|_k^2 + \|w\|_\ell^2.$$

Let us now see *when* (9) *results in a sequence of approximations* $u^N = u_d^N$ *to* $u = u_d$ *which are uniformly accurate in* d *as* $d \to 0$. This will lead to a FE method *free of locking* with *uniform boundary layer approximation.* We consider the approximation error in the $\|\cdot\|_{1,1}$ norm. For our analysis, we will use the space $W^N \subset Z^N$ defined by $W^N = \{w \in Z^N : \nabla w \in Y^N\}$. Also, for any space $X \subset H$, let $X_0 := \{u \in X : C_0 u = 0\}$. Moreover, for $X = Y^N$ or Z^N, we define for any integer $k \ge 1$,

$$\mathcal{D}(k, X) = \sup_{\|\psi\|_k \le 1} \inf_{v\in X} \|\psi - v\|_1,$$

where the supremum is over all ψ in $(H^k(\Omega))^2$ or $H^k(\Omega)$, respectively.

Theorem 3.1. *For $d > 0$, let $u_d = (\vec{\phi}_d, w_d)$ be the solution of* (2), *so that the decomposition* (5) *holds. Then, for any $k \ge 1$,*

$$\|u_d - u_d^N\|_{1,1} \le K\{\mathcal{F}_0(w_0, N) + \mathcal{F}_1(u_{d,k}^s, N) + \mathcal{F}_2(\chi\vec{\phi}_{d,k}^b, N) + \mathcal{F}_3(k, N)\}, \tag{11}$$

where the constant K is independent of N and d and
$\mathcal{F}_0(w, N) = \inf_{z\in W^N} \|w - z\|_2$ *for any* $w \in H^2(\Omega)$,
$\mathcal{F}_1(u, N) = \inf_{v\in H^N} \|u - v\|_{1,1}$ *for any* $u \in H$,
$\mathcal{F}_2(\vec{\tau}, N) = \inf_{\vec{\psi}\in Y^N}\{d\|\vec{\tau} - \vec{\psi}\|_1 + \|\vec{\tau} - \vec{\psi}\|_0\}$ *for any boundary layer term* $\vec{\tau}$,
$\mathcal{F}_3(k, N) = C(g)\{d^{1/2}\mathcal{D}(k, Y^N) + d^k\mathcal{D}(k, Z^N)\}$.

Proof. By (10),(9),(5) we have

$$(12)\quad \|u_d - u_d^N\|_{1,1} \le K\|u_d - u_d^N\|_{E,d} = K \inf_{v\in H^N} \|u_d - v\|_{E,d}$$

$$\le K\{\inf_{v_0\in H^N} \|u_0 - v_0\|_{E,d} + \mathcal{F}_1(u^s_{d,k}, N) + \mathcal{F}_2(\chi\vec{\phi}^b_{d,k}, N) + \inf_{v_1\in H^N} \|u^r_{d,k} - v_1\|_{E,d}\}.$$

Now since $C_0 u_0 = 0$,

$$\inf_{v_0\in H^N} \|u_0 - v_0\|_{E,d} \le K \inf_{v_o\in H_0^N} \|u_0 - v_0\|_{1,1} \le K\mathcal{F}_0(w_0, N)$$

as shown in [3]. Also, using (10), (6), we have

$$\begin{aligned}\inf_{v_1\in H^N} \|u^r_{d,k} - v_1\|_{E,d} &\le Kd^{-1}\{\inf_{\vec{\psi}\in Y^N} \|\vec{\phi}^r_{d,k} - \vec{\psi}\|_1 + \inf_{\xi\in Z^N} \|w^r_{d,k} - \xi\|_1\} \\ &\le KC(g)d^{-1}\{d^{\min(k+1,k+\frac{3}{2}-k)}\mathcal{D}(k, Y^N) + d^{k+1}\mathcal{D}(k, Z^N)\},\end{aligned}$$

from which (11) follows. □

We see, therefore, that to obtain uniform approximations in d, the spaces $\{V^N\}$ should be designed so that each of the approximation rates in (11) have the same order. Since $\mathcal{F}_1$ represents the usual approximation rate for the space V^N, and $\mathcal{F}_3$ is in general bounded by $\mathcal{F}_1$ for k large enough, we must essentially be able to bound both $\mathcal{F}_0$ and $\mathcal{F}_2$ by $\mathcal{F}_1$. In particular, if $\mathcal{F}_0 \le \mathcal{F}_1$, then our method will be *locking free*, while if $\mathcal{F}_2 \le \mathcal{F}_1$ (for all d), then we achieve *uniform boundary layer approximation.*

The conditions $\mathcal{F}_0 \le \mathcal{F}_1$, $\mathcal{F}_2 \le \mathcal{F}_1$ can be established (with minor modifications) for other boundary conditions as well, for the case that $\partial\Omega$ is smooth. If $\partial\Omega$ is nonsmooth (e.g. Ω is a polygon), then the presence of corner singularities complicates the analysis. Nevertheless, the conditions $\mathcal{F}_0 \le \mathcal{F}_1$ and $\mathcal{F}_2 \le \mathcal{F}_1$ will again provide important guidelines for constructing locking-free and uniformly boundary layer approximating spaces, respectively. For example, in [3], the necessity and sufficiency of $\mathcal{F}_0 \le \mathcal{F}_1$ for locking-free approximation in the case of a square plate with periodic boundary conditions has been rigorously established. Let us therefore briefly comment on these conditions $\mathcal{F}_0 \le \mathcal{F}_1$, $\mathcal{F}_2 \le \mathcal{F}_1$ for the case of polygonal domains, to see what the corresponding finite element spaces should look like.

1. *Locking-free spaces* ($\mathcal{F}_0 \le \mathcal{F}_1$). The construction of such spaces has been discussed in detail in [3]. For the *h version*, it is shown, for instance, that locking can be avoided in the case of triangular meshes by choosing polynomials of degree $p \ge 4$ for Y^N and degree $q \ge p+1$ for Z^N. For the *p version*, it is shown that the method is *locking free.* Various other results (e.g. for rectangular meshes) may be found in [3].

2. *Boundary layer approximation* ($\mathcal{F}_2 \le \mathcal{F}_1$). Let γ be a portion of the boundary, parameterized by θ, that is straight (and excludes vertices), and consider the region $A = \gamma \times t \in \Omega_0$ where $t = \{\rho : 0 < \rho < \rho_0\}$. Let Δ_θ, Δ_ρ be one-dimensional meshes on γ and t, respectively, and consider the tensor product space on A given by $Y^N = [S^q(\Delta_\theta) \otimes S^p(\Delta_\rho)]^2$, where $S^r(\Delta)$ denotes continuous piecewise polynomials of degree $\le r$ on Δ. Suppose we consider the problem of ensuring that $\mathcal{F}_2 \le \mathcal{F}_1$ on A. In order to make $\mathcal{F}_2$ small enough, Y^N should be such that the boundary layer terms $\vec{\phi}^b_i(\rho/d, \theta)$ given by (4) are well approximated. These functions are smooth in the θ-direction, so that $S^q(\Delta_\theta)$ only needs to have some general approximability

properties (independent of d). By a simple tensor product argument, the problem then reduces to picking appropriate spaces $S^p(\Delta_\rho)$ in the ρ-direction to approximate the $e^{-\alpha\rho/d}$ functions well, which is a one-dimensional problem. This problem has been considered in detail in [2], where, among other results, it is shown that for the h version FEM, exponentially graded meshes Δ_ρ are optimal. Moreover, the p version, when the first element of Δ_ρ has a length of $O(dp)$, can result in an *exponential* convergence rate. These results can then be used to construct appropriate tensor product spaces in the two-dimensional case, in an $O(d)$ neighborhood of the boundary. Obviously, a subdivision of such rectangular meshes into triangles will also lead to similar convergence rates for the boundary layers.

REFERENCES

1. Arnold, D. N. and Falk, R. S., *Asymptotic analysis of the boundary layer for the Reissner-Mindlin plate model*, Technical Report, Rutgers University, New Brunswick, NJ (1993).
2. Schwab, C. and Suri, M., *The h, p and h−p versions of the finite element method for problems with boundary layers I: One-dimensional problems.* (To appear).
3. Suri, M., Babuška, I., Schwab, C., *Locking effects in the finite element approximation of plate models*, Technical Report BN-1132, IPST, University of Maryland, College Park (1992).

DEPARTMENT OF MATHEMATICS AND STATISTICS, UNIVERSITY OF MARYLAND BALTIMORE COUNTY, BALTIMORE, MD 20910

E-mail address: Schwab@math.umbc.edu, Suri@umbc2.umbc.edu

Proceedings of Symposia in Applied Mathematics
Volume **48**, 1994

Efficient Spectral Galerkin Methods for Some Elliptic Problems

JIE SHEN

ABSTRACT. We present some efficient direct and iterative solvers, based on the spectral Galerkin approximations, for some typical elliptic equations. The key to the efficiency is to choose appropriate basis functions which lead to sparse or special matrices that can be effectively inverted.

1. Introduction

The spectral method employs global polynomials as the trial functions for the discretization of partial differential equations. It gives very accurate approximations with a relatively small number of unknowns. On the other hand, it usually leads to linear systems with full matrices, which are often severely ill-conditioned. Although many effective techniques have been made available (see for instance [**1, 8**]), solving these linear systems could still be quite expensive. Thus, it is particularly interesting if a spectral discretization would lead to sparse or special matrices which can be efficiently inverted. We shall see that this is possible, at least for problems with constant coefficients, by choosing a set of proper basis functions for the spectral Galerkin approximations.

2. Helmholtz equation and Stokes equation

We consider first the spectral Galerkin approximations to the two-dimensional Helmholtz equation

$$\alpha u - \Delta u = f \text{ in } \Omega = [-1,1]^2, \quad u|_{\partial\Omega} = 0. \tag{1}$$

1991 *Mathematics Subject Classification.* Primary 65N35, 65N22; Secondary 65F05, 35J05.
This work was supported in part by NSF Grant #9205300.
This paper is in final form, and no version of it will be submitted for publication elsewhere.

© 1994 American Mathematical Society
0160-7634/94 $1.00 + $.25 per page

Let $L_n(x)$ be the nth-degree Legendre polynomial, and let

$$S_N = \text{span}\{L_i(x)L_j(y) : i,j = 0,1,\ldots,N\}, \quad X_N = \{v \in S_N : v|_{\partial\Omega} = 0\}.$$

Then the standard Legendre-Galerkin approximation to (1) is to find $u_N \in X_N$ such that

$$\alpha(u_N, v) + (\nabla u_N, \nabla v) = (f, v), \ \forall\, v \in X_N, \tag{2}$$

where $(u,v) = \int_\Omega uv\,dxdy$ is the scalar product in $L^2(\Omega)$.

The efficiency of our algorithm relies on the following simple observation:

$$X_N = \text{span}\{\phi_i(x)\phi_j(y) : i,j = 0,1,\cdots,N-2\},$$

with $\phi_k(\cdot) = \frac{1}{\sqrt{4k+6}}(L_k(\cdot) - L_{k+2}(\cdot))$.

We shall see that the system (2) becomes extremely simple using the above basis for X_N. In fact, one can readily check that $\int_{-1}^{1} \phi_k'(x)\phi_j'(x)dx = \delta_{kj}$ and

$$b_{jk} \equiv \int_{-1}^{1} \phi_k(x)\phi_j(x)dx \neq 0 \ \text{ only for } \ j = k-2, k, k+2. \tag{3}$$

Hence, setting

$$\begin{aligned} &u_N = \sum_{k,j=0}^{N-2} u_{kj}\phi_k(x)\phi_j(y), \quad f_{kj} = (f, \phi_k(x)\phi_j(y)), \\ &U = (u_{kj})_{k,j=0,1,\ldots,N-2}, \ F = (f_{kj})_{k,j=0,1,\ldots,N-2}, \ B = (b_{kj})_{k,j=0,1,\ldots,N-2}, \end{aligned} \tag{4}$$

we find that (2) is equivalent to the following matrix equation:

$$\alpha BUB + UB + BU = F, \tag{5}$$

which can also be written in the tensor product form

$$(\alpha B \otimes B + I \otimes B + B \otimes I)\boldsymbol{u} = \boldsymbol{f}, \tag{6}$$

where $\boldsymbol{f}$ and $\boldsymbol{u}$ are respectively F and U written in the form of a column vector, and $\otimes$ denotes the tensor product of matrices, i.e., $A \otimes B = (Ab_{ij})_{i,j=0,1,\ldots,N-2}$.

The equation (5) can be solved, in particular, by the matrix decomposition method, similar to the technique used in [**2**]. However, compared to the spectral-tau approximation used in [**2**], the above treatment leads to more efficient and more accurate algorithms (cf. [**4, 5**]).

We can also naturally extend the above treatment to the Chebyshev Galerkin approximations, to the general second-order equation with constant coefficients and to the three-dimensional case. In summary (see [**4, 5**] for details), *the discrete systems arising from the Legendre or Chebyshev Galerkin approximations to the second-order equations with constant coefficients can be solved in about* $2(d-1)N^{d+1}$ *operations* ($d = 2$ or 3 is the dimension of the domain).

Furthermore, in the special case of the Legendre-Galerkin approximation applied to the 2-D Helmholtz equation, a more efficient algorithm can be constructed. In fact, since the matrix B (cf. (3)) can be split into two tridiagonal matrices, we can split the system (6) into four subsystems with block-tridiagonal matrices for, respectively, the even-even, even-odd, odd-even and odd-odd components of U. In case $\alpha = 0$, the subsystems have exactly the same structure as the linear system generated by the standard five-point finite difference discretization of separable second-order elliptic equations with variable coefficients (cf. [**7**]). Hence in particular, the four subsystems can be solved by the extended cyclic reduction method developed by Swarztrauber [**7**] in $O(N^2 \log_2 N)$ operations. In fact, the subroutine blktri.f in FISHPACK (written by J. Adams, P. N. Swarztrauber and R. Sweet, available via netlib) can be directly used to solve the above subsystems. We note that the case $\alpha \neq 0$ can be handled similarly [**6**].

We now consider the Stokes equation

$$\begin{aligned} &-\Delta \boldsymbol{u} + \nabla p = \boldsymbol{f},\ \nabla \cdot \boldsymbol{u} = 0 \ \text{ in } \ \Omega = [-1,1]^2, \\ &u|_{\partial\Omega} = 0,\quad \int_\Omega p\, dxdy = 0, \end{aligned} \tag{7}$$

and a sequence ($m = N, N-1, \dots$) of Legendre-Galerkin approximations: Find $(\boldsymbol{u}_{N,m}, p_{N,m}) \in X_N^2 \times S_m \cap L_0^2(\Omega)$ such that

$$\begin{aligned} &(\nabla \boldsymbol{u}_{N,m}, \nabla \boldsymbol{v}) - (\nabla \cdot \boldsymbol{v}, p_{N,m}) = (\boldsymbol{f}, \boldsymbol{v})\ ,\ \forall\, \boldsymbol{v} \in X_N^2; \\ &(\nabla \cdot \boldsymbol{u}_{N,m}, q) = 0\ ,\ \forall\, q \in S_m \cap L_0^2(\Omega). \end{aligned} \tag{8}$$

It is well known that the pressure approximation $p_{N,m}$ in (8) can be decoupled from the velocity approximation $\boldsymbol{u}_{N,m}$. In fact, by using $\phi_i(x)\phi_j(y)$ (resp. $L_i(x)L_j(y)$) as basis functions for X_N^2 (resp. S_m), one can derive from (8) that

$$D_{N,m} p_{N,m} \equiv (B_{N,m} A_N^{-1} B_{N,m}^T) p_{N,m} = B_{N,m} A_N^{-1} \tilde{f}_N, \tag{9}$$

where $A_N, B_{N,m}$ and $B_{N,m}^T$ are respectively the discrete "$-\Delta$", discrete "$-\nabla\cdot$" and discrete "∇" operators (see [6] for details). We note that the operator $D_{N,m}$ is positive definite for $m \le N-2$, and it has been determined numerically [**6**] and theoretically [**3**] that $\text{cond}(D_{N,m}) \sim CN$. Hence, the number of conjugate gradient iterations needed for solving (9) is of the order $N^{\frac{1}{2}}$. Thanks to the fast Poisson solver presented above, the matrix-vector product $D_{N,m} p_{N,m}$ can be performed in $O(N^2 \log_2 N)$ operations. We conclude that *the discrete Stokes systems for $m \le N-2$ can be solved in $O(N^{5/2} \log_2 N)$ operations.*

3. Biharmonic equation

Let $W_N = \{v \in S_N : v|_{\partial\Omega} = \frac{\partial v}{\partial \boldsymbol{n}}|_{\partial\Omega} = 0\}$. Then the Legendre-Galerkin approximation to the 2-D biharmonic equation

$$\Delta^2 u = f \text{ in } \Omega,\ v|_{\partial\Omega} = \frac{\partial v}{\partial \boldsymbol{n}}|_{\partial\Omega} = 0 \tag{10}$$

is to find $u_N \in W_N$ such that

$$(\Delta u_N, \Delta v) = (f, v)\ ,\ \forall\, v \in W_N. \tag{11}$$

One can check that $W_N = \text{span}\{\psi_i(x)\psi_j(y) : i,j = 0,1,\dots,N-4\}$, where

$$\psi_k(\cdot) = \frac{1}{\sqrt{2(2k+3)^2(2k+5)}}\left(L_k(\cdot) - \frac{2(2k+5)}{2k+7}L_{k+2}(\cdot) + \frac{2k+3}{2k+7}L_{k+4}(\cdot)\right).$$

Furthermore, $\int_{-1}^{1}\psi_i''(x)\psi_j''(x)dx = \delta_{ij}$. Letting $b_{kj} = \int_{-1}^{1}\psi_j(x)\psi_k(x)dx$ and $c_{kj} = \int_{-1}^{1}\psi_j'(x)\psi_k'(x)dx$, using similar notations as in (4) (replacing ϕ and $N-2$ by respectively ψ and $N-4$), we find that (11) is equivalent to the following matrix equation:

$$BU + 2CUC + UB = F. \tag{12}$$

We note that if $BC = CB$, the equation (12) could be efficiently solved by using the matrix decomposition method as above. But unfortunately $BC \neq CB$, owing to the fact that the equation (10) is not separable. However, $BC - CB$ has only eight nonzero entries whose positions have been identified in [**4**]. Therefore, we can modify B at the eight entries to form a new matrix $\tilde{B}$ such that $\tilde{B}C = C\tilde{B}$. Consequently, the modified equation

$$\tilde{B}U + 2CUC + UB = F \tag{13}$$

can be efficiently solved by the matrix decomposition method. Then the solution to the system (12) can be recovered by using the method of capacitance matrix (see [**4**] for details). In summary, *excluding the $O(N^3)$ operations needed in the preprocessing stage for the construction of the capacitance matrix, the system* (12) *can be solved in about $4N^3$ operations.* Note that general 2-D fourth-order equations with constant coefficients can be handled similarly.

References

1. C. Canuto, M. Y. Hussaini, A. Quarteroni and T. A. Zang, *Spectral Methods in Fluid Dynamics*, Springer-Verlag, 1987.
2. D. B. Haidvogel and T. A. Zang, *The accurate solution of Poisson's equation by expansion in Chebyshev polynomials*, J. Comput. Phys. **30** (1979), 167-180.
3. Y. Maday, A. T. Patera and E. M. Ronquist, *The $P_N \times P_{N-2}$ method for the approximation of the Stokes problem*, preprint (1992).
4. J. Shen, *Efficient spectral-Galerkin method I. Direct solvers for second and fourth order equations by using Legendre polynomials*, to appear in SIAM J. Sci. Comput.
5. J. Shen, *Efficient spectral-Galerkin method II. Direct solvers for second and fourth order equations by using Chebyshev polynomials*, to appear in SIAM J. Sci. Comput.
6. J. Shen, *On fast Poisson solver, inf-sup constant and iterative Stokes solver by Legendre Galerkin method*, to appear in J. Comput. Phys.
7. P. N. Swarztrauber, *A direct method for the discrete solution of separable elliptic equations*, SIAM J. Numer. Anal. **11** (1974), 1136-1150.
8. S. Zhang, *On the convergence of spectral multigrid methods for periodic problems*, Calcolo **28** (1991), 185-203.

Department of Mathematics, Penn State University, University Park, PA 16802

E-mail address: shen_j@math.psu.edu

Proceedings of Symposia in Applied Mathematics
Volume **48**, 1994

Periodic Solutions of Higher-Order Difference Equations in Two Independent Variables

QIN SHENG AND RAVI P. AGARWAL

ABSTRACT. Nonlinear difference equations of order $n+m$ in two independent variables, together with periodic boundary conditions, are considered. Comparison results are established. The inequalities obtained are then used to provide sufficient conditions for the convergence of monotone iterative methods for solving difference equation problems.

1. Introduction

Let $\mathbb{Z}^+$ be the set of positive integers and $\mathbb{Z}$ the set of nonnegative integers. Let $p,\ q \in \mathbb{Z}^+$, and let I_p, I_q and I_{pq} be the sets $\{x : \ x \in \mathbb{Z},\ 0 \le x \le p-1\}$, $\{y : \ y \in \mathbb{Z},\ 0 \le y \le q-1\}$ and $\{x : \ x \in \mathbb{Z},\ 0 \le x \le p-1\} \times \{y : \ y \in \mathbb{Z},\ 0 \le y \le q-1\}$, respectively. We consider the following higher-order nonlinear difference equation:

$$\Delta_x^n \Delta_y^m u(x,y) = g(x,y,< u >),\ (x,y) \in I_{pq}, \tag{1.1}$$

together with the periodic conditions

$$\Delta_x^i u(0,y) \ = \ \Delta_x^i u(p,y), \qquad y \in I_q,\ 0 \le i \le n-1; \tag{1.2}$$

$$\Delta_y^j u(x,0) \ = \ \Delta_y^j u(x,q), \qquad x \in I_p,\ 0 \le j \le m-1, \tag{1.3}$$

where Δ_x, Δ_y are the usual forward difference operators, and $< u >$ denotes

$$\big(u, \Delta_x u, \dots, \Delta_x^n u, \dots, \Delta_y^j u, \dots, \Delta_x^n \Delta_y^j u, \dots, \\ \Delta_y^{m-1} u, \dots, \Delta_x^n \Delta_y^{m-1} u, \Delta_y^m u, \dots, \Delta_x^{n-1} \Delta_y^m u\big)\,.$$

The following theorem holds (for a proof, see [**1**, **7**]).

1991 *Mathematics Subject Classification.* Primary 35H05; Secondary 35A35.
The detailed version of this paper will be published elsewhere.

© 1994 American Mathematical Society
0160-7634/94 $1.00 + $.25 per page

THEOREM 1. *Let $\mu(x,y)$, $\nu(x,y)$ be defined in $I_{p+n-1,q+m-1}$ and $g(x,y,< u >)$ be nondecreasing in $< u >$. If for $x \in I_p,\ y \in I_q$,*

$$\Delta_x^i\Delta_y^j\mu(0,y) \geq \Delta_x^i\Delta_y^j\nu(0,y), \qquad 0\leq i\leq n-1,\ 0\leq j\leq m, \tag{1.4}$$

$$\Delta_x^i\Delta_y^j\mu(x,0) \geq \Delta_x^i\Delta_y^j\nu(x,0), \qquad 0\leq i\leq n,\ 0\leq j\leq m-1, \tag{1.5}$$

$$\Delta_x^n\Delta_y^m\mu(x,y) \geq g(x,y,<\mu>),\ \Delta_x^n\Delta_y^m\nu(x,y) \leq g(x,y,<\nu>), \tag{1.6}$$

then we have

(1.7)

$$\Delta_x^i\Delta_y^j\mu(x,y) \geq \Delta_x^i\Delta_y^j\nu(x,y),\ 0\leq i\leq n, 0\leq j\leq m,\ i+j<n+m,\ (x,y)\in I_{pq}.$$

Further, strict inequalities in (1.4)–(1.6) imply strict inequalities in (1.7).

DEFINITION 1. Let $u(x,y)$ be a solution of the problem (1.1)–(1.3). Functions $w(x,y)$, $v(x,y)$, $(x,y)\in I_{pq}$, are said to be upper and lower solutions of the problem, if for $x\in I_p,\ y\in I_q$,

$$\Delta_x^n\Delta_y^m w(x,y) \geq g(x,y,<w>),\ \Delta_x^n\Delta_y^m v(x,y) \leq g(x,y,<v>), \tag{1.8}$$

$$\Delta_x^i\Delta_y^j w(0,y) \geq \Delta_x^i\Delta_y^j w(p,y), \qquad 0\leq i\leq n-1,\ 0\leq j\leq m, \tag{1.9}$$

$$\Delta_x^i\Delta_y^j v(0,y) \leq \Delta_x^i\Delta_y^j v(p,y), \qquad 0\leq i\leq n-1,\ 0\leq j\leq m, \tag{1.10}$$

$$\Delta_x^i\Delta_y^j w(x,0) \geq \Delta_x^i\Delta_y^j w(x,q), \qquad 0\leq i\leq n,\ 0\leq j\leq m-1, \tag{1.11}$$

$$\Delta_x^i\Delta_y^j v(x,0) \leq \Delta_x^i\Delta_y^j v(x,q), \qquad 0\leq i\leq n,\ 0\leq j\leq m-1. \tag{1.12}$$

2. Variation of Parameter Formula

Let $N_1 \neq -1$, $N_2 \neq -1$ be real numbers and let $'$ mean that the choice $i+j=0$ in the summation is deleted. For the linear partial difference equation

$$\begin{aligned}\Delta_x^n\Delta_y^m u(x,y) &= -\sum_{i=0}^{n}\sum_{j=0}^{m}{}' \binom{n}{i}\binom{m}{j}(-N_1)^i(-N_2)^j\Delta_x^{n-i}\Delta_y^{m-j}u(x,y)\\ &\quad +r(x,y), \qquad (x,y)\in I_{pq},\end{aligned} \tag{2.1}$$

subject to the periodic boundary conditions (1.2), (1.3), we have (cf. [**1, 7**]):

THEOREM 2. *Any solution $u(x,y)$ of the equation (2.1) and its differences are given by*

$$\begin{aligned}\Delta_x^i\Delta_y^j u(x,y) &= (1+N_1)^{x-n+i}(1+N_2)^{y-m+j}\\ &\quad \times\sum_{k=0}^{i}\sum_{\ell=0}^{j}\binom{i}{k}\binom{j}{\ell}\left(\frac{N_1}{1+N_1}\right)^k\left(\frac{N_2}{1+N_2}\right)^\ell\\ &\quad \times\left(\Phi_{i-k,j-\ell}(x,y)+\Psi_{i-k,j-\ell}(x,y)\right),\\ &\quad 0\leq i\leq n,\ 0\leq j\leq m,\ i+j\leq n+m-1,\ (x,y)\in I_{pq},\end{aligned} \tag{2.2}$$

where

$$(2.3)\quad \Phi_{i,j}(x,y) = \sum_{s=j}^{m-1} \frac{y^{(s-j)}}{(s-j)!}\Delta_x^i a_s(x) + \sum_{t=i}^{n-1} \frac{x^{(t-i)}}{(t-i)!}\frac{1}{(m-j-1)!} \times \sum_{\tau=0}^{y-m+j} (y-\tau-1)^{(m-j-1)}\Delta_\tau^m b_t(\tau),$$

$$(2.4)\quad \Psi_{i,j}(x,y) = \frac{1}{(n-i-1)!(m-j-1)!} \times \sum_{s=0}^{x-n+i}\sum_{t=0}^{y-m+j} (x-s-1)^{(n-i-1)}(y-t-1)^{(m-j-1)} \times (1+N_1)^{-s}(1+N_2)^{-t}r(s,t),$$

and $a_s(x)$, $b_t(y)$, $x \in I_p$, $y \in I_q$, $s = 0,1,\dots,m-1$; $t = 0,1,\dots,n-1$, *are arbitrary functions.*

THEOREM 3. *For the periodic boundary value problem* (2.1), (1.2), (1.3), *the functions* $a_s(x)$, $b_t(y)$, $x \in I_p$, $y \in I_q$, $s = 0,1,\dots,m-1$; $t = 0,1,\dots,n-1$, *in* (2.3) *are determined by*

$$(2.5)\quad (1+N_1)^p(\Phi_{i,0}(p,y) + \Psi_{i,0}(p,y)) = \Phi_{i,0}(0,y), \qquad 0 \le y \le q-1,$$

$$(2.6)\quad (1+N_2)^q(\Phi_{0,j}(x,q) + \Psi_{0,j}(x,q)) = \Phi_{0,j}(x,0), \qquad 0 \le x \le p-1,$$

$$0 \le i \le n-1,\ 0 \le j \le m-1.$$

3. Monotone Iterative Scheme

We consider

$$(3.1)\quad \Delta_x^n\Delta_y^m u_{k+1}(x,y) = -\sum_{i=0}^{n}\sum_{j=0}^{m}{}' \binom{n}{i}\binom{m}{j}(-N_1)^i(-N_2)^j \times\Delta_x^{n-i}\Delta_y^{m-j}(u_{k+1}-u_k)(x,y) + g(x,y,< u_k >),$$

$$(3.2)\quad \Delta_x^i u_{k+1}(0,y) = \Delta_x^i u_{k+1}(p,y), \qquad 0 \le i \le n-1,$$

$$(3.3)\quad \Delta_y^j u_{k+1}(x,0) = \Delta_y^j u_{k+1}(x,q), \qquad 0 \le j \le m-1,$$

$$k = 0,1,2,\dots,\ x \in I_p,\ y \in I_q.$$

Now we introduce the vector space $V = \{u : u = (u(0,0), u(0,1), \dots, u(0,q-1), u(1,0), u(1,1), \dots, u(1,q-1), \dots, u(p-1,0), u(p-1,1), \dots, u(p-1,q-1))^T\}$. In the space, we define a partial ordering in the following way: given any ν, $\mu \in V$, we say $\nu \preceq \mu$ if and only if

$$\Delta_x^i\Delta_y^j\nu \le \Delta_x^i\Delta_y^j\mu,\ 0 \le i \le n,\ 0 \le j \le m,\ n+m \le n+m-1.$$

The interval $\{u \in V : \nu \preceq u \preceq \mu\}$ is denoted by $[\nu,\mu]$.

THEOREM 4. *Let $N_1 \geq 0$, $N_2 \geq 0$ be two fixed parameters, v_0, w_0, $\eta \in V$, η being any function given by* (2.2), (2.5), (2.6), *and $v_0 \preceq \eta \preceq w_0$. Further, let* (1.8)–(1.12) *be satisfied for v_0, w_0. If*

$$(3.4) \quad g(x,y,<\mu>) - g(x,y,<\nu>) \geq \\ -\sum_{i=0}^{n}\sum_{j=0}^{m}{}' \binom{n}{i}\binom{m}{j}(-N_1)^i(-N_2)^j \Delta_x^{n-i}\Delta_y^{m-j}(\mu-\nu)(x,y),$$

whenever $<w_0> \geq <\mu> \geq <\nu> \geq <v_0>$, where μ, $\nu \in V$, then the sequences $\{v_k\}$, $\{w_k\}$ generated by the iterative scheme (3.1)–(3.3), *with the initial vectors v_0, w_0, converge monotonically to $v, w \in V$, respectively, that is,*

$$v_0 \preceq v_1 \preceq v_2 \preceq \cdots \preceq v_k \preceq \cdots \preceq v \preceq w \preceq \cdots \preceq w_k \preceq \cdots \preceq w_2 \preceq w_1 \preceq w_0,$$

and for any solution u of the problem (1.1)–(1.3) *satisfying $v_0 \preceq u \preceq w_0$ we have*

$$v \preceq u \preceq w.$$

Proof. Let

$$(3.5) \quad r(x,y) = g(x,y,<\eta>) \\ +\sum_{i=0}^{n}\sum_{j=0}^{m}{}' \binom{n}{i}\binom{m}{j}(-N_1)^i(-N_2)^j \Delta_x^{n-i}\Delta_y^{m-j}\eta(x,y), \ (x,y) \in I_{pq},$$

where $\eta(x,y) \in V$ and $<w_0> \geq <\eta> \geq <v_0>$. We consider (2.1) together with conditions (1.2), (1.3). The solution u of the problem (2.1), (1.2), (1.3) is given in (2.2). We can first show that the operator $T: \ T(\eta) = u$ is isotone from $[v_0, w_0]$ to U, where $U \subseteq V$ is the solution space of the problem (2.1), (1.2), (1.3).

Next, we may prove the monotonicity of the sequences $\{v_k\}$, $\{w_k\}$. Note that v_k, w_k for each k are solutions of (3.1)–(3.3). Using the monotonicity of the functions, and applying the Arzelà-Ascoli theorem, we find immediately that there exist v, w, with $v \preceq w$ such that

$$\lim_{k\to\infty} v_k(x,y) = v(x,y); \ \lim_{k\to\infty} w_k(x,y) = w(x,y), \ (x,y) \in I_{pq},$$

and this leads to the relations

$$\lim_{k\to\infty} \Delta_x^i\Delta_y^j v_k(x,y) = \Delta_x^i\Delta_y^j v(x,y), \\ \lim_{k\to\infty} \Delta_x^i\Delta_y^j w_k(x,y) = \Delta_x^i\Delta_y^j w(x,y), \\ 0 \leq i \leq n, \ 0 \leq j \leq m, \ i+j \leq n+m-1, \ (x,y) \in I_{pq}.$$

Further, by means of the continuity of T, we observe that

$$\lim_{k\to\infty} T^{k+1}(v_0) = T\left(\lim_{x\to\infty} T^k(v_0)\right) = T(v) = v, \ (x,y) \in I_{pq}, \\ \lim_{k\to\infty} T^{k+1}(w_0) = T\left(\lim_{x\to\infty} T^k(w_0)\right) = T(w) = w, \ (x,y) \in I_{pq}.$$

Thus, the conclusion follows immediately (cf. [**1**]). $\square$

COROLLARY 1. *If in Theorem 4 we further assume that for all* $\mu,\ \nu \in V$, *the function* g *satisfies the Lipschitz condition*

$$|g(x,y,<\mu>) - g(x,y,<\nu>)| \le \sum_{i=0}^{n}\sum_{j=0}^{m} \lambda_{i,j}|\Delta_x^i \Delta_y^j(\mu-\nu)(x,y)|,\ (x,y)\in I_{pq},$$

where $\lambda_{i,j} > 0$ *are constants, then* $v(x,y) \equiv w(x,y) \equiv u(x,y)$ *on* I_{pq}. *Consequently, problem* (1.1)–(1.3) *has a unique solution* $u(x,y) \in [v_0,\ w_0]$, *which is also the sole limit of the iterates* $v_k,\ w_k$.

REFERENCES

1. R. P. Agarwal, *The method of upper, lower solutions and monotone iterative scheme for higher order hyperbolic partial differential equations*, J. Austral. Math. Soc. (Ser. A) **47** (1989), 153–170.
2. R. P. Agarwal, *Difference Equations and Inequalities: Theory, Methods, and Applications*, Marcel Dekker, New York (1992).
3. S. R. Bernfeld and V. Lakshmikantham, *Monotone methods for nonlinear boundary value problems in Banach spaces*, Nonlinear Anal. **3** (1979), 303–316.
4. P. W. Eloe, *Difference equations and multipoint boundary value problems*, Proc. Amer. Math. Soc. **86** (1982), 253–259.
5. P. W. Eloe, *A boundary value problem for a system of difference equations*, Nonlinear Anal. **7** (1983), 813–820.
6. V. Lakshmikantham and S. G. Pandit, *Periodic solutions of hyperbolic partial differential equations*, Comput. Math. Appl. **11** (1985), 249–259.
7. Q. Sheng and R. P. Agarwal, *Periodic solutions of higher order hyperbolic partial differential equations*, Panamer. Math. J. **2** (1992), 1–22.

DEPARTMENT OF MATHEMATICS, NATIONAL UNIVERSITY OF SINGAPORE, KENT RIDGE, SINGAPORE 0511, SINGAPORE

E-mail address: matsheng@leonis.nus.sg

DEPARTMENT OF MATHEMATICS, NATIONAL UNIVERSITY OF SINGAPORE, KENT RIDGE, SINGAPORE 0511, SINGAPORE

E-mail address: matravip@nusvm.bitnet

Proceedings of Symposia in Applied Mathematics
Volume 48, 1994

Front Tracking Based on High-Resolution Wave Propagation Methods

KEH-MING SHYUE

1. Introduction

In previous work by LeVeque and the author [**6**],[**7**], a simple front-tracking approach has been developed and studied that models the propagation of discontinuous solutions for nonlinear hyperbolic systems of conservation laws in both one and two space dimensions. In this approach, we use a conservative finite volume method based on the *large time step* wave propagation approach [**4**],[**5**] to overcome a major difficulty associated with the limit on the time step when small cells are created by the tracked front cutting through the grid, while maintaining conservation of the algorithm. A variety of problems have been solved to validate the method for problems involving shock waves arising in gas dynamics. In this paper, we apply this method to the simulation of compressible Rayleigh-Taylor unstable interfaces, and to an oil reservoir simulation in porous media, in which a hyperbolic conservation equation is coupled with an elliptic PDE. Some preliminary results for both problems will be shown. Further details can be found in [**7**],[**10**].

2. Method

The basic idea of our front-tracking algorithm is quite simple. We choose a uniform underlying grid with some grid cells subdivided by tracked interfaces made up of moving points in 1D and curves in 2D, approximately aligned with discontinuities in the flow field. In each time step, we introduce a new set of

---

1991 *Mathematics Subject Classification.* Primary 65M06, 65M50; Secondary 76L05, 76M20.

This work was supported in part by NSF Grants DMS-8657319, DMS-9204329 and was conducted while the author was a graduate student in the Applied Mathematics Department at the University of Washington.

The final version of this paper will be submitted for publication elsewhere.

© 1994 American Mathematical Society
0160-7634/94 $1.00 + $.25 per page

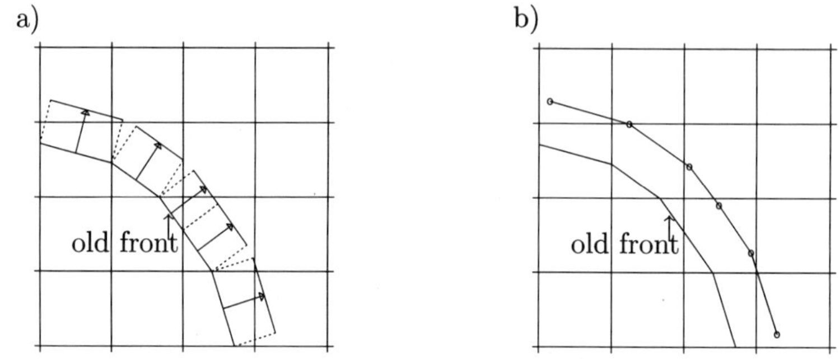

FIGURE 1. Two-dimensional front propagation. a) Tracked interfaces after advancing the original interfaces using the strong wave speeds obtained from solving the normal Riemann problems under the current time step. b) New tracked interfaces after taking an average of two neighboring points on the new interface location shown in a), and connecting the resulting points (indicated by large dots) consecutively by straight lines.

interfaces that are approximately aligned with the expected locations of discontinuities in the solution at the end of the time step. A high-resolution finite volume method is then applied on the resulting nonuniform grid to update the cell values. If we have chosen the new interface locations well, the resulting solution will remain sharp and be smooth away from these new interfaces. The old interfaces can then be eliminated by recombining the adjacent cells; see Figure 1.

To advance the tracked interfaces from the current time step to the next, we solve a one-dimensional Riemann problem at the tracked interface using the values from the adjacent cells as data, and follow strong waves to determine the new locations at the end of a time step. Since each tracked interface is the boundary between two cells, we can use the values from the adjacent cells as data to solve a one-dimensional Riemann problem in the direction normal to the interface. We expect the solution to this Riemann problem to consist of only one strong wave, corresponding to the shock or interface (contact discontinuity or slip line) being tracked, and other weaker waves.

Since we want the new tracked interfaces to form a continuous piecewise linear curve, as shown in Figure 1, we need to use the solutions of neighboring Riemann problems in some coordinated manner to determine the new interfaces. There are various ways that this can be done *via* some sort of curve fitting through points determined by the strong waves from the Riemann solutions; see Figure 1, for example. Our basic philosophy of tracking, however, gives us some flexibility on this score – we do not view the interface we introduce as being the definitive location of a tracked front, but rather as a grid interface that is sufficiently well located and aligned that the solution can be well captured on the resulting grid.

Note that for interface problems, the tracked front should be advanced not only in the normal direction to the front, as illustrated in Figure 1a, but also in the tangential direction. This can be done quite easily by moving points tangential to the front using, for example, an average tangential velocity from the data of the normal Riemann problem. It is easy to show that doing so gives exact front propagation for linear advection equations and for simultaneously diagonalizable linear hyperbolic systems. For general linear hyperbolic systems, or for nonlinear equations, this front-moving procedure gives a good approximation to the front motion. (See [**7**] for some test results.)

Once the new grid is constructed, the solution is then advanced using a fully conservative shock-capturing method. This method must be able to deal with the irregular cells near the tracked interface. In particular, we must maintain stability even if some of these cells are very small relative to the underlying mesh size used to determine the time step. We also hope to maintain second-order accuracy in the smooth flow on either side. This is accomplished using a high-resolution method based on the *large time step* wave propagation approach, developed by LeVeque [**4**],[**5**]. The main idea is that waves arising from the solution of Riemann problems at the cell boundaries are propagated the appropriate distance determined by the wave speed and time step, and used to update cell averages in any grid cell they encounter. The wave may affect more than one cell if the neighboring cell is very small. In this manner the stencil of the method adjusts automatically so that the CFL condition is always satisfied regardless of the configuration of the grid.

The high-resolution algorithm used for the calculations below is based on using rotated data to solve Riemann problems in the directions normal and tangential to the front, tangential splitting of the resulting waves to increase stability and resolution, and propagation of piecewise linear "correction waves" to increase the order of accuracy.

3. Numerical Results

We now show some preliminary results of interface tracking. As a first example, we consider a Rayleigh-Taylor unstable interface arising from the Euler equations of gas dynamics. (See [**9**] for an overview of this problem.) For comparison purposes, we choose the same initial setup as used in Gardner *et al.* [**1**], namely, we introduce a small single-mode perturbation of an isothermal equilibrium with a flat interface separating exponentially stratified flow above and below the interface. As in the laboratory experiments, the heavy fluid lies *below* the light fluid and gravity is directed *upwards*. For simplicity, we take the same gas with $\gamma = 1.4$ for both the heavy and light fluids. We use the following parameters: density ratio $D = 10$, $M^2 = 0.5$ (this parameter measures the compressibility of the fluids), $\varepsilon = 0.015$ (the amplitude of the perturbation). Figure 2 shows results at time $t = 6$ on a 40×160 grid. The growth of the interface and

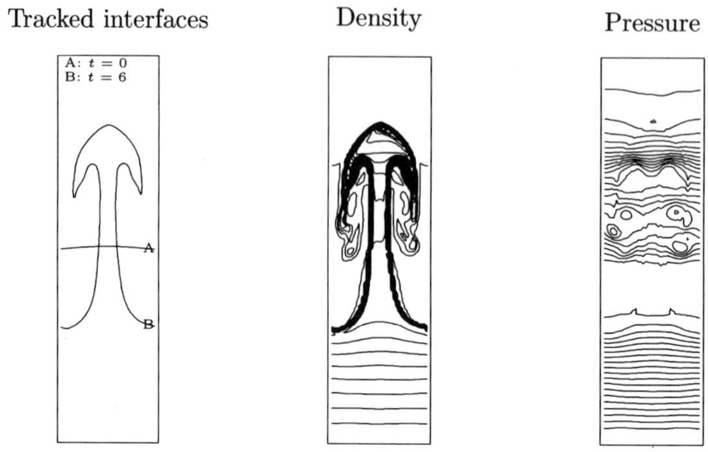

FIGURE 2. Results for the Rayleigh-Taylor instability at time $t = 6$ on a 40×160 grid. Parameters $D = 10$, $M^2 = 0.5$, and $\varepsilon = 0.015$ were used in the run.

the formation of a rising bubble and falling spike can be easily seen. Qualitatively, as far as the dominant unstable mode is concerned, our result agrees with the result shown in [**1**]. The results of [**1**] also show a secondary instability below the rising bubble. In our case, we do not observe such an instability. It is not clear which result is correct and work is continuing to clear up this discrepancy.

Our next example of interface tracking concerns a model problem for oil reservoir simulation in which the fluids are oil and solvent (such as CO_2). The governing equations for this model consist of a hyperbolic saturation equation, Darcy's law, and an equation enforcing the incompressibility of the fluids; see, *e.g.*, [**8**]. To solve this system of equations, we use the popular IMPES procedure, in which the hyperbolic saturation equation and the elliptic pressure equation are dealt with separately in two sequential steps. Here the hyperbolic saturation equation is solved using the front-tracking algorithm, and the elliptic pressure equation is solved using a standard five-point finite difference method on a uniform grid.

We consider a test problem in which the geometry is a standard five-spot pattern with fluid being injected into the center of a unit square domain and the oil being recovered from the four corners. The initial saturation we use is $s = 1$ (pure solvent) inside a perturbed circular interface and $s = 0$ (pure oil) outside the perturbed interface. In this model there is a dimensionless parameter called the frontal mobility ratio M, which characterizes the stability of the interface (for $M \leq 1$ the interface is stable, while for $M > 1$ the interface is unstable). Figure 3 shows results using a 50×50 grid with two different mobility ratios $M = 1$ and $M = 10$. For similar calculations, see [**2**],[**3**].

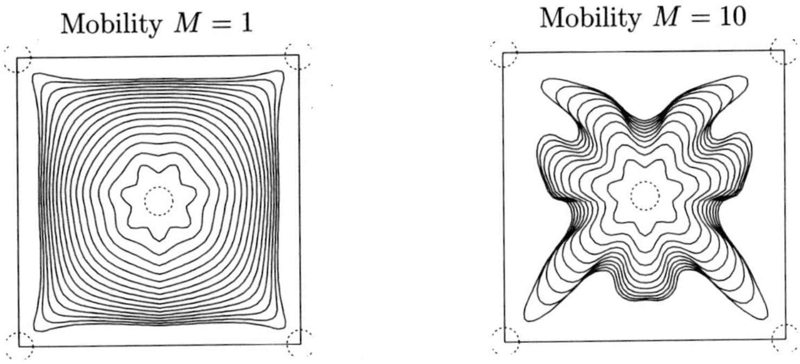

FIGURE 3. Evolution of the tracked interfaces for the miscible displacement oil reservoir simulation with mobility ratios $M = 1$ and $M = 10$, plotted every 10 time steps.

REFERENCES

1. C. L. Gardner, J. Glimm, O. McBryan, R. Menikoff, D. H. Sharp, and Q. Zhang, *The dynamics of bubble growth for Rayleigh-Taylor unstable interfaces*, Phys. Fluids, **31** (1988), 447-465.
2. J. Glimm, E. Isaacson, D. Marchesin, and O. A. McBryan, *Front tracking for hyperbolic systems*, Adv. in Appl. Math., **2** (1981), 91-119.
3. P. Lötstedt, *A front tracking method applied to Burgers' equation and two-phase porous flow*, J. Comput. Phys., **47** (1982), 211-228.
4. R. J. LeVeque, *A large time step generalization of Godunov's method for systems of conservation laws*, SIAM J. Numer. Anal., **22** (1985), 1051-1073.
5. ______, *High resolution finite volume methods on arbitrary grids via wave propagation*, J. Comput. Phys., **78** (1988), 36-63.
6. R. J. LeVeque and K.-M. Shyue, *One-dimensional front tracking based on high resolution wave propagation methods*, submitted for publication (1993).
7. ______, *Two-dimensional front tracking based on high resolution wave propagation methods*, in preparation (1993).
8. D. W. Peaceman, *Fundamentals of numerical reservoir simulation*, Elsevier, 1977.
9. D. H. Sharp, *An overview of Rayleigh-Taylor instability*, Physica D, **12** (1984), 3-18.
10. K.-M. Shyue, *Front tracking methods based on wave propagation*, Ph.D. thesis, University of Washington, June, 1993.

DEPARTMENT OF APPLIED MATHEMATICS AND STATISTICS, SUNY AT STONY BROOK, STONY BROOK, NY 11794

E-mail address: shyue@ams.sunysb.edu

Proceedings of Symposia in Applied Mathematics
Volume 48, 1994

Time-Splitting Methods for Nonhomogeneous Conservation Laws

TAO TANG AND ZHEN-HUAN TENG

ABSTRACT. In this work we show that time-splitting methods for conservation laws with source terms always converge to the unique solution satisfying the entropy condition. It is proved that the L^1 errors in the splitting methods are bounded by $O(\sqrt{\Delta t})$, where Δt is the splitting time step.

1. Introduction

We consider the initial value problem for the nonhomogeneous scalar conservation law

$$u_t + (f(u))_x = g(u), \qquad (x,t) \in R \times [0,T], \tag{1.1}$$

$$u(x,0) = u_0(x), \qquad x \in R. \tag{1.2}$$

For simplicity we discuss the one-dimensional equation (1.1) although the results generalize easily to the multi-dimensional case. Also, we assume that $g = g(u)$, but the results of this work can be easily extended to the more general case $g = g(x,u)$. We shall consider time-splitting methods for the numerical solutions of (1.1). In the simplest case the first step is to use the method known as operator splitting to remove the nonhomogeneous term $g(u)$ from (1.1). That is, we solve the homogeneous scalar conservation law

$$u_t + (f(u))_x = 0. \tag{1.3}$$

The second step is to solve the ordinary differential equation

$$u_t = g(u). \tag{1.4}$$

1991 *Mathematics Subject Classification.* Primary 65M15; Secondary 35L65.

The first author was supported in part by NSERC Canada Grant OGP0105545. The second author was supported in part by the National Natural Science Foundation of China.

The detailed version of this paper has been submitted for publication elsewhere.

© 1994 American Mathematical Society
0160-7634/94 $1.00 + $.25 per page

Let $S(t)$ denote the exact solution operator of (1.1) which satisfies the entropy conditions. Therefore, the solution of (1.1)–(1.2) can be expressed in the form $u(x,t) = S(t)u_0$. Similarly, let $S_1(t)$ and $S_2(t)$ denote the solution operators of (1.3) and (1.4), respectively. The first-order fractional step method ([**2**]) is based on the approximation

$$S(t_n)u_0 \approx (S_2(\Delta t)S_1(\Delta t))^n u_0, \qquad t_n = n\Delta t \in [0,T], \tag{1.5}$$

or on the one with the roles of S_2 and S_1 reversed, where Δt is the splitting time step. To maintain second-order accuracy, the Strang splitting [**4**] can be used, in which the solution $S(t_n)u_0$ is approximated by

$$S(t_n)u_0 \approx (S_2(\Delta t/2)S_1(\Delta t)S_2(\Delta t/2))^n u_0, \tag{1.6}$$

or by the one with the roles of S_2 and S_1 reversed. It should be pointed out that first-order accuracy and second-order accuracy are based on the truncation errors for smooth solutions. For discontinuous solutions of conservation laws, it is not difficult to show that both approximations (1.5) and (1.6) are at most first-order accurate ([**1**]). To analyze the principal properties of fractional step methods for discontinuous solutions, we shall concentrate on the splitting method (1.5). The main results for (1.5) in this work can be easily extended to the scheme (1.6).

THEOREM 1. *Let $u_0 \in BV(R)\cap L^\infty(R)\cap L^1(R)$, $f \in C^1(R)$ and assume that g satisfies a Lipschitz condition and $g(0) = 0$. Let $S(t)u_0$ denote the unique weak solution of* (1.1)–(1.2) *satisfying the entropy condition; then the L^1 convergence rate of the semidiscrete fractional step algorithm* (1.5) *is* 1/2. *More precisely, for any $t_n = n\Delta t \in [0,T]$, the following estimate holds:*

$$\left\| S(t_n)u_0 - \Big(S_2(\Delta t)S_1(\Delta t)\Big)^n u_0 \right\|_{L^1(R)} \le C\sqrt{\Delta t},$$

where C is a constant independent of Δt. Similar results are valid with the roles of S_2 and S_1 reversed.

2. Proof of Theorem 1

We first introduce a nonnegative function $\theta \in C^\infty$ satisfying $\theta(\xi) = \theta(-\xi)$, $\theta(\xi) = 0$ for $|\xi| \ge 1$, $\mathrm{sgn}(\xi)\theta'(\xi) \le 0$, and $\int_R \theta(\xi)d\xi = 1$. For $\epsilon > 0$, let

$$\tilde{\omega}(x,t) = \theta^\epsilon(t)\theta^\epsilon(x), \qquad \theta^\epsilon(t) = \frac{1}{\epsilon}\theta(t/\epsilon), \quad \theta^\epsilon(x) = \frac{1}{\epsilon}\theta(x/\epsilon).$$

Assume functions $p, q \in L^\infty(R\times[0,T])$ satisfying

$$\|p(\cdot+h,\tau_2) - p(\cdot,\tau_1)\|_{L^1(R)} \le A\max\{|h|, |\tau_2-\tau_1|\}, \tag{2.1}$$

$$\|q(\cdot+h,\tau_2) - q(\cdot,\tau_1)\|_{L^1(R)} \le B\max\{|h|, |\tau_2-\tau_1|\}, \tag{2.2}$$

for any $h \in R$ and $\tau_2, \tau_1 \in [0,T]$, where A and B are positive constants independent of h, τ_1 and τ_2. For a given $t \in (0,T]$, let $S = R\times[0,t]$, $ds = dxd\tau$ and

$ds' = dx'd\tau'$. For $p, q \in L^\infty(R \times [0,T])$ satisfying (2.1) and (2.2), we define the functional Λ as follows:

$$\begin{aligned}\Lambda(p,q,t) \quad &:= \quad -\int_S\int_S\Big(|p-q'|\omega_\tau + \mathrm{sgn}(p-q')(f(p)-f(q'))\omega_x\Big)dsds' \\ &\qquad + \int_S\int_R \{|p-q'|\omega\}\Big|_{\tau=0}^{\tau=t} dxds',\end{aligned} \tag{2.3}$$

where $S = R\times[0,t], p = p(x,\tau), q' = q(x',\tau')$ and $\omega = \omega(x,x',\tau,\tau') = \tilde{\omega}(x-x',\tau-\tau')$. Our global error bound result, Lemma 2.1, is based on Kuznetsov's lemma [**3**], which states that if $p, q \in L^\infty(R\times[0,T])$ satisfy (2.1) and (2.2), respectively, then for any $t \in [\epsilon, T]$,

$$\begin{aligned}&\|p(\cdot,t)-q(\cdot,t)\|_{L^1(R)} \\ &\le \|p(\cdot,0)-q(\cdot,0)\|_{L^1(R)} + \Lambda(p,q,t) + \Lambda(q,p,t) + (A+B)\epsilon.\end{aligned} \tag{2.4}$$

Following the notations in [**1**], we define $u_\Delta(x,t)$ as follows:

$$u_\Delta = \begin{cases} S_1(2(t-t_n))(S_2(\Delta t)S_1(\Delta t))^n u_0, & t\in[t_n, t_{n+1/2}),\\ S_2(2(t-t_{n+1/2}))S_1(\Delta t)(S_2(\Delta t)S_1(\Delta t))^n u_0, & t\in[t_{n+1/2}, t_{n+1}),\end{cases}$$

where $t_n = n\Delta t$ and $t_{n+1/2} = (n+1/2)\Delta t$. We can show that both u and u_Δ belong to $L^\infty(R\times[0,T]) \cap L^1(R\times[0,T])$ and satisfy (2.1) and (2.2).

Throughout this paper, C denotes a positive constant independent of $\epsilon, \Delta t$ and Δx, but possibly with different values at different places. Moreover, u' and u'_Δ denote $u(x',\tau')$ and $u_\Delta(x',\tau')$, respectively; and u, u_Δ and ω should be understood as $u(x,\tau), u_\Delta(x,\tau)$ and $\omega(x,x',\tau,\tau')$, respectively.

LEMMA 2.1. *Let u_0, f and g be subject to the conditions stated in Theorem* 1. *Moreover, if u, the entropy solution of* (1.1)–(1.2), *and u_Δ satisfy the condition*

$$\begin{aligned}&\Lambda(u_\Delta, u, t_n) \\ &\le -2\sum_{m=0}^{n-1}\int_{t_{m+1/2}}^{t_{m+1}}\int_R\int_S \mathrm{sgn}(u-u'_\Delta)g(u'_\Delta)\omega dsdx'd\tau' + \frac{C\Delta t}{\epsilon},\end{aligned} \tag{2.5}$$

where $S = R\times[0,t_n]$, then the following result holds:

$$\begin{aligned}&\|u(\cdot,t_n)-u_\Delta(\cdot,t_n)\|_{L^1(R)} \\ &\le \|u(\cdot,0)-u_\Delta(\cdot,0)\|_{L^1(R)} + C\left(\epsilon+\Delta t+\frac{\Delta t}{\epsilon}\right).\end{aligned} \tag{2.6}$$

In order to obtain Lemma 2.1, we need to use the following inequality ([**5**]):

$$\Lambda(u,u_\Delta,t_n) \le 2\sum_{m=0}^{n-1}\int_{t_{m+1/2}}^{t_{m+1}}\int_R\int_S \mathrm{sgn}(u-u'_\Delta)g(u)\omega dsdx'd\tau' + C\left(\Delta t+\frac{\Delta t}{\epsilon}\right).$$

Note that g satisfies a Lipschitz condition. The above inequality, together with (2.4) and (2.5), yields, for all $t_n \in [\epsilon, T]$, that

$$\begin{aligned}
&\|u(\cdot,t_n) - u_\Delta(\cdot,t_n)\|_{L^1(R)} \\
\leq\ & 2\sum_{m=0}^{n-1}\int_{t_{m+1/2}}^{t_{m+1}}\int_R\int_S \operatorname{sgn}(u-u'_\Delta)(g(u)-g(u'_\Delta))\omega ds dx' d\tau' \\
&+\|u(\cdot,0)-u_\Delta(\cdot,0)\|_{L^1(R)} + C\left(\epsilon+\Delta t+\frac{\Delta t}{\epsilon}\right) \\
\leq\ & 2\int_S\int_S |g(u)-g(u'_\Delta)|\omega ds ds' + \|u(\cdot,0)-u_\Delta(\cdot,0)\|_{L^1(R)} \\
&+C\left(\epsilon+\Delta t+\frac{\Delta t}{\epsilon}\right) \\
\leq\ & 2L\int_S\int_S |u-u'_\Delta|\omega ds ds' + \|u(\cdot,0)-u_\Delta(\cdot,0)\|_{L^1(R)} + C\left(\epsilon+\Delta t+\frac{\Delta t}{\epsilon}\right),
\end{aligned}$$

where L is the Lipschitz constant of g. Using the properties of ω, we can easily show that

$$\int_S\int_R \left|u(x,\tau)-u_\Delta(x',\tau')\right|\omega dx ds' \leq \|u(\cdot,\tau)-u_\Delta(\cdot,\tau)\|_{L^1(R)} + C\epsilon.$$

From the above two results, we obtain that for all $t_n \in [\epsilon, T]$

$$\begin{aligned}
\|u(\cdot,t_n) - u_\Delta(\cdot,t_n)\|_{L^1(R)} &\leq \|u(\cdot,0)-u_\Delta(\cdot,0)\|_{L^1(R)} \\
&+\ C\int_0^{t_n}\|u(\cdot,\tau)-u_\Delta(\cdot,\tau)\|_{L^1(R)}d\tau + C\left(\epsilon+\Delta t+\frac{\Delta t}{\epsilon}\right).
\end{aligned} \tag{2.7}$$

Note that $\|u(\cdot,t_n)-u_\Delta(\cdot,t_n)\|_{L^1(R)} \leq \|u(\cdot,0)-u_\Delta(\cdot,0)\|_{L^1(R)} + C\epsilon$ in the case $t_n \in [0,\epsilon]$. This, together with the Gronwall-type inequality (2.7), yields (2.6).

In order to obtain Theorem 1, we now only need to prove (2.5). To this end, we need the following results; their proofs can be found in [**5**]:

$$\begin{aligned}
&-\sum_{m=0}^{n-1}\int_S\int_R\int_{t_m}^{t_{m+1/2}}\Big(|u_\Delta-u'|\omega_\tau \\
&\qquad\qquad +2\operatorname{sgn}(u_\Delta-u')(f(u_\Delta)-f(u'))\omega_x\Big)d\tau dx ds' \\
\leq\ & -\sum_{m=0}^{n-1}\int_S\int_R\Big(|u_\Delta-u'|\omega\Big)\Big|_{t_m}^{t_{m+1/2}} dx ds'; \\
&-\sum_{m=0}^{n-1}\int_S\int_R\int_{t_{m+1/2}}^{t_{m+1}}\Big(|u_\Delta-u'|\omega_\tau + 2\operatorname{sgn}(u_\Delta-u')g(u_\Delta)\omega\Big)d\tau dx ds' \\
=\ & -\sum_{m=0}^{n-1}\int_S\int_R\Big(|u_\Delta-u'|\omega\Big)\Big|_{t_{m+1/2}}^{t_{m+1}} dx ds';
\end{aligned}$$

$$
\begin{aligned}
&-\sum_{m=0}^{n-1}\int_S\int_R\int_{t_{m+1/2}}^{t_{m+1}} \operatorname{sgn}(u_\Delta - u')(f(u_\Delta) - f(u'))\omega_x d\tau dx ds' \\
&\quad +\sum_{m=0}^{n-1}\int_S\int_R\int_{t_m}^{t_{m+1/2}} \operatorname{sgn}(u_\Delta - u')(f(u_\Delta) - f(u'))\omega_x d\tau dx ds' \\
\leq\ & C\Delta t/\epsilon.
\end{aligned}
$$

Adding the above three inequalities yields

$$
\Lambda(u_\Delta, u, t_n) \leq 2\sum_{m=0}^{n-1}\int_S\int_R\int_{t_{m+1/2}}^{t_{m+1}} \operatorname{sgn}(u_\Delta - u')g(u_\Delta)\omega d\tau dx ds' + \frac{C\Delta t}{\epsilon}.
$$

By changing notations, $\tau \leftrightarrow \tau'$, $x \leftrightarrow x'$, and noting that

$$
\omega(x, x', \tau, \tau') = \omega(x', x, \tau', \tau),
$$

we get

$$
\Lambda(u_\Delta, u, t_n) \leq -2\sum_{m=0}^{n-1}\int_{t_{m+1/2}}^{t_{m+1}}\int_R\int_S \operatorname{sgn}(u - u'_\Delta)g(u'_\Delta)\omega ds dx' d\tau' + \frac{C\Delta t}{\epsilon}.
$$

This completes the proof of (2.5). Finally, noting that $u_\Delta(\cdot, 0) = u(\cdot, 0)$, we obtain by using Lemma 2.1 that

$$
\|u(\cdot, t_n) - u_\Delta(\cdot, t_n)\|_{L^1(R)} \leq C\left(\epsilon + \Delta t + \frac{\Delta t}{\epsilon}\right). \tag{2.8}
$$

Hence, setting $\epsilon = \sqrt{\Delta t}$ in (2.8) yields Theorem 1.

REFERENCES

1. M. Crandall and A. Majda, *The method of fractional steps for conservation laws*, Numer. Math. **34** (1980), pp. 285–314.
2. S. K. Godunov, *Finite difference methods for numerical computations of discontinuous solutions of the equations of fluid dynamics*, Mat. Sb. **47** (1959), pp. 271–295 (in Russian).
3. N. N. Kuznetsov, *Accuracy of some approximate methods for computing the weak solutions of a first-order quasi-linear equation*, USSR Comput. Math. and Math. Phys. **16** (1976), pp. 105–119.
4. G. Strang, *On the construction and comparison of difference schemes*, SIAM J. Numer. Anal. **5** (1968), pp. 506–517.
5. T. Tang and Z.-H. Teng, *Error bounds for fractional step methods for conservation laws with source terms*, Research report 92–16, Dept. of Mathematics and Statistics, Simon Fraser University, Canada.

DEPARTMENT OF MATHEMATICS AND STATISTICS, SIMON FRASER UNIVERSITY, BURNABY, BRITISH COLUMBIA, CANADA V5A 1S6

E-mail address: tangtao@cs.sfu.ca

DEPARTMENT OF MATHEMATICS, PEKING UNIVERSITY, BEIJING 100871, CHINA

E-mail address: tengzh%bepc2@scs.slac.stanford.edu

Proceedings of Symposia in Applied Mathematics
Volume 48, 1994

Numerical Aspects of Uniform Airy-type Asymptotic Expansions

N. M. TEMME

ABSTRACT. An algorithm for evaluating Airy-type asymptotic representations of a class of special functions is discussed. The method is quite general, but the paper concentrates on Bessel functions. The algorithm is based on the differential equation of the Bessel functions, which has a turning point when order and argument of the Bessel functions are equal.

1. INTRODUCTION

From a computational point of view Airy-type uniform asymptotic expansions are unattractive. This is mainly due to the complexity of the coefficients in the expansions. In all known cases, for example Bessel functions, the coefficients are difficult to compute in the neighborhood of the turning point. Although in that case there are closed expressions for the coefficients in terms of elementary functions, there is massive cancellation in evaluating these expressions near the turning point.

In this paper we discuss a new numerical algorithm for uniform asymptotic representations of a class of functions. The method is quite general, but we only treat the case of Bessel functions. We use differential equations for functions representing (in an exact sense) the asymptotic series, and we base a numerical algorithm directly on these differential equations. Extra features are that we deal with *convergent* expansions, that we need only a small number of precomputed tabulated numbers, and that the method is applicable for quite small values of the large parameter. In some sense the method is similar to the one described for the computation of incomplete gamma functions in [4]. In that case the error function is the main approximant.

2. AIRY-TYPE ASYMPTOTICS OF ORDINARY BESSEL FUNCTIONS

The ordinary Bessel functions $J_\nu(z)$ and $Y_\nu(z)$ can be expanded in terms of Airy functions. From [3, p. 425], we obtain

$$
\begin{aligned}
J_\nu(\nu z) &\sim \frac{\phi(\zeta)}{\nu^{1/3}}\left[\mathrm{Ai}(\nu^{2/3}\zeta)\sum_{s=0}^{\infty}\frac{A_s(\zeta)}{\nu^{2s}}+\nu^{-4/3}\mathrm{Ai}'(\nu^{2/3}\zeta)\sum_{s=0}^{\infty}\frac{B_s(\zeta)}{\nu^{2s}}\right],\\
Y_\nu(\nu z) &\sim -\frac{\phi(\zeta)}{\nu^{1/3}}\left[\mathrm{Bi}(\nu^{2/3}\zeta)\sum_{s=0}^{\infty}\frac{A_s(\zeta)}{\nu^{2s}}+\nu^{-4/3}\mathrm{Bi}'(\nu^{2/3}\zeta)\sum_{s=0}^{\infty}\frac{B_s(\zeta)}{\nu^{2s}}\right],
\end{aligned}
\tag{2.1}
$$

as $\nu \to \infty$, uniformly with respect to $z \in [0, \infty)$. The expansions are valid for complex values of ν and z, but here we concentrate on real values of the parameters.

1991 *Mathematics Subject Classification.* Primary 65D20; Secondary 41A60, 33C10, 34E20.
Key words and phrases. Uniform asymptotic expansions, computation of special functions, turning points, Airy-type expansions, Bessel functions.

The final version of this paper will be submitted for publication elsewhere.

© 1994 American Mathematical Society
0160-7634/94 \$1.00 + \$.25 per page

The parameter ζ is defined by

$$\begin{aligned} \frac{2}{3}\zeta^{3/2} &= \ln\frac{1+\sqrt{1-z^2}}{z} - \sqrt{1-z^2}, \quad 0 \le z \le 1, \\ \frac{2}{3}(-\zeta)^{3/2} &= \sqrt{z^2-1} - \operatorname{arccos}\frac{1}{z}, \quad z \ge 1. \end{aligned} \tag{2.2}$$

Furthermore,

$$\phi(\zeta) = \left(\frac{4\zeta}{1-z^2}\right)^{1/4}. \tag{2.3}$$

The coefficients A_s, B_s are given in [1, p. 368], and [3, p. 425], with

$$A_0(\zeta) = 1, \quad B_0(\zeta) = -\frac{5}{48\zeta^2} + \frac{\phi^2(\zeta)}{2\zeta}\left[\frac{5}{24(1-z^2)} - \frac{1}{8}\right].$$

The coefficients A_s, B_s in (2.1) are complicated expressions. In particular, they are regular but rather difficult to compute near the *turning point* $z = 1$, or equivalently, near $\zeta = 0$. In [2] all needed coefficients A_s, B_s are expanded in Maclaurin series at the turning point, the power series being in terms of the variable $w^2 = 1 - z^2$. We propose a different approach, in which the direct computation of A_s, B_s can be circumvented. We define two functions F_ν, G_ν which represent the asymptotic series in the above representations of the Bessel functions:

$$\begin{aligned} J_\nu(\nu z) &= \phi(\zeta)\nu^{-1/3}\left[\operatorname{Ai}(\nu^{2/3}\zeta)\, F_\nu(\zeta) + \nu^{-4/3}\operatorname{Ai}'(\nu^{2/3}\zeta)\, G_\nu(\zeta)\right], \\ Y_\nu(\nu z) &= -\phi(\zeta)\nu^{-1/3}\left[\operatorname{Bi}(\nu^{2/3}\zeta)\, F_\nu(\zeta) + \nu^{-4/3}\operatorname{Bi}'(\nu^{2/3}\zeta)\, G_\nu(\zeta)\right]. \end{aligned} \tag{2.4}$$

Olver's approach for deriving Airy-type expansions for the Bessel functions is based on the differential equation

$$\frac{d^2W}{d\zeta^2} = [\nu^2\zeta + \psi(\zeta)]W, \tag{2.5}$$

where

$$\psi(\zeta) = \frac{5}{16\zeta^2} + \frac{\zeta z^2(z^2+4)}{4(z^2-1)^3}, \tag{2.6}$$

the relation between ζ and z being given in (2.2). The quantities within the square brackets in (2.4) are two solutions of equation (2.5).

By using equation (2.5), we can derive the following system of differential equations for the functions $F_\nu(\zeta), G_\nu(\zeta)$:

$$\begin{aligned} F'' + 2\zeta G' + G - \psi(\zeta)\, F &= 0, \\ G'' + 2\nu^2 F' - \psi(\zeta)\, G &= 0, \end{aligned} \tag{2.7}$$

where primes denote differentiation with respect to ζ. A Wronskian for the system (2.7) follows by eliminating the terms $\psi(\zeta)F_\nu(z)$ and $\psi(\zeta)G_\nu(z)$. This gives

$$F'G - FG' + \zeta G^2 - \nu^2 F^2 = -\nu^2. \tag{2.8}$$

The constant on the right-hand side follows by taking $\zeta = 0$ and by using the known behavior of the Bessel functions at the turning point; see [1, pp. 368 and 446].

3. Evaluation of the functions $F_\nu(\zeta)$, $G_\nu(\zeta)$

We solve the system of differential equations (2.7) by using analytical techniques. We expand the functions $F_\nu(\zeta)$, $G_\nu(\zeta)$ in Maclaurin series. The singular points of the function $\psi(\zeta)$, and hence those of the differential equations, occur at

$$\zeta^{\pm} = \left(\tfrac{3}{2}\pi\right)^{2/3} e^{\pm i\pi/3}$$

(see [3, p. 421]). It follows that the radius of convergence of the series of $F_\nu(\zeta)$ and $G_\nu(\zeta)$ in powers of ζ equals $2.81\ldots$.

We expand

$$F_\nu(\zeta) = \sum_{n=0}^{\infty} f_n(\nu)\zeta^n, \quad G_\nu(\zeta) = \sum_{n=0}^{\infty} g_n(\nu)\zeta^n, \quad \psi(\zeta) = \sum_{n=0}^{\infty} \psi_n\zeta^n. \tag{3.1}$$

The coefficients $f_0, f_1, \ldots, g_0, g_1, \ldots$ are to be determined, while the coefficients ψ_n are known. The first few ψ_n are

$$\psi_0 = \tfrac{1}{70}2^{1/3}, \quad \psi_1 = \tfrac{2}{75}, \quad \psi_2 = \tfrac{69}{13475}2^{2/3}, \quad \psi_3 = -\tfrac{148}{73125}2^{1/3}. \tag{3.2}$$

Upon substituting the expansions into (2.7), we obtain for $n = 0, 1, 2, \ldots$ the system of recursion relations

$$\begin{aligned} &(n+2)\,(n+1)\,f_{n+2} + (2n+1)\,g_n = \phi_n, \qquad &\phi_n = \sum_{k=0}^{n} f_k\psi_{n-k}, \\ &(n+2)\,(n+1)\,g_{n+2} + 2\nu^2(n+1)\,f_{n+1} = \chi_n, \qquad &\chi_n = \sum_{k=0}^{n} g_k\psi_{n-k}. \end{aligned} \tag{3.3}$$

The system of recursions in (3.3) is not stable in forward direction (especially when ν is large). When we try to use (3.3) in backward direction, for instance with false starting values f_N, g_N for some large integer N, a complication arises because of the terms ϕ_n, χ_n on the right-hand sides of (3.3). All terms ϕ_n, χ_n contain f_k, g_k for $k = 0, 1, 2, \ldots, n$. Hence, recursion in backward direction is not possible at all. A way out is to consider ϕ_n, χ_n as known quantities, and to treat (3.3) as a system of inhomogeneous difference equations.

A first step in this approach will be to solve the system (2.7) by iteration. That is, we choose an appropriate pair of functions F_0, G_0, and define two sequences of functions $\{F_m\}$, $\{G_m\}$ by writing for $m = 1, 2, 3, \ldots$:

$$\begin{aligned} F''_m + 2\zeta G'_m + G_m &= \psi(\zeta)\,F_{m-1}, \\ G''_m + 2\nu^2 F'_m &= \psi(\zeta)\,G_{m-1}. \end{aligned} \tag{3.4}$$

We rewrite (3.3) in backward form:

$$\begin{aligned} f_n &= \tfrac{1}{2\nu^2}\left[\tfrac{1}{n}\chi_{n-1} - (n+1)g_{n+1}\right], \\ g_{n-1} &= \tfrac{1}{2n-1}\left[\phi_{n-1} - n(n+1)f_{n+1}\right], \end{aligned} \tag{3.5}$$

where $n \geq 1$. The coefficients are assumed to belong to the functions $F_m(\zeta)$, $G_m(\zeta)$ of the iteration process described by (3.4), while the coefficients ϕ_{n-1}, χ_{n-1} are assumed to be known, and contain Maclaurin coefficients of $F_{m-1}(\zeta)$, $G_{m-1}(\zeta)$ and $\psi(\zeta)$. Observe that (3.5) does not define f_0. After having computed

$$f_1, f_2, \ldots, g_0, g_1, g_2, \ldots$$

by the backward recursion process, we compute f_0 from the Wronskian (2.8):

$$f_0 = \frac{-g_1 + \sqrt{g_1^2 + 4\nu^2(\nu^2 + f_1 g_0)}}{2\nu^2}. \tag{3.6}$$

$\nu = 25$

i	$\lvert f_0 - f_0^a\rvert$	$\lvert g_0 - g_0^a\rvert$	$\lvert f_5 - f_5^a\rvert$	$\lvert g_5 - g_5^a\rvert$	$\lvert f_{10} - f_{10}^a\rvert$	$\lvert g_{10} - g_{10}^a\rvert$
1	1.12e−11	5.29e−11	4.33e−12	1.55e−09	1.45e−13	4.09e−11
2	3.66e−16	4.00e−13	2.85e−14	8.86e−14	1.18e−15	5.32e−15
3	0.00e−00	2.52e−17	2.66e−18	1.55e−16	1.25e−18	4.63e−16
4	0.00e−00	0.00e−00	5.01e−20	1.46e−19	1.14e−18	4.48e−16
5	0.00e−00	1.69e−21	5.53e−20	1.47e−19	1.14e−18	4.48e−16

Table 1. Absolute errors during five iterations (i) of $f_0, g_0, f_5, g_5, f_{10}, g_{10}$ compared with more accurate values f_0^a, etc.

4. Numerical experiments

We have computed successive iterates of Maclaurin coefficients f_n, g_n defined in (3.1) for different values of ν. We have used simple starting values of the first set $\{f_n, g_n\}$, using the known behavior of the Maclaurin coefficients. During each iteration we start the backward recursions with $f_n = g_{n-1} = 0, n \geq 16$, and we compute $f_{15}, g_{14}, f_{14}, g_{13}, \ldots$ by using (3.5). We recompute the coefficients $\phi_k, \chi_k, k = 0, 1, 2, \ldots, 14$, using (3.3) with values f_k, g_k obtained in the previous iteration. For $\nu = 25$, Table 1 gives the errors in the values $f_0, g_0, f_5, g_5, f_{10}, g_{10}$, when compared with more accurate values f_0^a, etc.

From the table we conclude that for $\nu = 25$ we can already obtain an accuracy of 10^{-13} after two iterations. The algorithm is very efficient, especially when ν becomes larger.

References

1. M. Abramowitz and I.A. Stegun, *Handbook of mathematical functions*, Washington D.C., 1964.
2. D. Amos, *Algorithm* 644, *A portable package for Bessel functions of a complex argument and nonnegative order*, ACM Trans. Math. Software **12** (1986), 265–273.
3. F.W.J. Olver, *Asymptotics and Special Functions*, Academic Press, New York, 1974.
4. N.M. Temme, *On the computation of the incomplete gamma functions for large values of the parameters*, in *Algorithms for approximation*, Shrivenham, July 1985 (J. C. Mason and M.G. Cox, eds.), Clarendon, Oxford, 1987, pp. 479–489.
5. F.G. Tricomi, *Repertorium der Theorie der Differentialgleichungen*, Springer, Berlin, 1968.

CWI, P.O. Box 94079, 1090 GB Amsterdam, The Netherlands
E-mail address: nicot@cwi.nl

Proceedings of Symposia in Applied Mathematics
Volume 48, 1994

LOCAL DYNAMICS AND BIFURCATION CONSISTENCIES OF CONTINUOUS-TIME DYNAMICAL SYSTEMS AND THEIR NUMERICAL DISCRETIZATIONS

XIN WANG EDWARD K. BLUM QINGNAN LI

1. INTRODUCTION

Let X be an open subset of $\mathbb{R}^n$ and M be an open interval of $\mathbb{R}$. Given an autonomous one-parameter family of ordinary differential equations,

$$\dot{x}(t) = F(x(t); \mu), \; x(t) \in X, \; \mu \in M, \; t \geq 0, \tag{1.1}$$

where $F \in C^1(X \times M, \mathbb{R}^n)$, consider a one-step numerical method of order $p \geq 1$ [2, 6] for solving the system (1.1) with constant time step size $h > 0$,

$$\begin{aligned} x_h(k+1) &= x_h(k) + hF_h(x_h(k); \mu) \equiv f_h(x_h(k); \mu), \\ &\quad x_h(k) \in X, \; \mu \in M, \; k = 0, 1, 2, ..., \end{aligned} \tag{1.2}$$

where $F_h \in C^1(X \times M, X)$. Assume that $\Phi : \mathbb{R}_+ \times X \times M \to X$ is the continuous-time flow of the dynamical system defined by (1.1), namely,

$$x(t) = \Phi(t, x; \mu), \; x(0) = x, \; t \geq 0, \tag{1.3}$$

and let $\phi_h : \mathbb{N} \times X \times M \to X$ be the discrete-time flow of the dynamical system defined by (1.2), namely,

$$x_h(k) = \phi_h(k, x; \mu) = f_h^k(x; \mu), \; x_h(0) = x, \; k = 0, 1, 2, ... \tag{1.4}$$

The consistency problems we consider in this paper are as follows:

(a) Near a fixed point $x^* = x^*(\mu)$ of Φ, does ϕ_h have a fixed point $x_h^* = x_h^*(\mu)$?
(b) If so, does ϕ_h have the same dynamics near x_h^* as Φ has near x^*?
(c) When Φ undergoes a bifurcation near x^* as μ is varied, does ϕ_h go through the same bifurcation near x_h^*?

Our study of these consistency problems is motivated by our use of numerical methods for analyzing and simulating continuous-time dynamical systems such as Φ; the consistency problems address the reliability of the numerical methods for studying dynamical and bifurcation behavior of these systems. A simple example [3] showing that a numerical method may have different behavior is a damped nonlinear oscillator given by

$$\dot{x}_1 = x_2 - \mu x_1^3/3, \; \dot{x}_2 = -x_1. \tag{1.5}$$

1991 *Mathematics Subject Classification.* Primary 58F10, 58F14, 58F25, 58F30, 65L05, 65L07.
The detailed version of this paper will be submitted for publication elsewhere.

© 1994 American Mathematical Society
0160-7634/94 $1.00 + $.25 per page

Here, $(0, 0)$ is the unique fixed point, which is asymptotically stable when $\mu > 0$ and asymptotically unstable when $\mu < 0$ [9]; $\mu^* = 0$ is the bifurcation point. However, in the Euler method for solving (1.5),

(1.6)
$$x_{1h}(k+1) = x_{1h}(k) + h(x_{2h}(k) - \mu x_{1h}^3(k)/3), x_{2h}(k+1) = x_{2h}(k) - hx_{1h}(k),$$

the fixed point $(0, 0)$ is always asymptotically unstable and no local bifurcation occurs, no matter how small the step size h is. The reason for this is that the fixed point $(0, 0)$ is not *hyperbolic* for any μ in the dynamical system defined by (1.5).

In this paper, we formulate the concepts of (local) dynamics and bifurcation consistencies of the dynamical systems Φ in (1.3) and ϕ_h in (1.4) in terms of dynamical system theory and present general results for the consistencies of dynamics near hyperbolic fixed points and bifurcation due to change of the hyperbolicity of fixed points.

Owing to lack of space, we refer the reader to [2, 6, 9, 11] for basic concepts and results in one-step numerical integration methods and dynamical systems.

2. Dynamics Consistency

We first give a dynamical system definition of dynamics consistency.

Definition 1. Let x^* be a fixed point of Φ. Then ϕ_h is *dynamically consistent* with Φ at x^*, if there is a fixed point x_h^* of ϕ_h satisfying $x_h^* = x^* + O(h)$ such that ϕ_h and Φ have the same dimensions of local stable, unstable and center manifolds in some neighborhoods of x^* and x_h^*.

Considering (1.2) as the Euler method for solving $\dot{x}(t) = F_h(x(t); \mu)$, which is in turn a perturbation of (1.1), we prove the following result using the results in [10] on dynamical consistency for the Euler method and in [9] on topological conjugacy for perturbed dynamical systems; a similar result was given by Beyn in [1] using a different proof.

Theorem 1. *For a hyperbolic fixed point x^* of Φ, there exists an $h^* > 0$ such that, for each h, $0 < h < h^*$, ϕ_h is dynamically consistent with Φ at x^*.*

In many cases, upper bounds for h^* can be found. For example, we have obtained in [3, 10] the following result:

Theorem 2. *For the Euler and backward Euler methods, ϕ_h is dynamically consistent with Φ at a hyperbolic fixed point if and only if $h < 2|\mathrm{Re}(\Lambda)|/|\Lambda|^2$ for all eigenvalues Λ of $DF(x^*; \mu)$ with* $\mathrm{Re}\Lambda > 0$.

Another characterization of dynamics consistency can be had by comparing the dynamics of the h-time map $\Phi(h, \cdot\,; \mu)$ and the 1-step map $\phi_h(1, \cdot\,; \mu)$ of X into itself. Fečkan showed in [5] that the topological conjugacy of these two maps holds for the Euler method. We generalize this to any one-step method of order $p \geq 1$ as follows.

Theorem 3. *Let x^* be a hyperbolic fixed point of Φ. Then there exists an h^{**} such that for each $h < h^{**}$, there is a (hyperbolic) fixed point x_h^* of ϕ_h, and $\phi_h(1, \cdot\,; \mu)$ is topologically conjugate to $\Phi(h, \cdot\,; \mu)$ in some respective neighborhoods of x_h^* and x^*.*

3. Bifurcation Consistency

We now assume that Φ in (1.3) has a family of fixed points $x^*(\mu)$ ($\mu \in M$) and a local bifurcation at $\mu = \mu^*$. Further, assume that ϕ_h in (1.4) also has a family of fixed points $x_h^*(\mu)$ for μ in a neighborhood of μ^*. The following definition explains the notion of ϕ_h in (1.4) having a bifurcation of the same type and orientation.

Definition 2. ϕ_h is *bifurcationally consistent* with Φ at μ^* if ϕ_h has a bifurcation at some $\mu_h^* = \mu^* + O(h)$, and if there is a continuous and increasing map $\mu \mapsto \mu_h$ from a neighborhood M^* of μ^* onto a neighborhood M_h^* of μ_h^* such that, for every $\mu \in M^*$ (except possibly at $\mu = \mu^*$), $\phi_h(\cdot, \cdot\ ; \mu_h)$ is dynamically consistent with $\Phi(\cdot, \cdot\ ; \mu)$ at $x^*(\mu)$.

To obtain our result on bifurcation consistency, we need some generic conditions on bifurcations.

Definition 3. Assume Φ and ϕ_h have bifurcations at $\mu = \mu^*$ and $\mu = \mu_h^*$, respectively. Call an eigenvalue $\Lambda(\mu^*)$ of the Jacobian $DF(x^*; \mu^*)$ (respectively, $\lambda(\mu_h^*)$ of $Df_h(x_h^*; \mu_h^*)$) *neutral* if $\mathrm{Re}\Lambda(\mu^*) = 0$ ($|\lambda(\mu_h^*)| = 1$).

(i) The bifurcation of Φ (or ϕ_h) is *simple* if all neutral eigenvalues of the Jacobian $DF(x^*; \mu^*)$ (or $Df_h(x^*; \mu^*)$) are simple, i.e., have multiplicity one.

(ii) The *Hopf conditions* of the bifurcations of Φ and ϕ_h are

$$\left.\frac{\partial \mathrm{Re}\Lambda(\mu)}{\partial \mu}\right|_{\mu=\mu^*} \neq 0 \text{ and } \left.\frac{\partial |\lambda(\mu)|}{\partial \mu}\right|_{\mu=\mu^*} \neq 0, \tag{3.7}$$

respectively, for all neutral eigenvalues $\Lambda(\mu^*)$ and $\lambda(\mu_h^*)$.

Theorem 4. *Suppose that the bifurcation of Φ at μ^* is simple and satisfies the Hopf condition. If the bifurcation is of codimension one, then there exists an $h^* > 0$ such that for each h, $0 < h < h^*$, ϕ_h is bifurcationally consistent with Φ at μ^*. Moreover,*

(i) *the bifurcation of ϕ_h at μ_h^* is simple, satisfies the Hopf condition and is of codimension one; and*

(ii) *the shifting quantity and direction of μ_h^* away from μ^* are given by*

$$\mu_h^* = \mu^* + O(h^p) \text{ and } (\mu_h^* - \mu^*) \left.\frac{\partial \mathrm{Re}\Lambda(\mu)}{\partial \mu}\right|_{\mu=\mu^*} \leq 0.$$

We remark that bifurcations of codimension one [7, 8, 11] include the saddle-node, transcritical, pitchfork and Hopf bifurcations, which, in their normal forms, are simple and satisfy the Hopf condition. In this regard, and considering that it covers all one-step methods of order $p \geq 1$, Theorem 4 is much more general and precise than a result given by Brezzi, Ushiki and Fujii in [4] on how the Euler method "inherits" the Hopf bifurcation. The reason for requiring bifurcations of codimension one is because (1.1) is assumed to be a one-parameter system. Evidently, all the results in this paper are readily generalized to multiparameter and/or infinite-dimensional dynamical systems.

REFERENCES

1. W.-J. Beyn, *Numerical methods for dynamical systems*, Advances in Numerical Analysis, vol. 1, Oxford University Press, 1991, pp. 175–236.
2. E. K. Blum, *Numerical Analysis and Computation: Theory and Practice*, Addison-Wesley, 1972.
3. E. K. Blum, *Abstracts of AMS Joint Mathematical Conference, San Antonio*, January, 1993.
4. F. Brezzi, S. Ushiki, and H. Fujii, *"Real" and "ghost" bifurcation dynamics in difference schemes for ODEs*, in *Numerical Methods for Bifurcation Problems* (T. Küpper, H. D. Mittelmann, and H. Weber, eds.), Internat. Ser. Numer. Math., vol. 70, Birkhäuser, Basel, 1984, pp. 79–104.
5. M. Fečkan, *The relation between a flow and its discretization*, Mathematica Slovaca **42** (1992), 123–127.
6. C. W. Gear, *Numerical Initial Value Problems in Ordinary Differential Equations*, Prentice-Hall, New York, 1971.
7. M. Golubitsky and D. G. Schaeffer, *Singularities and Groups in Bifurcation Theory*, vol. 1, Springer-Verlag, New York, 1985.
8. J. K. Hale and H. Koçak, *Dynamics and Bifurcations*, Texts in Applied Mathematics, vol. 3, Springer-Verlag, New York, 1991.
9. M. W. Hirsch and S. Smale, *Differential Equations, Dynamical Systems, and Linear Algebra*, Academic Press, 1974.
10. X. Wang and E. K. Blum, *Discrete-time versus continuous-time neural networks*, J. Comput. System Sci. **49** (1992), 1–17.
11. S. Wiggins, *Introduction to Applied Nonlinear Dynamical Systems and Chaos*, Texts in Applied Mathematics, vol. 2, Springer-Verlag, New York, 1990.

DEPARTMENT OF MATHEMATICS, UNIVERSITY OF SOUTHERN CALIFORNIA, LOS ANGELES, CA 90089

E-mail: {wang,eblum,qingnanl}@cajal.usc.edu

Proceedings of Symposia in Applied Mathematics
Volume 48, 1994

Computing Integrals of the Complex Error Function

J.A.C. WEIDEMAN

ABSTRACT. We present an efficient expansion for computing the integral of the complex error function. Special cases include the integrals of the Dawson and complementary error functions.

1. Introduction

The error function of a complex variable is defined by [**1, 2, 3**]

$$w(z) = e^{-z^2}\operatorname{erfc}(-iz), \tag{1}$$

where

$$\operatorname{erfc}(z) = \frac{2}{\sqrt{\pi}}\int_z^\infty e^{-t^2}\,dt.$$

Of particular interest are the cases z real, and z pure imaginary. For real $z = x$, $w(z)$ is related to the Dawson function

$$\operatorname{Im}(w(x)) = \frac{2}{\sqrt{\pi}}e^{-x^2}\int_0^x e^{t^2}\,dt \equiv \frac{2}{\sqrt{\pi}}\operatorname{daw}(x). \tag{2}$$

For $z = iy$, with y real, $w(z)$ is related to the complementary error function

$$w(iy) = e^{y^2}\operatorname{erfc}(y). \tag{3}$$

In certain applications, it becomes necessary to compute integrals of $w(z)$. In particular, the integrals of (2) and (3) appear to be of some interest (see [**3**]):

$$I_1(r) = \int_0^r e^{-x^2}\int_0^x e^{t^2}\,dt\,dx, \quad I_2(r) = \int_0^r e^{y^2}\int_y^\infty e^{-t^2}\,dt\,dy. \tag{4}$$

In this note we derive a series expansion for the computation of integrals such as these.

1991 *Mathematics Subject Classification.* Primary 65D20; Secondary 33B20.
This paper is in final form and no version of it will be published elsewhere.

© 1994 American Mathematical Society
0160-7634/94 $1.00 + $.25 per page

2. Series Expansion

Various expansions for computing the complex error function exist; for a survey see [**4**]. Not all of these methods are readily amenable for computing integrals of $w(z)$ however. By contrast, the series expansion derived in [**4**] can be easily adapted for the computation of integrals. This series is

$$w(z) = \frac{1}{\sqrt{\pi}(L - iz)} + \frac{2}{L^2 + z^2} \sum_{n=1}^{\infty} a_n \left(\frac{L + iz}{L - iz} \right)^n, \quad \text{Im}(z) \geq 0, \tag{5}$$

where

$$a_n = \frac{L}{\pi} \int_{-\infty}^{\infty} e^{-t^2} \left(\frac{L - it}{L + it} \right)^n dt, \tag{6}$$

and L is any positive, real constant. Here, and below, attention is restricted to z in the upper half-plane, since elementary relations allow one to deal with the lower half-plane (see [**2**]).

In [**4**] it was demonstrated that the expansion (5) has primarily two attractive properties. First, it provides high accuracy uniformly in the complex plane, provided the parameter L is selected optimally. Second, the algorithm is very easy to implement; a recapitulation follows below.

We present here a third attractive property of the expansion (5): it can easily be integrated term by term to compute the integral of $w(z)$. This follows from the fact that (5) defines an analytic function in the upper half-plane. Thus the antiderivative relation

$$-\frac{i}{2Ln} \frac{d}{dz} \left(\frac{L + iz}{L - iz} \right)^n = \frac{1}{L^2 + z^2} \left(\frac{L + iz}{L - iz} \right)^n$$

can be invoked to obtain

$$\int_0^z w(\zeta)\, d\zeta = \frac{i}{\sqrt{\pi}} \log \left(\frac{L - iz}{L} \right) + \frac{i}{L} g(L) - \frac{i}{L} \sum_{n=1}^{\infty} \frac{a_n}{n} \left(\frac{L + iz}{L - iz} \right)^n, \tag{7}$$

valid for $\text{Im}(z) \geq 0$. Here,

$$g(L) = \sum_{n=1}^{\infty} \frac{a_n}{n} = L \left(\int_0^L e^{y^2} \text{erfc}(y)\, dy - \frac{1}{\sqrt{\pi}} \log(2) \right),$$

which follows from substituting $z = iL$ in (7).

Regarding the convergence of this series, we quote the following results from [**4**]. A steepest descent analysis of (6) estimates that for fixed L and $n \to \infty$

$$a_n = O\left(\exp\left[-(3/2)(Ln)^{2/3} \right] \right). \tag{8}$$

If the series is truncated at $n = N$, the choice $L = 2^{-1/4}N^{1/2}$ is optimal in the sense that a_N decays most rapidly as $N \to \infty$. This optimal decay is

$$a_N = O\left(N^{\frac{1}{2}}(\sqrt{2}-1)^N\right). \tag{9}$$

One can now estimate the truncation error in (7) as

$$E_N(z) \equiv \left|\frac{i}{L}\sum_{n=N}^{\infty}\frac{a_n}{n}\left(\frac{L+iz}{L-iz}\right)^n\right| \le \frac{1}{LN}\left|\frac{L+iz}{L-iz}\right|^N \sum_{n=N}^{\infty}|a_n|,$$

where we have used the fact that $|(L+iz)/(L-iz)| \le 1$ for z in the upper half-plane. The sum on the right is estimated to be $O(a_N)$. By (9) one can thus expect the truncation error to be

$$E_N(z) = O(N^{-1}(\sqrt{2}-1)^N), \tag{10}$$

uniformly for all z in the upper half-plane. However, we expect the accuracy to be higher near $z = iL$ than elsewhere, as may be confirmed in Fig. 1 below.

3. Implementation and Tests

In order to implement (7), one needs to find the expansion coefficients a_n, and one needs to compute $g(L)$. The rest is just polynomial evaluation in terms of the variable $Z = (L+iz)/(L-iz)$. To compute the a_n, one makes the change of variable $t = L\tan\frac{1}{2}\theta$ in (6). This yields a Fourier integral which can be evaluated rapidly and efficiently with the Fast Fourier Transform; for details, see [**4**]. For a given N, the a_n can be computed once and for all. The same remark applies to $g(L)$; a few values are listed in Table 1.

N	$g(L)$
8	1.602998832197634
16	2.889105710334415
24	3.994097690869298
32	4.988816722025917

TABLE 1. Values of $g(L)$, with $L = 2^{-1/4}N^{1/2}$.

It should be pointed out that evaluating the series (7) may be unstable for small $|z|$. This is equivalent to computing $1 - Z^n$ for $Z \approx 1$. One remedy is to use the identity $1 - Z^n = (1-Z)\sum_{j=0}^{n-1} Z^j$ to rearrange the series as

$$\int_0^z w(\zeta)\,d\zeta = \frac{i}{\sqrt{\pi}}\log\left(\frac{L-iz}{L}\right) + \frac{2z}{L(L-iz)}\sum_{n=0}^{\infty} b_n\left(\frac{L+iz}{L-iz}\right)^n, \tag{11}$$

where

$$b_n = \sum_{j=n+1}^{\infty}\frac{a_j}{j}.$$

This expansion is preferable when $|z| < 0.1$, roughly. However, when truncated it is not as accurate as the original series for larger $|z|$.

In [**4**], we presented an eight-line Matlab implementation of the series (5). This program requires but a simple modification to compute also the series (7) or (11). As a first test, we consider the table in [**3**, p. 190], which lists values of the integrals $I_1(r)$ and $I_2(r)$ over the interval $0 \leq r \leq 6$. The series (7), with 24 terms, reproduces those tables to all ten decimal places listed.

More comprehensive tests are offered in Fig. 1, which shows, on a log-scale, the relative errors in $\int_0^z w(\zeta)\, d\zeta$, for $z = re^{i\theta}$. The "true" values were computed to about 14 digits using the symbolic packages Mathematica and Maple. However, for large $|z|$ these packages had difficulty in achieving the desired accuracy and we resorted to the asymptotic expansion given in [**3**, p. 176]. The three curves in each window represent values $N = 8, 16, 24$, where N is the number of terms in the series (7). (The error curve corresponding to $N = 24$ is hardly visible in the top two pictures on the left.)

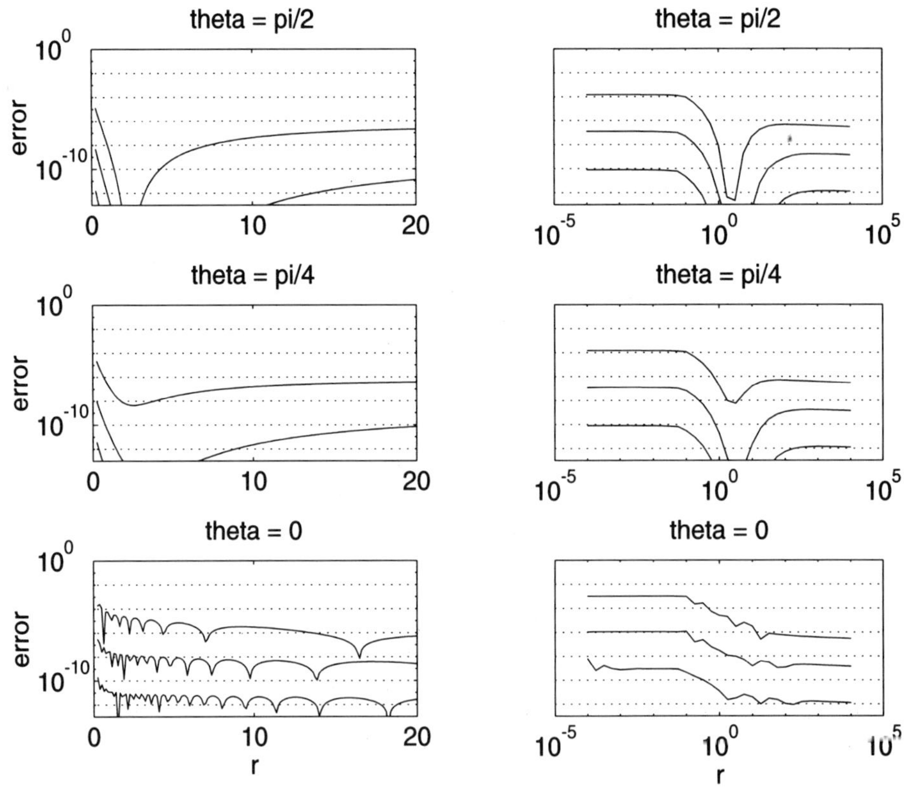

FIGURE 1. Relative error in the expansion (7), with $z = re^{i\theta}$, when truncated to $N = 8, 16, 24$ terms. The dotted lines represent values $10^{-2}, 10^{-4}, \ldots, 10^{-12}$.

The three rows in Fig. 1 represent values of $\theta = \pi/2, \pi/4$, and 0. Note that $\theta = \pi/2$ corresponds to the special integral I_2 in (4). For $\theta = 0$ we calculated the error in the imaginary part only, since this is related to the integral I_1 in (4). We also show two columns of results, the first represents values $r = 0.25(0.125)20$, and the second represents $r = 10^l$, where $l = -4(0.25)4$. In the right column, we have used the modified series (11) rather than the original series for values in the range $|z| < 0.1$. To aid the eye, the dotted lines indicate where the error $= 10^{-l}$, with $l = 2(2)12$. Since we have plotted relative errors, each such line represents roughly two digits of accuracy.

The results of Fig. 1 confirm that the series (7) provides high accuracy uniformly in the complex plane with relatively few terms. By retaining an additional eight terms in the series one increases the number of correct digits by more than three, for most z, in accordance with (10). Moreover, a single expansion is used (provided, of course, $|z|$ is sufficiently large to avoid the instability described below Table 1). In certain regions of the complex plane other expansions may be more efficient, like Taylor series near the origin, or the asymptotic approximation for large $|z|$. However, the fact that our method is based on a single expansion makes the method attractive for vectorized implementations such as the Matlab program alluded to above. This approach should be particularly efficient when $\int_0^z w(\zeta)\,d\zeta$ needs to be evaluated for a huge number of z's of intermediate magnitude.

Acknowledgment. The author is indebted to an anonymous referee for some constructive remarks.

References

1. V.N. Faddeeva and N.N. Terent'ev, *Tables of values of the function* $w(z) = e^{-z^2}\left(1 + \frac{2i}{\sqrt{\pi}}\int_0^z e^{t^2}\,dt\right)$ *for complex argument*, Gosud. Izdat. Teh.-Teor. Lit., Moscow, 1954; English transl., Pergamon Press, New York, 1961.
2. W. Gautschi, *Efficient computation of the complex error function*, SIAM J. Numer. Anal. **7** (1970), 187–198.
3. B. Rosser, *Theory and application of* $\int_0^z e^{-x^2}\,dx$ *and* $\int_0^z e^{-p^2y^2}\,dy \int_0^y e^{-x^2}\,dx$, Mapleton House, Brooklyn N.Y., 1948.
4. J.A.C. Weideman, *Computation of the complex error function*, to appear in SIAM J. Numer. Anal.

Department of Mathematics, Oregon State University, Corvallis, OR 97331-4605
E-mail address: weideman@math.orst.edu

Proceedings of Symposia in Applied Mathematics
Volume **48**, 1994

Quadratures for Improper Integrals and Their Applications in Integral Equations

YUESHENG XU AND YUNHE ZHAO

ABSTRACT. Quadrature formulas with high convergence orders for classes of improper integrals are derived. As an application, a product integration method and its discrete version for weakly singular Fredholm integral equations of the second kind is developed and analyzed.

1. Introduction

To construct a quadrature formula with a high convergence order for an improper integral, one needs to overcome difficulties caused by singularities of the integrand. Sinc quadratures are efficient methods for improper integrals (cf. [**5**, **8**]). For some classes of improper integrals, a suitable change of variables may smooth the integrands. This idea has been demonstrated in the literature ([**2**, p. 174] and the references cited there) in constructing quadrature formulas for certain singular integrals.

The purpose of this paper is to develop quadrature formulas by using the idea of change of variables for large classes of improper integrals and to supply a complete analysis for their convergence order. A product integration method for Fredholm integral equations of the second kind with weakly singular kernels is developed, based upon change of variables. A discrete product integration method is also derived. Convergence order of the methods is analyzed.

2. Quadrature Formulas

Let $0 \le s_1 < \cdots < s_k \le 1$ be a set of points and let $w_j = \int_0^1 \prod_{l \ne j} \frac{x - s_l}{s_j - s_l} dx$, $j = 1, 2, \ldots, k$. Denote by $[x]$ the largest integer $\le x$. In the following theorems, we assume $t_{ij} = ih + s_j h$, $j = 1, 2, \ldots, k$, $i = 0, 1, \ldots, n-1$, where n is an integer and h is to be defined specifically.

1991 *Mathematics Subject Classification*. Primary 65D30; Secondary 65D32, 45B05.

The work is in part supported by NASA grant NAG 3-1312.

The final version of this paper will be submitted for publication elsewhere.

© 1994 American Mathematical Society
0160-7634/94 \$1.00 + \$.25 per page

THEOREM 2.1. *Let $f \in C^k(0,1]$ satisfy $\left|f^{(m)}(x)\right| \le Cx^{\alpha-m}$ as $x \to 0^+$, $m = 0, 1, \dots, k$, where $C > 0$ is a constant and $\alpha > -1$. Let $\beta = \left[\frac{k+1}{\alpha+1}\right] + 1$ and $h = \frac{1}{n}$. Then*

$$\int_0^1 f(x)dx = \frac{\beta}{n}\sum_{j=1}^{k}\left(\sum_{i=0}^{n-1} f\left(t_{ij}^{\beta}\right) t_{ij}^{\beta-1}\right) w_j + O(n^{-k}).$$

PROOF. Let $x = t^\beta$. Then $\int_0^1 f(x)dx = \int_0^1 F(t)dt$, where $F(t) = f(t^\beta)(t^\beta)'$. We show that $F^{(m)}(t)$ can be continuously extended from $(0,1]$ onto $[0,1]$ with $F^{(m)}(0) = 0$ for $m = 0, 1, \dots, k$. Write $F^{(m)}(t) = \sum_{j=0}^{m} u_j^m(t) f^{(j)}(t^\beta)$. Then

$$F^{(m+1)}(t) = \sum_{j=0}^{m}\left[u_j^m(t)(t^\beta)' f^{(j+1)}(t^\beta) + \frac{du_j^m(t)}{dt} f^{(j)}(t^\beta)\right].$$

Thus, we have a recursion formula

$$u_j^{m+1}(t) = \begin{cases} \frac{du_0^m(t)}{dt}, & j = 0; \\ u_{j-1}^m(t)(t^\beta)' + \frac{du_j^m(t)}{dt}, & j = 1, 2, \dots, m; \\ u_m^m(t)(t^\beta)', & j = m+1. \end{cases}$$

In particular, $u_0^m = \left[(t^\beta)'\right]^{(m)} = c_0^m t^{\beta-m-1}$, $u_m^m(t) = \left[(t^\beta)'\right]^{m+1} = c_m^m t^{(m+1)(\beta-1)}$. By induction from the recursion formula, we have that for $m = 0, 1, \dots, k$ and $j = 0, 1, \dots, m$, there holds $u_j^m = c_j^m t^{(j+1)\beta-m-1}$, where the c_j^m's are constants. It follows that $\left|u_j^m(t) f^{(j)}(t^\beta)\right| \le C\left|c_j^m\right| t^{(1+\alpha)\beta-m-1}$ as $t \to 0$. Since $\beta > \frac{k+1}{1+\alpha}$ and $m \le k$, we have $(1+\alpha)\beta - m - 1 > 0$. Thus, $\lim_{t\to 0} u_j^m f^{(j)}(t^\beta) = 0$, $j = 0, 1, \dots, m$ and $m = 0, 1, \dots, k$. This implies that $\lim_{t\to 0} F^{(m)}(t) = 0$.

Denote by $\hat{F}$ the extension of F to $[0,1]$. Then $\hat{F} \in C^k[0,1]$ and $\int_0^1 f(x)dx = \int_0^1 \hat{F}(t)dt$. Now, applying the usual compound Newton-Cotes formula to the integral $\int_0^1 \hat{F}(t)dt$, one obtains the desired quadrature formula with convergence as stated in the theorem. □

A quadrature formula with an optimal convergence rate for the type of integrals considered in Theorem 2.1 was given in [**3**]. In a similar manner, one can prove the following three theorems.

THEOREM 2.2. *Let $f \in C^k(0,b]$ satisfy $\left|f^{(m)}(x)\right| \le x^{-m}\left|\sum_{j=0}^{p_m} c_j^m (\log x)^{\alpha_j^m}\right|$ as $x \to 0^+, m = 0, 1, \dots, k$, where $0 < b < 1$, p_m are nonnegative integers, α_j^m integers, and c_j^m constants. Let $\beta = [-\log(-\log b)] + 1$ and $h = \frac{1}{n}e^{-\frac{\log(-\log b)}{\beta}}$. Then*

$$\int_0^b f(x)dx = \beta h \sum_{j=1}^{k}\left(\sum_{i=0}^{n-1} f\left(e^{-t_{ij}^{-\beta}}\right) e^{-t_{ij}^{-\beta}} t_{ij}^{-\beta-1}\right) w_j + O(n^{-k}).$$

THEOREM 2.3. *Let $f \in C^k[1,\infty)$ satisfy $\left|f^{(m)}(x)\right| \le C\left|p_m(x)\right|e^{-\alpha x}$ as $x \to \infty$, $m = 0, 1, \dots, k$, where $p_m(x)$ are polynomials, C is a positive constant and $\alpha > 0$. Let $h = \frac{1}{n}$. Then*

$$\int_1^\infty f(x)dx = \frac{1}{n}\sum_{j=1}^{k}\left(\sum_{i=0}^{n-1} f(t_{ij}^{-1})t_{ij}^{-2}\right)w_j + O(n^{-k}).$$

THEOREM 2.4. *Let $f \in C^k[1,\infty)$ satisfy $\left|f^{(m)}(x)\right| \le Cx^{-\alpha-m}$ as $x \to \infty$, $m = 0, 1, \dots, k$, where $C > 0$ is a constant and $\alpha > 1$. Let $\beta = \left[\frac{k+1}{\alpha-1}\right] + 1$ and $h = \frac{1}{n}$. Then*

$$\int_1^\infty f(x)dx = \frac{\beta}{n}\sum_{j=1}^{k}\left(\sum_{i=0}^{n-1} f(t_{ij}^{-\beta})t_{ij}^{-\beta-1}\right)w_j + O(n^{-k}).$$

The theorems above have been applied to evaluate $\int_0^1 x^{-0.5}dx$, $\int_0^{0.5}\log x dx$, $\int_1^\infty e^{-x}\sin x dx$, $\int_0^\infty \frac{x^{-0.8}}{1+x}dx$, and $\int_1^\infty x^{-2}dx$. All numerical results appear to be in accordance with the theoretical estimates given in Theorems 2.1 – 2.4.

3. A New Product Integration Method

Consider integral equations of the second kind with weakly singular kernels:

$$\lambda\phi(s) - \int_0^1 g_\alpha(|s-t|)k(s,t)\phi(t)dt = y(s), \quad s \in [0,1], \tag{3.1}$$

where $g_\alpha(s) = s^{\alpha-1}$ for $0 < \alpha < 1$ and $g_1(s) = \log s$, $0 \ne \lambda \in R$, $y \in C^p[0,1]$, and $k \in C^{p+1}([0,1]\times[0,1])$. In operator notation, Equation (3.1) can be rewritten as $(\lambda I - K)\phi = y$. The operator K is compact on $C[0,1]$, and a unique continuous solution of (3.1) exists if and only if λ is not an eigenvalue of K [**1, 7**]. A product integration method for Equation (3.1) was constructed in [**7**] by using special nonuniform knots introduced in [**6**]. In this section, we employ the idea of change of variables to develop a new product integration method for (3.1) which has optimal convergence rate but does not require special nonuniform knots.

Let the function $w : [0,1] \to [0,1]$ be bijective, strictly monotonically increasing and infinitely differentiable. Assume that w and $1-w$ have zeros of order $q > \frac{p}{\alpha}$ at $s = 0$ and $s = 1$, respectively, that is, $w^{(j)}(0) = w^{(j)}(1) = 0, j = 0, 1, \dots, q-1$, but $w^{(q)}(0) \ne 0$ and $w^{(q)}(1) \ne 0$. One example is $w(s) = \frac{s^q}{s^q+(1-s)^q}$, $0 \le s \le 1$. Another example appears in [**4**].

The method presented here does not require equally spaced knots in either piecewise polynomials or interpolation. We use them only to simplify the presentation. Let $S_{n,p}$ be the set of piecewise polynomials of degree $p-1$ on the grid $0 = u_0 < \cdots < u_n = 1$, where $u_l = lh, h = \frac{1}{n}$. Let $\{0 \le \tau_1 < \cdots < \tau_p \le 1\}$ be a fixed set of points in $[0,1]$. Let $u_{lk} = u_l + \tau_k h, s_{lk} = w(u_{lk})$ for $k = 1, 2, \dots, p$ and $l = 0, 1, \dots, n-1$. Let $P_n : C[0,1] \to S_{n,p}$ be the interpolatory projection defined so that $P_n v$ interpolates $v \in C[0,1]$ at the np points u_{lk}. Then we define $K_n : C[0,1] \to C[0,1]$ in the following way:

(3.2)

$$(K_n\phi)(s) = \int_0^1 g_\alpha(|s-w(u)|)w'(u)[P_n(k(s,w(\cdot))\phi(w(\cdot))](u)du, \quad s \in [0,1].$$

Thus, $(K_n\phi)(s) = \sum_{i=0}^{n-1}\sum_{k=1}^{p} w_{ik}(s)k(s,s_{ik})\phi(s_{ik})$, where

$$w_{ik}(s) = \int_{u_i}^{u_{i+1}} g_\alpha(|s-w(u)|)w'(u)\prod_{l\neq k}\frac{u-u_{il}}{u_{ik}-u_{il}}du.$$

Then, a product integration solution to (3.1) is a function ϕ_n satisfying

$$\lambda\phi_n(s) - \sum_{i=0}^{n-1}\sum_{k=1}^{p} w_{ik}(s)k(s,s_{ik})\phi_n(s_{ik}) = y(s), \quad s \in [0,1]. \tag{3.3}$$

LEMMA 3.1. *Let ϕ be the exact solution of* (3.1). *Let $F(u) = \phi(w(u))$. Then $F^{(m)}(u)$ can be continuously extended from $(0,1)$ onto $[0,1]$ with $F^{(m)}(0) = 0$ and $F^{(m)}(1) = 0$ for $m = 1,2,\ldots,p$.*

THEOREM 3.1. *For sufficiently large n,* (3.3) *has a unique solution ϕ_n and $\|\phi-\phi_n\| = O(n^{-p})$.*

If $w(u) = \frac{u^q}{u^q+(1-u)^q}$, where $q \geq \frac{p+1}{\alpha}+1$ for $0<\alpha<1$ and $q = p+3$ for $\alpha = 1$, we evaluate $w_{ik}(s)$ in the following way: Let w^{-1} be the inverse of w, and $f(u)$ be the integrand of $w_{ik}(s)$. Denote $z = w^{-1}(s)$. If $z \notin [u_i, u_{i+1}]$, we substitute the variable u in $w_{ik}(s) = \int_{u_i}^{u_{i+1}} f(u)du$ by $z - t^\beta$ if $z > u_{i+1}$ and by $z + t^\beta$ if $z < u_i$, where $\beta = \left[\frac{p+2}{\alpha}\right]+1$, and then apply the usual compound Newton-Cotes formula with convergence order $O(n^{-(p+1)})$ to compute the integral so obtained. If $z \in [u_i, u_{i+1}]$, we split $w_{ik}(s)$ into $\int_{u_i}^{z} f(u)du + \int_{z}^{u_{i+1}} f(u)du$, transform the interval of integration in each of the above integrals into $[0,1]$, and then apply the quadrature formula in Theorem 2.1 with $k = p+1$ and $\beta = \left[\frac{p+2}{\alpha}\right]+1$ to each of the integrals obtained. Denote by $w_{ik}^*(s)$ the approximation to $w_{ik}(s)$ obtained by using the quadrature formula described above. We call the product integration method using $w_{ik}^*(s)$ the discrete product integration method. Then we have:

THEOREM 3.2. *For sufficiently large n, the discrete product integration method has a unique solution ϕ_n^* and $\|\phi-\phi_n^*\| = O(n^{-p})$.*

EXAMPLE 3.1. Consider the equation $2\phi(s) - \int_0^1 |s-t|^{-0.5}\phi(t)dt = y(s)$, $s \in [0,1]$, where $y(s)$ is chosen such that $\phi(s) = 2\sqrt{1-s}+2\sqrt{s}$ is the unique solution. Numerical results obtained by the discrete product integration method are shown below.

n	max. error ($p = 2$)	decay exp.	max. error ($p = 3$)	decay exp.
20	2.5889D-02		8.1223D-03	
40	6.7230D-03	1.95	8.7064D-04	3.22
60	3.0167D-03	1.98	2.1708D-04	3.43
80	1.7205D-03	1.95	8.7265D-05	3.17

REFERENCES

1. P. M. Anselone, *Collectively compact operator approximation theory and applications to integral equations*, Prentice Hall, Englewood Cliffs, NJ, 1971.
2. P. J. Davis and P. Rabinowitz, *Methods of numerical integration*, second edition, Academic Press, Orlando, FL, 1984.
3. K. Kaneko and Y. Xu, *Gaussian-type quadratures for weakly singular integrals and their applications to the Fredholm integral equation of the second kind*, Math. Comp. **62** (1994), to appear.
4. R. Kress, *A Nyström method for boundary integral equations in domains with corners*, Numer. Math. **58** (1990), 145–161.
5. J. Lund and K. L. Bowers, *Sinc methods for quadrature and differential equations*, SIAM, Philadelphia, 1992.
6. J. R. Rice, *On the degree of convergence of nonlinear spline approximation*, Approximation with Special Emphasis on Spline Functions (I. J. Schoenberg, ed.), Academic Press, New York, 1969, pp. 349–365.
7. C. Schneider, *Product integration for weakly singular integral equations*, Math. Comp. **36** (1981), 207–213.
8. F. Stenger, *Numerical methods based on sinc and analytic functions*, Springer-Verlag, New York, 1993.

DEPARTMENT OF MATHEMATICS, NORTH DAKOTA STATE UNIVERSITY, FARGO, ND 58105-5075

E-mail address: xu@plains.nodak.edu

DEPARTMENT OF MATHEMATICS, NORTH DAKOTA STATE UNIVERSITY, FARGO, ND 58105-5075

E-mail address: yunhe@plains.nodak.edu

Proceedings of Symposia in Applied Mathematics
Volume 48, 1994

Spline Harmonic Analysis and Wavelet Bases

VALERY A. ZHELUDEV

ABSTRACT. We present here a short description of a new computational technique, Spline Harmonic Analysis, based on periodic splines, and its application to constructing wavelet schemes. This technique provides a promising tool for solving numerous problems connected with differential and integral equations, digital signal processing, geometric design, and so on.

1. Spline Harmonic Analysis

We first introduce some notation. Throughout, $N = 2^j$ and $\sum_k^j$ stands for $\sum_{k=0}^{2^j-1}$. Denote $\omega = \exp(2\pi i/N)$. The Discrete Fourier Transform (DFT) of a vector $\mathbf{a} = \{a_k\}_0^{N-1}$ is

$$T_n^j(\mathbf{a}) = \frac{1}{N}\sum_k^j \omega^{-nk} a_k, \qquad a_k = \sum_n^j \omega^{nk} T_n^j(\mathbf{a}).$$

The function ${}_pB^j(x) = N^p \nabla_j^p (x_+^{p-1}/(p-1)!)$, where $x_+ = 0.5(x+|x|)$, is the B-spline of degree $p-1$ with knots at the points $\{k/2^j\}$. The symbol ∇_j denotes a backward difference with step 2^{-j}. The symbol ${}_pM^j(x)$ denotes the 1-periodic B-spline of degree $p-1$: ${}_pM^j(x) = \sum_{l=-\infty}^{\infty} {}_pB^j(x+l)$.

Throughout, ${}_p\mathfrak{B}^j$ will denote the space of 1-periodic splines of degree $p-1$ and of defect 1 with knots at the points $\{k/2^j\}$. Any spline ${}_pS^j \in {}_p\mathfrak{B}^j$ can be represented as follows: ${}_pS^j(x) = \frac{1}{N}\sum_k^j q_k^j {}_pM^j(x - k/N)$. Note that this representation allows one to compute the spline ${}_pS^j(x)$ immediately.

Denoting $\mathbf{q}^j = \{q_k^j\}_0^{N-1}$ and exploiting the relations of DFT, we write the

---

1991 *Mathematics Subject Classification*. Primary 41A15, 41A30; Secondary 65T20.

The work was supported by the Russian Foundation for Fundamental Researches, Grant No. 93-012-49.

The final version of this paper will be submitted for publication elsewhere.

© 1994 American Mathematical Society
0160-7634/94 $1.00 + $.25 per page

spline as ${}_pS^j(x) = \frac{1}{N}\sum_k^j {}_pM^j(x-k/N)\sum_r \omega^{rk} T_r^j(\mathfrak{q}^j) = \sum_r^j \xi_r^j {}_pm_r^j(x)$, where

$$
\begin{aligned}
{}_pm_r^j(x) &= \frac{1}{N}\sum_k^j {}_pM^j(x-k/N)\omega^{rk} \\
&= \frac{1}{N}\sum_k^j {}_pM^j(x+k/N)\omega^{-rk}, \qquad \xi_r^j = T_r^j(\mathfrak{q}^j).
\end{aligned}
$$

Hence, we see that the splines ${}_pm_r^j(x)$, $r = 0, 1, \ldots, 2^j - 1$, form a basis of the space ${}_p\mathfrak{B}^j$, and $\{\xi_r^j\}$ are the coordinates of the spline ${}_pS^j(x)$ with respect to this basis. We mention some properties of the splines ${}_pm_r^j$.

1. There holds $\langle {}_pm_r^j, {}_pm_k^j\rangle = \int_0^1 {}_pm_r^j(y)\,{}_p\overline{m}_k^j(y)\,dy = \delta_k^r\, {}_{2p}u_r^j$, where δ_k^r is the Kronecker delta, ${}_pu_r^j = \frac{1}{N}\sum_k^j \omega^{-rk}\, {}_pM^j(k/N + p/2N) > 0$. This property implies that the splines ${}_pm_r^j$ form an orthogonal basis of ${}_p\mathfrak{B}^j$, and we name these splines *ortsplines*.
2. As $p \to \infty$ one has

$$
{}_pm_r^j(x)/({}_{2p}u_r^j)^{1/2} \to \mu_r(x) = \exp(2\pi i r x), \qquad r = 0, \ldots, N-1. \tag{1.1}
$$

3. There holds ${}_pM^j(x + k/N) = \sum_r^j {}_pm_r^j(x)\omega^{rk}$.
4. One has ${}_pm_r^j(x)^{(s)} = N^s(1-\omega^{-r})^s\, {}_{p-s}m_r^j(x) \in {}_{p-s}\mathfrak{B}^j$.
5. If a spline ${}_lS^j(x) \in {}_l\mathfrak{B}^j$ is given as ${}_lS^j(x) = \sum_k^j \eta_k^j\, {}_lm_k^j(x)$, then the convolution is ${}_pm_r^j * {}_lS^j(x) = {}_{l+p}m_r^j(x)\eta_r^j \in {}_{p+l}\mathfrak{B}^j$.

The Properties 4 and 5 imply that the splines ${}_pm_r^j(x)$ are generalized eigenvectors of operators of convolution with any fixed spline and of differentiation. Therefore, the expansion of a spline ${}_pS^j(x)$ with respect to the orthogonal basis $\{{}_pm_r^j(x)\}$ can be treated as a kind of harmonic analysis of a spline ${}_pS^j(x)$ and, if the spline ${}_pS^j(x)$ approximates a function f, it can be looked upon as an approximate harmonic analysis of f. We name this *Spline Harmonic Analysis* (SHA). DFT is a special case of SHA in the space ${}_1\mathfrak{B}^j$, whereas the common Fourier analysis is the limit case of SHA in the spaces ${}_p\mathfrak{B}^j$ as $p \to \infty$. So, loosely speaking, SHA bridges the gap between the continuous and the discrete Fourier analysis. The natural fields of application of SHA are one- and multidimensional problems in which various forms of convolution appear, including differential equations with constant coefficients. The remarkable effectiveness of SHA methods can be demonstrated by solving some ill-posed problems. We will not discuss this here (see, however, [4]) but turn to a recent and very promising application, namely, to the wavelet transforms of periodic splines.

2. Wavelets based on periodic splines

The spaces ${}_p\mathfrak{B}^j$, $j=0, 1, \dots$, generate a multiresolution analysis (MRA) of the space $L^2(T)$ of 1-periodic square integrable functions [3]. Using the basis of ortsplines $\{{}_pm_r^j(x)\}$ yields a very feasible approach to constructing wavelets and to establishing wavelet relations. The outline of our technique is as follows:

1. The so-called two-scale relations are basic for wavelet construction. We establish these relations for ortsplines; the corresponding relations for B-splines follow from Theorem 1 below. Analogous relations may be established for other spline bases.

2. The two-scale relations for ortsplines enable us to construct a family of orthogonal bases of the space of wavelets ${}_p\boldsymbol{\mathfrak{B}}^{j-1}$, which is the orthogonal complement of ${}_p\mathfrak{B}^{j-1}$ in ${}_p\mathfrak{B}^j$ (Theorem 2). These bases involve the so-called ortwavelets, which generate a family of periodic wavelets in a conventional sense (Theorem 3). The shifts of each of these wavelets form a basis of the space ${}_p\boldsymbol{\mathfrak{B}}^{j-1}$. To change from the basis of ortwavelets to the basis of shifts of a wavelet, one can use the Fast Fourier Transform (FFT).

3. By means of ortwavelets we construct, along with a wavelet basis, the dual basis in the sense of Chui [2] (Theorem 4). The availability of the family of wavelet bases and their duals makes it possible to choose an appropriate basis for the problem to be solved.

4. We emphasize that the choice of a basis determines the procedures of decomposition of a spline into a wavelet representation and the inverse operation of reconstructing a spline from its wavelet representation. Once we use the SHA technique, these procedures become extremely simple. We cite a version of the decomposition procedure (Theorem 5).

We suggest the following scheme of wavelet processing of a function.

a) The function is projected onto a spline space ${}_p\mathfrak{B}^j$.

b) The spline obtained is represented via an ortspline basis by means of DFT.

c) The decomposition of the spline into an ortwavelet representation is carried out.

d) By means of DFT we change to the wavelet representation, whereupon we implement the processing of the spline.

e) The next step is the reciprocal one of changing to the ortwavelet representation, which is followed by the reconstruction of the spline via the ortspline basis of ${}_p\mathfrak{B}^j$.

f) The final step is changing by means of DFT to the B-spline basis of ${}_p\mathfrak{B}^j$ and computing the spline thus processed.

This technique may be extended immediately to the multidimensional case. To be specific, we cite here a few formulas illustrating our approach. For more details see [5].

THEOREM 1. *The following two-scale relations hold for* $r = 0, 1, \dots, 2^{j-1}-1$:

$$ {}_pm_r^{j-1}(x) = {}_pb_r^j\,{}_pm_r^j(x) + {}_pb_{r+N/2}^j\,{}_pm_{r+N/2}^j(x),\ {}_pb_r^j = 2^{-p}(1+\omega^{-r})^p $$
$$ \Rightarrow {}_pM^{j-1}(x) = 2^{-p}\sum_{l=0}^{p}\binom{p}{l}{}_pM^j(x-l/N). $$

THEOREM 2. *There exists a family of orthogonal bases*

$$ \{{}_p^\nu w_r^{j-1}(x)\}_0^{N/2-1} \quad of\ {}_p\mathfrak{W}^{j-1} \subset {}_p\mathfrak{V}^j, $$
$$ {}_p^\nu w_r^{j-1}(x) = {}_p^\nu a_r^j\,{}_pm_r^j(x) + {}_p^\nu a_{r+N/2}^j\,{}_pm_{r+N/2}^j(x), \qquad r = 0, 1, \dots, 2^{j-1}-1, $$
$$ {}_p^\nu a_r^j = 2^{-p}\omega^r\,{}_p^\nu\tau_r^{j-1}\,{}_p\overline{b}_{r+N/2}^j = 2^{-p}\omega^r(1-\omega^r)^p\,{}_p^\nu\tau_r^{j-1}({}_{2p}u_r^j)^{-1}, $$

where $\{{}_p^\nu\tau_r^{j-1}\}$ *is an arbitrary nonzero* 2^{j-1}*-periodic sequence.*

Choosing various sequences $\{{}_p^\nu\tau_r^{j-1}\}$, we obtain various bases $\{{}_p^\nu w_r^{j-1}(x)\}$ of the wavelet space ${}_p\mathfrak{W}^{j-1}$. The family of ortwavelets $\{{}_p^\nu w_r^{j-1}(x)\}$ generates the family of periodic wavelets in a conventional sense. To be precise, denote ${}_p^\nu A_k^{j-1} = \sum_r^j \omega^{rk}\,{}_p^\nu a_r^j$ and define the splines

$$ {}_p^\nu\psi^{j-1}(x) = \frac{1}{N}\sum_k^j {}_p^\nu A_k^{j-1}\,{}_pM^j(x-k/N) = \sum_r^j {}_p^\nu a_r^j\,{}_pm_r^j(x) \in {}_p\mathfrak{W}^{j-1}. $$

There hold the dual relations

$$ {}_p^\nu\psi^{j-1}(x+2l/N) = \sum_r^{j-1}\omega^{2rl}\,{}_p^\nu w_r^{j-1}(x), $$
$$ {}_p^\nu w_r^{j-1}(x) = \frac{2}{N}\sum_l^{j-1}\omega^{-2rl}\,{}_p^\nu\psi^{j-1}(x+2l/N). $$

THEOREM 3. *The splines* $\{{}_p^\nu\psi^{j-1}(x-2l/N)\}_{l=0}^{N/2-1}$ *form a family of bases of the space* ${}_p\mathfrak{W}^{j-1}$; *any spline* ${}_pW^{j-1}(x) \in {}_p\mathfrak{W}^{j-1}$ *can be written as*

$$ {}_pW^{j-1}(x) = \sum_r^{j-1} {}_p^\nu\eta_r^{j-1}\,{}_p^\nu w_r^{j-1}(x) = \frac{2}{N}\sum_l^{j-1} {}_p^\nu t_l^{j-1}\,{}_p^\nu\psi^{j-1}(x-2l/N). $$

Moreover, the following relations hold:

$$ {}_p^\nu t_l^{j-1} = \sum_r^{j-1}\omega^{2rl}\,{}_p^\nu\eta_r^{j-1} \Leftrightarrow {}_p^\nu\eta_r^{j-1} = \frac{2}{N}\sum_l^{j-1}\omega^{-2rl}\,{}_p^\nu t_l^{j-1}. $$

The theorem enables us to affirm that the splines ${}_p^\nu\psi^{j-1}(x)$ appear as wavelets in the sense of [3]. We emphasize that the shifts of any wavelet ${}_p^\nu\psi^{j-1}(x)$ form a basis of the wavelet space ${}_p\mathfrak{W}^{j-1}$. This family of wavelets contains the compactly supported (up to periodicity) spline wavelet (compare [3]) as well as the spline wavelet whose shifts form an orthonormal basis of ${}_p\mathfrak{B}^{j-1}$ (compare [1]). Together with a basis, this family contains its dual basis. Namely, the following proposition holds.

THEOREM 4. *The relation* $\langle {}^{\nu}_{p}\psi^{j-1}(x-2l/N), {}^{\mu}_{p}\psi^{j-1}(x-2s/N)\rangle = \delta^l_s$ *is true if and only if* ${}^{\nu}_{p}\tau^{j-1}_{r}\,{}^{\mu}_{p}\bar{\tau}^{j-1}_{r}\,({}_{2p}v^{j}_{r}\,{}_{2p}u^{j}_{r})^{-1}\,{}_{2p}u^{j-1}_{r} = 1/2N$.

We now establish formulae of decomposition.

THEOREM 5. *The following representation holds for* $r = 0, 1, \ldots, 2^j - 1$:

$$ {}_pm^j_r(x) = {}_ph^j_r\,{}_pm^{j-1}_r(x) + {}^{\nu}_{p}g^j_r\,{}^{\nu}_{p}w^{j-1}_r(x), $$

$$ {}_ph^j_r = \omega^r(1+\omega^r)^p\,{}_{2p}u^j_r/2^p\,{}_{2p}u^{j-1}_r, $$

$$ {}^{\nu}_{p}g^j_r = (1-\omega^{-r})^p\omega^{-r}\,{}_{2p}v^j_r\,{}_{2p}u^j_r/2^p\,{}_{2p}u^{j-1}_r\,{}^{\nu}_{p}\tau^{j-1}_r. $$

Any spline ${}_pS^j(x) = \frac{1}{N}\sum^j_k q^j_k\,{}_pM^j(x-k/N) = \sum^j_r \xi^j_r\,{}_pm^j_r(x) \in {}_p\mathfrak{B}^j$, $\mathfrak{q} = \{q_k\}_0^{N-1}$, $\xi^j_r = T^j_r(\mathfrak{q})$, *can be written as the following orthogonal sum:*

$$ {}_pS^j(x) = {}_pS^{j-1}(x) + {}^{\nu}_{p}W^{j-1}(x), $$

$$ {}_pS^{j-1}(x) = \sum_r^{j-1} {}_pm^{j-1}_r(x)\xi^{j-1}_r = \frac{2}{N}\sum_k^{j-1} q^{j-1}_k\,{}_pM^{j-1}(x-2k/N) \in {}_p\mathfrak{B}^{j-1}, $$

$$ {}^{\nu}_{p}W^{j-1}(x) = \sum_r^{j-1} {}^{\nu}_{p}w^{j-1}_r(x)\,{}^{\nu}\eta^{j-1}_r = \frac{2}{N}\sum_k^{j-1} {}^{\nu}t^{j-1}_k\,{}^{\nu}_{p}\psi^{j-1}(x-2k/N) \in {}_p\mathfrak{B}^{j-1}, $$

$$ \xi^{j-1}_r = \xi^j_r\,{}_ph^j_r + \xi^j_{r+N/2}\,{}_ph^j_{r+N/2}, \quad {}^{\nu}\eta^{j-1}_r = \xi^j_r\,{}^{\nu}_{p}g^j_r + \xi^j_{r+N/2}\,{}^{\nu}_{p}g^j_{r+N/2}, $$

$$ q^{j-1}_k = \sum_r^{j-1}\omega^{2kr}\xi^{j-1}_r, \quad {}^{\nu}t^{j-1}_k = \sum_k^{j-1}{}^{\nu}\eta^{j-1}_r. $$

REFERENCES

1. G. Battle, *A block spin construction of ondelettes. Part* I: *Lemarié functions*, Comm. Math. Phys. **110** (1987), 601–615.
2. C. K. Chui and J. Z. Wang, *On compactly supported spline wavelets and a duality principle*, Trans. Amer. Math. Soc. **330** (1992), 903–915.
3. Y. Meyer, *Ondelettes et opérateurs*. I, Hermann, Paris, 1990.
4. V. A. Zheludev, *Operational calculus connected with periodic splines*, Soviet. Math. Dokl. **42** (1991), 162–167.
5. ——, *Wavelets based on periodic splines*, Dokl. Akad. Nauk (1994) (to appear).

TATARSKY PER. 4–17, ST. PETERSBURG 197198, RUSSIA
E-mail address: adar@macro.lgu.spb.su

PART II

Minisymposium on Computational Number Theory

Dedicated to the memory of
Derrick Henry Lehmer

Invited Papers

Proceedings of Symposia in Applied Mathematics
Volume 48, 1994

Algorithms for quadratic orders

INGRID BIEHL AND JOHANNES BUCHMANN

ABSTRACT. We describe deterministic algorithms for solving the following algorithmic problems in quadratic orders: Computing fundamental unit and regulator, principal ideal testing, solving prime norm equations, computing the structure of the class group, computing the order of an ideal class and determining discrete logarithms in the class group. We also prove upper bounds for the time and space complexity of the algorithms.

1. Introduction

A *quadratic discriminant* is an integer $\Delta \in \mathbb{Z}$ which is not a square in $\mathbb{Z}$ such that $\Delta \equiv 0, 1 \mod 4$. By $\mathcal{D}$ we denote the set of all quadratic discriminants.

Let $\Delta \in \mathcal{D}$. By $\sqrt{\Delta}$ we denote the square root of Δ in $\mathbb{C}$ which for $\Delta > 0$ is positive and for $\Delta < 0$ has positive imaginary part. Then

$$\mathcal{O}_\Delta = \mathbb{Z} + \frac{\Delta + \sqrt{\Delta}}{2}\mathbb{Z}$$

is the *quadratic order* of discriminant Δ.

Important computational problems for $\mathcal{O}_\Delta$ are the following:

- For $\Delta > 0$ find the fundamental unit and the regulator R_Δ of $\mathcal{O}_\Delta$.
- Decide whether an ideal of $\mathcal{O}_\Delta$ is principal. If it is, find a generator.
- Decide whether a prime number p is a norm in $\mathcal{O}_\Delta$. If it is, find a number in $\mathcal{O}_\Delta$ of norm p.
- Find the ideal class group of $\mathcal{O}_\Delta$.
- Solve the discrete logarithm problem in the class group of $\mathcal{O}_\Delta$.

In this paper we describe deterministic algorithms for solving these problems and we prove upper bounds for their time and space bit complexity. To give a

1991 *Mathematics Subject Classification.* 11Y16.
This paper is in final form and no version of it will be submitted for publication elsewhere.

© 1994 American Mathematical Society
0160-7634/94 $1.00 + $.25 per page

simpler form for these upper bounds, we write $T = O^*(f)$ for functions $f, T : \mathbb{Z}_{>0} \to \mathbb{Z}_{>0}$ if $T \leq f \cdot p$ with some polynomial function $p : \mathbb{Z}_{>0} \to \mathbb{Z}_{>0}$. Also, for $\Delta < 0$ we let $\mathrm{R}_\Delta = 1$. We prove the following upper bounds both for time and space complexity:

Problem	complexity
Fundamental unit	$O^*(\mathrm{R}_\Delta^{1/2})$
Principal ideal testing	$O^*(\mathrm{R}_\Delta^{1/2})$
Norm p problem	$O^*(p^{1/2} + \mathrm{R}_\Delta^{1/2})$
Class group	$O^*(\|\Delta\|^{1/2})$
Discrete logarithm	$O^*(\|\Delta\|^{1/4})$

Algorithms for solving all the computational problems mentioned above were already developed by Lagrange, Gauß and others in the 18th century. While the running time of these classical methods is roughly the square of the running time of the algorithms described in this paper, they only require space $O^*(1)$ (see [**8**]). So they are more useful for hand computations.

With the invention of computers it became possible to handle large sets of data. It was Shanks [**16**], [**17**] who discovered that the running times of all the classical methods could be reduced to their square root at the expense of using a lot more storage. Some of the ideas of Shanks are explained and analyzed in papers of Lenstra [**11**] and Schoof [**15**]. There, however, the following problem was not solved. If for $\Delta > 0$ one represents the fundamental unit, the generator of an ideal, or a number of norm p in the standard way as $(x + y\sqrt{\Delta})/2$ with $x, y \in \mathbb{Z}$, then the number of bits required to write down this representation may be of the order of magnitude R_Δ. It is therefore not possible to prove the asserted complexity results using that representation. One can only prove upper running time bounds for computing an approximation to the regulator and for deciding principality or the existence of a number of norm p. The complexity results can only be shown with a new *binary representation* of numbers in the field of fractions K_Δ of $\mathcal{O}_\Delta$. This representation was introduced in [**2**]. The binary length of the binary representation for the fundamental unit, for example, is $O^*(1)$. Most of the operations that one wants to perform with numbers in K_Δ can be carried out in polynomial time with numbers given in binary representation. One can, for example, verify in polynomial time that a unit given in binary representation is a unit, i.e., generates the principal ideal $\mathcal{O}_\Delta$. Similar ideas were used in [**4**] and [**7**]. While Shanks relied on using probabilistic algorithms for finding a generating system for the class group, we restrict ourselves in this paper to deterministic methods.

Generalizations of our complexity results to arbitrary number fields can be found in [**1**]. Stronger results can be obtained if one is willing to assume the truth of the generalized Riemann hypothesis or if one applies randomized algorithms. There are various results in this direction. A systematic presentation will be given in a forthcoming paper.

The paper is organized as follows. After a few preliminary remarks we present algorithms for finite abelian groups which are given in such a way that multiplication, inversion and equality testing is possible. In particular, we describe algorithms for finding the structure of the group and for computing discrete logarithms. All these algorithms are based on ideas of Shanks. These algorithms are then applied to the class group. We define that group, and we describe how elements are represented using reduced ideals. We show how multiplication and inversion can be performed in polynomial time. We also show how equality testing is possible in polynomial time if $\Delta < 0$. We then discuss in great detail the problem of deciding whether an ideal class belongs to a given set of ideal classes for $\Delta > 0$. Again, we use an idea of Shanks to develop an algorithm for solving that problem. At the same time we are able to present algorithms for determining the fundamental unit, for principal ideal testing and for solving the norm p problem. At the end of the paper we present an algorithm of H.W. Lenstra [**13**] for computing the class group.

2. Preliminaries

2.1. Representation of rational numbers. With a few exceptions, which will be explicitly mentioned, we use for *integers* the binary representation. The *binary length* of an integer $z \neq 0$ is $\operatorname{size}(z) = \lfloor \log |z| \rfloor + 2$, where log denotes the logarithm to base 2. The extra bit encodes the sign of z. We also set $\operatorname{size}(0) = 2$.

Each mathematical object which we consider in this paper is represented by a finite sequence of integers. The binary length of such a representation is the sum of the binary lengths of the sequence elements. A rational number q, for example, is represented as $q = u/v$ with $u, v \in \mathbb{Z}$, $v > 0$.

2.2. Approximations to complex numbers. Let c be a complex number, and let q be an integer. An *approximation of precision* q to c is a number $\hat{c} \in 2^{-(q+1)}\mathbb{Z}[i]$ such that $|c - \hat{c}| < 2^{-q}$.

2.3. Ordered sets. An *ordered set* is a pair $G = (S(G), <_G)$, where $S(G)$ is a set and $<_G$ is a total ordering of $S(G)$. We also consider $\emptyset$ as an ordered set. Instead of explicitly defining $S(G)$ and $<_G$, we often just write the series of elements of $S(G)$ in increasing order with respect to $<_G$. So, for example, (g, h, k) is the ordered set $(S(G), <_G)$ with $S(G) = \{g, h, k\}$ and $g <_G h$, $h <_G k$ and $g <_G k$.

Let G be an ordered set. We write $\sharp G$ for cardinality of $S(G)$, and if $h \in S(G)$, we write $h \in G$. A *subset* of G is an ordered set G' where $S(G') \subset S(G)$ and $<_{G'}$ is $<_G$ restricted to $S(G')$. If G' is another ordered set which is disjoint from G, i.e., $S(G)$ and $S(G')$ are disjoint, then $G \circ G' = (S(G) \cup S(G'), <_{G \circ G'})$, where on G and G' the ordering $<_{G \circ G'}$ agrees with $<_G$ and $<_{G'}$, respectively, and where $g <_{G \circ G'} g'$ for every $g \in G$, $g' \in G'$.

The set of all maps from $S(G)$ to $\mathbb{Z}$ is denoted by $\mathbb{Z}^G$. For $v, w \in \mathbb{Z}^G$ we write $v < w$ if $v(g) < w(g)$ for all $g \in G$. Moreover, we use the *lexicographic ordering* on $\mathbb{Z}^G$: v is less than w with respect to the lexicographic ordering if $v(g) < w(g)$ for the smallest element $g \in G$ on which v and w disagree.

A map $f \in \mathbb{Z}^G$ can be interpreted as a vector $(f(g))_{g \in G}$ and vice versa. By 1 we denote the vector in $\mathbb{Z}^G$ whose first component is 1 and whose other components are 0. Let G' be a subset of G. Then each $y \in \mathbb{Z}^{G'}$ can be *extended* to a vector in $\mathbb{Z}^G$ by setting $y(g) = 0$ for $g \in G$, $g \notin G'$. For a set T the system of all ordered subsets of T is denoted by $\mathcal{G}(T)$.

3. Finite abelian groups

In order to describe the computation of the ideal class group, we present an algorithm for finding the structure of a finite abelian group H. We also solve the problem of computing orders of elements and determining discrete logarithms.

3.1. Group operations. We assume that H is a multiplicatively written finite abelian group. The elements of H are represented by finite sequences of integers. That representation is not necessarily unique. But we assume that we can carry out the following *group operations*: multiplication, inversion and deciding equality. If we can decide equality of two elements, then we can also decide whether an element belongs to a subset T of H. Such an operation is called a *containment test*.

3.2. Relation lattice. Let G be an ordered subset of H. For $e \in \mathbb{Z}^G$ we set

$$G^e = \prod_{g \in G} g^{e(g)}.$$

The *subgroup of H generated by the elements of G* is

$$\langle G \rangle = \{G^e : e \in \mathbb{Z}^G\}.$$

If $\langle G \rangle = H$, then G is called a *generating system* for H.

The map

$$\mathbb{Z}^G \to \langle G \rangle, \quad (e(g))_{g \in G} \mapsto \prod_{g \in G} g^{e(g)}$$

is a surjective homomorphism. Its kernel $L(G)$ is a $\sharp G$-dimensional sublattice of $\mathbb{Z}^G$ and

$$\langle G \rangle \cong \mathbb{Z}^G / L(G).$$

The lattice $L(G)$ is called the *relation lattice* for G. Its elements are called *relations.* Lattices are represented by bases.

3.3. Structure. Let $s_1, \dots, s_k$ be the *elementary divisors* of H, i.e., the s_i are positive integers for $1 \leq i \leq k$, s_i divides s_{i+1} for $1 \leq i \leq k-1$ and

$$H \cong \bigoplus_{i=1}^{k} \mathbb{Z}/s_i\mathbb{Z}. \tag{1}$$

If we know a generating system G and the relation lattice $L(G)$, then the elementary divisors of H, the isomorphism (1) and its inverse can be determined in polynomial time using the algorithm of [**6**]. Hence, finding $(G, L(G))$ means determining the structure of G.

3.4. The bounded relation problem. The *bounded relation problem* is the following. Given an ordered set G in H and an upper bound $u \in \mathbb{Z}^G$, $u > 1$, decide whether there is a relation y on G with $1 \leq y < u$. If there is one, find the minimal one with respect to the lexicographic ordering.

The problems of computing the structure of H, determining the order of an element in H and finding discrete logarithms in H can be reduced to solving the bounded relation problem.

The solution of the bounded relation problem is based on the following statement.

Proposition 3.1. *Let G be an ordered subset of H, $u \in \mathbb{Z}^G$, $u > 1$. Let y be the minimal such relation on G with $y \geq 1$. If $y < u$, then y can be written as $y = q\lfloor\sqrt{u}\rfloor + r$ with $q, r \in \mathbb{N}^G$, $0 \leq q^2 < u$ and $1 \leq r^2 < u$.*

This statement leads immediately to the following algorithm RELATION which is based on [**16**] and in which we use NEXT which, given G and $x \in \mathbb{Z}^G$, computes the successor of x according to the lexicographical ordering.

For RELATION we get the next theorem.

Theorem 3.2. *RELATION is correct and requires $O(\sqrt{k})$ operations and tests for containment in a subset of H with $O(\sqrt{k})$ elements, where $k = \prod_{g \in G} u(g)$. RELATION also requires space for $O(\sqrt{k})$ elements of H.*

Algorithm 3.3 (RELATION)

Input: H, $G \in \mathcal{G}(H)$, $u \in \mathbb{Z}^G$, $u > 1$.
Output: *If there is* $y \in L(G)$ *with* $1 \leq y < u$, *then the minimal such* y, *else* $y = 0$.

$S := \emptyset$; $r := 1$; $y := 0$
REPEAT
  $h := G^r$
  $S := S \cup \{(h, r)\}$
  $r := NEXT(G, r)$
UNTIL $r^2 \geq u$ *or* $h = 1$
$q := 0$
REPEAT
  $h := G^{-q\lfloor\sqrt{u}\rfloor}$
  IF $(h, r) \in S$ *for some* r
  THEN $y := q\lfloor\sqrt{u}\rfloor + r$
  ELSE $q := NEXT(G, q)$
UNTIL $y \neq 0$ *or* $q^2 \geq u$

3.5. Computing relation lattices. The algorithm BASIS which finds the relation lattice $L(G)$ for an ordered set G in H and which is described in this section is based on the following statement.

Proposition 3.4. *Let G be an ordered subset of H. For $h \in G$ let $e(h) = (e(g, h))_{g\in G}$ be the extension of the minimal relation on $\{g \in G : g \geq h\}$ to G. Then $e = (e(h))_{h\in G}$ is a basis of $L(G)$ with $e(g, h) \geq 0$ and*

$$e(g,h) \begin{cases} = 0 & \text{if } g <_G h, \\ > 0 & \text{if } g = h, \\ < e(g,g) & \text{if } h <_G g, \end{cases}$$

for all $g, h \in G$. Also, the basis e is uniquely determined by these conditions.

We say that the basis from Proposition 3.4 is in *Hermite normal form* (HNF).

Theorem 3.5. *BASIS is correct and requires $O(\sharp G(\log k)\sqrt{k})$ operations and tests for containment in each of $O(\sharp G \log k)$ subsets of H with $O(\sqrt{k})$ elements, where $k = \sharp\langle G\rangle$. BASIS also requires space for $O(\sqrt{k})$ elements in G.*

Proof. The number of iterations of the FOR loop is $\sharp G$. Let $e \in \mathbb{Z}^{G\times G}$ be the HNF basis of $L(G)$. Also, let $k = \sharp\langle G\rangle$. Suppose that the algorithm is executing an iteration of the FOR loop. Then $u(g) = e(g,g)$ for every g with $g >_G h$. We also always have $u(h) \leq 2e(h,h)$. Therefore, the number of iterations in the REPEAT loop is $O(\log e(h,h))$. Since $k = \prod_{g\in G} e(g,g)$, it follows from Theorem 3.2 that the iteration of the FOR loop requires $O((\log k)\sqrt{k})$ operations and tests for containment in each of $\log k$ subsets of H with $O(\sqrt{k})$ elements. The space required is $O(\sqrt{k})$. □

Algorithm 3.6 (BASIS)

Input: H, $G \in \mathcal{G}(H)$.
Output: *The HNF basis* e *of* $L(G)$.

```
G' := ∅
FOR all h ∈ G in descending order
    G' := {h} ∘ G'; u(h) := 2
    REPEAT
        e(h) := RELATION(H, G', u)
        u(h) := 2 * u(h)
    UNTIL e(h) ≠ 0
    Extend e(h) to G; u(h) := e(h, h)
```

3.6. Discrete logarithms. For $g, h \in H$ with $g \in \langle h \rangle$, the *discrete logarithm* of g to base h is the minimal positive integer y such that $g = h^y$. It is denoted by $\log_h g$.

Algorithm 3.7 (DISCRETE LOG)

Input: H, $g, h \in H$.
Output: $y = \log_h g$, *if it exists, and* $y = 0$ *otherwise.*

```
u(g) := u(h) := 2
REPEAT
    y := RELATION(H, (g, h), u)
    o := RELATION(H, (h), u(h))
    u(h) := 2u(h)
UNTIL y + o ≠ 0
```

The algorithm for computing discrete logarithms is based on the following observation.

Proposition 3.8. *Let $g, h \in H$. If $g \neq h^y$ for all y with $1 \leq y \leq$ order h, the discrete logarithm of g to base h does not exist.*

Theorem 3.9. *DISCRETE LOG is correct. The algorithm requires $O((\log k)\sqrt{k})$ operations and tests for containment in each of $O(\log k)$ subsets of H with $O(\sqrt{k})$ elements as well as space for $O(\sqrt{k})$ elements of H. If the discrete logarithm of g to base h exists, then $k = \log_h g$. Otherwise, $k =$ order h.*

4. Quadratic numbers

Let Δ be a quadratic discriminant. The *field of fractions* of $\mathcal{O}_\Delta$ is $\mathrm{K}_\Delta = \mathbb{Q}(\sqrt{\Delta})$. The *standard representation* of a number $\alpha \in \mathrm{K}_\Delta$ is $\alpha = (x+y\sqrt{\Delta})/(2z)$ with $x, y, z \in \mathbb{Z}$, $z > 0$ and $\gcd(x, y, z) = 1$. The integer z is called the *denominator* of α with respect to Δ; we write $z = d(\alpha)$. The *conjugate* of α is $\overline{\alpha} = (x - y\sqrt{\Delta})/(2z)$. The *height* of α is $\mathrm{H}(\alpha) = \max\{|\alpha|, |\overline{\alpha}|\}$. The *norm* of α is $\mathrm{N}(\alpha) = \alpha\overline{\alpha}$. If S is a set of numbers in K_Δ, then we write $\overline{S} = \{\overline{\alpha} : \alpha \in S\}$.

5. Ideals

We wish to use the algorithms from the previous section in class groups. In order to explain what they are, and how operations in those groups are performed, we first talk about ideals.

Let Δ be a quadratic discriminant. By an *ideal* of $\mathcal{O}_\Delta$ we always mean an *invertible ideal* of $\mathcal{O}_\Delta$.

5.1. Representation. A subset A of $\mathcal{O}_\Delta$ is an ideal of $\mathcal{O}_\Delta$ if and only if

$$\mathrm{A} = q\left(\mathbb{Z} + \frac{b+\sqrt{\Delta}}{2a}\mathbb{Z}\right), \tag{2}$$

where $a, b \in \mathbb{Z}$, $q \in \mathbb{Q}$, $a, q > 0$, $c = (b^2 - \Delta)/(4a) \in \mathbb{Z}$ and $\gcd(a, b, c) = 1$. In this representation, the numbers a, q are uniquely determined and the integer b is unique modulo $2a$. We write $a = a(\mathrm{A})$ and $q = q(\mathrm{A})$. A representation (2) is called *standard representation* if $|b| \leq \max\{a, \sqrt{|\Delta|}\}$. Ideals are always given in some standard representation. The *norm* of A is q^2a. The ideal A is called *normal* if $q = 1$. The *denominator* $d(\mathrm{A})$ of A is the smallest positive integer d such that $d\mathrm{A} \subset \mathcal{O}_\Delta$. If A is normal, then $d(\mathrm{A}) = a(\mathrm{A})$. The set I_Δ of invertible ideals of $\mathcal{O}_\Delta$ is an abelian group.

5.2. Inverse. The *inverse* of an ideal A of $\mathcal{O}_\Delta$ is $\mathrm{A}^{-1} = (1/\mathrm{N}(\mathrm{A}))\overline{\mathrm{A}}$.

5.3. Product. The *product* of two ideals A, B can be determined as follows. Let $b(\mathrm{A})$, $b(\mathrm{B})$ be the numbers b appearing in some standard representation of A and B (see § 5.1). Compute $m = \gcd(a(\mathrm{A}), a(\mathrm{B}), (b(\mathrm{A})+b(\mathrm{B}))/2)$. Let $j, k, \ell \in \mathbb{Z}$

with $ja(\mathrm{A}) + ka(\mathrm{B}) + \ell(b(\mathrm{A}) + b(\mathrm{B}))/2 = m$. Then $a(\mathrm{AB}) = a(\mathrm{A})a(\mathrm{B})/(m^2)$, $b(\mathrm{AB}) \equiv (ja(\mathrm{A})b(\mathrm{B}) + ka(\mathrm{B})b(\mathrm{A}) + \ell(b(\mathrm{A})b(\mathrm{B}) + \Delta)/2)/m \mod a(\mathrm{AB})$ and $q(\mathrm{AB}) = q(\mathrm{A})q(\mathrm{B})m$.

5.4. Principal ideals. By K_Δ^* we denote the set of nonzero elements of K_Δ. A *principal ideal* of $\mathcal{O}_\Delta$ is an ideal of the form $\alpha\mathcal{O}_\Delta$ with $\alpha \in \mathrm{K}_\Delta^*$. The number α is called a *generator* of $\mathcal{O}_\Delta$. Let $\alpha \in \mathrm{K}_\Delta^*$, $\alpha = (x+y\sqrt{\Delta})/(2z)$, be its standard representation. The *principal ideal* $\alpha\mathcal{O}_\Delta$ can be found as follows. Compute $m = \gcd(y, (x+y\Delta)/2)$ and let $k, \ell \in \mathbb{Z}$ such that $ky + \ell(x+y\Delta)/2 = m$. Then $a(\alpha\mathcal{O}_\Delta) = a = |x^2 - \Delta y^2|/(4m^2)$, $b(\alpha\mathcal{O}_\Delta) \equiv b = (kx + \ell(x+y)\Delta/2)/m \mod 2a$ and $q(\alpha\mathcal{O}_\Delta) = ma/z$.

5.5. Prime ideals. We denote by $\mathbb{P}_\Delta$ the set of all prime numbers p such that there is an invertible prime ideal of norm p in $\mathcal{O}_\Delta$. For a prime number $p \in \mathbb{P}_\Delta$ there is a prime ideal

$$\mathrm{A}_p = \mathbb{Z} + \frac{b_p + \sqrt{\Delta}}{2p}\mathbb{Z},$$

where b_p is the uniquely determined integer with $0 < b_p < p$, and $b_p^2 \equiv \Delta \mod 4p$. The only possible other prime ideal of norm p is $\overline{\mathrm{A}_p}$.

On input of Δ and a prime number p, the procedure PRIME IDEAL returns A_p if $p \in \mathbb{P}_\Delta$ and 0 otherwise. In PRIME IDEAL the square root of Δ mod p must be found. The fastest deterministic algorithm for factoring polynomials mod p requires time and space $O^*(\sqrt{p})$ (see [**18**]). Therefore, also PRIME IDEAL requires time and space $O^*(\sqrt{p})$.

6. Ideal class group

In this section we introduce the ideal class group and we explain how to perform the basic operations in these groups.

6.1. Definition. The set P_Δ of all principal ideals of $\mathcal{O}_\Delta$ is a subgroup of the group I_Δ of all ideals of $\mathcal{O}_\Delta$. The elements of the *ideal class group* $\mathrm{Cl}_\Delta = \mathrm{I}_\Delta/\mathrm{P}_\Delta$ are called *ideal classes* of $\mathcal{O}_\Delta$. The number $\mathrm{h}_\Delta = \sharp\mathrm{Cl}_\Delta$ is called *class number* of Δ. Ideals $\mathrm{A}, \mathrm{B} \in \mathcal{O}_\Delta$ are called *equivalent* if they belong to the same ideal class. The class of an ideal A is denoted by [A].

6.2. Representation. For $a, b \in \mathbb{Z}$, $a > 0$, we denote by $\tau(a, b)$ the uniquely determined number b' such that $b' \equiv b \mod 2a$ and

$$\begin{array}{ll} -2a + \sqrt{\Delta} < b' < \sqrt{\Delta} & \text{for } \Delta > 0 \text{ and } a < \sqrt{\Delta}, \\ -a < b' \leq a & \text{for } \Delta < 0 \text{ or } \Delta > 0 \text{ and } a > \sqrt{\Delta}. \end{array}$$

For an ideal A of $\mathcal{O}_\Delta$ we define

$$b(\mathrm{A}) = \tau(b, a), \quad c(\mathrm{A}) = |b(\mathrm{A})^2 - \Delta|/(4a(\mathrm{A}))$$

and

$$\gamma(\mathrm{A}) = \frac{b(\mathrm{A}) + \sqrt{\Delta}}{2a(\mathrm{A})}.$$

An ideal A of $\mathcal{O}_\Delta$ is called *reduced* if it is normal and

$$\begin{array}{ll} b(\mathrm{A}) \geq \max\{0, 2a(\mathrm{A}) - \sqrt{\Delta}\} & \text{for } \Delta > 0, \\ |b(\mathrm{A})| \leq a(\mathrm{A}) \leq c(\mathrm{A}) & \text{for } \Delta < 0. \end{array}$$

Each ideal class contains a reduced ideal. Hence, we can in fact represent each ideal class by a reduced ideal. For $\Delta > 0$, however, this representation is far from being unique.

Here are a few properties of reduced ideals. If A is reduced, then $a(\mathrm{A}), |b(\mathrm{A})| < \sqrt{|\Delta|}$, and if $\Delta > 0$, then $\gamma(\mathrm{A}) > 1$ and $-1 < \overline{\gamma(\mathrm{A})} < 0$. Also, a normal ideal A is reduced if $a(\mathrm{A}) < \sqrt{|\Delta|}/2$.

6.3. Multiplication, inversion. If we want to determine the inverse of an ideal class or the product of two ideal classes, then we compute the product or the inverse of their reduced representatives. The resulting ideals are, in general, not reduced. Therefore, in this section we describe a polynomial-time algorithm REDUCE which on input of an ideal A computes a reduced ideal in the ideal class of A. Thus, computing inverses and products of ideal classes can be effected in polynomial time.

Let A be a normal ideal. Define

$$\rho(\mathrm{A}) = (1/\gamma(\mathrm{A}))\mathrm{A} = \mathbb{Z} + \frac{1}{\gamma(\mathrm{A})}\mathbb{Z}.$$

Then $\rho(\mathrm{A})$ is normal,

$$a(\rho(\mathrm{A})) = c(\mathrm{A}), \quad b(\rho(\mathrm{A})) = \tau(-b(\mathrm{A}), c(\mathrm{A})).$$

Here is the reduction algorithm which works in polynomial time.

Algorithm 6.1 (REDUCE)

Input: $\Delta \in \mathcal{D}$, $A \in I_\Delta$.
Output: *A reduced ideal B in the class of A,* $\alpha \in A$ *with* $A = \alpha B$ *and* $H(\alpha) \leq q(A)$.

$\alpha := 1;\ B := A$		*(1)*
REPEAT		*(2)*
	$\alpha := \alpha\gamma(B)$	*(3)*
	$B := \rho(B)$	*(4)*
UNTIL B is reduced		*(5)*

The output α of the algorithm REDUCE is called the reducing number of A.

6.4. Containment problem. Let A and B be ideals of $\mathcal{O}_\Delta$ in the same ideal class. Then there is $\alpha \in \mathrm{K}_\Delta^*$ such that $\mathrm{A} = \alpha\mathrm{B}$. The number α is called a *generator of A relative to B.* For $\Delta > 0$ we assume without loss of generality that $\alpha > q(\mathrm{A})$. So in this case, the *minimal generator* is well defined. For $\Delta < 0$ each generator of A relative to B is called minimal.

The *containment problem* for Δ is the following. Given a quadratic discriminant Δ, a set T of reduced ideals in $\mathcal{O}_\Delta$ and a reduced ideal A of $\mathcal{O}_\Delta$, decide whether A is equivalent to a reduced ideal B in T. If so, find the minimal generator of A relative to B.

6.5. Solving the containment problem for $\Delta < 0$. If $\Delta < 0$, then there are at most two reduced ideals in each ideal class. More precisely, let A be a reduced ideal of $\mathcal{O}_\Delta$; then the only other possible reduced ideal of the same class is $\rho(\mathrm{A})$.

Hence, for $\Delta < 0$ the containment problem can be solved as follows. Let T be a set of reduced ideals and A be a reduced ideal of $\mathcal{O}_\Delta$. We order the elements of T such that binary search is possible. Then we determine the reduced ideals contained in [A], and using binary search, we check whether one of these ideals belongs to T. If so, the generator of A relative to that ideal is either 1 or $\gamma(\mathrm{A})$. Thus, we have proved the following statement.

Proposition 6.2. *For* $\Delta \in \mathcal{D}$, $\Delta < 0$ *the containment problem can be solved using time and space* $O^*(\sharp T)$.

6.6. The equivalence problem for $\Delta > 0$. If $\Delta > 0$, then there can be more than $\sqrt{\Delta}$ reduced ideals in one ideal class. Hence in this case, the containment problem is more difficult to solve. We first consider a special case.

The *bounded equivalence problem* is the following. Given a positive quadratic discriminant Δ, reduced ideals A, B of $\mathcal{O}_\Delta$ and a positive integer u, decide whether A has a generator α relative to B with $\alpha < u$. If so, find the minimal α.

This problem is very similar to the discrete logarithm problem in an abelian group. It is, however, not clear whether the decision problem belongs to the complexity class NP. In the sequel, we will prove that this is the case. We will also present an algorithm to solve the equivalence problem. This algorithm is very similar to the discrete logarithm procedure from above.

6.7. Minima. Let $\Delta > 0$. To show that the equivalence decision problem belongs to NP, we study the representation of relative generators. Let A be an ideal of $\mathcal{O}_\Delta$. A generator of A relative to a reduced ideal of $\mathcal{O}_\Delta$ is called a *minimum* of A. The logarithm of a minimum of A can be considered as some sort of discrete logarithm. These logarithms form a discrete subset of the real line. Each minimum has a right and a left neighbor which is easily computable. This will now be explained in more detail.

The ideal A is reduced if and only if 1 is a minimum of A. Let α be a minimum of A; then

$$|\mathrm{N}(\alpha)| < \mathrm{N}(\mathrm{A})\sqrt{\Delta}.$$

The *right neighbor* of a minimum α in A is the uniquely determined minimum β of A with minimal $\beta > \alpha$. It can be computed via

$$\mathrm{R\ MIN}(\Delta, \mathrm{A}, \alpha) = \alpha\gamma((1/\alpha)\mathrm{A}).$$

In particular, if A is reduced, then $\gamma(\mathrm{A})$ is the right neighbor of 1 in A. The *left neighbor* of a minimum α in A is the uniquely determined minimum β of A with minimal $|\overline{\beta}| > |\overline{\alpha}|$. It can be computed via

$$\mathrm{L\ MIN}(\Delta, \mathrm{A}, \alpha) = \alpha\overline{\gamma(\overline{(1/\alpha)\mathrm{A}})}.$$

Let α, β be minima of A. Then β is the right neighbor of α if and only if α is the left neighbor of β. Also $|\overline{\alpha}|, |\overline{\beta}|$ are minima of $\overline{\mathrm{A}}$, and $|\overline{\beta}|$ is the left neighbor of $\overline{\alpha}$ if β is the right neighbor of α.

Fix some minimum α_0 of A. Let $(\alpha_i)_{i\in\mathbb{Z}}$ be the sequence of minima in A with the property that α_i is the right neighbor of α_{i-1} for $i \in \mathbb{Z}$. Then we have

$$\log\alpha_{i+2} - \log\alpha_i \geq 1, \quad i \in \mathbb{Z}, \tag{3}$$

and

$$0 < \log\alpha_{i+1} - \log\alpha_i < (\log\Delta)/2, \quad i \in \mathbb{Z}. \tag{4}$$

The previous statements show that minima have a fairly regular distribution on the real line.

The sequence $(\alpha_i)_{i\in\mathbb{Z}}$ is periodic in the following sense. There is $k = k(\mathrm{A}) \in \mathbb{Z}_{>0}$ such that α_i/α_j is a unit in $\mathcal{O}_\Delta$ if and only if $i \equiv j \mod k$ for any $i, j \in \mathbb{Z}$. The number $\varepsilon = \alpha_k/\alpha_0$ is the fundamental unit of $\mathcal{O}_\Delta$. For every $i, l \in \mathbb{Z}$ we have $\alpha_{lk+i}/\alpha_i = \varepsilon^l$. The number $k(\mathrm{A})$ is an invariant of the ideal class of A. It is called the *period length* of [A]. If $k(\mathrm{A}) = 1$, then

$$\log \alpha_{i+1} - \log \alpha_i > 1/2, \quad i \in \mathbb{Z}. \tag{5}$$

6.8. Reduced ideals for $\Delta > 0$. Let A be an invertible ideal of $\mathcal{O}_\Delta$. The reduced ideals in [A] are exactly the ideals $(1/\alpha)\mathrm{A}$ where α is a minimum in A.

If A is reduced, then the *right neighbor* of A is

$$\mathrm{R\ IDEAL}(\Delta, \mathrm{A}) = \rho(\mathrm{A})$$

and the *left neighbor* of A is

$$\mathrm{L\ IDEAL}(\Delta, \mathrm{A}) = \overline{\rho(\overline{\mathrm{A}})}.$$

If B and C are reduced ideals in the class of A, $\mathrm{B} = (1/\beta)\mathrm{A}$, $\mathrm{C} = (1/\gamma)\mathrm{A}$ with minima β, γ of A, then C is the right neighbor of B if and only if γ is the right neighbor of β. The graph of reduced ideals in the class of A is a cycle, the *cycle of reduced ideals* in the class of A. The number of elements of that cycle is $k(\mathrm{A})$.

6.9. Binary representation of minima. Let A be a reduced ideal of $\mathcal{O}_\Delta$ and α a minimum of A. A *binary representation* of α in A is of the form

$$\alpha = \prod_{i=1}^{n} \beta_i^{2^{n-i}},$$

where $n \in \mathbb{Z}_{\geq 1}$, $\beta_i \in \mathrm{K}_\Delta^*$, $1 \leq i \leq n$, satisfy

- $n \leq \log\log \alpha + 2$,
- $\mathrm{H}(\beta_i) < 16\Delta^{5/4}$ for $1 \leq i \leq n$,
- $\alpha_j = \prod_{i=1}^{j} \beta_i^{2^{j-i}}$ is a minimum in A for $1 \leq j \leq n$.

We write

$$\alpha = (\beta_1, \dots, \beta_n).$$

It follows that the binary length of the standard representation of each β_i is $O^*(1)$, and the binary length of a binary representation of α is $O^*(\log\log \mathrm{H}(\alpha))$. We will show below that all minima have binary representations. Minima in reduced ideals will be always given in binary representation. Note that

$$\log \alpha = \sum (\log \beta_i) 2^{n-i}.$$

This looks like a real binary representation for the "discrete logarithm" $\log \alpha$ except that the $\log \beta_i$ are not bits but "small" numbers.

It is easy to see that there is a polynomial-time algorithm IDEAL that on input of Δ, A and a minimum α of A in binary representation determines the ideal $(1/\alpha)$A. It is also easy to see that there is a polynomial-time algorithm APPROX that on input of Δ, A, a minimum α of A in binary representation and a positive integer q in unary representation determines an approximation of precision q to $\log\alpha$.

6.10. Finding minima of a given size. Let A be a reduced ideal of $\mathcal{O}_\Delta$ and $s \in \mathbb{Q}_{>0}$. A minimum α in A is called *close to s* if

$$|s - \log\beta| > |s - \log\alpha| - 3/4 \tag{6}$$

for every minimum β in A. From (3) and (4) one can deduce

$$|s - \log\alpha| < (\log\Delta + 3)/4, \tag{7}$$

and

$$\begin{aligned} &\max\{2^{s-3/4}\Delta^{-1/4}, 2^{-s-3/4}\mathrm{N}(\mathrm{A})\Delta^{-1/4}\} \\ &\leq \mathrm{H}(\alpha) \\ &\leq \max\{2^{s+3/4}\Delta^{1/4}, 2^{-s+3/4}\mathrm{N}(\mathrm{A})\Delta^{3/4}\}. \end{aligned} \tag{8}$$

We now present the algorithm CLOSE which on input of Δ, A and s finds in polynomial time a minimum α of A which is close to s. The main tool that is used in CLOSE is the procedure DOUBLE, which on input of $t \in \mathbb{Q}_{>0}$ and a minimum $(\alpha_1, \dots, \alpha_{k-1})$ of A which is close to t determines in polynomial time a minimum $(\alpha_1, \dots, \alpha_{k-1}, \alpha_k)$ of A which is close to $2t$. Using DOUBLE as a subroutine, CLOSE works as follows.

Algorithm 6.3 (CLOSE)

Input: $\Delta \in \mathcal{D}$, $\Delta > 0$, *a reduced ideal* A *of* $\mathcal{O}_\Delta$, $s \in \mathbb{Q}_{\geq 0}$.
Output: *A minimum* $\alpha = (\alpha_1, \dots, \alpha_n)$ *of* A *close to* s, *where* $n = \lceil\log s\rceil + 1$ *and* $H(\alpha_i) \leq 4\Delta^{5/4}$ *for* $1 \leq i \leq n$ *if* $s > 0$, *otherwise 1.*

```
IF  s = 0
THEN  α = 1
ELSE  n := ⌈log s⌉ + 1
      α := 1
      t := s/2^n
      FOR  i = 1, 2, ..., n
           α := DOUBLE(A, α, t)
           t := 2t;
```

Proposition 6.4. *CLOSE is correct and has running time $O^*(1)$.*

Proof. The assertion follows from the fact that DOUBLE has polynomial running time and that we have initially $t < 1/2$. □

Next we present the procedure DOUBLE. DOUBLE uses the procedure NEAREST, which on input of an ideal B and a rational number u finds in polynomial time a minimum β in B *very close to* u, i.e., for every minimum γ of B we have

$$|u - \log \gamma| > |u - \log \beta| - 1/4.$$

Algorithm 6.5 (DOUBLE)

Input:	$\Delta \in \mathcal{D}$, $\Delta > 0$, *a reduced ideal* A *of* $\mathcal{O}_\Delta$, $t \in \mathbb{Q}_{>0}$, *a minimum* $(\alpha_1, \dots, \alpha_{k-1})$ *of* A.
Output:	*A minimum* $(\alpha_1, \dots, \alpha_{k-1}, \alpha_k)$ *of* A *close to* $2t$ *with* $H(\alpha_k) \leq 4\Delta^{5/4}$.

$B := (1/\alpha^2)A$	*(1)*
$u := 2(t - \mathit{APPROX}(\Delta, \alpha, 3))$	*(2)*
$\alpha_k := \mathit{NEAREST}(\Delta, B, u)$	*(3)*

Proposition 6.6. *DOUBLE is correct and has running time $O^*(1)$.*

Proof. NEAREST determines a minimum α_k in B. Hence, $\alpha' = (\alpha_1, \dots, \alpha_{k-1}, \alpha_k) = \alpha_k\alpha^2$ is a minimum in A. Let β be a minimum in A. Then $\gamma = \beta/\alpha^2$ is a minimum in B. It follows that

$$\begin{aligned} |2t - \log \beta| \geq& |u - \log \gamma| - |2 \cdot \mathrm{APPROX}(\Delta, \alpha, 3) - 2\log \alpha| \\ \geq& |u - \log \gamma| - 1/4 \geq |u - \log \alpha_k| - 1/2 \\ \geq& |2t - \log \alpha'| - 3/4. \end{aligned}$$

Thus, α' is close to $2t$.

Next we estimate the parameter u that is input for NEAREST. Since initially the minimum α is close to t, it follows that

$$|u| = 2|t - \mathrm{APPROX}(\Delta, \alpha, 3)| < (\log \Delta)/2 + 7/4. \tag{9}$$

It is easy to verify that $(1/\alpha^2)$A can be determined in polynomial time. Since B is the square of a reduced ideal, it follows that the running time of NEAREST is polynomial in $\log \Delta$. Also, it follows from (9) and (8) that $\mathrm{H}(\alpha_k) \leq 8\Delta^{5/4}$. □

Finally, we present NEAREST.

Algorithm 6.7 (NEAREST)

Input: $\Delta \in \mathcal{D}$, $\Delta > 0$, *an ideal* B *of* $\mathcal{O}_\Delta$, $u \in \mathbb{Q}_{>0}$.
Output: *A minimum* $\beta \in B$ *very close to* u.

```
(C, γ) := REDUCE(Δ, B); c := APPROX(Δ, γ, 2)
IF  c ≤ u
THEN  REPEAT
          β := γ
          γ := R MIN(R MIN(Δ, B, γ))
          c := APPROX(Δ, γ, 2)
      UNTIL  c > u
ELSE  REPEAT
          γ := L MIN(L MIN(Δ, B, γ))
          c := APPROX(Δ, γ, 2)
          β := γ
      UNTIL  c < u
β0 := L MIN(Δ, B, β)
β1 := β
β2 := R MIN(Δ, B, β)
β3 := R MIN(Δ, B, β2)
β4 := R MIN(Δ, B, β3)
For β choose that element from {β0, β1, β2, β3, β4} such that
|u − APPROX(Δ, β, 2)| is minimal
```

Proposition 6.8. *NEAREST is correct and has running time* $O^*(1)$.

Proof. We show that NEAREST works correctly by proving that the *optimal minimum* of B, i.e., the minimum β' with $|u - \log\gamma| \geq |u - \log\beta'|$ for every minimum γ of B, belongs to the set $\{\beta_i : 0 \leq i \leq 4\}$ from which the output is chosen.

NEAREST starts by computing the reducing number γ for B. Let $\gamma_0 = \gamma$, and let $(\gamma_i)_{i\in\mathbb{Z}}$ be the sequence of minima as in § 6.7. In the sequence $S = (\dots, \gamma_{-4}, \gamma_{-2}, \gamma_0, \gamma_2, \gamma_4, \dots)$ NEAREST determines the last minimum $\beta = \gamma_k$ such that $\text{APPROX}(\Delta, \beta, 2) \leq u$. That minimum is found in the two REPEAT loops since by (3), $\text{APPROX}(\Delta, \gamma_{i+2}, 2) - \text{APPROX}(\Delta, \gamma_i, 2) \geq 1/2$.

If $\log\gamma_k > u$, then for $i \geq 1$ the minimum γ_{k+i} cannot be optimal. Also, since

$\log \gamma_k < u + 1/4$, it follows from (3) that for $i \geq 2$ the minimum $\log \gamma_{k-i}$ cannot be optimal. If $\log \gamma_k < u$, then for $i \geq 1$ the minimum γ_{k-i} cannot be optimal. Also, since $\log \gamma_{k+2} > u - 1/4$, it follows from (3) that for $i \geq 4$ the minimum $\log \gamma_{k+i}$ cannot be optimal.

Next, we must show that the running time of NEAREST is polynomial. In NEAREST we have initially $\mathrm{N(B)} \leq \mathrm{H}(\beta) \leq q(\mathrm{B})$. It follows from this inequality and from (4) for all minima γ computed in the algorithm that

$$\min\{\log(\mathrm{N(B)}), u - \log \Delta - 1/4\} \leq \log \gamma \leq \max\{0, u + \log \Delta + 1/4\}$$

for every minimum γ computed in the algorithm. This shows that the sizes of the input to the procedures used in NEAREST are $O^*(1)$. □

6.11. Computing binary representations for minima. The procedure CLOSE is used in BINARY, which determines a binary representation for any minimum α of a reduced ideal A of $\mathcal{O}_\Delta$.

Algorithm 6.9 (BINARY)

Input: *$\Delta \in \mathcal{D}$, $\Delta > 0$, a reduced ideal A of $\mathcal{O}_\Delta$, for some minimum α of A the ideal $B = (1/\alpha)A$ and an approximation a of $\log \alpha$ of precision 3.*

Output: *A binary representation of α in A.*

$\alpha_1 :=$ *CLOSE*(Δ, A, a)	*(1)*
$\alpha_0 :=$ *L MIN*(Δ, B, α_1); $\alpha_2 :=$ *R MIN*(Δ, B, α_1)	*(2)*
For α choose the element in $\{\alpha_0, \alpha_1, \alpha_2\}$ such that $\vert$*APPROX*$(\Delta, \alpha, 3) - a\vert \leq 1/2$ *and* $(1/\alpha)B = A$	*(3)*

BINARY is based on the following statement.

Proposition 6.10. *Let α be a minimum in the reduced ideal A. Let a be an approximation to $\log \alpha$ of precision 3. Let α_1 be a minimum of A which is close to a, α_0 the left neighbor of α_1, α_2 the right neighbor of α_1, $M = \{\alpha_0, \alpha_1, \alpha_2\}$. Then $\alpha \in M$, and if $\alpha' \in M$, a' an approximation of precision 3 to $\log \alpha'$, then $\alpha' = \alpha$ if and only if $|a' - a| < 1/4$ and $(1/\alpha')B = (1/\alpha)B$.*

Proof. Since a is an approximation of precision 3 to $\log \alpha$, we have $|a - \log \alpha| < 1/8$. Since α_1 is close to a, it follows from (6) that $|a - \log \alpha_1| < 3/4 + 1/8$, and thus $|\log \alpha - \log \alpha_1| < 1$. By (3) we have $|\log \beta - \log \alpha_1| > 1$ for every minimum β in the sequence of right neighbors of a_2 and also for every minimum β in the sequence of left neighbors of α_0. Hence, α belongs to the set M.

Let $\alpha' \in M$ and a' an approximation of precision 3 to $\log \alpha'$. If $\alpha = \alpha'$, then we obviously have $|a' - a| < 1/4$ and $(1/\alpha')\mathrm{A} = (1/\alpha)\mathrm{A}$. On the other hand, assume that $\alpha \neq \alpha'$. If $k(A)$ exceeds 1, then the ideals $(1/\alpha_0)A \neq (1/\alpha_1)A$ and $(1/\alpha_2)A \neq (1/\alpha_1)A$. Also, $\log \alpha_2 - \log \alpha_0 > 1$, so $(1/\alpha')A \neq (1/\alpha)A$ or $|a - a'| > 1/4$. If $k(A) = 1$, then by (5), $|\log \alpha' - \log \alpha| > 1/2$ and so $|a' - a| > 1/4$. □

As an immediate consequence of Proposition 6.4 we obtain the following running time estimate.

Proposition 6.11. *BINARY is correct and has running time $O^*(1)$.*

6.12. Solving the bounded equivalence problem. Let $\Delta > 0$. The algorithm BOUNDED EQUIVALENCE is based on the following statement.

Proposition 6.12. *Let $\Delta \in \mathcal{D}$, $\Delta > 0$. Let A, B be reduced ideals, $u \in \mathbb{Z}_{>0}$, $q = \max\{0, \lceil (\log \alpha - ((\log \Delta)/4 + 1))/\sqrt{u} - 1 \rceil\}$. Let β be a minimum in A which is close to some s with $q\sqrt{u} \leq s < q\sqrt{u} + 1/4$. For $q = 0$ let $\beta = 1$. If there is a generator α of A relative to B with $0 < \log \alpha < u$, then for $\gamma = \alpha/\beta$ the inverse $1/\gamma$ is a minimum in B and $0 < \log \gamma < \sqrt{u} + (\log \Delta)/2 + 2$. Also, $q < \sqrt{u}$.*

Proof. Let α be a generator of A relative to B with $\log \alpha < u$. Let q be as above. Suppose that $q = 0$. Then $\gamma = \alpha$ and $\log \gamma = \log \alpha \leq \sqrt{u} + (\log \Delta)/4 + 1$.

Now assume that $q > 0$. Then

$$(q-1)\sqrt{u} < \log \alpha - ((\log \Delta)/4 + 1) - \sqrt{u} \leq q\sqrt{u}$$

and therefore

$$(\log \Delta)/4 + 1 < \log \alpha - q\sqrt{u} \leq \sqrt{u} + (\log \Delta)/4 + 1. \tag{10}$$

Since $\log \alpha < u$, this implies that $q < \sqrt{u}$. By (7) we have

$$|\log \beta - q\sqrt{u}| < (\log \Delta)/4 + 1. \tag{11}$$

Since $\gamma = \alpha/\beta$, the inverse $1/\gamma$ is a minimum in B, and it follows from (10) and (11) that $0 < \log \gamma < \sqrt{u} + (\log \Delta)/2 + 2$. □

BOUNDED EQUIVALENCE first determines all the numbers γ such that $1/\gamma$ is a minimum in B and $\log \gamma$ is below $\sqrt{u}$ plus some error, which is of technical relevance. Next, the algorithm constructs minima in A close to $q\sqrt{u}$ for $q = 1, 2, \ldots$ until $\beta\gamma$ is a generator of A relative to B. All the minima are stored in binary representation. To multiply two minima, one can determine the sum of approximations of precision 4 to the log of the factors and then use BINARY to construct the product. In the algorithm, we use the procedure APP, which determines approximations to values of elementary functions.

Algorithm 6.13 (BOUNDED EQUIVALENCE)

Input: $\Delta \in \mathcal{D}$, $\Delta > 0$, *reduced ideals* A, B *of* $\mathcal{O}_\Delta$, $u \in \mathbb{N}$.

Output: *The minimal generator α of A relative to B, if it exists, and* $\log \alpha < u$.

$\gamma := 1;\ S := \emptyset;\ \alpha := 0;\ C := B;\ d = \mathit{APP}((\log \Delta)/2, 2)$		
REPEAT		
	$\gamma := \gamma(L\ \mathit{MIN}(\Delta, C, 1))$	
	$c := \mathit{APPROX}(\Delta, \gamma, 2)$	
	$C := L\ \mathit{IDEAL}(\Delta, C)$	
	$S := S \cup \{(C, \gamma)\}$	
UNTIL $(c - d - 5/2)^2 \geq u$ *or* $C = A$		
Sort S *according to first components*		
$q := 0$		
REPEAT		
	$s = \mathit{APP}(q\sqrt{u} + 1/8, 3)$	
	$\beta := \mathit{CLOSE}(\Delta, A, s)$	
	$C := (1/\beta)A$	
	IF $(C, \gamma) \in S$ *for some* γ	
	THEN	$\alpha := \beta\gamma$
	ELSE	$q := q + 1;$
UNTIL $q^2 \geq u$ *or* $\alpha \neq 0$		

Proposition 6.14. *BOUNDED EQUIVALENCE is correct and requires time and space* $O^*(\sqrt{u})$.

Proof. Clearly, the algorithm terminates, and if the output is $\alpha \neq 0$, then α is a generator of A relative to B.

Suppose that there is a generator α of A relative to B and that $\log \alpha < u$. Let q be as in Proposition 6.12. Then $q < \sqrt{u}$. Let β be the minimum found in the second REPEAT loop, $\gamma = \alpha/\beta$. Then by Proposition 6.12 we have $0 < \log \gamma < \sqrt{u} + (\log \Delta)/2 + 2$ and therefore, $(c - d - 5/2)^2 < u$ for any approximation of precision 2 to $\log \gamma$. This implies that S contains the pair $(\gamma \mathrm{B}, \gamma)$ and thus, the algorithm finds a generator of A relative to B.

Next we show that this generator is the minimal one. Let $\alpha' = \beta\gamma'$ be the generator found in the algorithm. Let c' be the approximation of $\log \gamma'$ computed in the first REPEAT loop. Assume that $\alpha \neq \alpha'$. Since $(1/\alpha)\mathrm{A} = (1/\alpha')\mathrm{A} = \mathrm{B}$, it follows from the results of § 6.7 that $\log \alpha' - \log \alpha > 1/2$ and thus, $\log \gamma' - \log \gamma > 1/2$. This implies that $c < c'$ for every approximation c of precision 2 to $\log \gamma$.

Hence, S must contain the pairs $(\gamma\mathrm{B}, \gamma)$ and $(\gamma'\mathrm{B}, \gamma')$, and this is impossible since the first components of all pairs in S are pairwise distinct.

Finally, we prove the running time estimate. The number of elements in S is $O^*(\sqrt{u})$, so the time and space spent in that loop is $O^*(1)$. Since for the IF decision we can use binary search, the time and space required for the second REPEAT loop is also $O^*(\sqrt{u})$. $\square$

6.13. Principal ideal testing. Let B be a principal ideal of $\mathcal{O}_\Delta$. Let β be the reducing number for B and let α be the minimal generator of $\mathrm{A} = (1/\beta)\mathrm{B}$ relative to $\mathcal{O}_\Delta$. Then $\alpha\beta$ is a generator of B and it is called the *minimal generator* of B. It is represented as the pair (α, β). That representation is of binary length $O^*(1)$.

Algorithm 6.15 (PRINCIPAL)

Input: $\Delta \in \mathcal{D}$, $B \in I_\Delta$.
Output: *If B is principal, then the minimal generator (α, β) of B, else $(0, \beta)$ with some β.*

```
(A, β) := REDUCE(B); α := 0
IF  Δ < 0
THEN  IF  A = O_Δ
      THEN  α := 1
      ELSE  α := 0
ELSE  u := 1; ε := 0
      REPEAT
            ε := BOUNDED EQUIVALENCE(Δ, A, A, u)
            α := BOUNDED EQUIVALENCE(Δ, A, B, u)
            u := 2u
      UNTIL  α ≠ 0 or ε ≠ 0
      IF  ε ≠ 0 and α = 0
      THEN  R = APPROX(Δ, log ε, 4)
            α := BOUNDED EQUIVALENCE(Δ, A, B, R)
```

Proposition 6.16. *For $\Delta < 0$, PRINCIPAL has polynomial running time.*

Let $\Delta > 0$. If B is a principal ideal, then PRINCIPAL finds its minimal generator (α, β) using time and space $O^*((\log|\alpha|)^{1/2})$. Otherwise, PRINCIPAL decides in time and space $O^*(R_\Delta^{1/2})$ that B is not principal.

Proof. If u is such that $\mathrm{R}_\Delta < u$ and A is principal, then $\log \alpha < u$ for the minimal generator α of A. This proves the correctness. The assertions concerning time and space follow from § 6.5 and Proposition 6.14. □

6.14. Computing the fundamental unit. Let $\Delta > 0$; then the fundamental unit ε of $\mathcal{O}_\Delta$ can be computed via $(\varepsilon, 1) :=$ PRINCIPAL$(\Delta, \mathcal{O}_\Delta)$.

Proposition 6.17. *For $\Delta > 0$ the fundamental unit of $\mathcal{O}_\Delta$ can be determined in time and space $O^*(R_\Delta^{1/2})$.*

6.15. Solving norm equations. Another application of PRINCIPAL is the solution of norm equations. We consider only the case of prime norms. By means of combinatorial techniques, the method can be extended to arbitrary norms.

Algorithm 6.18 (NORM EQUATION)

Input: $\Delta \in \mathcal{D}$, *and a prime number* p.
Output: *A number* α *in* $\mathcal{O}_\Delta$ *of norm* p, *if it exists, and* $\alpha = 0$ *otherwise.*

IF $\left(\frac{\Delta}{p}\right) = 1$		*(1)*
THEN	$\alpha :=$ *PRINCIPAL*$(\Delta,$ *PRIME IDEAL*$(\Delta, p))$	*(2)*
ELSE	$\alpha := 0$	*(3)*

Proposition 6.19. *NORM EQUATION is correct. If there is a number of norm p in $\mathcal{O}_\Delta$, then NORM EQUATION requires time and space $O^*(p^{1/2} + (\log|\alpha|)^{1/2})$, where α is the minimal such number with $|\alpha| > 1$. Otherwise, the algorithm requires time and space $O^*(p^{1/2} + R_\Delta^{1/2})$.*

6.16. Solving the containment problem. Let T be a set of reduced ideals of $\mathcal{O}_\Delta$, and let A be a reduced ideal of $\mathcal{O}_\Delta$. We wish to find out whether A is equivalent to some $\mathrm{B} \in T$. If so, we wish to find the minimal generator of A relative to B.

For $\Delta < 0$ that problem was already solved in § 6.5. So assume that $\Delta > 0$. Since we might want to solve the containment problem for a fixed set T, but for many different reduced ideals A, we proceed in two steps. In a precomputation we determine the regulator of $\mathcal{O}_\Delta$ and we compute an *expansion* of T. As it was necessary in BOUNDED EQUIVALENCE to determine the set S of pairs (C, γ), it is now necessary to determine a similar set for each element of T. The union of all these sets is the expansion S of T.

Algorithm 6.20 (BABY STEPS)

Input: $\Delta \in \mathcal{D}$, *a set* T *of reduced ideals.*
Output: *The expansion* S *of* T, $u = \lceil\sqrt{R_\Delta}\rceil$.

```
S := {(C, 0) : C ∈ T}
IF Δ > 0
THEN ε := PRINCIPAL(Δ, O_Δ); u := ⌈(log ε)^{1/2}⌉
     FOR all C ∈ T
         γ := 1
         REPEAT
             γ := γ(L MIN(Δ, C, 1))
             C := L IDEAL(Δ, C)
             c := APPROX(Δ, γ, 4)
             S := S ∪ {(C, c)}
         UNTIL c ≥ u + (log Δ)/2 + 1 or C = A
```

Algorithm 6.21 (CONTAINMENT)

Input: *$\Delta \in \mathcal{D}$, The expansion S of set T of reduced ideals of $\mathcal{O}_\Delta$, a reduced ideal A of $\mathcal{O}_\Delta$.*

Output: *If A is equivalent to some $B \in T$, then (B, α), where α is the minimal generator of A relative to B, else $(A, 0)$.*

```
B := A; α := 0
IF Δ < 0
THEN IF (A, 0) ∈ S
     THEN α := 1
     ELSE IF (ρ(A), 0) ∈ S
          THEN B := ρ(A); α := γ(A)
ELSE q := 0
     REPEAT
         β := CLOSE(Δ, A, qu)
         C := (1/β)A
         IF (C, c) ∈ S for some c
         THEN b := APPROX(Δ, β, 4)
              α := BINARY(Δ, A, B, b + c)
         ELSE q := q + 1;
     UNTIL q = u or α ≠ 0
```

Proposition 6.22. *BABY STEPS is correct. For $\Delta < 0$ it requires time and space $O^*(\sharp T)$. For $\Delta > 0$ it requires time and space $O^*(\sharp T(R_\Delta)^{1/2})$.*

Proof. Each individual operation in BABY STEPS can be carried out in polynomial time. For $\Delta < 0$ the number of operations is $O^*(\sharp T)$. For $\Delta > 0$ it follows from (3) that the number of operations is $O^*(\sharp T(\mathrm{R}_\Delta)^{1/2})$. □

Proposition 6.23. *CONTAINMENT is correct. For $\Delta < 0$ it requires polynomial time. For $\Delta > 0$ it requires time and space $O^*((R_\Delta)^{1/2})$.*

7. Orders and discrete logarithms in the class group

Proposition 7.1. *Any discrete logarithm problem in Cl_Δ can be solved in time and space $O^*(|\Delta|^{1/4})$.*

Proof. Let $\Delta < 0$. From [**14**, pp. 400-401] there follows $\mathrm{h}_\Delta = O^*(|\Delta|^{1/2})$.

Multiplication, inversion in the class group can be performed in polynomial time. By Proposition 6.2, containment tests in the class group for a set with $O^*((\mathrm{h}_\Delta)^{1/2})$ elements requires a precomputation which uses time and space $O^*(|\Delta|^{1/4})$. Each individual containment test can then be effected in polynomial time. Hence, the result follows from § 3.6.

Let $\Delta > 0$. We have $\mathrm{h}_\Delta\mathrm{R}_\Delta = O^*(\Delta^{1/2})$. Again, multiplication and inversion in the class group can be carried out in polynomial time. By Proposition 6.2, containment tests in the class group for a set with $O^*((\mathrm{h}_\Delta)^{1/2})$ elements requires a precomputation which uses time and space $O^*((\mathrm{h}_\Delta\mathrm{R}_\Delta)^{1/2}) = O^*(|\Delta|^{1/4})$. Each individual containment test can then be effected using time and space $O^*(\mathrm{R}_\Delta^{1/2})$. Hence, the result follows from § 3.6. □

8. Computing the class group

We now sketch an algorithm CLASS GROUP, which finds the class group in time $O^*(|\Delta|^{1/2})$. CLASS GROUP is based on an idea of H.W. Lenstra, Jr. [**13**]. The idea is the following. By a theorem of Minkowski, each ideal class of $\mathcal{O}_\Delta$ contains an ideal of norm $\leq c\sqrt{|\Delta|}$, where $c = 1/2$ for $\Delta > 0$ and $c = 2/\pi$ for $\Delta < 0$. Therefore, the ideals of prime norm $p \leq c\sqrt{|\Delta|}$ generate the class group. The best known upper bound for the time for determining that generating system is $O^*(|\Delta|^{3/4})$. Hence, we do not determine that system but we only write down the list P of all prime numbers $p \leq c\sqrt{|\Delta|}$ such that there is an ideal of $\mathcal{O}_\Delta$ of norm p. We delete its first element p and determine the ideal A_p of norm p. Then we compute all the elements in the subgroup H generated by $[\mathrm{A}_p]$, and all the ideals of prime norm in P which, when input to REDUCE, yield a reduced ideal in a class of H. These prime ideals cannot contribute anything new to the class group. These norms are deleted from P. Then we pick the first prime number q in P that has survived. We compute the elements in $H = \langle [\mathrm{A}_p], [\mathrm{A}_q] \rangle$ and again we eliminate in P all norms of ideals which, when reduced, belong to a class of H. Iterating that process, we find the whole class group. Each time we determine a new prime ideal, the number of elements of H grows by a factor ≥ 2. Hence the total number of prime ideals that we must compute is only $O^*(1)$.

CLASS GROUP uses PRIME NORMS. On input of a reduced ideal A, the PRIME NORMS yields the set of all prime numbers p such that $\mathrm{A} = \mathrm{REDUCE}(\mathrm{A}_p)$. It is based on the following statement.

Proposition 8.1. *Let p be a prime number with $p < c\sqrt{|D|}$. If $\Delta > 0$, then A_p is reduced. If $\Delta < 0$, then A_p is reduced, or $A = \rho(A_p)$ is reduced and $b(A_p) = \sqrt{4a(A)p + \Delta}$.*

CLASS GROUP is a modification of BASIS. The details are left to the reader.

Proposition 8.2. *A generating system G for Cl_Δ with $O^*(1)$ elements and its relation lattice $L(G)$ can be determined in time and space $O^*(\sqrt{|\Delta|})$.*

References

1. J. Buchmann, *Zur Komplexität der Berechnung von Einheiten und Klassenzahlen algebraischer Zahlkörper*, Habilitationsschrift, Düsseldorf, 1987.
2. J. Buchmann, C. Thiel, H. C. Williams, *Short representation of quadratic integers*, to appear in Proceedings of CANT 1992.
3. J. Buchmann, S. Paulus, *Algorithms for finite abelian groups*, in preparation.
4. J. Buchmann, H. C. Williams, *On the existence of a short proof for the value of the class number and regulator of a real quadratic field*, Number Theory and Applications, in A. Molin (ed.), NATO ASI Series, Kluwer Academic Press, 1989, pp. 327-345.
5. G. W. Fung, H. C. Williams, *Compact representation of the fundamental unit in a complex cubic field*, unpublished manuscript, 1991.
6. J. L. Hafner, K. S. McCurley, *Asymptotically fast triangularization of matrices over rings*, SIAM J. Comput. **20** (1991), 1068-1083.
7. J. C. Lagarias, *Succinct certificates for the solvability of binary quadratic polynomials*, Proc. 20th IEEE Conference on Foundations of Computer Science, 1979, pp. 47-54.
8. J. C. Lagarias, *Worst-case complexity bounds in the theory of integral quadratic forms*, J. Algorithms **1** (1980), 142-186.
9. J. C. Lagarias, *Succinct certificates for the solvability of binary quadratic Diophantine equations*, in preparation.
10. A.K. Lenstra, H.W. Lenstra, Jr., *Algorithms in number theory*, in J. van Leeuwen (ed.), *Handbook of theoretical computer science*, Elsevier Science Publishers B.V., 1990, pp. 673–715.
11. H. W. Lenstra, Jr., *On the computation of regulators and class numbers of quadratic fields*, Lond. Math. Soc. Lect. Note Ser. **56** (1982), 123-150.
12. H. W. Lenstra, Jr., *Algorithms in algebraic number theory*, Bull. Amer. Math. Soc. (N.S.) **26** (1992), 211-244.
13. H. W. Lenstra, Jr., Private communication 1992.
14. W. Narkiewicz, *Elementary and analytic theory of algebraic numbers*, Polish Scientific Publishers, Warszawa, 1974.
15. R.J. Schoof, *Quadratic fields and factorization*, in H.W. Lenstra. Jr., R. Tijdeman (eds.), Computational methods in number theory, 1982, vol. 2, pp. 235–286.
16. D. Shanks, *Class Number, A Theory of Factorization and Genera*, Proc. Sympos. Pure Math. **20**, American Mathematical Society, 1970, pp. 415–440.

17. D. Shanks, *The infrastructure of a real quadratic field and its applications*, Proc. 1972 Number Theory Conference, Boulder, 1972, pp. 217-224.
18. V. Shoup, *On the deterministic complexity of factoring polynomials over finite fields*, Inform. Process. Lett., 1990, no. 33, pp. 261–267.
19. C. Thiel, *Short proofs using compact representations of algebraic integers*, to appear, 1993.

FACHBEREICH INFORMATIK, UNIVERSITÄT DES SAARLANDES, SAARBRÜCKEN, GERMANY

E-mail address: ingi@cs.uni-sb.de

FACHBEREICH INFORMATIK, UNIVERSITÄT DES SAARLANDES, SAARBRÜCKEN, GERMANY

E-mail address: buchmann@cs.uni-sb.de

Proceedings of Symposia in Applied Mathematics
Volume 48, 1994

Analytic Computations in Number Theory

ANDREW M. ODLYZKO

ABSTRACT. Number theory has been intimately associated with computation since ancient times. The early calculations were all with integers or rationals. However, starting with Riemann's discovery of the connection between distribution of primes and zeros of the Riemann zeta function, there has been intensive work on computation of analytic functions, concentrating on their zeros. The most extensive computations were those designed to test the truth of the Riemann Hypothesis as well as some other conjectures about the distribution of zeros of the Riemann zeta function and some related L-functions. Large computations of zeros have also been carried out to disprove conjectures about distributions of arithmetical functions, and to prove explicit bounds for various functions. This paper surveys the general area of computations of zeros of Dirichlet series and their applications, as well as of other computations in number theory that deal with analytic functions.

1. Introduction

Number theory and computation have been intimately connected since ancient times. The computations were usually those on the integers, involving factorizations into primes or solving diophantine equations. However, there is also a large area of computational number theory that involves analytic functions. Its origins are often dated to Riemann, who was the first one to discover the close connection between the distribution of primes and zeros of $\zeta(s)$, the Riemann zeta function. This function is defined for $s \in \mathbb{C}$, $\mathrm{Re}(s) > 1$, by

$$\zeta(s) = \sum_{n=1}^{\infty} \frac{1}{n^s} \,. \tag{1}$$

It is possible to trace this history even further back, about two centuries before Riemann. In 1650 Pietro Mengoli, an Italian mathematician, published a book

1991 *Mathematics Subject Classification.* Primary 11Y35, 11E45, 11M26. Secondary 11N05.

This paper is in final form and no version of it will be submitted for publication elsewhere.

© 1994 American Mathematical Society
0160-7634/94 $1.00 + $.25 per page

on summation of series. In it he raised the question of evaluating

$$\sum_{n=1}^{\infty} \frac{1}{n^2} = \zeta(2) \ . \tag{2}$$

Mengoli's question stimulated substantial research, with the most significant contributions coming from Euler. (See [**2**] for a beautiful survey of Euler's contributions in this area.) In attempting to evaluate $\zeta(2)$ numerically, Euler invented what is today called the Euler-Maclaurin summation formula, which we will discuss further in §2. This established an early connection between numerical analysis and number theory. Later, Euler discovered the exact formula

$$\zeta(2) = \sum_{n=1}^{\infty} \frac{1}{n^2} = \frac{\pi^2}{6} \tag{3}$$

(as well as similar formulas for $\zeta(4), \zeta(6), \dots$). His argument was not rigorous by modern standards, but can be made rigorous using the theory of analytic functions that was developed in the 19th century.

Euler's contributions to the theory of the zeta function that are mentioned above were all to numerical or closed-form summations. Perhaps the most important observation that Euler made, though, was that the zeta function has connections to number theory. He obtained the Euler product formula

$$\zeta(s) = \prod_{p} (1 - p^{-s})^{-1}, \ \ \mathrm{Re}(s) > 1 \ , \tag{4}$$

where p runs over all the primes. This shows that $\zeta(s)$ determines the distribution of primes. The product formula (4) allowed Euler to deduce, for example, that the sum of the reciprocals of the primes diverges.

Euler's work did not deal with the behavior of $\zeta(s)$ as an analytic function, and neither did that of various other mathematicians who followed him during the next century. Dirichlet, for example, introduced the Dirichlet L-functions,

$$L(s, \chi) = \sum_{n=1}^{\infty} \chi(n) n^{-s}, \ \ \mathrm{Re}(s) > 1 \ , \tag{5}$$

where χ is a Dirichlet character modulo an integer q. He used them in his proof that there are infinitely many primes in any arithmetic progression a, $a+q$, $a+2q, \dots$ for which $(a, q) = 1$. While he did have to show that these L-functions do not have zeros at $s = 1$, he did not consider their behavior far away from the real axis.

The importance of the analytic behavior of $\zeta(s)$ for number theory was pointed out by Riemann in his path-breaking paper [**54**]. The definition (1) applies only for $\mathrm{Re}(s) > 1$ and shows easily that $\zeta(s)$ is analytic there. Riemann showed that $\zeta(s)$ can be continued analytically to the entire complex plane with the exception of $s = 1$, where it has a first-order pole. Further, Riemann observed that the zeros of $\zeta(s)$ determine the distribution of primes. His formulas relating primes to zeros were only proved in the 1890s, after the necessary analytic machinery was developed by Hadamard, Weierstrass, and others. However, he did outline the path that was eventually followed in the study of the distribution of primes, and eventually yielded the proof of the Prime Number Theorem (PNT),

$$\pi(x) \sim \frac{x}{\log x} \quad \text{as} \quad x \to \infty \,, \tag{6}$$

which had been conjectured by Gauss and Legendre. (Throughout the paper, $\log x$ denotes the natural logarithm of x.) Moreover, this approach gave a more precise form of the PNT, namely

$$|\pi(x) - \mathrm{li}\,(x)| \le c_1 \exp(-c_2(\log x)^{1/2}) \tag{7}$$

for some constants $c_1, c_2 > 0$, where

$$\mathrm{li}\,(x) = \int_0^x \frac{du}{\log u} \,, \tag{8}$$

and the integral is the Cauchy principal value. We note that

$$\mathrm{li}\,(x) \sim \frac{x}{\log x} \quad \text{as} \quad x \to \infty \,. \tag{9}$$

The best bounds for the error term in the PNT that are known are not much smaller than the one in (7). On the other hand, if the Riemann Hypothesis (RH) is true, so that all the zeros of $\zeta(s)$ in $\mathrm{Re}(s) > 0$ lie on the critical line $\mathrm{Re}(s) = 1/2$, then

$$|\pi(x) - \mathrm{li}\,(x)| \le c_3 x^{1/2} \log x \tag{10}$$

for some $c_3 > 0$. Hence if the RH is true, the error term in the PNT is only about the square root of the main term.

It is not known what led Riemann to conjecture the RH. It was thought by some mathematicians, such as Felix Klein, that Riemann was motivated by a sense of general beauty and symmetry in mathematics. However, as a result of C. L. Siegel's study [**62**] of Riemann's unpublished notes, we now know that Riemann did obtain extensive numerical data about $\zeta(s)$ and its zeros, and that to do so he derived advanced computational techniques. We will discuss this in §2.

The importance of the RH has led to numerous attempts to prove it. At first it was not realized just how difficult a problem this is. Eventually, however, the perception of the difficulty of the RH changed, and a series of attempts were made to verify it numerically for initial sets of zeros. Section 2 describes these computations.

While computations designed to verify the RH are the most widely known, there have been many others as well, dealing with the zeta function as well as other functions in analytic number theory. Many, especially the large-scale ones, were designed to test conjectures such as the RH and its generalizations.

However, there have also been many computations aimed at indirect disproofs of number-theoretic conjectures, explicit bounds for arithmetical functions, and others. The purpose of this paper is to present a brief survey of this wide variety of different analytic computations. An excellent introduction to analytic number theory is given by [**12**], and readers not familiar with this subject can find definitions and motivation there.

2. Algorithms and computations

We first consider zeros of the Riemann zeta function. (For general information about $\zeta(s)$, see the books [**27, 50, 68**].) Since $\zeta(s) \in \mathbb{R}$ for $s \in \mathbb{R}$, $s > 1$, we find that $\zeta(\overline{\rho}) = 0$ whenever $\zeta(\rho) = 0$. Therefore we need to consider only the zeros ρ with $\text{Im}(\rho) \geq 0$. The Euler product (4) shows $\zeta(s)$ has no zeros in $\text{Re}(s) > 1$. The functional equation of the zeta function,

$$\xi(s) = \xi(1 - s) \ , \tag{11}$$

where

$$\xi(s) = \frac{1}{2} s(s-1)\pi^{-s/2}\Gamma(s/2)\zeta(s) \ , \tag{12}$$

then shows that $\zeta(s)$ has no zeros in $\text{Re}(s) < 0$ except for the *trivial zeros* at $s = -2, -4, -6, \ldots$. All the *nontrivial zeros* (i.e., those in the *critical strip* $0 \leq \text{Re}(s) \leq 1$) are strictly inside the critical strip, and none are real. The first 5 zeros in the upper half-plane $1/2 + i\gamma_n$, $1 \leq n \leq 5$ (first in order of distance from the real axis) are listed in Table 1. They all lie on the *critical line* $\text{Re}(s) = 1/2$, and thus satisfy the RH.

n	γ_n
1	14.13472514173469379045...
2	21.02203963877155499262...
3	25.01085758014568876321...
4	30.42487612585951321031...
5	32.93506158773918969066...

TABLE 1. Ordinates of first 5 zeros of $\zeta(s)$.

Table 2 lists the successive published records in proving that the first n zeros of the zeta function satisfy the RH. Thus the first such result was that of Gram in 1903, who showed that the RH holds for the first 15 zeros. The latest result is that of van de Lune, te Riele, and Winter [**38**], who showed that the RH is valid for the first 1.5×10^9 zeros.

It should be emphasized that the verifications that the RH holds for a given number of zeros are in principle completely rigorous. The functional equation (11) and the relation $\zeta(s) = \overline{\zeta(\overline{s})}$ imply that $\xi(1/2 + it) \in \mathbb{R}$ for $t \in \mathbb{R}$.

Investigator	n
Gram (1903)	15
Backlund (1914)	79
Hutchinson (1925)	138
Titchmarsh et al. (1936)	1,041
Turing (1953)	1,104
Lehmer (1956)	25,000
Meller (1958)	35,337
Lehman (1966)	250,000
Rosser et al. (1969)	3,500,000
Brent (1979)	81,000,001
van de Lune et al. (1986)	1,500,000,000

TABLE 2. Numerical verification of the Riemann Hypothesis for the first n zeros.

Therefore, sign changes of $\xi(1/2+it)$ correspond to zeros of $\zeta(s)$ that are exactly on the critical line, not just close to it. If we find n sign changes in the range $0 < t < T$, and then establish using the principle of the argument that there are exactly n zeros of $\zeta(s)$ in $0 < \text{Im}(s) < T$, then all the zeros in that region have to lie on the critical line. (In practice one does not even have to use the principle of the argument, as there is a beautiful method of Turing [**17, 46**], based on a theorem of Littlewood, which gives the same conclusion from just the calculations on the critical line.) What is required for this approach to the verification of the RH for a certain number of zeros to work is an effective algorithm for computing $\zeta(s)$ with a guaranteed error term. It is also necessary that all the zeros be simple and lie on the critical line. This has been shown to hold for all the zeros that have been examined.

The increase by a factor of 10^8 between 1903 and 1986 in the numbers of zeros for which the RH has been verified was due to a large extent to advances in technology, with the van de Lune et al. [**38**] calculations taking around two months on a large supercomputer. However, just as important have been advances in algorithms. The early computations, namely those of Gram, Backlund, and Hutchinson, were all carried out using the Euler-Maclaurin summation formula. This method is effective and is still used today for high-precision computations of low-order zeros, since it is simple to implement. However, it is not efficient for only moderately accurate calculations of high zeros, since it requires on the order of t steps to compute $\zeta(1/2+it)$. Zero number 1.5×10^9 of $\zeta(s)$ is at height approximately 5×10^8, so extensive computations at that height would be impractical with the Euler-Maclaurin formula. Fortunately, there are better methods. All the computations listed in Table 2, starting with that of Titchmarsh in the 1930s, have used the Riemann-Siegel formula, which allows one to compute $\zeta(1/2+it)$ in about $t^{1/2}$ steps. The Riemann-Siegel formula is not a general-purpose tool like the Euler-Maclaurin one. The latter can be used for a variety of computations of sums where the terms vary smoothly with the index of summation. The Riemann-Siegel formula applies only to the zeta function (although there are generalizations to some similar Dirichlet series, as we will mention later).

A remarkable fact about the Riemann-Siegel formula is that it was known to Riemann in the 1850s! It was discovered in Riemann's unpublished notes by Siegel [**62**]. Siegel's name deserves to be associated with it, since it required extensive work and deep insight to figure out what Riemann had done. Riemann had computed several zeros of the zeta function and possessed a deep understanding of its analytic behavior. It is not certain how large an influence this had in his proposal of the RH. However, it does prove that he did not make that conjecture in the absence of any numerical data, as Klein had thought.

Recently, an even faster method for simultaneous computation of large sets of zeros of the zeta function was invented by A. Schönhage and the author [**44, 47, 49**]. It makes possible the computation of the approximately $T^{1/2}$ zeros of $\zeta(1/2+it)$ in an interval $T \le t \le T + T^{1/2}$ in time on the order of $T^{1/2}$. This algorithm is based on the Riemann-Siegel formula, but has several new features, including use of the Fast Fourier Transform, of band-limited function interpolation, and of an algorithm similar to the multipole expansion of Greengard and Rokhlin [**21**]. It has been implemented and used to compute 1.75×10^8 zeros near zero number 10^{20}, for example [**47**]. It can also be adapted to check the validity of the RH beyond the 1.5×10^9 zeros that were treated in [**38**].

An important stimulus for the invention and implementation of the algorithm of [**49**] was the desire to check conjectures that in many ways go beyond the RH. A proof of the RH would still leave open many questions about the distribution of primes, such as those about gaps between consecutive primes. Answers to many of these problems depend on the vertical distribution of zeros of the zeta function. There are conjectures, originating with the work of Montgomery on the pair-correlation function [**41**], that connect the Hilbert-Pólya conjecture (which says that the RH is true because zeros of $\zeta(s)$ correspond to eigenvalues of a positive operator) together with the extensive work in physics on quantum chaos and random matrix theories [**3, 4, 6, 8, 40**]. For further details on the number-theoretic aspects of this work, see [**46**]. However, it seems safe to say that there will be additional work in this area, and new algorithms are likely to be invented [**47, 60**].

The Dirichlet L-function $L(s,\chi)$ for a character χ modulo q is defined by

$$L(s,\chi) = \sum_{n=1}^{\infty} \chi(n) n^{-s} \quad \text{for} \quad \mathrm{Re}(s) > 1 \; . \tag{13}$$

These functions play an important role in number theory, especially in the study of the distribution of primes in arithmetic progressions. They have many properties in common with the zeta function, especially an Euler product and a functional equation. The Generalized Riemann Hypothesis says that all their nontrivial zeros are on the critical line. All the evidence so far supports this conjecture, although it is not anywhere near as extensive as the data for the zeta function.

The Euler-Maclaurin formula can be used to compute zeros of $L(s,\chi)$. That was the method used in the early computations. The drawback to this approach is that at large heights the number of operations that are required becomes prohibitive.

Much more extensive computations of zeros of Dirichlet L-functions were carried out recently by Rumely [**58**]. He used a network of PCs to compute low zeros of many $L(s,\chi)$ with small conductors. The method Rumely used was also based on the Euler-Maclaurin formula, but he used also some ingredients inspired by [**47, 49**] that involve precomputations and reusing intermediate results from one L-function to another. He computed accurate values of zeros of $L(s,\chi)$ up to height 10^4 for all χ with conductors ≤ 13, up to 2.5×10^3 for all χ with conductors ≤ 72, and several other interesting collections of characters.

There do exist analogs of the Riemann-Siegel formula for Dirichlet L-functions, derived by Davies [**14**] and Deuring [**16**]. They were used by Hejhal [**23**] to compute high zeros of certain L-functions. For larger computations one could adapt the methods of [**47, 49**].

Since Dedekind zeta functions of abelian number fields are essentially products of Dirichlet L-functions, there is no difficulty in computing their zeros. (Some computations of this type are reported in [**23**]. The goal of that investigation was to determine how the distribution of spacings of the Dedekind zeta function compared to the distribution of spacings of individual L-functions.) The situation is quite different when one considers zeros of Dedekind zeta functions of nonabelian number fields. There the only computation that has been carried out so far is that of J. Lagarias and the author [**30**], and it deals only with pure cubic fields and only with low zeros. There are some other methods that might be used in the future, though, such as those of [**5, 20**].

If $Q(m,n) = am^2 + bmn + cn^2$ is a positive definite quadratic form, then the Epstein zeta function $Z(s)$ of Q is defined by

$$Z(s) = \frac{1}{2} \sum_{m,n}{}' Q(m,n)^{-s} \ , \ \operatorname{Re}(s) > 1 \ , \tag{14}$$

where $\sum'$ means the summation runs over all pairs of integers m and n except for the term $m = n = 0$. The function $Z(s)$ can be continued analytically to the entire complex plane except for $s = 1$, where it has a first-order pole. Epstein zeta functions play an important role in the study of quadratic forms and number fields. They do satisfy functional equations similar to those of the zeta function and Dirichlet L-functions, but with few exceptions they do not have Euler products and therefore are not expected to satisfy the RH. It is even known that most of them have many zeros away from the critical line.

Extensive computations have been done by Hejhal to investigate how many of their zeros are on the critical line [**23**]. There are also interesting questions about the behavior of zeros of parametrized families of Epstein zeta functions as the parameter passes through values for which these functions are expected to satisfy the RH. Some initial results in this area have been obtained by Arenstorf and Brewer [**1**]. (For other results on tracking of zeros of parametric families of functions that include the zeta function as a special case, see [**19**].)

Computations have been carried out of the zeros of some number-theoretic functions even more exotic than the ones mentioned above. For example, L-functions with Grössencharakteren were investigated in [**13**]. Zeros of L-functions associated with elliptic curves were studied by Fermigier [**18**], and zeros of the L-function with the Ramanujan τ-function as coefficients were computed by Keiper [**28**].

Extensive work has been carried out in recent years on Selberg zeta functions and related topics. Some references in this area are [**24, 25, 26**].

3. Indirect disproofs of conjectures

Values of zeros of $\zeta(s)$ and related functions have been used in many ways other than just to study the RH and its generalizations. For example, Stark [**66**] used high-precision values of the first few zeros of $\zeta(s)$ to obtain large lower bounds for the class number one problem. This result helped in completing some of the proofs of Gauss' conjecture that all imaginary quadratic fields with class number one have discriminants with absolute value ≤ 163. In a similar vein, the class number two problem was solved with the help of several high-precision values of low zeros of the zeta function [**67**] and independently by computing zeros of Dirichlet L-functions [**42, 69**].

Zeros of the zeta function have also been used in disproofs of various conjectures, or in obtaining effective bounds for least counterexamples to conjectures. For example, Littlewood showed around 1920 that the famous conjecture that $\pi(x) < \mathrm{li}\,(x)$ for all $x \geq 2$ is false for infinitely many integer values of x. The first bound for the smallest counterexample was found by Skewes, and was huge. However, Lehman [**33**] used precise values for large numbers of zeros of $\zeta(s)$ to bring the bound down to 10^{1166}. Later, te Riele [**53**] used Lehman's method together with more extensive computations to lower this bound to 10^{371}. (It is known there are no counterexamples below 10^{13}, and probably there are none below 10^{30}.) The disproof of the Mertens conjecture [**48**] involved computation of the first 2000 zeros of $\zeta(s)$ to 100 decimal places.

Consider the function

$$H_t(z) = \int_0^\infty e^{tu^2} \Phi(u) \cos(zu) du, \; t \in \mathbb{R}, \; z \in \mathbb{C} \, , \tag{15}$$

where

$$\Phi(u)=\sum_{n=1}^{\infty}(2\pi^2n^4e^{9u}-3\pi n^2e^{5u})\exp(-\pi n^2e^{4u})\ . \tag{16}$$

Then

$$H_0(z)=\xi(1/2+iz/2)/8\ , \tag{17}$$

and so the RH is equivalent to the statement that all the zeros of $H_0(z)$ are real. Pólya raised the question of what happens to the zeros of $H_t(z)$ for $t\neq 0$. De Bruijn [**9**] and Newman [**43**] showed there is a real constant $\Lambda\in(-\infty,1/2]$ such that for $t\geq\Lambda$, $H_t(z)$ has only real zeros, and for every $t<\Lambda$, $H_t(z)$ has some nonreal zeros. This constant Λ is now called the de Bruijn-Newman constant. The RH is equivalent to the statement that $\Lambda\leq 0$. On the other hand, Newman [**43**] conjectured that $\Lambda=0$. If true, this would be another piece of evidence that if the RH is true, it is barely true. Because of this connection with the RH, considerable work has been put into investigating Λ by Csordas, Norfolk, te Riele, Ruttan, Smith, and Varga. The latest result [**10**], which uses the techniques of Csordas, Smith, and Varga [**11**] is that

$$\Lambda>-5.895\times 10^{-9}\ . \tag{18}$$

The proof uses extensive mathematical analysis of the behavior of zeros of $H_t(z)$ and numerical values for a pair of unusually close zeros of $\zeta(s)$.

4. Explicit bounds for number-theoretic functions

An important use of values of zeros or even just of the knowledge that many initial zeros lie on the critical line is in proving precise estimates of number theoretic functions. Rosser and Schoenfeld [**55**] have used the results of their verification of the RH for the first 3.5×10^6 zeros of $\zeta(s)$ to prove a variety of estimates that have been used widely in number theory, combinatorics, and algebra. For example, the Prime Number Theorem says that for every $\epsilon>0$,

$$(1-\epsilon)\frac{x}{\log x}<\pi(x)<(1+\epsilon)\frac{x}{\log x}\quad\text{for }\ x\geq x_0(\epsilon)\ , \tag{19}$$

but does not immediately say what $x_0(\epsilon)$ is. Even forms of the prime number theorem with explicit remainder terms typically do not give good estimates for $x_0(\epsilon)$. Rosser and Schoenfeld have shown, for example, that

$$\frac{x}{\log x}\left(1+\frac{1}{2\log x}\right)<\pi(x)\ \text{ for }\ x\geq 59 \tag{20}$$

and

$$\pi(x)<\frac{x}{\log x}\left(1+\frac{3}{2\log x}\right)\quad\text{for }\ x>1\ . \tag{21}$$

These estimates are explicit and easy to use. For improvements of some of the results in [**55**], see [**56, 59**]. Based on the knowledge that the first 1.5×10^9 zeros of $\zeta(s)$ lie on the critical line, it should be possible to obtain substantial improvements of many of these estimates.

Ramaré and Rumely [**52**] have used Rumely's results [**58**] to obtain estimates for primes in arithmetic progressions. For example, they show that if $1 \le q \le 72$, $(a,q) = 1$, and $x \ge \exp(30)$, then

$$\max_{y \le x} \left| \psi(y;q,a) - \frac{y}{\phi(q)} \right| \le 0.012 \frac{x}{\phi(q)} , \tag{22}$$

where $\phi(q)$ is the Euler ϕ-function,

$$\psi(y;q,a) = \sum_{\substack{n \equiv a (\mathrm{mod}\ q) \\ n \le y}} \Lambda(n) , \tag{23}$$

and $\Lambda(n)$ is the von Mangoldt function, $\Lambda(p^r) = \log p$ if p is a prime and $r \in \mathbf{Z}^+$, $\Lambda(n) = 0$ otherwise. Estimates of this type played a crucial role in Ramaré's proof [**51**] that every integer $n \ge 2$ is a sum of at most 7 primes.

5. Computation of arithmetic functions

Analytic functions, such as the Riemann zeta function $\zeta(s)$, can be used to compute exactly certain arithmetic functions such as $\pi(x)$. There are exact formulas expressing $\pi(x)$ as a sum over the zeros of $\zeta(s)$, but they converge slowly and so cannot be used directly for efficient computation. (Formulas for $\pi(x)$ and related ones can be used for primality testing, since p is prime if and only if $\pi(p) - \pi(p-1) > 0$, but again there does not seem to be any way to make them efficient.) However, there is a method that uses analytic computations of $\zeta(s)$ (but does not use zeros of $\zeta(s)$) to compute $\pi(x)$ exactly and efficiently [**31, 32**]. One can write down formulas of the form

$$\sum_{p \le x} c(p) + \sum_{\substack{p^m \le x \\ m \ge 2}} \frac{1}{m} c(p^m) = \frac{1}{2\pi i} \int_{2-i\infty}^{2+i\infty} F(s) \log \zeta(s) ds , \tag{24}$$

where p ranges over primes and

$$c(u) = \frac{1}{2\pi i} \int_{2-i\infty}^{2+i\infty} F(s) u^{-s} du , \tag{25}$$

and $F(s)$ belongs to a certain class of functions. It turns out (see [**32**]) that one can choose $F(s)$ so that $c(p^m) = 1$ for $2 \le p^m \le x - y$, where $y = x^{1/2+o(1)}$, $c(p^m) = 0$ for $p^m > x$, with individual values of $c(u)$ and of $F(s)$ easy to compute, and so that $F(2+it)$ is rapidly decreasing for $|t| > x^{1/2}$. The integral in (24) can then be evaluated to within $1/10$ in $x^{1/2+o(1)}$ steps. The sum on the left of (24) differs from $\pi(x)$ by the contribution of primes $p \in [x-y, x]$ and proper prime powers $\le x$, and those terms can be estimated in $x^{1/2+o(1)}$ steps to within

1/10. Since $\pi(x)$ is an integer, this yields a value of $\pi(x)$ in a total of $x^{1/2+o(1)}$ steps as $x \to \infty$.

In contrast to the analytic method of computing $\pi(x)$ that is sketched above, the best combinatorial techniques that are known require time $x^{2/3+o(1)}$ as $x \to \infty$ [**29, 31**]. However, those methods are much easier to implement, and are the ones that have been used in the calculations of the largest values of $\pi(x)$ that are known, namely for $x = 4 \times 10^{16}$ [**29**] and more recently for $x = 10^{18}$ by M. Deleglise and J. Rivat [**15**].

Both the combinatorial and analytic methods mentioned above can be extended to the computation of other arithmetic functions [**29, 32, 61**].

Acknowledgements. Richard Brent, Keith Conrad, Walter Gautschi, and Hugh Montgomery provided corrections and helpful comments on an earlier version of this paper.

References

1. R. F. Arenstorf and L. L. Brewer, A study of the motion of zeros of the Epstein zeta function associated to $m^2 + y^2 n^2$ as y varies from 1 to $\sqrt{6}$, *Comput. Math. Appl.* **26**, no. 5, (1993), 57–69.
2. R. Ayoub, Euler and the zeta function, *Amer. Math. Monthly* **81** (1974), 1067–1086.
3. M. V. Berry, Riemann's zeta function: A model for quantum chaos?, pp. 1–17 in *Quantum Chaos and Statistical Nuclear Physics*, T. Seligman and H. Nishioka, eds., Lecture Notes in Phys. **263**, Springer, Berlin, 1986.
4. M. V. Berry, Quantum chaology, *Proc. Royal Soc. London A* **413** (1987), 183–198.
5. M. V. Berry and J. P. Keating, A new asymptotic representation for $\zeta(\frac{1}{2}+it)$ and quantum spectral determinants, *Proc. Royal Soc. London A* **437** (1992), 151–173.
6. O. Bohigas and M.-J. Giannoni, Chaotic motion and random matrix theories, pp. 1–99 in *Mathematical and Computational Methods in Nuclear Physics*, J. S. Dehesa, J. M. G. Gomez, and A. Polls, eds., Lecture Notes in Phys. **209**, Springer, Berlin, 1984.
7. E. Bombieri and D. A. Hejhal, Sur les zéros des fonctions zêta d'Epstein, *C. R. Acad. Sci. Paris Sér. I* **304** (1987), 213–217.
8. T. A. Brody, J. Flores, J. P. French, P. A. Mello, A. Pandey, and S. S. M. Wong, Random-matrix physics: spectrum and strength fluctuations, *Rev. Modern Phys.* **53** (1981), 385–479.
9. N. G. de Bruijn, The roots of trigonometric integrals, *Duke Math. J.* **17** (1950), 197–226.
10. G. Csordas, A. M. Odlyzko, W. Smith, and R. S. Varga, A new Lehmer pair of zeros and a new lower bound for the de Bruijn-Newman constant Λ, *Electr. Trans. Numer. Anal.* **1** (1993), 104–111.
11. G. Csordas, W. Smith, and R. S. Varga, Lehmer pairs of zeros, the de Bruijn-Newman constant Λ, and the Riemann Hypothesis, *Constr. Approx.* **10** (1994), 107–129.
12. H. Davenport, Multiplicative Number Theory, 2nd ed., revised by H. L. Montgomery, Springer, New York, 1980.
13. D. Davies, The computation of the zeros of Hecke zeta functions in the Gaussian field, *Proc. Royal Soc. Ser. A* **264** (1961), 496–502.
14. D. Davies, An approximate functional equation for Dirichlet L-functions, *Proc. Royal Soc. Ser. A* **284** (1965), 224–236.
15. M. Deleglise and J. Rivat, Computing $\pi(x)$: the Meissel, Lehmer, Lagarias, Miller, Odlyzko method, to be published.
16. M. Deuring, Asymptotische Entwicklungen der Dirichletschen L-Reihen, *Math. Ann.* **168** (1967), 1–30.
17. H. M. Edwards, *Riemann's Zeta Function*, Academic Press, New York, 1974.

18. S. Fermigier, Zéros des fonctions L de courbes elliptiques, *Experimental Math.* **1** (1992), 167–173.
19. B. Fornberg and K. S. Kölbig, Complex zeros of the Jonquière or polylogarithm function, *Math. Comp.* **29** (1975), 582–599.
20. E. Friedman, Hecke's integral formula, Sem. Théorie des Nombres Bordeaux, 1987–99, No. 5.
21. L. Greengard and V. Rokhlin, A fast algorithm for particle simulations, *J. Computational Phys.* **73** (1987), 325–348.
22. C. B. Haselgrove (in collaboration with J. C. P. Miller), *Tables of the Riemann Zeta Function,* Royal Soc. Math. Tables No. 6, Cambridge Univ. Press, 1960.
23. D. A. Hejhal, Zeros of Epstein zeta functions and supercomputers, pp. 1362–1384 in *Proc. Internat. Congress Math. 1986*, Amer. Math. Soc., Providence, 1987.
24. D. A. Hejhal, Eigenvalues of the Laplacian for Hecke triangle groups, *Memoirs Amer. Math. Soc.* **469** (1992).
25. D. A. Hejhal and S. Arno, On Fourier coefficients of Maass waveforms for $PSL(2, \mathbb{Z})$, *Math. Comp.* **61** (1993), 245–267 and S11–S16.
26. D. A. Hejhal and B. N. Rackner, On the topography of Maass waveforms for $PSL(2, Z)$, *Experimental Math.* **1** (1992), 275–305.
27. A. Ivić, *The Riemann Zeta-function: The Theory of the Riemann Zeta-function with Applications*, Wiley, New York, 1985.
28. J. B. Keiper, On the zeros of the Ramanujan τ-Dirichlet series in the critical strip, to be published.
29. J. C. Lagarias, V. S. Miller, and A. M. Odlyzko, Computing $\pi(x)$: The Meissel-Lehmer method, *Math. Comp.* **44** (1985), 537–560.
30. J. C. Lagarias and A. M. Odlyzko, On computing Artin L-functions in the critical strip, *Math. Comp.* **33** (1979), 1081–1095.
31. J. C. Lagarias and A. M. Odlyzko, New algorithms for computing $\pi(x)$, pp. 176–193 in *Number Theory: New York 1982*, D. V. Chudnovsky, G. V. Chudnovsky, H. Cohn, and M. B. Nathanson, eds., Springer-Verlag, Lecture Notes in Math. **1052**, 1984.
32. J. C. Lagarias and A. M. Odlyzko, Computing $\pi(x)$: An analytic method, *J. Algorithms* **8** (1987), 173–191.
33. R. S. Lehman, On the difference $\pi(x) - \mathrm{li}\,(x)$, *Acta Arith.* **11** (1966), 397–410.
34. D. H. Lehmer, Extended computation of the Riemann zeta-function, *Mathematika* **3** (1956), 102–108.
35. D. H. Lehmer, On the roots of the Riemann zeta-function, *Acta Math.* **95** (1956), 291–298.
36. M. E. Low, Real zeros of the Dedekind zeta function of an imaginary quadratic field, *Acta Arith.* **14** (1967/68), 117–140.
37. J. van de Lune and H. J. J. te Riele, Numerical computation of special zeros of partial sums of Riemann's zeta function, pp. 371–387 in *Computational Methods in Number Theory,* Part II, H. W. Lenstra, Jr., and R. Tijdeman, eds., Mathematisch Centrum, Amsterdam, 1982.
38. J. van de Lune, H. J. J. te Riele, and D. T. Winter, On the zeros of the Riemann zeta function in the critical strip. IV., *Math. Comp.* **46** (1986), 667–681.
39. K. S. McCurley, Explicit estimates for the error term in the prime number theorem in arithmetic progressions, *Math. Comp.* **42** (1984), 265–285.
40. M. L. Mehta, *Random Matrices*, 2nd revised and enlarged ed., Academic Press, Boston, 1991.
41. H. L. Montgomery, The pair correlation of zeros of the zeta function, pp. 181–193 in *Analytic Number Theory,* H. G. Diamond, ed., Proc. Symp. Pure Math. **24**, Amer. Math. Soc., Providence, 1973.
42. H. L. Montgomery and P. J. Weinberger, Notes on small class numbers, *Acta Arith.* **24** (1973/74), 529–542.
43. C. M. Newman, Fourier transforms with only real zeros, *Proc. Amer. Math. Soc.* **61** (1976), 245–251.
44. A. M. Odlyzko, New analytic algorithms in number theory, pp. 466–475 in *Proc. Internat.*

Congress Math. 1986, Amer. Math. Soc., Providence, 1987.
45. A. M. Odlyzko, On the distribution of spacings between zeros of the zeta function, *Math. Comp.* **48** (1987), 273–308.
46. A. M. Odlyzko, Zeros of the Riemann zeta function: Conjectures and computations, manuscript in preparation.
47. A. M. Odlyzko, *The* 10^{20}*-th Zero of the Riemann Zeta Function and 175 Million of its Neighbors*, manuscript in preparation.
48. A. M. Odlyzko and H. J. J. te Riele, Disproof of the Mertens conjecture, *J. reine angew. Math.* **357** (1985); 138–160.
49. A. M. Odlyzko and A. Schönhage, Fast algorithms for multiple evaluations of the Riemann zeta function, *Trans. Amer. Math. Soc.* **309** (1988), 797–809.
50. S. J. Patterson, *An Introduction to the Theory of the Riemann Zeta-Function,* Cambridge Univ. Press, 1988.
51. O. Ramaré, On Šnirel'man's constant, to be published.
52. O. Ramaré and R. Rumely, Primes in arithmetic progressions, to be published.
53. H. J. J. te Riele, On the sign of the difference $\pi(x) - \mathrm{li}\,(x)$, *Math. Comp.* **48** (1987), 323–328.
54. B. Riemann, Über die Anzahl der Primzahlen unter einer gegebenen Grösse, *Monatsberichte der Berliner Akad.* (1858/60), 671–680. Reprinted in *Gesammelte Mathematische Werke*, Teubner, Leipzig, 1892, and later by Dover Publications. English translation by H. M. Edwards, pp. 299–305 in [**17**].
55. J. B. Rosser and L. Schoenfeld, Approximate formulas for some functions of prime numbers, *Illinois J. Math.* **6** (1962), 64–94.
56. J. B. Rosser and L. Schoenfeld, Sharper bounds for the Chebyshev functions $\theta(x)$ and $\psi(x)$, *Math. Comp.* **29** (1975), 243–269.
57. J. B. Rosser, J. M. Yohe, and L. Schoenfeld, Rigorous computation and the zeros of the Riemann zeta function, pp. 70–76 in *Information Processing 68, Proc. IFIP Congress, Edinburgh, 1968*, vol. 1, North-Holland, Amsterdam, 1969. Errata in *Math. Comp.* **29** (1975), 243.
58. R. Rumely, Numerical computations concerning the ERH, *Math. Comp.* **61** (1993), 415–440 and S17–S23.
59. L. Schoenfeld, Sharper bounds for the Chebyshev functions $\theta(x)$ and $\psi(x)$. II, *Math. Comp.* **30** (1976), 337–360.
60. A. Schönhage, Numerik analytischer Funktionen und Komplexität, *Jahresber. Deutsch. Math.-Verein.* **92** (1990), 1–20.
61. P. Shiu, Counting sums of two squares: The Meissel-Lehmer method, *Math. Comp.* **47** (1986), 351–360.
62. C. L. Siegel, Über Riemanns Nachlass zur analytischen Zahlentheorie, *Quellen und Studien zur Geschichte der Math. Astr. Phys.* **2** (1932), 45–80. Reprinted in C. L. Siegel, *Gesammelte Abhandlungen*, Springer, Berlin, 1966, Vol. 1, pp. 275–310.
63. C. L. Siegel, Contributions to the theory of the Dirichlet L-series and the Epstein zeta functions, *Ann. Math.* **44** (1943), 143–172. Reprinted in C. L. Siegel, *Gesammelte Abhandlungen*, Springer, Berlin, 1966, Vol. 2, pp. 360–389.
64. R. Spira, Zeros of approximate functional approximations, *Math. Comp.* **21** (1967), 41–48.
65. R. Spira, Zeros of Hurwitz zeta functions, *Math. Comp.* **30** (1976), 863–866.
66. H. M. Stark, On complex quadratic fields with class number equal to one, *Trans. Amer. Math. Soc.* **122** (1968), 112–119.
67. H. M. Stark, On complex quadratic fields with class number two, *Math. Comp.* **29** (1975), 289–302.
68. E. C. Titchmarsh, *The Theory of the Riemann Zeta-function*, 2nd ed. (revised by D. R. Heath-Brown), Oxford Univ. Press, 1986.
69. P. J. Weinberger, On small zeros of Dirichlet L-functions, *Math. Comp.* **29** (1975), 319–328.

AT&T Bell Laboratories, Murray Hill, New Jersey 07974
E-mail address: amo@research.att.com

Proceedings of Symposia in Applied Mathematics
Volume 48, 1994

THE NUMBER FIELD SIEVE

CARL POMERANCE*

Dedicated to the memory of D. H. Lehmer

ABSTRACT. The most exciting recent development in the integer factorization problem is the number field sieve. It has had some spectacular successes with integers in certain special forms, most notably the factorization in 1990 of the 155 decimal digit number $2^{512}+1$. For arbitrary hard integers, it now appears to threaten the quadratic sieve as the algorithm of choice. In this paper the number field sieve, and the ideas behind it, are described.

1. Introduction

The problem of factoring integers is a good one to test our mettle as mathematicians. First, it is as fundamental as a problem can be. Second, while having the patina of centuries of history, the problem has taken on a new urgency for its connection with public-key cryptography. Third, it is a very *hard* problem, but not so hard that we do not occasionally gain an insight and make an advance. In this paper I would like to describe the background and main ideas of one of these advances, the number field sieve of John Pollard.

Twenty years ago, at the dawn of the continued fraction factoring algorithm of Brillhart and Morrison, factoring hard 50-digit numbers seemed barely possible, while 100-digit numbers could not even be dreamed about. Ten years ago, when my quadratic sieve factoring algorithm first began to enjoy some success, we indeed did dream of 100-digit numbers, and within a few more years, they were falling regularly. Today, with the number field sieve, our dreams have moved on to the hard 150-digit numbers.

Other advances have enabled us to factor broad classes of formerly unattackable numbers, and in the process have changed our thinking on which integers

1991 *Mathematics Subject Classification.* Primary 11Y05.

Key words and phrases. Number field sieve, quadratic sieve, smooth numbers, exponent vectors.

*Supported in part by an NSF grant.

This paper is in final form and no version of it will be submitted for publication elsewhere.

© 1994 American Mathematical Society
0160-7634/94 $1.00 + $.25 per page

should be considered as hard to factor. Among these developments are the elliptic curve method of Lenstra and certain optimal cases of the number field sieve itself. One of these optimal cases is the Fermat number $F_9 = 2^{512} + 1$. Despite the fact that it has 155 decimal digits, it is particularly well suited to the number field sieve. It was factored in 1990 into three prime factors – see [15].

To be sure, some part of our success in factoring has little to do with new algorithms and theory, but rather with dramatic improvements in computers and their availability. On the other hand, these advances in computing resources helped to influence and shape the kinds of algorithms that were being developed. For example, almost all of the newer methods are easily distributable to scores or more of unextraordinary computers and most of our greatest successes were accomplished by doing just this.

Several of the key papers on the development of the number field sieve are collected together in the book [13]. Included there is Pollard's original note whose informal distribution started the whole ball rolling. The number field sieve has also been suggested for use in computing discrete logarithms in the multiplicative group of a prime finite field (and perhaps more generally). See [11, 25] for more on this.

Thanks are due to Joe Buhler, Arjen Lenstra, Hendrik Lenstra and Walter Gautschi for their critical comments on an earlier version of this paper.

2. Congruent squares

For which odd numbers $n > 1$ does the congruence $x^2 \equiv y^2 \bmod n$ for integers x, y force us to conclude that we must have $x \equiv \pm y \bmod n$? Since $x^2 \equiv y^2 \bmod n$ is equivalent to n dividing $(x-y)(x+y)$, clearly it is just odd primes n with this property. Perhaps not so interesting as a primality test, this observation can be quite useful in factoring n. For if $x^2 \equiv y^2 \bmod n$ and $x \not\equiv \pm y \bmod n$, then the greatest common divisor of $x - y$ and n is a nontrivial factor of n.

It is a simple matter to compute the greatest common divisor of two given integers by means of Euclid's algorithm. This has us replace the larger integer of the pair with its remainder upon division by the smaller integer. If this remainder is 0, then the smaller integer is the greatest common divisor, while if it is not 0, we may repeat the process.

A wide class of factoring algorithms thus set out to find two squares x^2, y^2 that are congruent mod n. It seems difficult to force the condition $x \not\equiv \pm y$ mod n, which is necessary for splitting n. Thus, the factoring algorithms in this class either use randomness or "pseudorandomness" to count on at least a fair proportion of the congruent pairs of squares produced to lead to a factorization.

Thus, perhaps we should revise the question above and ask for which odd numbers $n > 1$ do we have at least as many pairs of integers x, y with $x^2 \equiv y^2 \bmod n$ and $x \not\equiv \pm y \bmod n$ as we have pairs of integers x, y with $x \equiv \pm y$ mod n. Let us restrict our attention to pairs of integers x, y with $(xy, n) = 1$. The problem is thus reduced to counting solutions to $x^2 \equiv 1 \bmod n$. This congruence has 2^l solutions if n has l distinct prime factors, so that the answer to our revised question is: for odd numbers n divisible by at least two distinct primes.

Hence, finding pairs of congruent squares may not be a good way to factor n if n is a power of a prime. However, perfect powers are easy to factor. For example, to factor a square, we can first find a real approximation to the square root, round to nearby integers and see if one of them squared is our number. The same idea works for higher powers, and for a given number n, we only have to check exponents up to $\log_2 n$, since if $n = m^k$ where $n > m > 1$ are integers, then $k = \log_m n \le \log_2 n$.

In summary, the problem of factoring n for n odd and composite can be reduced to checking if n is a nontrivial power and, if not, finding solutions to $x^2 \equiv y^2 \bmod n$ in a random or pseudorandom fashion.

3. Multiplying congruences

How then are we to find solution pairs x, y to $x^2 \equiv y^2 \bmod n$? We exploit the ideas that being a square is a multiplicative condition and that congruences mod n may be multiplied. Thus, if we have integers s_i, t_i for $i = 1, \dots, k$ where $\prod s_i = x^2$, $\prod t_i = y^2$ and each $s_i \equiv t_i \bmod n$, then $x^2 \equiv y^2 \bmod n$.

Suppose we have a very large set of pairs s_i, t_i where each pair has $s_i \equiv t_i \bmod n$. Is there a nonempty subset $\mathcal{I}$ of subscripts i such that $\prod_{i\in\mathcal{I}} s_i$ and $\prod_{i\in\mathcal{I}} t_i$ are both squares? And if there is such a set $\mathcal{I}$, is there an efficient way to find it?

To make matters a little simpler, let us ask these questions just for one side of the congruence. Given integers $s_1, \dots, s_k$, is there a nonempty set $\mathcal{I} \subset \{1, \dots, k\}$ such that $\prod_{i\in\mathcal{I}} s_i$ is a square? And can we find such a set $\mathcal{I}$?

For any integer $s \ne 0$, consider the "exponent vector" for s. This is the vector $v(s) = (v_{-1}(s), v_2(s), v_3(s), \dots)$ where the subscripts are -1 and the primes, and where

$$s = (-1)^{v_{-1}(s)} \prod_p p^{v_p(s)}.$$

The exponent $v_{-1}(s)$ is taken to be in $\{0, 1\}$. For a given integer s, all but finitely many of the exponents $v_p(s)$ are 0. Note that $\prod_{i\in\mathcal{I}} s_i$ is a square if and only if

$$\sum_{i\in\mathcal{I}} v_{-1}(s_i), \quad \sum_{i\in\mathcal{I}} v_2(s_i), \quad \sum_{i\in\mathcal{I}} v_3(s_i), \dots$$

are all even numbers.

With this observation, we can reduce our question about squares to one on linear algebra over the field $\mathbb{F}_2$ of two numbers. Thus, given $s_1, \dots, s_k$, we first find the exponent vectors $v(s_1), \dots, v(s_k)$, next we reduce them mod 2 so that we have a sequence of vectors $v(s_1) \bmod 2, \dots, v(s_k) \bmod 2$ over $\mathbb{F}_2$. Our problem is to determine if these vectors are linearly dependent and, if so, to find a nontrivial relation. Since there is just one nonzero scalar, a linear dependency takes the form $\sum_{i\in\mathcal{I}} v(s_i) \bmod 2 = 0$. This corresponds exactly to $\prod_{i\in\mathcal{I}} s_i$ being a square.

4. Smooth numbers

We have been a little vague about the *dimension* of the vector space in the preceding discussion. Clearly, the exponent vectors $v(s) \bmod 2$, as s runs over the positive integers, span an infinite-dimensional vector space. Why then should we expect $v(s_1) \bmod 2, \ldots, v(s_k) \bmod 2$ to be linearly dependent?

In general, we should not expect this to occur. However, if the integers $s_1, \ldots, s_k$ are chosen randomly from among the integers of a certain size, and if k is sufficiently large, then it will be highly likely that the vectors $v(s_i) \bmod 2$ will be linearly dependent over $\mathbb{F}_2$; that is, that some nonempty set $\mathcal{I}$ exists with $\prod_{i \in \mathcal{I}} s_i$ being a square.

Let us formulate a precise problem. For x, k positive real numbers, say we choose random integers $s_1, s_2, \ldots, s_{[k]}$ independently and with the uniform distribution from the interval $[1, x]$. Let $\Pr(x, k)$ denote the probability that some nonempty set $\mathcal{I} \subset \{1, \ldots, [k]\}$ exists with $\prod_{i \in \mathcal{I}} s_i$ a square. How large should k be so that $\Pr(x, k)$ is greater than $1/2$?

Let us prove the following result.

Proposition 4.1. *For each $\varepsilon > 0$ we have*

$$\lim_{x \to \infty} \Pr(x, \exp((1+\varepsilon)\sqrt{2 \log x \log\log x})) = 1.$$

Proof. We shall say a positive integer n is z-smooth if no prime factor of n exceeds z. Let $\psi(x, z)$ denote the number of z-smooth integers n in $[1, x]$. Let

$$z = \exp(\sqrt{(1/2) \log x \log\log x}), \quad k = \exp((1+\varepsilon)\sqrt{2 \log x \log\log x}). \tag{4.2}$$

Say we choose $s_1, s_2, \ldots, s_{[k]}$ from among the integers in $[1, x]$ independently and with the uniform distribution. Then the expected number of $i \le k$ with s_i being z-smooth is

$$\frac{[k]}{[x]} \psi(x, z). \tag{4.3}$$

From [5], we have that

$$\psi(x, z) = x / u^{(1+o(1))u} \quad \text{as} \quad x \to \infty,$$

where $u = \log x / \log z = \sqrt{2 \log x / \log\log x}$. Thus,

$$\psi(x, z) = x \cdot \exp(-(1+o(1))\sqrt{(1/2) \log x \log\log x}) = x / z^{1+o(1)}$$

as $x \to \infty$. (That is, the particular choice of z above satisfies $\psi(x, z) \approx x/z$.)

Thus, the expression in (4.3) exceeds

$$\frac{k}{x} \cdot \frac{x}{z^{1+\varepsilon}} = z^{1+\varepsilon}$$

for all large x. In particular, from the binomial distribution, the probability that there are at least z values of $i \le k$ with s_i being z-smooth tends to 1 as $x \to \infty$.

So let us assume that there are at least z values of $i \le k$ with s_i being z-smooth. Clearly, for a z-smooth number s, the coordinates of the exponent vector for s are all 0 for primes larger than z. By truncating these 0's, we have the exponent vectors lying in a $(\pi(z)+1)$-dimensional space, where $\pi(z)$ denotes the number of primes up to z. But we have at least z such vectors and $z > \pi(z)+1$ for all $z > 3$. Thus, the exponent vectors (reduced mod 2) are linearly dependent over $\mathbb{F}_2$ and, as we saw in the preceding section, this leads us to a nonempty set of indices i where the product of the s_i's is a square.

This simple proposition (or rather, its proof) is crucial for understanding the role of smooth numbers in combination of congruences factorization algorithms. We are presented in some manner with random or pseudorandom integers s_i in some range, and we wish to find a nonempty subsequence whose product is a square. Proposition 4.1 not only tells us we are likely to be successful by a certain point, it suggests an efficient way to *find* such a subsequence. First, if the numbers s_i all lie below x, we take z by (4.2) and *discard* any s_i which is not z-smooth. When we have found sufficiently many values of i with s_i being z-smooth, a linear algebra step over $\mathbb{F}_2$ on a system of size about z allows us to find the sought-after subsequence.

I conjecture that Proposition 4.1 is best possible in that if "$1+\varepsilon$" is replaced with "$1-\varepsilon$", the limit in the Proposition is 0.

5. How to recognize smooth numbers

If we are presented with many randomly chosen integers $s \le x$, and if we choose z by (4.2), then only a few such numbers s will be z-smooth. In fact, as we saw in the last section, the probability that s is z-smooth is about $1/z$. If we are to use the above ideas in an efficient factorization algorithm, we shall need to have an efficient "smoothness test," that is, an algorithm that can quickly recognize the z-smooth numbers among many random (or pseudorandom) integers.

The first method that comes to mind is trial division, taking about z steps to recognize whether s is z-smooth. This is the smoothness test used in the continued fraction factoring algorithm. The "early abort strategy" (see [21]) does not complete the trial division on a particular number s if at increasing points a factored portion of s does not exceed an increasing threshold. On average, this method takes about $z^{1/2}$ steps as a smoothness test, though some smooth numbers may be lost. If trial division in the early abort strategy is replaced with the "fast factorials" method of Pollard and Strassen (see [21]), the combined method takes about $z^{1/4}$ steps on average.

The elliptic curve method factors smooth numbers much sooner than numbers with several large prime factors. Used as a smoothness test, it takes z^{ε} steps to recognize a z-smooth number. More precisely, the elliptic curve method is expected to take $\exp((1+o(1))\sqrt{2\log z \log\log z})$ steps as $z \to \infty$. This is what we conjecture. What can be proved is that it recognizes most z-smooth numbers in $\exp((\log z)^{5/6+\varepsilon})$ steps (see [18, 23]). Recently, Lenstra, Pila and I showed that a certain *hyperelliptic* curve method is expected to factor any z-smooth number in $\exp((\log z)^{2/3+\varepsilon})$ steps (see [17] for the beginning of the proof).

The elliptic and hyperelliptic curve methods are not practical subroutines in current combination of congruences algorithms, though they are of interest in a rigorous treatment of factoring. It is not inconceivable, though, that the elliptic curve method, perhaps augmented with an early abort strategy, could be of practical use in a combination of congruences algorithm. And perhaps the algorithm of [6] is a candidate.

If the numbers s are truly random, then we know of no better smoothness tests. This is true even for certain pseudorandom sequences, such as in the continued fraction factoring algorithm. However, there are other pseudorandom sequences which quite nicely lend themselves to a very simple smoothness test.

The sieve of Eratosthenes is well understood as an efficient method of finding all the primes to some point. In this sieve composite numbers are crossed off until only primes remain. Let us look at these numbers that are crossed off. They are marked as many times as they have prime factors used in the sieve. Thus, smooth numbers will be crossed off many times over, and we might use this phenomenon as a way of recognizing them.

To make this idea a little more precise, one should not count the number of times a number is crossed off, which treats each prime factor of the number with equal weight, but rather higher primes (below z) should carry more weight since they make up a larger portion of the numbers that they divide. We weight each prime in the sieve (the primes up to z) with its logarithm (say to base 2 and rounded to the nearest integer). Marking a location with a prime amounts to adding the approximate logarithm of the prime to a counter at the location. When the count at a particular location exceeds some threshold, the number corresponding to this location is reported as a candidate z-smooth number. Since reports are few, we can afford to use a slow method, such as trial division, to check the candidates.

This is all well and good if the numbers s are consecutive integers up to some point. But what about more general sequences? A moment's reflection reveals that we can use this sieve idea whenever the numbers s are consecutive values of a polynomial with integer coefficients. If we consider at least z consecutive values of this polynomial, then the time spent sieving is only about $O(\log\log z)$ steps per value, the O-constant depending on the choice of polynomial. (For z tending to infinity, this O-constant tends to the number of distinct irreducible factors of the polynomial.) Moreover, as we have seen, these steps are particularly easy computer instructions such as adding low precision integers.

The comparison between z steps per candidate with trial division and $\log\log z$ steps with sieving is striking. Our problem has us searching for z-smooth needles in a huge haystack. Trial division would have us examine every straw by hand to see if it is a needle. In contrast, when sieving is appropriate, it is as if we could pass a strong magnet over the haystack fed over a rapidly moving conveyor belt, with the needles jumping to the magnet as they pass by.

6. The quadratic sieve

In §3 we described a general factoring scheme where one has at hand many congruences $s_i \equiv t_i \bmod n$ and somehow finds a nonempty set $\mathcal{I}$ of indices i where $\prod_{i\in\mathcal{I}} s_i$ and $\prod_{i\in\mathcal{I}} t_i$ are both squares, say x^2 and y^2. Then $x^2 \equiv y^2 \bmod n$ and, as we saw in §2, this may lead to a nontrivial factorization of n via the greatest common divisor of $x - y$ and n.

Consider the congruences $t^2 \equiv t^2 - n \bmod n$, where t runs over consecutive integers. Since one side of these congruences is already a square, the problem is reduced to finding a set $\mathcal{T}$ of values of t where $\prod_{t\in\mathcal{T}}(t^2 - n)$ is a square. The search for $\mathcal{T}$ is reduced, via the ideas of §4, to a search for values of t with $t^2 - n$ smooth. Since this function of t is a polynomial with integer coefficients, we can use a sieve, as described in §5, to locate rapidly those numbers t with $t^2 - n$ smooth.

This simple algorithm is the quadratic sieve factorization algorithm. As we saw in §4, the number k of values of t that must be examined depends on x, a bound for the size of the numbers $t^2 - n$ that we hope are smooth. By taking t in an interval near $\sqrt{n}$, the numbers $t^2 - n$ are $O(n^{1/2+\varepsilon})$, so that the number k in §4 is about $\exp(\sqrt{\log n \log\log n})$.

This expression is in fact the approximate heuristic complexity of the quadratic sieve algorithm. To be sure, one must also account for the complexity of finding a linear dependency among the exponent vectors mod 2. By using the sparse matrix methods in [7, 8, 12, 20, 24, 26], this complexity is the same as the sieving time; that is, about $\exp(\sqrt{\log n \log\log n})$.

A problem with the quadratic sieve is that the numbers $t^2 - n$ grow as t moves away from $\sqrt{n}$. In particular, if T values of t are considered, all we can say about the size of the numbers $t^2 - n$ is that they are $O(T\sqrt{n})$. The penalty for large values of $t^2 - n$ is that they are less likely to be smooth.

This problem is neatly mitigated by an idea of Peter Montgomery (see [22]). He has us replace t above with $at + b$, where a, b are integers satisfying $b^2 \equiv n \bmod a$, $|b| \le a/2$. Then the polynomial $(at + b)^2 - n$ has all of its values divisible by a. If also a is a square, then we are only concerned with whether

$$f_{a,b}(t) = \frac{1}{a}((at + b)^2 - n)$$

has smooth values. The advantage is that there are many choices for a, b, and thus many polynomials that one may use. If one wishes to sieve $2M$ values per polynomial, one chooses a near $\sqrt{2n}/M$ and sieves over an interval centered at 0. A simple computation shows that $f_{a,b}(t) = O(M\sqrt{n})$, so that the numbers we hope are smooth do not grow as we sieve over more and more numbers. We just take more and more polynomials $f_{a,b}(t)$. This is the multiple polynomial variation of the quadratic sieve. Another version of this idea was proposed earlier by Davis and Holdridge [10].

The size of the numbers we wish to find smooth in the basic quadratic sieve is about $\sqrt{n}\exp(\sqrt{\log n \log\log n})$. This drops to about $\sqrt{n}\exp\left(\frac{1}{2}\sqrt{\log n \log\log n}\right)$ in the multiple polynomial version. Both expressions are of the form $n^{1/2+\varepsilon}$, so both versions of the quadratic sieve have about the same heuristic complexity. However, the drop to a smaller "ε" makes a big difference in practice.

This thought underscores an important principle in combination of congruences factorization algorithms: The smaller are the auxiliary numbers that you hope to find smooth, the faster the algorithm will be. As we will see, in the number field sieve, the auxiliary numbers are about $\exp((\log n)^{2/3})$; that is, they are much smaller than $\sqrt{n}$. This is why the method is so exciting and has so much potential.

7. The number field sieve

Suppose $f(t)$ is a monic polynomial irreducible over $\mathbb{Z}$, suppose m is an integer with $f(m) \equiv 0 \bmod n$ and $\alpha \in \mathbb{C}$ is a root of f. There is thus a natural homomorphism φ from the ring of algebraic integers $\mathbb{Z}[\alpha]$ to $\mathbb{Z}/n\mathbb{Z}$, where $\varphi(\alpha) = m \bmod n$. That is, if $g(t)$ is any polynomial over $\mathbb{Z}$, then $\varphi(g(\alpha)) = g(m) \bmod n$.

Of what interest to us are f, α, m and φ? Well, suppose we could find a set $\mathcal{S}$ of polynomials g over $\mathbb{Z}$ such that $\prod_{g\in\mathcal{S}} g(\alpha)$ is a square, say β^2, in $\mathbb{Z}[\alpha]$ and $\prod_{g\in\mathcal{S}} g(m)$ is a square, say y^2, in $\mathbb{Z}$. Let x be an integer such that $\varphi(\beta) = x \bmod n$. Then

$$x^2 \equiv \varphi(\beta)^2 \equiv \varphi(\beta^2) \equiv \varphi\left(\prod_{g\in\mathcal{S}} g(\alpha)\right) \equiv \prod_{g\in\mathcal{S}} g(m) \equiv y^2 \bmod n.$$

That is, we have found a pair of squares that are congruent mod n and we may attempt to factor n by computing $(x-y, n)$.

The reader is an expert now and may be a bit suspicious. This scenario will be promising only if we can find choices for $g \in \mathbb{Z}[t]$ such that $g(m)$ is small and so likely to be smooth. And how in the world are we to combine various numbers $g(\alpha)$ to get a square in $\mathbb{Z}[\alpha]$? Do we have a notion of smoothness in $\mathbb{Z}[\alpha]$, and can we factor $g(\alpha)$ into primes to get an exponent vector?

Let us first attack the question on $g(m)$, for it is much easier. Suppose we wish our polynomial $f(t)$ to have degree d. The size of d as a function of n will be discussed momentarily, but for now we assume that $d > 1$ and $n > 2^{d^2}$. Before the polynomial f is chosen, we do the somewhat surprising thing of first choosing the integer m. We let $m = \lfloor n^{1/d} \rfloor$. We write n in the base m, so we find integers $c_0, c_1, \ldots, c_d$ in $\{0, 1, \ldots, m-1\}$ with

$$n = c_d m^d + c_{d-1} m^{d-1} + \cdots + c_0.$$

It is easy to show we must have $c_d = 1$. Let

$$f(t) = t^d + c_{d-1}t^{d-1} + \cdots + c_0.$$

Then f is a monic polynomial in $\mathbb{Z}[t]$ and $f(m) = n$. Is f irreducible? It probably is since most primitive polynomials over $\mathbb{Z}$ are. In fact, though, if f is not irreducible then we are in luck. Using the algorithm of [14], we can factor f into irreducibles in time polynomial in $\log n$, and if the factorization is nontrivial, by substituting m for t, we get a nontrivial factorization of n (see [3]). In this event we cancel plans to spend a lot of effort splitting n, for we have just succeeded!

So let us assume that f is irreducible over $\mathbb{Z}$. Suppose we take for our polynomials $g(t)$ the linear polynomials $a - bt$, where a, b run over small coprime numbers with $b > 0$, say $0 < b \leq B$, $0 \leq |a| \leq B$. Since m is about $n^{1/d}$, we see that the integers $a - bm$ can be rather small compared with n, especially if we do not have to take B large and we can choose d with some size. Further, it is easy to see how a sieve might be used to pick out pairs a, b with $a - bm$ being smooth: we fix b and sieve over a, then choose the next b and sieve again, continuing until we exhaust the choices for b.

Well, I have put off the questions on how to produce squares in $\mathbb{Z}[\alpha]$ long enough. If $\mathbb{Z}[\alpha]$ happens to be a unique factorization domain, we can get a glimmer of what it means for $a - b\alpha$ to be a smooth element, namely it factors into "small" primes. But we cannot count on any special properties of $\mathbb{Z}[\alpha]$.

Consider the norm map N from the field $\mathbb{Q}(\alpha)$ to $\mathbb{Q}$. If $\alpha = \alpha_1, \alpha_2, \ldots, \alpha_d$ are the conjugates in $\mathbb{C}$ of α, and $g \in \mathbb{Q}[t]$, then $N(g(\alpha)) = \prod_{i=1}^{d} g(\alpha_i)$. The norm map is multiplicative and it sends algebraic integers to $\mathbb{Z}$. In particular, $N(\mathbb{Z}[\alpha]) \subset \mathbb{Z}$. We thus have the following simple result: If $\mathcal{S}$ is a set of coprime pairs (a, b) of integers and $\prod_{(a,b)\in\mathcal{S}}(a - b\alpha)$ is a square in $\mathbb{Z}[\alpha]$, then $\prod_{(a,b)\in\mathcal{S}} N(a - b\alpha)$ is a square in $\mathbb{Z}$.

Can we at least arrange for this necessary condition to hold? Note that for any numbers $a, b \in \mathbb{Q}$ with $b \neq 0$, we have

$$\begin{aligned} N(a - b\alpha) &= \prod_{i=1}^{d}(a - b\alpha_i) = b^d \prod_{i=1}^{d}\left(\frac{a}{b} - \alpha_i\right) = b^d f\left(\frac{a}{b}\right) \\ &= a^d + c_{d-1}a^{d-1}b + \cdots + c_1 ab^{d-1} + c_0 b^d. \end{aligned}$$

Thus, the norm of $a - b\alpha$ is a polynomial with integral coefficients in the two variables a, b.

For integers a, b, let us say that $a - b\alpha$ is z-smooth if $N(a - b\alpha)$ is z-smooth. So now it should be clear how we may find a set $\mathcal{S}$ of coprime integer pairs (a, b) such that $\prod_{(a,b)\in\mathcal{S}} N(a - b\alpha)$ is a square in $\mathbb{Z}$. Namely, we use a sieve to detect pairs a, b where $a - b\alpha$ is z-smooth, create exponent vectors from the prime factorizations of the corresponding norms, and use linear algebra over $\mathbb{F}_2$ to create a square.

But is it good enough for our purposes that the product of the norms is a square? Consider the following example. In $\mathbb{Z}[i]$ we have the elements $2 \pm i$. They both have norm 5, so their product has norm 5^2. However, their product is 5, which is not a square in $\mathbb{Z}[i]$. It should be clear what it is we are throwing away when we look at norms. A particular prime number p in $\mathbb{Z}$ may factor into several prime ideals in an algebraic extension, each having norm a power of p. The norm map lumps all of these prime ideals together. Is there any way for us to tease them apart?

Let us look a little more closely at how we would sieve the integers $N(a - b\alpha)$ as a, b run over integers. If we are trying to detect pairs a, b where $a - b\alpha$ is z-smooth, then we only consider prime numbers p with $p \leq z$. Let us denote by $R(p)$ the set of integers $r \in \{0, 1, \ldots, p-1\}$ with $f(r) \equiv 0 \bmod p$. Then for a, b coprime, p is a divisor of $N(a - b\alpha)$ if and only if $a \equiv br \bmod p$ for some $r \in R(p)$ and $b \not\equiv 0 \bmod p$. Thus, the prime factor p of $N(a - b\alpha)$ comes with a unique *signature*, a particular number r of $R(p)$.

This idea suggests that we may create somewhat more complicated exponent vectors when $a - b\alpha$ is z-smooth and a, b are coprime. For each prime $p \leq z$ and each $r \in R(p)$, define the function $v_{p,r}(a - b\alpha)$ to be the exponent on p in the prime factorization of $N(a - b\alpha)$ if $a \equiv br \bmod p$, and define $v_{p,r}(a - b\alpha)$ to be 0 otherwise. We also include a coordinate for the sign of $N(a - b\alpha)$, so our exponent vectors $v(a - b\alpha)$ will have $1 + \sum_{p \leq z} \#R(p)$ coordinates.

Is it true that if $\prod_{(a,b)\in\mathcal{S}}(a - b\alpha)$ is a square in $\mathbb{Z}[\alpha]$, then $\sum_{(a,b)\in\mathcal{S}} v(a - b\alpha) \bmod 2$ is the zero vector over $\mathbb{F}_2$? Yes it is true, but the proof is not trivial. With some effort, it is shown in [4] that the exponent vector map is well defined on the subgroup $\mathcal{H}$ of $\mathbb{Q}(\alpha)^*$ generated by elements $a - b\alpha$ with a, b coprime rational integers, and in fact this map is a homomorphism. Thus squares have even exponent vectors.

Is the converse true? That is, if $\sum_{(a,b)\in\mathcal{S}} v(a - b\alpha) \bmod 2$ is the zero vector, must $\prod_{(a,b)\in\mathcal{S}}(a - b\alpha)$ be a square in $\mathbb{Z}[\alpha]$? Unfortunately no, as the following two examples show. In $\mathbb{Z}[i]$, consider the associate elements $2 + i$ and $-1 + 2i$. Since they both have positive norm, they have the same exponent vectors: all 0, except $v_{5,3} = 1$. The sum of their exponent vectors has all even coordinates, but $(2 + i)(-1 + 2i) = i(2 + i)^2$ is not a square in $\mathbb{Z}[i]$. Or consider $\mathbb{Z}[5i]$. If the element $5i$ is written as $a - b \cdot 5i$, then $a = 0$, $b = -1$. Thus $0 \in R(5)$ and $v_{5,0}(5i) = 2$ (since $N(5i) = 25 = 5^2$). But $5i$ is not a square in $\mathbb{Z}[5i]$ (nor is it the associate of a square).

Although the answer to our last question was no, in some sense it should not be considered an emphatic no. The condition that $\sum_{(a,b)\in\mathcal{S}} v(a - b\alpha) \bmod 2$ is the zero vector takes us a long way towards concluding that $\prod_{(a,b)\in\mathcal{S}}(a - b\alpha)$ is a square in $\mathbb{Z}[\alpha]$. There are just a few small obstructions keeping us from this promised land, and in the next section we shall show how a certain simple device allows us to overcome these obstructions. For now though, let us assume that this problem is solved and try to put the above ideas together to form a factorization algorithm.

Recall that we wish to construct *two* squares, not just one. We accomplish this with still more complicated exponent vectors. For a coprime pair a, b with *both* $a - bm$ and $a - b\alpha$ being z-smooth, we consider the vector $v(a, b)$ which has the usual exponent vector $v(a - bm)$ in its first $1 + \pi(z)$ coordinates and the exponent vector $v(a - b\alpha)$ in its next $1 + \sum_{p \le z} \#R(p)$ coordinates. If we find a set $\mathcal{S}$ of coprime integer pairs (a, b) with $\sum_{(a,b) \in \mathcal{S}} v(a, b) \bmod 2$ being the zero vector, then both $\prod_{(a,b)\in\mathcal{S}}(a - bm)$ will be a square in $\mathbb{Z}$ and $\prod_{(a,b)\in\mathcal{S}}(a - b\alpha)$ will be a square in $\mathbb{Z}[\alpha]$ (modulo the few obstructions still to be discussed), and we have fulfilled our main goal.

What is the size of the numbers we wish to find smooth? These numbers are the integers $(a - bm)N(a - b\alpha)$ where a, b run over coprime integer pairs with $0 < b \le B$, $|a| \le B$. We have already seen that we would like to have d, the degree of our polynomial, be large to make m small and so make $|a - bm|$ small. The norm form $N(a - b\alpha)$ contains the coefficients of our polynomial f, which are bounded by m, so it may seem that we should like to make d large to keep $N(a - b\alpha)$ small as well. However, $N(a - b\alpha)$ is a homogeneous polynomial in the variables a, b of degree d. Thus, when a, b are at the upper end of their range, $N(a - b\alpha)$ may well be as large as $(d + 1)B^d n^{1/d}$. When multiplied by $a - bm$, the absolute value of the product may well be as large as $(d + 1)B^{d+1} n^{2/d}$. This shows that we should not choose d *too* large.

There is a delicate aspect to our problem concerning the parameter B. We will be sieving over approximately B^2 pairs (a, b), so surely this will be a lower bound for the running time. Thus, we would like to take B small. But if we do so, we may not have found enough pairs (a, b) with $(a - bm)N(a - b\alpha)$ being z-smooth. Further, the larger we take B the larger the numbers $(a - bm)N(a - b\alpha)$ get, and so the less likely they are to be z-smooth.

Nevertheless, it is more or less routine to solve this optimization problem. Let $\varepsilon > 0$ be arbitrary but fixed, let $\beta = (8/9)^{1/3} + \varepsilon$ and let

$$B = \exp(\beta(\log n)^{1/3}(\log\log n)^{2/3}), \quad d = [(2/\beta)^{1/2}(\log n/\log\log n)^{1/3}].$$

Then the expression $x = (d + 1)B^{d+1}n^{2/d}$, which is a bound for the numbers we wish to find smooth, is less than $\exp((8\beta + \varepsilon)^{1/2}(\log n)^{2/3}(\log\log n)^{1/3})$. By Proposition 4.1, it will likely suffice if we have $\exp((1 + \delta)\sqrt{2\log x \log\log x})$ such auxiliary numbers, for some fixed $\delta > 0$. But this expression is less than $\exp((128\beta/9 + 2\varepsilon)^{1/4}(\log n)^{1/3}(\log\log n)^{2/3})$, if we choose δ as an appropriate function of ε. Note that $(128\beta/9 + 2\varepsilon)^{1/4} < 2\beta$, so the B^2 auxiliary numbers we have should indeed be sufficient for the task at hand. Thus, an upper bound for the heuristic running time of the algorithm is about B^2, that is, about $\exp((64/9)^{1/3}(\log n)^{1/3}(\log\log n)^{2/3})$. (Note that the number $(8/9)^{1/3}$ in the expression for β is chosen as the real root v of $(128v/9)^{1/4} = 2v$ and that d is chosen so that $B^d \approx n^{2/d}$. The smoothness bound z is that given by (4.2); it is about B. Some of the inequalities given in the above argument hold only when

n is sufficiently large depending on the choice of ε. For a fuller treatment of the heuristic complexity of the number field sieve, see [4].)

It should be noted that the exponent $(64/9)^{1/3}$ in the heuristic complexity of the number field sieve can be reduced using an idea of Coppersmith [6]. However, this improvement is not likely to prove practical until the numbers factored become *very* large.

8. Obstructions

As we just saw, in the number field sieve we hope that numbers $(a-bm)N(a-b\alpha)$ are smooth, where a, b run over small integers. A bound x for the size of these numbers is about $\exp((\log n)^{2/3})$, where n is the number we are trying to factor. The smallness of this bound, when compared to the analogous bound of about $\sqrt{n}$ with the quadratic sieve algorithm, is so encouraging that it gives us great motivation to overcome any remaining obstructions.

To overcome these obstructions, however, we should at least have an idea of what they are. One problem comes from the fact that the ring $\mathbb{Z}[\alpha]$ may not be the full ring $\mathcal{O}$ of integers in $\mathbb{Q}(\alpha)$. Thus, even if $\sum_{(a,b)\in\mathcal{S}} v(a-b\alpha) \bmod 2$ is the zero vector, we may not have $\prod_{(a,b)\in\mathcal{S}}(a-b\alpha)\mathcal{O}$ being the square of an ideal in $\mathcal{O}$. Even if it is the square of an ideal I in $\mathcal{O}$, it may be that $\mathcal{O}$ is not a principal ideal domain, so I may not be principal. Even if $I = (\gamma)$ is principal, because $\mathcal{O}$ may have a complicated unit group, it may be that $\prod_{(a,b)\in\mathcal{S}}(a-b\alpha) \neq \gamma^2$. There are two more problems, but let us deal with these three first.

These three obstructions deal with three algebraic objects: $\mathcal{O}/\mathbb{Z}[\alpha]$, the class group of $\mathcal{O}$, and the unit group of $\mathcal{O}$. These three groups are each related to a different aspect of the "obstruction group" $\mathcal{G}$. To describe $\mathcal{G}$, first take the subgroup $\mathcal{H}$ of $\mathbb{Q}(\alpha)^*$ mentioned above. Then $\mathcal{G}$ is the subgroup of $\mathcal{H}$ of elements with even exponent vectors modulo the group of squares in $\mathcal{H}$. In the obstruction group every element other than the identity has order 2, and so it may be thought of as an $\mathbb{F}_2$ vector space. In [4] it is shown that the dimension of this vector space is less than $\log n/\log 2$. Surely, such a low-dimensional vector space should not be hard to deal with!

In [1], Adleman suggested a particularly neat device for dealing with this vector space. Suppose l is large compared with $\log n/\log 2$, say $l \approx 3\log n/\log 2$. And say we have l "random" multiplicative maps $\chi_1, \dots, \chi_l$ from the z-smooth members of $\mathbb{Z}[\alpha]$ to $\{1, -1\}$. A necessary condition for $\prod_{(a,b)\in\mathcal{S}}(a-b\alpha)$ to be the square of an element in $\mathbb{Z}[\alpha]$ is that each $\prod_{(a,b)\in\mathcal{S}}\chi_i(a-b\alpha)$ is equal to 1 for $i = 1, \dots, l$. If we write $v_i(a-b\alpha)$ for the member of $\{0,1\}$ with $\chi_i(a-b\alpha) = (-1)^{v_i(a-b\alpha)}$, then we can enlarge our exponent vectors by these l coordinates and get an even stronger necessary condition for $\prod_{(a,b)\in\mathcal{S}}(a-b\alpha)$ to be a square. Since the three obstructions described above have dimension so much smaller than l, and since the maps $\chi_1, \dots, \chi_l$ are random, it should be highly likely that we overcome these three obstructions.

So where are we to find these multiplicative maps $\chi_1, \dots, \chi_l$? We can take these as quadratic characters modulo primes exceeding our smoothness bound z. In particular, if q_i is a prime exceeding z, and s_i is an integer satisfying $f(s_i) \equiv 0 \bmod q_i$, $f'(s_i) \not\equiv 0 \bmod q_i$, then we may take for $\chi_i(a - b\alpha)$ the Legendre symbol $\left(\frac{a-bs_i}{q_i}\right)$ (provided that $a \not\equiv bs_i \bmod q_i$, which will be the case if $a - b\alpha$ is z-smooth). We may consider these choices of χ_i as pseudorandom, and in fact with these characters we can overcome the three mentioned obstructions.

There are just two problems left, one minor and one major. The minor problem is that even if $\prod_{(a,b)\in\mathcal{S}}(a-b\alpha) = \beta^2$ for some $\beta \in \mathcal{O}$, it may be that $\beta \notin \mathbb{Z}[\alpha]$. We solve this by multiplying the square $\prod_{(a,b)\in\mathcal{S}}(a-bm)$ with the square $f'(m)^2$ and multiplying β^2 by $f'(\alpha)^2$. It is not hard to show that for any $\beta \in \mathcal{O}$, we have $f'(\alpha)\beta \in \mathbb{Z}[\alpha]$, so $f'(\alpha)^2\beta^2$ is indeed the square of an element of $\mathbb{Z}[\alpha]$.

So now assume that $f'(\alpha)^2 \prod_{(a,b)\in\mathcal{S}}(a-b\alpha) = \gamma^2$ for some $\gamma \in \mathbb{Z}[\alpha]$. We have worked long and hard to get to this point. But we cannot attempt to factor n unless we can actually find the element γ of $\mathbb{Z}[\alpha]$. This is the major problem mentioned above. Note that this problem also occurs with the quadratic sieve when we locate a set of integers $\mathcal{T}$ with $\prod_{t\in\mathcal{T}}(t^2 - n)$ being a square. But in that case, our exponent vectors give us the prime factorization of the square and so the prime factorization of the square root. Moreover, the square root need only be computed modulo n, so this is not a difficult problem.

We still have exponent vectors for our numbers $a - b\alpha$, but unless we are working in a unique factorization domain it is not so easy to use them. A brute-force method is to multiply out the huge product and somehow take its square root. More subtle methods for solving the "square root problem" are discussed in [4, 9, 19]. Suffice it to say that a subroutine for doing this can be fashioned so that asymptotically this phase of the algorithm takes a negligible amount of time compared with sieving and the linear algebra with the exponent vectors. In practice though, this problem is not trivial, though it should not be considered as a barrier to running the algorithm – see [2].

9. The special number field sieve

For some numbers n it may be possible to find a polynomial $f(t)$ for use in the number field sieve where f has especially small coefficients. For example, in the case $n = F_9 = 2^{512} + 1$, we may take $f(t) = t^5 + 8$ and $m = 2^{103}$. Then $f(m) = 8n \equiv 0 \bmod n$.

When f has such small coefficients, the complexity of the algorithm is reduced. In particular, when the coefficients of f have negligible size, the bound $(d + 1)B^{d+1}n^{2/d}$ on the numbers we wish to find smooth, where d is the degree of f and B^2 is the size of the sieve, is reduced to $(d+1)B^{d+1}n^{1/d}$. This in turn reduces the complexity of the number field sieve to $\exp((32/9)^{1/3}(\log n)^{1/3}(\log\log n)^{2/3})$.

There can be other advantages to working with special polynomials such as $t^5 + 8$. In particular, if we are working in a unique factorization domain, explicit factorizations of the auxiliary numbers $a - b\alpha$ into prime elements and units can allow us to overcome the obstructions of the previous section without characters.

In addition, such factorizations allow for a simple solution of the square root problem.

It is naturally with the special number field sieve that we have seen the most spectacular factorizations. For more on the experience with this, see [16].

One lesson that the special number field sieve has for the general case is that it is very desirable to find a polynomial with small coefficients. An averaging argument shows that we cannot hope to do very much better than the "base m method" of §7, but that there is room for improvement. One of the simpler improvements we have is to choose m as $\lceil n^{1/(d+1)} \rceil$ and write n in the base m as before. The resulting polynomial is not monic, but this is not a big problem to overcome. No one has yet come up with a substantially better way of choosing a polynomial in the general case.

10. Conclusion

It seems clear that there is a correlation between algorithmic developments and the evolution of computer hardware. Though Lehmer and Powers in 1931 had almost all of the ingredients of the continued fraction factoring algorithm, they did not make the leap to smooth numbers and exponent vectors reduced modulo 2 as did Brillhart and Morrison, almost surely because these ideas are not well suited for the hand computations that Lehmer and Powers did. Similarly, Kraitchik, 50 years before the introduction of the quadratic sieve, had suggested finding a set of integers $\mathcal{T}$ with $\prod_{t \in \mathcal{T}} (t^2 - n)$ a square as a way of factoring n. Because he did not have a large computer to work on, the idea of sieving to discover smooth values of $t^2 - n$ and then combining them via linear algebra mod 2 did not occur to him.

In some sense the number field sieve for general integers may be a little ahead of its time. Although heuristically it is the *asymptotic* champion, it has not yet factored the largest composite number of no special form and with no small prime factor ever factored. This honor still belongs to the quadratic sieve. We believe we are close to the crossover point now, namely 120 to 130 decimal digits. But to factor such numbers takes an enormous amount of computing power, despite our clever algorithms. Given perhaps an order of magnitude improvement of the speed (and size) of computers, I believe the number field sieve would emerge as the clear method of choice for the hardest numbers.

It is perhaps a little too self-serving, though, to say that others should first improve their aspects of the problem. Rather than sitting on our hands, let us instead find ways to improve the number field sieve now, or perhaps even find a wholly new factorization method. At one point some suggested that $\exp(\sqrt{\log n \log\log n})$ might be the true complexity of factoring, only because we had several different methods with this complexity and none faster. Now we have the number field sieve and it has heuristic complexity of the shape $\exp(c(\log n)^{1/3}(\log\log n)^{2/3})$. Is this now the limit of what we can do? It may be, but it is unlikely an advance will be made by people who think they cannot succeed.

REFERENCES

1. L. M. Adleman, *Factoring numbers using singular integers*, Proc. 23rd Ann. ACM Symp. on Theory of Computing (STOC) (1991), 64-71.
2. D. J. Bernstein and A. K. Lenstra, *A general number field sieve implementation,* in [13], 103-126.
3. J. Brillhart, M. Filaseta and A. Odlyzko, *On an irreducibility theorem of A. Cohn*, Canad. J. Math. **33** (1981), 1055-1059.
4. J. P. Buhler, H. W. Lenstra, Jr. and C. Pomerance, *Factoring integers with the number field sieve,* in [13], 50-94.
5. E. R. Canfield, P. Erdős and C. Pomerance, *On a problem of Oppenheim concerning "factorisatio numerorum"*, J. Number Theory **17** (1983), 1-28.
6. D. Coppersmith, *Modifications to the number field sieve*, J. Cryptology (to appear); also see IBM Research Report #RC 16264, Yorktown Heights, NY, 16 pages, November 1990.
7. ———, *Solving linear equations over* $GF(2)$*: block Lanczos algorithm*, Linear Algebra and its Applications **192** (1993), 33-60; also see IBM Research Report #RC 16997, July 1991, and see Proceedings of Essen Workshop on Linear Algebra, July 1992.
8. ———, *Solving homogeneous linear equations over* $GF(2)$ *via block Wiedemann algorithm*, Math. Comp. **62** (1994), 333-350.
9. J.-M. Couveignes, *Computing a square root for the number field sieve,* in [13], 95-102.
10. J. A. Davis and D. B. Holdridge, *Factorization using the quadratic sieve algorithm*, Sandia Report Sand 83-1346, Sandia National Laboratories, Albuquerque, NM (1983).
11. D. Gordon, *Discrete logarithms in* $GF(p)$ *using the number field sieve*, SIAM J. Discrete Math. **6** (1993), 124-138.
12. B. A. LaMacchia and A. M. Odlyzko, *Solving large sparse linear systems over finite fields*, in Advances in Cryptology – CRYPTO 90 (A. J. Menezes and S. A. Vanstone, eds.), Lecture Notes in Computer Science **537**, Springer-Verlag, Berlin, 1991, 109-133.
13. A. K. Lenstra and H. W. Lenstra, Jr. (eds.), *The development of the number field sieve,* Lecture Notes in Mathematics **1554**, Springer-Verlag, Berlin, 1993.
14. A. K. Lenstra, H. W. Lenstra, Jr. and L. Lovász, *Factoring polynomials with rational coefficients*, Math. Ann. **261** (1982), 515-534.
15. A. K. Lenstra, H. W. Lenstra, Jr., M. S. Manasse and J. M. Pollard, *The factorization of the ninth Fermat number*, Math. Comp. **61** (1993), 319-349.
16. ———, *The number field sieve,* in [13], 11-42, Extended abstract: Proc. 22nd Ann. ACM Symp. on Theory of Computing (STOC) (1990), 564-572.
17. H. W. Lenstra, Jr., J. Pila and C. Pomerance, *A hyperelliptic smoothness test,* I, Phil. Trans. Royal Soc. Ser. A (to appear).
18. H. W. Lenstra, Jr. and C. Pomerance, *A rigorous time bound for factoring integers*, J. Amer. Math. Soc. **5** (1992), 483-516.
19. P. L. Montgomery, *Square roots of products of algebraic numbers*, preprint, August 6, 1993.
20. A. M. Odlyzko, *Discrete logarithms in finite fields and their cryptographic significance*, in Advances in Cryptology – Proceedings of EUROCRYPT 84 (T. Beth et al., eds.), Lecture Notes in Computer Science **209**, Springer-Verlag, Berlin, 1985, 224-314.
21. C. Pomerance, *Analysis and comparison of some integer factoring algorithms*, in Computational methods in number theory (H. W. Lenstra, Jr. and R. Tijdeman, eds.), Mathematical Centre Tracts **154/155**, Mathematisch Centrum, Amsterdam, 1982, 89-139.
22. ———, *The quadratic sieve factoring algorithm*, in Advances in Cryptology – Proceedings of EUROCRYPT 84 (T. Beth et al., eds.), Lecture Notes in Computer Science **209**, Springer-Verlag, Berlin, 1985, 169-182.
23. ———, *Fast, rigorous factorization and discrete logarithm algorithms*, in Discrete algorithms and complexity (D. S. Johnson et al., eds.), Academic Press, Orlando, 1987, 119-143.

24. C. Pomerance and J. W. Smith, *Reduction of huge, sparse matrices over a finite field via created catastrophes*, Experimental Math. **1** (1992), 90-94.
25. O. Schirokauer, *On pro-finite groups and on discrete logarithms*, Ph.D. thesis, University of California, Berkeley, May 1992.
26. D. Wiedemann, *Solving sparse linear equations over finite fields*, IEEE Trans. Inform. Theory **32** (1986), 54-62.

Department of Mathematics
University of Georgia
Athens, Georgia 30602 U.S.A.
e-mail: carl@ada.math.uga.edu

Proceedings of Symposia in Applied Mathematics
Volume 48, 1994

Factoring Integers Before Computers

H. C. WILLIAMS AND J. O. SHALLIT

Dedicated to the memory of our friend, D. H. Lehmer (1905–1991)

ABSTRACT. We examine some aspects of the history of computational number theory, with emphasis on the problems of integer factorization and primality testing.

1. Introduction

We revel in the details of history because they are the source of our being.
Stephen Jay Gould

During the last twenty years, the field of computational number theory has achieved the status of a discipline in its own right. This development is due, in large part, to the invention of cryptosystems (methods for encoding and decoding messages) whose security is based on the presumed computational difficulty of certain number-theoretic problems such as integer factorization. Also quite influential was the development of the theory of computational complexity, which enabled the comparison of different algorithms to be placed on a firmer mathematical footing. Because of the influence of these two areas, one may legitimately ask whether computational number theory is properly a part of mathematics or computer science.

But, of course, the most profound influence upon the establishment of computational number theory is the development of the computer. From the very beginning of computer technology (and, as we will see, even before then), number theorists were using whatever "hardware" was available to aid them in their research.

1991 *Mathematics Subject Classification.* Primary 11Y05; Secondary 11Y11, 11A51, 11Y16, 01A55, 01A60.

Key words and phrases. Primality testing; factoring.

The research of both authors is supported in part by grants from NSERC of Canada.

This paper is in final form, and no version of it will be submitted for publication elsewhere.

© 1994 American Mathematical Society
0160-7634/94 $1.00 + $.25 per page

Computational number theory, unlike many other areas of mathematics, is properly a science: experimental method is a significant component in practice. Those who develop number-theoretic algorithms are often concerned with their eventual implementation, and a great deal can be learned from such experimentation.

The release from the immense drudgery of hand calculation afforded by electronic computers leads to a much more thorough development and testing of algorithms, and gives us new insight into the theorems on which those algorithms are based. Furthermore, it becomes possible to use algorithms that are extremely complicated: many powerful modern methods would be completely unusable if only hand computation were available.

Given this, why should we even consider computational number theory before the age of electronic computers? The answer is that the subject was studied for centuries before computers were available, and it is in this work that the modern subject claims its origins. It is important to understand and appreciate the ingenuity and tireless energy of the pioneers of our discipline. It is easy to be condescending toward these people; after all, they did not express their ideas using modern terms such as "polynomial time" or "succinct certificate" that we casually use today. Although they had not formulated such language, they often understood the ideas implicitly.[1] Since they did their calculations by hand or with the aid of simple mechanical tools, they had to be concerned with efficiency and accuracy.

Why did the early workers in computational number theory not use machines? The answer is that they did: not only did they make use of whatever calculating device was available at the time, on occasion they even developed new machines, as we will see later. One great difficulty that frequently plagued these researchers was that the technology of the day was simply not equal to the demands of the mechanisms they envisaged. For example, according to some historians, Babbage's failures with the differential engine were due in part to the inadequacy of the British machine-tool industry at the time (but for a different view, see [**176**]). A sieving device described by Lawrence in 1896 was only constructed successfully by Carissan in 1919. Furthermore, it was not until 1932 that Lehmer was able to make a sieve mechanism run at a fairly rapid rate without human intervention. Another problem, easy to appreciate by researchers of today, was the difficulty in obtaining research funding for these projects.

We do not intend in this paper to provide anything like a complete history of the development of computational number theory. For our purposes, it will suffice to illustrate our points by concentrating on one particular problem which is still of great interest today: the integer factoring problem. Indeed, the only "proof" that this problem is intractible (a very important issue in modern cryptography) is based on its history. More precisely, we will study the computational

[1] In making this observation, we hope, of course, to avoid the fallacy of interpreting historical events solely in the light of modern understanding.

history of both factoring and primality testing. These two problems really do belong together, because if we wish to obtain the complete factorization of an integer N as a product of prime powers, we must be able to show that the primes involved in this representation are indeed primes. Also, as we will see below, a partial factorization of numbers like $N-1$ and $N+1$ proves useful if one wants to prove that N is prime.

These problems have a very long history, but, with the exception of the invention of the Eratosthenian sieve, nothing truly significant was done until the time of Fermat. From then on, progress has been steady (if not always spectacular) up to the present day.

We emphasize here that early researchers in computational number theory were often uninterested in general methods for factorization or primality testing; rather, particular classes of numbers (such as $2^n \pm 1$ and $10^n \pm 1$) served as stimuli for their efforts. The reader is no doubt familiar with the stories of Mersenne's discussion of the primality of $2^n - 1$, and Fermat's incorrect belief that all numbers of the form $2^{2^n}+1$ are prime. In 1856, Reuschle [**159**] published (among other items) a table of factors of $2^n - 1$ and $10^n - 1$ for $n \leq 42$, and in 1869, Landry [**63**] published a table of factors of $2^n \pm 1$ for $n \leq 64$. The "holes" in these and other tables provided a source of challenges in factoring and primality testing that occupied the energies of many workers over several decades, as do the holes in the Cunningham Table [**13**] even today.

Our main focus in this paper is on the computational history of factoring and primality testing from 1750 to about 1950 (although we discuss modern work in §14). (For citations before 1750, the reader may consult Volume 1 of Dickson's History of the Theory of Numbers [**20**].) Nevertheless, it may be of some interest for comparison's sake to mention what can currently be done. The largest prime currently known is $2^{756839} - 1$, a number of 227832 decimal digits. This was discovered by Slowinski and Gage on 19 February 1992; see Ewing [**23**]. One of the largest integers ever proved prime by a general method[2] is $(2^{3539} + 1)/3$, a number with 1065 decimal digits; see Atkin and Morain [**3**] and Bosma and van der Hulst [**10**]. D. Bernstein and A. K. Lenstra announced the factorization of $2^{523} - 1$, a 158-digit number, on 29 October 1992; they used the lattice sieve, a version of the number-field sieve. The largest integer ever factored by a general method is a 116-digit divisor of $10^{142}+1$, by A. K. Lenstra, M. Manasse, and the assistance of many others; see [**98**]. For further information on this subject, we refer the reader to the papers of Williams [**186**], Silverman [**172**], and the books of Riesel [**161**], Bressoud [**11**], and the Cunningham project book of Brillhart et al. [**13**].

---

[2]By a *general method* we understand an algorithm that does not depend on any special properties of its input. For example, a primality testing algorithm that applies only to numbers of the form $2^n - 1$ would not be a general method.

2. Primality testing

We begin our discussion of the history of computational number theory with the problem of testing a positive integer for primality.

From a modern point of view, the definition of the term "primality test" is not universally agreed upon. (Indeed, the authors of this paper are not in complete agreement!) In general usage, a *test* can mean a "procedure for determining the presence or absence of some property"; consider, for example, "syphilis test". As the term is generally understood, a test may not be accurate with 100% confidence. It is also possible for a test to return the result "inconclusive".

This intuitive usage of the term "test" is what is generally meant by most complexity theorists. By a "primality test", most complexity theorists understand any randomized algorithm[3] that recognizes either the set of prime numbers or the set of composite numbers. More precisely, any Atlantic City algorithm is regarded as satisfactory; here, Atlantic City refers to an algorithm with the property that the output may be incorrect, but the probability of error is no more than $1/2 - \epsilon$, for some fixed positive constant ϵ. It is important to note that the probability of error does not depend on the input N; rather it is computed by considering all possible sequences of coin flips used by the algorithm. If a set is accepted by an Atlantic City algorithm running in polynomial time, then we say that the set is in the complexity class BPP (bounded away from zero probabilistic polynomial time). It is known that the set of prime numbers are in this class; thus, complexity theorists and cryptographers are generally happy with the algorithms of Solovay-Strassen [**173**] and Miller-Rabin [**156**].

However, many number theorists find this rather general definition unsatisfactory, since Atlantic City primality tests do not, in general, provide a proof that a prime number really is prime - only good evidence that this is the case. (However, the Solovay-Strassen and Miller-Rabin algorithms actually have one-sided error, and hence if they reply "composite" on input N, then a proof that N is composite is provided by the details of the computation.) In contrast, most number theorists do not object on philosophical grounds to the so-called Las Vegas tests: these use a source of random numbers, but their output does provide a proof that the number in question is either prime or composite. Only the running time, not the conclusion, of a Las Vegas algorithm is probabilistic. If a set is accepted by a Las Vegas algorithm running in expected polynomial time, then we say that the set is in the complexity class ZPP (zero error probabilistic polynomial time).

As most of the early workers tended to regard a primality test as being a technique which would yield a rigorous proof of primality, we will adopt for use throughout this paper the following more narrow definition of "primality test":

DEFINITION 2.1. A primality test on a given integer N is the computational

[3] A *randomized algorithm* is one that has access to a sequence of random, unbiased flips of a fair coin.

verification of the hypothesis of a theorem whose conclusion is that N is prime. Some have also called such a test a "primality proof".

Many such tests can be described. For example, consider Wilson's theorem (which was probably known to Leibniz; see Vacca [**185**] and Mahnke [**127**]): $N > 1$ is a prime if and only if N divides $(N-1)! + 1$. However, this test seems to be computationally impractical, since there is no known way to compute $(N-1)! + 1 \pmod{N}$ in any reasonable length of time for large N. (Note that Wilson's Theorem actually gives a necessary and sufficient test for primality, something not explicitly required by our definition.)

Having described what we mean by a primality test, we now give four features that are desirable: the test should be efficient, the test should be universally applicable, the test should be robust, and the test should produce an efficiently verifiable certificate of primality.

The term "efficient" is, of course rather vague and subject to interpretation, but most would agree that the obvious implementation of Wilson's theorem is not efficient. Today we might say that an algorithm is efficient if it runs in polynomial time, that is, if its running time is bounded by a polynomial in $\log N$, where N is the input. This definition has its problems in real life, however, since the polynomial referred to may have high degree or very large coefficients. For example, although it is a great theoretical advance, the Las Vegas primality test of Adleman and Huang [**1**] is probably not useful in practice for precisely these reasons.

The second feature, universal applicability, means that the test should work for all prime numbers. This is a desirable goal, but one that not all otherwise useful tests satisfy. As we will see below, many of the early practical primality tests only worked for numbers of special forms: Lucas' test for numbers of the form $2^p - 1$, for example.

The third desirable feature of a primality test is robustness. Robustness is a vague notion which is difficult to express precisely; we use it to mean that errors that take place during the course of the computation will be detectable. Here is an example: suppose we are trying to determine if a number a is a quadratic residue modulo some given odd prime p. By Euler's criterion, it suffices to compute $a^{(p-1)/2} \pmod{p}$. The result, if computed correctly, will be $+1$ or -1, according to whether a is a residue or nonresidue. Nearly every mistake made in this computation will be detectable, since most will result in an answer which is neither $+1$ nor -1.

Robustness was especially important in the days before computers, when hand computation was very error-prone, and checking a computation was not simply a matter of running a program again on another machine. Lucas' necessary and sufficient test for primality of numbers of the form $2^p - 1$ depends on obtaining the result 0 after a long series of hand calculations involving numbers with p bits. If the result is 0, then one may be reasonably confident that it is correct, and hence the number in question is prime, since the probability of obtaining 0 after a random series of multiplications and additions of numbers this large is quite small. On the other hand, if the result is not 0, the theory implies that the number in question is composite. But since almost any error introduced into the computation would also give a nonzero result, one's confidence in this result is necessarily smaller.

The last criterion, that of producing an easily verifiable certificate of primality, was explicitly pointed out for the first time by Pratt in 1975 [**150**]. He showed that, for any given prime p, it is possible to obtain a "succinct certificate" that proves p prime and which can be verified in polynomial time. It is, of course, quite conceivable that actually *finding* such a certificate may require more than polynomial time.

Note that we have not discussed what may happen if the input to a primality test is composite. There are several possibilities: the algorithm may diverge (run forever); the algorithm may reply that the test is inconclusive; or the algorithm may provide a proof that the input is indeed composite. While this behaviour was a legitimate concern in the past, today a proof of compositeness may be given in polynomial expected time using the strong pseudoprime test (as in Miller-Rabin).

Having discussed some abstract principles, we now move on to the history of primality testing. In 1588, Cataldi proved that $N_1 = 524287 = 2^{19} - 1$ is a prime by trial division by all the primes less than the square root of N_1. (The method of trial division was known to Fibonacci in 1202.) Until 1772, this seems to have been the largest prime known. At that time, Euler proved that $N_2 = 2147483647 = 2^{31} - 1$ is a prime. He did this by improving upon the algorithm used by Cataldi.

Euler was aware that if any prime q is a divisor of $2^p - 1$, where p is an odd prime, then $q \equiv 1 \pmod{2p}$, a result known to Fermat. He also realized that since $2N_2 = (2^{16})^2 - 2$, then 2 must be a quadratic residue of q; hence $q \equiv \pm 1 \pmod 8$. By combining these observations, Euler could assert that if q is a prime divisor of N_2, then $q = 248k + 1$ or $q = 248k + 63$ for some integer $k \geq 0$. If $q < \sqrt{N_2}$, then there are only at most 186 values of k to check. Thus, after performing at most 372 trial divisions (fewer if one eliminates composite trial values of q), Euler was able to show that N_2 is a prime. Thus, Euler's test was more efficient than Cataldi's.

In 1859, F. Landry [**61**], who at that time believed that $2^{31} - 1$ was still the largest known prime[4], attempted to give a shorter proof of its primality than Euler's. By using a number of clever tricks he was able to do this; in fact, by 1867 he was able to claim [**62**] that he had so reduced the calculations necessary to effect the verification of the primality of N_2 that they would fit on a single slip of paper. Furthermore, in this same paper Landry goes on to state that he had proved the primality of the thirteen-digit divisor 1133836730401 of $2^{75} + 1$.

[4]In fact, according to Reuschle [**159**, p. 3], Looff had proved in 1851 that 999999000001, the 12-digit factor of $10^{18} + 1$, is prime.

What he said next we quote here, in translation from the French:

> At this point we are, if not uneasy, then at least somewhat embarrassed.
>
> Indeed, when one has succeeded in factoring a number, and has given its factors, this can be verified immediately. But it is a different matter when the methods used fail to discover any factor, and one then asserts that the number is prime. How could one then transmit to another such a totally personal conviction? Who would be convinced, without having redone all the calculations, and without having understood the principles on which those calculations were based?
>
> We understand well that our claim is valid only as an assertion, worthwhile until someone proves the contrary, or until we make known our methods and enable others to apply them.

Today, we would say that Landry was struggling with the difficulty of providing a *succinct certificate.* Landry clearly anticipated the basic idea of what would become famous a hundred years later in computational complexity theory: the class NP. Even today, providing such a certificate is a difficult problem, and those who have been involved in primality testing can easily understand and appreciate Landry's concern.

In 1869, Landry [**63**] went on to state that he had proved that the 14-digit number $(2^{53}+1)/(3\cdot 107)$ is prime. This was the largest prime known until 1876, when Lucas made his breakthrough. Incidentally, Landry did not describe in [**62, 63**] how he had carried out his impressive factorizations. It was not until 1880, when he wrote a letter [**64**] to Charles Henry, that he revealed his method. Landry's technique had been known to Fermat and will be described later. (It was this convergence of techniques that particularly struck Landry and caused him to write his letter.) After Landry's work in 1869, only the following four numbers of the form $2^n \pm 1$ $(n \leq 64)$ remained unfactored:

$$2^{59}-1, 2^{61}-1, 2^{61}+1, 2^{64}+1.$$

By 1878, Landry (according to Lucas [**114**, pp. 239–240]) had factored $2^{59}-1$; we will discuss the investigation of the others later. It is interesting to note in passing that Landry suspected that $2^{61}-1, (2^{61}+1)/3$, and $2^{64}+1$ were all primes, but he never seems to have said that he had proved the primality of any of them. Indeed, in [**63**] he stated that these had resisted his efforts, and in 1880 he said [**64**] that the first two still remain to be investigated (the third is composite).

2.1. Babbage. Charles Babbage, the English pioneer of the modern computer, was one of the first to apply computational ideas to number theory.

Legendre and Euler had both noticed that the polynomials $x^2 \pm x + 41$ had the unusual property of generating nothing but primes for small values of x. The relationship between this curiosity and the class number of $\mathbb{Q}(\sqrt{-163})$ was not explained until 1912 by Frobenius and Rabinovitch. Babbage knew about these polynomials, and one of the first tasks he set his Difference Engine to was the computation of values of $x^2 + x + 41$. In June 1822, he wrote:

> With this machine I have constructed ... a table from the singular formula $x^2 + x + 41$, which comprises amongst its terms so many prime numbers.

His machine could only display results, and not print them. Babbage employed a friend to write down the results. In a letter to Sir Humphrey Davy the next month, Babbage discussed the speed of his machine:

> [The Difference Engine] proceeded to make a table from the formula x^2+x+41. In the earlier numbers my friend, in writing quickly, rather more than kept pace with the engine; but as soon as four figures were required, the machine was at least equal in speed to the writer.

The letter also mentioned a possibility for a future machine (perhaps the Analytical Engine):

> I have also certain principles by which, if it should be desirable, a table of prime numbers might be made, extending from 0 to ten millions.

See Babbage's collected letters and papers [**4**] for more details.

3. Lucas' discovery

In order to discuss the work of Edouard Lucas, it is first necessary to introduce the functions that today carry his name. The Lucas functions U_n and V_n (or $U_n(P,Q)$ and $V_n(P,Q)$) are defined by

$$U_n = U_n(P,Q) = (\alpha^n - \beta^n)/(\alpha - \beta),$$

$$V_n = V_n(P,Q) = \alpha^n + \beta^n,$$

where α, β are the zeros of the polynomial $x^2 - Px + Q$, and P, Q are coprime integers.

These functions did not originate with Lucas; they (or functions very similar to them) had been studied by Euler, Lagrange, Legendre, and Gauss. Several combinatorial and number-theoretic properties of the function U_n had been discovered by Siebeck [**171**] almost thirty years previous to Lucas' work. Also, Genocchi [**33**] had used properties of functions very similar to U_n and V_n in order to demonstrate the infinitude of primes of the forms $ax + 1$ and $ax - 1$, for any given a. Nevertheless, in our opinion the functions are properly called the Lucas functions because of Lucas' singular contribution to their study, only a part of which we will discuss here.

In early January of 1876, it appears that Lucas [**102**] had considered the functions U_n and V_n as we see them above, but he was mainly interested in the particular case of $U_n(P, Q)$ where $P = 1$ and $Q = -1$. Today, these numbers are called Fibonacci numbers, but at the time it was common to call the sequence

$$0, 1, 1, 2, 3, 5, 8, 13, 21, 34, \ldots \tag{3.1}$$

the series of Lamé. That was because Lamé **[60]** in 1844 had used these numbers to investigate the number of steps performed by the Euclidean algorithm to compute the greatest common divisor of two numbers. Lamé had shown that the worst case of the Euclidean algorithm occurred when the inputs were successive Fibonacci numbers, a fact that had been given without proof by Léger **[69]** in 1837. This is one of the earliest results in the analysis of algorithms. Lamé further showed that the number of division steps in the Euclidean algorithm was always bounded by five times the number of decimal digits in the smaller of the two inputs. It was Lucas, however (see **[103]** and **[112]**), who discovered that these numbers had been known to Fibonacci six centuries earlier, and in Lucas' next paper **[103]** dated May of 1876, he calls (3.1) the Fibonacci series.

There are several results mentioned in **[102]**, all given without proof, but the most striking of these are results such as the following.

THEOREM 3.1. *If* $N \equiv \pm 3 \pmod{10}$, $N | U_{N+1}$, *and* $N \nmid U_d$ *for all divisors* d *of* $N + 1$ *[with* $d < N + 1$*], then* N *is a prime.*

The reason for the importance of this theorem will be fully discussed later. Lucas pointed out that it permits one to know that a number is prime or composite without the need of a table of primes, i.e., without employing trial division. Notice that this is somewhat similar in spirit to Wilson's theorem in that to show that N is a prime it is necessary to show that it divides a certain number. However, Lucas was the first individual to devise an efficient test for primality or compositeness, which did not necessarily have to reveal the factors, if any, of the number being tested. He then goes on to say:

> It is with the aid of this theorem that I think that I have shown that the number $N = 2^{127} - 1$ is a prime.

This $N = 170141183460469231731687303715884105727$ is a number of 39 digits, far greater than the largest prime known at the time. (Lucas at this time was unaware of the work of Looff and Landry and thought that $2^{31} - 1$ was the largest known prime.)

Lucas had clearly made a most remarkable discovery, and when we think about his statement today, there are at least four important questions that we are led to ask:

(i) *How* did Lucas discover this test?
(ii) *Why* did he test $2^{127} - 1$, and not some other number?
(iii) *How* did he test $2^{127} - 1$?
(iv) *Why* was he unsure of the result?

We will attempt to answer the first two of these questions in this section, leaving for later sections the answers to the others. Previous to his paper **[102]**,

Lucas had written only one other paper [**101**] on factoring. In this paper he attempts to factor various numbers of the form $a^n \pm 1$ by determining the possible linear forms of the prime divisors (as did Euler) and then performing trial divisions by taking logarithms of the numbers involved and subtracting. This work, which appeared in November of 1875, seems much less sophisticated than the work of the following January. What happened? In [**118**, p. 14] Lucas himself provides us with the answer. In commenting on [**102**] he says:

> This note is the point of departure for new investigations by the author in the theory of prime numbers. After having decomposed by means of a table of prime divisors the early terms [of the Fibonacci numbers], the observations of the forms of these divisors led the author to the demonstration of a certain number of theorems which allowed him to develop some new and general considerations concerning the theory of prime numbers. One finds here for the first time an indication of a new method for finding prime numbers, which is independent of a previously constructed table of prime numbers.

As to why Lucas selected $2^{127} - 1$ to test, we can only speculate. Probably he wanted to find a prime of the form $M_n = 2^n - 1$; after all, Euler himself had been interested in such primes and Euclid had shown that $2^{n-1}M_n$ is a perfect number when M_n is a prime. Also, it is easy to find all of the divisors of $M_n + 1 = 2^n$; thus, the theorem of Lucas quoted earlier is easy to apply (see §4 below). Lucas knew that any prime value of M_n would require that n be prime and he also knew all the primes of the form M_n that had been identified by 1876:

$$2^2 - 1, 2^3 - 1, 2^5 - 1, 2^7 - 1, 2^{13} - 1, 2^{17} - 1, 2^{19} - 1, 2^{31} - 1.$$

Notice that some of these values of n (i.e., 3, 7, 31) are also of the form M_k. The next possible prime value for such an n is 127 and $M_{127} \equiv -3 \pmod{10}$; thus, his theorem could be used. We don't really know that this was how Lucas was thinking, but in a later letter [**111**] he speculates on the existence of a sequence of primes of the form $2^n - 1, 2^{n_1} - 1, 2^{n_2} - 1, \dots$, where $n_1 = 2^n - 1$, $n_2 = 2^{n_1} - 1$, etc.; thus, it's quite likely that he was thinking along the lines of our speculation. According to [**118**], Lucas wrote a thesis for presentation to the Faculty of Science at Paris. This thesis, dated September 24, 1877, was a development of the paper [**102**] and could perhaps cast some further light on these problems. Unfortunately, all attempts by the authors to locate a copy of it have so far failed.

4. Implementation of Lucas' test

In order to discuss how Lucas actually used his theorem to test M_{127}, we provide the following simple identities satisfied by U_n and V_n:

$$U_{2n} = U_n V_n;$$

$$V_{2n} = V_n^2 - 2Q_n.$$

On examining these identities and referring back to Lucas' theorem, it is easy to see that if $N = 2^n - 1$ and $N|U_{N+1}$ but $N\nmid U_d$ for any proper divisor d of $N+1$, then $N|V_{(N+1)/2}$. Thus, if we define $S_k = V_{2^k}(1,-1)$, then $S_k = S_{k-1}^2 - 2$ for $k \geq 2$. We have the following test:

If $M_n \equiv \pm 3 \pmod{10}$, then M_n is a prime if $M_n|S_{n-1}$, where $S_1 = 3$ and $S_k \equiv S_{k-1}^2 - 2 \pmod{M_n}$ $(k = 2, 3, ..., n-1)$.

Notice that this test for primality certainly possesses the features of efficiency and robustness discussed in §2. As he mentions in [**112**], this is the form of the test that Lucas used. By December of 1876 he [**107**] has modified this result as follows:

THEOREM 4.1. *Let $p = 2^{4m+3} - 1$, where the exponent is a prime. If one forms the sequence*

$$3, 7, 47, 2207, \ldots$$

with $r_1 = 3$ and $r_{n+1} = r_n^2 - 2$, then p is a prime when the least value of k such that $p|r_k$ is $4m+2$. The number is composite if none of the first $4m+2$ values of r_n is divisible by p. Finally, if α denotes the first value of k such that $p|r_k$, then the [prime] divisors of p have the form $2^\alpha t \pm 1$ combined with those forms which can divide $x^2 - 2y^2$.

So we know what sort of test Lucas used, but how did he actually perform the arithmetic? Remember M_{127} is a number of 39 digits; the kind of arithmetic needed here would involve numbers larger than any used previously. In several papers Lucas [**103, 106, 108, 114**] alludes to the use of binary arithmetic for doing this, but he does not really describe how he actually did it. Finally, in [**112**, p. 152 ff] he is more explicit. The solution to this problem is very characteristic of him; he made the performance of this tedious arithmetic into something like a game. He said he used a 127×127 chessboard to effect the computations. To illustrate his idea he used a simple example which we will discuss here.

Consider (instead of M_{127}) $M_7 = 2^7 - 1 = 127$. We have $S_1 = 3$, $S_2 = 7$, $S_3 = 47$, $S_4 = 47^2 - 2 \equiv 48 \pmod{127}$, $S_5 \equiv 48^2 - 2 \equiv 16 \pmod{127}$, $S_6 \equiv 16^2 - 2 \equiv 0 \pmod{127}$; hence, we have a proof of the primality of 127. Lucas notes that if $M_n = 2^n - 1$, then

$$2^{n+m} \equiv 2^m \pmod{M_n}. \tag{4.1}$$

The main operation of testing involves squaring, subtracting 2, and reduction modulo M_n. If we try to perform this on S_3 to obtain S_4 (using M_7) we first consider the binary representation of S_3 which is 101111. We can then begin to compute its square by simply multiplying as usual

$$
\begin{array}{ccccccccccc}
 & & & & & 1 & 0 & 1 & 1 & 1 & 1 \\
 & & & & & 1 & 0 & 1 & 1 & 1 & 1 \\
\hline
 & & & & & 1 & 0 & 1 & 1 & 1 & 1 \\
 & & & & 1 & 0 & 1 & 1 & 1 & 1 & \\
 & & & 1 & 0 & 1 & 1 & 1 & 1 & & \\
 & & 1 & 0 & 1 & 1 & 1 & 1 & & & \\
 & 0 & 0 & 0 & 0 & 0 & 0 & & & & \\
1 & 0 & 1 & 1 & 1 & 1 & & & & & \\
\hline
\end{array}
$$

However, because we only need to find the answer modulo $2^7 - 1$, we can use (4.1) to put the information between the horizontal lines into a neat 7×7 square because each row wraps around cyclically.

7	6	5	4	3	2	1
0	1	0	1	1	1	1
1	0	1	1	1	1	0
0	1	1	1	1	0	1
1	1	1	1	0	1	0
1	1	1	0	1	0	1
0	0	0	0	0	0	0
1	1	0	1	0	1	1

In order to avoid a lot of writing, Lucas suggests the use of a 7×7 chessboard with pawns in squares representing the positions of the ones and empty squares representing the positions of the zeros.

Once we have this array of pawns on our chessboard we make use of two rules:

(i) Take (when possible but only once) a pawn away from column 2. This corresponds to the subtraction of 2 from the square. If a pawn never appears in column 2, then 2 must be subtracted from the final answer.

(ii) For each pair of pawns in any column remove one from the board and move the other into the column to the left. Remember that the column to the left of the left-most column (column 7 here) is column 1 because of (4.1).

Continue performing these operations until the only pawns remaining are in the first row. In our case we get the first row to be

7	6	5	4	3	2	1
	•	•				

which represents 0 1 1 0 0 0 0 or 48 in decimal.

Lucas said that with a little training one can succeed in becoming quite quick at the manoeuvre (ii) above. By making use of a 127×127 chessboard, this was how Lucas performed his test on M_{127}. The advantage of not having to write anything down (contrary to Lehmer's criticism in [**71**]) is to some degree offset by having nothing left at the end of the process but an answer. If an error were made, it would be difficult to detect. Depending on how quickly Lucas could execute the operations needed to perform this test, we estimate that it must have taken between 170–300 hours for him to complete it. This explains why he only performed this test of M_{127} a single time (see [**112**, p. 152]), and this may be why he was a little unsure of his result, but we'll discuss this more fully in §5.

One thing that is very striking about Lucas' technique is how easy it should be to mechanize (and how much more accurate the results would be if it were). Certainly, Lucas thought so. In several of his papers [**103, 105, 106, 107**]; and [**114**, p. 305] he mentions the possibility of constructing a machine for doing this. For example in [**103**] he says:

> The construction of this mechanism allows for the rapid calculation in the binary number system of residues of the V_n with respect to the number whose decomposition we are seeking. It is based on the preceding theorems and the mathematical laws of the geometry of weaving.

In [**106**] he goes so far as to say:

> I have conceived ... the plan of a mechanism which allows ... for the determination of prime numbers having a thousand digits in the decimal system.

The reference to the geometry of weaving is not as outlandish as it may seem. Lucas knew a great deal about weaving and wrote or presented several papers on this subject. Also, there were mechanisms in existence at that time which did allow for the construction of a fabric with the properties of the chessboard representation of the square of an integer (mod M_n) discussed above. In this representation a 1 might be denoted by a vertical thread passing over a horizontal one, whereas a zero would then be represented by a vertical thread passing under a horizontal one. Nevertheless, by the time Lucas published in 1878 his main work [**114**] on the Lucas Functions, his claims are much more subdued. Also, in his major paper on the geometry of weaving [**119**] (see [**126**] for a French translation) he makes no mention of using his ideas for constructing a machine. However, he never forgot these ideas, and in 1887 he says [**121**]:

> I will add that Mr. Genaille is currently constructing for me a machine that will enable us to [find all primes of the form $2^n - 1$] which have no more than 150 digits.

Henri Genaille was a very prolific inventor of calculating mechanisms. Lucas praises him very warmly in his paper on calculating devices, which was reprinted in Vol. III of *Récréations* [**125**]. He is credited with constructing over 28 such devices in [**129**], all of which were donated in 1886–1888 to the Conservatoire National des Arts et Métiers in Paris. Also, he and Lucas worked together on a number of various computing devices (see [**2**]), and in 1891 we learn in an all-too-brief mention [**32**] that he had constructed an Arithmetic Piano, a device which was to be used for determining primes of the form $2^n - 1$. We quote the article [**32**] in full:

> The Arithmetic piano allows one to put into practice the method formulated by Mr. E. Lucas, at the Congress of Clermont-Ferrand, for the verification of large prime numbers. By the simple manoeuvering of some pegs, the verification of prime numbers of the form $2^n - 1$ is reduced in the greater number of cases to a labour of a few hours. This machine, which can manage to carry out automatically calculations of the greatest importance, will effect one day the realization of a calculating machine which performs arithmetic operations by itself.

We have currently no idea what happened to this machine. Unfortunately, Lucas died at the early age of 49 a few weeks after the meeting at which the machine was exhibited and nothing seems to have been heard about it since.

Gérardin [**40**], writing in 1912, claimed that Genaille had not constructed anything, but this is certainly not the sense of [**31**], where it is explicitly stated that the machine was admired by all the members of the session. Indeed, in view of the facts that 1) Genaille was a skilled fabricator of calculation machines and 2) that he had been working on this device for a period of about 4 years, it is difficult to believe that he would have discussed a nonexistent mechanism. However, it is possible that the device was not in a completely finished state. Very likely the machine was in the possession of Genaille when Lucas died, and there being no further use for it, Genaille probably abandoned the project. We wonder if somewhere in France this machine might still exist.

It is curious to note that both Fauquembergue [**25**] in 1912 and Mason [**128**] in 1914 described devices based on Lucas' ideas for testing the primality of M_n, devices which were based on the same multiplication technique of Cauchy [**18**], even though it appears that Mason was unaware of [**18**]. In the case of Fauquembergue, a device resembling a rather large slide rule was actually constructed and used. Mason's device, which was more sophisticated in concept than that of Fauquembergue, seems never to have been constructed.

5. Lucas and Mersenne

The most difficult-to-answer of all the questions posed in §3 is (iv). This is because it does not appear that Lucas had ever settled this question in his own mind. What is particularly remarkable is his failure to mention the primality of M_{127} in several of his papers where it would have been appropriate, especially in his memoir [**114**], the notice of his works [**118**], or his book [**122**]. In many ways his doubts or caveats, expressed in one way or another in [**102, 112, 120, 121, 123, 124**] seem strange to us today. For, after going through all the labour necessary to find S_{126} (mod M_{127}) and finding it to be the particular residue zero, and considering all the possibilities for error, he would almost certainly have had the correct answer. The probability that it would be wrong is very remote indeed. However, this is not necessarily the way that Lucas thought about these matters. Remember that he had just discovered this test and did not have all the experience in these matters that we have today.

He was evidently of two minds concerning his result; nowhere is this better illustrated than in his paper [**120**], written in 1886. This note seems to have been in response to one of Seelhoff [**165**] (see also [**164**]) in which, among other matters, it was stated that there were only eight perfect numbers known. Lucas replied that there were *nine* (his italics) such numbers and listed $2^{126}M_{127}$ as one. Since from the time of Euler, it has been known that the only even perfect numbers are those of the form $2^{n-1}M_n$, where M_n is a prime, it is clear at this point that Lucas considered M_{127} a prime. However, on the next page of this same note he gave a list of 24 values of n up to 257 which includes the number 127. Of this list of values he stated that it remains to study (meaning in this context to determine the prime or composite character of) the values of $2^n - 1$, for these n-values, a statement he also made in [**122**, p. 376].

The likely cause of Lucas' difficulties was a statement of Mersenne. In 1644, Mersenne asserted (without giving any reason) that of the 55 prime values of $n \leq 257$, M_n is a prime for only the following 11 values:

$$n = 2, 3, 5, 7, 13, 17, 19, 31, 67, 127, 257.$$

Mersenne was a correspondent of many of the great mathematical thinkers of his time such as Descartes, Fermat, and Frenicle; thus, this statement carried a lot of weight. The attentive reader might well ask why it was not this list that caused Lucas to attempt to prove M_{127} a prime? (The symbol M here commemorates Mersenne and numbers of the form M_n, where n is a prime, are called Mersenne numbers today.) The answer to that question is that Lucas was unaware of Mersenne's list when he tested M_{127}. He was informed of it later in a paper [**34**] of Genocchi, where the list is mentioned in a lengthy footnote.

While it is hard for us to understand this today, it is important to realize that throughout most of his academic life, Lucas attached considerable importance to Mersenne's list. We now know that it contains five errors (M_{67} and M_{257} are composite and M_{61}, M_{89}, M_{107} are primes). In fact, in [**113**] Lucas declared that Mersenne was in possession of arithmetical methods that are now lost. In [**114**, p. 237] he went on to say that Mersenne's method would probably not deviate from the principles of Fermat, and as a consequence, would not differ essentially from the method he (Lucas) had deduced. Another part of the reason he believed in this list is that until 1887 every factoring result concerning Mersenne numbers, of which Lucas was aware, confirmed the truth of it. Indeed, the 24 values of n mentioned earlier were all that remained to be tested of the original 55. His faith in Mersenne's list was so strong that he seems to have refused to believe the results of his own calculations on M_{67}. In [**103**] he states that he does not think that 67 belongs in the list because he had already applied his method to it. But later in [**114**, p. 307] he says:

> One could thus construct diagrams for prime numbers of the form $2^{4q+3}-1\ldots$ we hope ultimately to give these for the numbers $2^{67}-1$ and $2^{127}-1$.

These diagrams were depictions of $n \times n$ chessboards whose rows represented in binary the residues S_k modulo M_n for $k = 0$ $(S_0 = 0)$, $1, 2, \ldots, n-1$. For example, the diagram for M_7 would appear as

6	5	4	3	2	1	0	
						●	0
					●	●	1
				●	●	●	2
	●		●	●	●	●	3
	●	●					4
		●					5
							6

In fact, Lucas [**112**] went to some trouble to compute such a diagram for M_{31}. This represented his attempt at producing a succinct certificate for its primality. He [**104**] also issued, as a challenge, the problem of proving M_{31} a prime without the use of a table of primes.

In 1887, Lucas [**121**] learned that M_{61} is a prime. Seelhoff [**168**] had shown this in 1886, and Hudelot, after 54 hours of labour, in 1887 confirmed Seelhoff's result. This confirmation was very much needed, as Seelhoff's proof is certainly not complete. He appears to have done little more than verify that

$$3^{N-1} \equiv 1 \pmod{N}$$

for $N = M_{61}$. This is certainly a necessary condition for primality, but it is not sufficient, as Lucas himself well knew. Hudelot made use of the following test of Lucas [**124**] (see also [**110**]).

THEOREM 5.1. *If $p = 2^{4m+1} - 1$, where $4m+1$ is a prime, then p is a prime if $p|S_{4m}$, where $S_1 = 4$ and $S_k \equiv S_{k-1}^2 - 2 \pmod{p}$ $(k = 2, 3, \ldots, 4m)$.*

Hudelot's result is correct, but in fact Pervouchine (see Imchenetzki and Bouniakowsky [**49**] or Raik [**157**]), by using the same test, showed that M_{61} is prime in 1883. So Mersenne's list was known to Lucas in 1887 to be wrong; this should have discouraged Lucas' belief in it, but according to Tannery [**177**], Lucas looked at more of what Mersenne had written on this topic and attributed the following "proposition" to Fermat:

CONJECTURE 5.1. *For $2^p - 1$ to be prime, it is necessary and sufficient that p be a prime of one of the forms $2^{2n}+1$, $2^{2n} \pm 3$, $2^{2n+1}-1$.*

This is very close (without the attribution to Fermat) to the modern conclusion of Drake [**21**] concerning the numbers in Mersenne's list. Of course, under this rule, 61 should be included in the list. Possibly, then, Lucas rationalized that Mersenne's failure to include 61 was just an oversight. At any event the discovery of the primality of M_{61} did not appear to shake Lucas' belief in Mersenne (Fermat), but it did seem to affect how he felt about M_{127}; for, he says in [**121**] that M_{61} is the largest prime currently known, and in none of his subsequent publications does he ever mention that M_{127} is a prime; he does however reaffirm in the year of his death (see [**122**, p. 376]) his belief that Mersenne was in possession of an important method. Today we believe, as did Tannery, that Mersenne's list is most likely an empirical result (see [**21**] or Heyworth [**47**]).

To confuse further this picture, we mention that in [**122**, p. 376] Lucas stated that he thought he had shown by very lengthy calculations that M_{67} and M_{89} are not primes. But according to Tannery [**178**], Lucas thought to the end of his life that M_{67} was nevertheless a prime. As to why he included this statement about M_{67} and M_{89}, we can again only speculate. Lucas had a correspondent named Reuss, and according to Fauquembergue [**25**], Reuss had written Lucas in January of 1877, saying that he had shown by using Lucas' test that M_{89} is not a prime and that his (Reuss') computations required about 240 hours of work. In September of 1888, Reuss [**160**] wrote Lucas and informed him that he had shown that M_{67} is not a prime. Fortunately, he lists his values of $S_1, S_2, \ldots, S_{66} \pmod{M_{67}}$ and we are able to see that he made an error in determining the value of S_{22}, and that all subsequent values are also incorrect. Thus, for the wrong reason, Reuss arrived at the correct conclusion. Possibly it was these computations that in some way confirmed Lucas' results and caused him to include the remark about M_{67} and M_{89} in his book.

The only explanation we can offer for Lucas' behaviour is that he did not believe the results of his own computations. Since he did not have faith in these calculations, he was unable to state anything consistent about the primality of M_{127} or M_{67}. He praises Hudelot's work on proving M_{61} a prime very highly, saying it is one of the most beautiful numerical memoirs that he has ever seen. Unfortunately, he seems to have little praise for his own calculations. This is a pity because he had almost certainly proved M_{127} a prime and may very well have proved that M_{67} is composite.

As a postscript to this section, we note that in 1894 Fauquembergue [**24**], by a then unpublished process of computation, declared that M_{67} is composite. Later in 1912 he says in [**25**] that he used his device (see §4) to prove that M_{67} is composite and it required only 20 hours of work to do this. Unfortunately, he did not publish his calculations as he did later for other numbers. Furthermore, in 1914 [**42**], he used his device to confirm that M_{127} is a prime. In 1911, Powers [**148**] proved that M_{89} is a prime (a result confirmed in 35 hours by Fauquembergue [**25**]) and in 1914 Fauquembergue also determined that M_{107} is a prime. In fact, according to [**26**], he had proved M_{107} a prime in February of

1914. Gérardin [**41**] states that Fauquembergue let him know of his result on June 7 of that year and that Powers [**149**] had communicated the same result (as determined by him) to the London Mathematical Society on June 11. However, in [**42**] Gérardin points out that Powers had cabled his result to Bromwich on June 1. Several of these computations were published in [**42**]. It is somewhat ironic that Fauquembergue should be the one to confirm Lucas' result on M_{127}. We do not have his calculations for M_{67} and M_{89}, but of the six numbers that he investigated (M_{101}, M_{103}, M_{107}, M_{109}, M_{127}, M_{137} in [**42**] and [**27**]) for which we do have his calculations, he gets the wrong answer for S_{n-1} for four of them: M_{101}, M_{103}, M_{109}, M_{137}. (See Robinson [**162**]).

6. Lucas and necessity

In 1877, Lucas [**109**] turned his attention from tests for the primality of Mersenne numbers to those for the primality of the Fermat numbers F_n. In 1640 Fermat had conjectured that any number of the form $F_n = 2^{2^n}+1$ is prime. This is certainly the case for $n = 0, 1, 2, 3, 4$, but when $n = 5$, Euler showed in 1747 that 641 is a divisor of F_n. This is where matters stood when Lucas began to consider the question of the primality of F_n. In [**109**] he presented the following test (with some modifications and two misprints corrected):

THEOREM 6.1. *Let $F_n = 2^r+1$ $(r = 2^n)$ and $T_1 = 3$. If we define the sequence $\{T_i\}$ by $T_{i+1} = 2T_i^2 - 1$ $(i = 1, 2, 3, \dots)$, then F_n is a prime if the first term of this sequence which is divisible by F_n is T_{r-1}. Also, F_n is composite if none of these terms up to T_{r-1} is divisible by F_n. Finally, if k denotes the rank of the first term which is divisible by F_n, the prime divisors of F_n must have the form $2^{k+1}q+1$.*

Later in [**114**], Lucas gives this result in a somewhat different form (again with our modifications and corrections):

THEOREM 6.2. *Let $F_n = 2^r+1$ $(r = 2^n)$ and $S_1 = 6 = V_2(2, -1)$. If we define the sequence $\{S_i\}$ by $S_{i+1} = S_i^2 - 2$, then F_n is a prime when $F_n | S_k$ for some k such that $r/2 \le k \le r-1$. Also, F_n is composite if $F_n \nmid S_k$ for any $k \le r-1$. Finally, if $F_n | S_k$ with $k < r/2$, then any prime divisor of F_n must have the form $2^{k+1}q+1$.*

One simple difference here from the earlier version is that $S_i = 2T_i$. But the more important difference is that primality is determined as soon as $F_n | S_k$ for $k \ge r/2$. We see, of course, that this easily follows from the statement of the previous version. For, if $k \ge r/2$, then

$$2^{k+1}q+1 > 2^{2^{n-1}}+1 > \sqrt{F_n},$$

which means that F_n must be prime.

A few weeks after the appearance of [**109**], Pepin [**138**] noted that Lucas' test was only sufficient for the primality of F_n, but not necessary. He then proved that F_n is a prime if and only if $5^{(F_n-1)/2} \equiv -1 \pmod{F_n}$. Pepin also noted that this could be made into a simple Lucas-like test by defining $T_1 = 5^2$, and $T_{i+1} = T_i^2$ $(i = 1, 2, \dots)$. In this case, F_n is a prime if and only if $F_n | T_{r-1} + 1$. Pepin also mentioned that the number 10 could also be used here instead of 5. In this case the values of the T_i's could be very easily evaluated as powers of 10. (In this remark Pepin demonstrates his ignorance of the actual use of these tests; for in usage, as noted by Lucas [**109**], we want to keep the values of our intermediate terms as small as possible. Hence, we compute them modulo the number being tested. Given this, no real advantage accrues on the use of 10 as opposed to 5.)

We should point out here that the test, often called "Pepin's test", today has a 3 in place of the 5 used above. Indeed, shortly after the appearance of Pepin's paper, Lucas ([**110**], [**114**, p. 313]) showed that any integer a can be used for the 5 in Pepin's test as long as the Jacobi symbol $\left(\frac{a}{F_n}\right) = -1$. Actually, though, it was Proth [**153**] in 1878 who mentioned the use of the 3. He seemed, however, to be unable to give a complete proof of his result (in spite of the existence of [**110**]), but Lucas [**116**] later supplied him with one. Thus, the test that we call Pepin's test is actually Proth's test with a proof supplied by Lucas.

It appears that the first time Lucas ever came to grips with the problem of the necessity of his tests was when he read Pepin's paper. With respect to Lucas and necessity, Lehmer [**82**] stated the following:

> A great deal of confusion exists in Lucas's writings about the exact enunciation and actual proofs of these tests for primality. Nevertheless, it is evident that Lucas was in possession of the facts needed to prove the sufficiency of his tests. The confusion arose from the fact that he was unable or unwilling to consider the necessity also.

As we shall see, Lucas was certainly able to consider the problem of necessity, but he was rather strangely unwilling to do so in his tests. On this subject we allow Lucas to speak (with some minor errors corrected) for himself [**110**, p. 165]:

> Meanwhile one should observe that if the method indicated by Father Pepin leads to a form more clear and precise to state, which thus becomes similar to that of the Theorem of Wilson, it is preferable to keep for application to the form which we have adopted. For, the application of these theorems rests on a hypothesis, that of considering as prime a number of certain form taken arbitrarily; it is more likely to suppose, on the contrary, that the number is composite, as seems to be indicated by the assertion of Father Mersenne. As a consequence, instead of postponing the verification to the extreme limit by the use of quadratic nonresidues, it would be more practical to use one of the $\varphi(2^{2^{n-1}})$ numbers which belong to the exponent $2^{2^{n-1}}$ for

> the modulus F_n considered as a prime. This cuts the process short by one half; but a direct search [for these numbers] is most difficult. Meanwhile one can assure oneself that by our previous procedure it suffices to demonstrate that (F_2, F_3, F_4) are primes by executing respectively 3, 6, 12 operations instead of the maximum number of 4, 8, 16 which would be required by the other method; as to the numbers F_5 and F_6, they are composite.

Lucas seems to be a little confused here. He claims, rightly, that numbers of the forms under consideration are more likely to be composite rather than prime; but his test, like that of Pepin, must go to the extreme limit to verify this. This could be improved if one could solve a very difficult problem, but he doesn't know how to do it. Anyway, he concludes that his test works better on F_2, F_3, F_4 than does Pepin's. Although this all seems somewhat self-serving, it is curious to note that by using a modern result we can offer some support to his point of view. For, since

$$F_n = (2^{2^{n-1}})^2 + 1 = (2^{2^{n-1}} - 1)^2 + 2(2^{2^{n-2}})^2,$$

we can use a result of E. Lehmer [**97**] to show that if F_n is a prime, then $F_n|S_t$, where $t \leq r - 5$ $(n \geq 4)$. Thus, if F_n is composite, Lucas' test could be modified to be possibly a little shorter than Pepin's. (But see the implementation of Pepin's test by Morehead and Western [**132**].)

We should add here that Lucas [**110**] used his test to show that F_6 is composite. Note that $F_6 = 2^{64} + 1$, one of the numbers Landry thought to be a prime. On learning of the composite nature of F_6, Landry set out to work on its factorization. After a labour of several months, and at the age of 82, he discovered by a rather clever method (see Williams [**187**]) adapted from his technique [**61**] for showing that M_{31} is a prime, that F_6 is divisible by 274177 and that the other factor is a prime (see [**124**]). In 1905, Morehead and Western [**131**] independently used Proth's test $(a = 3)$ to show that F_7 is composite and in 1909, Morehead and Western [**132**], working together in one of the great feats of hand calculation (and perhaps the first example of a distributed computation in number theory!), showed that F_8 is composite by using a modification of this same test. Robinson [**162**] has verified by computer the correctness of this incredible piece of numerical work.

In spite of Lucas' lack of interest in the idea of necessity in practical applications, he did provide some tests that were both necessary and sufficient for the primality of Mersenne numbers. We mention two of these taken from [**115**].

THEOREM 6.3. *For $p = M_{4q+3}$ to be a prime, it is necessary and sufficient that*

$$\left(\frac{1+\sqrt{5}}{2}\right)^{\frac{p+1}{2}} + \left(\frac{1-\sqrt{5}}{2}\right)^{\frac{p+1}{2}} \equiv 0 \pmod{p}.$$

THEOREM 6.4. *For $p = M_{4nq+2n+1}$ to be a prime, it is necessary and sufficient that*

$$\left(2^n + \sqrt{2^{2n}+1}\right)^{\frac{p+1}{2}} + \left(2^n - \sqrt{2^{2n}+1}\right)^{\frac{p+1}{2}} \equiv 0 \pmod{p}.$$

We can only assume that Lucas left his tests in these forms because he was not really very interested in using them in any application. For, if one examines the first of these, one sees that

$$S_{4q+2} = V_{(p+1)/2}(1,-1) = \left(\frac{1+\sqrt{5}}{2}\right)^{\frac{p+1}{2}} + \left(\frac{1-\sqrt{5}}{2}\right)^{\frac{p+1}{2}};$$

that is, the test given in §3 is both sufficient and necessary for primality, contrary to a remark of Lehmer [**76**, p. 444]. It is difficult to believe that this simple fact escaped the notice of Lucas, but he never mentioned it.

The second of these tests is even more interesting; for, if we put $q = 0$, $\alpha = 2^n + \sqrt{2^{2n}+1}$, $\beta = 2^n - \sqrt{2^{2n}+1}$, we see that the theorem states that $p = M_{2n+1}$ is a prime if and only if

$$M_{2n+1} | V_{(p+1)/2}(2^{n+1}, -1).$$

If we put $S_i = V_{2^i}(2^{n+1}, -1)$ (a trick Lucas often uses), then

$$S_1 = V_2(2^{n+1}, -1) = 2^{2n+2} + 2 \equiv 4 \pmod{M_{2n+1}}.$$

Thus, if we put $S_1 = 4$ and define $S_k \equiv S_{k-1}^2 - 2 \pmod{M_{2n+1}}$ $(k = 2, 3, \ldots)$, we see that M_{2n+1} is a prime if and only if $M_{2n+1} | S_{2n}$. This is the celebrated Lucas-Lehmer test for the primality of Mersenne numbers. Lucas seems to have known this test only as a sufficiency test (see [**124**]); it was Lehmer [**76**] who in 1930 showed that it was also necessary. Lucas had already done that in 1878, but his feelings about necessity blinded him to the result. This test has achieved a well-deserved fame. It is simple, elegant, and the means by which all of the large Mersenne primes have been verified. Many different proofs have been given for it (see [**186**] for references), the most recent being that of Rosen [**163**].

7. General primality tests

In [**106**] and [**114**, p. 302], Lucas presented the following theorem for primality testing.

THEOREM 7.1. *If $|\epsilon| = 1$, $p | U_{p+\epsilon}$, and $p \nmid U_k$ for all divisors k of $p+\epsilon$ such that $k < p+\epsilon$, then p is a prime.*

Lucas did not seem to be greatly interested in general primality tests, and did not, therefore, attempt to take this result much further. He did present in [**107, 110, 115**] and [**114**, pp. 309–311] a number of tests for numbers of the particular forms $N = Ar^n \pm 1$ $(r = 2, 3, 5)$, but these, of course, were such that $N + \epsilon$ could be easily factored. In spite of this favourable circumstance, several of his tests are not properly stated (in several of them he should have said something about limiting the size of A to $A < \sqrt{N}$, but did not), and they often contain other minor errors.

Lucas has been criticized for his many errors and omissions, especially by Carmichael [**17**]; however, there can be no doubt that Lucas understood very well the basic principles behind his results. It is true that his statements and proofs thereof often leave something to be desired; but, on the other hand, these deficits are usually very easily repaired and do not require the machinery that Carmichael thought necessary. Indeed, it is the authors' opinion that Carmichael muddied the waters rather than clarified them; certainly his presentation lacks the charm and infectious enthusiasm that so characterizes Lucas' work. It should be kept in mind that during the brief time during which Lucas was developing his seminal work on primality testing (according to his publication list [**118**] and that of Harkin [**45**], this is represented by 13 papers published between January of 1876 and January of 1878) he wrote at least 70 papers on many other subjects. Considering this immense outpouring of activity during such a brief time period, it is easy to forgive him for the few easily correctable deficiencies that we have mentioned.

The Theorem of Fermat states that if a is an integer and p is a prime such that $p \nmid a$, then $a^{p-1} \equiv 1 \pmod{p}$. Lucas knew that the strict converse of this theorem is false; for in [**110**] he gives the example $2^{37\cdot 73-1} \equiv 1 \pmod{37 \cdot 73}$. In fact, in [**110**] he singles out the following particular case of the theorem given at the beginning of this section as the proper converse of Fermat's theorem.

THEOREM 7.2. *If $a^k - 1$ is divisible by N for $k = N - 1$, and is not divisible by N for any k [$< N - 1$ and] a divisor of $N - 1$, then N is a prime.*

It is strange that later in [**114**, p. 302] he presented this result in a weaker form.

THEOREM 7.3. *If $a^k - 1$ is divisible by N when $k = N - 1$ and is not divisible by N for $k < N - 1$, then N is a prime.*

However, by the time he published his book, he had returned to the earlier result as the converse of Fermat's theorem (see [**122**, p. 441]). There is no evidence that Lucas ever used this result (even when $N - 1$ is easy to factor) to prove any numbers prime. This was possibly because he was unaware of a fast method for exponentiation (mod N) when the exponent is large, but this is difficult to believe because the technique is mentioned in Legendre's *Théorie des Nombres*, a book that Lucas must have read. Nevertheless, there is no indication in any of his published work that he knew of this device.

The next important contribution that was made to the problem of primality testing for a general N was a remarkable half page of theorems (without proof) given by Proth [**152**] in December of 1878. We state the two most interesting of these below.

THEOREM 7.4. *N is a prime if $N|a^n - 1$ for $n = N - 1$ and* $\gcd(a^n - 1, N) = 1$ *for every value of n such that $n|N - 1$ and $n < \sqrt{N}$.*

THEOREM 7.5. *Let $N = m2^k + 1$, where $2\nmid m$ and $2^k > m$. If the Jacobi symbol $\left(\frac{a}{N}\right) = -1$, then N is a prime if and only if $a^{(N-1)/2} \equiv -1 \pmod{N}$.*

Notice that the first of these is an improvement on Lucas' converse of Fermat's theorem, and the second is a generalization of the result of Pepin mentioned in §6. While no proofs of these results were offered, it is almost certain that Proth must have had correct proofs. He says in a letter to a Dr. Pein (see [**155**, p. 156]) that his proofs were long and that he did not have the time to copy them. The statements are correct as given, and it is difficult to believe that Proth would have had these correct statements if he did not have proofs, especially in the case of the first result which, as noted by Poulet in [**147**], even Dickson [**20**, p. 92] states incorrectly.

It is unfortunate that this work was never expanded upon; Proth certainly had intentions of doing this, but this remarkable, self-educated farmer was dead in January of 1879 at the age of 27 [**38**]. Also, because Proth was often guilty of making unsubstantiated claims (see, for example, [**151**]), it is likely that these results were ignored by many that might otherwise have taken up this work. As an interesting example of one of Proth's claims see [**154**] where it appears that he was aware of Gilbreath's conjecture [**50**] eighty years before Gilbreath. It seems that he thought he had a proof of it, something that still eludes us today.

The next important observation on this subject came from Pocklington [**144**] in 1914. He proves a result which can be stated as follows:

THEOREM 7.6. *If $N - 1 = q^n R$, where q is a prime, $q\nmid R$ and for some a we have $a^{N-1} \equiv 1 \pmod{N}$ and $\gcd(a^{(N-1)/q} - 1, N) = 1$, then each prime factor p of N is of the form $p = 1 + kq^n$.*

Notice that, although Pocklington was apparently ignorant of the work of both Lucas and Proth, he produced a result which is a generalization of their work. He was also aware that if one needed to evaluate $a^b \pmod{m}$ $(0 < a, b < m)$, then this could be done in $O((\log m)^3)$ elementary arithmetic operations; consequently, his test was computationally practical.

The computational model used by most number theorists today is that of "naive bit complexity". In this model, one can add two numbers of n bits using $O(n)$ operations, multiply two n-bit numbers in $O(n^2)$ operations, etc. Pocklington seems to have been the first to analyze a number-theoretic algorithm using this model. In his 1910 paper [**143**], he gave an algorithm to find $\operatorname{ord}_p a$, the least positive e such that $a^e \equiv 1 \pmod{p}$, when the prime factorization of $p - 1$ is known. He remarked:

> We notice that the labour required here is proportional to a power of the logarithm of the modulus, not to the modulus itself or its square root as in the indirect processes, and hence see that

> in the case of a large modulus the direct process will be much quicker than the indirect.

Perhaps Pocklington also deserves credit as the inventor of the randomized algorithm. In a 1917 paper [**145**], he gave an algorithm for computing square roots modulo a prime p which used $O((\log p)^3)$ steps and observed "the labour varies roughly as the cube of the number of digits in the modulus, and so remains moderate for large moduli." His algorithm depended on finding a quadratic non-residue of a special form by trial and error. He must have viewed this use of randomness as unsatisfactory, for he explained in a footnote:

> We have to do this [find the nonresidue] by trial, using the Law of Quadratic Reciprocity, which is a defect in the method. But as for each value of u half the values of t are suitable, there should be no difficulty in finding one.

In modern language, we would say that his method is a random polynomial-time algorithm, with each iteration having failure probability 1/2.

Pocklington's primality test was ignored for several years until D. H. Lehmer [**72**] realized its significance. Lehmer [**73**] also extended Pocklington's result to give the forms of the factors of N/δ in the case when $\delta = \gcd(a^{(N-1)/q}-1, N) > 1$, but his most important contribution was that of showing how these tests could be used in the actual practice of primality testing. In the course of doing this, he was able to demonstrate that large numbers which are not of simple forms like $Ar^n \pm 1$ could still be tested for primality fairly expeditiously. As an example we mention that in [**72**] Lehmer proved that $(10^{31}+1)/11$ is a prime. Lehmer's ideas will be discussed more fully in later sections, but in order to do this, we must now begin our discussion of the factoring problem.

8. Factoring

As has been mentioned earlier, the verification of a partial factorization of an integer N into $N = ab$ with $a, b > 1$ is a much simpler process than the verification that a prime value of N is indeed that. Nevertheless, the problem of factorization seems, at least empirically, to be much more difficult than that of primality testing. Also, of course, if a factorization algorithm is guaranteed, within a fixed bound on the time required, to find a nontrivial factor of N and one is not found, then we have a primality test for N. Thus, progress on the factorization problem has always seemed to be somewhat less dramatic than on the primality testing problem. For example, we have already seen that Lucas was able to verify the primality of a 39-digit number in 1876, but it was only in 1970 (and by use of a computer) that it was finally possible to factor a 39-digit number (see Morrison and Brillhart [**133**]). It should be emphasized here that when we speak, as above, of factorization of a number N, the assumption being made is that the value of N under consideration is not trivially factorable, i.e., N has at least two large prime factors.

The first interesting factorization idea is that of Fermat [**28**] in 1643. (See also Henry [**46**].) We note that if N is composite and odd, then $N = rs$ with r, s odd and $r < \sqrt{N}$. If we put $a = (r+s)/2$, $b = (s-r)/2$, then $\sqrt{N} < a < (N+1)/2$ and $N = a^2 - b^2$. Thus, in order to factor N, one need only search values of $x = a^2 - N$ as $a = \lfloor\sqrt{N}\rfloor + 1, \lfloor\sqrt{N}\rfloor + 2, \dots, (N-1)/2$ until one finds a perfect square. This process can be simplified by using differences; that is, if we define $x_i = (\lfloor\sqrt{N}\rfloor + i)^2 - N$; $y_i = 2\lfloor\sqrt{N}\rfloor + 1 + 2i$, then $x_{i+1} = x_i + y_i$, $y_{i+1} = y_i + 2$.

Fermat also points out that one can shorten the search somewhat by rejecting any of the values of x whose last few digits could not be those of a perfect square; for example, if the value of x is 46619, then because the last two digits are 19 and there is no b such that $b^2 \equiv 19 \pmod{100}$, then x cannot be a perfect square. Thus, Fermat was aware of an idea that we will later call modular exclusion. It is this technique that Landry and Le Lasseur (see [**64**]) rediscovered and used to effect their factorizations. We emphasize that the method as presented here is really only going to be of some value when r and s are relatively close in value. If s is very much larger than r, then the technique is very ineffective.

In 1798, Legendre proposed finding square-free values of a such that there exist integers x, y, z for which

$$x^2 - kNz^2 = ay^2. \tag{8.1}$$

Notice that regardless of the value of k, if p is any prime divisor of N, then the Legendre symbol $\left(\frac{a}{p}\right) = 1$. For a given a such values of p can only have certain linear forms; for example, if $a = 2$, then $p \equiv \pm 1 \pmod 8$; if $a = 3$, then $p \equiv \pm 1 \pmod{12}$; and if $a = 5$, then $p \equiv \pm 1 \pmod 5$, etc. Legendre tabulated these forms for various values of a.

If, for example, we could solve (8.1) for $a = 2, 3, 5$, then if $p | N$, we must have $p \equiv 1, 49, 71, 119 \pmod{120}$. As any prime ($\neq 2, 3, 5$) can be congruent modulo 120 to $\varphi(120) = 32$ possible values, we see that this knowledge allows us to reduce the amount of possible trial division by a factor of 8. In general, if we have r such independent conditions, the number of trial divisions of N can be reduced to about $2^{-r}\pi(\sqrt{N})$ trials, where by $\pi(x)$ we mean the number of primes which do not exceed x.

In 1929, D. N. Lehmer [**94**] and D. H. Lehmer (see [**89**]) produced a set of factor stencils which could be used to factor numbers up to $48593^2 = 2361279649$. For each value of a up to 250 in absolute value, a stencil was created as a matrix of 100 rows and 50 columns. A hole was punched in any spot corresponding to those of the first 5000 primes for which a is a quadratic residue; that is, $\left(\frac{a}{p}\right) = 1$ for these particular primes p. (D. N. Lehmer considered 1 to be a prime; thus, his first 5000 primes begin at 1 and end at 48593). The stencils were approximately 20cm by 43cm, and were provided in a wooden box with a glass cover. To use the stencils (see [**95**]), one first discovers the values of a such that $|a| < 250$ and (8.1) is solvable. Then the corresponding stencils are stacked on top of each other and placed on top of the glass cover. An electric light is then introduced into the box to shine through the holes (an early example of a user interface!). These holes correspond to the only possible primes that could divide N. In 1939, Elder [**22**] put these stencils on 2000 Hollerith cards (i.e., IBM punch cards).

There may be a question about the use of k in (8.1). Normally one would think that k could simply be put equal to one, but it is often convenient to use other values of k, especially if $\left(\frac{N}{p}\right) = -1$ for many of the smaller primes, in order to solve (8.1). Today we call these values of k 'multipliers'. They are used rather extensively in certain factoring techniques.

We are now left with the problem of how to solve (8.1). Legendre suggested using the continued fraction expansion of $\sqrt{kN}$. For if

$$\sqrt{kN} = q_0 + \cfrac{1}{q_1 + \cfrac{1}{q_2 + \cfrac{1}{\ddots + \cfrac{1}{q_{n-1} + \cfrac{1}{\theta_n}}}}} \tag{8.2}$$

where $\theta_n = (P_n + \sqrt{kN})/Q_n$ $(P_n, Q_n \in \mathbb{Z})$, then $0 < Q_n < 2\sqrt{kN}$ $(n > 0)$, and we can use the values of $q_0, q_1, \dots, q_{n-1}$ to compute easily integers A_{n-1}, B_{n-1} such that

$$A_{n-1}^2 - kNB_{n-1}^2 = (-1)^n Q_n. \tag{8.3}$$

Thus, a can be taken as the square-free part of $(-1)^n Q_n$. For the purpose of this technique, however, we shall see that there is no need for the coefficient of kN in (8.1) to be a perfect square. We would be content to know that there exists a value of x such that

$$x^2 \equiv a \pmod{N}. \tag{8.4}$$

Such values of a are called quadratic residues of N. The determination of small (in absolute value) quadratic residues of N is still considered to be a very difficult task. We shall discuss this further in what follows.

We conclude this section by mentioning a very important observation of Gauss [**30**, Art. 320]. Suppose $f(x), g(x) \in \mathbb{Z}[x]$, and we wish to solve the Diophantine equation

$$f(x) = g(y) \tag{8.5}$$

for integers x and y. (Gauss only treats the equation $a + mx = y^2$, but the idea applies more generally.) Select different (exclusion) moduli $E_1, E_2, \dots, E_r$. We must have $f(x) \equiv g(y) \pmod{E_i}$ for $1 \le i \le r$ for any solution (x, y) of (8.5). For each value, then, of $y = 0, 1, 2, \dots, E_i - 1$, find the possible residues classes for x modulo E_i. Combine these to find the permissible residue classes for x modulo the least common multiple of the E_i's.

To use Gauss' example, suppose we want to solve $22+97x=y^2$. We consider $E_1=3$, $E_2=4$, $E_3=5$, $E_4=7$. Permissible values of x modulo E_1 are $0,2$; modulo E_2 are $2,3$; modulo E_3 are $1,2,4$; and modulo E_4 are $0,1,4,6$. It is easy to see that the least integer satisfying all of these is 11, and in fact $22+97\cdot 11=33^2$. Notice that with one exception the E_i-values here are distinct primes; thus, on using a new E_i-value, we eliminate about half of the previous candidates for n in this problem. In subsequent sections we shall illustrate the importance of Gauss' idea of exclusion. We should mention that Gauss had several other ideas for factorization techniques (see [**30**, Arts. 329–334]), but only the exclusion idea has turned out to be of real practical importance.

9. Seelhoff and Pepin

In this section we discuss some of the important work on factorization techniques done by two individuals whose contributions appear to have been largely ignored. We use the term 'appear' here because it is not entirely clear whether or not their work was taken over and expanded by others. In any event their ideas were important and merit some attention.

Between the years 1885 and 1887 Paul Seelhoff published a number of papers on factorization and primality testing. Some of this work was not of very high quality (we have already mentioned his less than adequate proof of the primality of M_{61}), but at least one of his papers contained some interesting ideas.

In [**166**] he presented a method of factorization which first requires that for a given N we find a number of different values of a such that (8.4) holds. He attacked this problem by putting $N=\omega^2+r$, where $\omega=\lfloor\sqrt{N}\rfloor$. If p is any prime such that $\left(\frac{N}{p}\right)=1$, solve $\rho^2\equiv N \pmod{p^k}$ and put $\alpha=\omega\pm(\rho+yp^k)$. We then have $N=(\omega-\alpha)^2+b$, where $b=\alpha(2\omega-\alpha)+r$ and $p^k|b$. The point here is that, if the $|\alpha|$-values are near p^k, then the corresponding b-values will tend to be of size about $2p^k\sqrt{N}$, and since they are already divisible by p^k, they should be somewhat easy to factor. On putting $b=at^2$, we have a value of a satisfying (8.4).

Seelhoff suggested that for values of N of about 15 digits, the values of p^k to try are the primes from 7 to 97 with $k\le 2$; 5 with $k\le 4$; 3 with $k\le 6$; and 2 with $k\le 10$. (Of course, only those primes p for which $\left(\frac{N}{p}\right)=1$ would be used.)

His next step is best illustrated by using an example. He selected for that purpose $N=7\cdot 2^{34}+1$ and took as his collection of possible values of p the set

$$\{2,5,7,11,19,31,37,47,53,67,71,93,127\}.$$

We have $N=346783^2+635200$.

After several trials he found that for

$$\alpha=1950 \quad (\alpha\equiv 0 \pmod 5, \alpha\equiv 581 \pmod{37^2});$$

$$\alpha=143432 \quad (\alpha\equiv 3 \pmod{11}, \alpha\equiv 14400 \pmod{127^2});$$

$$\alpha = -3836 \quad (\alpha \equiv 3 \pmod{11}, \alpha \equiv 271 \pmod{37^2})$$

one has

$$\begin{aligned} N &= 344833^2 + 2 \cdot 7 \cdot 11 \cdot 2960^2 \\ &= 203351^2 + 7 \cdot 106172^2 \\ &= 350619^2 - 2 \cdot 11 \cdot 11026^2. \end{aligned}$$

From the first two of these he obtained

$$11 \cdot 832082029^2 \equiv 2 \cdot 150479740^2 \pmod{N}.$$

He combined this with the third and found

$$5045995048467^2 \equiv 26380527979530^2 \pmod{N}$$

or

$$N | (5045995048467 - 26380527979530) \times (5045995048467 + 26380527979530).$$

In fact, $\gcd(5045995048467 - 26380527979530, N) = 317306291$ and $N = 379 \cdot 317306291$. Those who are familiar with modern factoring techniques will be struck by the similarity of this method to those which make use of the idea of combining congruences, such as the Quadratic Sieve and CFRAC (see below). Of course, we have more sophisticated methods now for effecting the divisions needed to determine the a-values (indeed, no divisions need be performed if the techniques of the Quadratic Sieve are employed) and for assembling the dependencies[5] in order to get values of x and y such that $x^2 \equiv y^2 \pmod{N}$, but Seelhoff seems to have understood the basic idea behind some of our modern methods back in 1886. He used his method [**167, 170**] to factor some numbers of the form $A2^n + 1$, as he was interested in these as possible factors of Fermat numbers.

Father Théophile Pepin S.J. is best known today for the primality test which bears his name. However, Pepin did a lot of work on factoring. His most useful techniques, one of which we will discuss here, was pretty well developed in 1890 (see [**139**]), but in a series of papers [**140, 141, 142**] (see Statuti [**174**] for a partial bibliography of Pepin's work) he continued to refine and apply it to various numbers of the form $(a^n - 1)/(a - 1)$. Numbers of this type were of particular interest to Pepin because of their importance in the study of perfect numbers.

To factor $N = (a^n - 1)/(a - 1)$ with n odd, we first notice that N is odd and that if $p | N$, then $p \equiv 1 \pmod{2n}$. We now consider the more general problem of factoring any odd N whose factors must be of the form $2nx + 1$ with n odd. We first demonstrate that N has no prime factor less than or equal to m (by trial division, say).

[5]By a *dependency* we mean a collection of possible values for a whose product is a perfect integral square. In the previous example the product of $-2 \cdot 7 \cdot 11$, -7, $2 \cdot 11$ is the perfect square $(2 \cdot 7 \cdot 11)^2$.

Suppose $N = s_1 s_2$; since $s_2 > m$, we have $N/(ms_1) > 1$; hence $0 < s_1 - m < (s_1 - m)N/(ms_1)$ and $s_1 + s_2 = s_1 + N/s_1 < m + N/m$.

Furthermore, by using the results at the beginning of §8, we also see that $s_1 + s_2 > 2\sqrt{N}$. If we put $L = m + N/m$, $s_1 = 2nx + 1$, $s_2 = 2ny + 1$, we get

$$\frac{\sqrt{N} - 1}{n} < x + y < \frac{L - 2}{2n}. \tag{9.1}$$

Also, since $4n^2xy + 2n(x+y) = N - 1 = 2nN'$, it follows that if $N' = 2nN'' + r$, then

$$x + y = r + 2nz. \tag{9.2}$$

Putting $x - y = u$ and noting that $4xy = (x+y)^2 - (x-y)^2$, we get

$$(2nz + r)^2 - u^2 + 4z = 4N''.$$

If $2|r$, we can replace u by $2u'$ and divide by 4. In any event, we get an expression of the form

$$az^2 + 2bz + c = u^2, \tag{9.3}$$

where a, b, c are known integers and the value of z is bounded by using (9.1) and (9.2). So far, all this really represents is a more sophisticated version of the difference of squares method of Fermat. The equation (9.3) is a Diophantine equation which can be solved by using Gauss' method of modular exclusion.

As one of several examples of his method, Pepin considered $N = (5^{13} - 1)/4 = 305175781$. Since any prime divisor of N is of the form $26x + 1$ and since there is no such prime divisor less than 1000, he was able to say that if $N = (26x + 1)(26y + 1)$, then $1343 < x + y < 5946$. Furthermore, $26xy + x + y = (N - 1)/26 = 11737530$; hence, $x + y = 26z + 12$ and

$$169z^2 + 157z - 451407 = u^2 \tag{9.4}$$

with

$$51 < z < 229. \tag{9.5}$$

Using $8, 9, 5, 7, 11, 13$ as exclusion moduli, he found, by using a technique like the Sieve of Eratosthenes, that there are no candidates for z satisfying (9.4) and (9.5). Thus, N is a prime.

Pepin went on to factor many numbers by using this technique or some variant thereof. If one examines [**62**] in the light of [**64**] it seems that Landry must have used a technique very like this to effect his factorizations in [**63**]. Indeed, in 1880 Landry [**65**] essentially proved that if p is any prime and $N \equiv a^2 - b^2 \pmod{p}$, then there are precisely $(p + (\frac{N}{p}))/2$ possible values of a modulo p. This, of course, is why the exclusion principle works; each additional prime exclusion modulus cuts the number of possible values for a to one half their former number. It is possible that Pepin's ideas did not receive the currency they merited because of the rather obscure journals in which they were published. The method, however, became the most important factorization technique until well after the arrival of computers.

10. An example of Cole

In §5 we mentioned that Fauquembergue (and possibly Lucas) had shown by 1891 that M_{67} is composite. In 1903, Cole [**19**] succeeded in finding the factors of M_{67}. This marvelous feat of hand calculation has been justly celebrated through the story related by Bell [**9**].

> I should like here to preserve a small bit of history before all the American mathematicians of the first half of the twentieth century are gone. When I asked Cole in 1911 how long it had taken him to crack M_{67}, he said "three years of Sundays." But this, though interesting, is not the history. At the October, 1903, meeting in New York of the American Mathematical Society, Cole had a paper on the program with the modest title *On the factorization of large numbers.* When the chairman called on him for his paper, Cole — who was always a man of very few words — walked to the board and, saying nothing, proceeded to chalk up the arithmetic for raising 2 to the sixty-seventh power. Then he carefully subtracted 1. Without a word he moved over to a clear space on the board and multiplied out, by longhand, $193,707,721 \times 761,838,257,287$. The two calculations agreed. Mersenne's conjecture — if such it was — vanished into the limbo of mathematical mythology. For the first and only time on record, an audience of the American Mathematical Society vigorously applauded the author of a paper delivered before it. Cole took his seat without having uttered a word. Nobody asked him a question.

(Given Bell's accuracy on other matters, aspects of this story may be apocryphal.) What seems to be much less known is how Cole actually achieved this factorization. He appears to have been unaware of Fauquembergue's work, but he did take seriously Lucas' comment in [**122**] concerning the composite character of M_{67}. It was this that caused him to seek its factors. He began by finding a collection of quadratic residues of M_{67}. These he listed as

$$2, -3, -7, 13, -23 \cdot 53, 37, 41, 61, -67, -71, 23 \cdot 83, 89, 97, 101, \ldots$$

etc. He did not say exactly how he discovered these, but as he refers to the work of Seelhoff mentioned in §9, he likely used Seelhoff's technique. He made the important observation (refined later by Kraitchik) that if a prime r (taken with the proper sign) is a quadratic residue of N, then there can be at most $(r+3)/4$ possible values of a modulo r satisfying $N \equiv a^2 - b^2 \pmod{r}$. Notice that this is an improvement on the upper bound of $(r+1)/2$ mentioned in §9.

Next, by using the same reasoning as Euler concerning M_{31}, he noted that any prime factor of M_{67} must be of the form $536k + 1$ or $536k + 135$. Since -3 is a quadratic residue of M_{67}, these forms can be further refined to $1608k + 1$ or $1608k + 1207$. By using these forms and some of his other quadratic residues of N, he eliminated (but not entirely convincingly) all possible prime factors < 16000000.

He then attempted to solve $N = a^2 - b^2$. He found that

$$\begin{aligned} a &\equiv 671 \pmod{67^2}, \\ a &\equiv 0 \pmod 8, \\ a &\equiv 1, 4^* \pmod 5, \\ a &\equiv 1, 3^* \pmod 7, \\ a &\equiv 0, 1^*, 12 \pmod{13}, \\ a &\equiv 10, 37, 46^*, 64 \pmod{81}. \end{aligned}$$

There are 48 possible residue classes here for a modulo $1323536760 = 8 \cdot 5 \cdot 7 \cdot 13 \cdot 81 \cdot 67^2$, but only one is the correct one. After what must have been a great deal of work, Cole selected that class indicated by the asterisks; that is, $a = 1323536760x + 1160932384$. This was used to find an equation like (9.3). The exclusion moduli $23, 37, 41, 53, 61$ were then used to narrow x down to 287 from which Cole obtained $a = 381015982504, b = 380822274783$ and his factorization.

11. The sieve

During the years 1894–1897, Lawrence [**66, 67, 68**] rediscovered the difference of squares technique and some of Pepin's refinements to it. In fact in [**68**], he was able to use his method to prove that the factors of $10^{29} - 1$ and $10^{25} - 1$ found by Looff [**100**] are primes. However, the most important contribution of Lawrence occurs in [**67**], where he discussed a means of mechanizing the process of solving (9.3) by the use of exclusion moduli. He first described a technique for doing this that made use of moveable paper strips, but it is his second idea which is really of interest to us here. He suggested the construction of a machine in which gears of m teeth would be used for each exclusion modulus m; thus, each such gear would represent (as it rotates) an endless paper strip. The gears (each with the same tooth size) would be driven by the same driving gear, and, as they would be of differing diameters, would have to be mounted on different shafts. The teeth on each m-toothed gear would be numbered $0, 1, 2, \dots, m-1$, and a brass stud would penetrate through it at the point(s) of an acceptable (mod m) residue (one for which (9.3) could hold for $z \pmod m$)). When studs from each of the gears were all in contact a circuit would be completed and a bell would ring or the machine would stop, indicating to the operator that a solution of (9.3), modulo the least common multiple of the exclusion moduli, had been detected. Of course, in order to determine the z-value, a count would have to be recorded of the number of rotations of the driving gear.

Several other details for the construction of this machine are also provided, but it seems that Lawrence never actually built such a device, being instead content simply to describe it in principle. His description is not complete, as he admitted, but it is sufficient to convince anyone that the basic idea of mechanizing the modular exclusion principle for solving (9.3) is a real possibility.

This idea of mechanizing the solution of (9.3) seems to have been forgotten until 1912, when a flurry of activity, apparently inspired by Andre Gérardin, began to produce some results. This may have originated through Gérardin's publishing in 1910 a French translation of Lawrence's paper [**67**] in his peculiar journal/newsletter called *Sphinx-Oedipe.* The first machine announced was that of Maurice Kraitchik in February of 1912 [**37**]. Kraitchik had read the version of Lawrence's paper which was published in *Sphinx-Oedipe* because he [**52**] refers to it in an earlier issue. Furthermore, the machine, when it is described in [**39**] (or in somewhat less detail in [**53**, pp. 43–44]), is rather similar to that of Lawrence, except that the gears representing the various exclusion moduli are each mounted on the same shaft. Also, he mentions the possibility of putting holes in the gears (instead of studs) and shining light through onto a screen. When a spot of light appears on the screen a solution has been detected. (Holes that did not represent a possible solution, of course, would have to be plugged.) Curiously, however, Kraitchik never refers to Lawrence's work in this connection.

In March of 1912, Gérardin [**37**] refers to machines of Pierre Carissan, Kraitchik, and himself. Kraitchik's machine is described in some detail in April of 1912, but the descriptions of Carissan's (built by his brother Eugène-Olivier) and Gérardin's machines, given respectively in [**16**] and [**40**], are rather vague. It appears that each of these machines existed only as a prototype and that none ever produced any important result. Later in 1913–14, E.-O. Carissan built another model of such a mechanism and was so encouraged by its performance that he decided to have a precision device constructed. Unfortunately, World War I intervened, and his device was not completed until 1919. This seems to have been the first automatic sieve mechanism to have ever been successfully constructed. A full and convincing description of it is given in [**16**]. It is also mentioned again by d'Ocagne [**135**] in his lengthy paper on calculating machines. Indeed it is rather a telling point that neither Kraitchik's nor Gérardin's machines are described by d'Ocagne.

Carissan's machine was made up of 14 concentric metallic rings representing the exclusion moduli: $19, 21, 23, 26, 29, 31, 34, 37, 41, 43, 47, 53, 55, 59$. A pinion gear whose axis was parallel to the plane of the rings drove all of them at the same rate. Each ring had a number of equidistant studs on the upper side. The number of such studs was equal to the appropriate exclusion modulus. Those studs representing a possible solution were covered by a nonconducting cap. When these passed under a 'line of investigation', a circuit would be completed when a solution (each stud under the line was capped) appeared and a sound would be made in a telephonic receiver placed on the ear of the operator, alerting him to the solution.

While this device was hand-driven, plans were also made to construct a machine-driven model; however, this seems never to have been done. The hand-driven device ran through the numbers at the rate of 35-40 per second. A counter (up to 1000000) would keep track of how many numbers had been sieved. As an example, we mention that the machine required only 15 minutes to prove that M_{31} is a prime.

Unfortunately, Pierre Carissan died in 1923, and E.-O. Carissan died in 1925, events that likely contributed greatly to their pioneering sieve not receiving the attention that it deserved. The machine itself languished in a drawer in the Observatoire de Floirac (near Bordeaux) for many years. It was inspected there by the authors several times. Recently, France Silber, a daughter of E.-O. Carissan, took possession of the machine, and there is hope that it will soon be donated to the collection of the Conservatoire National des Arts et Métiers. We are extremely grateful to Marie-Josèphe Salefran-Carissan and France Silber for informing us of the sieve's location.

The next individual to look at the problem of constructing number sieves was D. H. Lehmer [**74**]. Lehmer was unaware of the work of the Carissans and remained so until 1989. In fact in his paper [**81**, p. 663] on sieves he states the following:

> The literature contains at least two descriptions of such machines which, though impractical in their design, are theoretically interesting. As far as the writer knows, the first successful machine of this type was constructed by him in 1927.

Lehmer was referring to the machines of Lawrence and Kraitchik here. In fact it is rather strange that he gives so very little credit to them, especially since his photoelectric sieve of 1932 represents in many ways the fruition of their ideas; however, Lehmer had determined that neither of these earlier mechanisms, as described, would actually work and therefore decided to maintain a discrete silence concerning them and their creators. His device, described in [**79, 81**], would search through numbers at the rate of 5000 per second. In constructing this remarkable machine, Lehmer pushed contemporary technology about as far as he could. For a charming description of the function of it we refer the reader to the paper [**96**] of Lehmer's father, D. N. Lehmer.

As it would take considerable space to discuss all of Lehmer's work on sieves, we simply provide below a list of all sieves known to the authors at present. All, but those of Carissan and Gérardin (see [**43**]) and the last four are Lehmer's. For a more detailed account on sieves and their history we refer the reader to the paper of Stephens and Williams [**175**] and Patterson [**137**].

Sieve Systems

Machine	Year	Rings	Trials/Sec
E. Carissan	1919	14	35-40
Bicycle Chain	1926	19	50
Optical Gears	1932	30	5,000
16 mm Movie Film	1936	18	50
A. Gérardin	1937	?	< 10
SWAC†	1950's	?	1,450
IBM 7094†	1960's	21 or 22	100,000
DLS-127	1966	31	1,000,000
DLS-157	1969(?)	37	1,000,000
ILLIAC IV†	1960's	64	15,000,000
Shift Register	1975	42	20,000,000
UMSU	1983	32	133,000,000
OASiS	1989	16 variable	215,000,000
Sun4/280†	1989	32	2,000,000
SSU	1991	30	200,000,000

† Denotes software running on a general purpose computer.

It may interest the reader to learn that up to 1970 these sieve methods were still the fastest techniques known for factoring. In fact the DLS-127 was used to factor a 33-digit divisor of $2^{136}+1$ in 2600 hours (see [**13**, p. 1vi]). In spite of the successes of these cheap, easy-to-construct, and useful mechanisms, their history has been characterized by a lack of interest or neglect. Indeed, in speaking about the sieve problem in his paper [**88**, p. 456] on the history of the sieve process, Lehmer stated:

> It's very esoteric, of course, and since I am practically the only man working in this field you can see how widespread the interest in it is.

The situation is not much different today, and that is a great pity; there's a lot you can do with a sieve. Possibly the current easy availability of off-the-shelf computer chips and components will cause modern researchers to begin to investigate the problem of construction and utilization of number sieves. We certainly hope so.

12. Kraitchik and Lehmer

In 1911, only one number from Landry's original table of factorizations of $2^n \pm 1$ ($n \leq 64$) remained to be factored; this was $N = (2^{61}+1)/3$. Gérardin [**36**] posed the factorization of this number as a contest in May of that year. A then new contributor to *Sphinx-Oedipe*, Kraitchik, first suggested the following conjecture:

CONJECTURE 12.1. *If $2^n - 1$ is prime, then $(2^n + 1)/3$ is also prime.*

Later [**51**] he found that there were no factors of N between $\lfloor\sqrt{N}\rfloor = 876706528$ and 61003051. Finally, Poulet [**146**] proved that N could have no prime factor less than 68200400, and thus N is a prime. Poulet also pointed out that Kraitchik's conjecture is false, as it fails for $n = 89$. Much more recently a revised form of this conjecture has been suggested by Bateman, Selfridge, and Wagstaff [**6**]. Their conjecture reads as follows:

CONJECTURE 12.2. *If two of the following statements about an odd positive integer p are true, then so is the third one:*

- $p = 2^k \pm 1$ *or* $p = 2^{2k} \pm 3$.
- $M_p = 2^p - 1$ *is a prime.*
- $(2^p + 1)/3$ *is a prime.*

This has been verified for all primes p up to 100000. Incidentally, it was in [**6**] that the number $(2^{2539} + 1)/3$, mentioned in §1, was identified as a probable prime.

This debut of Kraitchik into the factoring business was followed over ten years later by the publication of several books containing a great deal of information about number theory in general and factoring in particular. In [**53**, Chapter 6], he described in detail how he factored the 19-digit number $2^{61} + 2^{31} + 1$. He used a method which is little more than a minor refinement of the technique of Pepin and Lawrence, but does not attribute the technique to either of them. This is peculiar because, as we have already seen, he was aware of Lawrence's work, and he must have known about Pepin's work through a description of it by Gérardin [**35**] published in *Sphinx-Oedipe*. Furthermore, Cole's paper [**19**] had also appeared in French translation in 1910 in *Sphinx-Oedipe*.

Kraitchik also stated in [**53**] that he proved that $(10^{19} - 1)/9$ is a prime, but this had already been done earlier by Hoppe [**48**] in 1918. Kraitchik became quite adept at using a modified version of Lawrence's technique of moveable strips in order to effect his factorizations. As a result of this skill, he was able to factor several large numbers such as: $2^{68} + 1, 2^{53} + 2^{27} + 1, 2^{122} + 1, 2^{77} - 1$.

In 1851, Looff [**100**] published a table of factors of numbers of the form $10^n - 1$. These same factorizations for $n \leq 42$ and with some addenda were reported later by Reuschle. There were several holes in Reuschle's tables, namely for $n = 17, 19, 23, 25, 27, 29, 31, 33, 34$ (because of 17), $35, 37, 38, 39, 40, 41$, where the factorizations were incomplete. Le Lasseur (see Lucas [**117**], Brocard [**14**]) factored $(10^{17} - 1)/9$ in or before 1879; we have already seen that Lawrence and Hoppe had found $(10^{19} - 1)/9$ to be prime in 1918. This was how matters stood until 1926 when Lehmer [**72**] turned his attention to the factorization of the 15-digit value of $N = (10^{27} - 1)/(3^5 \cdot 37 \cdot 757 \cdot 333667)$. He was able to use his primality testing ideas to prove N a prime. He did this by first factoring

$N-1 = 2 \cdot 5 \cdot 3^3 \cdot 31249 \cdot 52189481$ and then showing that $3^{N-1} \equiv 1 \pmod{N}$ and $\gcd(3^m - 1, N) = 1$, where $m = (N-1)/52189481$. Hence any prime divisor of N must be of the form $52189481k + 1$. Since $\sqrt{N} < 20984153$, N must be a prime. He also showed that $(10^{19}+1)/11$ is a prime and computed the value of 3^{N-1} for $N = (10^{20}+1)/(10^4+1)$ and $N = (10^{23}-1)/9$. In both cases he obtained values of $3^{N-1} \pmod{N}$ different from 1 and concluded that both of these numbers must be composite. Later, using his bicycle chain sieve, he [**74**] was able to factor $(10^{20}+1)/(10^4+1)$.

Lehmer's most impressive computation [**78**], however, was rather modestly announced in June of 1932. Kraitchik [**55**, p. 142] stated in 1926 that he had shown in 1922 (by use of the Lucas-Lehmer test) that the last number in Mersenne's list, M_{257}, is composite. However, he provided no details of this Herculean labour except to say that it required the months from August to December of 1922 to complete the work. Kraitchik's book appeared after July of 1926; in June of that same year Lehmer (unaware of Kraitchik's work) began the task of investigating the 78-digit M_{257}. He made use of a 10-digit calculating machine and his cross division process [**70**], a modification of an idea of his father [**91, 93**], to effect the computations. He also checked each value of S_i by repeating the calculations modulo 1001. Here we use $S_1 = 4$, $S_i \equiv S_{i-1}^2 - 2 \pmod{M_{257}}$, for $0 \le i < 257$.

On March 6, 1927 he found that $S_{256} \neq 0$. Lehmer did not publish this finding, although he did publish in [**71**] his discovery that M_{139} is composite; indeed, he began his work on M_{257} less than a month after he completed his work on M_{139} and he used the same techniques. Possibly, he was forestalled by the appearance of [**55**] or he was unsure of the accuracy of his calculations. We know nothing more about this until we read in Kraitchik [**56**, p. 83] that Lehmer found that he had made an error in the 47th term (in fact, it was S_{48} that was wrong) of his calculations. According to the dates Lehmer provides in his journal, this meant that all the work from August of 1926 to March of 1927 had been in vain. Lehmer had also asked Kraitchik for a copy of his computations; however, Kraitchik said that he had sent them to Gérardin without keeping a copy for himself. On being requested to return them, Gérardin declared that owing to the state of his health, he could not at that time recover Kraitchik's calculations from among the thousands of kilograms of archives in his possession. These calculations have still not been located to this day. Later, when Lehmer visited Gérardin, he discovered why a search of Gérardin's papers might prove hazardous to anyone's health. The entire office was full of papers piled up everywhere and in no apparent filing system.

Actually, it was Lehmer's wife Emma who had detected the error in S_{48}. At Lehmer's request she began in 1930, starting at S_{47}, to redo the calculations. The values of S_i were now checked modulo 10^8+1 and modulo 10^9+1. Unfortunately, this work in Lehmer's journal is not dated as frequently as the earlier work; we only know that S_{128} had been computed on September 17, 1930. Both Lehmer and his wife shared the computational chores until S_{256} was computed. This time, as noted later by Robinson [**162**], the calculation of S_{256} was correct. Lehmer was able to announce in June of 1932 that M_{257} is composite. According to Reid [**158**] this computational feat required over 700 hours of work, whereas the determination of the composite character of M_{149}, also announced in [**78**], had required only 70 hours.

Kraitchik and Lehmer went on to improve their techniques and produce other factorizations. A glance at the tables of factors of $2^n \pm 1$ in Kraitchik [**56**, pp. 84–88] (but see Lehmer [**84**, pp. 29–30]), [**58**, pp. 12–13; 38–39], [**59**] and [**85**] will give the reader some idea of the strength of both their methods and their endurance. By using his sieves, Lehmer was able to achieve his factorizations more rapidly than Kraitchik; for example, see [**80**], where Lehmer used his photoelectric sieve to factor 5 difficult numbers, one of 21 digits, in a matter of a few hours. Naturally, both of these authors made errors in their calculations from time to time; this is only to be expected, but Kraitchik made a serious mathematical error which, taken with some simple calculation mistakes, led him to the belief that certain numbers such as $(2^8+1)(2^{120}+1)/((2^{24}+1)(2^{40}+1))$ were prime when they are not.

13. Quadratic Residues and a "Fallacious Principle"

In many of his attempts to factor (or primality test) an integer N, Kraitchik was first very concerned about determining a collection of quadratic residues of N. His techniques for finding these quadratic residues are fully described in [**55**, pp. 123–132] and again in [**58**, pp. 134–138]:

For a given N find small integers (say primes ≤ 50) $r_1, r_2, r_3, \ldots$ for which $\left(\frac{N}{r_i}\right) = 1$. Then find values of x such that the difference $R = \pm(x^2 - N)$ is divisible by $r_1, r_1^2, r_1^3, \ldots, r_2, r_2^2, r_2^3, \ldots, \ldots r_1^2r_2^2, r_1^2r_3^2, \ldots, r_i^2r_j^2, r_1^2r_2^2r_3^2, \ldots$.

Also, take for the values of x those that are close in value to $\sqrt{N}$. Thus, we will know in advance certain factors of R and likely not have much trouble in putting R in the form ay^2, where a is square-free. According to Kraitchik the factors of R will fall into three categories:

(i) small factors of about the same size as the r_i, but which were not included among the r_i-values;
(ii) factors of value about $\sqrt[4]{N}$;
(iii) large factors.

The factors in the first category can be added to the list of possible r_i-values; those of the second category can be used to search for residues of x such that $\pm(x^2 - N)$ is divisible by the squares of those factors; and those of the third category can be used for finding values of x such that $\pm(x^2 - N)$ is divisible by these factors, but not their squares.

He provides the following example. Consider $N = 535 \cdot 2^{30} + 1$. One finds $N = 1054929^2 - 2 \cdot 2909 \cdot 9620^2$; one then finds $N = 327679^2 + 2 \cdot 2909 \cdot 8960^2$. (Note that $1054924 - 327679 = 2909 \cdot 250$.) Since both $-2 \cdot 2909$ and $2 \cdot 2909$ are quadratic residues of N, their product must be, and since this is -5818^2, we see that -1 is a quadratic residue of N. One finds further that

$$\begin{aligned} N &= 718721^2 + 41 \cdot 353 \cdot 2000^2, \\ N &= 714485^2 + 2 \cdot 13 \cdot 37 \cdot 353 \cdot 434^2, \\ N &= 757671^2 + 13 \cdot 37 \cdot 41 \cdot 140^2. \end{aligned}$$

(Note $718721 - 714485 = 12 \cdot 353$, $757671 - 718721 = 41 \cdot 950$.) Thus $-41 \cdot 353$, $-2 \cdot 13 \cdot 37 \cdot 353$, $-13 \cdot 37 \cdot 41$ are each quadratic residues of N. Their product is -2 times a perfect square; hence -2 (or, in this case, 2) is also a quadratic residue of N.

Kraitchik became quite skilled at this sort of trial and error process for finding quadratic residues. On seeing these ideas, one is very much reminded of the work of Seelhoff [**166**]. This is not too surprising when it is pointed out that a French translation of [**166**] appeared in 1912 in *Sphinx-Oedipe*.

Notice that, although Kraitchik was completely familiar with Legendre's idea of using continued fractions to find quadratic residues (see §8), he did not, in general, advocate using this method. This is because he could not control the finding of a second factorization of some Q_i having some of the same prime factors as another Q_j that he had already factored. Given that a prime p divides some Q_j, it seems to be rather difficult to predict another place in the continued fraction where p divides the corresponding Q_i.

Kraitchik seems to have had at least three reasons for wanting to find these quadratic residues. First, if p is an odd prime such that $\left(\frac{N}{p}\right) = 1$, then if $(-1)^{(p-1)/2}p$ is a quadratic residue of N, he showed in [**55**, pp. 150–153] that

$$a^2 - b^2 \equiv N \pmod{p} \tag{13.1}$$

has exactly $(p+1)/4$ solutions for a modulo p when $p \equiv -1 \pmod 4$, and either $(p-1)/4$ or $(p+3)/4$ solutions (depending on whether $\left(\frac{N}{p}\right)_4 = 1$ or -1) when $p \equiv 1 \pmod 4$. This is a more precise version of Cole's observation in [**19**], and, of course, it is very useful to use such a p as an exclusion modulus when we want to set up our sieving algorithm for solving $a^2 - b^2 = N$, as we have cut in half the usual number of solutions of (13.1).

Second, Kraitchik [**55**, pp. 195–208] had developed a method of factoring an integer N which made use of what he called 'cycles'. Suppose we have a set of congruences $a_i x_i^2 \equiv b_i y_i^2 \pmod N$ $(i = 1, 2, \dots, k)$. This is what Kraitchik called a cycle. If

$$\prod_{1 \le i \le k} a_i = dA^2; \quad \prod_{1 \le i \le k} b_i = dB^2,$$

$$\prod_{1 \le i \le k} x_i = X; \quad \prod_{1 \le i \le k} y_i = Y,$$

we get $(AX)^2 \equiv (BY)^2 \pmod{N}$ (if $\gcd(d, N) = 1$). If $AX \equiv \pm BY \pmod{N}$, the cycle was called *primitive*; if this were not the case, the cycle was called *derived*. Of course, a derived cycle leads to a factorization of N, since $N > \gcd(AX - BY, N) > 1$. Kraitchik used this idea, in connection with his means of finding quadratic residues, to factor several values of N. For example, let $N = 409 \cdot 2^{30} + 1$. Kraitchik always kept his examples to less than 15 digits; that way, he could use D. N. Lehmer's [**92**] factor tables up to 10^7 in order to factor his values of $\pm(x^2 - N)$. By using his technique to find values of x, y, a such that $N = x^2 \pm ay^2$, he found

$$\begin{aligned} 976169^2 &\equiv 31^2 \cdot 23^2 \cdot 7^2 \cdot 2^4 \cdot 1289, \\ 47^2 \cdot 11^2 \cdot 2^3 \cdot 7 \cdot 1289 &\equiv -647971^2, \\ 661761^2 &\equiv -29^2 \cdot 2^9 \cdot 7 \cdot 409, \\ 709823^2 &\equiv 47^2 \cdot 7^2 \cdot 2^7 \cdot 7 \cdot 23 \cdot 29, \\ 41^2 \cdot 7^2 \cdot 2^7 \cdot 7 \cdot 23 \cdot 29 &\equiv -624447^2, \\ 409 \cdot 2^{30} &\equiv -1. \end{aligned}$$

Taking the products of both sides and deleting common factors, we get

$$(2^{10} \cdot 11 \cdot 41 \cdot 976169 \cdot 661761 \cdot 709823)^2 \equiv (7 \cdot 23 \cdot 29 \cdot 31 \cdot 647971 \cdot 624447)^2 \pmod{N}$$

or

$$50511214193^2 \equiv 132469785291^2 \pmod{N},$$

from which we get

$$N = 14621 \cdot 30036277.$$

This technique is little more than a slight variation of that of Seelhoff [**166**], and, as mentioned earlier, can be regarded as an early ancestor of modern methods like the Quadratic Sieve Algorithm and CFRAC.

In [**90**] Lehmer and Powers described a method of factorization which utilizes continued fractions to obtain cycles. From (8.3) we note that if we put $Q_n^* = (-1)^n Q_n$, we get

$$A_{n-1}^2 \equiv Q_n^* \pmod{N}.$$

Thus, if we find a set $\{Q_{n_1}^*, Q_{n_2}^*, \dots, Q_{n_k}^*\}$ of some of these Q_n^*-values such that

$$\prod_{1 \le i \le k} Q_{n_i}^* = B^2 \quad (B \in \mathbb{Z}),$$

then putting $a_i = 1$, $x_i = A_{n_i - 1}$, $b_i = Q_{n_i}^*$, $y_i = 1$, we would get a cycle, and this might lead to a factorization of N. When this idea was published in 1931, it was not considered by Lehmer to be practical; indeed, his own experience with it had been most frustrating. It was not until 1970, that a practical version of this idea, developed by Morrison and Brillhart [**133**], succeeded in factoring F_7. Their method, now commonly referred to as CFRAC (<u>C</u>ontinued <u>Frac</u>tion), is discussed at length in [**134**], and the central idea in it is at the heart of most modern general-purpose factoring techniques. This is the technique that ended the dominance of the number sieve in factorization.

Kraitchik's third use for quadratic residues was, unfortunately, not as scientific as the first two. He may have developed this idea from a fallacious principle of Seelhoff [**169**], which seems to have been accepted by Cole [**19**]. In Seelhoff's form, the principle seems to assert that if at least r primes such that $2^r > \pi(\sqrt{N})$ can be found, which when taken with the proper sign (see discussion below) are quadratic residues of N, then N is a prime. Now as Lehmer [**77**] rightly asserts, this principle allows us to get rid of the hardest part of attempting to construct a primality test, namely the problem of combining all the acceptable residues or sieving. No proof of this principle is offered by Seelhoff, and his example is wrong. He considers the factor $N = 204084568497$ of $2^{43} - 1$ and shows that $-1, 2, 7, 11, 17, 19, 23, 31, 43, 53, 61, 67, 83, 97, 113, 131$ are all quadratic residues of N. Since $2^{15} > \pi(\sqrt{N})$, N must be a prime. However, Landry had found 17 years earlier that $N = 9719 \cdot 2099863$. Clearly, -1 is not a quadratic residue of N; Seelhoff must have made an arithmetic error. (-1 is not a prime, but the knowledge that -1 is a quadratic residue is used in determining other quadratic residues.)

This principle occurs again in a somewhat different guise in Kraitchik's [**56**], an entire book practically devoted to it. He bases his idea on the following (unproved) theorem: If every prime p such that $\left(\frac{N}{p}\right) = 1$ is such that p, taken with the proper sign, is a quadratic residue of N, then N is a prime. Although he does not say this explicitly, the proper sign is $(-1)^{(p-1)/2}$ when p is odd, -1 when $p = 2$ and $N \equiv 3, 5 \pmod 8$, $+1$ when $p = 2$ and $N \equiv 1, 7 \pmod 8$ (see §14.) In fact, Hall [**44**] showed later that this theorem is true when we replace the conclusion with the words: then N is a prime or a prime power. Of course, this is not a useful primality test, so Kraitchik simply asserts that N is a prime if a sufficient number of such p can be found. This number is not defined, but seems, rather, to be up to the discretion of the user as long as it is large enough. This is hardly scientific, and it was vigorously criticized by Lehmer [**77**] and Beeger [**7**]. Kraitchik's unsatisfactory response in [**57**] did not really address the theoretic aspects of his principle but instead concentrated, and this only briefly, on what he claimed was the question of practicality.

He used his principle in [**56**] to 'prove' that

$$(2^{96} + 1)/(2^{32} + 1), (2^{93} - 1)/(17(2^{31} - 1))$$

are primes (they are) and that

$$(2^{120} + 1)(2^8 + 1)/((2^{24} + 1)(2^{40} + 1)), N = 4(3^{55} + 1)/((3^{11} + 1)(3^5 + 1))$$

are primes (they aren't). Indeed he devotes twenty pages of [**56**] to showing that this latter number is a prime. He finds after much labour that

$$-1, 2, 3, 5, 7, 11, 13, 17, 19, 37, 41, 47, 53, 59, 67, 71, 73, 79, 89, 101, 103, 127, 137,$$

$$139, 193, 337, 419, 421, 487, 547, 643, 677, 761, 1279, 1877, 5711$$

are all quadratic residues of N and concludes that N must be prime (with "moral certainty"). In fact, according to Brillhart [**12**], Riesel found that N is divisible by 65967. Kraitchik, as in the case of the factor of $2^{120}+1$ (see Lehmer [**83**]), must have made one or more simple arithmetic errors. For example, we see that -1 cannot be a quadratic residue of N.

Kraitchik was more successful when he used his criterion to provide evidence that N is not a prime. He does this several times in [**56**], and when he becomes sure that the number he is testing is not a prime, he then attempts to factor it. In one interesting instance of identifying an integer to be prime, Kraitchik did turn out to be correct.

Recall that Lehmer had determined that $3^{N-1} \pmod N$ for $N=(10^{23}-1)/9$ was not 1. According to [**56**, p. 39], he asked Kraitchik to factor N. Kraitchik found some quadratic residues of N and then used them to find the forms of the divisors $< 10^8$ that would divide N. He found no factor; he then used his sieving process and still found no factor. As he points out in [**56**, pp. 47–48], this would constitute a proof of the primality of N except that the large number (4367532788) of possible values for a in (13.1) rendered this proof very indirect. He then obtained more quadratic residues of N and this convinced him that N must be a prime.

He communicated his conviction to Lehmer on September 23, 1928. Lehmer then re-evaluated $3^{N-1} \pmod N$ and again did not get 1, but he also didn't get the answer he got before. Kraitchik then attempted to evaluate $3^{N-1} \pmod N$ and also made an error. Finally, in a letter of December 25, 1928 Lehmer could assert that N is a prime. The proof given in [**75**] is very simple and is a beautiful representative of Lehmer's methods.

Lehmer evaluated $3^{N-1} \pmod N$ and finally got 1; he then computed $r_1 \equiv 3^{(N-1)/11} \pmod N$ and $r_2 \equiv 3^{(N-1)/4093} \pmod N$. Since $\gcd(r_1-1, N) = \gcd(r_2-1, N) = 1$, any prime factor of N must have the form $11 \cdot 23 \cdot 4093n + 1 = 11390819n + 1$. Thus, the smallest possible prime factor of N must exceed $2 \cdot 11390819 = 22781638$.

If we try to express $N = a^2 - b^2$, then $a = 11390819^2 n + 115222895547343$. If we restrict a modulo 12 and 25, the least possible value for a is 5435003952668544. But $a < \frac{1}{2}(m + N/m)$ when $m > 22781638$; hence, $a < 243861122499491 < 5435003952668544$ and N must be a prime.

14. A modern version of Kraitchik's principle

A few years after the appearance of [**56**], Hall [**44**] attempted to put Kraitchik's principle on a more solid mathematical footing. He first defined what he called apparent residues and nonresidues of an integer N. If p is an odd prime and $p' = (-1)^{(p-1)/2}p$, then p' is an apparent residue or nonresidue of N according as the value of the Legendre symbol $\left(\frac{N}{p}\right) = +1$ or -1, respectively. Also, -1 is an apparent residue or nonresidue of N when the Jacobi symbol $\left(\frac{-1}{N}\right) = 1$ or -1, and the apparent residue or nonresidue character of 2 is defined similarly.

He then went on to consider integers L_p which he defined as the least non-square positive integer congruent to 1 modulo 8, such that $\left(\frac{L_p}{q}\right) = 1$ for all odd primes $q \leq p$. Curiously, Kraitchik had considered these numbers previously in [**54**, pp. 41–46], where he tabulated them up to L_{47} by making use of his moveable paper strips method of sieving. This table was later extended by Lehmer in [**74, 86**]; also, in [**86**] Lehmer introduced the term *pseudosquare* for any such L_p. In Table 1 below we give all the pseudosquares that are currently known.

The main theorem of [**44**] is the following:

THEOREM 14.1. *If all the factors of N are less than L_p and if $-1, 2, \ldots, (-1)^{(p-1)/2}p$ can be divided into two classes, $A = \{a_1, a_2, \ldots, a_r\}$ (the apparent residues of N), and $B = \{b_1, b_2, \ldots, b_s\}$ (the apparent nonresidues of N) such that every member of A is a true quadratic residue of N and the product of every pair of elements from B is also a true quadratic residue of N, then N is either a prime or a prime power.*

This test was actually used by Beeger [**8**] to show that the large factor $N = 9298142299081$ of $12^{45} + 1$ is a prime. He noted that it had previously been shown that any prime divisor of N must exceed 10^5; hence, any prime divisor of N must be less than $N/10^5 < L_{67}$. Further,

$$-1, 2, -3, 5, -11, 13, 17, -23, 29, -31, -43, -47, 53, -59, 61$$

are all apparent residues of N and $-7, -19, 37, 41, -67$ are apparent nonresidues of N. By using Kraitchik's method to find quadratic residues of N, he was able to show that $-1, 2, 3, 5, 11, 13, 17, 23, 29, 31, 43, 47, 53, 59, 61, 7\cdot19, 7\cdot37, 7\cdot41, 7\cdot67$ are all quadratic residues of N; hence, N must be a prime or prime power. Since $N < 10^{15}$, and N is not a perfect square, N is a prime.

The main difficulty with this technique is that determining the quadratic residues of N is a rather slow, trial and error process. This problem can be solved by making use of an idea of Selfridge and Weinberger mentioned in [**186**]. When this is done, we get the following theorem:

THEOREM 14.2. *If*

(i) *No prime factor of an odd N is $\leq B$;*
(ii) *$N/B < L_p$;*
(iii) *$p_i^{(N-1)/2} \equiv \pm 1 \pmod{N}$ for all primes $p_i \leq p$;*
(iv) *$p_j^{(N-1)/2} \equiv -1 \pmod{N}$ for some odd $p_j \leq p$ if $N \equiv 1 \pmod{8}$; or $2^{(N-1)/2} \equiv -1 \pmod{N}$ if $N \equiv 5 \pmod{8}$,*

then N is a prime or prime power.

p	L_p	Comments
3	73	Kraitchik
5	241	(1924) [**54**]
7	1009	Moveable Strips
11	2641	
13	8089	
17	18001	
19	53881	
23	87481	
29	117049	
31	515761	
37	1083289	
41	3206641	
43	3818929	
47	9257329	
53	22000801	Lehmer (1928) [**74**]
59	48473881	Bicycle chains
61	48473881	
67	175244281	Lehmer (1954) [**86**]
71	427733329	SWAC
73	427733329	
79	898716289	
83	2805544681	Lehmer, Lehmer,
89	2805544681	Shanks (1970) [**87**]
97	2805544681	DLS-127
101	10310263441	
103	23616331489	
107	85157610409	
109	85157610409	
113	196265095009	
127	196265095009	
131	2871842842801	Lehmer
137	2871842842801	(Unpublished)
139	2871842842801	DLS-57
149	26250887023729	
151	26250887023729	
157	112434732901969	Williams (1988)
163	112434732901969	(Unpublished)
167	112434732901969	UMSU
173	178936222537081	
179	178936222537081	
181	696161110209049	
191	696161110209049	
193	2854909648103881	Stephens and Williams [**175**]
197	6450045516630769	OASiS (1989)
199	6450045516630769	
211	11641399247947921	
223	11641399247947921	

TABLE 1. Table of Pseudosquares

Notice that if $B = 1$ and N is a prime, there must exist some odd $q \le p$ such that $\left(\frac{N}{q}\right) = -1$ when $N \equiv 1 \pmod 8$. If $N \equiv 5 \pmod 8$, then $\left(\frac{2}{N}\right) = -1$. Thus, both conditions (iii) and (iv) must hold when N is a prime and $B = 1$.

The next problem we encounter on using this test is the growth rate of L_p. Clearly, if we could show that L_p grows rapidly, we would have an efficient primality test. In fact, Bach [**5**] has shown that if G is a nontrivial subgroup of R, the group of reduced residues modulo m, such that $n \in G$ for all $0 < n < x$, then $x < 2(\log m)^2$. (The value 2 can be replaced by a somewhat smaller constant for sufficiently large values of m.) If we put $m \equiv 1 \pmod 8$ and define $\left(\frac{a}{m}\right)$ to be the Jacobi symbol, then if G is the group of reduced residues of a modulo m where $\left(\frac{a}{m}\right) = 1$, G is a subgroup of R. Also, if m is not a perfect square, then $p^a \,||\, m$ with a odd for some odd prime p. If we let t be some integer such that $\left(\frac{t}{p}\right) = -1$, and put

$$\begin{aligned} b &\equiv t \pmod{p^a}, \\ b &\equiv 1 \pmod{m/p^\alpha}, \end{aligned}$$

then $\left(\frac{b}{m}\right) = \left(\frac{b}{p^\alpha}\right) = \left(\frac{b}{p}\right) = \left(\frac{t}{p}\right) = -1$, $\gcd(b, m) = 1$ and $1 < b < m$. Hence, b is a reduced residue which is not in G, and therefore G is a proper subgroup of R.

If we put $m = L_p$, then $\left(\frac{n}{m}\right) = 1$ for all $1 \le n \le p$ and we get $p < 2(\log L_p)^2$. It follows that

$$L_p > e^{\sqrt{p/2}}. \tag{14.1}$$

Unfortunately, Bach's result is conditional on the truth of the Extended Riemann Hypothesis (ERH), a hypothesis which has not yet been proved; indeed, there are some who believe that it is false. Certainly, the values of L_p computed so far satisfy (14.1). If the ERH is ever proved, or the restriction of the ERH could be removed from Bach's theorem, we would have a primality test which satisfies all the desiderata of §2.

15. Conclusion

We have concluded this paper in much the same way as we began it, by discussing primality tests. We should in this connection mention that just at the beginning of the development of computers, the prime character of all the Mersenne numbers M_p for $p \le 257$ had been determined without the use of a computer. This achievement was due greatly to the immense efforts of H.S. Uhler [**180, 181, 182, 183, 184**], who (correctly) showed that M_{157}, M_{167}, M_{199}, M_{227} and M_{193} are all composite by using the Lucas-Lehmer test. Briefly put, Uhler's [**179, 180**] modification to the test technique of Lehmer consisted of replacing the division by M_p by the less time-consuming process of multiplication by an approximation to $1/M_p$.

The largest integer ever proved prime without the use of a computer is the 44-digit $(2^{148}+1)/17$. This was done by Ferrier [**29**] in 1952 by using essentially the same technique that Lehmer used to prove $(10^{23}-1)/9$ a prime.

We realize that there are many whose names we have not mentioned in this history and many other developments that we have left undescribed. In spite of the length of the paper, we have not made any attempt at completeness — that would require a book. Rather, we have attempted only to outline the main developments of our subject. It is our hope that the reader has gained some appreciation for all the work that was done before computers appeared on the scene and now understands why the introduction of these machines caused such a sudden acceleration in the further development of Computational Number Theory. Indeed, this is one of the few papers on the subject that will not soon be out of date.

One problem that is of great importance to us is the determination of the growth rate of L_p. This, naturally, can be approached in several directions. One of us (HCW) is pursuing this through further computation of the L_p-values. This will require the construction of another sieve device which has recently been described by Patterson [**136, 137**]. This machine is planned to be able to sieve through the integers in order to find special numbers like pseudosquares at the rate of 5.38×10^{11} per second. We recognize on saying this that one of us (at least) is promising to build a new machine. Perhaps, in view of several of our themes in this paper, it is fitting to close with the French proverb: Plus ça change, plus c'est la même chose.

16. Acknowledgements

The authors would like to thank Richard Mollin, Bruce Reznick and Michael Rosen for helping them to obtain various obscure library materials. Thanks are also due to Brown University for permission to use its excellent mathematics collection. We are pleased to acknowledge assistance and suggestions we received from Eric Bach, John Brillhart, Hendrik W. Lenstra, Jr. and Andrew Odlyzko. François Morain was extremely helpful in obtaining certain bibliographic information and, especially, in locating information about Carissan and his wonderful sieve. We are most grateful to Emma Lehmer for making Dick's journals, papers, etc. available to us and for reviewing a preliminary version of this paper. Finally, we thank the referee for a careful reading and useful suggestions.

References

1. L. M. Adleman and M.-D. Huang, *Primality Testing and Abelian Varieties over Finite Fields*, Lecture Notes in Mathematics, vol. 1512, Springer-Verlag, 1992.
2. Anonymous, *Les appareils à calculs exacts et instantanés pour simplifier la multiplication et la division, inventés par M. Henri Genaille, et perfectionnés par M. Edouard Lucas*, Nouvelles Annales de Mathématiques (3) **4** (1885), 516–519.
3. A. O. L. Atkin and F. Morain, *Elliptic curves and primality proving*, Math. Comp. **61** (1993), 29–68.
4. H. P. Babbage, *Babbage's Calculating Engines*, Spon & Company, London, 1889. Reprinted by Tomash Publishers, Los Angeles, 1982.

5. Eric Bach, *Explicit bounds for primality testing and related problems*, Math. Comp. **55** (1990), 355–380.
6. P. T. Bateman, J. L. Selfridge and S. S. Wagstaff, Jr., *The new Mersenne conjecture*, Amer. Math. Monthly **96** (1989), 125–128.
7. N. G. W. H. Beeger, *Sur la decomposition de grands nombres*, Nieuw Archief voor Wiskunde (2) **16** (1929/30), 37–42.
8. N. G. W. H. Beeger, *Note sur la factorisation de quelques grands nombres*, Institut Grand-Ducal de Luxembourg, Sect. d. Sc. nat. phys. math. Archives **16** (1946), 93–95.
9. E. T. Bell, *Mathematics: Queen and Servant of Science*, McGraw-Hill, NY, 1951, p. 228.
10. W. Bosma and M.-P. van der Hulst, *Primality Proving with Cyclotomy*, University of Amsterdam, 1990.
11. D. M. Bressoud, *Factorization and Primality Testing*, Springer-Verlag, New York, 1989.
12. J. Brillhart, *Fermat's factoring method and its variants,* Cong. Numerantium **32** (1981), 29–48.
13. J. Brillhart, D. H. Lehmer, J. L. Selfridge, B. Tuckerman, and S. S. Wagstaff, Jr., *Factorizations of* $b^n \pm 1$*,* $b = 2, 3, 5, 6, 7, 10, 11, 12$ *Up to High Powers*, vol. 22 of Contemporary Mathematics, American Mathematical Society, 1988, 2nd Edition.
14. H. Brocard, *Question 528*, Mathesis **7** (1887), 73–75.
15. J. Brown, L. C. Noll, B. K. Parady, J. F. Smith, G. W. Smith, and E. S. Zarantonello, *Letter to the Editor*, Amer. Math. Monthly **97** (1990), 214.
16. E. Carissan, *Machine à resoudre les congruences*, Bulletin de la Société d'Encouragement pour l'Industrie Nationale **132** (1920), 600–607.
17. R. D. Carmichael, *On the numerical factors of the arithmetic forms* $\alpha^n \pm \beta^n$, Ann. of Math. **15** (1913), 30–70.
18. A. Cauchy, *Sur les moyens d'éviter les erreurs dans les calculs numériques*, C. R. Acad. Sci. Paris **9** (1840), 789–798.
19. F. N. Cole, *On the factoring of large numbers*, Bull. Amer. Math. Soc. **10** (1903), 134–137. French translation in Sphinx-Oedipe **5** (1910), 122–124.
20. L. E. Dickson, *History of the Theory of Numbers*, Vol. 1: Divisibility and Primality. Carnegie Institution of Washington, Publication No. 256, 1919. Reprinted by Chelsea Books, New York, 1971.
21. S. Drake, *The rule behind 'Mersenne's numbers'*, Physis-Riv. Internaz. Storia Sci. **13** (1971), 421–424.
22. J. D. Elder, *Factor Stencils*, Carnegie Institution of Washington, D.C., 1939.
23. J. Ewing, *The latest Mersenne prime*, Amer. Math. Monthly **99** (1992), 360.
24. E. Fauquembergue, *Question 266*, L'Intermédiaire des Mathématiciens **1** (1894), 148.
25. E. Fauquembergue, *Vérification des nombres de Mersenne*, Sphinx-Oedipe **7** (1912), 20–23.
26. E. Fauquembergue, *Réponse*, L'Intermédiaire des Mathématiciens **24** (1917), 33.
27. E. Fauquembergue, *Nombres de Mersenne*, Sphinx-Oedipe **15** (1920), 17–18.
28. P. Fermat, *Fragment d'une lettre de Fermat*, Oeuvres de Fermat **2** (1894), 256–258.
29. A. Ferrier, *The determination of a large prime*, Math. Tables and other Aids to Computation **6** (1952), 256.
30. C. F. Gauss, *Disquisitiones Arithmeticae*, G. Fleischer, Leipzig, 1801. English translation by A. A. Clarke, Springer-Verlag, New York, 1986.
31. H. Genaille, *Calculateur ou table de multiplication*, Assoc. Française pour l'Avancement des Sciences. Comptes Rendus **20** (1891), I partie, 159.
32. H. Genaille, *Piano arithmétique pour la vérification des grands nombres premiers*, Assoc. Française pour l'Avancement des Sciences. Comptes Rendus **20** (1891), I partie, 159.
33. A. Genocchi, *Intorno ad alcune forme di numeri primi*, Annali Mat. Pura Appl. (2) **2** (1868/69), 256–267.
34. A. Genocchi, *Intorno a tre problemi aritmetici*, Atti della Reale Accademia delle Scienze di Torino **11** (1876), 811–829.
35. A. Gérardin, *Le Père Theophile Pepin S.J.*, Sphinx-Oedipe **5**, 1er trimestre (1910), 1–13.
36. A. Gérardin, *Question 278*, Sphinx-Oedipe **6** (1911), 70.

37. A. Gérardin, *Question 335*, Sphinx-Oedipe **7** (1912), 47–48. Also see p. 31.
38. A. Gérardin, *F. Proth*, Sphinx-Oedipe **7** (1912), 50–51.
39. A. Gérardin, *Réponse 335*, Sphinx-Oedipe **7** (1912), 61–64.
40. A. Gérardin, *Rapport sur diverses méthodes de solutions employées en théorie pour la décomposition des nombres en facteurs*, Assoc. Française pour l'Avancement des Sciences. Comptes Rendus **41** (1912), II partie, 54–57.
41. A. Gérardin, *Dernière heure*, Sphinx-Oedipe **9** (1914), 85.
42. A. Gérardin, *Nombres de Mersenne*, Sphinx-Oedipe **9** (1914), 103–108.
43. A. Gérardin, *Machine à congruences (Modèle 1937)*, 70e Congrès des Sociétés Savantes de Paris et des Départements, Section des Sciences, Gauthier-Villars, Paris, 1937, p. 14; II, p. 37.
44. M. Hall, *Quadratic residues in factorization*, Bull. Amer. Math. Soc. **39** (1933), 758–763.
45. D. Harkin, *On the mathematical work of François-Édouard-Anatole Lucas*, Enseign. Math. **3** (1957), 276–288.
46. Ch. Henry, *Sur divers points de la théorie des nombres. Remarques historiques*, Assoc. Française pour l'Avancement des Sciences. Comptes Rendus **9** (1880), 201–207.
47. M. Heyworth, *A conjecture on Mersenne's conjecture*, New Zealand Math. Mag. **19** (1982), 147–151.
48. O. Hoppe, *The primality of* $(10^{19}-1)/9$, Proc. London Math. Soc. (2) **17** (1918), xix.
49. V. G. Imchenetzki and V. Bouniakowsky, *Sur un nouveau nombre premier, annoncé par le père Pervouchine*, Bull. Acad. Imp. Sci. St. Petersbourg **31** (1887), 532–533. Also in Mélanges Mathématiques et Astronomiques tirés du Bulletin de l'Académie Imperiale des Sciences de St. Petersbourg **6** (1883), 553–554.
50. R. B. Killgrove and K. E. Ralston, *On a conjecture concerning the primes*, Math. Comp. **13** (1959), 121–122.
51. M. Kraitchik, *Réponse 278*, Sphinx-Oedipe **6** (1911), 95.
52. M. Kraitchik, *Sur l'équation* $x^2 - y^2 = N$, Sphinx-Oedipe **7** (1912), 23–29.
53. M. Kraitchik, *Théorie des Nombres*, Gauthier-Villars, Paris, 1922.
54. M. Kraitchik, *Recherches sur la Théorie des Nombres*, Gauthier-Villars, Paris, 1924.
55. M. Kraitchik, *Théorie des Nombres*, T. II, Gauthier-Villars, Paris, 1926.
56. M. Kraitchik, *Recherches sur la Théorie des Nombres*, T. II, Gauthier-Villars, Paris, 1929.
57. M. Kraitchik, *Factorisation des grands nombres*, Sphinx **1** (1931), 35–37.
58. M. Kraitchik, *Introduction à la Théorie des Nombres*, Gauthier-Villars, Paris, 1952.
59. M. Kraitchik, *On the factorization of* $2^n \pm 1$, Scripta. Math. **18** (1952), 39–52.
60. G. Lamé, *Note sur la limite du nombre des divisions dans la recherche du plus grand commun diviseur entre deux nombres entiers*, C. R. Acad. Sci. Paris **19** (1844), 867–870.
61. F. Landry, *Procédés Nouveaux pour Démontrer que le Nombre 2147483647 est Premier*, Librairie Hachette, Paris, 1859.
62. F. Landry, *Aux Mathématiciens de Toutes les Parties du Monde. Communication sur la Décomposition des Nombres en Leurs Facteurs Simples*, Librairie Hachette, Paris, 1867.
63. F. Landry, *Decompositions des Nombres* $2^n \pm 1$ *en Leurs Facteurs Premiers de* $n = 1$ *à* $n = 64$ *(Moins Quatre)*, Librairie Hachette, Paris, 1869.
64. F. Landry, *Letter to Charles Henry*, Bollettino di Bibliografia e di Storia delle Scienze Matematiche e Fisiche **13** (1880), 469–470.
65. F. Landry, *Méthode de décomposition des nombres en facteurs premiers*, Assoc. Française pour l'Avancement des Sciences. Comptes Rendus **9** (1880), 185–189.
66. F. W. Lawrence, *Factorisation of numbers*, Messenger of Math. **24** (1894/95), 100–109.
67. F. W. Lawrence, *Factorisation of numbers*, Quart. J. Pure Appl. Math. **28** (1896), 285–311. French translation in Sphinx-Oedipe 5 (1910), 98–121.
68. F. W. Lawrence, *Determination of certain primes*, Proc. Lond. Math. Soc. **28** (1896/97), 465–475.
69. E. Léger, *Note sur le partage d'une droite en moyenne et extrême, et sur un problème d'arithmétique*, Corresp. Math. Phys. **9** (1837), 483–485.
70. D. H. Lehmer, *A cross-division process and its application to the extraction of roots*,

Amer. Math. Monthly **33** (1926), 198–206.
71. D. H. Lehmer, *Note on the Mersenne number* $2^{139}-1$, Bull. Amer. Math. Soc. **32** (1926), 522.
72. D. H. Lehmer, *Tests for primality by the converse of Fermat's theorem*, Bull. Amer. Math. Soc. **33** (1927), 327–340.
73. D. H. Lehmer, *A further note on the converse of Fermat's theorem*, Bull. Amer. Math. Soc. **34** (1928), 54–56.
74. D. H. Lehmer, *The mechanical combination of linear forms*, Amer. Math. Monthly **35** (1928), 114–121.
75. D. H. Lehmer, *On the number* $(10^{23}-1)/9$, Bull. Amer. Math. Soc. **35** (1929), 349–350.
76. D. H. Lehmer, *An extended theory of Lucas' functions*, Ann. Math. **31** (1930), 419–448.
77. D. H. Lehmer, *A fallacious principle in the theory of numbers*, Bull. Amer. Math. Soc. **36** (1930), 847–850.
78. D. H. Lehmer, *Note on Mersenne numbers*, Bull. Amer. Math. Soc. **38** (1932), 383–384.
79. D. H. Lehmer, *A photo-electric number sieve*, Amer. Math. Monthly **40** (1933), 401–406.
80. D. H. Lehmer, *Some new factorizations of* $2^n \pm 1$, Bull. Amer. Math. Soc. **39** (1933), 105–108.
81. D. H. Lehmer, *A machine for combining sets of linear congruences*, Math. Annalen **109** (1934), 661–667.
82. D. H. Lehmer, *On Lucas's test for the primality of Mersenne's numbers*, J. Lond. Math. Soc. **10** (1935), 162–165.
83. D. H. Lehmer, *Sur les essais directs de primalité*, Sphinx **8** (1938), 87–88.
84. D. H. Lehmer, *Guide to Tables in the Theory of Numbers*, National Research Council, Washington, D.C., 1941.
85. D. H. Lehmer, *On the factors of* $2^n \pm 1$, Bull. Amer. Math. Soc. **53** (1947), 164–167.
86. D. H. Lehmer, *A sieve problem on "pseudo-squares"*, Math. Tables and Other Aids to Computation **8** (1954), 241–242.
87. D. H. Lehmer, E. Lehmer and D. Shanks, *Integer sequences having prescribed quadratic character*, Math. Comp. **24** (1970), 433–451.
88. D. H. Lehmer, *A history of the sieve process*, in *A History of Computing in the Twentieth Century* (N. Metropolis, J. Howlett and G.-C. Rota, eds.), Academic Press, New York, 1980, pp. 445–456.
89. D. H. Lehmer, *Factorization then and now*, Computers in Mathematics, Lecture Notes in Pure and Applied Mathematics, vol. 125, Marcel Dekker, New York, 1990, 311–320.
90. D. H. Lehmer and R. E. Powers, *On factoring large numbers*, Bull. Amer. Math. Soc. **37** (1931), 770–776.
91. D. N. Lehmer, *A new short method of multiplication*, Science **16** (1902), 71–74.
92. D. N. Lehmer, *Factor Table for the First Ten Millions containing the Smallest Factor of Every Number Not Divisible by 2, 3, 5, or 7 between 0 and 10017000.* Carnegie Institute Publication 105, Washington, D.C., 1909.
93. D. N. Lehmer, *On the multiplication of large numbers*, Amer. Math. Monthly **30** (1923), 67–68.
94. D. N. Lehmer, *Factor Stencils*, Carnegie Institution of Washington, D.C., 1929.
95. D. N. Lehmer, *On the factorization of large numbers*, University of California Chronicle **32** (1930), 326–341.
96. D. N. Lehmer, *Hunting big game in the theory of numbers*, Scripta Mathematica **1** (1932/33), 229–235.
97. E. Lehmer, *On the quartic character of some quadratic units*, J. Reine Angew. Math. **268/269** (1974), 294–301.
98. A. K. Lenstra and M. S. Manasse, *Factoring by electronic mail*, Advances in Cryptology, Eurocrypt '89, Lecture Notes in Computer Science, vol. 434, Springer, 1990, pp. 355–371.
99. A. K. Lenstra, H. W. Lenstra, Jr., M. S. Manasse, and J. M. Pollard, *The factorization of the ninth Fermat number*, Math. Comp. **61** (1993), 319–349.
100. W. Looff, *Über die Periodicität der Decimalbrüche*, Archiv der Mathematik und Physik **16** (1851), 54–57.

101. Ed. Lucas, *Sur la décomposition des nombres en facteurs premiers*, Nouvelles Annales de Mathématiques (2) **14** (1875), 523–525.
102. Ed. Lucas, *Note sur l'application des séries récurrentes à la recherche de la loi de distribution des nombres premiers*, C. R. Acad. Sci. Paris **82** (1876), 165–167.
103. Ed. Lucas, *Sur la théorie des nombres premiers*, Atti della Reale Accademia delle Scienze di Torino **11** (1875/76), 928–937.
104. Ed. Lucas, *Questions nouvelles d'arithmétique supérieure*, Nouvelles Annales de Math. (2) **15** (1876), 82–83.
105. Ed. Lucas, *Sur les rapports qui existent entre la théorie des nombres et le calcul intégral*, C. R. Acad. Sci. Paris **82** (1876), 1303–1305.
106. Ed. Lucas, *Sur la recherche des grands nombres premiers*, Assoc. Française pour l'Avancement des Sciences. Comptes Rendus **5** (1876), 61–68.
107. Ed. Lucas, *Nouveaux théorèmes d'arithmétique supérieure*, C. R. Acad. Sci. Paris **83** (1876), 1286–1288.
108. Ed. Lucas, *Sur l'extension du théorème de Fermat généralisé, et du Canon arithmeticus*, C. R. Acad. Sci. Paris **84** (1877), 439–442.
109. Ed. Lucas, *Sur la division de la circonférence en parties égales*, C. R. Acad. Sci. Paris **85** (1877), 136–139.
110. Ed. Lucas, *Considérations nouvelles sur la théorie des nombres premiers et sur la division géométrique de la circonférence en parties égales*, Assoc. Française pour l'Avancement des Sciences. Comptes Rendus **6** (1877), 159–167.
111. Ed. Lucas, *Extrait d'une lettre de M. Édouard Lucas*, Nouvelle Correspondance Mathématique **3** (1877), 315–316.
112. Ed. Lucas, *Recherches sur plusieurs ouvrages de Leonard de Pise et sur diverses questions d'arithmétique supérieure*, Bollettino di Bibliografia e di Storia delle Scienze Matematiche e Fisiche **10** (1877), 129–193, 239–293.
113. Ed. Lucas, *On the interpretation of a passage in Mersenne's works*, Messenger of Math. **7** (1878), 185–187.
114. Ed. Lucas, *Théorie des fonctions numériques simplement périodiques*, Amer. Journ. Math. **1** (1878), 184–240, 289–321.
115. Ed. Lucas, *Sur la série récurrente de Fermat*, Bollettino di Bibliografia e di Storia delle Scienze Matematiche e Fisiche **11** (1878), 783–798.
116. Ed. Lucas, *Question 453*, Nouvelle Correspondance Mathématique **5** (1879), 137.
117. Ed. Lucas, *Question 454*, Nouvelle Correspondance Mathématique **5** (1879), 138–139.
118. Ed. Lucas, *Notice sur les Titres et Travaux Scientifiques de M. Édouard Lucas*, D. Jouaust, Paris, 1880.
119. Ed. Lucas, *Principi fondamentali della geometria dei tessuti*, L'Ingegneria Civile e le Arti Industriali **6** (1880), 104–111, 113–115.
120. Ed. Lucas, *Sur les nombres parfaits*, Mathesis **6** (1886), 145–146.
121. Ed. Lucas, *Sur le neuvième nombre parfait*, Mathesis **7** (1887), 45–46.
122. Ed. Lucas, *Théorie des Nombres*, Gauthier-Villars, Paris, 1891.
123. Ed. Lucas, *Récréations Mathématiques*, T. I, 2nd edition, Gauthier-Villars, Paris, 1891, pp. 235–236.
124. Ed. Lucas, *Récréations Mathématiques*, T. II, 2nd edition, Gauthier-Villars, Paris, 1893, pp. 230–235.
125. Ed. Lucas, *Le calcul et les machines à calculer*, *Récréations Mathématiques*, T. III, Gauthier-Villars, Paris, 1893, pp. 27–86.
126. Ed. Lucas, *Les principes fondamentaux de la géométrie des tissus*, Assoc. Française pour l'Avancement des Sciences. Comptes Rendus **40** (1911), 72–88.
127. D. Mahnke, *Leibniz auf der Suche nach einer allgemeinen Primzahlgleichung*, Bibliotheca Math. **13** (1912–13), 29–61.
128. T. E. Mason, *Mechanical device for testing Mersenne numbers for primes*, Proc. Indiana Acad. Sci. (1914), 429–432.
129. Ministère du Commerce, de l'Industrie, des Postes et du Travail, *Catalogue Officiel des Collections du Conservatoire National des Arts et Métiers*, E. Bernard, Paris, 1906,

pp. 169–206.

130. F. Morain, *Distributed primality proving and the primality of* $(2^{3539}+1)/3$, Advances in Cryptology - EUROCRYPT '90, Springer-Verlag, Berlin, 1991, pp. 110–123.
131. J. C. Morehead, *Note on Fermat's numbers*, Bull. Amer. Math. Soc. **11** (1905), 543–545.
132. J. C. Morehead and A. E. Western, *Note on Fermat's numbers*, Bull. Amer. Math. Soc. **16** (1909), 1–6.
133. M. A. Morrison and J. Brillhart, *The factorization of* F_7, Bull. Amer. Math. Soc. **77** (1971), 264.
134. M. A. Morrison and J. Brillhart, *A method of factoring and the factorization of* F_7, Math. Comp. **29** (1975), 183–205.
135. M. d'Ocagne, *Vue d'ensemble sur les machines à calculer*, Bulletin des Sciences Mathématiques **46** (1922), 102–144.
136. C. Patterson, *A 538 Billion Integer per Second Sieve*, Proc. 1991 Canad. Conf. on Electrical and Computer Engineering, 1991, pp. 13.1.1-13.1.9.
137. C. Patterson, *The Derivation of a High Speed Sieve Device*, Ph. D. Thesis, Dept. of Computer Science, University of Calgary, Calgary, Canada, 1991.
138. T. Pepin, *Sur la formule* $2^{2^n}+1$, C. R. Acad. Sci. Paris **85** (1877), 329–331.
139. T. Pepin, *Sur la décomposition des grands nombres en facteurs premiers*, Atti della Accademia Pontificia dei Nuovi Lincei **43** (1889/90), 163–191.
140. T. Pepin, *Extension de la méthod d'Euler pour la décomposition des grands nombres en facteurs premiers*, Memorie della Accademia Pontificia dei Nuovi Lincei **9** (1893), 47–76.
141. T. Pepin, *Sur la decomposition des grands nombres en facteurs premiers*, Memorie della Accademia Pontificia dei Nuovi Lincei **17** (1900), 321–344.
142. T. Pepin, *Décomposition en facteurs premiers du nombre* $N = \frac{(151)^5-1}{5\cdot 150} = 104670701$ [sic], Atti della Accademia Pontificia dei Nuovi Lincei **54** (1901), 89–93.
143. H. C. Pocklington, *The determination of the exponent to which a number belongs, the practical solution of certain congruences, and the law of quadratic reciprocity*, Proc. Cambridge Phil. Soc. **16** (1910), 1–5.
144. H. C. Pocklington, *The determination of the prime or composite nature of large numbers by Fermat's theorem*, Proc. Cambridge Phil. Soc. **18** (1914–16), 29–30.
145. H. C. Pocklington, *The direct solution of the quadratic and cubic binomial congruences with prime moduli*, Proc. Cambridge Phil. Soc. **19** (1917), 57–59.
146. P. Poulet, *Réponse 278*, Sphinx-Oedipe **15** (1920), 74–75.
147. P. Poulet, *Note sur un Théorème de Proth*, Sphinx-Oedipe **15** (1920), 53–55, 97–100, 113–119.
148. R. E. Powers, *The tenth perfect number*, Amer. Math. Monthly **18** (1911), 195–197.
149. R. E. Powers, *On Mersenne's numbers*, Proc. London Math. Soc. (2) **13** (1914), xxxix.
150. V. R. Pratt, *Every prime has a succinct certificate*, SIAM J. Computing **4** (1975), 214–220.
151. F. Proth, *Énoncés de divers théorèmes sur les nombres*, C. R. Acad. Sci. Paris **83** (1876), 1288–1289.
152. F. Proth, *Mémoires présentés*, C. R. Acad. Sci. Paris **87** (1878), 374.
153. F. Proth, *Correspondance*, Nouvelle Correspondance Mathématique **4** (1878), 210–211.
154. F. Proth, *Sur la série des nombres premiers*, Nouvelle Correspondance Mathématique **4** (1878), 236–240.
155. F. Proth, *Lettres*, Sphinx-Oedipe **6** (1911), 36–37, 151–156.
156. M. O. Rabin, *Probabilistic algorithm for testing primality*, J. Number Theory **12** (1980), 128–138.
157. A. E. Raik, *Ural'ski matematik Ivan Mikheevich Pervoushin*, Istoriko-Matematicheskie Issledovaniya **6** (1953), 535–572.
158. C. Reid, *Perfect numbers*, Scientific American **188** (1953), 84–86.
159. K. G. Reuschle, *Mathematische Abhandlung, enthaltend: Neue zahlentheoretische Tabellen*, Stuttgart, 1856.
160. Em. Reuss, *Lettre à Lucas*, Sphinx-Oedipe **20** (1925), 6–9.
161. H. Riesel, *Prime Numbers and Computer Methods for Factorization*, Birkhäuser, Boston,

1985.

162. R. M. Robinson, *Mersenne and Fermat numbers*, Proc. Amer. Math. Soc. **5** (1954), 842–846.
163. M. Rosen, *A proof of the Lucas-Lehmer test*, Amer. Math. Monthly **95** (1988), 855–856.
164. P. Seelhoff, *Über die vollkommenen Zahlen, insbesondere über die bis jetzt zweifelhaften Fälle* $2^{40}(2^{41}-1), 2^{46}(2^{47}-1), 2^{52}(2^{53}-1)$, Archiv der Mathematik und Physik **2** (1885), 327–329.
165. P. Seelhoff, *Notes mathématiques 6*, Mathesis **6** (1886), 100–101.
166. P. Seelhoff, *Die Auflösung grosser Zahlen in ihre Factoren*, Zeitschrift für Mathematik und Physik **31** (1886), 166–172. French translation in Sphinx-Oedipe **7** (1912), 84–88.
167. P. Seelhoff, *Die Theilbarkeit des Binoms* $2^{2^n}+1$, Zeitschrift für Mathematik und Physik **31** (1886), 172–174.
168. P. Seelhoff, *Die neunte vollkommene Zahl*, Zeitschrift für Mathematik und Physik **31** (1886), 174–178.
169. P. Seelhoff, *Ein neues Kennzeichen für die Primzahlen*, Zeitschrift für Mathematik und Physik **31** (1886), 306–310.
170. P. Seelhoff, *Zur Analyse sehr grosser Zahlen*, Archiv der Mathematik und Physik **3** (1886), 325–329.
171. H. Siebeck, *Die recurrenten Reihen vom Standpuncte der Zahlentheorie aus betrachtet*, J. Reine Angew. Math. **33** (1846), 71–76.
172. R. D. Silverman, *The multiple polynomial quadratic sieve*, Math. Comp. **48** (1987), 329–339.
173. R. Solovay and V. Strassen, *A fast Monte-Carlo test for primality*, SIAM J. Comput. **6** (1977), 84–85. Erratum **7** (1978), 118.
174. A. Statuti, *Cenno necrologico del Rev. Prof. P. Teofilo Pepin*, Atti della Accademia Pontificia dei Nuovi Lincei **58** (1904–05), 210–215.
175. A. J. Stephens and H. C. Williams, *An open architecture number sieve*, London Math. Soc. Lecture Note Series **154** (1990), 38–75.
176. Doron D. Swade, *Redeeming Charles Babbage's Mechanical Computer*, Scientific American **268** (2) (February 1993), 86–91.
177. P. Tannery, *Question 660*, L'Intermédiaire des Mathématiciens **2** (1895), 317.
178. P. Tannery, *Réponse*, L'Intermédiaire des Mathématiciens **5** (1898), 166.
179. H. S. Uhler, *Multiplication of large numbers*, Amer. Math. Monthly **28** (1921), 447–448.
180. H. S. Uhler, *First proof that the Mersenne number M_{157} is composite*, Proc. Nat. Acad. Sci. **30** (1944), 314–316.
181. H. S. Uhler, *Note on the Mersenne numbers M_{152} and M_{167}*, Bull. Amer. Math. Soc. **52** (1946), 178.
182. H. S. Uhler, *On Mersenne's number M_{199} and Lucas's sequences*, Bull. Amer. Math. Soc. **53** (1947), 163–164.
183. H. S. Uhler, *On Mersenne's number M_{227} and cognate data*, Bull. Amer. Math. Soc. **54** (1948), 378–380.
184. H. S. Uhler, *On all of Mersenne's numbers particularly M_{193}*, Proc. Nat. Acad. Sci. **34** (1948), 102–103.
185. G. Vacca, *Sui manoscritti inediti di Leibniz*, Boll. Bib. Storia Sc. Mat. **2** (1899), 113–116.
186. H. C. Williams, *Primality testing on a computer*, Ars Combinatoria **5** (1978), 127–185.
187. H. C. Williams, *How was F_6 factored?*, Math. Comp. **61** (1993), 463–474.

Department of Computer Science, University of Manitoba, Winnipeg, Manitoba R3T 2N2, Canada

E-mail address: hugh_williams@csmail.cs.umanitoba.ca

Department of Computer Science, University of Waterloo, Waterloo, Ontario N2L 3G1, Canada

E-mail address: shallit@graceland.uwaterloo.ca

Contributed Papers

Proceedings of Symposia in Applied Mathematics
Volume **48**, 1994

Explicit Bounds for Primes in Residue Classes

ERIC BACH AND JONATHAN SORENSON

ABSTRACT. Let E/K be an abelian extension of number fields, with $E \neq \mathbb{Q}$. Let Δ and n denote the absolute discriminant and degree of E. Let σ denote an element of the Galois group of E/K. We prove the following theorems, assuming the Extended Riemann Hypothesis:

1) There is a degree-1 prime ideal $\mathfrak{p}$ of K such that $\left(\frac{\mathfrak{p}}{E/K}\right) = \sigma$ that satisfies $N\mathfrak{p} \leq (1 + o(1))(\log \Delta + 2n)^2$.
2) There is a degree-1 prime ideal $\mathfrak{p}$ of K such that $\left(\frac{\mathfrak{p}}{E/K}\right)$ generates the same group as σ that satisfies $N\mathfrak{p} \leq (1 + o(1))(\log \Delta)^2$.

We can in fact take $\mathfrak{p}$ to be unramified in $K/\mathbb{Q}$. A special case of this result is the following.

3) If $\gcd(m, q) = 1$, then the least prime $p \equiv m \pmod q$ satisfies $p \leq (1 + o(1))(\varphi(q) \log q)^2$.

It follows from our proof that 1) and 2) also hold for arbitrary Galois extensions, provided we replace σ by its conjugacy class $<\sigma>$. Our theorems lead to explicit versions of 1) – 3), including the following: the least prime $p \equiv m \pmod q$ is at most $2(q \log q)^2$.

1. Introduction

Several algorithms require the use of a prime in a given residue class or a prime ideal with a given Artin symbol, and the running time analyses of these algorithms rely on the Extended Riemann Hypothesis (ERH) to bound this prime or prime ideal (see, for example, Adleman and Lenstra [**1**], Bach and Shallit [**3**], and Bach and Sorenson [**4**]). In this paper, we present explicit versions of the theorems that provide these bounds.

Dirichlet's theorem [**7, 8**] states that if $\gcd(m, q) = 1$, then there are infinitely many primes p satisfying $p \equiv m \pmod q$. Linnik [**10**] proved that the least such p satisfies $p \leq q^{O(1)}$; the sharpest known estimate of the exponent is due

---

1991 *Mathematics Subject Classification.* Primary 11N13, 11M26; Secondary 11R44, 11Y35 .

This research was supported in part by NSF Grants CCR-9208639 and CCR-9204414, a Butler University Fellowship, and the Wisconsin Alumni Research Foundation.

The detailed version of this paper will be submitted for publication elsewhere.

© 1994 American Mathematical Society
0160-7634/94 $1.00 + $.25 per page

to Heath-Brown [**9**]: $p = O(q^{11/2})$. However, the available data on primes in progressions [**5, 17**] suggest this exponent is too large. In an attempt to obtain realistic estimates, several authors have invoked the ERH (see, for example, Chowla [**6**], Titchmarsh [**15**], Turán [**16**], Wang et al. [**18**]).

In algebraic number theory, Dirichlet's theorem generalizes to the theorem of Chebotarev [**14**], which states that there are infinitely many prime ideals with each possible Artin symbol, and estimates their density. Assuming the ERH, estimates for the least prime ideal with a given Artin symbol were given by Lagarias and Odlyzko [**12**], and Lagarias, Montgomery, and Odlyzko [**11**]. Oesterlé [**13**] stated that if E/K is a Galois extension of number fields, then there is a prime ideal $\mathfrak{p}$ of K with a given Artin symbol that satisfies $N\mathfrak{p} \leq 70(\log|\Delta|)^2$, where Δ is the discriminant of E. We improve on this result in two ways: our constant factor is smaller, and we show that $\mathfrak{p}$ can be taken to have degree 1, a property that is important for applications.

In the design of algorithms, one frequently uses a prime ideal with a given Artin symbol not for its own sake, but because it has some group-theoretic property. For this reason, we say that two elements of a group are *equivalent* if they generate the same subgroup. Our results are slightly sharper for the relaxed problem of finding a prime ideal with Artin symbol equivalent to a given one.

We present our asymptotic results in §2 and our explicit results in §3.

2. Asymptotic Bounds

In this section, we give asymptotic versions of our estimates. Our proofs rely upon analytic lemmas that are omitted owing to space restrictions. All of our results rely on the ERH, which states that the zeta functions of E, K, and $\mathbb{Q}$ are zero-free in $\Re(s) > 1/2$.

We refer the reader to [**2**] for undefined notation and terminology.

THEOREM 2.1 (ERH). *Let E/K be an abelian extension of number fields, with $E \neq \mathbb{Q}$. Let Δ denote the absolute value of E's discriminant. Let n denote the degree of E. Let $\sigma \in G$, the Galois group of E/K.*

(i) *There is a degree-1 prime ideal $\mathfrak{p}$ of K with $\left(\frac{\mathfrak{p}}{E/K}\right) = \sigma$ that satisfies $N\mathfrak{p} \leq (1+o(1))(\log\Delta + 2n)^2$.*

(ii) *There is a degree-1 prime ideal $\mathfrak{p}$ of K such that $\left(\frac{\mathfrak{p}}{E/K}\right)$ is equivalent to σ that satisfies $N\mathfrak{p} \leq (1+o(1))(\log\Delta)^2$.*

In both cases we can take $\mathfrak{p}$ to be unramified in $K/\mathbb{Q}$.

Before sketching the proof we note two facts (proofs omitted).

THEOREM 2.2. *If Theorem 2.1 holds for cyclic extensions, then it holds for arbitrary abelian extensions.*

COROLLARY 2.3 (ERH). *Theorem 2.1 holds for Galois extensions E/K, provided we replace σ by its conjugacy class $<\sigma>$.*

Proof of Theorem 2.1. We sketch a proof of (i). A proof for (ii) is obtained by substituting zero for $p(x)$.

By Theorem 2.2, we may assume that E/K has a cyclic Galois group G.

If χ is a character of G, the Artin symbol induces a function on the integral ideals of K; for simplicity, we use the symbol χ for both. We also write $\hat{\chi}$ for the primitive character induced by χ.

As in [**2**], we consider

$$S(x,\chi) = \sum_{N\mathfrak{a}<x} \Lambda(\mathfrak{a})\chi(\mathfrak{a}) \left(\frac{N\mathfrak{a}}{x}\right)^a \log\left(\frac{x}{N\mathfrak{a}}\right), \tag{2.1}$$

where $0 < a < 1$. If $\mu = \sigma^{-1}$, we may sum this over all characters χ of G to obtain

$$\sum_\chi \chi(\mu) S(x,\chi) = |G| \sum_{\substack{N\mathfrak{a}<x \\ \chi(\mathfrak{a})=\chi(\sigma)}} \Lambda(\mathfrak{a}) \left(\frac{N\mathfrak{a}}{x}\right)^a \log\left(\frac{x}{N\mathfrak{a}}\right). \tag{2.2}$$

Denote the contributions of proper prime powers, ramified primes, and powers of primes of degree greater than 1 to the right-hand side above by $p(x)$, $r(x)$, and $d(x)$, respectively. Also let $i(x)$ denote the absolute error incurred by using primitive characters in the left-hand sum. If no primes $< x$ meet the conditions of the theorem, then

$$\sum_\chi \chi(\mu) S(x,\hat{\chi}) \le i(x) + p(x) + r(x) + d(x). \tag{2.3}$$

Let $a = 1/\log\log\Delta$; for sufficiently large Δ we will have $0 < a < 1$.

By a residue computation and Minkowski's Theorem we have

$$\sum_\chi \chi(\mu) S(x,\hat{\chi}) \quad \sim \quad x - \sqrt{x}(\log\Delta - 2n) + \log x (\log\Delta)^{1+o(1)}.$$

We also have the following bounds:

$$\begin{aligned} i(x), r(x) &\le \log x(\log\Delta)^{1+o(1)}, \\ d(x), p(x) &\le 2n\sqrt{x}(1+o(1)). \end{aligned}$$

Thus,

$$x(1+o(1)) \quad \le \quad \sqrt{x}(\log\Delta + 2n) + \log x(\log\Delta)^{1+o(1)}.$$

Dividing by $\sqrt{x}$, noting that we may assume $x \ge (\log\Delta)^2$, and then squaring both sides completes the proof. $\square$

We give two applications of this result (proofs omitted).

COROLLARY 2.4 (ERH). *Let m and q be integers, with* $\gcd(m,q) = 1$. *There is a prime* $p \equiv m \pmod{q}$ *that satisfies* $p \le (1+o(1))(\varphi(q)\log q)^2$.

COROLLARY 2.5 (ERH). *Let K be a quadratic field, with discriminant D and class number h. Each ideal class of K contains an unramified degree-1 prime $\mathfrak{p}$ that satisfies $N\mathfrak{p} \le (1 + o(1))(h \log |D|)^2$.*

We do not know if the $(2n)$-term in Theorem 2.1 (i) can be eliminated. Our proof does show that the coefficient 2 can be replaced by any number larger than $\psi(1) - \log 2\pi - 4 = 1.584907...$ (here ψ is the digamma function).

3. Explicit Bounds

In this final section, we give explicit versions of Theorem 2.1 and Corollary 2.4.

THEOREM 3.1 (ERH). *Under the hypotheses of Theorem 2.1, there is an unramified degree-1 prime $\mathfrak{p}$ of K with $\left(\frac{\mathfrak{p}}{E/K}\right) = \sigma$ that satisfies $N\mathfrak{p} \le (4 \log \Delta + 2.5n + 5)^2$.*

THEOREM 3.2 (ERH). *Let m and q be integers, with $\gcd(m, q) = 1$. There is a prime $p \equiv m \pmod q$ that satisfies $p < 2(q \log q)^2$.*

The following tables provide more precise bounds. In each table, across the top are the ranges for n, and along the left side the ranges for $\log \Delta$. Each triple of the form (A, B, C) in the table corresponds to the bound $N\mathfrak{p} \le (A \log \Delta + Bn + C)^2$.

Bounds for $N\mathfrak{p}$, $\deg \mathfrak{p} = 1$, $\left(\frac{\mathfrak{p}}{E/K}\right) = \sigma$

$\log \Delta_E$	$n = \deg(E/\mathbb{Q})$ 2–4	5–14	15+
1–5	$(3.913, 2.57, 4.7)$	—	—
5–10	$(3.298, 2.07, 4.6)$	—	—
10–100	$(2.811, 1.94, 4.8)$	$(2.966, 1.82, 4.7)$	$(2.249, 1.6, 4.7)$
100–10000	$(1.644, 1.83, 6.1)$	$(1.759, 1.74, 5.9)$	$(1.94, 1.5, 4.9)$
10000+	$(1.048, 3.03, 19.8)$	$(1.058, 3.16, 18.1)$	$(1.073, 3.34, 16.1)$

Bounds for $N\mathfrak{p}$, $\deg \mathfrak{p} = 1$, $\left(\frac{\mathfrak{p}}{E/K}\right)$ equivalent to σ

$\log \Delta_E$	$n = \deg(E/\mathbb{Q})$ 2–4	5–14	15+
1–5	$(4.381, 0.55, 4.9)$	—	—
5–10	$(3.503, -0.05, 4.6)$	—	—
10–100	$(2.929, -0.17, 4.8)$	$(3.257, -0.21, 4.7)$	$(2.509, -0.29, 5)$
100–10000	$(1.653, -0.24, 6.1)$	$(1.787, -0.27, 5.8)$	$(1.437, -0.19, 6.9)$
10000+	$(1.048, 0.37, 18.5)$	$(1.057, 0.45, 16.9)$	$(1.073, 0.55, 15.1)$

Bounds for p, $K = \mathbb{Q}$, $\left(\frac{p}{E/\mathbb{Q}}\right) = \sigma$

$\log \Delta_E$	$n = \deg(E/\mathbb{Q})$ 2–4	5–14	15+
1–5	$(3.43, 1.44, 4.9)$	—	—
5–10	$(2.967, 0.67, 4.7)$	—	—
10–100	$(2.533, 0.47, 5)$	$(2.838, 0.35, 4.8)$	$(2.228, 0.1, 4.9)$
100–10000	$(1.527, 0.17, 6.5)$	$(1.668, 0.09, 6.1)$	$(1.885, 0, 5.6)$
10000+	$(1.038, 0.43, 20.6)$	$(1.049, 0.51, 18.3)$	$(1.065, 0.63, 15.9)$

The dashes in the tables above indicate combinations of Δ and n that are not possible by Minkowski's Theorem.

REFERENCES

1. L. M. Adleman and H. W. Lenstra, Jr., *Finding irreducible polynomials over finite fields*, Proc. 18th Ann. ACM Symp. Theory of Computing, ACM, New York, 1987, 350–355.
2. E. Bach, *Explicit bounds for primality testing and related problems*, Math. Comp. **55** (1990), 355–380.
3. E. Bach and J. Shallit, *Factoring with cyclotomic polynomials*, Math. Comp. **52** (1989), 201–219.
4. E. Bach and J. Sorenson, *Sieve algorithms for perfect power testing*, Algorithmica **9** (1993), 313–328.
5. E. Bach and L. Huelsbergen, *Statistical evidence for small generating sets*, Math. Comp. **61** (1993), 69–82.
6. S. Chowla, *On the least prime in an arithmetical progression*, J. Indian Math. Soc. **2** (1934), 1–3.
7. P. G. Lejeune Dirichlet, *Beweis eines Satzes über die arithmetische Progression*, Ber. Akad. Wiss. Berlin (1837), 108–110. Reprinted in *Werke*, **1**, 307–312.
8. ———, *Beweis des Satzes, daß jede unbegrenzte arithmetische Progression, deren erstes Glied und Differenz ganze Zahlen ohne gemeinschaftlichen Factor sind, unendlich viele Primzahlen enthält*, Abh. Akad. Wiss. Berlin (1837–9), 45–81. Reprinted in *Werke* **1**, 313–342.
9. R. Heath-Brown, *Zero-free regions for Dirichlet L-functions and the least prime in an arithmetic progression*, Proc. London Math. Soc. **64** (1991), 265–338.
10. Ju. V. Linnik, *On the least prime in an arithmetic progression, I. The basic theorem*, Mat. Sbornik **15** (1944), 139–178.
11. J. Lagarias, H. Montgomery, and A. Odlyzko, *A bound for the least prime ideal in the Chebotarev density theorem*, Invent. Math. **54** (1979), 271–296.
12. J. Lagarias and A. Odlyzko, *Effective versions of the Chebotarev density theorem*, Algebraic Number Fields (A. Fröhlich, ed.), Academic Press, London, 1977, pp. 409–464.
13. J. Oesterlé, *Versions effectives du théorème de Chebotarev sous l'hypothèse de Riemann generalisée*, Soc. Math. France Astérisque **61** (1979), 165–167.
14. N. Tschebotareff, *Die Bestimmung der Dichtigkeit einer Menge von Primzahlen, welche zu einer gegebenen Substitutionsklasse gehören*, Math. Ann. **95** (1926), 191–228.
15. E. Titchmarsh, *A divisor problem*, Rend. Circ. Mat. Palermo **54** (1930), 414–429.
16. P. Turán, *Über die Primzahlen der arithmetischen Progression*, Acta Sci. Math. **8** (1936/37), 226–235.
17. S. S. Wagstaff, Jr, *Greatest of the least primes in arithmetic progressions having a given modulus*, Math. Comp. **33** (1979), 1073–1080.
18. Y. Wang, S.-K. Hsieh, and K.-J. Yu, *Two results on the distribution of prime numbers*, Zhongguo Kexue Jishu Daxue Xuebao **1** (1965), 32–38. In Chinese. Reviewed in Math. Rev. **34** (1964), #7482.

COMPUTER SCIENCES DEPARTMENT, UNIVERSITY OF WISCONSIN-MADISON, 1210 WEST DAYTON STREET, MADISON, WI 53706
E-mail address: bach@cs.wisc.edu

MATHEMATICS AND COMPUTER SCIENCE DEPARTMENT, BUTLER UNIVERSITY, 4600 SUNSET AVENUE, INDIANAPOLIS, IN 46208
E-mail address: sorenson@butler.edu

Proceedings of Symposia in Applied Mathematics
Volume **48**, 1994

Ramanujan and Euler's Constant

RICHARD P. BRENT

ABSTRACT. We consider Ramanujan's contribution to formulas for Euler's constant γ. For example, in his second notebook Ramanujan states that (in modern notation)

$$\sum_{k=1}^{\infty} \frac{(-1)^{k-1}}{nk} \left(\frac{x^k}{k!}\right)^n = \ln x + \gamma + o(1)$$

as $x \to \infty$. This is known to be correct for the case $n = 1$, but incorrect for $n > 2$. We consider the case $n = 2$. We also suggest a different, correct generalization of the case $n = 1$.

1. Introduction

Ramanujan gave many beautiful formulas for π and $1/\pi$. Euler's constant $\gamma = -\Gamma'(1) = 0.57721566\ldots$, which occurs in many well-known formulas involving the Gamma function, the Riemann zeta function, the divisor function $d(n)$, etc. [**15**], seems to be more mysterious and more difficult to compute than π. For example, quadratically convergent iterations are known [**6, 8, 21**] for π, but none are known for γ. Also, π is transcendental, but it is not even known if γ is irrational [**3**]. If $\gamma = p/q$ is rational, then $q > 10^{15000}$. This result follows from a computation [**10**] of the regular continued fraction expansion for γ.

There may be an analogy with $\zeta(3)$. Apéry [**2, 17**] proved $\zeta(3)$ irrational, using the series

$$\zeta(3) = \frac{5}{2} \sum_{k=1}^{\infty} \frac{(-1)^{k-1} k! k!}{(2k)! k^3},$$

and, in Chapter 9 of his Notebooks, Ramanujan gives several similar series, some

---

1991 *Mathematics Subject Classification*. Primary 01A60, 33C10, 41A60; Secondary 33-01, 33-03, 33E99, 40A25, 65D20.

This paper is in final form and no version of it will be submitted for publication elsewhere.

© 1994 American Mathematical Society
0160-7634/94 \$1.00 + \$.25 per page

involving $\zeta(3)$. Ramanujan [**4**, I, p. 252] rediscovered Euler's formula

$$\zeta(3) = \sum_{k=1}^{\infty} \frac{H_k}{(k+1)^2},$$

where $H_k = \sum_{j=1}^{k} 1/j$ is a Harmonic number. Harmonic numbers also occur in formulas involving γ. For example, the well-known result

$$H_n = \ln n + \gamma + O(1/n) \tag{1}$$

as $n \to \infty$ is often used to give an alternative definition of γ.

2. Ramanujan's Papers and Notebooks

Ramanujan published one paper [**18**] specifically on γ. In it he generalizes an interesting series of Glaisher:

$$\gamma = 1 - \sum_{k=1}^{\infty} \frac{\zeta(2k+1)}{(k+1)(2k+1)}.$$

Because the generalizations all involve the Riemann zeta function or related functions, they are not convenient for computational purposes.

Much of Ramanujan's work was not published during his lifetime, but was summarized in his Notebooks. These were first printed in facsimile [**20**], and edited editions have since been published by Berndt [**4**]. Scanning the Notebooks, we find many occurrences of γ. Owing to space limitations, we concentrate on Chapter 4, Entry 9, Corollaries 1–2 [**4**, I, p. 98], because these are potentially useful for computing γ. Corollary 1 is (in modern notation)

$$\sum_{k=1}^{\infty} \frac{(-1)^{k-1} x^k}{k!k} = \ln x + \gamma + o(1) \tag{2}$$

as $x \to \infty$. In fact, Euler showed the more precise result [**4**, II, p. 167],

$$\sum_{k=1}^{\infty} \frac{(-1)^{k-1} x^k}{k!k} - \ln x - \gamma = \int_x^{\infty} \frac{e^{-t}}{t}\,dt = O\left(\frac{e^{-x}}{x}\right), \tag{3}$$

and this has been used by Sweeney [**22**] and others [**5, 9**] to compute Euler's constant (one has to be careful because of cancellation in the series). In Ch. 12, Entry 44(ii) [**4**, II, p. 168], Ramanujan correctly states that the error term $O(e^{-x}/x)$ in (3) is between $e^{-x}/(1+x)$ and e^{-x}/x.

2.1. A Generalization. Ramanujan's Corollary 2 [**4**, I, p. 98] is that

$$\sum_{k=1}^{\infty} \frac{(-1)^{k-1}}{nk} \left(\frac{x^k}{k!}\right)^n = \ln x + \gamma + o(1) \tag{4}$$

for $n > 0$. We assume that n is a fixed positive integer. Clearly, (4) generalizes (2), which is just the case $n = 1$.

Berndt [**4**, I, p. 98], using a result of Olver [**16**, Ex. 8.4, p. 309], shows that (4) is false for $n > 2$, because the function defined by the left side of (4) changes sign infinitely often and grows exponentially large as $x \to \infty$. Berndt leaves the case $n = 2$ open. In fact, (4) is true if $n = 2$. Theorem 1 below gives an exact expression for the error in (4) as an integral involving the Bessel function $J_0(x)$, and Corollary 1 deduces an asymptotic expansion.

The exact expression for $n = 2$ is a special case of a formula given by Luke [**13**, p. 48] and [**1**, formula 11.1.20]. However, the connection with Ramanujan does not seem to have been noticed before.

2.2. Avoiding Cancellation. In Chapter 3, Entry 2, Corollary 2, Ramanujan states that the sum on the left side of (2) can be written as

$$e^{-x} \sum_{k=0}^{\infty} H_k \frac{x^k}{k!}.$$

This is easy to prove [**4**, I, pp. 46–47]. Thus, (2) gives

$$\sum_{k=0}^{\infty} H_k \frac{x^k}{k!} \Big/ \sum_{k=0}^{\infty} \frac{x^k}{k!} = \ln x + \gamma + o(1). \tag{5}$$

This is more convenient than (2) for computation, because there is no cancellation in the series when $x > 0$. In § 4 we indicate how Ramanujan might have generalized (5) in much the same way that he attempted to generalize (2).

3. Ramanujan's Corollary for $n = 2$

The following result [**11**] shows that (4) is valid for $n = 2$. Recall that $J_0(x)$ is a Bessel function of the first kind and order zero.

THEOREM 1. *Let*

$$e(x) = \sum_{k=1}^{\infty} \frac{(-1)^{k-1}}{2k} \left(\frac{x^k}{k!}\right)^2 - \ln x - \gamma.$$

Then, for real positive x,

$$e(x) = \int_{2x}^{\infty} \frac{J_0(t)}{t} dt.$$

We omit details of the proof of Theorem 1. However, the reader should be able to construct a proof by proceeding as in [**4**, I, p. 99], using the fact that

$$\int_0^{\infty} \left(\frac{e^{-t} - J_0(2t)}{t}\right) dt = 0. \tag{6}$$

A slightly more general result than (6) is given in [**12**, equation 6.622.1] and is attributed to Nielsen [**14**]. An independent proof is given in [**11**].

COROLLARY 1. *Let $e(x)$ be as in Theorem 1. Then, for large positive x, we have*

$$e(x) = \frac{1}{2\pi^{1/2}x^{3/2}}\left(\cos\left(2x + \frac{\pi}{4}\right) + \frac{13\sin\left(2x + \frac{\pi}{4}\right)}{16x} + O\left(\frac{1}{x^2}\right)\right).$$

For computational purposes, it is much better to take $n = 1$ than $n = 2$ in (4), because the error for $n = 1$ is $O(e^{-x}/x)$.

4. A Different Generalization

Equation (4) may be obtained from (2) by replacing $x^k/k!$ by $(x^k/k!)^n/n$. An analogous generalization of (5) is

(7) $$\sum_{k=0}^{\infty} H_k \left(\frac{x^k}{k!}\right)^n \Big/ \sum_{k=0}^{\infty} \left(\frac{x^k}{k!}\right)^n = \ln x + \gamma + o(1)$$

as $x \to \infty$. Relation (5) is just the case $n = 1$.

It is easy to show that (7) is valid for all positive integers n. An essential difference between (4) and (7) is that there is a large amount of cancellation between terms of size $\Omega_\pm(e^{nx}x^{-(n+2)/2})$ on the left side of (4), but there is no cancellation in the numerator and denominator on the left side of (7). The function $(x^k/k!)^n$ acts as a smoothing kernel with a peak at $k \simeq x - \frac{1}{2}$. In view of (1), the result (7) is not surprising, but the speed of convergence may be surprising. Brent and McMillan [**10**] show that

(8) $$\sum_{k=0}^{\infty} H_k \left(\frac{x^k}{k!}\right)^n \Big/ \sum_{k=0}^{\infty} \left(\frac{x^k}{k!}\right)^n = \ln x + \gamma + O(e^{-c_n x})$$

as $x \to \infty$, where $c_n = \begin{cases} 1 & \text{if } n = 1, \\ 2n\sin^2(\pi/n) & \text{if } n \geq 2. \end{cases}$

In the case $n = 2$, the formula (8) has error $O(e^{-4x})$. Brent and McMillan [**10**] used this case with $x \simeq 17{,}400$ to compute γ to more than 30,000 decimal places. From Corollary 1, the same value of x in (4) would give less than 8-decimal place accuracy. Also, more than 15,000 decimal places would have to be used in the computation to compensate for cancellation of terms $\Omega_\pm(e^{2x}/x^2)$ in (4).

The case $n = 3$ of (8) is interesting because $\max_{n=1,2,\ldots} c_n = c_3 = 4.5$. However, no one seems to have used $n > 2$ in a serious computation of γ.

It would be interesting to consider the behavior of the functions occurring in (4) and (8) for positive but nonintegral values of n. Certainly, (7) is valid for all positive n, but we do not know if (8) holds when n is positive but not an integer (assuming a suitable extension of the definition of c_n).

REFERENCES

1. M. Abramowitz and I. A. Stegun (eds.), *Handbook of Mathematical Functions with Formulas, Graphs, and Mathematical Tables*, National Bureau of Standards, Washington, 1964 (reprinted by Dover, 1965). (Chapter 11 was written by Y. L. Luke.)
2. R. Apéry, *Irrationalité de $\zeta(2)$ et $\zeta(3)$,* Journées arithmétiques Luminy, Astérisque **61** (1979), 11–13.
3. D. Bailey, *Numerical results on the transcendence of constants involving π, e, and Euler's constant,* Math. Comp. **50** (1988), 275–281.
4. B. C. Berndt, *Ramanujan's Notebooks,* Parts I–III, Springer-Verlag, New York, 1985–1991.
5. W. A. Beyer and M. S. Waterman, *Error analysis of a computation of Euler's constant,* Math. Comp. **28** (1974), 599–604.
6. J. M. Borwein and P. B. Borwein, *Pi and the AGM,* John Wiley and Sons, New York, 1987.
7. J. M. Borwein, P. B. Borwein and D. H. Bailey, *Ramanujan, modular equations, and approximations to pi or how to compute one billion digits of pi,* Amer. Math. Monthly **96** (1989), 201–219.
8. R. P. Brent, *Multiple-precision zero-finding methods and the complexity of elementary function evaluation,* Analytic Computational Complexity (J. F. Traub, ed.), Academic Press, New York, 1975, 151–176.
9. R. P. Brent, *Computation of the regular continued fraction for Euler's constant,* Math. Comp. **31** (1977), 771-777.
10. R. P. Brent and E. M. McMillan, *Some new algorithms for high-precision computation of Euler's constant,* Math. Comp. **34** (1980), 305–312. (There is an error on page 310: in the definition of $V_p(z)$, "$z/k!$" should be "$z^k/k!$".)
11. R. P. Brent, *An asymptotic expansion inspired by Ramanujan,* Austral. Math. Soc. Gaz. (to appear). Also Report CMA-MR02-93, ANU, Feb. 1993 (available by ftp from `dcssoft.anu.edu.au` in the directory `pub/Brent`).
12. I. S. Gradshteyn and I. M. Ryzhik, *Tables of Integrals, Series, and Products*, fourth edition (trans. Alan Jeffrey), Academic Press, New York, 1965.
13. Y. L. Luke, *Integrals of Bessel Functions*, McGraw-Hill, New York, 1962.
14. N. Nielsen, *Theorie des Integrallogarithmus und verwandter Transzendenten*, Teubner, Leipzig, 1906 (reprinted by Chelsea, New York, 1965).
15. J. Nunemacher, *On computing Euler's constant,* Math. Mag. **65** (1992), 313–322.
16. F. W. J. Olver, *Asymptotics and Special Functions*, Academic Press, New York, 1974.
17. A. van der Poorten, *A proof that Euler missed ... Apéry's proof of the irrationality of $\zeta(3)$,* Math. Intelligencer **1** (1979), 195–203.
18. S. Ramanujan, *A series for Euler's constant γ,* Messenger of Mathematics **46** (1917), 73–80 (reprinted in [**19**]).
19. S. Ramanujan, *Collected Papers of Srinivasa Ramanujan* (G. H. Hardy, P. V. Seshu Aiyar and B. M. Wilson, eds.), Cambridge University Press, Cambridge, 1927. Reprinted by Chelsea, New York, 1962.
20. S. Ramanujan, *Notebooks*, two volumes, Tata Institute of Fundamental Research, Bombay, 1957.
21. E. Salamin, *Computation of π using arithmetic-geometric mean,* Math. Comp. **30**, 1976, 565–570.
22. D. Sweeney, *On the computation of Euler's constant,* Math. Comp. **17** (1963), 170–178.

COMPUTER SCIENCES LAB., AUSTRALIAN NATIONAL UNIVERSITY, CANBERRA, ACT 0200, AUSTRALIA

E-mail address: rpb@cslab.anu.edu.au

Proceedings of Symposia in Applied Mathematics
Volume **48**, 1994

Congruential Sieves on FPGA Computers

NATHAN D. BRONSON AND DUNCAN A. BUELL

ABSTRACT. Congruential sieves have been built as special-purpose processors since at least the 1920s. In this paper we describe the efficient design of a sieve for the Splash FPGA-based attached processor board. The combination of a host workstation and the programmable Splash board achieves the highest performance to date of any sieve on a programmable computer. We illustrate the design with a search for negative pseudosquares.

1. Introduction

Congruential sieves have been used for many years to solve problems in number theory [**LEHM70, STEP90**]. One such problem is that of finding negative pseudosquares. We shall refer to an integer $N_i \equiv 7 \pmod 8$ as a *negative pseudosquare for the* ith *prime* p_i if $-N_i$ is a nonzero quadratic residue of all odd primes $q \le p_i$. Negative pseudosquares are related to class numbers of imaginary quadratic fields [**BUEL77, SHAN73**] and are also relevant in a primality testing method [**WILL78**].

In searching for negative pseudosquares, we may use a congruential sieve. In its simplest form, a sieve comprises a set of arrays $R_{p_i}[j]$, $0 \le j \le p_i - 1$, of length p_i for some chosen primes p_i. Initially stored in the arrays are values 0 and 1, $R_{p_i}[j] = 1$ usually indicating that it is permissible to have N for which $N \equiv j \pmod{p_i}$. For negative pseudosquares, the arrays $R_{p_i}[j]$ contain 1 for those j for which $-j$ is a nonzero quadratic residue modulo p_i, and 0 for those for which $-j$ is zero or a nonresidue.

Having initialized the arrays and begun a counter at an appropriate starting value, the counter is incremented by 1, the arrays are circularly shifted one position, and the leading values (subscript 0) are ANDed together. The AND

1991 *Mathematics Subject Classification.* Primary 11-04, 11A07; Secondary 11Y55.

The first author was partially supported by the Research Science Institute of the Center for Excellence in Education, McLean, Virginia.

This paper is in final form, and no version of it will be submitted for publication elsewhere.

© 1994 American Mathematical Society
0160-7634/94 $1.00 + $.25 per page

is a 1 if and only if the current value N of the counter satisfies the appropriate residue conditions for each prime p_i. If a solution is found, the value N is output.

Special computing devices have been built since the 1920s to implement sieves, all of them either logically or physically using the above basic idea [**LEHM70, STEP90**]. One problem in obtaining high speed for such a computing problem is the need for circular shift registers of prime length; these computational steps are ill suited for general-purpose computers with intrinsic word lengths. Further, the simplicity of the design and the ability to perform the shifts for each prime in parallel make this an easy problem for which to build fast, special-purpose processors. These same characteristics—the need for multiple word lengths, the simplicity of the design, and the intrinsic parallel nature of the problem—also make congruential sieves well suited for a computing engine based on Field-Programmable Gate Arrays (FPGAs), of which Splash is one example of several [**GOKH91, MOOR91, SHBV90**]. FPGAs, especially of the reprogrammable sort (Xilinx 3090) used in Splash, allow the programmer to design an Arithmetic-Logic Unit (ALU) tailored to the application, in this case prime-length shift registers with some additional Boolean logic. The design executes at hardware speeds and with the inherent parallelism of true circuitry but can be reprogrammed as often as desired for different applications.

2. The Splash Hardware

Splash [**GOKH91**] is a linear array of 32 Xilinx 3090 FPGAs [**XILI91**]. Each FPGA contains a 16×20 rectangular grid of Configurable Logic Blocks (CLBs) with extensive signal routing resources throughout the chip. Each CLB can be used as two flip-flops, each storing one bit. Up to 70 data lines can be connected from Xilinx chip to Xilinx chip on Splash, providing high interchip bandwidth. The board itself is driven by a clock asynchronously with the Sun host at speeds of 32, 16, 8, ... MHz. Even though the theoretical maximum speed is 32 MHz, communication with the host is possible only at speeds of 16 MHz or less. Input from, and output to, the host workstation can be accomplished either through a (slow) direct read/write of the entire board by the host, or by a (much faster) set of input/output FIFOs that mitigate the performance loss of direct communication and control from the two computing systems operating at different speeds on different clocks.

3. Sieve Implementation

As described above, the sieve requires ANDing the 0th bit of the R_{p_i}-arrays, recording a hit if the result is 1, incrementing a counter by 1, and circularly shifting the arrays one location. We can increase the speed of the sieve by replicating the R_{p_i}-arrays until they are each at least K bits long, ANDing bits $R_{p_i}[j]$ together for each j, $0 \leq j \leq p_i - 1$, incrementing the counters by K, and circularly shifting the arrays by K. This permits a K-fold speed-up with only a moderate cost in hardware complexity and a slight overhead in controlling the sieve. In this Splash sieve implementation we do 64 tests at once, at a rate of 16 MHz, for a total of 1024 million tests per second.

The sieve itself on Splash is pipelined; each chip contains the R_{p_i}-array for a different prime p_i. Chip 0, for example, has the array for the prime 3, whose 64-bit array S_3 (the leading 64 bits of the extended R_3-array) is passed to Chip 1 (for the prime 5), ANDed with the Chip 1/prime 5 array S_5, passed on to Chip 2, and so forth. At present there are 22 chips used for 22 distinct primes. The signal propagation of an array S_{p_i} completely across a Xilinx chip cannot be done in one time step at 16 MHz, so the array is staged through four registers on each chip. This means that results appearing at the end of the Splash processor array lag behind the current counter by a fixed pipeline delay of 124 (each chip, including the unused chips, in the 32-chip array contributes a 4-tick propagation delay), which is corrected for by the host computer. We also note that we are, in essence, counting down and not up, since we are interested in negative pseudosquares.

In order to maximize performance, communication of answers between the Splash board and the Sun host must be done in one clock tick through the 32-bit FIFO. However, there are 94 bits of information (30 bits of counter and 64 bits that are the ANDed S_i-arrays). This problem is handled by passing the 30-bit counter, a bit indicating if any of the solution bits 0 through 31 is set, and a bit indicating if any of the solution bits 32 through 63 is set. Communication with the host thus occurs either on a counter rollover or on a pseudosquare hit. The host Sun then sieves all 32 (or 64) possible numbers for that counter value, applying both the primes implemented on Splash and those extra primes necessary for the computation. This is not computationally burdensome because 32 possibilities can be checked with a single 32-bit arithmetic operation. That is, the 32 bits of a word on the host can be ANDed in one instruction with an appropriately aligned R_{p_i}-array for each prime. The pseudosquare hit N then can be determined from the counter value and the offset into the word of any nonzero bits surviving the ANDing.

The sieve itself, implementing one prime per Xilinx chip, logically requires a circular shift register of length at least 64, shifted by 64 bits at each time step. On an FPGA like the Xilinx chip, however, it is important to minimize the physical distance that signals must travel, so the implementation of the logical shift register was somewhat more elaborate. The diagram for the prime 5 is shown in Figure 1. Initially, the integers 0-63 (0-7 are shown) labelling the boxes below them are congruent to 0, 1, 2, 3, 4, 0, ... modulo 5, and -1 and -4 are squares modulo 5, but -2 and -3 are not, and 0 is not allowed. (The value i/j in a box indicates the congruence class $-i \pmod 5$, which is or is not a square according as $j = 1$ or 0.)

Since $64 \equiv -1 \pmod 5$, incrementing the counter by 64 causes the pattern of 0's and 1's to shift by one position, the data movement indicated by the arrows in Figure 1. The long end-around shift can be avoided by using the short cycle from 4 back to 0.

In a similar but often more complicated manner, further chips hold arrays for the remaining primes between 3 and 71 inclusive, except for 53, and the primes 97, 127, and 131. Some primes have been left out of the Splash sieve because

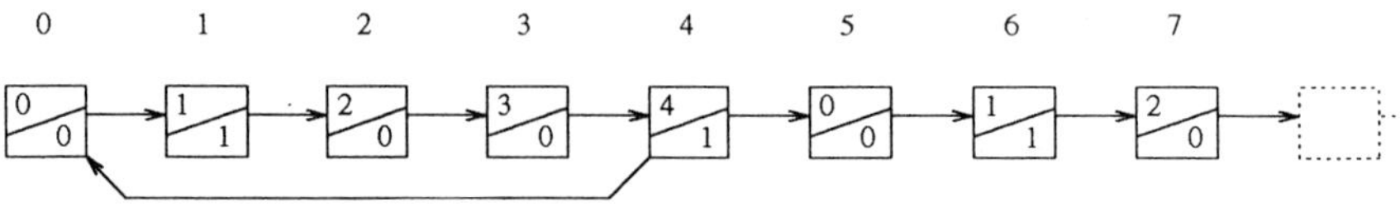

FIGURE 1.

Prime	*Least N*	*Least Prime N*	*Prime*	*Least N*	*Least Prime N*
3	23	23	101	2570169839	33857579279
5	71	71	103	2570169839	33857579279
7	311	311	107	2570169839	89206899239
11	479	479	109	2570169839	121560956039
13	1559	1559	113	328878692999	328878692999
17	5711	5711	127	328878692999	328878692999
19	10559	10559	131	513928659191	513928659191
23	18191	18191	137	844276851239	4306732833311
29	31391	31391	139	1043702750999	4306732833311
31	118271	366791	149	4306732833311	4306732833311
37	366791	366791	151	8402847753431	8402847753431
41	366791	366791	157	47375970146951	70864718555231
43	2155919	4080359	163	52717232543951	317398900373231
47	2155919	12537719	167	100535431791791	501108392233679
53	2155919	30706079	173	251109340045079	501108392233679
59	6077111	36415991	179	493092541684679	501108392233679
61	6077111	82636319	181	493092541684679	501108392233679
67	98538359	120293879	191	493092541684679	5551185799073591
71	120293879	120293879	193	1088144332169831	5551185799073591
73	131486759	131486759	197	1088144332169831	5551185799073591
79	131486759	131486759	199	1088144332169831	7832488789769159
83	508095719	2929911599	211	1088144332169831	7832488789769159
89	2570169839	2929911599	223	71608584429428591	102097158739597271
97	2570169839	7979490791	227	88163809868323439	102097158739597271

TABLE 1. Least Negative Pseudosquares

they did not lend themselves well to an efficient Xilinx array. In each case, the pattern analogous to Figure 1 was determined optimally by a case-by-case examination.

Sieving for the "inconvenient" primes is left to the host; by including a sufficient number of larger "convenient" primes on the Splash board, the host's workload is kept to a manageable level.

4. Computational Results

The sieve has been implemented, as stated, for the primes on Splash mentioned above and optionally for any combination of primes less than 500 on the host workstation. At the 16 MHz operating speed, testing 64 numbers per clock tick, a sieving rate of 1024 million tests per second has been sustained over long periods of time. Since only one residue is present modulo 8 and modulo 3, only numbers of the form $24k + 23$ need to be tested, yielding a 24-fold increase in the effective sieve rate: we are, in fact, testing $24 \cdot 1024 \cdot 10^6 = 2.4576 \cdot 10^{10}$ integers per second. In a single overnight run of 14 hours, then, we can test integers in a block of size $14 \cdot 3600 \cdot 2.4576 \cdot 10^{10} = 1.2386304 \cdot 10^{15}$. This has allowed us to extend Table 3 of Stephens and Williams [**STEP90**] to Table 1. We can also correct an error in [**LEHM70**] propagated in that table: the least negative pseudosquare for the prime 31 is 118271, not 307271.

References

[ARNO92] Jeffrey M. Arnold, Duncan A. Buell, and Elaine Davis, *Splash 2*, Proceedings, ACM Symposium on Parallel Algorithms and Architectures (1992), 316-322.

[BRIL83] John Brillhart, D. H. Lehmer, J. L. Selfridge, B. Tuckerman, and S. S. Wagstaff, *Factorizations of $b^n \pm 1$, $b = 2, 3, 5, 6, 7, 10, 11, 12$, up to High Powers*, American Mathematical Society, Providence, R. I., 1983.

[BUEL77] Duncan A. Buell, *Small class numbers and extreme values of L-functions of quadratic fields*, Math. Comp. **31** (1977), 786-796.

[GOKH91] Maya Gokhale, William Holmes, Andrew Kopser, Sara Lucas, Ronald Minnich, Douglas Sweely, and Daniel Lopresti, *Building and using a highly parallel programmable logic array*, IEEE Computer **24** (1991), 81-89.

[GOKH90] Maya Gokhale, Andrew Kopser, Sara Lucas, and Ronald Minnich, *The Logic Description Generator*, SRC Technical Report SRC-TR-90-011 (1990).

[LEHM70] D. H. Lehmer, Emma Lehmer, and Daniel Shanks, *Integer sequences having prescribed quadratic character*, Math. Comp. **24** (1970), 433-451.

[MOOR91] Will B. Moore and Wayne Luk (eds.), *FPGAs*, Abingdon EE & CS Books, Abingdon, England, 1991.

[SHBV90] M. Shand, P. Bertin, and J. Vuillemin, *Hardware speedups for long integer multiplication*, Proceedings, ACM Symposium on Parallel Algorithms and Architectures (1990), 138-145.

[SHAN73] Daniel Shanks, *Systematic examination of Littlewood's bounds on $L(1, \chi)$*, Proceedings, Symposia in Pure Mathematics, vol. 24, American Mathematical Society, Providence, Rhode Island, 1973, pp. 267-283.

[STEP90] A. J. Stephens and H. C. Williams, *An open architecture number sieve*, London Mathematical Society Lecture Note Series **154** (1990), 38-75.

[WILL78] H. C. Williams, *Primality testing on a computer*, Ars Comb. **5** (1978), 127-185.

[XILI91] Xilinx, Inc., *The Programmable Gate Array Data Book*, Xilinx, Inc., San Jose, California, 1991.

Supercomputing Research Center, 17100 Science Drive, Bowie, Maryland 20715

E-mail address: ndbrons@super.org

Supercomputing Research Center, 17100 Science Drive, Bowie, Maryland 20715

E-mail address: duncan@super.org

Proceedings of Symposia in Applied Mathematics
Volume **48**, 1994

Lehmer Pairs of Zeros and the Riemann ξ-Function

GEORGE CSORDAS, WAYNE SMITH, AND RICHARD S. VARGA

Dedicated to the memory of Professor Derrick H. Lehmer

ABSTRACT. A rigorous formulation is given for a pair of consecutive simple zeros of the Riemann ξ-function to be a "Lehmer pair" of zeros. It is shown that Lehmer pairs exist, and that each such pair gives a lower bound for the de Bruijn-Newman constant Λ (where the Riemann Hypothesis is equivalent to $\Lambda \leq 0$). This is applied to provide the following new lower bound for Λ:

$$-4.379 \cdot 10^{-6} < \Lambda.$$

The theoretical contribution of this note is a proof (based on a result of Littlewood) that if infinitely many Lehmer pairs of zeros exist, then $0 \leq \Lambda$, which is an improved version of a recent result of ours.

Let

$$H_t(x) := \int_0^\infty e^{tu^2}\, \Phi(u) \cos(xu) du \qquad (t \in \mathbb{R}; x \in \mathbb{C}), \tag{1}$$

where

$$\Phi(u) := \sum_{n=1}^{\infty} \left(2\pi^2 n^4 e^{9u} - 3\pi n^2 e^{5u}\right) \exp\left(-\pi n^2 e^{4u}\right) \qquad (0 \leq u < \infty). \tag{2}$$

Then H_0 is essentially Riemann's ξ-function, and the *Riemann Hypothesis is equivalent to the statement that all zeros of H_0 are real* (cf. p. 255 of [**T**]).

In 1950, de Bruijn [**B**] established that $H_t(x)$ has only real zeros for $t \geq 1/2$, and that if $H_t(x)$ has only real zeros for some real t, then $H_{t'}(x)$ also has only real zeros for any $t' \geq t$. Subsequently, C.M. Newman [**N**] showed in 1976 that there is a real constant $\Lambda \leq 1/2$ such that H_t has only real zeros if and only if

---

1991 *Mathematics Subject Classification.* Primary 30D10; Secondary 11M26.
Support by the National Science Foundation for this research is gratefully acknowledged.
The detailed version of this paper will appear elsewhere.

© 1994 American Mathematical Society
0160-7634/94 \$1.00 + \$.25 per page

$t \geq \Lambda$. This constant Λ is now called the *de Bruijn-Newman constant*, and the Riemann Hypothesis is equivalent to the statement that $\Lambda \leq 0$.

In 1956, Lehmer [**L**] found two simple zeros of the Riemann ζ-function on the critical line which were exceptionally close to one another. This has been referred to in the literature as "*Lehmer's near counterexample*" to the Riemann Hypothesis (cf. p. 177 of [**E**]). The zeros of H_0 corresponding to these zeros are

$$x_{6709}(0) = 14,010.125\,732\ldots, \qquad x_{6710}(0) = 14,010.201\,129\ \ldots. \tag{3}$$

Here, and below, we find it convenient to use the above numbering system where the zeros of H_t in Re $z > 0$ are numbered according to increasing modulus, and where, from the evenness of H_t, one has $x_{-j}(t) := -x_j(t)$ $(j = 1, 2, \ldots)$.

Differentiation under the integral sign in equation (1) shows that $H_t(x)$ satisfies the backward heat equation:

$$\frac{\partial H_t(x)}{\partial t} = -\frac{\partial^2 H_t(x)}{\partial x^2} \qquad (x, t \text{ real}).$$

From this, it is clear that if H_{t_0} has a local extremum at x_0 with $H_{t_0}(x_0) = 0$, and if $\epsilon > 0$ is sufficiently small, then H_t has no real zeros near x_0 for $t_0 - \epsilon < t < t_0$. But H_t must continue to have at least two zeros near x_0, so it follows that $t_0 \leq \Lambda$ since these zeros are necessarily nonreal. Similar considerations apply when H_{t_0} has a multiple zero of odd multiplicity, proving the following proposition.

PROPOSITION 1. *Suppose that H_{t_0} has a multiple real zero. Then $t_0 \leq \Lambda$. In particular, if $t > \Lambda$, then the zeros of H_t are real and simple.*

We next give a rigorous formulation of the notion of a "close pair" of zeros.

DEFINITION 1. With k a positive integer, let $x_k(0)$ and $x_{k+1}(0)$ (with $0 < x_k(0) < x_{k+1}(0)$) be two consecutive simple positive zeros of H_0, and set

$$\Delta_k := x_{k+1}(0) - x_k(0). \tag{4}$$

Then, $\{x_k(0); x_{k+1}(0)\}$ is a *Lehmer pair of zeros of H_0* if

$$\Delta_k^2 \cdot g_k(0) < 4/5, \tag{5}$$

where

$$g_k(0) := \sum_{j \neq k, k+1}{}' \left\{ \frac{1}{(x_k(0) - x_j(0))^2} + \frac{1}{(x_{k+1}(0) - x_j(0))^2} \right\}; \tag{6}$$

here (and below), the prime in the summation means that $j \neq 0$, and the summation in (6) extends over all positive and negative integers j with $j \neq k, k+1$.

From density considerations, consecutive pairs $\{x_{k_i}(0); x_{k_i+1}(0)\}$ of real zeros of H_0 can be found for which $\lim\limits_{i \to \infty} \Delta_{k_i} = 0$. We remark however that inequality (5) is not solely dependent on Δ_k, so that a Lehmer pair of zeros of H_0 requires more than just close pairs of zeros.

With the above notations, our main result can be stated as

THEOREM 1. *Let* $\{x_k(0); x_{k+1}(0)\}$ *be a Lehmer pair of zeros of* H_0. *If* $g_k(0) \leq 0$, *then* $\Lambda > 0$. *If* $g_k(0) > 0$, *set*

$$\lambda_k := \frac{(1 - \frac{5}{4}\Delta_k^2 \cdot g_k(0))^{4/5} - 1}{8g_k(0)}, \tag{7}$$

so that $-1/[8g_k(0)] < \lambda_k < 0$. *Then, the de Bruijn-Newman constant* Λ *satisfies*

$$\lambda_k \leq \Lambda. \tag{8}$$

The details of the proof can be found in [**CSV**]; here we just indicate the method. The first step is to show that if $t_0 \geq t > \Lambda$ and if $x_k(t)$ and $x_{k+1}(t)$ are two consecutive simple positive zeros of H_t, then

$$x'_{k+1}(t) - x'_k(t) = \frac{4}{[x_{k+1}(t) - x_k(t)]} - f_k(t) \cdot [x_{k+1}(t) - x_k(t)], \tag{9}$$

where

$$f_k(t) := \sum_{j \neq k, k+1}{}' \frac{2}{[x_k(t) - x_j(t)][x_{k+1}(t) - x_j(t)]}. \tag{10}$$

If the second term on the right side of (9) were zero, then its solution would be

$$(x_{k+1}(t) - x_k(t))^2 - (x_{k+1}(t_0) - x_k(t_0))^2 = 8(t - t_0). \tag{11}$$

It would follow that H_t has a multiple real zero for $\hat{t} := t_0 - (x_{k+1}(t_0) - x_k(t_0))^2/8$, and thus $\hat{t} \leq \Lambda$ by Proposition 1. This method of establishing a lower bound for Λ remains valid if f_k is not "too large". The assumption that $\{x_k(0); x_{k+1}(0)\}$ is a Lehmer pair is what is required to provide a suitable estimate for f_k.

COROLLARY 1. *If* H_0 *has infinitely many Lehmer pairs of zeros, then* $0 \leq \Lambda$.

PROOF. If H_0 has any nonreal zeros, then $\Lambda > 0$ by definition. Hence, we may assume that H_0 has only real zeros. This implies that $g_k(0) > 0$. Now it follows from a result of Littlewood (see p. 224 of [**T**]) that the separation between the real parts of consecutive zeros of H_0 tends to zero, and so $\lim_{k \to \infty} \Delta_k = 0$, since we are assuming H_0 has only real zeros. Suppose that $\{x_{k_i}(0); x_{k_i+1}(0)\}$ are Lehmer pairs of zeros of H_0. A straightforward calculation using (7) yields that

$$-\frac{5}{32}\,\Delta_{k_i}^2 < \lambda_{k_i} < -\frac{1}{8}\,\Delta_{k_i}^2, \tag{12}$$

and consequently $\lim_{i \to \infty} \lambda_{k_i} = 0$. Hence $0 \leq \Lambda$, by Theorem 1. □

Corollary 1 above is a strengthened version of Corollary 1 of [**CSV**]. It gives credence to the conjecture of Newman [**N**] that $0 \leq \Lambda$. Note that the existence of infinitely many Lehmer pairs, as in Corollary 1, would *not* disprove the Riemann Hypothesis, but would (paraphrasing Newman [**N**]) yield a "quantitative version of the dictum that the Riemann Hypothesis, if true, is only barely so".

Theorem 1 might seem to be a purely *theoretical* result, since its application requires first showing that a given pair $\{x_k(0); x_{k+1}(0)\}$ of zeros of H_0 is a *Lehmer pair* by showing that inequality (5) is satisfied. But this inequality involves *all* (real or complex) zeros of H_0 (cf. (5) and (6)), and these zeros are certainly not all known. Fortunately, given accurate values for a sufficient number of zeros of H_0 on either side of the given pair $\{x_k(0); x_{k+1}(0)\}$, we can provide an upper bound for $g_k(0)$ which may be sufficient for satisfying (5).

Indeed, this is the case for Lehmer's pair of zeros of H_0 given in (3). Our calculations, using the tabulations of te Riele [**R**], show (see §4 of [**CSV**]) that

$$\Delta^2_{6709} \cdot g_{6709}(0) < 6.936 \cdot 10^{-3} < 4/5, \tag{13}$$

so that this pair of zeros is a Lehmer pair of zeros. Applying Theorem 1 then yields the following lower bound for Λ:

$$-7.113 \cdot 10^{-4} < \Lambda, \tag{14}$$

which improves all previously known lower bounds for Λ (cf. [**CSV**]).

From A.M. Odlyzko's tabulation of the zeros of $\zeta(z)$ on the critical line [**O**], the minimum value of Δ_k for $1 \le k \le 2 \cdot 10^6$ occurs when $k = 1,115,578$. In the same manner, it is shown in §5 of [**CSV**] that

$$\Delta^2_{1,115,578} \cdot g_{1,115,578}(0) < 3.001 \cdot 10^{-4} < 4/5, \tag{15}$$

and so $\{x_{1,115,578}(0); x_{1,115,579}(0)\}$ is a Lehmer pair. Applying Theorem 1 to this pair yields the lower bound $-4.379 \cdot 10^{-6} < \Lambda$, as claimed in the Abstract.

Acknowledgment. We heartily thank Dr. A.M. Odlyzko for his kindness in providing us the zeros of $\zeta(z)$ used in the computation above.

References

[B] N.G. de Bruijn, *The roots of trigonometric integrals*, Duke J. Math. **17** (1950), 197-226.

[CSV] G. Csordas, W. Smith, and R.S. Varga, *Lehmer pairs of zeros, the de Bruijn-Newman constant Λ, and the Riemann Hypothesis*, Constr. Approx. (to appear).

[E] H.M. Edwards, Riemann's Zeta Function, Academic Press, New York, 1974.

[L] D.H. Lehmer, *On the roots of the Riemann zeta-function*, Acta Math. **95** (1956), 291-298.

[N] C.M. Newman, *Fourier transforms with only real zeros*, Proc. Amer. Math. Soc. **61** (1976), 245-251.

[O] A.M. Odlyzko, personal communication.

[R] H.J.J. te Riele, *Table of the first 15,000 zeros of the Riemann zeta function to 28 significant digits and related quantities*, Report Number 67/79 of the Mathematisch Centrum, Amsterdam, 1979.

[T] E.C. Titchmarsh, The Theory of the Riemann Zeta-function, 2nd ed. (revised by D.R. Heath-Brown), Oxford University Press, Oxford, 1986.

Department of Mathematics, University of Hawaii, Honolulu, Hawaii 96826
E-mail address: george@math.hawaii.edu, wayne@math.hawaii.edu

Inst. for Comp. Math., Kent State University, Kent, Ohio 44242
E-mail address: varga@mcs.kent.edu

Proceedings of Symposia in Applied Mathematics
Volume **48**, 1994

A Record Aliquot Sequence

Andrew W. P. Guy & Richard K. Guy

In memory of Dick Lehmer

ABSTRACT. The aliquot sequence for 840 terminates after reaching a maximum with 49 decimal digits, just larger than the square of the previous record maximum for a terminating sequence.

The **aliquot sequence** for n, $\{s^k(n)\}$, is generated from $s^0(n) = n$ by iterating the sum of aliquot parts function, $s(n) = \sigma(n) - n$, where $\sigma(n)$ is the sum of the divisors of n. The Catalan-Dickson conjecture [**1, 3**] is that all such sequences **terminate** at 1, the previous term having been prime, or **cycle** on reaching a perfect number, amicable pair or larger cycle [**4, 5, 6, 8, 9, 10**]. Selfridge and the second author and others [**2, 5, 6, 7, 11**] have given heuristic and experimental evidence for the contrary conjecture that many sequences, perhaps almost all with n even, are **unbounded**.

The main sequences with $n < 1000$ for which there was doubt were the 'Lehmer six' with $n = 276, 552, 564, 660, 840, 966$. These are now the 'Lehmer five' since a program using PARI has found that $s^{746}(840) = 601$, $s^{747}(840) = 1$, after reaching a record

$$\begin{aligned} s^{287}(840) &= 3\,463982\,260143\,725017\,429794\,136098\,072146\,586526\,240388 \\ &= 2^2 \cdot 64970467217 \cdot 6237379309797547 \cdot 2136965558478112990003 \end{aligned}$$

for the maximum of a terminating sequence.

Our program has extended the knowledge of the Lehmer five, of the 'Godwin fourteen' (main sequences starting between 1000 and 2000 whose status was unknown) and of other main sequences. In the following table the third row is the number of decimal digits in $s^k(n)$ and the fourth row shows the **driver** or **guide** [**5**] in action at that stage.

1991 *Mathematics Subject Classification.* Primary 11A25, 11Y55.
This paper is in final form and no version of it will be submitted for publication elsewhere.

© 1994 American Mathematical Society
0160-7634/94 $1.00 + $.25 per page

n	276	552	564	660	966	1074	1134	1248	1464
k	616	242	551	229	243	889	176	497	1533
	63	53	53	53	51	50	51	50	56
	2^6	$2^2 \cdot 7$	$2 \cdot 3$	$2^5 \cdot 3 \cdot 7$	$2^3 \cdot 3 \cdot 5$	2^2	$2^5 \cdot 3^2 \cdot 7$	$2^2 \cdot 7$	$2^4 \cdot 31$
n	1476	1488	1512	1560	1578	1632	1734	1920	1992
k	710	461	949	273	441	331	1061	855	851
	51	55	50	50	51	52	50	51	54
	$2^5 \cdot 3$	$2^2 \cdot 7$	$2^2 \cdot 7$	$2^2 \cdot 7$	$2^2 \cdot 7$	$2^3 \cdot 3$	$2^2 \cdot 7$	$2^4 \cdot 31$	2^4
n	2232	2340	2360	2484	2514	2664	2712	2982	3270
k	197	327	690	474	204	263	548	549	205
	50	50	50	51	50	51	50	53	53
	$2^3 \cdot 3 \cdot 5$	$2^3 \cdot 3 \cdot 5$	$2 \cdot 3$	$2^2 \cdot 7$	$2^3 \cdot 3 \cdot 5$	$2^5 \cdot 3 \cdot 7$	$2^2 \cdot 7$	$2^4 \cdot 31$	2^4
n	3366	3408	3432	3556	3564	3630	3678	3774	3876
k	899	453	793	288	500	923	385	781	527
	55	50	56	50	53	54	51	50	50
	2^6	$2^3 \cdot 3 \cdot 5$	$2^3 \cdot 3 \cdot 5$	$2^2 \cdot 7$	$2^5 \cdot 3^2$	$2^7 \cdot 3 \cdot 5$	$2^2 \cdot 7$	$2^2 \cdot 7$	$2^2 \cdot 7$
n	3906	4116	4170	4224	4290	4350	4380	4788	4800
k	421	481	152	378	626	245	698	737	571
	52	50	54	50	51	53	54	57	50
	$2^2 \cdot 7$	$2^3 \cdot 3 \cdot 5$	2^6	$2^3 \cdot 3 \cdot 5$	$2^2 \cdot 7$	$2^5 \cdot 3 \cdot 7$	$2^2 \cdot 7$	2^4	2^6
n	4842	5148	5208	5250	5352	5400	5448	5736	5748
k	195	835	351	146	371	567	1079	619	698
	51	50	51	54	50	50	61	51	53
	$2^2 \cdot 7$	2^4	$2^5 \cdot 3 \cdot 7$	2^2	$2 \cdot 3$	2^6	$2^3 \cdot 3 \cdot 5$	$2^2 \cdot 7$	$2^2 \cdot 7$
n	5778	6160	6396	6552	6680	6822	6832	6984	7044
k	501	1078	236	536	405	496	397	257	638
	50	55	51	53	50	50	50	50	50
	$2^4 \cdot 31$	$2^6 \cdot 127$	2^3	2^4	2^4	$2 \cdot 3$	$2^2 \cdot 7$	$2^2 \cdot 7$	$2^3 \cdot 5$

Seven other sequences which terminate are those for 1848 (discovered by Godwin around 1980), whose maximum is $s^{165}(1848) =$
$71189\,813234\,834347\,691364\,750868 = 2^2 \cdot 13 \cdot 149 \cdot 151 \cdot 1327 \cdot 16227697 \cdot 2825683276589$
with final prime 43 and length 672; – for 2580, whose maximum is $s^{157}(2580) =$
$16241\,338435\,391662\,496599\,610156 = 2^2 \cdot 11 \cdot 263 \cdot 1403503148582065545852023$
with final prime 1511 and length 704; – for 2850, whose maximum is $s^{260}(2850) =$
$1\,524837\,594903\,346209\,713807\,945264 = 2^4 \cdot 2085447209 \cdot 45698759129538405931$
with final prime 43 and length 600; – for 4488, whose maximum is $s^{459}(4488) =$

$$1\,807814\,653481\,873196\,164108 = 2^2 \cdot 13 \cdot 991 \cdot 146819 \cdot 3201581 \cdot 74632871$$

with final prime 43 and length 801; – for 4830, whose maximum is $s^{286}(4830) =$

$$2\,977531\,567363\,875104\,232820\,996315\,575241\,838615\,517076 = 4p,$$

close to the record, with final prime 31 and length 562; – for 6792, whose maximum is $s^{659}(6792) = 89463\,834268\,303248\,322465\,650004\,897188\,841494\,754216 =$

$$2^3 \cdot 83 \cdot 5851 \cdot 13001 \cdot 17712202615378967889295490511935762 69$$

with final prime 59 and length 1341; – and for 7752, whose maximum is $s^{178}(7752)$ $= 497\,004760\,443101\,287799\,450014 = 2 \cdot 7^2 \cdot 563 \cdot 249749 \cdot 36068021511207089,$ with final prime 601 and length 359.

Contemporaneously, Stephen Wolfram is running the 276 sequence, using one line of code from Mathematica©, and has confirmed most, and anticipated many, of our calculations for that sequence. Cameron Patterson has recently acquired Arjen Lenstra's ECM implementation and is now helping with recalcitrant factorizations.

Details are available from the authors, who are indebted to Graham Birtwistle for arranging for considerable CPU time, and to the Natural Sciences and Engineering Research Council of Canada for financial support.

Stop press: In a 94-01-05 email, Mitchell Dickerman reports finding in the previous night that the 1248 sequence terminates after reaching a maximum

$$s^{583}(1248) = 1231\,636691\,923602\,991963\,829388\,638861\,714770\,651073\,275257\,065104 = 2^4 p$$

of 58 digits; the length of the sequence is 1075.

He also announces that $s^{628}(276)$, $s^{352}(552)$, $s^{659}(554)$, $s^{262}(660)$, $s^{280}(966)$, $s^{976}(1074)$ and $s^{712}(1134)$ have respectively 65, 63, 63, 64, 63, 62 and 64 decimal digits, and are under the guidance of 2^4, $2^3 \cdot 5$, $2 \cdot 3$, $2^5 \cdot 3 \cdot 7$, $2^3 \cdot 3 \cdot 5$, $2^3 \cdot 5$ and $2^5 \cdot 3$. I am grateful for his permission to state his results.

REFERENCES

1. E. Catalan, *Propositions et questions diverses*, Bull. Soc. Math. France **16** (1887–88), 128–129.

2. J. S. Devitt, R. K. Guy and J. L. Selfridge, *Third report on aliquot sequences*, Congr. Numer. XVIII, Proc. 6th Manitoba Conf. Numer. Math., 1976, 177–204.

3. L. E. Dickson, *Theorems and tables on the sum of the divisors of a number*, Quart. J. Math. **44** (1913), 264–296.

4. Achim Flammenkamp, *New sociable numbers*, Math. Comp. **56** (1991), 871–873.

5. Richard K. Guy and J. L. Selfridge, *What drives an aliquot sequence?* Math. Comp. **29** (1975), 101–107. Corrigendum, ibid. **34** (1980), 319–321.

6. Richard K. Guy and M. R. Williams, *Aliquot sequences near* 10^{12}, Congr. Numer. XII, Proc. 4th Manitoba Conf. Numer. Math., 1974, 387–406.

7. H. W. Lenstra Jr., *Problem* 6064, Amer. Math. Monthly, **82** (1975), 1016; *solution*, ibid. **84** (1977), 580.

8. David Moews and Paul C. Moews, *A search for aliquot cycles below* 10^{10}, Math. Comp. **57** (1991), 849–855.

9. David Moews and Paul C. Moews, *A search for aliquot cycles and amicable pairs*, Math. Comp. **61** (1993), 935–938.

10. P. Poulet, *Nouvelles suites arithmétiques*, Sphinx, Deuxième Année (1932), 53–54.

11. H. J. J. te Riele, *A note on the Catalan-Dickson conjecture*, Math. Comp. **27** (1973), 189–192.

guya@cpsc.ucalgary.ca
Department of Computer Science,
The University of Calgary,
Calgary, Alberta, Canada T2N 1N4.

rkg@cpsc.ucalgary.ca
Department of Mathematics & Statistics,

Proceedings of Symposia in Applied Mathematics
Volume 48, 1994

Implications of Computational Mathematics for the Philosophy of Mathematics

ANDREW J. LAZARUS

Electronic machinery continues to play an indispensable role in mathematical research, and there is no reason to expect this phenomenon to abate. One sometimes hears that the proof of the Four Color Theorem was the first example of a proof in which the computer made a contribution that could not have been done by humans unassisted. Recall that Haken and Appel first showed that it would suffice to establish the theorem for a certain finite set of graphs; then they used a computer to four-color each of these. Mark Kac [**KRS86**, p. 93f] wrote of this proof, "The latter task was relegated to the computer which came through with flying colors, completing the tedious and relatively uninteresting part of the proof."

I believe that most mathematicians who care about the ramifications of the use of a computer agree with Kac, as do I. There is, however, a school—at least a trend—amongst philosophers of mathematics to describe computer-assisted proof as part of a "paradigm shift" (cf., of course, [**Kuh70**]) that makes the meaning of the term 'proof' different from what it was before. The *new* tradition in philosophy of mathematics is philosophical analysis of mathematical praxis, over against the *old* traditional multifurcated arguments of the formalists, the logicists, the intuitionists, and the constructivists. I respect the new tradition as a valuable, even overdue, endeavor, but within its framework what I am espousing—and what I think most computational mathematicians believe—is a conservative position.

1. History

Computer-assisted proofs have grown in both number and breadth since the Four Color Theorem, even discounting results which are calculations of the prop-

---

1991 *Mathematics Subject Classification.* Primary 00A30; Secondary 03–06.
The author expects to treat the questions in this paper in more depth later.

© 1994 American Mathematical Society
0160-7634/94 $1.00 + $.25 per page

erties of individual specific numbers, such as primality testing. Besides the Four Color Theorem, we may give partial credit to machines for the disproof of Mertens's Conjecture, the nonexistence of a certain projective plane, new results on high-dimensional polyhedra, and so on. Computational mathematics has come so much to the fore that it seemed remarkable to read that Andrew Wiles did not have any computer in the room where he apparently solved Fermat's Last Theorem, not even for word processing and e-mail.

However, the novelty of computation in pure mathematics has been greatly exaggerated. One "new philosopher" of mathematics attributed the first machine-assisted proofs to D. H. Lehmer.[1] In his Gibbs Lecture of 1965, published in [**Leh66**], Lehmer himself attributed the first electronic mathematics to a harmonic analyzer developed in the 1890s. According to Lehmer, its graphical output stumped Michelson and led Gibbs to his eponymous phenomenon. Thus, original computerized proofs are some 30 years old, while computerized inspiration has been around almost a century. Inspiration from calculation has been around even longer. We have Gauss's statement that he was led to conjecture the Prime Number Theorem on the basis of his own tables. In thinking of the electronic computer as a novelty, we forget the human calculators who were on occasion hired by *theoretical* mathematicians. These pre-electronic computers were persons who did not necessarily have high-level mathematics training but were unusually efficient at mental arithmetic [**Bec89**, p. 104]. The history of computation as heuristic is old indeed.

2. Aesthetics

Having settled that computational mathematics is not just recent mathematics, we can ask, "Is it attractive mathematics?" There is no accounting for taste, but to answer this question in the negative is to repudiate, not to affirm, the traditional standard. Gauss both used and enjoyed computations. Boring, trivial mathematics can and will be created more rapidly with computers. These results, even if published, are destined for the same obscurity as the tedious hand computations of determinants that fascinated some British mathematicians of the last century. I suggest that one adopt a holistic approach when evaluating results which include calculations.[2] To denigrate the Four Color Theorem as uninteresting by reflecting onto it a judgment of a proof (e.g. [**DH81**, p. 384]) is a dangerous practice. The paradoxical implication is that if another proof

[1]Possibly [**LLMS62**], which was brought to my attention by its fourth-named author, where the statement of the theorem was inspired by computer-generated data, and an iterated calculation in which the computer was programmed to recognize parameter values indicating success was used to finish off its proof. The Lehmers made a special point of remarking that their result was a statement in full generality about an *infinite* family of primes.

[2]My own ranking would note that the Four Color Theorem required considerable human ingenuity to reduce an infinite problem to an effective algorithm, while the disproof of Mertens's Conjecture involved a search not known in advance to succeed, and the nonexistence of the projective plane involved a finite calculation too large to do by hand.

of the Four Color Theorem without a computer appears, four-coloring becomes interesting again. Heretofore problems became *less* interesting after being solved once.

Similar criticism is voiced by Jerry King [**Kin92**, p. 89]: computerized proof will lead to a "future without elegance, a world of disfigured mathematics." Let us assume that King's inelegance is restricted to the computational components of the proofs, and not to the theorems, to avoid the paradox above. How does one distinguish the beauty of a computation by the identity of the computing agent? One could, theoretically, argue that computation will now be less elegant because the computer's speed will lead to a loss of interest in beautiful algorithms—for example, that our computer programs will sum arithmetic series by looping, not by formula. This undesirable outcome is exactly the opposite of the actual situation in computational mathematics: efficiency and complexity are being analyzed and quantified as never before. For a fixed algorithm, I see no meaningful aesthetic distinction between a hand computation and a machine computation. Criticism like King's should be understood as applying to both. Such anti-computational sentiments consign to aesthetic inferiority some of number theory's most fertile techniques, such as sieves. I doubt if this view would survive widely if someone presented a sieve-theoretic proof of the Riemann Hypothesis.

3. Epistemological Warrant

The most controversial aspect of computer-assisted proof is whether it truly entails new concepts of proof and of mathematics itself. The novel position is that the computer has turned mathematics into an experimental, empirical science, closer to (say) physics.

That mathematics is an experimental science I readily concede. Imre Lakatos showed in his *Proofs and Refutations* [**Lak77**] that mathematical knowledge increases through a process of conjecture-formation, testing, and modification: this process describes the behavior of the mathematical community much better than the sausage-grinder model of mathematicians extruding new, securely-encased theorems.

I have some difficulty, however, with the label "empirical" and with the assertion that mathematical truth is similar to physical truth. I am willing to accept Lakatos's label of "quasi-empirical" not because it means "we won't insist that mathematics is exactly like an empirical science if you'll acknowledge it's quasi-empirical"—as alleged, perhaps in jest, by the philosopher Thomas Tymoczko [**Tym91**, p. 202]—but because I think that the way in which we apprehend abstract objects is through empirical manifestations or avatars—a stronger concept than 'example.'

Tymoczko stated that the proof of the Four Color Theorem relegated mathematics to the status of physics because it introduced black-box testing into the proof [**Tym79**]. He conjured up Martian mathematicians who began to consult a Martian named Simon of super-human (or super-Martian!) computational capabilities. Eventually the other Martians, who could no longer understand how Simon arrived at his results, filled their proofs with "Simon says." I find this a dubious analogy to the process of computer programming and debugging. The computer is a tool, not an oracle: the reasons a computer programmer has for believing the output of his own program, even if unjustified, are not the same the Martians have for believing Simon. Mathematical knowledge, viewed as a collection of beliefs, has long featured reliance on others. The epistemic consequences of reliance are complicated, but in my view irrelevant to a discussion of *incremental* changes introduced by the computer.

I reiterate that it is dangerous to classify the Four Color Theorem *itself* on the basis of its known proof, which may be supplanted. But I think it is an additional error here to classify the *proof* on the basis of how much time an effective algorithm takes. Had Haken and Appel the use of a billion slaves, they could have produced the same proof without a computer. I assert that the computer and slave versions of the proof are epistemically although not morally equivalent.

The point of Tymoczko's argument was to adduce the Four Color Theorem as an example of a necessary a posteriori proposition. To the extent that he considers mathematical results as necessary, I agree with him. However, *if* the a priori/a posteriori dichotomy is coherent—by no means settled—I think the distinction must be between statements that are known *only* a posteriori and those *potentially* known a priori, say by an agent with superhuman, but not omniscient, computational abilities (cf. Krakowski [**Kra80**]). Otherwise it is possible that some of the more complicated proofs done without a computer would likewise qualify.[3]

[3] The canonical example is the complete enumeration of finite simple groups, whose proof, in original form, ran several thousand pages. To see the problem inherent in Tymoczko's classification, assume that the only person with enough experience in the field to apprehend its validity without reliance on others was D. Gorenstein. What is the status of the a priority of this theorem now that Gorenstein has died?

An even more radical attack on the status of computer-assisted proof was launched by the logician Nicolas Goodman, who claimed that such a proof is empirical because conjoined to its explicit hypotheses is the truth of the laws of physics [**Goo91**, p. 122].Goodman misinterprets the technical evidence for his claim in [**Goo91**] that mathematics is a natural science. Take, for example, his assertion that phlogistic chemistry and Euler's manipulation of divergent series are both "completely dead." The vocabulary of modern chemistry is so unsuited to describing phlogiston theory, and conversely, that it is used as an exemplar of incommensurability of competing paradigms, while Euler's *Algebra* is still perfectly readable. Goodman labels "absurd" Euler's statement $-1 = 1 + 2 + 4 + \dots$ and says it was discarded after Cauchy's development of convergence. Even if so, it was resurrected, unlike phlogiston: Goodman ignores the 2-adic integers in which Euler's series converges and his equation is true. Euclid, *pace* Goodman, was not "refuted" by Einstein but I have no space here for similar analysis. See also Koetsier [**Koe91**, p. 288].

We might as well say that a proof that depends on published articles relies not only on their correctness, but also on the empirical laws of optics. This extremely skeptical approach

Finally, I suggest that mathematics differs from physics in respect to revisability. Since Einstein it has been recognized that the truth of physics grows by both expansion and revision, i.e., our understanding about physical objects is inherently incomplete and this is reflected in the "laws of physics." Our knowledge of mathematical objects is also incomplete but this is *not* reflected in theorems. We write theorems and if we are correct, in my view their validity is permanent. It is true that later mathematicians may have to refine or re-contextualize our work, as happened when the concept of convergence was improved in the nineteenth century, but, unlike the physicist, most mathematicians do not work with the expectation revision is inevitable. This belief I share and find well justified.

REFERENCES

[Bec89] Petr Beckmann, *A history of π*, Dorset, New York, 1989.

[Dav72] Philip J. Davis, *Fidelity in mathematical discourse: Is one and one really two?*, Amer. Math. Monthly **79** (1972), 252–263.

[DH81] Philip J. Davis and Reuben Hersh, *The mathematical experience*, Birkhäuser, Boston, 1981.

[Goo91] Nicolas D. Goodman, *Modernizing the philosophy of mathematics*, Synthese **88** (1991), no. 2, 119–126.

[Kin92] Jerry P. King, *The art of mathematics*, Plenum Press, New York and London, 1992.

[Koe91] Teun Koetsier, *Lakatos' philosophy of mathematics*, Studies in the history and philosophy of mathematics, vol. 3, North-Holland, Amsterdam, 1991.

[Kra80] Israel Krakowski, *The Four-Color problem revisited*, Philos. Studies **38** (1980), 91–96.

[KRS86] Mark Kac, Gian-Carlo Rota, and Jacob T. Schwartz, *Discrete thoughts*, Birkhäuser, Boston, 1986.

[Kuh70] Thomas S. Kuhn, *The structure of scientific revolutions*, Second revised ed., University of Chicago, Chicago, 1970.

[Lak77] Imre Lakatos, *Proofs and refutations*, Cambridge University Press, Cambridge, 1977, Reprinted with corrections.

[Leh66] D. H. Lehmer, *Mechanized mathematics*, Bull. Amer. Math. Soc. **72** (1966), no. 5, 739–750.

[LLMS62] D. H. Lehmer, E. Lehmer, W. H. Mills, and J. L. Selfridge, *Machine proof of a theorem on cubic residues*, Math. Comp. **16** (1962), 407–415.

[Tym79] Thomas Tymoczko, *The Four-Color problem and its philosophical significance*, J. Philos. **76** (1979), 57–83.

[Tym91] Thomas Tymoczko, *Mathematics, science and ontology*, Synthese **88** (1991), no. 2, 201–228.

NATURAL LANGUAGE INC., ALAMEDA, CA

Current address: 2745 Elmwood Avenue, Berkeley, CA 94705–2311

E-mail address: lazarus@nli.com

has indeed been promulgated by Philip J. Davis [**Dav72**]. In Davis's concept of mathematical truth even hand computations have no exact, truthful answer, but merely one answer much more probable than the others. History has not been kind to this article, in which (1) Davis asks why mathematicians do not seriously consider the possibility that Fermat's Last Theorem is undecidable and (2) Davis adduces the failures of computers at handwriting recognition as evidence for his stand against symbolic fidelity. I can not imagine how to assign 'probabilities' in this context but have no space here to discuss all of the problems I see with Davis's conception of mathematics.

Proceedings of Symposia in Applied Mathematics
Volume **48**, 1994

Square Roots of Products of Algebraic Numbers

PETER L. MONTGOMERY

ABSTRACT. Let α be an algebraic number and $g_i(\alpha) \in \mathbb{Q}(\alpha)$ for $1 \le i \le n$. Suppose we suspect $\gamma(\alpha) = \prod_i g_i(\alpha)$ is a nonzero square in $\mathbb{Q}(\alpha)$. We assume that the prime ideal factorization of each $(g_i(\alpha))$ is known. Using this ideal factorization, we construct a square root of $\gamma(\alpha)$, if it exists. The algorithm uses lattice basis reduction to estimate a square root, repeatedly simplifying the problem until it can be done directly.

1. Introduction

The Number Field Sieve (NFS) algorithm [**5, 2, 1**] is asymptotically the fastest known algorithm for factoring large integers. One phase of NFS requires finding the square root of an algebraic number.

Let $f(X) = \sum_{j=0}^{d} c_j X^j \in \mathbb{Z}[X]$ be an irreducible polynomial of degree d. Let α be a root of f. To simplify the presentation, we assume that f is monic (and α is an algebraic integer), although this assumption is not satisfied in practice.

NFS constructs a nonempty finite set S of integer pairs (a_i, b_i) such that $\prod_{i \in S}(a_i - b_i\alpha)$ is believed to be a nonzero square in $\mathbb{Q}(\alpha)$. The product might have 10^6 terms, with each a_i and b_i around 10^7. This product is an example of the more general product

$$\gamma = \gamma(\alpha) = \prod_{i \in S} g_i(\alpha), \tag{1.1}$$

where each $g_i(\alpha) \in \mathbb{Q}(\alpha)^*$ and where $\sqrt{\gamma} \in \mathbb{Q}(\alpha)^*$. We assume that the prime factorization of each norm $N(g_i(\alpha))$ is known.

Couveignes [**3**] gives an algorithm for constructing $\sqrt{\gamma}$ when the degree d is odd and $\sqrt{\gamma} \in \mathbb{Z}[\gamma]$. He observes that exactly one of $\pm\sqrt{\gamma}$ has positive norm, and uses the Chinese Remainder Theorem after determining the coefficients of $\sqrt{\gamma}$ modulo several single-precision inert primes.

1991 *Mathematics Subject Classification.* Primary 11Y40; Secondary 11Y05.

Supported in part by NSF Grant #DMS–9012989.

The final version of this paper will be submitted for publication elsewhere.

© 1994 American Mathematical Society
0160-7634/94 $1.00 + $.25 per page

Couveignes's time is at least $\mathcal{O}(|S|^2)$, since at least $\mathcal{O}(|S|)$ inert primes are needed and it takes time $\mathcal{O}(|S|)$ to evaluate (1.1) modulo any prime. We give an algorithm for $\sqrt{\gamma}$ whose time is linear in $|S|$ (ignoring the implicit growth of a_i and b_i as $|S| \to \infty$) and which works for all values of d. Like the original $\gamma(\alpha)$, its constructed square root will have a product form.

2. Some algebraic number theory

Let α have d conjugates $\{\alpha_j\}_{j=1}^d$. Given $\beta = \beta(\alpha) \in \mathbb{Q}(\alpha)$, define its *norm*

$$N(\beta) = N_{\mathbb{Q}(\alpha)/\mathbb{Q}}(\beta) = \prod_{j=1}^{d} \beta(\alpha_j)$$

to be the product of all algebraic conjugates of β. This norm is a rational number.

We denote the set of algebraic integers in $\mathbb{Q}(\alpha)$ by $\mathcal{O}_\alpha$. The set $\mathcal{O}_\alpha$ is closed under addition and multiplication.

A *fractional ideal* of $\mathcal{O}_\alpha$ is a subset $I \subseteq \mathbb{Q}(\alpha)$ such that

(1) There exists $r \in \mathbb{Q}(\alpha)$ such that $I \subseteq r\mathcal{O}_\alpha$;
(2) $I \neq \emptyset$ and $I \neq \{0\}$;
(3) I is closed under addition;
(4) I is closed under multiplication by elements of $\mathcal{O}_\alpha$.

If equality holds in (1), then I is said to be a *principal ideal* with generator r; this ideal is written (r).

If I_1 and I_2 are fractional ideals, then their product $I_1 I_2$ is the smallest subset of $\mathbb{Q}(\alpha)$ which is closed under addition and which contains all products $i_1 i_2$ where $i_1 \in I_1$ and $i_2 \in I_2$. The inverse of I is $I^{-1} = \{i \in \mathbb{Q}(\alpha) : iI \subseteq \mathcal{O}_\alpha\}$.

A fractional ideal I of $\mathcal{O}_\alpha$ is said to be an *integral ideal* if $I \subseteq \mathcal{O}_\alpha$ (i.e., if $r = 1$ works in (1)).

A *prime ideal* is an integral ideal I such that if I_1 and I_2 are integral ideals with $I_1 I_2 \subseteq I$, then $I_1 \subseteq I$ or $I_2 \subseteq I$.

THEOREM 2.1. *The fractional ideals of $\mathcal{O}_\alpha$ form an abelian group under multiplication, with identity* $(1) = \mathcal{O}_\alpha$. *Any fractional ideal I can be uniquely expressed as a product*

$$I = \mathfrak{P}_1^{e_1} \cdot \mathfrak{P}_2^{e_2} \cdots \mathfrak{P}_k^{e_k}, \tag{2.1}$$

where the $\mathfrak{P}_i$'s are distinct prime ideals and $e_i \in \mathbb{Z}$.

DEFINITION 2.1. If I is a fractional ideal of $\mathcal{O}_\alpha$, then its *norm* $N(I)$ is the least positive rational q such that $N(r)/q \in \mathbb{Z}$ for all $r \in I$.

The norm is multiplicative: $N(I_1 I_2) = N(I_1)N(I_2)$ and $N(I^{-1}) = N(I)^{-1}$.

DEFINITION 2.2. Let I be a fractional ideal of $\mathcal{O}_\alpha$, with factorization (2.1). Define the *numerator* and *denominator* of I to be numer(I) and denom(I), respectively, where

$$\operatorname{numer}(I) = \mathfrak{P}_1^{\max(e_1,0)} \cdot \mathfrak{P}_2^{\max(e_2,0)} \cdots \mathfrak{P}_k^{\max(e_k,0)},$$
$$\operatorname{denom}(I) = \mathfrak{P}_1^{\max(-e_1,0)} \cdot \mathfrak{P}_2^{\max(-e_2,0)} \cdots \mathfrak{P}_k^{\max(-e_k,0)}.$$

Equivalently, $\operatorname{numer}(I) = I \cap \mathcal{O}_\alpha$ and $\operatorname{denom}(I) = I^{-1} \cap \mathcal{O}_\alpha$.

THEOREM 2.2. *If I is a fractional ideal of $\mathcal{O}_\alpha$, then* numer(I) *and* denom(I) *are integral ideals. Also*

$$I = \operatorname{numer}(I)/\operatorname{denom}(I) \qquad \textit{and} \qquad N(I) = N(\operatorname{numer}(I))/N(\operatorname{denom}(I)).$$

3. Lattice basis reduction

Fix an integral basis $\{\mathbf{o}_1, \mathbf{o}_2, \ldots, \mathbf{o}_d\}$ of the d-dimensional lattice $\mathcal{O}_\alpha$ over $\mathbb{Z}$. If $\mathbf{v} = \sum_{i=1}^d v_i \mathbf{o}_i \in \mathcal{O}_\alpha$, define $\|\mathbf{v}\| = (v_1^2 + \cdots + v_d^2)^{1/2}$.

Suppose L is a d-dimensional sublattice of $\mathcal{O}_\alpha$. Lovász et al. [**4**, §1] give a polynomial-time algorithm (called *LLL reduction*) for finding a basis of short vectors, given an arbitrary basis. Their algorithm finds $\mathbf{v} \in L$ such that

$$0 < \|\mathbf{v}\| \leq 2^{(d-1)/4} \det(L)^{1/d}, \tag{3.1}$$

where $\det(L)$ is the determinant of the lattice. If L is an ideal, then $\det(L) = N(L)$.

4. Square roots via successive approximations

We want a square root of (1.1). Define $\gamma_1 = \gamma$. Initialize $\ell = 1$. If $\ell \geq 1$, then step ℓ constructs $\delta_\ell = \delta_\ell(\alpha)$ from γ_ℓ such that $\gamma_{\ell+1}$ is in some sense "smaller" than γ_ℓ, where

$$\gamma_{\ell+1} = \gamma_\ell \delta_\ell^{-2s_\ell}, \qquad s_\ell \in \{\pm 1\}.$$

The identity

$$\gamma(\alpha) = \gamma_\ell(\alpha) \left[\prod_{k=1}^{\ell-1} \delta_k(\alpha)^{s_k}\right]^2 \tag{4.1}$$

will hold for $\ell \geq 1$. Eventually, some γ_ℓ will be sufficiently small that its coefficients can be computed explicitly and a square root found by another method.

Start by evaluating the roots of f numerically and using (1.1) to approximate each embedding $|\gamma_1(\alpha_j)|$. Determine the prime ideal factorization of the fractional ideal $(\gamma_1(\alpha))$. Unless all exponents are even, reject $\gamma(\alpha)$ as a nonsquare.

Let $\ell \geq 1$. At the start of step ℓ we will have a product formula for $\gamma_\ell(\alpha)$ obtained from (4.1). We will know approximations to $|\gamma_\ell(\alpha_j)|$ for $1 \leq j \leq d$. We will know the prime ideal factorization of $(\gamma_\ell(\alpha))$, with each ideal having an even exponent. The algorithm selects $s_\ell = +1$ or $s_\ell = -1$ depending primarily upon whether the numerator or the denominator of γ_ℓ is larger (i.e., it chooses whether to simplify γ_ℓ or γ_ℓ^{-1}). If one assumes $s_\ell = +1$, it finds $\delta_\ell \in \mathcal{O}_\alpha$ such that (δ_ℓ) is divisible by several of the prime ideals dividing the numerator of $(\sqrt{\gamma_\ell})$. This δ_ℓ is an approximation to $\sqrt{\gamma_\ell}$. Although (δ_ℓ) may be divisible by other prime ideals as well, we can bound the norm of these additional ideals in terms of the coefficients of f. If we require enough ideals to divide (δ_ℓ), then the product of norms of the numerator and denominator of $(\gamma_\ell/\delta_\ell^2)$ will be smaller than the corresponding product for γ_ℓ. Once this product becomes sufficiently small, subsequent γ_ℓ will be algebraic integers (i.e., denominator norm 1). Then the algorithm concentrates on reducing the unit contribution by choosing δ_ℓ so as to minimize the absolute values $\delta_\ell(\alpha_j)/|\gamma_\ell(\alpha_j)|^{1/2}$ of the embeddings.

More precisely, if $s_\ell = +1$, choose an ideal $I = \mathfrak{P}_1^{e_1} \cdot \mathfrak{P}_2^{e_2} \cdots \mathfrak{P}_k^{e_k}$ dividing $\operatorname{numer}((\sqrt{\gamma_\ell}))$, where the $\mathfrak{P}_i$ are distinct prime ideals and $e_i > 0$, with $N(I)$ as large as computationally convenient. Find a basis for I in $\mathcal{O}_\alpha$. Use LLL reduction to find a small nonzero $\mathbf{v} \in I$ satisfying (3.1) (with $L = I$).

By the formula for the resultant, if $\mathbf{v} = \sum_{j=1}^{d} v_j \mathbf{o}_j = \sum_{j=0}^{d-1} v_j' \alpha^j$, then

$$N(\mathbf{v}) = \pm \begin{vmatrix} v'_{d-1} & v'_{d-2} & v'_{d-3} & \cdots & v'_0 & 0 & \cdots & 0 \\ 0 & v'_{d-1} & v'_{d-2} & \cdots & v'_1 & v'_0 & \cdots & 0 \\ \vdots & 0 & \ddots & \ddots & \ddots & \ddots & \cdots & 0 \\ 0 & 0 & 0 & 0 & v'_{d-1} & v'_{d-2} & \cdots & v'_0 \\ c_d & c_{d-1} & c_{d-2} & \cdots & c_1 & c_0 & & 0 \\ 0 & c_d & c_{d-1} & \cdots & c_2 & c_1 & \cdots & 0 \\ \vdots & \ddots & \ddots & \ddots & \ddots & \ddots & \cdots & 0 \\ 0 & 0 & \cdots & c_d & c_{d-1} & c_{d-2} & \cdots & c_0 \end{vmatrix}$$

(a $(2d-1) \times (2d-1)$ matrix). Hadamard's inequality gives

$$|N(\mathbf{v})| \le \left[\sum_{j=0}^{d-1} (v_j')^2\right]^{d/2} \left[\sum_{j=0}^{d} c_j^2\right]^{(d-1)/2} = \|\mathbf{v}'\|^d \, \|f\|^{d-1}, \tag{4.2}$$

where $\|f\| = \left(\sum_{j=0}^{d} c_j^2\right)^{1/2}$. This and $\|\mathbf{v}\|^d \le 2^{d(d-1)/4} N(I)$ imply

$$\left|\frac{N(\mathbf{v})}{N(I)}\right| \le 2^{d(d-1)/4} \, \|f\|^{d-1} \frac{\|\mathbf{v}'\|^d}{\|\mathbf{v}\|^d} \le C,$$

where C depends only on f and on the basis $\{\mathbf{o}_j\}$. If we choose $\delta_\ell = \mathbf{v}$, then

$$N(\gamma_{\ell+1}) = N\left(\frac{\gamma_\ell}{\delta_\ell^{2s_\ell}}\right) = \frac{N(\gamma_\ell)}{N(\mathbf{v})^2} = \frac{N(\gamma_\ell)/N(I)^2}{N(\mathbf{v})^2/N(I)^2}.$$

The numerator norm of $(\sqrt{\gamma_{\ell+1}})$ has dropped by a factor of $N(I)$ (compared to that of $(\sqrt{\gamma_\ell})$), while the denominator norm has increased at most by a factor of C. If $N(I) \gg C$, then repeated application of this construction will reduce the product of the numerator and denominator norms of $\sqrt{\gamma}$ until both are below C. Two more iterations (in which I is the entire remaining numerator or entire remaining denominator) will replace γ by an algebraic integer of norm at most C^2.

REMARK 4.1. The algorithm as stated requires the full factorization of (γ_ℓ) into prime ideals. The factorization of $(\gamma_{\ell+1})$ can be obtained from those of (γ_ℓ) and of $(\mathbf{v})$. Several ideals dividing $(\mathbf{v})$ are known since $\mathbf{v} \in I$, but we may be left with a nonprime ideal I' after dividing by known factors and not know how to factor the integer $N(I')$. Fortunately, the algorithm does not require the full factorization of I' if it is relatively prime to the other ideals in $(\gamma_{\ell+1})$.

REMARK 4.2. It may be computationally convenient to use $\{1, \alpha, \dots, \alpha^{d-1}\}$ rather than $\{\mathbf{o}_1, \dots, \mathbf{o}_d\}$. The powers of α might not generate all of $\mathcal{O}_\alpha$, but this only increases C in (4.2).

If γ_ℓ has small integer coefficients, then its norm must be small; alas this condition is not sufficient (e.g., γ_ℓ might be a power of a unit). But since we know the embeddings $|\gamma_\ell(\alpha_j)|$, we can bound its coefficients using Lagrange's interpolation formula. If no embedding is large, then we can compute γ_ℓ and its square root using modular or p-adic arithmetic. If some embedding $|\gamma_\ell(\alpha_j)|$ is too large, then we can choose δ_ℓ to be small near α_j, ensuring $\gamma_{\ell+1} = \gamma_\ell\delta_\ell^2$ is better behaved. Details will appear in a later version of this paper.

REFERENCES

1. Daniel J. Bernstein and A. K. Lenstra, *A general number field sieve implementation*, The Development of the Number Field Sieve (A.K. Lenstra and H. W. Lenstra, Jr., eds.), Lecture Notes in Mathematics, 1554, Springer–Verlag, Berlin, 1993, pp. 103–126.
2. J. P. Buhler, H. W. Lenstra, Jr., and Carl Pomerance, *Factoring integers with the number field sieve*, The Development of the Number Field Sieve (A.K. Lenstra and H. W. Lenstra, Jr., eds.), Lecture Notes in Mathematics, 1554, Springer–Verlag, Berlin, 1993, pp. 50–94.
3. Jean-Marc Couveignes, *Computing a square root for the number field sieve*, The Development of the Number Field Sieve (A.K. Lenstra and H. W. Lenstra, Jr., eds.), Lecture Notes in Mathematics, 1554, Springer–Verlag, Berlin, 1993, pp. 95–102.
4. A. K. Lenstra, H. W. Lenstra, Jr., and L. Lovász, *Factoring polynomials with rational coefficients*, Math. Ann. **261** (1982), 515–534.
5. A. K. Lenstra, H. W. Lenstra, Jr., M. S. Manasse, and J. M. Pollard, *The number field sieve*, Proceedings of the Twenty Second Annual ACM Symposium on Theory of Computing, Baltimore, May 14–16, 1990 (New York), ACM, pp. 564–572.

DEPARTMENT OF MATHEMATICS, OREGON STATE UNIVERSITY, CORVALLIS, OR 97331–4605
Current address: 780 Las Colindas Rd., San Rafael, CA 94903–2346
E-mail address: pmontgom@math.cwi.nl

Proceedings of Symposia in Applied Mathematics
Volume **48**, 1994

A Locally Parameterized Version of Lehmer's Problem

GARY A. RAY

ABSTRACT. Lehmer asked if there exist monic polynomials $p(x)$ with integer coefficients such that Mahler's measure $M(p)$ is arbitrarily close to 1. No reciprocal polynomial has been found whose measure is smaller than Lehmer's 10th-degree example. We parameterize the problem by looking at reciprocal $p_t(x) = p(x,t) \in R[x,t]$ of even degree, where p_0 is a product of cyclotomics. Hence, the parameterization "moves" $M(p_t)$ from 1 to values greater than one, and we are interested in small values on an integer lattice. We look at this problem locally by expanding the roots $\{x(t)\}$ of p_t and truncating the resulting fractional power series. We solve questions analogous to Lehmer's question as well as a related conjecture by Schinzel and Zassenhaus for this locally parameterized version for certain families of products of cyclotomics, using the LLL and/or Choleski square reduction algorithm. Some of these local solutions turn out to give Boyd's examples of small limit points of $\{M(p)\}$.

1. Introduction

In the study of certain sequences, Lehmer in [**5**] asked the following question:

> Are there polynomials p with integer coefficients such that $M(p)$ is arbitrarily close to 1?

Here, $M(p)$ is Mahler's measure of the polynomial $p = a_0 + a_1x + \cdots + a_nx^n = a_n \prod_{i=1}^n (x - r_i)$, which is the absolute value of a_n times the product of roots outside the unit circle, i.e., $M(p) = |a_n| \prod_{i=1}^n \max\{|r_i|, 1\}$. Lehmer's question is still unsolved but the answer is expected to be negative. Indeed, no smaller measure greater than 1 has been found than Lehmer's example:

$$M(x^{10} + x^9 - x^7 - x^6 - x^5 - x^4 - x^3 + x + 1) \approx 1.17628.$$

1991 *Mathematics Subject Classification.* Primary 12-04; Secondary 11Y40, 11H55.
The final version of this paper will be submitted for publication elsewhere.

© 1994 American Mathematical Society
0160-7634/94 \$1.00 + \$.25 per page

A related conjecture by Schinzel and Zassenhaus says that there should exist a constant c such that

$$\Omega(p) < 1 + c/n$$

implies p only has roots on the unit circle (and possibly z^m factors), where $\Omega = \max\{|r_i|\}$.

Now by a classical result of Kronecker, if $M(p) = 1$, then p must be "cyclotomic", i.e., a product of cyclotomic polynomials and z^m. Thus in some sense we are asking for integer polynomials which are "close" to cyclotomics. Because of a result of Smyth ([**7**]), we need only look at monic reciprocal polynomials of even degree $2n$.

Let $p_t(x) = r(x) + tq(x) \in \mathbf{R}[x,t]$ of degree $2n$, where $q = 0 + q_1 x + q_2 x + \cdots + q_n x^n + \cdots + q_1 x^{2n-1} + 0x^{2n}$ with r and q (thought of as having degree $2n$) being reciprocal. Then p_t is a family of monic reciprocal polynomials with real coefficients parameterized by t. Furthermore, if $r(x) = p_0(x)$ is cyclotomic, the parameterization moves $M(p_t)$ continuously from a value of 1 to values greater than 1 and we are interested in small values of $M(p_t)$ where the first half of the coefficients of q lie on an integer lattice in $\mathbf{R}^n$ and t is a nonzero integer (which is later chosen to be 1).

We can expand $M(p_t)$ in powers of t by expanding the algebraic curves defined by $p_t(x) = 0$ in powers of t. Therefore, let $p(x)$ be a reciprocal cyclotomic polynomial of degree $2n-2$ with distinct roots and such that $p(1) \neq 0$. Let

$$\begin{aligned} p_t(x) &= p(x)(x-1)^2 + tq(x) \\ &= (x - s(t))(x - 1/s(t)) \prod_{j=1}^{n-1} (x - r_j(t)) \prod_{j=1}^{n-1} (x - 1/r_j(t)). \end{aligned} \tag{1.1}$$

Here, $s(t), 1/s(t)$ are the curves which start at 1, and $r_j(t)$ is the curve which starts at the jth cyclotomic root u_j of $p(x)$ on the upper half of the unit circle. Then each $r_j(t)$ is locally analytic at 0, and can be expanded as

$$r_j(t) = u_j(1 + ib_j t + \cdots),$$

where b_j is real. Because 1 is a double root,

$$s(t) = 1 + a\sqrt{t} + \cdots.$$

For simplicity, we can translate the problem from fractional power series to regular power series by replacing t by t^2 in Eq. (1.1). If we then reduce Eq. (1.1) modulo t^3, we can look at an approximate localized problem which may be easier. After truncating the expansions, it follows that a simple matrix equation then relates $(a^2, b_1, \ldots, b_{n-1})$ to the integer vector $(q_1, q_2, \ldots, q_n)$ via

$$M_r(a^2, b_1, \ldots, b_{n-1})^t = (q_1, q_2, \ldots, q_n)^t.$$

It is easy to show the following.

THEOREM 1.1. *The integer* $-2\sum_{j=1}^{n-1} q_j - q_n$ *is a positive multiple of* a^2. *In fact,* $-2\sum_{j=1}^{n-1} q_j - q_n = a^2 \prod_{j=1}^{n-1}(1-u_j)(1-\overline{u_j}) \leq (2n-2)a^2$.

Thus, to minimize $a \neq 0$ and hence the first coefficient in the expansion of $M(p_t)$, set $q_n = -1 - 2\sum_{j=1}^{n-1} q_j$. Then the largest "root" $1+a > 1+\frac{1}{\sqrt{2n-2}}$, satisfying an analogue to the Schinzel-Zassenhaus conjecture[1].

The other coefficients b_j relate to movements of the roots along the unit circle. Hence, to minimize this movement, we choose to minimize the 2-norm of the vector $b = (b_1, b_2, \dots, b_{n-1})$, given this value of q_n. This amounts to solving for the minimum of an inhomogeneous quadratic form $(q-d)^t M_p (q-d)$ over $q \in Z^{n-1}$ for a certain matrix M_p derived from M_r and a certain vector d. Thus, the locally parameterized version of Lehmer's question amounts to the solution of families of inhomogeneous quadratic forms associated with perturbed cyclotomic polynomials. Heuristically speaking, if $(q-d)^t M_p (q-d)$ approached zero fast enough, the "roots" on the unit circle would not then join and leave the unit circle, and so would not contribute to Mahler's measure, thus keeping $M(p_1)$ close to 1.

2. Solutions for certain cyclotomic families

Case I. Let $p(x) = (x^{2n+1}-1)/(x-1)$. Then $(2n+1)M_p$ and $(2n+1)d$ have integer entries and can be determined explicitly. Furthermore, the Choleski factorization of M_p can also be determined, leading to the result that the minimum of $(q-d)^t M_p (q-d)$ is gotten by setting $q_j = 0, j \leq n$, and this unique minimum tends to $\frac{1}{24}$, which says heuristically that more and more roots are leaving the unit circle as $n \to \infty$. Therefore, $q_{n+1} = -1$ and the solution to Case I is $p_t = (x^{2n+1}-1)(x-1) - tx^{n+1}$. By setting $t=1$, this gives Boyd's family of polynomials with the second smallest known limit point (see [**1**]), namely, $M(p_1) \overset{n\to\infty}{\longrightarrow} 1.28573 = M(x+y+1+x^{-1}+y^{-1})$!

Case II. First consider $p(x) = (x^{2n+1}-1)(x^{2n-1}-1)/(x-1)^2$. It turns out that M_p has a special form as before; and via a permutation of the rows and columns, the Choleski factorization becomes very simple and the zero vector is the unique minimum again. Hence, $q_{2n} = -1$ and the solution to this case is $p_t = (x^{2n+1}-1)(x^{2n-1}-1) - tx^{2n}$. This gives the family of polynomials considered by Smyth when $t=1$ with the same limit point as before.

Second, consider another two-factor case, $p(x) = (x^{2n}+1)(x^{2n-1}-1)/(x-1)$. Then the result is more complicated. It turns out that there are $2^{\lfloor (n+1)/2 \rfloor}$ minimal vectors with the same minimum as the all-zero vector. In the case of the all-zero vector, $p_t = (x^{2n}+1)(x^{2n-1}-1)(x-1) - tx^{2n}$ and $M(p_1) \to 1.25542$, the smallest known limit point for Mahler's measure.

Third, consider the more general $p(x) = (x^{2n+1}-1)(x^{2m+1}-1)/(x-1)^2$. It might seem reasonable to guess that the all-zero vector minimizes the quadratic

[1]Note that the problem looks more analogous when $p_t(x) = p(x)(x-1) + tq(x)$, since in this case there is no square root in the denominator.

form as before. Indeed, computations show this is true for $1 \le m < n < 19$, but it is false for certain values beyond that point.

Example. Let $p(x) = (x^{39} - 1)(x^{35} - 1)/(x - 1)^2$. Then it turns out that $q_1(x) = -x^{72} + x^{37} - x^2$ gives a smaller norm than $q_2(x) = -x^{37}$, and in fact their Mahler measures have a similar relationship: $M(p(x) + q_1(x)) \approx 1.2472$ while $M(p(x) + q_2(x)) \approx 1.2883$.

Case III. Suppose $p(x) = (x^{2n+1} - 1)(x^{2m+1} - 1)(x^{2l+1} - 1)/(x - 1)^3$. Then a computation was done over small values of (n, m, l), and very few polynomials with small measure were generated. No further cases have yet been examined.

3. Comments about the algorithm

The algorithm to solve the locally parameterized version of Lehmer's problem is essentially as follows. Given a cyclotomic polynomial $p(x)$ of degree $2n - 2$ with distinct roots and such that $p(1) \neq 0$, form the matrix M_r, compute its inverse and solve for the inhomogeneous vector d. Then from these compute the matrix M_p and solve for the associated vector of shortest length, using the LLL lattice basis reduction algorithm to speed up the shortest-vector search, which gives the coefficents of $q(x)$. Note that one can choose appropriate integer factors so that the whole problem can be done with integer arithmetic.

We ran this algorithm, using random cyclotomic polynomials $p(x)$ with distinct factors and small degree, and many small Salem numbers in the list in [**2**] were found.

The computational complexity of the above algorithm can be estimated from results in [**6**] and [**4**], which give the number of operations necessary for LLL factorization and shortest-vector searches. It turns out that under certain conditions (satisfied for example by the cyclotomic families studied above), the algorithm is polynomial in n. However, it is not advisable to search by simply looping over all cyclotomic polynomials with distinct factors, owing to a result of [**3**], which shows that this number grows roughly like $\exp(\sqrt{n})$.

References

1. D. Boyd, *Speculations concerning the range of Mahler's measure*, Canad. Math. Bull. **24** (1981), 453–469.
2. D. Boyd, *Small Salem Numbers*, Duke Math. J. **44** (1977), 315–328.
3. D. Boyd and H. L. Montgomery, *Cyclotomic Partitions*, First Canadian Number Theory Conference, Walter de Gruyter, New York, 1990, pp. 7-25.
4. U. Fincke and M. Pohst, *Improved methods for calculating vectors of short length in a lattice, including a complexity analysis*, Math. Comp. **44** (1985), 463–471.
5. D. H. Lehmer, *Factorization of certain cyclotomic functions*, Ann. Math. (2) **34** (1933), 461–479.
6. M. Pohst and H. Zassenhaus, *Algorithmic Algebraic Number Theory*, Cambridge, UK: Cambridge University Press, 1989.
7. C. J. Smyth, *On the product of the conjugates outside the unit circle of an algebraic integer*, Bull. London Math. Soc. **3** (1971), 169–175.

16949 SE 144th, Renton, WA 98059

Proceedings of Symposia in Applied Mathematics
Volume **48**, 1994

A New Method for finding Amicable Pairs

H.J.J. TE RIELE

ABSTRACT. Let $\sigma(x)$ denote the sum of all divisors of the (positive) integer x. An *amicable pair* is a pair of integers (m, n) with $m < n$ such that $\sigma(m) = \sigma(n) = m + n$. A new method for finding amicable pairs is presented, based on the following observation of Erdős: For given s, let $x_1, x_2, \dots$ be solutions of the equation $\sigma(x) = s$; then any pair (x_i, x_j) for which $x_i + x_j = s$ is amicable. An algorithm is described and applied to yield 116 new amicable pairs (m, n) with $2 \times 10^{11} < m < 5 \times 10^{11}$.

1. Introduction

Let m be a positive integer and let $\sigma(m)$ be the sum of all divisors of m. An *amicable pair* is a pair of positive integers (m, n), $m < n$, which satisfies $\sigma(m) = \sigma(n) = m + n$. Let $g = \gcd(m, n)$ and write $m = gM$, $n = gN$. The amicable pair (m, n) is called *regular* if $\gcd(g, M) = \gcd(g, N) = 1$, and M and N are squarefree. Other pairs are called *irregular* or *exotic*. The smallest regular amicable pair is $(220, 284) = (2^2 5 \cdot 11, 2^2 71)$ and the smallest irregular amicable pair is $(1184, 1210) = (2^5 37, 2 \cdot 5 \cdot 11^2)$. Below 10^{10} there are 1082 regular and 345 irregular amicable pairs (sometimes abbreviated to 'AP') [**11**].

Three essentially different methods to find amicable pairs are known. The first is a numerical method, the second and third are algebraic methods. In the first, a number m is chosen, $n := \sigma(m) - m$ is computed and, if $n > m$, $t := \sigma(n) - n$ is computed. If $t = m$, (m, n) is an amicable pair. By letting m run through a given interval, we can trace *all* amicable pairs (m, n) with m in that interval. In the second method, an assumption is made about the prime structure of m and n, for example, $m = 2^k pq$, $n = 2^k r$, where p, q and r are mutually different prime numbers. This yields two equations which k, p, q and r must satisfy. In the third method, amicable pairs are *constructed* from special numbers called

1991 *Mathematics Subject Classification.* Primary 11A25; Secondary 11Y50.
Key words and phrases. Amicable pairs, sum of divisors.
This paper is in final form and no version of it will be submitted for publication elsewhere.

© 1994 American Mathematical Society
0160-7634/94 $1.00 + $.25 per page

breeders [**3**], which may be amicable numbers themselves [**10**]. The first method is exhaustive and CPU time-consuming (cf. [**11**]), whereas the second and the third are nonexhaustive and relatively inexpensive (cf. [**1**] and [**5**]). Thousands of amicable pairs are known [**1, 12**]. An exhaustive list of amicable pairs below 10^{10} was published in [**11**]. This bound was extended recently by Moews and Moews to 10^{11} [**8**] and to 2.01×10^{11} [**9**].

In this paper we present and apply a new method to find amicable pairs, the principle of which stems from Erdős: If $x_1, x_2, \ldots$ are solutions of the equation $\sigma(x) = s$, then any pair (x_i, x_j) for which $x_i + x_j = s$ is an amicable pair. Heuristically, values of s for which $\sigma(x) = s$ has *many* solutions have an increased chance to yield amicable pairs.

2. An algorithm to solve $\sigma(x) = s$ and derive amicable pairs

We are going to find solutions $x = x_1, x_2, \ldots$ of $\sigma(x) = s$ for suitable s. We restrict our search to amicable pairs with members which are divisible by 2 or by 3. We exclude the possibility that one member is divisible by 6 since in that case the other member must be *odd* [**6**]: no such amicable pair is known, and if it exists, it probably is very large. Only recently, amicable pairs have been found with members coprime to 6 [**2, 4**], but these are very large compared with the numbers in our search interval.

Algorithm A (*For given positive integer s, find solutions x of the equation $\sigma(x) = s$, where $2 \mid x$ or $3 \mid x$ but $6 \nmid x$ and find amicable pairs among pairs of solutions found*). If $\sigma(p^e) \mid s$ for a prime p and positive integer exponent e, and if the equation $\sigma(y) = s/\sigma(p^e)$ has a solution y which is coprime to p, then $x = p^e y$ is a solution of the equation $\sigma(x) = s$ by the multiplicativity of the σ-function. This suggests our recursive algorithm: it finds a divisor $\sigma(p^e)$ of s and solves $\sigma(y) = s/\sigma(p^e)$, with $\gcd(p, y) = 1$. Two tables T_1 and T_2 of triples $(p, e, \sigma(p^e))$ will be used (where $\sigma(p^e) = p^e + p^{e-1} + \cdots + 1$). The ith triple from table T_1 will be denoted by T_{1i}, and similarly for table T_2. Solutions found are stored in $x_1, x_2, \ldots$. Choose upper bounds B_1 and B_2 for the $\sigma(p^e)$-values admitted in tables T_1 and T_2, respectively.

A1 [Precomputation of tables of $\sigma(p^e)$-values]. Fill table T_1 with triples $(p, e, \sigma(p^e))$ for $p = 2, 3$ and for those integers $e = 1, 2, \ldots$ for which $\sigma(p^e) < B_1$. Similarly, fill table T_2 for all the primes $p = 5, 7, \ldots$ and integers $e = 1, 2, \ldots$ for which $\sigma(p^e) < B_2$, such that the $\sigma(p^e)$-values are in *increasing* order. Set $i_{\max}$ and $j_{\max}$ to the number of triples in tables T_1 and T_2, respectively.

A2 [Initialize]. Set $d \leftarrow 1$, $s_d \leftarrow s$, $i \leftarrow 0$, $n \leftarrow 0$. The current value of d indicates that the dth prime divisor of x is being looked for (so $d - 1$ prime divisors have been found so far). The integer s_d is the current value of s for which $\sigma(x) = s$ is being solved; j_d ($d \geq 2$) remembers the location in table T_2 where a prime power divisor of x has been found;

p_d and e_d are the prime and the corresponding exponent in that prime power.

A3 [Select next triple from table T_1]. Set $i \leftarrow i+1$. If $i > i_{\max}$, goto step A8, otherwise set $(p, e, \sigma) \leftarrow T_{1i}$. If $\sigma \nmid s_1$, repeat step A3, otherwise set

$$p_1 \leftarrow p, e_1 \leftarrow e, d \leftarrow 2, s_2 \leftarrow s_1/\sigma, sq \leftarrow \sqrt{s_2}, j \leftarrow 0.$$

A4 [Select next triple from table T_2]. Set $j \leftarrow j+1$. If $j > j_{\max}$, goto step A5, otherwise set $(p, e, \sigma) \leftarrow T_{2j}$. If $p = p_l$ for some $l \in \{2, 3, \ldots, d-1\}$, repeat step A4. If $\sigma > sq$, goto step A6. If $\sigma \mid s_d$, set

$$j_d \leftarrow j, p_d \leftarrow p, e_d \leftarrow e, s_{d+1} \leftarrow s_d/\sigma, sq \leftarrow \sqrt{s_{d+1}}, d = d+1.$$

Repeat step A4.

A5 [Check if $s_d - 1$ is prime]. If $s_d - 1$ is prime, set

$$n \leftarrow n+1, x_n \leftarrow (s_d - 1) \prod_{k=1}^{d-1} p_k^{e_k}.$$

Goto step A7.

A6 [Check if s_d occurs as σ-value in table T_2]. If $\exists l$ and $T_{2l} = (p, e, \sigma)$ with $\sigma = s_d$, set

$$n \leftarrow n+1, x_n \leftarrow p^e \prod_{k=1}^{d-1} p_k^{e_k}.$$

Goto step A7.

A7 [Decrease depth d]. Set $d \leftarrow d-1$. If $d = 1$, goto step A3, otherwise set $j \leftarrow j_d, sq \leftarrow \sqrt{s_d}$; return to step A4.

A8 [Sort solutions and check pair sums]. Sort the solutions $x_1, x_2, \ldots, x_n$ in increasing order and find amicable pairs, i.e., pairs (x_i, x_j) with $x_i + x_j = s$. Notice that this algorithm finds amicable pairs (m, n), if any, with $m < s/2$ and $n > s/2$, where m and n are in the neighborhood of $s/2$.

Application of this algorithm with $B_1 = 70$ and $B_2 = 100$ to $s = 504 = 2^3 3^2 7$ yields the five solutions $x_1 = 2 \cdot 11 \cdot 13$, $x_2 = 2 \cdot 167$, $x_3 = 2^2 5 \cdot 11$, $x_4 = 2^2 71$, $x_5 = 2^5 7$, in this order, and the amicable pair $(x_3, x_4) = (220, 284)$, since $x_3 + x_4 = s$. The equation $\sigma(x) = 504$ has five more solutions, viz., $x = 192, 246, 415, 451$, and 503, but the first two of them are divisible by 6, and the other three have a smallest prime divisor > 3, and such solutions are excluded from our algorithm.

3. Applying Algorithm A to suitable numbers s

Inspection of the pair sums of known amicable pairs reveals that in many cases these sums have only small prime divisors. In particular, among the 1427 amicable pairs below 10^{10} there are 37 *pairs* of amicable pairs with the same pair sum (and no such *triples*), and in these 37 pair sums the largest occurring prime is 37. Suggested by the frequencies of occurrence of these primes in the 37 pair sums, and by the exponent ranges of these primes (cf. Table 5 in [**11**]), we have generated two sets, S_1 and S_2, of s-values as input for Algorithm A, as follows. Let

$$S = \{s = 2^{i_1}3^{i_2}5^{i_3}7^{i_4}11^{i_5}13^{i_6}17^{i_7}19^{i_8}23^{i_9}31^{i_{10}} \mid$$

$$6 \le i_1 \le 14,\ 3 \le i_2 \le 6,\ 0 \le i_3 \le 3,\ 0 \le i_4 \le 2,\ 0 \le i_k \le 1 \text{ for } k = 5, \ldots, 10\};$$

then $S_1 := \{s \in S \mid 4 \times 10^{10} < s < 2 \times 10^{11}\}$ and $S_2 := \{s \in S \mid 4 \times 10^{11} < s < 10^{12}\}$. Algorithm A was applied to all elements of S_1 and S_2, with $B_1 = 260$ and $B_2 = 10^6$.

From the 3582 elements in S_1, Algorithm A found 200 numbers s which yielded one or more amicable pairs, viz., 2 each yielded three APs, 8 each yielded two APs, and 190 each yielded one AP, hence a total of 212 APs (m, n) with $m < 10^{11}$. Computing time was about 13 CPU hours on a 33 MHZ SGI workstation. All the 212 APs found occur in the list by Moews and Moews [**8**]. This list contains *four* triples of amicable pairs having the same pair sum, viz., $2^{14}3^65^27 \cdot 31$, $2^{12}3^45^47^211$, $2^{13}3^45 \cdot 11 \cdot 13^219$, and $2^{10}3^55^27 \cdot 11 \cdot 13 \cdot 19$. The first and fourth of these pair sums belong to our set S_1, and our algorithm found the two corresponding triples of APs.

In the set S_2, which contains 1850 elements, Algorithm A found, after about 30 CPU hours of computing time, 118 numbers s which yielded one or more APs, viz., 1 yielded three APs, 7 each yielded two APs, and 110 each yielded one AP, hence a total of 127 APs. They are in the interval $[1.88 \times 10^{11}, 5 \times 10^{11}]$. Of these, 102 (80%) are regular, and 25 (20%) are irregular. Moews and Moews [**9**] found that there are 970 amicable pairs in the interval $[10^{11}, 2 \times 10^{11}]$, 778 (80%) of them being regular and 192 (20%) being irregular, so it appears that our algorithm finds a selection of regular and irregular amicable pairs in a ratio which is representative for *all* the amicable pairs in the corresponding interval. The other known nonexhaustive methods usually only find regular amicable pairs and very few irregular pairs.

The first eight of the 127 APS found from S_2 have their smaller member $< 2.01 \times 10^{11}$ and, therefore, occur in the list by Moews and Moews [**9**]. Two others occur in [**12**], and one in [**7**]. This leaves a total of 116 APs (m, n) with $2.01 \times 10^{11} < m < 5 \times 10^{11}$, which seem to be new. A list of these APs is available from the author upon request. The new *triple* of APs having the same pair sum was found for $s = 2^{14}3^65^27 \cdot 11 \cdot 19$, and the three corresponding amicable pairs are: $(3^3 \cdot 5 \cdot 11 \cdot 29 \cdot 113 \cdot 167 \cdot 263, 3^3 \cdot 5 \cdot 11 \cdot 131 \cdot 839 \cdot 1367)$, $(3 \cdot 5 \cdot 7 \cdot 19 \cdot 29 \cdot 43 \cdot 86183, 3 \cdot 5 \cdot 7 \cdot 19 \cdot 89 \cdot 113 \cdot 11087)$, and $(2^3 \cdot 31 \cdot 43 \cdot 113 \cdot 419 \cdot 431, 2^3 \cdot 37 \cdot 47 \cdot 167 \cdot 197 \cdot 479)$.

Let N_s be the number of solutions x of $\sigma(x) = s$ found with our algorithm. If $N_s \approx \sqrt{s}$, and if solutions are "randomly" distributed in $[1, s]$, we have a reasonable chance to find a pair (x_1, x_2) among the N_s solutions for which $x_1 + x_2 = s$. For the numbers s in S_1 and S_2, we found values of $N_s/\sqrt{s}$ in the range $(0.01, 0.43)$. The maximum value we found was 0.4224, for $s = 2^{12}3^65^27^211 \cdot 13$ with $N_s = 305536$, but no amicable pair resulted from this s. The second-largest quotient was 0.388, for $s = 2^{13}3^55^27^211 \cdot 19$ with $N_s = 277222$. This s yielded two new amicable pairs, viz. $(3^35 \cdot 11^223 \cdot 223 \cdot 2969, 3^35 \cdot 11 \cdot 197 \cdot 839 \cdot 1063)$ and

$(2 \cdot 5 \cdot 13 \cdot 37 \cdot 109 \cdot 419 \cdot 1151, 2 \cdot 5 \cdot 29 \cdot 31 \cdot 43 \cdot 113 \cdot 5879)$. In order to get some feeling for the behavior of the quotient $N_s/\sqrt{s}$, we have computed it for $s = i!$, $i = 8, \ldots, 15$, and for a few other values of s. The results as given in Table 1 show an increasing tendency for the quotient, which gives some support to our expectation that many (but much larger) smooth numbers s (i.e., only with small prime factors) exist for which the quotient exceeds 1, and these numbers can be expected to yield many amicable numbers. For $s = 3 \cdot 14!$, $s = 4 \cdot 14!$, and $s = 15!$ Algorithm A yielded two APs, one AP, and one AP, respectively; for the other values of s in Table 1 no AP was found.

Table 1. N_s and $N_s/\sqrt{s}$ for various values of s

s	11!	12!	13!	14!	$2 \cdot 14!$	$3 \cdot 14!$	$4 \cdot 14!$	$8 \cdot 14!$	$12 \cdot 14!$	15!
N_s	1665	6999	25656	110137	163869	200965	236219	331105	449253	442439
$N_s/\sqrt{s}$	0.264	0.320	0.325	0.373	0.392	0.393	0.400	0.397	0.439	0.387

Acknowledgments. We thank Carl Pomerance for pointing out Erdős's idea to us, and an anonymous referee for pertinent remarks concerning the quotient $N_s/\sqrt{s}$.

References

1. S. Battiato, *Ueber die Produktion von* 37803 *neuen befreundeten Zahlenpaaren mit der Brütermethode*, Bergische Universität Gesamthochschule, Wuppertal, Master's Thesis, June 1988.
2. W. Borho and S. Battiato, *Are there odd amicable pairs not divisible by 3?*, Math. Comp. **50** (1988), 633–637.
3. W. Borho and H. Hoffmann, *Breeding amicable numbers in abundance*, Math. Comp. **46** (1986), 281–293.
4. Mariano Garcia, *Favorable conditions for amicability*, Hostos Community College Mathematics Journal (Spring 1989), New York, 20–25.
5. Elvin J. Lee, *Amicable numbers and the bilinear Diophantine equation*, Math. Comp. **22** (1968), 181–187.
6. Elvin J. Lee, *On divisibility by nine of the sums of even amicable pairs*, Math. Comp. **23** (1969), 545–548.
7. Elvin J. Lee, private communication, February, 1987.
8. David Moews and Paul C. Moews, *A search for aliquot cycles and amicable pairs*, Math. Comp. **61** (1993), 935–938.
9. David Moews and Paul C. Moews, private communication, January 8, 1993.
10. H.J.J. te Riele, *On generating new amicable pairs from given amicable pairs*, Math. Comp. **42** (1984), 219–223.
11. H.J.J. te Riele, *Computation of all the amicable pairs below* 10^{10}, Math. Comp. **47** (1986), 361–368, S9–S40.
12. H.J.J. te Riele, W. Borho, S. Battiato, H. Hoffmann and E.J. Lee, *Table of Amicable Pairs between* 10^{10} *and* 10^{52}, CWI, Amsterdam, Tech. Rept. NM-N8603, September 1986.
13. W. Sierpiński, *Elementary Theory of Numbers*, Second edition (edited by A. Schinzel), North-Holland, Amsterdam, 1988.

Centre for Mathematics and Computer Science (CWI), Kruislaan 413, 1098 SJ Amsterdam, The Netherlands

E-mail address: herman@cwi.nl

Proceedings of Symposia in Applied Mathematics
Volume 48, 1994

Generalized Fermat Numbers

HANS RIESEL AND ANDERS BJÖRN

ABSTRACT. A search for prime factors is currently being carried out for the numbers $F_n(a,b) = a^{2^n} + b^{2^n}$ with $a, b \leq 12$, $(a,b) = 1$, and $n < 1000$. Several thousand new factors have been found. For $7 \leq n \leq 14$, and for some numbers with $n = 15$ and $n = 16$, we are also establishing the primality or compositeness of the cofactors of the numbers in question. In addition, complete factorizations have been achieved for all numbers $a^{2^n} + b^{2^n}$ for $n \leq 6$, for $a^{2^7} + b^{2^7}$ with $a, b \leq 5$, and for some larger numbers. On those numbers with less than 60000 digits, for which no factor is found, we plan to perform a strong pseudoprimality test, and then try to prove the pseudoprimes found, if any, to be primes.

1. Why study Fermat numbers?

The interest in Fermat numbers stems from the fact that $F_n(a,b)$ lacks algebraic factors. This leaves the possibility open that F_n might be a prime, just as is the case for ordinary Fermat numbers for $n \leq 4$. Thus, e.g., $F_6(11,8)$ is a prime.

2. Possible prime factors of $F_n(a,b)$

The line of reasoning, leading to the well-known theorem that every prime factor of $2^{2^n} + 1$ is of the form $p = 2^m k + 1$, with k odd and $m \geq n+2$, can easily be extended to the general case. It depends on Legendre's theorem for the prime factors of $a^s \pm b^s$ and quadratic residuacity. We have the following

THEOREM. *Each prime factor of $F_n(a,b)$, with $(a,b) = 1$, except for a possible prime factor 2, is of the form $p = 2^m k + 1$, with k odd and $m \geq n+1$. One has $m = n+1$ if ab is a quadratic nonresidue* $\bmod p$, *otherwise $m > n+1$.*

PROOF. The proof, which is quite straightforward, will be published elsewhere.

1991 *Mathematics Subject Classification.* Primary 11-04, 11A51, 11Y05, 11Y11.
The final version of this paper will be submitted for publication elsewhere.

© 1994 American Mathematical Society
0160-7634/94 $1.00 + $.25 per page

3. A device to save computing time

All a, b with $12 \geq a > b \geq 1$ make 45 different pairs. We have omitted the pair $(2,1)$, which has been extensively studied by previous authors [1], and the pairs $(4,1)$, $(9,1)$, and $(9,4)$, which are simply $(2,1)$, $(3,1)$, and $(3,2)$, respectively, for the next value of n. This left us with 41 actual pairs of a and b to study.

Let, as usual, $p = 2^m k + 1$, with k odd. Instead of computing $F_n(a,b) \bmod p$ for each pair (a,b), we computed the 9 values of $a^{2^s} \bmod p$ for $a = 4$ to $a = 12$ for each value of s up to and including $s = m - 1$. The values of $a^{2^s} \bmod p$ for $a = 2$ and 3 were taken as the values of $(a^2)^{2^{s-1}} \bmod p$, which were computed in the previous step. Then, for each s, from the values obtained, we computed the different sums required to find $F_n(a,b) \bmod p$ and checked if some of these values were $\equiv 0 \bmod p$. Thus, at the cost of only 9 squarings $\bmod p$, and some minor computations, we were able to investigate 41 different numbers $F_n(a,b) \bmod p$.

To demonstrate our approach, we show $a^{2^s} \bmod 769$ for $s = 5, 6$, and 7:

$a =$	1	2	3	4	5	6	7	8	9	10	11	12
$s = 5$	1	19	360	361	62	−81	40	−62	−361	−360	−164	−1
$s = 6$	1	361	−361	360	−1	−360	62	−1	360	−361	−19	1
$s = 7$	1	360	360	−361	1	−361	−1	1	−361	360	361	1

From these values it is easy to figure out that the numbers $8^{2^5} + 5^{2^5}$, $9^{2^5} + 4^{2^5}$, $10^{2^5} + 3^{2^5}$, $12^{2^5} + 1$, $3^{2^6} + 2^{2^6}$, $5^{2^6} + 1$, $6^{2^6} + 4^{2^6}$, $8^{2^6} + 1$, $9^{2^6} + 6^{2^6}$, $10^{2^6} + 2^{2^6}$, $12^{2^6} + 5^{2^6}$, $12^{2^6} + 8^{2^6}$, $7^{2^7} + 1$, $7^{2^7} + 5^{2^7}$, $8^{2^7} + 7^{2^7}$, $11^{2^7} + 4^{2^7}$, $11^{2^7} + 6^{2^7}$, $11^{2^7} + 9^{2^7}$, $12^{2^7} + 7^{2^7}$, and no others of the given form, all are $\equiv 0 \bmod 769$.

4. Limits of k and m

Since we find the smaller numbers $F_n(a,b)$ more interesting than the larger ones, we are working with the limit of k in steps. Thus we have chosen

$$\begin{array}{lll} k \text{ odd and } < 10^9 & \text{for} & m \leq 100 \quad \text{and} \\ k \text{ odd and } < 10^4 & \text{for} & 100 < m \leq 1000. \end{array}$$

For each m, runs are in progress for n up to and including $n = m - 1$, which ensures that *all* generalized Fermat numbers with $a, b \leq 12$ and $n \leq 999$ will be checked for all prime factors of the form $p = 2^m k + 1$ in the given range.

When we submitted this paper for publication, our runs had reached $k = 732,000,000$.

5. Which numbers p should we use as candidates for factors?

For each number $p = 2^m k + 1$ we search for small factors up to a limit $p_{\max}$, in order to avoid testing composite numbers p as factors of F_n. The limit $p_{\max}$ is put to 12,500 for $m \leq 100$, and is raised to 500,000 for $m > 100$. These limits reflect the usual trade-off between the computing time such a test of small factors of p takes, and the time used up in testing p as a possible factor of all the numbers $F_n(a,b)$ in our study. The prime candidates remaining after the sieving have to pass a strong pseudoprimality test, before being accepted as trial divisors of the numbers F_n.

All factors found will be checked for multiplicity. The largest multiple factor found so far is 17 (in $9^8 + 7^8$ and $11^8 + 4^8$).

The numbers $F_n(8,1)$ are algebraically divisible by $F_n(2,1)$, the quotient being $Q_n = 4^{2^n} - 2^{2^n} + 1$. By removing the factors of $F_n(2,1)$ from those of $F_n(8,1)$, we are assembling a table of factors of the numbers Q_n.

6. The tables

For each pair (a,b) we are assembling a table of factors. The tables will start with $n = 7$. In addition to the factor search we are, for $7 \leq n \leq 14$, and for some numbers with $n = 15$ and $n = 16$, also testing the primality of the remaining cofactor, after the known factors have been removed. If no factor is found, the number F_n itself (or $F_n/2$ in case a and b are both odd) will be tested for primality. The result of these tests will be presented when the whole investigation has been finished. The largest strong pseudoprime we have found so far is the cofactor of $7^{2^{12}} + 3^{2^{12}}$ (3427 digits).

Tables 1a–d show the complete factorizations of the numbers F_n, for $a = 11, 12$ if $n = 4, 5$, for all a and b if $n = 6$, and for $a \leq 6$, if $n = 7$. We achieved the factorizations with the computer program *Mathematica*, which contains a standard package for factorization, based on Lenstra's Elliptic Curve Method, ECM. The twelve largest of the numbers in this table were, however, too large for the *Mathematica* package. They were kindly factored for us by Richard P. Brent, who used ECM to find the smaller factors and the Multiple Precision Quadratic Sieve, MPQS, to find the larger penultimate factors. Before a number was scrutinized by ECM, a pseudoprimality test was performed, also with *Mathematica*, to ensure that the number in question is composite. All pseudoprimes found will, if possible, subsequently be confirmed as primes.

The prime or composite character of numbers with more than about 300 digits was checked with a strong pseudoprimality test. The base b was chosen as the smallest prime ≥ 13 with Jacobi's symbol $(b/N) = -1$. This was programmed and run on a CRAY XMP and on various Sparc work stations, using David Bailey's Portable High Performance Multiprecision Package. As a spin-off effect we also happened to find the complete factorizations of some numbers larger than those given in Table 1. These factorizations will be published elsewhere.

7. Acknowledgements

We would like to thank Richard P. Brent for his kind and helpful assistance in filling in the holes in Table 1c–d for us. The computer time is graciously put at our disposal by the National Supercomputing Centre of Sweden, by the Linköping University and by the Royal Institute of Technology. We also acknowledge David H. Bailey and NASA Ames for allowing the use of the software package MPFUN.

References

1. John Brillhart, D.H. Lehmer, J.L. Selfridge, Bryant Tuckerman and S.S. Wagstaff, Jr., *Factorizations of* $b^n \pm 1, b = 2, 3, 5, 6, 7, 10, 11, 12$ *up to high powers*, 2nd ed., AMS, Providence R.I., 1988.
2. Hans Riesel, *Prime numbers and computer methods for factorization*, Birkhäuser, Boston, 1985.
3. Hans Riesel, *Common prime factors of the numbers* $A_n = a^{2^n} + 1$, Nord. Tidskr. Informationsbeh. **9** (1969), 264–269.

Table 1a–b.

a, b	Factorization of $a^{16} + b^{16}$	Prime factorization of $a^{32} + b^{32}$
11, 2	697409 · 65886344833	257 · 558205057 · 14717669124989680991233
11, 3	2 · 68074913 · 337493857	2 · 193 · 364417 · 15009975792943059177726401
11, 4	1249 · 59009 · 623450977	257 · 1153 · 1217 · 54804376513 · 106831093776257
11, 5	2 · 25633 · 896303250721	2 · 2113 · 18157889 · 27520769 · 999793963860481
11, 6	673 · 1153 · 39521 · 1498433	2839553 · 22259029358081 · 33404865330049
11, 7	2 · 22991481397070881	2 · 1055689389481464764733130779754561
11, 8	577 · 80123405269121	193 · 257 · 622337 · 685697 · 5727809 · 17415878657
11, 9	2 · 577 · 41423526908513	2 · 193 · 1153 · 47517657436908173721858111969
11, 10	97 · 3617 · 159469543489	22113776745352552855456615254209921
12, 5	184884411482927041	449 · 3385409 · 224875627431912786899654 41
12, 7	2693153 · 68662081889	257 · 13572412993 · 979963043811008532 9857
12, 11	929 · 2689 · 262369 · 352193	449 · 80831997464815383957493223207873

Table 1d.

a, b	Prime factorization of $a^{128} + b^{128}$
3, 2	257 · 4587620458264040144560783324427797511339173138865086722688 1
4, 3	257· 4505528764097906895476870145808213205219565995259113996796466003258919795 21
5, 2	257 · 115201 · 152833 · 79280897 · 207676203460222729984 1· 3944561699895512147807036336288996553199598277 13
5, 3	2 · 7681 · 36097 · 259841 · 21440257 · 26556161· 35821084158205420800845791659686744842677019632207505966274 57
5, 4	257 · 1719885157889· 664856597964527775591232190673009405184495378793489594446703567585557440257
6, 5	3329 · 8253953 · 3108109313 · 29759533890817 · 28232308132114188879656724422657· 559133441067962474012042631317191 69

Table 1c.

a, b	Prime factorization of $a^{64} + b^{64}$
3, 2	769 · 121899667073 · 36629538145348481
4, 3	337153 · 1009281751473729386230842057136129
5, 2	274568286337 · 1974376186976624976974230015921793
5, 3	2 · 257 · 290320001 · 3632789233881076244813676034342529
5, 4	1699132282369 · 319046040243900998188151306914049
6, 5	257 · 1189782673537 · 207149065298675818568214979118531969
7, 2	47047681 · 62177153 · 416970669270749472494126795751788515969
7, 3	2 · 271004161 · 2250446050597421823605463563798919148824618881
7, 4	400303811518337 · 3047086868869250345637754038506015544961
7, 5	2 · 257 · 6529 · 9345903233 · 6084515529914753 · 6391716288418607254529
7, 6	1441838081 · 45814876214017 · 18466054092468533735568034082561
8, 3	35201 · 13298174977 · 13409485752562918448342707397650793232918401
8, 5	7681 · 9857 · 82908039533579409153578534393535276660420426832513
8, 7	1153 · 257921 · 865537 · 186270593 · 130947618578460692408608101578574209
9, 2	257 · 32310529 · 480865793 · 2952701387950377265900522969510993211622913
9, 5	2 · 15539860481 · 5584960066874701094401 · 67924026214760827303058881153
9, 7	2 · 206697065730709487073032833 · 28520447921741769926385807851011457
9, 8	257 · 17921 · 21750529 · 124683908353 · 10517420781313 · 89798052356669480217601
10, 3	257 · 7800222337 · 4988384196692145582667808890329898418176842028878209
10, 7	257 · 143489 · 7354241 · 407083829648792831580161 · 90578799127308003578311937
10, 9	4993 · 1010193993337978111110606210433 · 1984930895961378122350344691969
11, 2	98561 · 235009 · 749633772534063781633 · 256739747797740287066020373380619521
11, 3	2 · 641 · 1409 · 2467929969100966926383385686050917127870777924886726183123969
11, 4	3730817 · 117071733905921 · 10206476795408368212421659203564483078337512321
11, 5	2 · 3602561 · 13347415255864961 · 46354653502784142190063289765715640124892673
11, 6	641 · 10369 · 11777 · 97624432793901757313· 583369537719534966217090312323137153
11, 7	2 · 610344589666960166969722938113 · 3651966249884858172122790877141138817
11, 8	4457915690803004131256192897205630962697827851093882159977969339137
11, 9	2 · 257 · 13313 · 2947336985726430593· 221036580703378954253624978864524844875777
11, 10	15361 · 137089 · 2518913 · 3736495361· 225426649400435020917315129910488452522753
12, 5	257 · 769 · 13186177 · 2877166849 · 15068584506047617· 10341508117354075765162470036737
12, 7	1153 · 6189697 · 46380768089432146049· 3529905780381047084953395405549974142593
12, 11	641 · 441738113 · 4142196236086054595120177167536933992184466613226792461569

Department of Numerical Analysis and Computing Science, Royal Institute of Technology, S-10044 Stockholm, Sweden

E-mail address: riesel@nada.kth.se

Department of Mathematics, Linköping University, S-58183 Linköping, Sweden

E-mail address: anbjo@math.liu.se

Proceedings of Symposia in Applied Mathematics
Volume **48**, 1994

Evaluation of $\zeta_K(2)$ For Some Totally Real Algebraic Number Fields K of Degree 9

KISAO TAKEUCHI

Dedicated to Professor Hideo Shimizu on his 60th birthday

ABSTRACT. We calculate the exact values $\zeta_K(2)$ for some totally real number fields K of degree 9 with small discriminant $d(K)$ containing cubic subfields F by using Public Domain Software UBASIC-86.

1. Introduction

Let K be a totally real algebraic number field of degree n and $d(K)$ be its discriminant. Let $\zeta_K(s)$ be the Dedekind zeta function of K. By definition we have

$$\zeta_K(s) = \sum_A \frac{1}{N(A)^s},$$

where A runs over all integral ideals of K and $N(\cdot)$ denotes the norm map of K to the rational number field $\mathbf{Q}$.

It is known that $\zeta_K(s)$ can be continued meromorphically to the whole complex s-plane and satisfies the functional equation

$$\zeta_K(1-s) = d(K)^{s-\frac{1}{2}}C(s)^n\zeta_K(s), \tag{1}$$

where $C(s) = 2^{1-s}\pi^{-s}\Gamma(s)\cos(\pi s/2)$.

The function $\zeta_K(s)$ has the Euler product

$$\zeta_K(s) = \prod_P \frac{1}{1-N(P)^{-s}}, \tag{2}$$

where P runs over all prime ideals of K.

The evaluation of $\zeta_K(2)$ has been done by many people for the various fields K of various degrees n. The explicit formula for $\zeta_K(2)$ is given by Siegel ([7]).

1991 *Mathematics Subject Classification.* Primary 11R04, 11R42, 11R80, 11Y40.
This paper is in final form and no version of it will be submitted for publication elsewhere.

© 1994 American Mathematical Society
0160-7634/94 $1.00 + $.25 per page

By using the results of Serre ([5]), Cartier and Roy calculated $\zeta_K(2)$ for many cases: $2 \leq n \leq 6$ ([1]).

For any prime number p we have the decomposition of the ideal (p) in K,

$$(p) = P_1^{e_1} \cdots P_g^{e_g}, \tag{3}$$

where P_i $(1 \leq i \leq g)$ is a prime ideal of K with $N(P_i) = p^{f_i}$.

We call $(p; f_1, \dots, f_g)$ the decomposition data of p. By (2) we have

$$\zeta_K(2) = \prod_p \prod_{i=1}^{g} \frac{1}{1 - p^{-2f_i}}, \tag{4}$$

where p runs over all prime numbers. Now let p_r be the rth prime number in order of magnitude. Then we have $p_1 = 2, p_2 = 3, p_3 = 5, \dots$. We put

$$\zeta_K(2, r) = \prod_{p \leq p_r} \prod_{i=1}^{g} \frac{1}{1 - p^{-2f_i}}. \tag{5}$$

Since $n = \sum_{i=1}^{g} e_i f_i$, we have the inequalities

$$\prod_{i=1}^{g} \frac{1}{1 - p^{-2f_i}} \leq \prod_{i=1}^{g} \frac{1}{1 - p^{-2}} \leq \left(\frac{1}{1 - p^{-2}}\right)^n. \tag{6}$$

Hence we have the estimation (cf. Cartier and Roy [1])

$$\zeta_K(2, r) < \zeta_K(2) < \zeta_K(2, r)((\zeta_{\mathbf{Q}}(2)/\zeta_{\mathbf{Q}}(2, r))^n, \tag{7}$$

where $\zeta_{\mathbf{Q}}(s)$ is the Riemann zeta function and

$$\zeta_{\mathbf{Q}}(s, r) = \prod_{p \leq p_r} \frac{1}{1 - p^{-s}}. \tag{8}$$

It is well known that $\zeta_{\mathbf{Q}}(2) = \pi^2/6$. The inequalities in (7) give an approximate value of $\zeta_K(2)$ for a sufficiently large integer r, but not the exact value.

The value $\zeta_K(2)$ appears in the volume formula for the fundamental domains of arithmetic discrete groups. Therefore, it is useful to have a simple procedure to determine the exact value of $\zeta_K(2)$.

In this article we calculate the exact value of $\zeta_K(2)$ for some fields K of degree 9 by using the results of Serre ([6]). In computing numerical data, we use the public domain software UBASIC-86, which was developed by Y. Kida (cf. [4]), running on a personal computer NEC PC-9800 series.

2. Evaluation of $\zeta_K(2)$

Let K be a totally real field of degree 9 with discriminant $d(K)$. Recently, H. Fujita has obtained the following result ([2, 3]).

THEOREM 1 (FUJITA). *Let K be a totally real algebraic number field of degree 9 containing cubic subfields F with the condition*

$$d(K) \leq 20000000000.$$

Let $f(x) = x^9 + a_1x^8 + a_2x^7 + a_3x^6 + a_4x^5 + a_5x^4 + a_6x^3 + a_7x^2 + a_8x + a_9$ be a defining equation for the field K. Then the complete list of such fields K is given as follows:

	a_1	a_2	a_3	a_4	a_5	a_6	a_7	a_8	a_9	$d(K)$
(*i*)	-1	-9	4	26	-2	-25	-1	7	1	16240385609
(*ii*)	3	-7	-14	17	18	-13	-8	3	1	16440305941
(*iii*)	-1	-8	7	21	-15	-20	10	5	-1	16983563041
(*iv*)	5	-2	-34	-24	26	29	3	-4	-1	17515230173
(*v*)	5	-2	-27	4	40	-13	-18	10	-1	19936446593

Each field K given above is uniquely determined by $d(K)$ up to $\mathbf{Q}$-isomorphism, and $d(K)$ is equal to the discriminant $d(f)$ of $f(x)$. Moreover, the class number $h(K)$ of each field K is equal to 1.

Now we shall calculate the exact value of $\zeta_K(2)$ for the fields K given above. By (1) we have

$$\zeta_K(-1) = -\frac{d(K)^{3/2}}{2^9\pi^{18}}\zeta_K(2). \tag{9}$$

It is known that $\zeta_K(-1)$ is a rational number ([7]). As to the denominator of $\zeta_K(-1)$, Serre proved theorems which are very efficient for the explicit evaluation ([5, 6]). Let ζ_n be the primitive nth root of unity for any natural number n. Let $\mathbf{Q}(\zeta_n)$ be the cyclotomic field generated by ζ_n over $\mathbf{Q}$ and $\mathbf{Q}^+(\zeta_n)$ be the maximal real subfield of $\mathbf{Q}(\zeta_n)$. For any prime number ℓ let $m(\ell)$ be the maximal natural number m such that $\mathbf{Q}^+(\zeta_{\ell^m})$ is a subfield of K. We define

$$w(K) = 2^{m(2)+1}\prod_{\ell\neq 2}\ell^{m(\ell)}. \tag{10}$$

We need the following theorem.

THEOREM 2 (SERRE [6]). *Let K be a totally real algebraic number field of degree n and $w(K)$ be as above. Then $w(K)\zeta_K(-1)$ is a rational integer which is divisible by 2^n.*

By this theorem we have the following expression:

$$\zeta_K(-1) = -2^9\frac{a(K)}{w(K)}, \tag{11}$$

where $a(K)$ is a positive integer. From (6), (7), (9) and (11) we have the inequalities

$$b(K,r) < a(K) < c(K,r), \tag{12}$$

where

$$(13) \qquad b(K,r) = \frac{w(K)d(K)^{3/2}}{(2\pi)^{18}}\zeta_K(2,r),$$

$$(14) \qquad c(K,r) = b(K,r)(\frac{\pi^2}{6}\prod_{p\le p_r}(1-1/p^2))^9.$$

In order to calculate the approximate values of $b(K,r)$ and $c(K,r)$, it is necessary to have the decomposition data $(p; f_1, \ldots, f_g)$ for each prime $p \le p_r$.

Let θ be an algebraic integer of K generating K over $\mathbf{Q}$, and $f(x)$ be the irreducible polynomial of θ over $\mathbf{Q}$. As to the discriminant $d(f)$ of $f(x)$, we have

$$(15) \qquad d(f) = m^2 d(K),$$

where m is a positive integer. Now we need the following theorem.

THEOREM 3 (DEDEKIND). *Let K be an algebraic number field of degree n and $f(x)$ be the irreducible polynomial for an algebraic integer θ of K generating K over $\mathbf{Q}$. For any prime p not dividing m in* (15), *if $f(x)$ has the following decomposition*

$$(16) \qquad f(x) \equiv F_1(x)^{e_1}\cdots F_g(x)^{e_g} \pmod{p},$$

where $F_i(x)$ is a monic polynomial in $\mathbf{Z}[x]$ which is irreducible modulo p, then the ideal (p) has the decomposition

$$(p) = P_1^{e_1}\cdots P_g^{e_g}$$

in the field K.

In the computer language UBASIC-86 we can use the function "prm(r)" which generates the rth prime number for $1 \le r \le 12251$. Moreover, the program "POLFACT2" is installed which can give a factorization modulo p of any integral polynomial $f(x)$ for any prime $p \le 65521$. By using the modified "POLFACT2" as a subroutine, we can write a program "ZETAK2" with which we can compute $b(K,r)$ and $c(K,r)$ in (13) and (14). The numerical data for each field K in Theorem 2 (Fujita) are as follows:

	(i)	*(ii)*	*(iii)*	*(iv)*	*(v)*
$w(K)$	2^3*3*7	2^3*3	2^3*3*19	2^3*3*7	2^3*3*7

Using the above data for $w(K)$, we have

	r	p_r	$b(K,r)$	$c(K,r)$
(i)	1553	13037	1513.98905236	1514.08904567
(ii)	341	2293	221.98829759	222.08829024
(iii)	3692	34549	4382.98875901	4383.08874670
(iv)	1714	14633	1709.98882145	1710.08878017
(v)	1992	17327	2059.98874919	2060.08874333

Hence we have the following table for $a(K)$:

	$a(K)$
(i)	$1514 = 2 * 757$
(ii)	$222 = 2 * 3 * 37$
(iii)	$4383 = 3 * 3 * 487$
(iv)	$1710 = 2 * 3 * 3 * 5 * 19$
(v)	$2060 = 2 * 2 * 5 * 103$

By (9) and (11) we have our main result:

THEOREM 4. *For each field K listed in Theorem* 1 (*Fujita*) *the exact values of* $\zeta_K(-1)$ *and* $\zeta_K(2)$ *are as follows:*

	$\zeta_K(-1)$	$\zeta_K(2)$
(i)	$-2^7 * 757/(3 * 7)$	$2^{16} * 757 * 3^{-1} * 7^{-1} * \pi^{18}/d(K)^{3/2}$
(ii)	$-2^7 * 37$	$2^{16} * 37 * \pi^{18}/d(K)^{3/2}$
(iii)	$-2^6 * 3 * 487/19$	$2^{15} * 3 * 487 * 19^{-1} * \pi^{18}/d(K)^{3/2}$
(iv)	$-2^7 * 3 * 5 * 19/7$	$2^{16} * 3 * 5 * 19 * 7^{-1} * \pi^{18}/d(K)^{3/2}$
(v)	$-2^8 * 5 * 103/(3 * 7)$	$2^{17} * 5 * 103 * 3^{-1} * 7^{-1} * \pi^{18}/d(K)^{3/2}$

REFERENCES

1. P. Cartier and Y. Roy, *Certains calculs numériques relatifs à l'interpolation p-adique des séries de Dirichlet,* in *Modular functions of one variable III* (W. Kuyk and J.-P. Serre, eds.), Lecture Notes Math. **350**, Springer, Berlin, 1973, pp. 269–307.
2. H. Fujita, *The minimum discriminant of totally real algebraic fields of degree 9 with cubic subfields*, Math. Comp. **60** (1993), 801–810.
3. ———, *The minimum discriminant of totally real algebraic fields of degree 9 with cubic subfields II*, Saitama Math. J. **9** (1991), 9–18.
4. W. D. Neumann, *UBASIC: A Public-Domain BASIC for Mathematics*, Notices Amer. Math. Soc. **36** (1989), 557–559.
5. J.-P. Serre, *Formes modulaires et fonctions zêta p-adiques,* in *Modular functions of one variable III* (W. Kuyk and J.-P. Serre, eds.), Lecture Notes Math. **350**, Springer, Berlin, 1973, pp. 191–268.
6. ———, *Cohomologie des groupes discrets*, Prospects in Mathematics, Ann. Math. Stud. **70** (1971), 77–169.
7. C. L. Siegel, *Berechnung von Zetafunktionen an ganzzahligen Stellen*, Göttingen Nachr. **10** (1969), 87–102.

DEPARTMENT OF MATHEMATICS, FACULTY OF SCIENCE, SAITAMA UNIVERSITY, URAWA, SAITAMA, 338, JAPAN

E-mail address: ktakeuch@rimath.saitama-u.ac.jp

Proceedings of Symposia in Applied Mathematics
Volume 48, 1994

The period of the Bell exponential integers modulo a prime

SAMUEL S. WAGSTAFF, JR.

ABSTRACT. We show that the minimum period of the Bell exponential integers reduced modulo p is $(p^p - 1)/(p - 1)$ for all primes $p < 82$ and several larger p. Our proof of this result requires the prime factorization of these periods. For about one-half of the primes p the factoring is aided by an algebraic formula.

The first-order Bell exponential integer $B(n)$ is the number of ways of placing n distinguishable objects into 1 to n indistinguishable cells so that no cell is empty. The Bell numbers may be expressed as a sum $B(n) = \sum_{r=1}^{n} S(n,r)$ of Stirling numbers of the second kind. See [**4**] and its references.

The first few Bell numbers may be computed easily from the difference formula $B(n) = \Delta^n B(1)$ of Cesàro [**2**]. The first few values are $B(0) = 1$ (by definition), $B(1) = 1$, $B(2) = 2$, $B(3) = 5$, $B(4) = 15$, $B(5) = 52$ and $B(6) = 203$.

Consider the sequence of Bell numbers reduced modulo a prime p. After one computes $B(n) \bmod p$ for $0 \le n < p$ by Cesàro's formula, one may compute further terms quickly by the congruence

$$B(n+p) \equiv B(n) + B(n+1) \pmod{p} \tag{1}$$

of Touchard [**7**]. It is clear from (1) that the sequence $\{B(n) \bmod p; n = 0, 1, \dots\}$ is eventually periodic. Williams [**8**] proved that for each prime p the sequence is periodic from the beginning and that the minimum period divides

$$N_p = \frac{p^p - 1}{p - 1}.$$

1991 *Mathematics Subject Classification.* Primary 11-04, 11B73; Secondary 11Y05, 12-04, 12E10, 12Y05.

Key words and phrases. Bell numbers, period modulo p, integer factorization, Aurifeuillian factorization.

Some of the computing reported in this work was performed on a MasPar computer at Purdue University which was supported in part by NSF Infrastructure Grant CDA-9015696.

The final version of this paper will be submitted for publication elsewhere.

© 1994 American Mathematical Society
0160-7634/94 $1.00 + $.25 per page

By hand compuation, he showed that the minimum period is precisely N_p for $p = 2$, 3 and 5. Levine and Dalton [**4**] used a computer to show that the minimum period is exactly N_p for $p = 7$, 11, 13 and 17. They also investigated the period for the other primes < 50. Using the same general technique, we show that the minimum period is exactly N_p for each prime < 82 and for several larger primes. Great advances in integer-factoring methods since 1962 allowed us to extend their work so far.

Given a prime p, to test whether the period of $\{B(n) \bmod p\}$ divides some factor N of N_p, it suffices because of (1) to compare $B(N+i) \bmod p$ with $B(i) \bmod p$ for $0 \leq i < p$. For primes $p < 180$, we factored N_p as much as possible, using techniques described below. The factorization of N_p was complete for all primes $p < 82$ and for the six larger primes mentioned in Theorem 1. For each prime $p < 180$ and each known prime divisor q of N_p we tested whether the period divides $N = N_p/q$. It never did, and we have proved

THEOREM 1. *The minimum period of the sequence* $\{B(n) \bmod p\}$ *is* N_p *when* p *is a prime* < 82 *and also when* $p = 89$, 97, 101, 163, 167 *or* 173.

We conjecture that the minimum period of the sequence $\{B(n) \bmod p\}$ is N_p for every prime p.

We computed $B(N) \bmod p$ for large N via the congruence $B(n + p^m) \equiv B(n+1) + mB(n) \pmod p$ of Touchard [**7**], which generalizes (1). Starting from the block $B(i) \bmod p$, $0 \leq i \leq p$, we computed successive blocks of length $p+1$, using the digits of N in radix p to direct our choice of the blocks towards the final block $B(N+i) \bmod p$, $0 \leq i \leq p$. See Levine and Dalton [**4**] for details.

We now describe our efforts to factor N_p for primes $p < 180$. The Table shows the factorization of those N_p which we could factor completely. We use Pxx in the Table to mean a prime of xx digits. Some trial division was done first, using the fact that all prime factors of N_p have the form $2kp + 1$ for some positive integer k. Most of the larger factors in the Table were found by the Elliptic Curve Method [**3**], using a program written by Peter Montgomery. This work was aided greatly by the use of Aurifeuillian factorizations. That is, when p is prime and $\equiv 1 \pmod 4$, N_p splits algebraically into two nearly equal factors (called pL and pM in the Table). We computed these two Aurifeuillian factors from Theorem 2.

We would be happy to send our partial factorizations of the N_p not shown in the Table to any reader. The first p for which we could not factor N_p completely was $p = 83$, which has a composite cofactor of 147 digits. The smallest remaining composite cofactor of an N_p was the 100-digit divisor of $113M$. For the primes $p < 180$ not listed in the Table, we checked that no known proper divisor of N_p can be a period.

For integers $n > 0$ let $\Phi_n(x)$ denote the cyclotomic polynomial. When p is an odd prime, $N_p = \Phi_p(p)$. Let (m, n) be the greatest common divisor of m and n. Let $\phi(n)$ denote Euler's totient function. Let $(m|n)$ be the Jacobi symbol. Theorem 2 follows from Theorem 1 of Schinzel [**6**].

Table. Factors of $N_p = (p^p - 1)/(p - 1)$ for some primes p in $10 < p < 180$

p	Prime factorization of N_p
11	$15797 \cdot 1806113$
$13L$	1803647
$13M$	$53 \cdot 264031$
$17L$	2699538733
$17M$	$10949 \cdot 1749233$
19	$1099122030922396643840221$
23	$461 \cdot 1289 \cdot 831603031789 \cdot 1920647391913$
$29L$	$84449 \cdot 2428577 \cdot 549334763$
$29M$	$59 \cdot 16763 \cdot 14111459 \cdot 58320973$
31	$568972471024107865287021434301977158534824481$
$37L$	$149 \cdot 41903425553544839998158239$
$37M$	$1999 \cdot 7993 \cdot 16651 \cdot 17317 \cdot 10192715656759$
$41L$	$1752341 \cdot 20567159 \cdot 1876859311090803007$
$41M$	$83 \cdot 5926187589691497537793497756719$
43	$173 \cdot 120401 \cdot P62$
47	$1693 \cdot 255742492896763511474638530188876017 \cdot P39$
$53L$	$107 \cdot 16505521259654533 \cdot 143470720478589313288313473$
$53M$	$141829 \cdot 1303396057963132488045544988140899439 2143$
59	$709 \cdot 141579233 \cdot P92$
$61L$	$977 \cdot 343625872243632312073 \cdot 39885328645607179260991799590 7$
$61M$	$1000403244183535565720394723140528028235711874491322863$
67	$269 \cdot 4021 \cdot 730837 \cdot 10960933 \cdot$ $\cdot 151495488509660402356228791573 0049 \cdot P69$
71	$105649 \cdot 3388409395214741 \cdot 17882954877203881 \cdot P93$
$73L$	$1414741 \cdot 1295720382587 \cdot 1192167517020392933 \cdot P31$
$73M$	$293 \cdot 439 \cdot 25239167 \cdot 56377463 \cdot 3611379501352361 \cdot P32$
79	$317 \cdot 1558537597 \cdot 17135507183050838 9477 \cdot$ $\cdot 5449313290804337826320291 3 \cdot P91$
$89L$	$179 \cdot 8009862103557709 \cdot 5964844210432006407836201 \cdot P43$
$89M$	$3730759891225349089330219913 3 \cdot P58$
$97L$	$P95$
$97M$	$389 \cdot 363751 \cdot 684640163 \cdot 119437287337412947643906021 53 \cdot P51$
$101L$	$1213 \cdot 9931988588681 \cdot 102208068907493 \cdot 393101595766008847 \cdot P53$
$101M$	$607 \cdot 5657 \cdot 157561 \cdot P89$
163	$653 \cdot 2609 \cdot 41729 \cdot 31943437 \cdot 3727539197017 \cdot 391683908074297 \cdot$ $\cdot 8224734227858383253 \cdot P294$
167	$16033 \cdot 1001953110409 \cdot 669806250678629514045626189 \cdot P326$
$173L$	$347 \cdot 685081 \cdot P184$
$173M$	$161297590410850151 \cdot P176$

THEOREM 2. *Let $p \equiv 1 \pmod 4$ be squarefree. Then there exist polynomials $C_p(x)$ and $D_p(x)$ with integer coefficients and degrees $\phi(p)/2$ and $\phi(p)/2 - 1$, respectively, with the following properties. For any odd positive integer h,*

$$\Phi_p(p^h) = (C_p(p^h) - p^{(h+1)/2} D_p(p^h))(C_p(p^h) + p^{(h+1)/2} D_p(p^h)).$$

The coefficients of $C_p(x)$ and $D_p(x)$ may be computed from the identity

$$C_p(x^2) - \sqrt{p}\, x D_p(x^2) = \prod_{\substack{s=1 \\ (s,p)=1}}^{(p-1)/2} (x^2 - 2(s|p) \cos \frac{2\pi s}{p} x + 1).$$

Brent [**1**] gives an algorithm for computing the coefficients of $C_p(x)$ and $D_p(x)$, which uses integer arithmetic throughout.

A table of coefficients of $C_p(x)$ and $D_p(x)$ for $p < 120$ may be found in Table 34 on page 453 ff. of Riesel [**5**].

To prove Theorem 1, we used Theorem 2 only when p is prime and $h = 1$.

Acknowledgements

I thank John W. Wrench, Jr. for suggesting this research and R. P. Brent for sending me a preprint of [**1**]. I am grateful to P. L. Montgomery for the use of his ECM program and for several factors which I missed. I thank A. K. Lenstra and B. Dixon for letting me use their ECM program for the MasPar computer. It found several factors in the Table.

REFERENCES

1. Richard P. Brent, *On computing factors of cyclotomic polynomials*, Math. Comp. **61** (1993), 131–149.
2. E. Cesàro, *Sur une équation aux différences mèlées*, Nouvelles Annales de Math. (3) **4** (1885), 36–40.
3. H. W. Lenstra, Jr., *Factoring integers with elliptic curves*, Ann. of Math. **126** (1987), 649–673.
4. Jack Levine and R. E. Dalton, Math. Comp. **16** (1962), 416–423.
5. Hans Riesel, *Prime Numbers and Computer Methods for Factorization*, Birkhäuser, Boston, 1985.
6. A. Schinzel, *On the primitive prime factors of $a^n - b^n$*, Proc. Cambridge Philos. Soc. **58** (1962), 555–562.
7. J. Touchard, *Propriétés arithmétiques de certains nombres récurrents*, Ann. Soc. Sci. Bruxelles **53A** (1933), 21–31.
8. G. T. Williams, *Numbers generated by the function e^{e^x-1}*, Amer. Math. Monthly **52** (1945), 323–327.

DEPARTMENT OF COMPUTER SCIENCES, PURDUE UNIVERSITY, WEST LAFAYETTE, IN 47907

E-mail address: ssw@cs.purdue.edu

Proceedings of Symposia in Applied Mathematics
Volume **48**, 1994

Computing Invariant Polynomials of p-adic Reflection Groups

CHANGSHENG XU

ABSTRACT. It is known that the ring of invariants of a p-adic reflection group is a polynomial algebra if p does not divide the order $|G|$ of G. But in modular cases (i.e., $p||G|$), the algebraic structure of the rings of invariants of some p-adic reflection groups is still unknown. Using Gröbner bases theory in conjunction with the computer package *MACAULAY*, we prove that the rings of invariants of groups number 29 and 31 (p=5) in the Shephard-Todd classification are polynomial algebras over the finite field F_5.

1. Introduction

Fix an odd prime p and let V be an n-dimensional vector space over the finite field F_p. Assume that G is a subgroup of the general linear group $GL_n(F_p)$. Let $S(V)$ be the symmetric algebra on V^*, the dual of V. Any basis $\{t_1, t_2, \ldots, t_n\}$ of V^* gives the identity $S(V) = F_p[t_1, t_2, \ldots, t_n]$. The action of G on V induces an action on V^*. This action extends naturally to $S(V)$. For an odd prime p, consider $S(V) = F_p[t_1, t_2, \ldots, t_n]$ as a graded algebra with generators $t_1, t_2, \ldots, t_n$ of degree 2. We use $S(V)^G = \{x \in S(V)| \ gx = x \text{ for all } g \in G\}$ to denote the algebra of invariants of the action. $S(V)^G$ is a finitely generated algebra over F_p. However, in general, $S(V)^G$ is not a polynomial algebra. In the nonmodular case (i.e., p does not divide the order $|G|$ of G), there is a correspondence between p-adic reflection groups and polynomial algebras, and Adams and Wilkerson [**1**] and Clark and Ewing [**3**] classified the unstable algebras over the Steenrod algebra which are polynomial algebras.

In this paper, we study two modular finite reflection groups, number 29 and number 31 in the Shephard-Todd classification [**5**]. Our main result is that the rings of invariants of these two groups are polynomial algebras over F_5.

1991 *Mathematics Subject Classification*. Primary 55P35, 13A50; Secondary 55-04.

The author was supported in part by NSF grant DMS-9207731.

The final version of this paper will be submitted for publication elsewhere.

© 1994 American Mathematical Society
0160-7634/94 $1.00 + $.25 per page

2. Invariant theory over F_p

PROPOSITION. *Let $w_1, w_2, \dots, w_n$ in $S(V)^G$ be homogeneous polynomials with $\prod_{i=1}^n |w_i|/2 = |G|$, where $|w_i|$ denotes the degree of w_i. Assume that $w_1, \dots, w_n$ are algebraically independent over F_p; then*

$$S(V)^G = F_p[w_1, w_2, \dots, w_n]$$

is a polynomial algebra.

LEMMA (Wilkerson [8, Lemma 3.2]). *If $R = F_p[w_1, \dots, w_n]$ is included in $S(V)$ such that $S(V)$ is integral over R, then the degree of the underlying field extension is $\prod_{i=1}^n |w_i|/2$.*

Proof of Proposition. Since $w_1, \dots, w_n$ are algebraically independent, $R = F_p[w_1, \dots, w_n]$ is a polynomial algebra with the same transcendence degree n as $S(V)$. Thus, $S(V)$ is integral over $R = F_p[w_1, \dots, w_n]$. We need only to show that $S(V)^G = F_p[w_1, \dots, w_n]$.

Let $K = F_p(w_1, \dots, w_n)$ and $L = F_p(t_1, \dots, t_n)$ be the field of fractions of $R = F_p[w_1, \dots, w_n]$ and $S(V) = F_p[t_1, \dots, t_n]$, respectively. Let E be the field of fractions of the invariants $S(V)^G$. For any u/v in L^G, the invariant subfield of L, the denominator can always be converted to one in $S(V)^G$, namely, $\prod_{g \in G} gv$ (if v is not in $S(V)^G$). So $L^G = E$. Now we have a tower of fields:

$$K = F_p(w_1, \dots, w_n) \subset E = L^G \subset L = F_p(t_1, \dots, t_n).$$

The degree $[L : E] = [L : L^G]$ is $|G|$. From the above lemma, the degree $[L : K]$ is $\prod_{i=1}^n |w_i|/2 = |G|$. So the quotient field K of $R = F_p[w_1, \dots, w_n]$ equals the quotient field E of $S(V)^G$. Since the polynomial ring $R = F_p[w_1, \dots, w_n]$ is integrally closed in its field of fractions, and $S(V)^G$ is integral over R, we get $R = S(V)^G$. □

3. The invariants of groups number 29, G_{29}, and number 31, G_{31}

From the complex representation, we easily get the representation of G_{31} over the finite field F_5:

$$r_1 = \begin{pmatrix} 3 & -3 & -3 & -3 \\ -3 & 3 & -3 & -3 \\ -3 & -3 & 3 & -3 \\ -3 & -3 & -3 & 3 \end{pmatrix}, \qquad r_2 = \begin{pmatrix} 0 & -2 & 0 & 0 \\ 2 & 0 & 0 & 0 \\ 0 & 0 & 1 & 0 \\ 0 & 0 & 0 & 1 \end{pmatrix},$$
$$r_3 = \begin{pmatrix} 0 & 1 & 0 & 0 \\ 1 & 0 & 0 & 0 \\ 0 & 0 & 1 & 0 \\ 0 & 0 & 0 & 1 \end{pmatrix}, \qquad r_4 = \begin{pmatrix} 1 & 0 & 0 & 0 \\ 0 & 0 & 1 & 0 \\ 0 & 1 & 0 & 0 \\ 0 & 0 & 0 & 1 \end{pmatrix},$$
$$r_5 = \begin{pmatrix} 1 & 0 & 0 & 0 \\ 0 & 1 & 0 & 0 \\ 0 & 0 & 1 & 0 \\ 0 & 0 & 0 & -1 \end{pmatrix}.$$

The group number 29, G_{29}, is generated by the first four unitary reflections. It is easily checked that

$$w_4 = \sum_{i=1}^{4} t_i^4 - 2t_1t_2t_3t_4$$

is invariant under G_{29}. But w_4 is not invariant under G_{31}. The reflection r_5 changes w_4 to $t_1^4t_2^4t_3^4t_4^4 + 2t_1t_2t_3t_4$. From Maschke [**4**, §3], and applying r_i to it, we see that the orbit of w_4 under the action of G_{31} consists of the following polynomials:

$\psi_1 = w_4 = (\sum_{i=1}^4 t_i^4) - 2t_1t_2t_3t_4$,
$\psi_2 = (\sum_{i=1}^4 t_i^4) + 2t_1t_2t_3t_4$,
$\psi_3 = 2(\sum_{i=1}^4 t_i^4) - 2(t_1^2t_2^2 + t_3^2t_4^2) - 2(t_1^2t_3^2 + t_2^2t_4^2) - 2(t_1^2t_4^2 + t_2^2t_3^2)$,
$\psi_4 = 2(\sum_{i=1}^4 t_i^4) - 2(t_1^2t_2^2 + t_3^2t_4^2) + 2(t_1^2t_3^2 + t_2^2t_4^2) + 2(t_1^2t_4^2 + t_2^2t_3^2)$,
$\psi_5 = 2(\sum_{i=1}^4 t_i^4) + 2(t_1^2t_2^2 + t_3^2t_4^2) - 2(t_1^2t_3^2 + t_2^2t_4^2) + 2(t_1^2t_4^2 + t_2^2t_3^2)$,
$\psi_6 = 2(\sum_{i=1}^4 t_i^4) + 2(t_1^2t_2^2 + t_3^2t_4^2) + 2(t_1^2t_3^2 + t_2^2t_4^2) - 2(t_1^2t_4^2 + t_2^2t_3^2)$.

The first six elementary symmetric functions of these homogeneous polynomials are

$W_4 = \sum \psi_i = 0$,
$W_8 = \sum \psi_i\psi_j = \sum t_i^8 - \sum t_i^4t_j^4 - 2t_1^2t_2^2t_3^2t_4^2$,
$W_{12} = -3\sum \psi_i\psi_j\psi_k = \sum t_i^{12} + 2\sum t_i^8t_j^4 + 2t_1^2t_2^2t_3^2t_4^2 \sum t_i^4$,
$W_{16} = \sum \psi_i\psi_j\psi_k\psi_l = -(W_8)^2$,
$W_{20} = -\sum \psi_i\psi_j\psi_k\psi_l\psi_m$,
$W_{24} = \psi_1\psi_2\psi_3\psi_4\psi_5\psi_6$.

Since G_{29} is a subgroup of G_{31}, we have that $w_4, w_8 = W_8, w_{12} = W_{12}$ and $w_{20} = W_{20}$ are invariant under G_{29}. Thus, we obtain invariants $\{w_4, w_8, w_{12}, w_{20}\}$ for the group G_{29} and $\{W_8, W_{12}, W_{20}, W_{24}\}$ for the group G_{31}.

4. Algebraic independence

We have obtained invariants $\{w_4, w_8, w_{12}, w_{20}\}$ for G_{29} and $\{W_8, W_{12}, W_{20}, W_{24}\}$ for G_{31}. In order to show that they are the fundamental invariants, we need to show that they are algebraically independent. For this purpose, we use computer algorithms. Buchberger's Gröbner bases theory gives

SUBROUTINE (Algebraic Dependence [**7**, Subroutine 3.7])

Input: $F := \{f_1, \dots, f_m\} \subset k[t_1, \dots, t_n]$, where $m \leq n$

Questions: Is F algebraically dependent over k? If so, find an m-variate polynomial P such that $P(f_1, \dots, f_m) = 0$ in $k(t_1, \dots, t_n)$.

Solution: Introduce m new "slack" variables $\{s_1, \dots, s_m\}$, and compute a Gröbner basis GB of $\{f_1 - s_1, \dots, f_m - s_m\}$ with respect to the reverse lexicographic order induced by $t_1 > \cdots > t_n > s_1 > \cdots > s_m$. The set F is algebraically independent if and only if $GB \bigcap k[s_1, \ldots, s_m] = \emptyset$. On the other hand, if $P \in GB \bigcap k[s_1, \ldots, s_m]$, then $P(f_1, \ldots, f_m) = 0$ in $k[t_1, \ldots, t_n]$.

In our cases, we consider $F = \{w_4, w_8, w_{12}, w_{20}\}$ for G_{29}. In the computer package *MACAULAY*, we define a ring $k[t_1, \ldots, t_4, s_1, \ldots, s_4]$ with the reverse lexicographic order induced by $t_1 > \cdots > t_4 > s_1 > \cdots > s_4$, where t_1, t_2, t_3, t_4 have weight 1 and s_1, s_2, s_3, s_4 have weights 4,8,12,20. After computing a Gröbner basis GB of $\{w_4 - s_1, w_8 - s_2, w_{12} - s_3, w_{20} - s_4\}$, we want to find a Gröbner basis for the intersection $GB \bigcap k[s_1, \ldots, s_4]$. In *MACAULAY*, this is accomplished by the command **elim**, which eliminates the variables t_1, t_2, t_3, t_4 from GB [**6**]. We get $GB \bigcap k[s_1, \ldots, s_4] = \emptyset$, i.e., $F = \{w_4, w_8, w_{12}, w_{20}\}$ is algebraically independent. Similarly, $\{W_8, W_{12}, W_{20}, W_{24}\}$ is algebraically independent. Since $4 \cdot 8 \cdot 12 \cdot 20 = |G_{29}|$ (and $8 \cdot 12 \cdot 20 \cdot 24 = |G_{31}|$), from the Proposition of §2 we obtain the following theorem.

THEOREM. *The ring of invariants of G_{29} is a polynomial algebra with generators of degrees* 8, 16, 24 *and* 40:

$$F_5[t_1, t_2, t_3, t_4]^{G_{29}} = F_5[w_4, w_8, w_{12}, w_{20}].$$

The ring of invariants of G_{31} is a polynomial algebra with generators of degrees 16, 24, 40 *and* 48:

$$F_5[t_1, t_2, t_3, t_4]^{G_{31}} = F_5[W_8, W_{12}, W_{20}, W_{24}].$$

REFERENCES

1. J.F. Adams and C.W. Wilkerson, *Finite H-spaces and algebras over the Steenrod algebra*, Ann. of Math. **111** (1980), 95-143.
2. B. Buchberger, *Gröbner bases - an algorithmic method in polynomial ideal theory*, Chapter 6 in *Multidimensional systems theory*, N.K. Bose (ed.), Reidel, Dordrecht, 1985.
3. A. Clark and J. Ewing, *The realization of polynomial algebras as cohomology rings*, Pacific J. Math **50** (1974), 425-434.
4. H. Maschke, *Ueber die quaternäre,endliche,lineare,Substitutionsgruppe der Borchardt'schen Moduln*, Math. Ann. **30** (1887), 496-515.
5. G.C. Shephard and J.A. Todd, *Finite unitary reflection groups*, Canad. J. Math. **6** (1954), 274-304.
6. M. Stillman and D. Bayer, *Macaulay User Manual* (preprint).
7. B. Sturmfels and N. White, *Computing combinatorial decompositions of rings*, Combinatorica **11** (1991), 275-293.
8. C. Wilkerson, *A primer on the Dickson invariants*, Contemp. Math. **19** (1983), 421-434.

DEPARTMENT OF MATHEMATICS, PURDUE UNIVERSITY, W. LAFAYETTE, IN 47907

E-mail address: xu@math.purdue.edu

Author Index

Subject Index

Recent Titles in This Series

(Continued from the front of this publication)

18 **H. Grad, editor,** Magneto-fluid and plasma dynamics (New York City, April 1965)

17 **R. Finn, editor,** Applications of nonlinear partial differential equations in mathematical physics (New York City, April 1964)

16 **R. Bellman, editor,** Stochastic processes in mathematical physics and engineering (New York City, April 1963)

15 **N. C. Metropolis, A. H. Taub, J. Todd, and C. B. Tompkins, editors,** Experimental arithmetic, high speed computing, and mathematics (Atlantic City and Chicago, April 1962)

14 **R. Bellman, editor,** Mathematical problems in the biological sciences (New York City, April 1961)

13 **R. Bellman, G. Birkhoff, and C. C. Lin, editors,** Hydrodynamic instability (New York City, April 1960)

12 **R. Jakobson, editor,** Structure of language and its mathematical aspects (New York City, April 1960)

11 **G. Birkhoff and E. P. Wigner, editors,** Nuclear reactor theory (New York City, April 1959)

10 **R. Bellman and M. Hall, Jr., editors,** Combinatorial analysis (New York University, April 1957)

9 **G. Birkhoff and R. E. Langer, editors,** Orbit theory (Columbia University, April 1958)

8 **L. M. Graves, editor,** Calculus of variations and its applications (University of Chicago, April 1956)

7 **L. A. MacColl, editor,** Applied probability (Polytechnic Institute of Brooklyn, April 1955)

6 **J. H. Curtiss, editor,** Numerical analysis (Santa Monica City College, August 1953)

5 **A. E. Heins, editor,** Wave motion and vibration theory (Carnegie Institute of Technology, June 1952)

4 **M. H. Martin, editor,** Fluid dynamics (University of Maryland, June 1951)

3 **R. V. Churchill, editor,** Elasticity (University of Michigan, June 1949)

2 **A. H. Taub, editor,** Electromagnetic theory (Massachusetts Institute of Technology, July 1948)

1 **E. Reissner, editor,** Non-linear problems in mechanics of continua (Brown University, August 1947)

ISBN 0-8218-0291-7

9 780821 802915